Accounting Fundamentals

Dr. Donald J. Guerrieri

Dr. F. Barry Haber

William B. Hoyt

Robert E. Turner

COVER: Thapana_Studio/Shutterstock

mheducation.com/prek-12

Copyright © 2025 McGraw Hill

All rights reserved. No part of this publication may be reproduced or distributed in any form or by any means, or stored in a database or retrieval system, without the prior written consent of McGraw Hill, including, but not limited to, network storage or transmission, or broadcast for distance learning.

Send all inquiries to:
McGraw Hill
8787 Orion Place
Columbus, OH 43240

ISBN: 978-1-26-500369-2
MHID: 1-26-500369-6

Printed in the United States of America.

1 2 3 4 5 6 7 8 9 LWI 28 27 26 25 24

Meet Our Authors

Donald J. Guerrieri
Donald J. Guerrieri has taught accounting at both the secondary (Accounting Instructor at Norwin High School, North Huntington, Pennsylvania) and college levels. His work for a firm of certified public accountants gave him hands-on experience in working with real-world clients. Dr. Guerrieri has written numerous articles about teaching accounting for a number of educational journals. He has been a program speaker at business education seminars and conferences.

F. Barry Haber
F. Barry Haber has taught accounting courses, including principles, cost, and advanced for more than 40 years. As a professor and Certified Public Accountant, he served on many local, state, and national accounting committees. Dr. Haber has authored three other textbooks in addition to writing accounting articles and cases.

William B. Hoyt
William B. Hoyt is an Accounting Instructor at Wilton High School, Wilton, Connecticut. He has been a force in accounting curriculum development for many years and served as a member of the writing task force for the National Standards for Business Education. The author of various articles in professional journals, Mr. Hoyt was named Connecticut's Outstanding Business Educator of the Year in 1996, and in 1997 the National Business Education Association named him National Secondary Teacher of the Year.

Robert E. Turner
Robert E. Turner is an Assistant Professor in the College of Business at Northwestern State University, Natchitoches, Louisiana. Previously, he served as Vice President for Business Affairs at McNeese State University, Lake Charles, Louisiana, and at the University of Louisiana at Monroe. Mr. Turner's broad experience includes teaching accounting and other business subjects at the high school and college levels. He is a frequent speaker at educational conferences, seminars, and workshops around the country.

Contributors

Cynthia A. Ash
Davenport University - Granger, Indiana

Louann Cummings
The University of Findlay - Findlay, Ohio

Patrick C. Hale
Angelina College - Lufkin, Texas

Sandra S. Lang
McKendree College Lebanon, Illinois

Alice Sineath
Forsyth Technical Community College
Winston-Salem, North Carolina

Kimberly D. Smith
County College of Morris - Randolph, New Jersey

Educational Reviewers

Jayne Abernathy
Accounting Teacher
Centerville High School
Centerville, Ohio

Barbara A. Berg
Vocational Business Teacher
Carey High School
Carey, Idaho

Sibyl Cole-Olfzewski
Computer Tech. Teacher
Medina High School
Medina, Ohio

Naomi Crane
Accounting Teacher
Southwest High School
Fort Worth, Texas

Johnsie Crawford
Accounting Teacher
Brandon High School
Brandon, Florida

Carrie Davis
Business Teacher
Yoncalla High School
Yoncalla, Oregon

Melinda G. Dibenedetto
Business Department Chair
Shades Valley High School
Birmingham, Alabama

Christine Fink
Business Teacher
Cocalico High School
Denver, Pennsylvania

Patricia Kay Fordham
Business Teacher
Central High School
Salt Lake City, Utah

Patsy A. Frost
Business Department Chair
Scituate High School
North Scituate, Rhode Island

Ruby Garr
Business Department Head
I.H. Kempner High School
Sugar Land, Texas

Karen King Gilliland
Business Department Chair
Winchester Community
High School
Winchester, Indiana

Norma C. Isassi
Accounting Teacher
H.M. King High School
Kingsville, Texas

Karen LaClair
Business Department Chair
Simsbury High School
Simsbury, Connecticut

Chris Mendez
Accounting Teacher
Bel Air High School
El Paso, Texas

Regena Morris
Business Education Teacher
Arlington Heights High School
Fort Worth, Texas

Jeff Noyes
Business Education Teacher
Minnetonka High School
Minnetonka, Minnesota

Karen Poidomani
Business Teacher
William Floyd High School
Mastic Beach, New York

Martha Ramirez
Business Teacher
McCallum High School
Austin, Texas

William Sinai
Computerized Accounting
Dept. Chair
West Valley Occupational Center
Woodland Hills, California

Violet Snell
Business Department Chair
MacArthur High School
San Antonio, Texas

Alfonso Vargas
Accounting Teacher
Hueneme High School
Oxnard, California

Karen Wood
Accounting Teacher
Centerville High School
Centerville, Ohio

Dianne L. Young
Dean of Students
Wilmington High School
Wilmington, Delaware

Educational Consultants

We wish to acknowledge the contributions of the following:

Jayne Abernathy, Xenia, OH
Jim Adams, New Albany, IN
G. Armstrong, Bath, NY
Ron Baldini, Pulaski, WI
Alan Balog, Ravenna, OH
B. A. Banks-Burke, Hudson, OH
Denise M. Barbaris, West Milford, NJ
E. Patricia Barnes, Evansville, IN
Lou Ellen Bass, Four Oaks, NC
Linda Becker, Cary, NC
Perry Beckerman, Brooklyn, NY
Kathleen Beetar, Houston, TX
Scott Behnke, Wales, WI
Len C. Bigler, Grand Rapids, MI
Jackie Billings, McAllen, TX
Lee Birnbaum, Farmingdale, NJ
Cheryl Birx, Canandaigua, NY
Linda J. Black, Mt. Airy, GA
Elbert F. Black, Hastings, MI
Cindi Blansfield, Auburn, WA
Josephine Bliss, Merced, CA
D. Bogataj, Flat Rock, MI
Kelly Brock, Columbus, GA
William A. Brogdon, German Town, TN
Rachelle Brown, Morocco, IN
Jessie Broxmeyer, Long Beach, NY
Margaret Broyles, Forest Park, OH
Gus Buchttolz, Wyoming, MI
Ronda K. Budd, Grove City, OH
Pete Bush, Leesburg, FL
Michelle Cardoza, Tulare, CA
Walter Carlson, Torrance, CA
Marisa Charles, Houston, TX
Bill Christianson, Spokane, WA
Jodi S. Clark, Warner Robins, GA
Anthony Paul Corbisiero, West New York, NJ
Ron Craig, Anaheim, CA
Lynn Crawford, Cudahy, WI
Claudia Cullari, Fair Lawn, NJ
Jo Damron, Abilene, TX
Jan Davis, Port Neches, TX
Toni DeFuria, Old Tappan, NJ
Denise Deltondo, Girard, OH
Carolyn Deshler, Moorpark, CA
Diane DeWitt, Freeport, IL
Wanda Drake, Jackson, TN
Buzz Drake, Barrington, IL
Ray Dunavant, Denison, TX
Tearle Dwiggins, Pendleton, IN
Susan España, Rocklin, CA

Morris P. Fair, Sr., Jackson, TN
Glenn Fong, Sacramento, CA
Julie Fox, Wisconsin Rapids, WI
Carolyn Francis, Baytown, TX
Cherilynn Frost, Livonia, MI
Patricia T. Gardella, Vineland, NJ
Robert D. Garrison, Belleview, FL
Kimberly Morris Gehres, Adrian, MI
JoAnn Gipson, Garland, TX
Nancy Glavic, New Concord, OH
Joyce Greemon, Gastonin, NC
Sheree Green, Teaneck, NJ
Andrea Gross, Vidor, TX
Andrew Gump, Waynesburg, PA
Guzzeta, Plant City, FL
Ronald L. Hainds, Jacksonville, IL
Kellie Hair, Kennesaw, GA
Sheryl J. Hane, Norco, CA
Janice Hanes, Fort Worth, TX
Marion Hanneld, Omaha, NE
Farolyn Hanscom, Riverside, CA
Wanda Harris, Eden, NC
Tammy Hatfield, Tullahoma, TN
Felicia S. Hatten, Detroit, MI
Debra R. Hauser, Shrub Oak, NY
Richard Heim, Latham, NY
Audrey M. Hendry, Huntington Beach, CA
John Hodgins, Marysville, WA
Helen Hogan, Lithonia, GA
Betty L. Holloway, Maplewood, NJ
Carol Huprich, Dover, OH
Antoinette Hutchings, Slate Hill, NY
David Ifkovic, Wilton, CT
Steve Ingmire, Indianapolis, IN
Waydene C. Jackson, Stone Mountain, GA
M. Jackson, Houston, TX
Irma D. James, Dayton, OH
Carol Jekel, Flushing, MI
Lori Jepson, Montclair, CA
Kay Jernigan, Vanceboro, NC
Glenn Johnson, Mukwonago, WI
Judy Jones, Amarillo, TX
Mary K. Jones, Neenah, WI
Ted Juske, Mundelein, IL
Beverly Kaeser, Appleton, WI
Linda Keats, Stevens Point, WI
Sue Kelley, Hubbard, OH
Mary Kay King, Dickson, TN
Carol Kinney, Eau Claire, WI

Educational Consultants   v

Educational Consultants

Ann Klingelsmith, Greenville, MI
Tammy Koch, Minocqua, WI
Patrick Kubeny, Rhinelander, WI
Cheryl M. Lagan, Ossining, NY
Amelia Lard, Memphis, TN
Marianna Larkin, Waxahachie, TX
Becky Larsen, Irvine, CA
Sharon Larson, Crystal Lake, IL
Inez Lauerman, Shingle Springs, CA
Annette Laughlin, Oswego, IL
Steve Leighton, Brooklyn, NY
Sue Lewis, Littlerock, CA
Stan Lewis, Champaign, IL
Donna Lewis, Morehead City, NC
Theresa Lueras, Daly City, CA
Sherry Lund, Port Angeles, WA
Linda Lupi, West Bloomfield, MI
A. Mallette, New York, NY
John Matera, Kenosha, WI
Alice Matthews, Artesia, CA
Sarah McBride, Hickory, NC
Rebecca McKnight, Gainesville, FL
Pam Meyer, The Woodlands, TX
J. Miller, Madison, OH
Gloria Morris, Fort Worth, TX
Karen Mozingo, Greenville, NC
Ralph A. Muro, Binghamton, NY
Dianne Murphy, Rancho Cucamonga, CA
Jill N. Murphy-Totten, Middletown, OH
John Murray, Fraser, MI
Karen Neuman, Parma, OH
William Nitch, Niles, OH
Bonnie N. Nute, Munford, TN
Betty O'Dell, Escondido, CA
Deborah Ottmers, San Antonio, TX
Donna Owens, Missouri City, TX
RaeNell Parker, Menomonie, WI
Ronda Patrick, Kirkland, WA
Diane Personett, Rushville, IN
Hank Petite, Bloomfield, NJ
Glenn Philpott, Belleville, IL
Kip Pichel, Auburn Hills, MI
Patti Pletcher, Bristol, IN
Mark G. Pretz, Harlingen, TX
Diane A. Prom, Oak Creek, WI
Lydia Quiñones, Dunwoody, GA
Mary C. Rabb, Winchester, TN
Margaret Ralph, Portland, IN
Glenda Razor, Santa Ana, CA
Tom Reach, Aliso Viejo, CA

C. Redmond, Lakewood, CA
Suzanne Reisert, Floral Park, NY
Edlyn Reneau, Corpus Christi, TX
Ruth Riley, Poland, OH
Melanie Rodges, Keller, TX
Rosalinda Rodriguez, McAllen, TX
Mary R. Roth, Warner Robins, GA
Frank Rudnesky, Pine Hill, NJ
Joyce Sadd, Hastings, NE
Lawrence Sakalas, Southgate, MI
Gary L. Schepf, Irving, TX
Donna M. Scheuerer, Farmingdale, NY
Jerry Schlecher, Hartland, MI
Sue Smith, Whittier, CA
Pete Smith, Mishawaka, IN
James Smith, Greensboro, NC
Linda Songer, Orange Park, FL
Linda Spellich, Belleville, MI
Carolyn R. Spencer, East St. Louis, IL
Camille Sroka, Toms River, NJ
Will Stauske, Cedarburg, WI
Beverly Stergeos, Fort Worth, TX
Linda Stevens, Houston, TX
Anne Stewart, Clearwater, FL
Geoffrey Strauss, Endicott, NY
Amy C. Strickland, Columbus, GA
Debbie Stuebbe, Bakersfield, CA
Ken Swiergosz, Toledo, OH
Bonnie Toone, Evansville, IN
Robert Torres, Anaheim, CA
Jeff VanArsdel, West Lafayette, IN
Jill Vicino, Darien, IL
Lina Volesky, West Bend, WI
Sandy Wainio, Durham, NC
David D. Weaver, Palmdale, CA
Irene Webb, Hackensack, NJ
Kathy Webb, Garland, TX
Gloria Wein, Aberdeen, NJ
Victoria Wenke, Burlington, WI
Donna Wheeler, Madison, IN
Patricia Williams, Tampa, FL
June Winkel, Plymouth, WI
Ann Wittmer, Waco, TX
Karen M. Wolf, West Carrollton, OH
William W. Wunrow, Green Bay, WI
Thomas F. YingLing, St. Marys, OH
Carol Young, Yorba Linda, CA
Cy Young, Garden Grove, CA
Cheri Zuccarelli, Phelan, CA

Table of Contents

Unit 1: Introduction to Accounting 1

Chapter 1 You and the World of Accounting 2

Reading Guide 2
Section 1.1 Exploring Careers 4
    Assessment 12
Section 1.2 Accounting Careers: The Possibilities are Endless! 13
    Assessment 19
Visual Summary 20
Review and Activities 21
Standardized Test Practice 22
Problems 23
Real-World Accounting Careers 25
Real-World Applications & Connections 26

Chapter 2 The World of Business and Accounting 28

Reading Guide 28
Section 2.1 Exploring the World of Business 30
    Assessment 36
Section 2.2 Accounting: The Universal Language of Business 37
    Assessment 41
Visual Summary 42
Review and Activities 43
Standardized Test Practice 44
Problems 45
Real-World Accounting Careers 48
Real-World Applications & Connections 49

Unit 2: The Basic Accounting Cycle 51

Chapter 3 Business Transactions and the Accounting Equation 52

Reading Guide 52
Section 3.1 Property and Financial Claims 54
    Assessment 57
Section 3.2 Transactions that Affect Owner's Investment, Cash, and Credit 59
    Assessment 65
Section 3.3 Transactions that Affect Revenue, Expense, and Withdrawals by the Owner 66
    Assessment 70
Visual Summary 71
Review and Activities 72
Standardized Test Practice 73
Computerized Accounting 74
Problems 75
Real-World Applications & Connections 82

| Chapter 4 | Transactions that Affect Assets, Liabilities, and Owner's Capital | 84 |

Reading Guide 84
Section 4.1 Accounts and the Double-Entry Accounting System 86
    Assessment 91
Section 4.2 Applying the Rules of Debit and Credit 92
    Assessment 98
Visual Summary 99
Review and Activities 100
Standardized Test Practice 101
Computerized Accounting 103
Problems 104
Real-World Accounting Careers 111
Real-World Applications & Connections 112

| Chapter 5 | Transactions That Affect Revenue, Expenses, and Withdrawals | 114 |

Reading Guide 114
Section 5.1 Relationship of Revenue, Expenses, and Withdrawals to Owner's Equity 116
    Assessment 122
Section 5.2 Applying the Rules of Debit and Credit to Revenue, Expense, and Withdrawals Transactions 124
    Assessment 130
Visual Summary 131
Review and Activities 132
Standardized Test Practice 133
Computerized Accounting 134
Problems 135
Real-World Accounting Careers 141
Real-World Accounting & Connections 142

| Chapter 6 | Recording Transactions in a General Journal | 144 |

Reading Guide 144
Section 6.1 The Accounting Cycle 146
    Assessment 149
Section 6.2 Recording Transactions in the General Journal 151
    Assessment 166
Visual Summary 168
Review and Activities 169
Standardized Test Practice 170
Computerized Accounting 171
Problems 172
Real-World Applications & Connections 180

Chapter 7   Posting Journal Entries to General Ledger Accounts   182

Reading Guide 182
- **Section 7.1** The General Ledger 184
  - **Assessment** 188
- **Section 7.2** The Posting Process 189
  - **Assessment** 196
- **Section 7.3** Preparing A Trial Balance 197
  - **Assessment** 202
- Visual Summary 203
- Review and Activities 204
- Standardized Test Practice 205
- Computerized Accounting 206
- Problems 207
- Real-World Accounting Careers 211
- Real-World Applications & Connections 212

Mini Practice   **Set 1:** Setting Up Accounting Records for a Sole Proprietorship   214

Chapter 8   The Six-column Work Sheet   216

Reading Guide 216
- **Section 8.1** Preparing the Work Sheet 218
  - **Assessment** 223
- **Section 8.2** Completing the Work Sheet 225
  - **Assessment** 231
- Visual Summary 233
- Review and Activities 234
- Standardized Test Practice 235
- Computerized Accounting 237
- Problems 238
- Real-World Applications & Connections 242

Chapter 9   Financial Statements for a Sole Proprietorship   244

Reading Guide 244
- **Section 9.1** The Income Statement 246
  - **Assessment** 251
- **Section 9.2** The Statement of Changes in Owner's Equity 252
  - **Assessment** 256
- **Section 9.3** The Balance Sheet and the Statement of Cash Flows 258
  - **Assessment** 265
- Visual Summary 266
- Review and Activities 267
- Standardized Test Practice 268
- Computerized Accounting 269
- Problems 270
- Real-World Accounting Careers 275
- Real-World Applications & Connections 276

Chapter 10 Completing the Accounting Cycle for a Sole Proprietorship — 278

Reading Guide — 278

Section 10.1 Preparing Closing Entries — 280
- Assessment — 290

Section 10.2 Posting Closing Entries and Preparing a Post-closing Trial Balance — 291
- Assessment — 295

Visual Summary — 296
Review and Activities — 297
Standardized Test Practice — 298
Computerized Accounting — 299
Problems — 300
Real-World Applications & Connections — 306

Chapter 11 Cash Controls and Banking Activities — 308

Reading Guide — 308

Section 11.1 Banking Procedures — 310
- Assessment — 316

Section 11.2 Reconciling the Bank Statement — 318
- Assessment — 327

Visual Summary — 328
Review and Activities — 329
Standardized Test Practice — 331
Computerized Accounting — 332
Problems — 333
Real-World Accounting Careers — 337
Real-World Applications & Connections — 338

Mini Practice Set 2: Completing the Accounting Cycle for a Sole Proprietorship — 340

Unit 3: Accounting for a Payroll System — 343

Chapter 12 Payroll Accounting — 344

Reading Guide — 344

Section 12.1 Calculating Gross Earnings — 346
- Assessment — 351

Section 12.2 Payroll Deductions — 353
- Assessment — 357

Section 12.3 Payroll Records — 359
- Assessment — 364

Visual Summary — 365
Review and Activities — 366
Standardized Test Practice — 367
Computerized Accounting — 368
Problems — 369
Real-World Applications & Connections — 376

Chapter 13   Payroll Liabilities and Tax Records   378

Reading Guide   378

Section 13.1 Journalizing and Posting the Payroll   380
  - Assessment   385

Section 13.2 Employer's Payroll Taxes   386
  - Assessment   390

Section 13.3 Tax Liability Payments and Tax Reports   391
  - Assessment   401

- Visual Summary   402
- Review and Activities   403
- Standardized Test Practice   404
- Computerized Accounting   405
- Problems   406
- Real-World Accounting Careers   410
- Real-World Accounting & Connections   411

Mini Practice Set 3: Payroll Accounting   413

Unit 4: The Accounting Cycle for a Merchandising Corporation   417

Chapter 14   Accounting for Sales and Cash Receipts   418

Reading Guide   418

Section 14.1 Accounting for a Merchandising Business   420
  - Assessment   424

Section 14.2 Analyzing Sales Transactions   425
  - Assessment   434

Section 14.3 Analyzing Cash Receipt Transactions   436
  - Assessment   445

- Visual Summary   446
- Review and Activities   447
- Standardized Test Practice   448
- Computerized Accounting   449
- Problems   450
- Real-World Applications & Connections   456

Chapter 15   Accounting for Purchases and Cash Payments   458

Reading Guide   458

Section 15.1 Purchasing Items Needed by a Business   460
  - Assessment   464

Section 15.2 Analyzing and Recording Purchases on Account   465
  - Assessment   472

Section 15.3 Analyzing and Recording Cash Payments   474
  - Assessment   481

- Visual Summary   482
- Review and Activities   483
- Standardized Test Practice   484
- Computerized Accounting   486
- Problems   487
- Real-World Accounting Careers   493

Chapter 16   Special Journals: Sales and Cash Receipts   496

Reading Guide 496
- **Section 16.1** The Sales Journal 498
  - **Assessment** 506
- **Section 16.2** The Cash Receipts Journal 508
  - **Assessment** 519
- Visual Summary 520
- Review and Activities 521
- Standardized Test Practice 522
- Computerized Accounting 523
- Problems 524
- Real-World Applications & Connections 528

Chapter 17   Special Journals: Purchases and Cash Payments   530

Reading Guide 530
- **Section 17.1** The Purchases Journal 532
  - **Assessment** 539
- **Section 17.2** The Cash Payments Journal 540
  - **Assessment** 554
- Visual Summary 555
- Review and Activities 556
- Standardized Test Practice 557
- Computerized Accounting 558
- Problems 559
- Real-World Accounting Careers 565
- Real-World Applications & Connections 566

Chapter 18   Adjustments and the Ten-Column Work Sheet   568

Reading Guide 568
- **Section 18.1** Identifying Accounts to be Adjusted and Adjusting Merchandise Inventory 570
  - **Assessment** 576
- **Section 18.2** Adjusting Supplies, Prepaid Insurance, and Federal Corporate Income Tax 577
  - **Assessment** 582
- **Section 18.3** Completing the Work Sheet and Journalizing and Posting the Adjusting Entries 583
  - **Assessment** 591
- Visual Summary 592
- Review and Activities 593
- Standardized Test Practice 594
- Computerized Accounting 596
- Problems 597
- Real-World Applications & Connections 602

Chapter 19 Financial Statements for a Corporation — 604

Reading Guide — 604

Section 19.1 The Ownership of a Corporation — 606
    Assessment — 611
Section 19.2 The Income Statement — 612
    Assessment — 620
Section 19.3 The Statement of Retained Earnings, Balance Sheet, and Statement of Cash Flows — 622
    Assessment — 628
Visual Summary — 629
Review and Activities — 630
Standardized Test Practice — 631
Computerized Accounting — 633
Problems — 634
Real-World Applications & Connections — 638

Mini Practice Set 4: Recording Business Transactions in Special Journals — 640

Chapter 20 Completing the Accounting Cycle for a Merchandising Corporation — 644

Reading Guide — 644

Section 20.1 Journalizing Closing Entries — 646
    Assessment — 650
Section 20.2 Posting Closing Entries — 651
    Assessment — 658
Visual Summary — 659
Review and Activities — 660
Standardized Test Practice — 661
Computerized Accounting — 663
Problems — 664
Real-World Applications & Connections — 670

Chapter 21 Accounting for Publicly Held Corporations — 672

Reading Guide — 672

Section 21.1 Publicly Held Corporations — 674
    Assessment — 679
Section 21.2 Distribution of Corporate Earnings — 680
    Assessment — 684
Section 21.3 Financial Reporting for a Publicly Held Corporation — 685
    Assessment — 688
Visual Summary — 689
Review and Activities — 690
Standardized Test Practice — 691
Computerized Accounting — 692
Problems — 693
Real-World Accounting Careers — 698
Real-World Applications & Connections — 699

Unit 5: Accounting for Special Procedures 701

Chapter 22 Cash Funds — 702

- Reading Guide 702
- **Section 22.1** The Change Fund 704
  - Assessment 708
- **Section 22.2** The Petty Cash Fund 709
  - Assessment 718
- Visual Summary 719
- Review and Activities 720
- Standardized Test Practice 721
- Computerized Accounting 723
- Problems 724
- Real-World Applications & Connections 730

Chapter 23 Plant Assets and Depreciation — 732

- Reading Guide 732
- **Section 23.1** Plant Assets and Equipment 734
  - Assessment 738
- **Section 23.2** Calculating Depreciation 739
  - Assessment 741
- **Section 23.3** Accounting for Depreciation Expense at the End of a Year 742
  - Assessment 750
- Visual Summary 752
- Review and Activities 753
- Standardized Test Practice 754
- Computerized Accounting 756
- Problems 757
- Real-World Accounting Careers 761
- Real-World Applications & Connections 762

Chapter 24 Uncollectible Accounts Receivable — 764

- Reading Guide 764
- **Section 24.1** The Direct Write-Off Method 766
  - Assessment 770
- **Section 24.2** The Allowance Method 771
  - Assessment 780
- **Section 24.3** Estimating Uncollectible Accounts Receivable 781
  - Assessment 785
- Visual Summary 786
- Review and Activities 787
- Standardized Test Practice 788
- Computerized Accounting 790
- Problems 791
- Real-World Applications & Connections 796

Chapter 25 Inventories — 798

- **Reading Guide** — 798
- **Section 25.1** Determining the Quantity of Inventories — 800
  - Assessment — 803
- **Section 25.2** Determining the Cost of Inventories — 804
  - Assessment — 809
- **Section 25.3** Choosing an Inventory Costing Method — 810
  - Assessment — 812
- Visual Summary — 813
- Review and Activities — 814
- Standardized Test Practice — 815
- Computerized Accounting — 817
- Problems — 818
- Real-World Accounting Careers — 823
- Real-World Applications & Connections — 824

Chapter 26 Notes Payable and Receivable — 826

- **Reading Guide** — 826
- **Section 26.1** Promissory Notes — 828
  - Assessment — 832
- **Section 26.2** Notes Payable — 833
  - Assessment — 841
- **Section 26.3** Notes Receivable — 842
  - Assessment — 844
- Visual Summary — 845
- Review and Activities — 846
- Standardized Test Practice — 847
- Computerized Accounting — 849
- Problems — 850
- Real-World Applications & Connections — 854

Mini Practice Set 5: Completing the Accounting Cycle for a Merchandising Corporation — 856

Unit 6: Additional Accounting Topics — 861

Chapter 27 Introduction to Partnerships — 862

- **Reading Guide** — 862
- **Section 27.1** Partnership Characteristics and Partners' Equity — 864
  - Assessment — 868
- **Section 27.2** Division of Income and Loss — 869
  - Assessment — 875
- Visual Summary — 877
- Review and Activities — 878
- Standardized Test Practice — 879
- Computerized Accounting — 881
- Problems — 882
- Real-World Applications & Connections — 886

Chapter 28   Financial Statements and Liquidation of a Partnership   888

Reading Guide 888
- **Section 28.1** Financial Statements for a Partnership 890
  - Assessment 892
- **Section 28.2** Liquidation of a Partnership 894
  - Assessment 898
- Visual Summary 899
- Review and Activities 900
- Standardized Test Practice 901
- Computerized Accounting 903
- Problems 904
- Real-World Applications & Connections 908

Mini Practice   Set 6: Completing the Accounting Cycle for a Partnership   910

Chapter 29   Ethics in Accounting   914

Reading Guide 914
- **Section 29.1** The Nature of Ethics 916
  - Assessment 920
- **Section 29.2** Ethics in the Accounting Profession 921
  - Assessment 925
- Visual Summary 926
- Review and Activities 927
- Problems 928
- Real-World Applications & Connections 930

APPENDIX A The Accrual Basis of Accounting A-2
APPENDIX B Federal Personal Income Tax A-12
APPENDIX C Using the Numeric Keypad A-18
APPENDIX D Advanced Accounting Concepts A-22
APPENDIX E Additional Reinforcement Problems A-27
APPENDIX F Answers to Section Assessment Problems A-45
CAREER SKILLS HANDBOOK C-1
GLOSSARY G-1
INDEX A I-1
INDEX B Real-World Applications and Connections I-12

UNIT 1

Introduction to Accounting

A Look Ahead

In Unit 1, you will learn about ways to explore careers and various careers in accounting.

Embark on a thrilling journey as you discover the diverse paths available, from forensic accountants to auditors, in the field of accounting and uncover the exciting opportunities that await you. This unit will provide valuable insights into the world of accounting careers, helping you align your passions and skills with your future aspirations. Get ready to uncover the perfect fit and set the foundation for a successful and fulfilling professional journey.

Keys to Success

Q Why are effective writing skills helpful when beginning a job search?

A Writing helps you develop ideas, express opinions, and learn about yourself. Developing lists and journalizing can help you organize your thoughts and goals. Effective writing skills are also essential when you apply for a job. When you give a prospective employer your résumé and cover letter, you are providing a list of your skills and work experience, and also a writing sample that shows your ability to communicate clearly.

English Language Arts

Writing Imagine yourself in five years. What type of job do you have? Where are you living? Write a one-page paper that describes how these questions might help you find a career that is right for you.

Focus on the Photo

By learning accounting, you can navigate the financial ups and downs of the business world, manage your personal finances, and even pursue an accounting career. *Have you ever seen financial information about a company? Did you understand what it reported?*

Chapter 1 • You and the World of Accounting   1

Chapter 1
You and the World of Accounting

Chapter Topics

1-1 Exploring Careers

1-2 Accounting Careers: The Possibilities are Endless!

    Visual Summary

    Review and Activities

    Standardized Test Practice

    Problems

    Real-World Accounting Careers

    Real-World Applications & Connections

Essential Question

As you read this chapter, keep this question in mind:
How can you set the career goal that is best for you?

Main Idea

A successful career choice begins with learning about yourself.

Chapter Objectives

Concepts	Analysis	Procedures
C1 Describe how personal skills, values, and lifestyle goals affect career decisions.	**A1** Find information about a variety of careers. **A2** Set career goals. **A3** Identify career opportunities in the accounting field.	**P1** Describe the types of businesses and organizations that hire consultants. **P2** Compare for-profit businesses and not-for-profit organizations.

Real-World Business Connection

McFarlane Companies

Inspiration and risk transformed former college baseball player Todd McFarlane from talented writer and graphic artist to iconic entrepreneur. After leaving Marvel Comics with other artists to form their own independent company, McFarlane earned a massive following with his *Spawn* and *Haunt* comic-book series, fueling plans to build a diverse line of products. Today, the McFarlane Companies is a multi-million-dollar business developing not just comics, but also toys, video games, films, and television shows.

Connect to the Business

The success of McFarlane Companies depends on more than just the work of artists and writers. All types of businesses extend opportunities to behind-the-scenes professionals, such as accountants, who ensure smooth business operations. Defining your skills, values, and career goals will help you learn about yourself and decide whether an accounting career is right for you.

Analyze

What type of job opportunities do you think exist at McFarlane Companies?

Focus on the Photo

Many people change careers throughout their lives, and others stick with one profession. Before Todd McFarlane created *Spawn* and owned companies that developed films, games, and toys, he was drawing comic book heroes in high school. **What jobs have you had? Were they in a field that interests you?**

SECTION 1.1
Exploring Careers

Film producer? Animator? Web site designer? History teacher? Pediatrician? Financial analyst? Environmental consultant? What career will you choose?

Choosing a Career

What Do You Want to Do?

Let's face it: Not many of us know what we want to do with the rest of our lives, especially when we're still in high school. You may be thinking, "I'll just take some liberal arts courses in college and something will come up" or "I'll just work for a while to make some money and then fall into a career." You know what? Something may come up, and you may just fall into a career; but if you ask people who really love what they do, odds are that they took some time to really get to know themselves and what turned them on.

Section Vocabulary
- skills
- values
- lifestyle
- personality
- personal interest tests
- networking
- foundation
- acknowledgment
- resources

Danielle and Steve have been friends since their freshman year. Like most students, they've been too wrapped up with classes, homework, and activities to think about what their lives will be like after high school. This all changed at the start of their senior year. Steve sees it as the end of a long haul, but Danielle sees it as the beginning of a whole new life.

Closing his locker door with his elbow, Steve turns to see Danielle smiling at him. "Hey Dani! How's it going?"

"Great," says Danielle, holding up her class schedule. "They say the last year in high school is always the best."

"What are you taking?"

"Let's see. I have chemistry, literature, algebra, speech, and accounting."

4 Chapter 1 • You and the World of Accounting

"Ouch, tough schedule! Did you say accounting? I would have never guessed you to be a number cruncher."

"I don't think I want to just 'crunch numbers.' I've always aced math, so I figured if I take accounting now, I'll know if I really like it enough to study it as a career. Besides, there are tons of ads in the paper for accounting jobs. If you're good at it, you can write your own ticket. Think about it: Companies need employees who know how to run their businesses. Plus it's not just sitting at a desk crunching numbers. My aunt is a financial advisor. She loves her job and travels all over the world."

"Slow down, Dani, it's only your senior year in high school. Don't you think it's a bit too soon to be thinking about careers?"

"The sooner the better, Steve!"

"Maybe you're onto something here."

Unlike a job, which is simply work for pay, a career is built on a **foundation** of interest, knowledge, training, and experience. Have you thought of what you may want your career to be? If not, you are not alone.

Assess Yourself in Terms of a Career Vision

Who Are You?

Before choosing a career, you'll want to do a little soul-searching. The more you know about yourself, the easier it will be to make career choices.

- What are your personal interests and skills?
- What are your values, and how will they affect your career?
- What lifestyle interests you?
- How will your personality affect a career choice?

You have the answers to the questions but just don't know it yet!

Chapter 1 • You and the World of Accounting 5

Skills and Traits	Career Example
Creative thinking	Fine arts and humanities careers, such as actor, artist, or musician
Knowing how to learn	Training or teaching careers, such as business consultant or training coordinator
Responsibility	Health-related careers, such as surgeon, dental hygienist, or home health aide
Friendliness	Hospitality and recreation careers, including cruise director, hotel manager, or park ranger
Honesty	Child-care workers, veterinarians, and other family and consumer science workers
Decision making	Financial planners, accountants, and other business and office careers
Analytical	Construction careers, including surveyors, general contractors, or electricians
Adaptability	Any of the public service careers, such as teaching, fire fighting, or serving in the armed forces
Self-control	Entrepreneur or manager
Self-esteem	Communications and media occupations, such as computer artist or book editor

Your Interests, Skills, and Traits

You're more likely to enjoy a career that uses your interests, **skills**, and traits. Skills are activities that you do well. Consider John King. When John was a kid, he built an elaborate fort in his yard. As a teen he studied major metropolitan architecture. When he finished high school, he worked at a graphics company making calendars. The pay was good, but John lost interest and enrolled in school to study his true passion, architecture. Today John is a partner in a successful Los Angeles architecture firm and loves his job.

Are you good at math? Do you write well? Are you creative? Do you like to meet lots of new people or enjoy building things? Are you good at solving problems and paying attention to detail? Everyone has different skills and abilities—it's the combination that gives you a unique selling point. The chart on page 8 lists some skills and traits that employers have identified as being valuable. Although these skills and traits are useful in various situations, the chart shows examples of careers that require them.

How many of these traits do you have? What other skills or competencies do you have? Make a list. It will come in handy when you begin to consider careers that interest you.

Values

One way to get to know yourself better is to examine your values. **Values** are the principles you live by and the beliefs that are important to you. Values are really about actions, not words. If you like to spend your free time volunteering at a local hospital or senior center, one of your values might be helping others.

What you value and believe may change as you get older. Most people, though, have a basic set of values that they follow throughout their life. As you read this section, think about your personal beliefs. Remember, values are actions, not aspirations. Which values are important to you? Can you think of careers that would benefit from these values?

Responsibility.

- Being responsible means being dependable and taking positive actions, such as showing up on time to take a friend to an appointment or honoring a commitment.
- If you value responsibility, you might think about a career as a supervisor or manager.

Achievement.

- You value achievement if you try to be successful in whatever you do.
- I know, you're thinking, "Who doesn't want to be a success?"
- The truth is that wanting and achieving are two different things.
- For example, if you take action to train outside of regular practices to make first string on the basketball team, you value achieving goals.

Relationships.

- If you especially value interacting with your friends and family, relationships are important to you.
- After all, sharing the joy of your accomplishments is half the fun.
- Those who value these types of connections might avoid occupations that require a lot of travel and might base their career decisions on the ability to live close to family and friends.

Compassion.

- Do you care deeply about people, animals, or special causes?
- For you, a career that allows you to show your compassion may outweigh all the money in the world.
- For example, if you love being around animals, you might enjoy a career in veterinary medicine or the marine sciences.

Courage.

- Courage is not just being brave in the face of physical danger.
- Courage is also about overcoming other fears.
- For example, it takes courage to make a speech to the whole school, even if you are fearful or nervous.
- If you can put your beliefs on the line, you may be headed for a career in politics or law.

Recognition.

- If receiving **acknowledgment** and appreciation of your work is important to you, then you value recognition.
- You might consider a career as a novelist or a television news reporter.
- Many people share the same values, but how they apply them is unique to each person.
- To some, courage may be accepting the challenge of a job they know little about.
- To others, it may be turning down the big money to do something they really love.
- Think about your values.
- What can you learn from your values that will help you narrow your career choices?

Lifestyle Goals

Your **lifestyle** is the way you use your time, energy, and **resources**. For example, many people devote themselves to work, earn lots of money, and put off the benefits of free time until they are older. Others accept smaller paychecks and work fewer hours to spend more time with family and friends now. If you want to work as a business manager or accountant for a professional sports team, there will be more opportunities if you live in a city that has such a team. For some positions, remote working arrangements can be negotiated allowing workers the flexibility to work from varied locations. If you want to be an actor, get ready for life on the road.

- What's really important to you?
- Do you want to go for the big bucks?
- Do you want to live in a big city with endless activities or a small town where everyone knows your name?
- Do you want to collaborate with a group of people or to work solo?

Make a list of how you would like to spend your time, energy, and resources. These are your lifestyle goals. Try to focus on careers that closely match them.

Personality Traits

Imagine what it would be like if all your friends had the same personality. What if they were all shy or serious? Even worse, what if they all had the same sense of humor? It's a good thing we each have our own **personality**—a set of unique qualities that makes us different from all other people.

What is your personality? Are you confident, dependable, funny, friendly, sympathetic? Be honest. Do you prefer being with people or spending your time with things, such as reading books or working with computers? You probably wonder what this has to do with accounting. Well, your personality affects your preferences for working with data, people, or things.

As you collect information about yourself, you'll want to complete a personal career profile such as the one shown in **Figure 1–1.** Use this profile to help you evaluate whether the careers you're considering match your skills, interests, values, and personality.

Reading Check

Recall

How can your career choice affect your lifestyle?

Making Career Decisions

How Can Networking Help You Develop Career Goals?

Once you have a clear vision of yourself and how you want to live, you are ready to research careers and set goals.

Research the Possibilities

How do you find the right career for you? Here are some places to start:

Guidance Counselors.

- School counselors do a lot more than just show you the quickest way to get from your homeroom to the cafeteria.
- They can help you identify the things you like to do.
- One way they accomplish this is through testing, called **personal interest tests**, which help you identify your preferences.

Contacts.

- **Networking** is making contacts with people to share information and advice.
- Find out about specific careers through networking.

Library.

- Print materials on every career imaginable can be found in your public library.
- Try books on careers or magazines that focus on your interests like *House & Garden, Metropolitan Home,* or *Forbes.*

Personal Career Profile Form

Name *Darla Johnson*   Date *December 12*   Career *Marketing Manager, Music Industry*

Career Profile

Your Values I believe in equal opportunities for all people. I like to be creative.	**Career Values** All kinds of people work in the music industry. As a marketing manager, I would be able to use my creativity, as well as work with other creative people.
Your Interests I have a large collection of jazz and blues CDs and keep up with up-and-coming artists that are featured on independent labels. I love getting together with friends and having parties.	**Career Duties and Responsibilities** As a marketing manager, I would make contacts with music stores and distributors, labels, and artists. I might send out press releases, arrange for artist appearances, and map out company marketing strategies to increase profits.
Your Personality I am very outgoing and get bored with sitting in class or reading. I have a great imagination and love group discussions.	**Personality Type Needed** A marketing manager must work well with people. Sharp communications skills and attention to details are important.
Skills and Aptitudes My best subject is history, and I am president of my school's debate team. I am not big on writing letters or grammar but love to communicate in person.	**Skills and Aptitudes Required** Good verbal communication skills are essential for a marketing professional. Although history may not be particularly important, good perceptions of what works and what doesn't work might be important.
Education/Training Acceptable I would be interested in learning more about getting a business degree, but also believe that if I could get in on the ground level in the music industry and learn the ropes, I could be successful as well.	**Education/Training Required** I suppose a degree in marketing might open a lot of doors for me. A knowledge of how marketing and accounting fits into the big picture of a music corporation would definitely help.

Figure 1–1 Personal Career Profile Form

Internet

- The Internet is a great source of educational and career information.
- Check out ideas for putting together the perfect résumé or browse through job opportunities at companies you are interested in.

Organizations.

- Professional organizations are groups of people who have common career interests.
- You can learn about interesting careers by getting to know the American Institute of Certified Public Accountants (AICPA) and the Institute of Management Accountants (IMA).

Set Career Goals

Once you have a clear vision of your interests and the types of careers you want to pursue, it's time to put a plan into action.

Map Out a Plan.
- For starters you will need to make a list of the careers you have researched and compare possibilities.
- The easiest way to do this is to make a chart. List the careers you are interested in and list your personal information.
- Then see which careers align with your personal career profile information.

After selecting the career choices that look the most promising, decide which one you will pursue. Reaching your ultimate career goal is not going to happen overnight. There are many intermediate goals to achieve along the way. Here are some steps to help you achieve your career goals:

1. Decide on a long-term goal. Learn as much as you can about careers that interest you. Visit job fairs to obtain information.
2. Identify actions that will lead to the long-term goal. What skills, education, or training will you need? Make plans early.
3. Take action! Make your plans and put them to work.
4. Diversify your skills. Experience different work environments.
5. Realize your long-term goal. Setting and implementing these steps will help you achieve a career that you desire.

Education.
- Most careers require some education or training beyond high school.
- Unfortunately, deciding to further your education and finding the cash to pay for it are two different things.
- For ideas on help in paying tuition, you can turn to many of the same places you turned to for career information: books, the Internet, and your friend, the guidance counselor.

On-the-Job Training.
- Imagine taking only your favorite school subject and getting paid for it! That's how on-the-job training works.
- Suppose you're thinking about eventually opening your own accounting firm.
- While in school, you might work as an office assistant at a local accounting firm to learn how accounting services are provided.
- This is a great way to find out if accounting is the career for you, and you'll be paid for your efforts.

Internships.
- Another way to obtain career experience is to work as an intern.
- Many companies offer summer or longer internships to students. Some offer modest pay.
- Successful interns are often offered positions in the company once their internships are completed.
- The key is to become so valuable that they miss you when you are gone!

SECTION 1.1
Assessment

After You Read

Reinforce the Main Idea

Draw a diagram like the one shown here. Show how people your age may benefit in their careers by first learning about themselves. Add answer circles as needed.

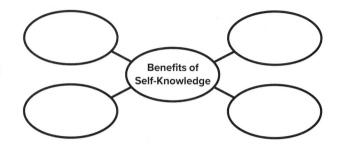

Problem 1 Studying Yourself

Instructions Think about the things you like to do and your particular skills. Make a list of at least five personal interests or skills. After you complete your list, use the career resource materials described in this section to identify one or more careers that match each interest or skill. Choose one career and write a one-page description of how your skills and interests fit this career.

Problem 2 Gathering Career Resources

Instructions Use the personal career profile form in **Figure 1–1** as a guide, and compare three careers that you find interesting. Use the resources mentioned in this section to gather information: guidance counselors, networking, print materials, the Internet, and professional organizations. You may find other references in your school or local public library. After completing your comparison, write a brief summary; identify which of your choices you prefer and explain why.

Math for Accounting

Congratulations! The good news is that you've decided to attend a local community college after high school. The challenge is that tuition for an 18-week semester is $1,820. You estimate travel expenses of $20 per week and $300 for books and supplies. You plan to work 20 hours a week, earning $7.75 per hour, and you will apply all your paychecks toward college expenses. Social Security and income taxes will take about 15 percent of your earnings.

1. What is the total cost for a semester?

2. Assume that you have already saved $300 toward your first semester's costs. Approximately how long will it take to save enough to cover your tuition and books for one semester?

SECTION 1.2
Accounting Careers: The Possibilities are Endless!

Some people believe accounting is boring. There, we said it. It's a pretty safe bet that you did not enroll in an accounting course because you thought it would lead to a career in the spotlight. How often do you see a film star starring in an action thriller about a jet-setting accountant? However, who do you think develops and approves the budget for films that go into production? Who do you think advises film stars about how to invest the hefty salary they make? You got it—someone just like you, who is good with accounting concepts and knows how to handle money.

The point is: Accounting can be a lot of things you would never have imagined. So, if an accounting career matches up with some of your interests and goals, hang in there. We promise there is a lot more to it than crunching numbers!

Section Vocabulary

- accountant
- accounting clerk
- for-profit business
- not-for-profit organization
- public accounting firm
- audit
- certified public accountant (CPA)
- dynamic

The Changing Horizon

How Is Accounting More Than Arithmetic?

Sure, some careers in accounting can be a little dry, but there are many more that are **dynamic** and exciting. Think of any business—Warner Music, Nike, the Hard Rock Cafe. They all look to accountants to help run their businesses. Think of your favorite celebrities—maybe Robert Pattinson or Beyoncé. Most have financial advisors. Movie producers hire accountants to track production costs. Publishers of magazines, such as PC World and Sports Illustrated, depend on accountants to work with national advertisers to keep things running smoothly. In this section you will identify career opportunities in the accounting field.

Accounting is not just adding and subtracting. An **accountant** handles a broad range of responsibilities, makes business decisions, and prepares and interprets financial reports. The activities of accountants affect business and industry. If you are good, the sky is the limit.

Types of Organizations

What Career Opportunities Exist in Accounting?

If you still picture accountants huddled over pages of numbers in back offices, the following scenarios should set you straight.

For-Profit Businesses

"Okay, everybody. Show time in five minutes." The announcement comes as the members of the band adjust guitar straps and prepare to go out on stage. A crowd of 30,000 fans, dropping thirty-five bucks per person, waits for their entrance. Drew Taylor jokes with the band just offstage and then watches as they head out to the fans' applause.

Drew gets a rush being backstage amid all the excitement of a live show. As the financial assistant to the band's business manager, he often spends time at the performances. For Drew, working with numbers came as naturally as his love for music. He studied accounting throughout high school and college, and in his spare time he dabbled with the guitar and hung around recording studios with friends who were in the music business.

It wasn't long before Drew started working as an accounting clerk at a studio. An **accounting clerk** is an entry-level job that can vary from specializing in one part of the system to doing a wide range of tasks. Drew's accounting clerk experience and networking efforts landed him a position with one of the music industry's top business managers. Drew still dreams of recording his own music one day, but for now, he's quite content helping orchestrate the financial security of his music idols.

Entertainers perform to earn a profit. **For-profit businesses** operate to earn money for their owners. Profit is money that is kept after all expenses of running a business have been deducted from the income. It is one way of measuring the company's success. The majority of businesses in the United States are for-profit businesses.

Today, computers handle much of the basic accounting work, freeing accountants to do more planning for future operations. Computer technology means that accountants are no longer tied to a desk or an office. If you worked for artists such as Taylor Swift, Jay-Z, or Radiohead, you would probably spend part of your time traveling. Laptop computers and wireless Internet connections allow you to work as you travel to other cities and countries.

In order to explore accounting career opportunities, you need to compare the various types of business organizations. Not all organizations are for-profit businesses.

Not-For-Profit Organizations

"Listen up, folks! Here's the draft of the news release we're sending out tomorrow," shouts Darin Korman, waving a stack of papers in the air. "We need to get our position before the public while Congress is still considering the environmental legislation. Questions?"

"What else are we doing to alert voters to the potential value of this legislation?" a team member asks from the back of the room.

Darin, the team leader, opens a folder. "Here's our total plan. We begin filming a TV spot tomorrow. Next week, the art for our magazine ads will be finished, and they'll run in six different magazines. We'll also post a call-to-action on our Web site asking visitors for their support. We can all thank Maya for putting together a budget for the media campaign."

Maya Cruz beams as her boss Darin describes a typical campaign put together for a group—like Sierra Club or Audubon Society—that works to protect and preserve the environment. Such groups operate as **not-for-profit organizations**, also known as *nonprofit organizations*. These organizations operate for purposes other than making a profit. When Maya was very young, her parents founded a similar group dedicated to cleaning up the local waterways. Maya helped her parents in the difficult task of soliciting donations within their community. As she grew older, Maya pursued the study of accounting with the sole purpose of continuing her involvement in environmental causes.

It is important to identify the various accounting functions involved with each type of business organization. Most not-for-profit organizations have the goal of balancing their income with expenses rather than earning a profit for owners or investors. Some not-for-profit organizations, like United Way or Boy Scouts of America, get their income from donations. Other organizations get income through tax dollars. Government agencies, such as your school or a federal agency, fall under this category.

There are thousands of opportunities in the not-for-profit arena. Like Maya did, you may prefer to combine your interest in accounting with a cause that you hold close to your heart.

Reading Check

Recall

What is profit?

Public Accounting Firms

"It seems like all the ducks are in a row." Jana Passeno takes another look at the Whitley Company accounting reports on her desk.

"Yes, we have identified the last few questions we need to go over with Whitley's accounting manager," Greg Hally says as he finalizes a short list.

Jana and Greg work for the public accounting firm Radcliff & Pratt. **Public accounting firms** provide clients a variety of accounting services including the independent audit. An **audit** is the review of a company's accounting systems and financial statements to confirm that they follow generally accepted accounting principles.

An *independent audit* is not done by company employees. For example, the accountants who work at Whitley cannot do an independent audit. Instead, it is done by certified public accountants who work for public accounting firms. A **certified public accountant (CPA)** is an accountant who has met certain education and experience requirements and passed a national test. Like doctors and lawyers, CPAs are licensed by the states.

Radcliff & Pratt is a small firm, but many certified public accountants work for the four largest accounting firms known as "The Big Four": Ernst & Young, Deloitte & Touche, KPMG, and PricewaterhouseCoopers.

Jana feels pride in her job. When companies such as Coca-Cola or General Electric sell their stock on a U.S. stock exchange, potential buyers depend on audited financial statements to make decisions. Once Jana signs off on the financial statements of a company, the public has greater assurance that the information is correct.

Public accounting firms have many job opportunities besides auditing. Possibilities include financial planning and preparing tax returns. You might be interested in a job as a forensic accountant. *Forensic accountants* take cases that involve issues like fraud or employee theft. Did you know that forensic accountants have played a major role in tracking down suspected terrorists?

See **Figure 1–2** on pages 17 and 18 for information about accounting careers. From entertainment to health care, conservation to entrepreneurship, the career possibilities in accounting are endless!

Figure 1–2 Careers in Accounting

Career	Education Requirements	Roles	Responsibilities
Certified Public Accountant	All states require the CPA candidate to pass the CPA exam to qualify for a certificate and state license. Most states require CPAs to complete a certain number of continuing professional education seminars, courses, or conferences before their license can be renewed.	Advise companies, board of directors, and compliance committees on establishing control systems and developing audit methodologies to detect potential noncompliance with accounting standards and procedures.	To exercise professional and moral judgment in all activities, maintain public trust, exhibit ethical behavior.
General Ledger Accountant	A college education is usually required depending on the size of the organization. For small organizations a two-year degree in accounting may be sufficient, although a four-year college degree is preferred.	Ensure the accuracy and completeness of the company's accounting records.	To ensure that all business transactions have been properly accounted for, accurately recorded, and completely reflected on the company's books. The main responsibility of the general ledger accountant is to analyze and maintain the general ledger for preparation of financial statements.
Management Accountant	Minimum of a bachelor's degree in accounting or a related field. Management accountants, unlike CPAs, do not serve the public directly, are not required to be licensed, and there are no state minimum requirements for education.	Record and analyze financial information <u>inside</u> the organization. Their role is to serve the needs of managers running the organization.	To provide cost and asset management, establishing production standards, performance evaluation, segment reporting, and planning and control.
Auditor	Minimum of a bachelor's degree in accounting or a related field. Auditors are usually CPAs and must meet educational standards set forth by state boards of accountancy and the American Institute of Public Accounting.	Provide statement and budget analysis, investment planning, technology advancement, and tax avoidance strategies.	To provide independent reports on clients' financial statements to ensure that statements follow Generally Accepted Accounting Principles (GAAP).

Career	Education Requirements	Roles	Responsibilities
Government Accountants	Minimum of a bachelor's degree in accounting or a related field. Some become a Certified Government Financial Manager (CGFM).	Create, maintain, examine, audit, and analyze the accounting records of government agencies and private businesses which are regulated by the government. They verify the proper collection and spending of taxpayer money.	To create budgets, examine revenue and expenses, and analyze publicly funded programs. They must make sure that money collected through taxes and other forms of revenue are accounted for. They must also examine all expenditures to verify the money is spent in accordance with budget allocations and all legal requirements are met.
International Accountants	Minimum of a bachelor's degree in accounting or a related field. Some acquire additional certifications such as the FAIA (Fellow) or AAIA (Associate) membership in the Association of International Accountants.	Provide the knowledge and skills in financial accounting, financial analysis, auditing, taxes, and cost accounting for global business activity.	To align accounting functions with the International Accounting Standards Board as well as the accounting standards of foreign countries. All accounting practices and procedures must meet the requirements of each country where business is being conducted.
Forensic Accountants	Higher education degree in accounting or a professional degree (masters degree with 150 college credit hours) in accounting Typically forensic accountants are Certified Public Accountants (CPA) and/or a Certified Fraud Examiner (CFE)	Investigate and analyze financial evidence; develop software to assist in the analysis and presentation of financial evidence; communicate findings in the form of reports, exhibits and collections of documents; and assist in legal proceedings, including testifying in court as an expert witness and preparing visual aids to support trial evidence.	To recover money earned through crime. To investigate and analyze financial evidence. To assist in legal proceedings, including testifying in court as an expert witness and prepare visual aids to support trial evidence.

Figure 1–2 (continued)

SECTION 1.2
Assessment

After You Read

Reinforce the Main Idea

Create an organizer like the one shown here. List at least three accounting career opportunities in each type of organization.

Career Opportunities		
For-Profit Businesses	Not-for-Profit Organizations	Public Accounting Firms

Problem 3 Checking Out Accounting Careers

Instructions Using the resources described in Section 1.1, research possible careers for people with accounting degrees. List at least five different careers and the formal training and work experience needed for each. Choose one career as your preference and write a paragraph describing why this career appeals to you.

Problem 4 Matching Interests and Careers

Instructions Using the three career examples described in this section, make a list of the personal interests and skills of the accountants described in each situation. Compare the list to your own interests and skills. Then think of three types of businesses (or actual companies) for which you might want to work. How would you learn about accounting career opportunities in those companies? Aside from pursuing needed training or education, what else would you do to prepare to work in that career? Share your ideas in class.

Problem 5 Researching Public Accounting Firms

Instructions Surf the Internet or conduct research in your library to find information about "The Big Four" accounting firms. Create a table of information about each firm and the services the firm provides.

Problem 6 Interviewing Accountants

Instructions Interview members of the accounting field to investigate entry-level job requirements, career tracks for the profession, and projected trends for the future. Write a short report about your findings.

Math for Accounting

Assume that the federal government pays new accountants without a master's degree an annual salary of approximately $23,500. The starting salary for new accountants with a master's degree is approximately $35,500. If you spent $21,000 to attend graduate school, how many months would you work at a new job in the federal government before the difference in salary pays for your graduate school expenses?

Chapter 1
Visual Summary

Concepts
Values that play a vital role in choosing a career.

- Responsibility
- Achievement
- Relationships

- Compassion
- Courage
- Recognition

Analysis

Set Career Goals
- Map out a plan.
- Consider the education required.
- Find a part-time job to get on-the-job training.
- Look for internships in your field of interest.

Procedures

Career Opportunities in Accounting
- Auditing
- Financial Planning
- Preparing Tax Returns
- Forensic Accounting

Chapter 1
Review and Activities

Answering the Essential Question

How can you set the career goal that is best for you?

Setting a career goal takes time, patience, exploration, and experimentation. Chances are the first goal you set will change as you find out more about yourself and the world of work. To help you explore your interests, list some careers that sound interesting to you. Under each item in your list, write the pros and cons that you see right now. Revisit the list occasionally to revise the pros and cons and add or erase careers. This can lead to a better understanding of career choices that might be right for you.

Vocabulary Check

1. **Vocabulary** Divide the vocabulary words into groups of related words. Label each group to explain why the words belong together.

   - foundation
   - skills
   - values
   - acknowledgment
   - lifestyle
   - personality
   - resources
   - personal interest tests
   - networking
   - dynamic
   - accountant
   - accounting clerk
   - for-profit business
   - not-for-profit organization
   - public accounting firm
   - audit
   - certified public accountant (CPA)

Concept Check

2. **Evaluate** Name the six values that most people share. Select the two that you feel are most important to a career and explain why you feel this way.
3. **Predict** List the five sources of career information. What kind of information would you expect to find in each?
4. Describe the five steps in mapping out a plan to achieve your career goals.
5. **Analyze** What does an accountant do? Explain why accountants are important to business and industry.
6. What task can be performed only by certified public accountants employed by public accounting firms?
7. Explain the differences between for-profit and not-for-profit organizations.
8. **Math** Corey Murphy earns $11.80 an hour for a regular 40-hour week at the pet store. His overtime pay is 1½ times his regular hourly rate. Last week Corey earned $613.60 in total pay. How many hours of overtime did he work?
9. **English Language Arts** Write a short report about yourself. Focus on your reasons for taking this course and how it fits with your goals and, if applicable, with your plan for a future career. Include one paragraph each on the following areas: 1) your interests, skills, and traits; 2) your values; 3) your lifestyle goals; and 4) your personality traits.

Chapter 1
Standardized Test Practice

Multiple Choice

1. Which of these is a good source for finding more information about making a good career choice?

    a. The library

    b. The Internet

    c. Professional organizations

    d. All of these

2. A _____ is an accountant who has met specific education and experience requirements and has passed a national test.

    a. accounting clerk

    b. business manager

    c. financial auditor

    d. CPA

True or False

3. Accountants mostly spend their work hours adding and subtracting numbers.

Short Answer

4. Accountants routinely _____ a company's accounting systems and financial statements to make sure they follow generally accepted accounting principles.

Extended Response

5. What kinds of information make up a person's value system?

6. As you read in this chapter, there are many different career opportunities in accounting, including those described in **Figure 1–2** on pages 17 and 18. In your own words, explain how your interests, skills, and traits would match with the roles and responsibilities of one of these accounting careers. Then explain the steps you might take to meet the career's educational requirements.

Chapter 1
Problems

Problem 7 Researching Careers in Your Library

It is never too early to begin researching careers.

Instructions Complete your research project using these steps.

1. Using a reference book (such as the *Occupational Outlook Handbook*), choose a career area that interests you.
2. Research the skills, education, and experience you would need to work in that career area. Your research can utilize any number of resources including books, magazines, and the Internet.
3. Write a brief profile of your chosen career, including a description of jobs in the field, education or training requirements, potential earnings, and working conditions.
4. Present your profile to the class.

Problem 8 Researching Careers in Your Local Newspaper

Instructions Follow these steps to organize employment information.

1. Review the employment ads in a local or regional newspaper.
2. Collect information on at least 10 job titles plus the skills and education required for each. Include at least two job titles related to the accounting field.
3. Present your information in table format. If you have access to a word processing program, use it to create and print your table.

Problem 9 Assessing Your Skills and Interests

Instructions Complete an assessment of your personal skills and interests by answering the questions in your working papers. Use the survey included in the working papers, or you can ask your guidance counselor to administer a personal interest test. Using the results of the test, find at least three careers that match your skills and interests.

Problem 10 Working with Others

Instructions As an accountant for a large business, you might be put in charge of training new hires in the accounting department. Or you might be asked to discuss project cost overruns with a department manager or to present operating results to senior managers. In these situations you need skills other than just your accounting knowledge. For each situation make a list of the skills needed; then decide whether you have those skills and, if not, how you plan to acquire them.

Chapter 1
Problems

Problem 11 Summarizing Personal Traits

Instructions Sometimes you can learn about yourself by asking other people how they see you. Ask at least 10 people to name three words they think describe you. Ask them to give you descriptive words such as these:

- Dependable
- Fun
- Quick-thinking
- Decisive

Do not just choose friends. Ask teachers, co-workers, relatives, and others who are willing to give you an honest opinion. Make a list of each person's descriptions; then summarize your findings by identifying the five characteristics or traits that were mentioned most often. Do these descriptions match your self-perception? Why or why not?

Problem 12 Gathering Career Information

Instructions Arrange to interview someone who currently works in a career area that interests you. Before the interview prepare a list of the questions you want to ask. You may wish to cover the following topics:

- What are the major tasks that you perform?
- What do you enjoy most about your job?
- Is there much variety in your work?
- What specific skills are involved?

After the interview write two or three paragraphs describing your interview and the information you learned. Explain how this information will help you choose a career.

Problem 13 Exploring Careers in Accounting

Instructions Choose a local company that interests you, and find out who works as an accountant for the company. Call and ask whether you can observe the person at work for part of a day. Write a summary of your observations, and share the information with the class.

Problem 14 Exploring Global Careers

Instructions Many U.S. businesses operate in the global economy, which means they need accountants who understand international business. Find a local company that imports or exports products to or from one or more countries. Interview the accounting manager, and find out the skills, formal study, and personal traits he or she looks for in an accountant who works in international accounting. Write a short report about your findings.

Real-World Accounting Careers

Samantha Roettker
Luxottica Retail

Q What do you do?

A I am responsible for making sure the inventory records of our accounting system are accurate. My team and I also work to build better methods and achieve greater efficiency in our processes.

Q What are your day-to-day responsibilities?

A I review and approve all journal entries and account reconciliations to ensure the financial statements are correct. I work with my managers to discuss new accounting pronouncements and determine how they will impact our financial statements. Then, we develop a plan to incorporate such changes. I also interact with other departments to help them forecast, understand variances and make plans for future inventory buying.

Q What factors have been key to your success?

A Dedication and the willingness to put in hard work in order to learn more and get results.

Q What do you like most about your job?

A I'm fortunate to work with talented people from whom I'm continuously learning. Accounting is a field that is always changing, so I'm constantly learning to adapt to changes that affect what we do.

Career Facts

Real-World Skills
Attention to detail; strong communication skills

Training And Education
Bachelor's degree in accounting is required; master's degree and/or CPA license is beneficial

Career Skills
Inventory accounting is a specialized area, so experience working for a manufacturer is important

Career Path
Start as a staff accountant, then move to senior accountant and accounting manager

Tips from Robert Half International

Before a job interview, learn as much as you can about the company by conducting research on the Internet and talking to members of your network. The information you uncover will enable you to identify specific ways you can contribute to the organization's success.

College and Career Readiness

Use the Internet to research the most common questions employers ask during job interviews. Choose five questions and think about how you might respond. Write your answers in a one-page paper.

Real-World Applications & Connections

CASE STUDY

Career Advice

Sean Smith is a senior in high school. He is taking an accounting course this year because he wants to be an accountant. Sean likes working with numbers, but he also likes working with people. He plans to go to college but doesn't know which ones offer accounting programs and what costs are involved.

ACTIVITY Take the role of Sean's career counselor.

Write a one-page report advising Sean. Include the following in your report:

- Information on how to find the resources he needs to choose a college.
- Details on how to set education and career goals.
- Steps Sean needs to take over the next several months.
- A list of resources Sean might use to learn more about accounting careers, colleges, and financial help.

Success Skills

Keys to Success

Locate and read a current magazine article featuring a successful entrepreneur. Web sites for periodicals like *Forbes* or *Inc.* are a good place to start. Consider what the entrepreneur you chose did that made his or her business a success, and what character traits and skills might have helped get the business off the ground. Write a summary of the article, and explain why you think the person succeeded.

Career Wise

Human Resources Director

Once a company grows to a certain size, it needs a dedicated person whose job is to take care of the needs of all the other employees. This person, the human resources director, is in charge of employee benefits such as insurance and vacation days; he or she is also involved in the hiring (and sometimes firing) of employees.

Research And Share

a. Use the Internet to learn more about the role of human resources in a business. How do human resources departments hire new employees and promote, transfer, and release current employees? Which laws and regulations affect human resources departments? Compare employee benefits that may be available through an employer.

b. As a class, discuss skills and characteristics that you think would be useful for a human resources director to have.

Spotlight on Personal Finance

Your Career

In this chapter you considered how interests, values, and lifestyle affect your career goals. You can learn more about what you want in a career from personal assessment tests. Meet with your guidance counselor or try an Internet search engine to learn more about these tests.

Activity

Imagine you have two job offers to work as an accountant. One is for a for-profit business. The work is not very exciting, but the company pays a high salary. The other job is at a not-for-profit organization dedicated to a cause that is important to you, but it pays about one-half as much. Based on what you discovered in your self-assessment, which job fits your goals? Why?

A Matter of Ethics

Padding a Résumé

Part of landing a great job is putting together a résumé that effectively represents your skills and background. Imagine that BMW has just opened a new regional headquarters in your area. You would like to work for BMW as a payroll clerk, but you're afraid you don't have the right qualifications. A friend suggests that you "change" your résumé to make yourself look better.

Activity

1. What are the ethical issues?
2. What are the alternatives?
3. Who are the affected parties?
4. How do the alternatives affect the parties?
5. What would you do?

Chapter 2

The World of Business and Accounting

Chapter Topics

2-1 Exploring the World of Business

2-2 Accounting: The Universal Language of Business

Visual Summary

Review and Activities

Standardized Test Practice

Problems

Real-World Accounting Careers

Real-World Applications & Connections

Essential Question

As you read this chapter, keep this question in mind:

What role does accounting play in the free enterprise system?

Main Idea

The accounting system produces information used by businesses to make decisions.

Chapter Objectives

Concepts

C1 Describe profit, risk-taking, and entrepreneurs.

C2 Describe service, merchandising, and manufacturing businesses.

C3 Compare the sole proprietorship, partnership, and corporate forms of business.

Analysis

A1 List the advantages and disadvantages of each form of business organization.

A2 Describe the purpose of accounting.

Procedures

P1 Explain financial and management accounting.

P2 Describe the three basic accounting assumptions.

Real-World Business Connection

Olympus

Olympus pretty much wrote the book on developing cutting-edge camera technology. Though most people know the company as a producer of high-end equipment for a select group of photo-hobbyists, Olympus actually began more than 90 years ago as a microscope manufacturer. It later expanded to design cameras with great success. With increased consumer demand, Olympus saw the need for lighter, more compact camera equipment. In 1959, Olympus focused on a truly innovative design, the Olympus Pen F, which was the smallest and lightest (and world's first) half-frame single lens reflex camera. Olympus later developed the first digital camera with live-view technology, which is now a feature in all digital cameras.

Connect to the Business

Olympus competes with companies like Nikon and Canon, the other leading camera manufacturers. Innovation is the edge it needs to earn the rights to lead the industry, and with that edge, Olympus zoomed in on the competition.

Analyze

In a free-enterprise system, competition is what drives the economy and the practices of companies within that system. What are some ways that you see businesses compete?

Focus on the Photo

Olympus and its competing companies offer many choices to consumers seeking quality digital cameras in a free enterprise system. ***How do you think accountants at Olympus use their knowledge to help manage the business?***

SECTION 2.1
Exploring the World of Business

In this chapter you will learn about different types of businesses and how they are organized. All businesses need information that is generated by accounting systems in the form of reports. You will learn about the different types of accounting and accounting reports.

The Environment of Business

What Makes Up the Business Environment?

When you hear the word *business*, what do you think of? Nike, Microsoft, General Motors, IBM, Coca-Cola, or American Airlines? Large businesses like these are certainly players in the world of business today. Your neighborhood convenience store, clothing boutique, video rental store, and grocer are also important contributors in our free enterprise system. In a **free enterprise system**, people are free to produce the goods and services they choose. Individuals are free to use their money as they wish: spend it, invest it, save it, or donate it. Business owners in this system must compete to attract the customers they need to continue operating.

One measure of success in a business operation is the amount of profit it earns. The amount of money earned over and above the amount spent to keep the business operating is called **profit**. Businesses that spend more money than they earn operate at a **loss**. Whatever its size, a business must do two things to survive. It must operate at a profit, and it must attract and keep an individual willing to take the risk to run it.

Section Vocabulary

- free enterprise system
- profit
- loss
- entrepreneur
- capital
- service business
- merchandising business
- manufacturing business
- sole proprietorship
- partnership
- corporation
- charter
- attitude
- incurs
- sufficient

The Need for Profit

The continued operation and success of a business requires profit. Your local pizza parlor must pay for raw materials (flour, sauce, cheese, toppings), equipment (mixers, ovens, freezers), employees, utilities, and rent. The selling price of the pizza must be high enough to cover all costs. Once the owner pays these costs, the money left over is profit.

The Need for a Risk-Taker

In a business many people play different roles in daily operations, planning, and management. Inventors create ideas for new products or services. Investors provide money to help businesses get started. Employees supply the labor needed to operate the business. Managers supervise and plan.

Role of Entrepreneurs.

- An **entrepreneur** is a person who transforms ideas for products or services into real-world businesses.
- Owning a business can offer flexible schedules, self-direction, and financial gain.
- Yet business ownership is not free from risk.
- See **Figure 2–1** on page 32 for some pros and cons of entrepreneurship.

Figure 2–1 Entrepreneurship: Pros and Cons

Pros	Cons
• You are your own boss. • You create opportunities for earning money. • You create and control your work schedule. • You choose the people you serve. • You select the people who work with you. • You benefit from the rewards of your own hard work. • You choose your own work hours.	• You probably need to work long hours. • You lose the security of steady wages and medical benefits an employer provides. • You market your own services or products. • You pay for your own operating expenses. • You must be motivated and energetic each day. • You face the possibility of losing money.

Traits of Entrepreneurs.

- Most entrepreneurs share certain behaviors and **attitudes**.
- They are motivated self-starters, have a strong work ethic, and are willing to take necessary risks to create profitable and useful businesses.
- Strong organizational skills, marketing knowledge, and accounting skills are three areas of expertise that contribute to successful business ownership.

Entrepreneur—Who, Me?

- Have you ever considered owning a business? If so, you must first inventory your skills and interests.
- Do you have good writing and speaking skills? Are you creative? Mechanical? A self-starter? Do you have accounting or marketing skills? If you have vision, great energy, and imagination, you may have the makings of an entrepreneur! See **Figure 2–2** for the types of decisions entrepreneurs face.

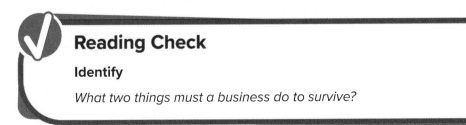

Reading Check

Identify

What two things must a business do to survive?

Traits of the Entrepreneur

1. Before

Entrepreneurs are aware of their environment.

- No bakery nearby
- Does the community want one?

2. In Between

Entrepreneurs evaluate alternatives.

- Customer preferences
- Accountant's guidance
- New tools and equipment are expensive. Use tools from home and buy new ones only when necessary.

Figure 2–2

Types and Forms of Businesses

What Do Businesses Do, and How Are They Organized?

Our free enterprise system allows an entrepreneur to choose the kind of business to operate as well as its organizational form.

Types of Business Operations

The three types of business operations are service, merchandising, and manufacturing. They are alike in many ways. Each sells products or services to customers and **incurs** expenses. They differ, however, in some basic ways. Each type needs money to begin and maintain operations; buy or make products; and cover operating costs like rent, utilities, and wages. Money that investors, banks, or business owners supply is called **capital**.

Service Businesses.

- A **service business** provides a needed service for a fee.
- Service businesses include travel agencies, salons like Fantastic Sam's, repair shops, real estate offices like Century 21, and medical centers.

Merchandising Businesses.

- A **merchandising business** buys finished products and resells them to individuals or other businesses.
- Examples are department stores, car dealers, supermarkets, drugstores, and hobby shops.

Entrepreneurs seek the best solution.
- Find commercial space for rent. Distribute fliers promoting the business.

3. After

Entrepreneurs turn ideas into actions.
- Donna's Bakery quickly becomes the community's top option for cupcakes and other baked goods.

Manufacturing Businesses.

- A **manufacturing business** buys raw materials, uses labor and machinery to transform them into finished products, and sells the finished products to individuals or other businesses.
- Examples are shipbuilders, bakeries, and tire factories.

Forms of Business Organization

To start a business, a potential owner must have a **sufficient** amount of capital and must choose an appropriate form of business organization. With a few exceptions, U.S. businesses are organized in one of three basic ways: sole proprietorship, partnership, or corporation, illustrated in **Figure 2–3**. Notice that each of the different types of operations can use any form of organization.

Figure 2–3 Forms of Business Organization

Form of Business Organization

Type of Business Operation	Sole Proprietorship	Partnership	Corporation
Service	X	X	X
Merchandising	X	X	X
Manufacturing	X	X	X

Sole Proprietorship.

- *Sole* means "single" or "one." *Proprietor* means "owner."
- A **sole proprietorship**, therefore, is a business owned by one person.
- It is sometimes simply called a *proprietorship*.
- Being a sole proprietor does not mean working alone.
- Based on the operation's size and scope, a sole proprietorship may have many managers and employees.

The oldest and most common form of business organization, the sole proprietorship is the easiest business form to start. Little or no legal paperwork (forms and documents) is required. The success or failure of the business depends heavily on the efforts and talent of the owner. The advantages and disadvantages of organizing as a sole proprietorship are shown in **Figure 2–4**.

Figure 2–4 Advantages and Disadvantages of a Sole Proprietorship

Sole Proprietorship

Advantages	Disadvantages
• Easy to set up	• Limited expertise
• All profits go to owner	• Hard to raise money
• Owner has total control	• Owner has all the risks
• Few regulations to follow	• Hard to attract talented employees

Partnership.

- A **partnership** is a business owned by two or more persons, called *partners*, who agree to operate the business as co-owners.
- Business partners usually enter into a written, legal agreement.
- This agreement specifies each partner's investment in money or property, responsibilities, and percent of profits and losses.
- Partnerships are often formed when a business needs more capital than one person can invest.
- Partnerships are not always small.
- For example, partnerships like the large accounting firm KPMG may have as many as 1,600 partners and more than 18,000 employees.
- **Figure 2–5** lists some partnership advantages and disadvantages.

Figure 2–5 Advantages and Disadvantages of a Partnership

Partnership

Advantages	Disadvantages
• Easy startup • Pooled skills and talents • More money available	• Risk of partner conflict • Shared profits • Shared risks

Corporation.

- A **corporation** is a business recognized by law to have a life of its own.
- Unlike a sole proprietorship and a partnership, a corporation must get permission from the state to operate.
- This legal permission, called a **charter**, gives a corporation certain rights and privileges.
- It also spells out the rules under which the corporation is to operate.

Corporations often start as sole proprietorships or partnerships. The business owner(s) may incorporate to obtain money needed to expand. To raise this money, organizers sell shares of stock to hundreds or even thousands of people. These *shareholders*, or *stockholders*, are the corporation's legal owners. **Figure 2–6** outlines a few advantages and disadvantages of a corporation.

Figure 2–6 Advantages and Disadvantages of a Corporation

Corporation

Advantages	Disadvantages
• Easier to raise money • Easy to expand • Easy to transfer ownership • Losses limited to investment	• More costs to start • More complex to organize • More regulations • Higher taxes

SECTION 2.1
Assessment

After You Read

Reinforce the Main Idea

Use a diagram like this one to show details that support the following main idea: The United States has a free enterprise system with various kinds of businesses.

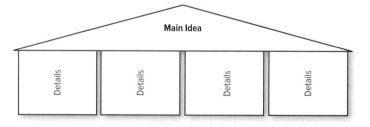

Problem 1 Assess Your Entrepreneurship Potential

Instructions The characteristics or traits needed to set up an owner-operated business and to run it successfully are listed below. Rank yourself, indicating whether this trait is most like you or least like you. Use the form provided in your working papers.

	Most Like Me				Least Like Me
Persistent	5	4	3	2	1
Creative	5	4	3	2	1
Responsible	5	4	3	2	1
Inquisitive	5	4	3	2	1
Goal-oriented	5	4	3	2	1
Independent	5	4	3	2	1
Demanding	5	4	3	2	1
Self-confident	5	4	3	2	1
Risk-taking	5	4	3	2	1
Restless	5	4	3	2	1
Strong work ethic	5	4	3	2	1

List the categories in which you ranked 3 or lower. Identify experiences, activities, and ways in which you could improve your entrepreneurial potential.

Math for Accounting

Suppose that you bought 500 shares of Pinewood Dairy Corporation for $39 per share several months ago. Yesterday you sold 200 of the shares at the current price of $45 per share. How much money did you gain or lose on the shares that you sold? What was the gain or loss per share? Based on the price of $45 per share, what was the market value of your remaining shares yesterday?

SECTION 2.2
Accounting: The Universal Language of Business

Have you ever wondered how a band determines which venues it will play, what a tour costs, how profitable CD sales have been, or how much ticket prices should be? All these issues are probably tackled by the band's financial team: accountants, clerks, and financial advisors.

The Accounting System

What Does the Accounting System Do?

Whether you are the accountant for a band or the financial manager for a chain of mountain bike stores, accounting methods are universal. The **accounting system** is designed to collect, document, and report on financial transactions affecting the business. In a **manual accounting system**, the accounting information is processed by hand. In a **computerized accounting system**, also known as an *automated accounting system*, the financial information is recorded by entering it into a computer. Owners and investors who risk their money depend on accountants to report the results of operations and the financial condition of the businesses in which they invest. **Figure 2–7** outlines the inputs, processing, and outputs of an accounting system.

> **Section Vocabulary**
> - accounting system
> - manual accounting system
> - computerized accounting system
> - GAAP
> - financial reports
> - financial accounting
> - management accounting
> - business entity
> - accounting period
> - going concern
> - fundamental
> - volume
> - specific

All accountants follow the same accounting framework, or set of rules to prepare financial reports. The Financial Accounting Standards Board (FASB) issues these rules. The rules are referred to as *generally accepted accounting principles* or **GAAP** (pronounced "gap").

Accounting principles provide a way to communicate financial information in a form understood by those interested in the operations and financial condition of a business.

Because it is so **fundamental** to the communication of financial information and has a significant impact on business and industry, accounting is often called the "language of business." **Financial reports** are summarized information about the financial status of a business.

Figure 2–7 How the Accounting System Works

Inputs	Processing	Outputs
Source Documents • Checks • Invoices • Sales Slips • Receipts	**Tasks** • Analyzing • Classifying • Recording	**Financial and Management Accounting Reports** • Financial condition • Results of operations • Investments by and distributions to owners

Using Accounting Reports for Making Business Decisions

Who Uses Accounting Reports?

The purpose of accounting is to provide financial information about a business or a not-for-profit organization. Individuals also need financial information in order to make decisions.

The information found in accounting reports has a wide audience. In general there are two groups that use accounting reports:

- individuals *outside* the business who have an interest in the business
- individuals *inside* the business

Financial Accounting

Financial accounting focuses on reporting information to *external* users. Financial accounting reports are prepared for individuals not directly involved in the day-to-day operations of the business.

Who uses financial accounting reports? The following examples describe just a few situations in which people use financial accounting reports.

- Suppose you wanted to invest in a business, such as Trek Mountain Bikes Inc. You would analyze financial accounting reports to estimate if an investment in the business would be profitable.
- Individuals or institutions, like banks, that loan money to a business use financial accounting reports to determine whether or not the business will be able to repay loans.
- Local, state, and federal governments review financial accounting reports. For example, the Internal Revenue Service may compare the tax return and financial reports of a business to determine whether the business is paying the proper amount of taxes.
- Workers, consumers, union leaders, and competitors are interested in the performance of businesses as presented in financial accounting reports.

Management Accounting

Management accounting, which focuses on reporting information to management, is often referred to as accounting for *internal* users of accounting information. Management accounting reports are prepared for managers involved in making the day-to-day operating decisions like purchasing, hiring, production, payments, sales, and collections.

Managers need accounting information so they can decide what to do, how to do it, when to do it, and whether or not the results match the plans for the future.

Reading Check

How is financial accounting similar to management accounting? How is it different?

Accounting Assumptions

What Are Three Accounting Assumptions?

An assumption is something taken for granted as true. When you attend a movie, you assume the **volume** will be at a reasonable level, there will be a place to sit, and the film will be in focus. If these assumptions are wrong, the movie will be a disappointing experience.

Each business sets up an accounting system for its specific needs, but all businesses follow GAAP. These generally accepted accounting principles include three important assumptions:

- business entity
- accounting period
- going concern

Recall that the corporation is the only form of business that is a separate *legal* entity. For *accounting* purposes, however, all businesses are separate entities from their owners. A **business entity** exists independently of its owner's personal holdings. The accounting records and reports are maintained separately and contain financial information related only to the business. The owner's personal financial activities or other investments are not included in the reports of the business.

For example, the personal residence of a flower shop owner, valued at $100,000, is not reported in the accounting records of the flower shop. Buildings owned by the business, however, are included in the financial records and reports of the flower shop.

For reporting purposes, the life of a business is divided into **specific** periods of time. An **accounting period**, also known as an *accounting cycle*, is a period of time covered by an accounting report. The accounting period can cover one month or three months (quarterly), but the most common period is one year. This assumption is necessary when accountants assign the cost of buildings and equipment over the estimated period of time they will be used. Comparison of reports from one period to the next also makes the accounting period concept necessary.

Unless there is evidence to the contrary, accountants assume that a business has the ability to survive and operate indefinitely. In other words a business is expected to continue as a **going concern**. Although many businesses fail within the first five years, the accountant assumes the business will continue to operate unless it is clear that it cannot survive.

SECTION 2.2
Assessment

After You Read

Reinforce the Main Idea

List at least three ways people use each type of accounting report. Use an organizer like the one shown here.

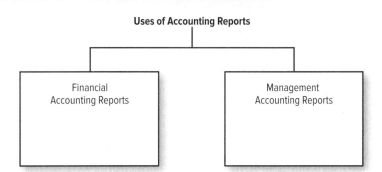

Problem 2 Using Financial Information

Instructions Imagine that you are the manager of Valley Skateboard Shop. The accountant for the business has delivered the news that sales are down by 15 percent compared to last month. She also reports that the rent on the building will be increasing next month by $100. You believe the shop needs a more experienced salesperson and that the building needs new paint and awnings. How might the information delivered by the accountant affect your decisions and course of action?

Problem 3 Identifying Accounting Assumptions

Suppose you work for a company called Greenwood Sky Divers as a sky-diving instructor. Since you have your pilot's license, you also give flying lessons on the side. You and the owner of Greenwood Sky Divers both belong to a local club, the Greenwood Eagles, whose members get together for sky dives on weekends. Greenwood Sky Divers is a sole proprietorship that has been in business for six years. The owner of the company regularly reviews financial reports, and the company's accountant prepares end-of-the-year reports to compare results from one year to the next.

Instructions Decide what information in this paragraph is relevant to accounting for Greenwood Sky Divers. Identify the accounting assumptions for Greenwood, and give an example of each assumption from the information given here.

Math for Accounting

Suppose that you are opening a 1,000-square-foot baseball card store in a shopping center. You have signed a lease that requires you to pay a monthly rent of $1.75 per square foot, plus 5 percent of your gross annual sales. What is the total amount of rent you will pay in one year if you have sales of $85,000?

Chapter 2
Visual Summary

Concepts
Compare sole proprietorship, partnership, and corporate forms of business

Ownership	Only one owner	Owned by two or more persons	Can be owned by shareholders or stockholders
Start-Up	Easy	Easy	Costly
Paperwork	Little to none	Written, legal partnership agreement	Lots of paperwork required

Analysis
List the advantages and disadvantages of each form of business organization

Sole Proprietorship	• Easy to set up • All profits go to owner • Owner has total control • Few regulations to follow	• Limited expertise • Hard to raise money • Owner has all the risks • Hard to attract talented employees
Partnership	• Easy startup • Pooled skills and talents • More money available	• Risk of partner conflict • Shared profits • Shared risks
Corporation	• Easier to raise money • Easy to expand • Easy to transfer ownership • Losses limited to investment	• More cost to start • More complex to organize • More regulations • Higher taxes

Procedures

THE THREE BASIC ACCOUNTING ASSUMPTIONS

Business Entity: A business exists separately from the personal finances of the owner.

Accounting Period: The life of the business is divided into specific periods of time covered by accounting reports.

Going Concern: A business has the ability to survive and operate indefinitely unless there is evidence otherwise.

Chapter 2
Review and Activities

Answering the Essential Question

What role does accounting play in the free enterprise system?

In a free enterprise system, people are free to produce the goods and services they choose. This freedom imposes on businesses the responsibility of managing and tracking profits and losses. How does the success of a business depend on accounting?

Vocabulary Check

1. **Vocabulary** Write your own definition of each content vocabulary term; include an example whenever possible.

- free enterprise system
- profit
- loss
- entrepreneur
- attitude
- incur
- sufficient
- capital
- service business
- merchandising business
- manufacturing business
- sole proprietorship
- partnership
- corporation
- charter
- fundamental
- accounting system
- manual accounting system
- computerized accounting system
- GAAP
- financial reports
- financial accounting
- management accounting
- volume
- specific
- business entity
- accounting period
- going concern

Concept Check

2. **Interpret** What is an entrepreneur? In your opinion, why would risk-taking be a beneficial trait for entrepreneurs to possess?
3. Name and describe the three types of business operations
4. Describe each of the three forms of business organization.
5. List two advantages and two disadvantages of each business form.
6. **Evaluate** Why is accounting called the "Universal Language of Business"? Discuss the impact of accounting on both business and industry.
7. **Analyze** Discuss the ways in which financial accounting and management accounting are alike and the ways in which they are different.
8. What are three important assumptions of accounting, and what does each mean?
9. **Math** Linh Lam works for a company that is opening a facility in England. She is being transferred later this year. At the time of her transfer, the exchange rate is expected to be 0.75. In other words, 1 U.S. dollar ($) will only buy 0.75 worth of English pounds (£). Linh earns $43,650 a year in the U.S. and will receive a 15% raise after her transfer. Calculate Linh's salary in English pounds after her transfer.

Chapter 2
Standardized Test Practice

Multiple Choice

1. Which of these is not an advantage to setting up a business as a sole proprietorship?
    a. The taxes are lower.
    b. There are fewer regulations.
    c. Losses are limited to the owner's investment.
    d. The owner controls the business decisions.

2. A merchandising business –
    a. provides a needed service for a fee.
    b. transforms raw materials into finished products.
    c. buys finished products and resells them for a profit.
    d. can be a shipyard, repair shop, or real estate office.

3. A business in which two or more persons combine their assets is called a(n)
    a. corporation.
    b. sole proprietorship.
    c. partnership.
    d. merchandising business.
    e. none of the above.

4. An example of an accounting assumption is a(n)
    a. accounting system.
    b. accounting statement.
    c. going concern
    d. computerized accounting.

Short Answer

5. Someone who transforms ideas for products or services into real-world business is called a(n) _____.

6. A _____ buys raw materials and transforms them into finished products.

Extended Response

7. Explain the difference between financial accounting and management accounting.

Chapter 2
Problems

Problem 4 — Identifying Types of Businesses

Instructions In your working papers, indicate whether each of the following businesses is a service business, a merchandising business, or a manufacturing business.

1. International Business Machines (IBM)
2. Gap
3. Glendale Memorial Hospital
4. Avis Rent A Car System
5. Ford Motor Company
6. Bank of America
7. Ace Hardware Stores
8. Michigan City Animal Hospital
9. Office Depot
10. Wal-Mart Stores
11. Allstate Insurance
12. U.S. Steel

Problem 5 — Understanding Accounting Assumptions

Instructions In your working papers, indicate the assumption from the list below that best matches each numbered statement.

Accounting Period

Business Entity

Going Concern

1. Accounting reports may cover a month, a quarter, or a year.
2. Accountants expect a business to last indefinitely.
3. The personal property of a business owner is not included in the accounting records of the business.
4. The business has been in operation for several years and is expected to continue.
5. An owner's personal activities or properties are not mixed with the business activities or properties.
6. The accountant's report shows how much profit the business earned for one month.

Chapter 2
Problems

Problem 6 | Understanding Business Operations

Mary Torres owns and operates a bakery. During the past week, she sold 500 loaves of bread at $2.25 per loaf. The raw materials for each loaf cost Ms. Torres $1.80.

Instructions In your working papers, write the answers to the following questions:

1. What type of business does Ms. Torres operate?
2. What was the profit on bread sales for the week?
3. Aside from the costs for raw materials, what other costs will Ms. Torres use to calculate her profit for the week?

Problem 7 | Categorizing Forms of Business Organizations

Instructions Match the letter next to each form of business with the appropriate items in the following list of advantages and disadvantages. You may use a letter more than once. Use the form provided in your working papers.

A. **Sole Proprietorship**
B. **Partnership**
C. **Corporation**

1. Easier to raise money
2. Limited expertise
3. Higher start-up costs
4. Owner has total control
5. Shared profits
6. Higher taxes
7. Fewer regulations to follow
8. Easy transfer of ownership
9. Risk of conflict between owners
10. Easy to expand

Chapter 2
Problems

Problem 8 Working as an Entrepreneur

Instructions Using the form in your working papers, identify each of the following as an advantage or disadvantage of being an entrepreneur. For each item you label a disadvantage, decide what actions you would take to overcome that disadvantage. Analyze and describe how your actions might affect the profit your business earns.

1. Risking the loss of your savings
2. Deciding what you and everyone else needs to do each day
3. Lacking steady wages and employee benefits
4. Choosing when and where to work
5. Keeping the financial benefits of your hard work
6. Choosing the people you want to work with
7. Paying all the expenses of a new business

Real-World Accounting Careers

Sheree Chapela
Hospital Corporation of America (HCA)

Q What do you do?

A I work in the finance unit of the company's construction group. I oversee the review and processing of invoices and charge orders. I also am responsible for the contracts for architects and general contractors who work on building projects for the company.

Q What are your day-to-day responsibilities?

A I audit contractors' invoices to make sure they comply with our company's requirements and the Sarbanes-Oxley Act of 2002, and I authorize payments. I also train the contractors' staff in terms of meeting all of our requirements.

Q What factors have been key to your success?

A I take pride in doing the absolute best job I can. I started out as an office support professional to get in the door with a construction firm. Then I became an assistant to the accounting manager and learned as much as I could about applying accounting principles in a construction setting.

Q What do you like most about your job?

A My job makes sense; it's not abstract. When I come to work, I get to see the accounting process from start to finish.

Career Facts

Real-World Skills
Strong work ethic; time management skills; dependability

Training And Education
Associate's degree with a focus in accounting

Career Skills
Strong knowledge of basic accounting principles; experience in accounting for a particular industry

Career Path
Start at a small firm to gain experience in accounts payable and receivable, invoicing, job contracts, and other areas

Tips from Robert Half International
Many job openings are filled by word of mouth or through referrals, so be sure to tell everyone you know about your job hunt. This includes your family, friends, neighbors and even teachers. You might be surprised by who is able to provide useful advice or a promising lead.

College and Career Readiness

Write a one-page article for a career newsletter. Describe the field of accounting and the types of businesses and organizations where accountants might work.

Real-World Applications & Connections

CASE STUDY

Accounting and Entrepreneurships

Robert's Creative Toys keeps all the books and records using a manual accounting system. The business has grown and the owner realizes that a computerized accounting system would be helpful. However, he knows little about the advantages of using a computer. What advantages would there be by using a computer accounting system over a manual accounting system?

ACTIVITY Discuss the advantages and disadvantages of integrating technology into the company's accounting system.

Success Skills

Time Management

Balancing your time between school, work, sports, hobbies, and other activities can be a challenge. Learning to manage your time as a student can help you prepare for the workplace. Multi-tasking and prioritizing are vital skills in the workplace. Developing a strong work ethic and the ability to manage workloads and deadlines can play a major role in being successful. Understanding and practicing prioritization skills now can prepare you to do so even more efficiently in years to come.

ACTIVITY

a. Make a list of your daily activities, identify the relative importance of these activities, then arrange them in order of importance, or priority. Think through which activities you should focus on first, and why.

b. Present your priority list to a classmate, explaining how you will manage your time. Do you see any common themes in how you each prioritized your activities?

Career Wise

Accountant

Accountants play a critical role in the economy. They analyze and communicate financial information for companies, individual clients, and government organizations. Accountants need strong math, analytical, communication, and computer skills. Specific job duties may vary widely between different accounting fields, but fundamental tasks are likely to include preparing, analyzing, and verifying financial documents, budget analysis, financial planning, and sometimes limited legal services.

Activity

a. Visit the web site of the U.S. Department of Labor's Bureau of Labor Statistics and find the section on "Accountants and Auditors."

b. Find the average annual salary for five different accounting professions, and compile them into a chart.

Spotlight on Personal Finance

Maintaining a Checking Account

A bank checking account is a safe and convenient place to hold money. It's designated to hold funds for a short period of time, from which you can write checks or withdraw money using a bank card. Managing your checking account online gives you 24-hour access to information about your account.

Activity

Visit local banks or research online to discover the various transactions that can be conducted through online banking at two or more banks. Using spreadsheet software, create a table summarizing this information. Note any fees charged for these transactions.

Global Accounting

International Accounting Standards

When two companies from different countries prepare financial statements, do you think they use the same accounting rules? If both companies use International Accounting Standards (IAS), the statements would be comparable and understandable. The International Accounting Standards Board (IASB) develops accounting rules so that investors can compare financial statements of publicly traded companies from any country. Since International Accounting Standards are not always the same as U.S. generally accepted accounting principles (GAAP), it is important to know which rules were used to prepare financial statements.

Instructions

Explain why a U.S. company might choose to prepare its financial statements using both GAAP and IAS.

UNIT 2

The Basic Accounting Cycle

A Look Back
Unit 1 described different types of accounting careers and the three forms of business organization.

A Look Ahead
In Unit 2, you will analyze transactions and prepare financial records for a business.

Keys to Success

Q Why are effective reading skills important for an accounting career?

A You may not think reading skills are important to an accountant who spends a lot of time crunching numbers and detailing costs. But without effective reading skills, an accountant would not be able to analyze important financial documents, make important financial recommendations, or stay up-to-date on current events and changes in laws that affect the accounting field. Whether you plan to go to a two- or four-year college or find a job after graduation, effective reading skills are among your most important keys to success.

English Language Arts

Reading Keeping up with current events will help you stay competitive in the workplace. Read two financial news articles online or in a newspaper or magazine. Compare the two texts in a one-page paper. *What companies are involved? Are they sole proprietorships, partnerships, or corporations? How can you tell?*

Focus on the Photo
The accounting cycle translates economic events into useful information. **What economic events can you think of that might occur in a bicycle shop like this one?**

Chapter 3 • Business Transactions and the Accounting Equation 51

Chapter 3

Business Transactions and the Accounting Equation

Chapter Topics

3-1 Property and Financial Claims

3-2 Transactions that Affect Owner's Investment, Cash, and Credit

3-3 Transactions that Affect Revenue, Expense, and Withdrawls by the Owner

Visual Summary

Review and Activities

Standardized Test Practice

Computerized Accounting

Problems

Real-World Accounting Careers

Essential Question

As you read this chapter, keep this question in mind:

Why is understanding the accounting equation crucial to understanding the condition of any business?

Main Idea

Any item of property has at least one financial claim against it. Accounts are used to analyze business transactions. Owner's equity is changed by revenue, expenses, and withdrawals.

Chapter Objectives

Concepts	Analysis	Procedures
C1 Describe the relationship between property and financial claims. **C2** Explain the meaning of the term *equities* as it is used in accounting. **C3** List and define each part of the accounting equation.	**A1** Learn how businesses use accounts. **A2** Demonstrate the effects of transactions on the accounting equation.	**P1** Check the balance of the accounting equation after a business transaction has been analyzed and recorded.

52 Chapter 3 • Business Transactions and the Accounting Equation

Real-World Business Connection

Hulu

Have you ever missed a crucial episode of your favorite TV show? Have you ever sat at the library with nothing to do? Then Hulu might be a site you want to check out. Hulu lets you watch television shows, clips, and movies online—anytime. You can also use Hulu to embed video on your blog or Web page, or just email a clip to your friends. Since Hulu was founded in 2007, it has become hugely successful. Within its first year of operation, it was one of the top five online video sites in the United States.

Connect to the Business

Hulu employs about 150 people, which is a much smaller staff than companies that generate similar revenues, and so is able to keep its payroll costs to a minimum. Lower operating cost is a big advantage in business because it helps increase profitability.

Analyze

How do technology companies like Hulu use technology to keep their operating costs down?

Focus on the Photo

Hulu's main assets are virtual in the world of streaming video, but producing its "product" requires technological equipment and real space in which staff develops Hulu's innovative media distribution. **What types of assets would a business like Hulu have?**

SECTION 3.1
Property and Financial Claims

In Chapter 2 you learned that accounting is the language of business. In this chapter you will learn how to apply basic accounting concepts and terminology. You will also learn how the accounting equation expresses the relationship between property and the rights, or claims, to the property.

United Parcel Service (UPS), a corporation that provides global delivery services, uses accounting reports to communicate with its managers, employees, and investors. The UPS financial reports identify the property used in the business, such as airplanes, trucks, and computers. The reports also show how the company obtained the property, either from loans or from **funds** provided by investors.

Section Vocabulary

- property
- financial claim
- credit
- creditor
- assets
- equities
- owner's equity
- liabilities
- accounting equation
- funds
- acquire

Property

What Is Property?

The right to own property is basic to a free enterprise system. **Property** is anything of value that a person or business owns and therefore controls. When you own an item of property, you have a legal right to that item. For example, suppose you paid $600 for a mountain bike. As a result of the payment, you own the bike. If you had rented the bike for the weekend instead of buying it, you would pay a much smaller amount of money, but you would have the bike for only a limited time. You would have the right to use the bike for the weekend, but you would not own it.

Businesses also own property. One of the purposes of accounting is to provide financial information about property and financial claims to that property. A **financial claim** is a legal right to property. In accounting, property and financial claims are measured in dollar amounts. Dollar amounts measure both the cost of the property and the financial claims to the property. In our mountain bike example, since you paid $600 cash to buy the bike, you have ownership and a financial claim of $600 to the bike. This relationship between property and financial claims is shown in the following equation.

Property (Cost)	=	Financial Claims
Bike	=	Your Claim to the Bike
$600	=	$600

When you buy property with cash, you **acquire** all of the financial claims to that property at the time of purchase. What happens to the financial claim, however, when you don't pay for the property right away?

When you buy something and agree to pay for it later, you are buying on **credit**. The business or person selling the item on credit is called a **creditor**. A creditor can be any person or business to which you owe money.

54 Chapter 3 • Business Transactions and the Accounting Equation

When you buy property on credit, you do not have the only financial claim to the property. You share the financial claim to that property with your creditor. For example, suppose you want to buy a $100 lock for the mountain bike, but you have only $60. You pay the store $60 and sign an agreement to pay the remaining $40 over the next two months. Since you owe the store (the creditor) $40, you share the financial claim to the lock with the creditor. The creditor's financial claim to the lock is $40 and your claim is $60. The combined claims equal the cost of the property. Your purchase of the lock can be expressed as an equation:

PROPERTY = FINANCIAL CLAIMS

Property	=	Financial Claims		
Bike Lock	=	Creditor's Financial Claim	+	Owner's Financial Claim
$100	=	$40	+	$60

As you can see, two (or more) people can have financial claims to the same property.

Only the property owner has control of the property. For example, suppose that you buy a used vehicle for $8,000. You pay $1,200 in cash and take out a loan from the credit union for the remaining $6,800.

Property	=	Financial Claims		
Vehicle	=	Creditor's Financial Claim	+	Owner's Financial Claim
$8,000	=	$6,800	+	$1,200

As the owner, you have control of the vehicle. However, if you don't make the payments to the credit union, the credit union can exercise its legal claim to the vehicle and you will lose ownership.

 Reading Check

Recall

When you buy an item on credit, with whom do you share a financial claim?

Financial Claims in Accounting

What Are the Two Types of Equities?

Property or items of value owned by a business are referred to as **assets**. Businesses can have various types of assets, such as:

- cash
- office equipment
- manufacturing equipment
- buildings
- land

The accounting term for the financial claims to these assets is **equities**. Let's explore the meaning of the term *equities* by introducing Crista Vargas and her new business, Zip Delivery Service, organized as a sole proprietorship.

Suppose Zip Delivery Service purchases a delivery truck for $10,000. Zip makes a cash down payment of $3,000 to the seller. A local bank loans Zip the remaining $7,000. Both Zip and the bank now have financial claims to the truck.

Property	=	Financial Claims		
Truck	=	Creditor's Financial Claim	+	Owner's Financial Claim
$10,000	=	$7,000	+	$3,000

Over the years as Zip repays the loan, its financial claim will increase. As less money is owed, the financial claim of the creditor (the bank) will decrease.

For example, after Zip pays one-half of the loan ($7,000 × ½ = $3,500), the financial claims to the property will change as follows:

Property	=	Financial Claims		
Truck	=	Creditor's Financial Claim	+	Owner's Financial Claim
$10,000	=	$3,500	+	$6,500

When the loan is completely repaid, the creditor's financial claim will be canceled. In other words, the owner's financial claim will then equal the cost of the truck.

In accounting there are separate terms for owner's claims and creditor's claims. The owner's claims to the assets of the business are called **owner's equity**. Owner's equity is measured by the dollar amount of the owner's claims to the total assets of the business.

The creditor's claims to the assets of the business are called **liabilities**. Liabilities are the debts of a business. They are measured by the amount of money owed by a business to its creditors. The relationship between assets and the two types of equities (liabilities and owner's equity) is shown in the **accounting equation:**

ASSETS = LIABILITIES + OWNER'S EQUITY

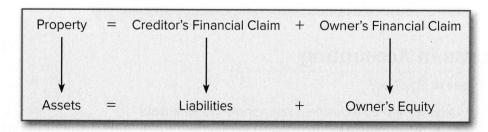

SECTION 3.1
Assessment

After You Read

Reinforce the Main Idea

Use a diagram like this one to describe the financial claims of creditors and owners.

Problem 1 Balancing the Accounting Equation

Instructions Determine the missing dollar amount indicated by the question mark in each equation. Write each missing amount in your working papers.

	ASSETS	=	LIABILITIES	+	OWNER'S EQUITY
1.	$17,000	=	$7,000	+	?
2.	?	=	$6,000	+	$20,000
3.	$10,000	=	?	+	$7,000
4.	?	=	$9,000	+	$17,000
5.	$8,000	=	$2,000	+	?
6.	$20,000	=	$7,000	+	?
7.	?	=	$12,000	+	$4,000
8.	$30,000	=	?	+	$22,000
9.	$22,000	=	$1,000	+	?
10.	$25,000	=	$5,000	+	?
11.	?	=	$10,000	+	$25,000
12.	$7,500	=	?	+	$3,000

continued

SECTION 3.1
Assessment

Math for Accounting

Owner's equity can be expressed as a fraction of the total equities. It can also be expressed as a decimal. Consider the following example:

Assets = Liabilities + Owner's Equity
$10,000 = $9,000 + $1,000

Owner's equity is equal to 1/10, or 10 percent, of the total equities:

Owner's Equity ÷ Total Equities
= $1,000 ÷ $10,000
= 1/10
= 0.10

Convert the following fractions into decimals.

1. $\frac{1}{2}$
2. $\frac{1}{5}$
3. $\frac{1}{3}$
4. $\frac{4}{5}$
5. $\frac{7}{8}$
6. $\frac{7}{10}$

Convert the following decimals into fractions.

7. 0.889
8. 0.60
9. 0.375
10. 0.25
11. 0.667
12. 0.75

SECTION 3.2
Transactions that Affect Owner's Investment, Cash, and Credit

When you purchase a new sweater, buy popcorn at the movies, or put cash in your savings account, you are participating in business transactions. Business transactions involve the purchase, sale, or exchange of goods and services.

Business Transactions

How Are Accounts Used?

A **business transaction** is an economic event that causes a change—either an increase or a decrease—in assets, liabilities, or owner's equity. The change is reflected in the accounting system of the business.

Section Vocabulary

- business transaction
- account
- accounts receivable
- accounts payable
- investment
- on account
- eventually
- transfer

When a business buys a computer with cash, its cash decreases, but its computer equipment increases. It records increases and decreases caused by business transactions in specific accounts. An **account** is a subdivision under assets, liabilities, or owner's equity. It shows the balance for a specific item and is a record of the increases or decreases for that item. Accounts represent things in the real world, such as money invested in a business or owed to a creditor. An account for office furniture represents the dollar cost of all office furniture the business owns.

Every business sets up its accounts and its accounting system to meet its needs. The number of accounts needed varies. Some businesses use only a few accounts, but others use hundreds. No matter how many accounts a business has, all of its accounts may be classified as either assets, liabilities, or owner's equity. Zip Delivery Service uses these accounts:

Assets	=	Liabilities	+	Owner's Equity
Cash in Bank		Accounts Payable		Crista Vargas, Capital
Accounts Receivable				
Computer Equipment				
Office Equipment				
Delivery Equipment				

The second asset account listed is **Accounts Receivable. Accounts receivable** is the total amount of money owed to a business—money to be received later because of the sale of goods or services on credit. The **Accounts Receivable** account is an asset because it represents a claim to the assets of other people or businesses. It represents a future value that **eventually** will bring cash into the business. When the business receives payment, it cancels the claim.

The liability account is **Accounts Payable. Accounts payable** is the amount owed, or payable, to the creditors of a business. The owner's equity account title is the owner's name, a comma, and then the word *Capital*.

Transactions and the Accounting Equation

How Do Transactions Affect the Accounting Equation?

When a business transaction occurs, an accounting clerk analyzes the transaction to see how it affects each part of the accounting equation. Analyzing business transactions is simple. Use the following steps.

Business Transaction

Analysis		
	Identify	1. Identify the accounts affected.
	Classify	2. Classify the accounts affected.
	+/−	3. Determine the amount of increase or decrease for each account affected.
	Balance	4. Make sure the accounting equation remains in balance.

Most businesses have the following types of transactions: investments by the owner, cash transactions, credit transactions, revenue and expense transactions, and withdrawals by the owner.

Investments by the Owner

An **investment** is money or other property paid out in order to produce profit. Owner Crista Vargas made two investments in her business, Zip Delivery Service. The first was a cash investment; the second was a **transfer** of property.

Business Transaction 1

Crista Vargas took $25,000 from personal savings and deposited that amount to open a business checking account in the name of Zip Delivery Service.

Analysis		
	Identify	1. Cash transactions are recorded in the account **Cash in Bank**. Crista Vargas is investing personal funds in the business. Her investment in the business is recorded in the account called **Crista Vargas, Capital**.
	Classify	2. **Cash in Bank** is an asset account. **Crista Vargas, Capital** is an owner's equity account.
	+/−	3. **Cash in Bank** is increased by $25,000. **Crista Vargas, Capital** is increased by $25,000.
	Balance	4. The accounting equation remains in balance.

	Assets	=	Liabilities	+	Owner's Equity
	Cash in Bank				Crista Vargas, Capital
Trans. 1	+$25,000	=	0	+	+$25,000

Business Transaction 2

Crista Vargas transferred two telephones valued at $200 each from her home to the business.

Analysis	Identify	1. Crista Vargas gave two telephones to the business. This affects the account **Office Equipment**. The investment of these assets affects the account **Crista Vargas, Capital**.
	Classify	2. **Office Equipment** is an asset account. **Crista Vargas, Capital** is an owner's equity account.
	+/−	3. **Office Equipment** is increased by $400. **Crista Vargas, Capital** is increased by $400.
	Balance	4. The accounting equation remains in balance.

	Assets		=	Liabilities	+	Owner's Equity
	Cash in Bank	Office Equipment				Crista Vargas, Capital
Prev. Bal.	$25,000	0		0		$25,000
Trans. 2		+ $400				+400
Balance	$25,000 +	$400	=	0	+	$25,400

Cash Payment Transactions

Transaction 3 is the cash purchase of an asset. Any asset purchased for cash is recorded this way, but the account name of the asset purchased may vary. Transaction 3 affects only the assets side of the equation. Zip exchanged one asset (cash) for another asset (computer equipment).

Business Transaction 3

Zip issued a $3,000 check to purchase a computer system.

Analysis	Identify	1. Transactions involving any type of computer equipment are recorded in the **Computer Equipment** account. The business paid cash for the computer system, so the account **Cash in Bank** is affected. Check payments are treated as cash payments and are recorded in **Cash in Bank**.
	Classify	2. **Computer Equipment** and **Cash in Bank** are both asset accounts.
	+/−	3. **Computer Equipment** is increased by $3,000. **Cash in Bank** is decreased by $3,000.
	Balance	4. The accounting equation remains in balance.

	Assets			= Liabilities +	Owner's Equity
	Cash in Bank	Computer Equipment	Office Equipment		Crista Vargas, Capital
Prev. Bal.	$25,000	0	$400	0	$25,400
Trans. 3	−3,000	+$3,000			
Balance	$22,000 +	$3,000 +	$400 =	0 +	$25,400

Credit Transactions

Now that you have learned about cash transactions, let's look at how the use of credit affects the accounting equation. When a business buys an item on credit, it is buying **on account**. In the next four transactions, you will learn about a purchase on account, a sale on account, a payment made on account, and a payment received on account.

Business Transaction 4

Zip bought a used truck on account from Coast to Coast Auto for $12,000.

Analysis

Identify
1. Zip purchased a truck to be used as a delivery vehicle, so the account **Delivery Equipment** is affected. The business promised to pay for the truck at a later time. The promise to pay is a liability; therefore, the **Accounts Payable** account is affected.

Classify
2. **Delivery Equipment** is an asset account. **Accounts Payable** is a liability account.

+/−
3. **Delivery Equipment** is increased by $12,000. **Accounts Payable** is also increased by $12,000.

Balance
4. The accounting equation remains in balance.

	Assets				= Liabilities +	Owner's Equity
	Cash in Bank	Computer Equipment	Office Equipment	Delivery Equipment	Accounts Payable	Crista Vargas, Capital
Prev. Bal.	$22,000	$3,000	$400	0	0	$25,400
Trans. 4				+$12,000	+$12,000	
Balance	$22,000 +	$3,000 +	$400 +	$12,000 =	+$12,000 +	$25,400

Business Transaction 5

Zip sold one telephone to Green Company for $200 on account.

Analysis

Identify
1. Since Zip has agreed to receive payment for the telephone at a later time, the **Accounts Receivable** account is affected. The business sold the telephone, so the account **Office Equipment** is also affected.

	Classify	2. Both **Accounts Receivable** and **Office Equipment** are asset accounts.
	+/−	3. **Accounts Receivable** is increased by $200. **Office Equipment** is decreased by $200.
	Balance	4. The accounting equation remains in balance.

		Assets				= Liabilities +	Owner's Equity
	Cash in Bank	Accounts Receivable	Computer Equipment	Office Equipment	Delivery Equipment	Accounts Payable	Crista Vargas, Capital
Prev. Bal.	$22,000	0	$3,000	$400	$12,000	$12,000	$25,400
Trans. 5		+$200		−200			
Balance	$22,000 +	$200 +	$3,000 +	$200 +	$12,000 =	$12,000 +	$25,400

Business Transaction 6

Zip issued a check for $350 in partial payment of the amount owed to its creditor, Coast to Coast Auto.

Analysis	Identify	1. The payment decreased the total amount owed to the creditor, so **Accounts Payable** is affected. Payment was made by check, so the account **Cash in Bank** is affected.
	Classify	2. **Accounts Payable** is a liability account. **Cash in Bank** is an asset account.
	+/−	3. **Accounts Payable** is decreased by $350. **Cash in Bank** is also decreased by $350.
	Balance	4. The accounting equation remains in balance.

		Assets				= Liabilities +	Owner's Equity
	Cash in Bank	Accounts Receivable	Computer Equipment	Office Equipment	Delivery Equipment	Accounts Payable	Crista Vargas, Capital
Prev. Bal.	$22,000	$200	$3,000	$200	$12,000	$12,000	$25,400
Trans. 6	−350					−350	
Balance	$21,650 +	$200 +	$3,000 +	$200 +	$12,000 =	$11,650 +	$25,400

Business Transaction 7

Zip received and deposited a check for $200 from Green Co. The check received is full payment for the telephone sold on account in Transaction 5.

Analysis	Identify	1. The check decreases the amount owed to Zip, so **Accounts Receivable** is affected. A check is given in payment, so **Cash in Bank** is affected.
	Classify	2. **Accounts Receivable** and **Cash in Bank** are asset accounts.
	+/−	3. **Accounts Receivable** is decreased by $200. **Cash in Bank** is increased by $200.
	Balance	4. The accounting equation remains in balance.

	Assets					=	Liabilities	+	Owner's Equity
	Cash in Bank	Accounts Receivable	Computer Equipment	Office Equipment	Delivery Equipment		Accounts Payable		Crista Vargas, Capital
Prev. Bal.	$21,650	$200	$3,000	$200	$12,000		$11,650		$25,400
Trans. 7	+$200	−200							
Balance	$21,850 +	$0 +	$3,000 +	$200 +	$12,000	=	$11,650 +		$25,400

As you can see, each business transaction causes a change in assets, liabilities, or owner's equity. Analyzing each transaction to see how it affects the accounting equation keeps everything in balance.

SECTION 3.2
Assessment

After You Read

Reinforce the Main Idea

Create a diagram like the one shown here. For the boxes marked "?", fill in the correct labels and give one example of a related account for each category.

| ? | = | ? | + | Owner's Equity |

Leo Jones, Capital

Problem 2 Determining the Effects of Transactions on the Accounting Equation

Instructions Use these accounts to analyze the business transactions of WordService.

Assets	=	Liabilities	+	Owner's Equity
Cash in Bank Accounts Receivable Computer Equipment Office Furniture		Accounts Payable		Jan Swift, Capital

On the form provided in your working papers, identify the accounts affected by each transaction and the amount of increase or decrease in each account. Make sure the accounting equation is in balance after each transaction.

1. Jan Swift, owner, deposited $30,000 in the business checking account.
2. The owner transferred to the business a desk and chair valued at $700.
3. WordService issued a check for $4,000 for the purchase of a computer.
4. The business bought office furniture on account for $5,000 from Eastern Furniture.
5. The desk and chair previously transferred to the business by the owner were sold on account for $700.
6. WordService wrote a check for $2,000 in partial payment of the amount owed to Eastern Furniture Company.

Math for Accounting

The basic accounting equation is in the form of $A = L + OE$

1. What is the algebra equation to find L?
2. What is the algebra equation to find OE?

Using the rules of algebra, determine the missing dollar amount in each equation.

Assets	=	Liabilities	+	Owners Equity
?		$9,000		$21,000
$25,000		?		$11,000
$10,000		$2,000		?

SECTION 3.3
Transactions that Affect Revenue, Expense, and Withdrawals by the Owner

United Parcel Service (UPS) has thousands of shareholders who expect a return on their investment in the business. The most common way for a business to provide a return is by selling goods or providing services. UPS earns revenue by providing a global delivery service. To provide the delivery service, UPS incurs expenses like salaries, transportation, and insurance. In this section you will learn about revenue and expense transactions as well as owner's withdrawals.

Section Vocabulary
- revenue
- expense
- withdrawal
- generate
- conduct

Revenue and Expense Transactions

What Are Revenue and Expenses?

Income earned from the sale of goods or services is called **revenue**. Examples of revenue are fees earned for services performed and cash received from the sale of merchandise. Revenue increases owner's equity because it increases the assets of the business.

Both revenues and investments by the owner increase owner's equity, but these represent very different transactions:

- Revenue is income from the sale of goods and services.
- Investment by the owner is the dollar amount contributed to the business by the owner.

To **generate** revenue most businesses must also incur expenses to buy goods, materials, and services. An **expense** is the cost of products or services used to operate a business. Examples of business expenses are

- rent,
- utilities, and
- advertising.

Expenses decrease owner's equity because they decrease the assets of the business or increase liabilities.

The effects of revenue and expenses are summarized as follows:

- Revenue increases assets and increases owner's equity.
- Expenses decrease assets and decrease owner's equity *or* increase liabilities and decrease owner's equity.

Reading Check

Contrast

What is the difference between revenue and investments by the owner?

Business Transaction 8

Zip received a check for $1,200 from a customer, Sims Corporation, for delivery services.

Analysis

Identify
1. Zip received cash, so **Cash in Bank** is affected. The payment received is revenue. Revenue increases owner's equity, so **Crista Vargas, Capital** is also affected.

Classify
2. **Cash in Bank** is an asset account. **Crista Vargas, Capital** is an owner's equity account.

+/−
3. **Cash in Bank** is increased by $1,200. **Crista Vargas, Capital** is also increased by $1,200.

Balance
4. The accounting equation remains in balance.

	Assets					= Liabilities	+ Owner's Equity
	Cash in Bank	Accounts Receivable	Computer Equipment	Office Equipment	Delivery Equipment	Accounts Payable	Crista Vargas, Capital
Prev. Bal.	$21,850	$0	$3,000	$200	$12,000	$11,650	$25,400
Trans. 8	+1,200						+1,200
Balance	$23,050 +	$0 +	$3,000 +	$200 +	$12,000 =	$11,650 +	$26,600

Business Transaction 9

Zip wrote a check for $700 to pay the rent for the month.

Analysis

Identify
1. Zip pays rent for use of building space. Rent is an expense. Expenses decrease owner's equity, so the account **Crista Vargas, Capital** is affected. The business is paying cash for the use of the building, so **Cash in Bank** is affected.

Classify
2. **Crista Vargas, Capital** is an owner's equity account. **Cash in Bank** is an asset account.

+/−
3. **Crista Vargas, Capital** is decreased by $700. **Cash in Bank** is decreased by $700.

Balance
4. The accounting equation remains in balance.

	Assets					= Liabilities +	Owner's Equity
	Cash in Bank	Accounts Receivable	Computer Equipment	Office Equipment	Delivery Equipment	Accounts Payable	Crista Vargas, Capital
Prev. Bal.	$23,050	$0	$3,000	$200	$12,000	$11,650	$25,600
Trans. 9	−700						−700
Balance	$22,350 +	$0 +	$3,000 +	$200 +	$12,000 =	$11,650 +	$25,900

Withdrawals by the Owner

What Is a Withdrawal?

If a business earns revenue, the owner will take cash or other assets from the business for personal use. This is called a **withdrawal**. Withdrawals and investments have opposite effects. A withdrawal decreases both assets and owner's equity.

A withdrawal is not the same as an expense. Both decrease owner's equity, but each represents a different transaction. An expense is the price paid for goods and services used to operate a business. For example, a gardening service purchases fertilizer and lawncare supplies to **conduct** daily operations. Withdrawals by the owner are cash or other assets taken from the business for the owner's personal use. Transaction 10 illustrates the impact of a withdrawal on the accounting equation.

Business Transaction 10

Crista Vargas withdrew $500 from the business for her personal use.

Analysis	Identify	1. A withdrawal decreases the owner's claim to the assets of the business, so **Crista Vargas, Capital** is affected. Cash is paid out, so the **Cash in Bank** account is affected.
	Classify	2. **Crista Vargas, Capital** is an owner's equity account. **Cash in Bank** is an asset account.
	+/−	3. **Crista Vargas, Capital** is decreased by $500. **Cash in Bank** is decreased by $500.
	Balance	4. The accounting equation remains in balance.

	Assets					= Liabilities +	Owner's Equity
	Cash in Bank	Accounts Receivable	Computer Equipment	Office Equipment	Delivery Equipment	Accounts Payable	Crista Vargas, Capital
Prev. Bal.	$22,350	$0	$3,000	$200	$12,000	$11,650	$25,900
Trans. 10	−500						−500
Balance	$21,850 +	$0 +	$3,000 +	$200 +	$12,000 =	$11,650 +	$25,400

The following summarizes the transactions of this chapter. Can you describe what is happening in each line?

	Cash in Bank	Accounts Receivable	Computer Equipment	Office Equipment	Delivery Equipment	=	Accounts Payable	+	Owner's Equity
Prev. Bal.	0	0	0	0	0		0		0
Trans. 1	+25,000								+25,000
Trans. 2				+400					+400
Trans. 3	−3,000		+3,000						
Trans. 4					+12,000		+12,000		
Trans. 5		+200		−200					
Trans. 6	−350						−350		
Trans. 7	+200	−200							
Trans. 8	+1,200								+1,200
Trans. 9	−700								−700
Trans. 10	−500								−500
Balance	$21,850 +	$0 +	$3,000 +	$200 +	$12,000	=	$11,650	+	$25,400

SECTION 3.3
Assessment

After You Read

Reinforce the Main Idea

Use a table like the one shown here to describe four transactions of a home decorating business. Indicate how each transaction affects owner's equity.

Type of Transaction	Transaction Description	Does Owner's Equity Increase or Decrease?
Revenue		
Expense		
Investment		
Withdrawal		

Problem 3 Determining the Effect of Transactions on the Accounting Equation

Instructions Use the accounts of WordService to analyze these business transactions. The beginning balance for each account is shown following the account name.

Assets	=	Liabilities	+	Owners Equity
Cash in Bank, $24,000		Accounts Payable		Jan Swift, Capital
Accounts Receivable, $700		$3,000		$30,700
Computer Equipment, $4,000				
Office Equipment, $5,000				

On the form provided in your working papers, identify the accounts affected by each transaction and the amount of the increase or decrease for each account. Make sure the accounting equation is in balance after each transaction.

1. Paid $50 for advertising in the local newspaper.
2. Received $1,000 as payment for preparing a report.
3. Wrote a $600 check for the month's rent.
4. Jan Swift withdrew $800 for her personal use.
5. Received $200 on account from the person who had purchased the old office furniture.

Math for Accounting

Determine the **Cash in Bank** balance for Wiemack Landscape Designs after the third transaction that follows. All three transactions occurred on the same day. The **Cash in Bank** balance before the first transaction was $10,000.

1. John Wiemack, the owner, withdrew $1,000 from personal savings and deposited that amount in the business checking account.
2. Purchased computer equipment for $5,000; issued a check for 20 percent of the price and agreed to pay the balance at a later date.
3. Issued a check for $100 to buy tools.

Chapter 3
Visual Summary

Concepts
Describe the relationship between property and financial claims

PROPERTY
Property is anything of value that a person or business owns. Property is measured in dollars.

=

FINANCIAL CLAIMS
Financial claims are the legal rights to property and are also measured in dollars.

 +

Analysis
List and define each part of the accounting equation.

PROPERTY = **FINANCIAL CLAIMS**

ASSETS — Items of value owned by the business

=

LIABILITIES — Financial claims of creditors

+

OWNERS EQUITY — Financial claims of owners

Procedures
Use the following steps to analyze a business transaction:

IDENTIFY — Identify the accounts affected.

CLASSIFY — Classify the accounts affected.

+/− — Determine the amount of increase or decrease for each account affected.

BALANCE — Make sure the accounting equation remains in balance.

Chapter 3
Review and Activities

Answering the Essential Question

Why is understanding the accounting equation crucial to understanding the condition of any business?

Suppose your cell phone (property/asset) is worth $150.00. You had to borrow $100.00 from your parents (creditor's financial claim/liability) to buy it. What part of the total value of the phone is yours (owner's financial claim/owner's equity)? Why would it be important for owners and investors to have that kind of information about a business?

Vocabulary Check

1. **Vocabulary** Arrange the vocabulary terms below into categories—groups of related words. Explain why you put the words together.

- acquire
- credit
- creditor
- financial claim
- funds
- property
- assets
- accounting equation
- equities
- liabilities
- owner's equity
- account
- accounts receivable
- business transaction
- eventually
- accounts payable
- investment
- transfer
- on account
- expense
- generate
- revenue
- withdrawal
- conduct

Concept Check

2. **Analyze** Define *property* and *financial claims*. What is the relationship between the two?
3. Name the two types of equities in a business. Give an example of an account that is used for each.
4. What is the accounting equation? Give an example of an account for each part of the equation.
5. **Evaluate** How does each of the following transactions affect the three parts of the accounting equation?
   A. Owner transfer of cash from personal savings account to the business
   B. Owner withdrawal of cash from the business for personal use
   C. Purchase of equipment for cash
   D. Purchase of equipment on credit
   E. Issued check for payment on Transaction D
   F. Received payment for design consulting
   G. Issued check to pay telephone bill
6. **Synthesize** At least how many accounts must be affected by each business transaction. Why?

72 Chapter 3 • Business Transactions and the Accounting Equation

Chapter 3
Standardized Test Practice

Multiple Choice

1. A creditor's claim to the property of an individual or business is called —

    a. an equity

    b. an asset

    c. a liability

    d. a credit

2. An business owner's claim to the property of the business is called —

    a. an equity

    b. an asset

    c. a liability

    d. a credit

3. The money a business earns is

    a. revenue

    b. a withdrawal

    c. an asset

    d. an account payable

True or False

4. The accounting equation is stated as "Assets + Liabilities = Owner's Equity."

5. Cash withdrawals decrease owner's equity by decreasing assets and the owner's financial claim.

6. Accounts payable is the amount of money owed, or payable, to the creditors of a business.

Short Answer

7. List three kinds of assets that a business can own.

Extended Response

8. Explain the difference between Accounts Receivable and Accounts Payable.

Chapter 3
Computerized Accounting

Exploring Electronic Spreadsheets

Computer spreadsheets are important tools for organizing and analyzing data. A spreadsheet is made up of rows and columns. The columns are identified by letters and the rows are identified by numbers. As you create a spreadsheet, you will enter numbers, labels, and formulas into cells. Microsoft® Excel is the most commonly used spreadsheet application in the business world. The following is an example of a basic electronic spreadsheet:

	A	B	C	D	E	F	G	H
1		Jan	Feb	March	April	May	June	July
2	Sales	2,500	5,000	1,650	10,000	3,100	8,200	12,500
3	Expenses	700	1,200	225	3,550	1,800	2,300	5,600
4	Net income	1,800	3,800	1,425	6,450	1,300	5,900	6,900
5								
6								

Before you create a computer spreadsheet, review the following spreadsheet terms.

	Spreadsheet Terms	Description
1	Row	Identified by numbers down the left side of the spreadsheet.
2	Column	Identified by letters along the top of the spreadsheet.
3	Cell address	Identified by a column letter and row number. For example, the cell address B4 indicates the cell where the number 1,800 is found.
4	Active cell	Indicated by a dark border.
5	Scroll arrows	Allows you to view other parts of the spreadsheet.
6	Labels	Text that identifies columns or rows of information; cannot be used for calculations.
7	Values	Numbers inserted in cells that can be used for calculations.
8	Formulas	Mathematical functions entered in a particular cell that tells the software to add, subtract, divide, or multiply values. For example, E2 − E3 represents 10,000 − 3,550, or 6,450

Chapter 3
Problems

Problem 4 Classifying Accounts

All accounts belong in one of the following classifications: Asset, Liability, Owner's Equity.

Instructions In your working papers, indicate the classification for each of the following accounts.

1. John Jones, Capital
2. Cash in Bank
3. Accounts Receivable
4. Accounts Payable
5. Office Equipment
6. Delivery Equipment
7. Camping Equipment
8. Building
9. Land
10. Computer Equipment

Analyze
Identify the accounts that represent financial claims to property.

Problem 5 Completing the Accounting Equation

A business owned and operated by Mike Murray uses these accounts.

Instructions Look at the following list of accounts, and determine the missing amount for each of the question marks.

Assets		=	Liabilities	+	Owners Equity
Cash in Bank	$4,500		Accounts Payable		Mike Murray, Capital $9,250
Accounts Receivable	1,350		?		
Office Equipment	5,000				
	?				

Analyze
Predict what would happen to owner's equity if this business paid all of its bills today.

Chapter 3 • Business Transactions and the Accounting Equation 75

Chapter 3
Problems

Problem 6 — Classifying Accounts Within the Accounting Equation

Listed here are the account names and balances for Wilderness Rentals.

Accounts Payable	$ 7,000	Cash in Bank	$ 5,000
Accounts Receivable	2,000	Office Equipment	3,000
Camping Equipment	12,000	Ronald Hicks, Capital	15,000

Instructions Using these account names and balances:

1. List and total the assets of the business.
2. Determine the amount owed by the business.
3. Give the amount of the owner's equity in the business.

Analyze

Design a diagram that shows the accounting equation for Wilderness Rentals.

Chapter 3
Problems

Problem 7 — Determining Increases and Decreases in Accounts

Hot Suds Car Wash uses the following accounts:

Assets	=	Liabilities	+	Owners Equity
Cash in Bank		Accounts Payable		Regina Delgado, Capital
Accounts Receivable				
Office Equipment				
Office Furniture				
Car Wash Equipment				

Instructions Use a form similar to the one that follows. For each transaction:

1. Identify the accounts affected.
2. Classify the accounts.
3. Determine the amount of the increase (+) or decrease (−) for each account affected.

The first transaction is completed as an example.

Trans.	Account Affected	Classification	Amount of Increase (+) or Decrease (−)
1	Cash in Bank	Asset	+$25,000
	Regina Delgado, Capital	Owner's Equity	+$25,000

Date	Transactions
Jan. 1	1. Regina Delgado, the owner, invested $25,000 cash in the business.
4	2. Bought car wash equipment with cash for $12,000.
5	3. Purchased, on account, $2,500 of office equipment.
10	4. Wrote a check for the monthly rent, $800.
12	5. Received cash for services performed, $1,000.
15	6. The owner withdrew $600 cash from the business for personal use.
20	7. Purchased a desk for $1,000, paying $200 cash and agreeing to pay the balance of $800 in 30 days.
25	8. Provided services worth $600 on account.

Analyze
Identify the transaction that affects the most accounts.

Chapter 3
Problems

Problem 8 — Determining the Effects of Transactions on the Accounting Equation

After graduating from college, Abe Shultz decided to start a pet grooming service called Kits & Pups Grooming.

Instructions Use a form similar to the one that follows. For each of the following transactions:

1. Identify the accounts affected, using the account names on the form.
2. Determine the amount of the increase or decrease for each account.
3. Write the amount of the increase (+) or decrease (−) in the space under each account affected.
4. On the following line, write the new balance for each account.
5. Transaction 1 is completed as an example.

Trans.	Assets				=	Liabilities	+	Owner's Equity
	Cash in Bank	Accts. Rec.	Office Equip.	Grooming Equip.	=	Accounts Payable	+	Abe Shultz, Capital
1	+$10,000							+$10,000

Date	Transactions
Jan. 2	1. Abe Shultz began the business by depositing $10,000 in a checking account at the Shoreline National Bank in the name of the business, Kits & Pups Grooming.
3	2. Bought grooming equipment for cash, $1,000.
8	3. Issued a check for $900 for the monthly rent.
9	4. Bought $6,000 worth of new office equipment on account for use in the business.
15	5. Received $700 cash for services performed for customers during the first week of business.
21	6. Issued a $2,000 check to the creditor as partial payment for the office equipment purchased on account.
29	7. Performed grooming services and agreed to be paid for them later, $500.

Analyze

Explain the difference between Transaction 5 and Transaction 7.

Chapter 3
Problems

Problem 9 Determining the Effects of Transactions on the Accounting Equation

Juanita Ortega is the owner of a professional guide service called Outback Guide Service.

Instructions Use a form similar to the one below. Complete these steps for each of the following transactions:

1. Identify the accounts affected.
2. Write the amount of the increase (+) or decrease (−) in the space provided on the form in your working papers.
3. Determine the new balance for each account.

Trans.	Assets					=	Liabilities	+	Owner's Equity
	Cash in Bank	Accts. Rec.	Hiking Equip.	Rafting Equip.	Office Equip.	=	Accounts Payable	+	Juanita Ortega, Capital

Date		Transactions
Jan. 3	1.	Ms. Ortega, the owner, opened a checking account for the business by depositing $60,000 of her personal funds.
6	2.	Paid by check the monthly rent of $3,000.
8	3.	Bought hiking equipment for the business by writing a check for $3,000.
9	4.	Purchased $24,000 of rafting equipment by writing a check.
11	5.	Purchased office equipment on account for $4,000.
15	6.	Received payment for guide services, $2,500.
18	7.	Ms. Ortega contributed a desk valued at $450 to the business.
21	8.	Withdrew $3,000 cash from the business for personal use.
26	9.	Wrote a check to a creditor as partial payment on account, $1,500.
30	10.	Took a group on a tour and agreed to accept payment later, $1,200.

Analyze

Calculate the amount owed to creditors after Transaction 10.

Chapter 3
Problems

Problem 10   Describing Business Transactions

Showbiz Video is a business owned by Greg Failla. The transactions that follow are shown as they would appear in the accounting equation.

Instructions In your working papers, describe what has happened in each transaction. Transaction 1 is completed as an example.

Example:

1. The owner invested $30,000 in the business.

Trans.	Assets				=	Liabilities	+	Owner's Equity
	Cash in Bank	Accts. Rec.	Office Equip.	Video Equip.	=	Accounts Payable	+	Greg Failla, Capital
1	+$30,000							+$30,000
2	−$ 2,000		+$ 2,000					
3				+$ 8,000		+$ 8,000		
4	+$ 700							+$ 700
5		+$ 500						+$ 500
6			+$ 200					+$ 200
7	−$ 3,000					−$ 3,000		
8		+$ 200	−$ 200					
9	+$ 500	−$ 500						
10	−$ 1,000							−$ 1,000

Analyze

Calculate the balance of the account, Greg Failla, Capital.

Chapter 3
Problems

Problem 11 Completing the Accounting Equation

The account names and balances for Job Connect are listed below.

Instructions Determine the missing amount for each of the question marks. Use the form in your working papers and write in the missing amounts.

Trans.	Assets			=	Liabilities	+	Owner's Equity
	Cash in Bank	Accounts Receivable	Business Equipment	=	Accounts Payable	+	Richard Tang, Capital
1	?	$ 2,000	$ 1,000		$ 500		$ 7,500
2	$ 3,000	$ 9,000	?		$2,000		$16,000
3	$ 8,000	$ 1,000	$ 10,000		?		$15,000
4	$ 4,000	?	$ 4,000		$ 1,000		$17,000
5	$ 9,000	$ 7,000	$ 6,000		$5,000		?
6	$ 10,000	$ 14,000	?		$6,000		$32,000
7	$ 6,000	$ 4,000	$ 10,000		?		$15,000
8	?	$ 5,000	$ 9,000		$ 1,000		?

Hint: In line 8, total assets are $18,000.

Analyze

Explain the mathematical operations used to solve for the accounting equation.

Real-World Applications & Connections

Career Wise

Certified Public Accountant (CPA)

A certified public accountant can be a great career for those who are interested in working with a variety of clients and doing a variety of accounting tasks, rather than working with only one organization or specializing in a certain type of task.

Education Requirements

Most states require a CPA to have 150 hours of college coursework, which means pursuing a master's degree in addition to a bachelor's degree. CPAs must also pass a special licensing exam and continue to take classes throughout their career.

Roles

The CPA acts as an accountant for the general public. The CPA advises the general public on tax and financial matters and advises companies on how to control their procedures to make sure they comply with accounting standards and procedures.

Responsibilities

CPAs must always exercise good judgment and conduct their work ethically. They must do a good job advising their clients and be a good representative of the accounting profession.

Activity

CPAs very often have their own businesses. Use the career information tools available to you (Internet, library, and your career center) to research the extra responsibilities that come with owning your own business. Write a paragraph summarizing what you discovered.

Success Skills

Information and Media Literacy

Making decisions, whether business or personal, requires being able to acquire and analyze information. Imagine you have just graduated and are ready to look for a job matching your skills and interests.

Instructions

1. Use the Internet to find local job postings and then to research the companies that are hiring. Would you be interested in working for any of these companies?

2. Write a summary of your findings, and describe the accounting career opportunities you think these organizations might offer.

A Matter of Ethics

Company Property

Many companies provide office supplies for their employees' use while on the job. Imagine that you work for a large department store like JC Penney. Several of your coworkers take company supplies home for their personal use, such as pens, bags, hangers, and boxes. You need boxes to store some items at home, so you consider taking them from the supply room.

Activity

Define the ethical issue involved. Then describe how you think ethics relates to taking supplies from the workplace. Who are the affected parties? What are the alternatives? How do the alternatives affect the parties? What would you do?

Spotlight on Personal Finance

Your Earning Power

Your future income will depend on various things including your career choice, your education, and the region where you live.

Activity

List three jobs that are in different fields and different regions in the United States. Using the Internet or library, research the education requirements and the salary ranges of the three jobs. Create a table and organize the results of your research.

H.O.T. Audit

Conducting an Audit

To audit means to examine and verify financial records and reports.

Instructions

Review the business transactions recorded in the accounting equation below for errors. On a separate sheet of paper, make any corrections necessary.

A. Purchased a delivery truck for $25,000 on account from A-1 Trucks Inc.

B. Wrote a check for $1,000 to A-1 Trucks Inc. for payment on account.

ASSETS		= LIABILITIES	+ OWNERS EQUITY
Cash in Bank	Delivery Equipment	Accounts Payable	J. Smith, Capital
A. 0	−$25,000	−$25,000	0
B. −$1,000	0	−$ 1,000	0

Chapter 4

Transactions that Affect Assets, Liabilities, and Owner's Capital

Chapter Topics

4-1 Accounts and The Double-Entry Accounting System

4-2 Applying The Rules of Debit and Credit

Visual Summary

Review and Activities

Standardized Test Practice

Problems

Real-World Accounting Careers

Real-World Applications & Connections

 Essential Question

As you read this chapter, keep this question in mind:

Why would a business need to balance the money it's earning with the money it's spending?

Main Idea

A business's transactions can be analyzed by using the double-entry accounting system, which recognizes the different sides of business transactions as debits and credits.

Chapter Objectives

Concepts

C1 Describe the chart of accounts.

C2 Explain the purpose of double-entry accounting.

Analysis

A1 Use T accounts to analyze transactions that affect assets, liabilities, and the owner's capital account.

Procedures

P1 Prepare a chart of accounts.

P2 Identify the normal balance of accounts.

P3 Use T accounts to illustrate the rules of debit and credit.

P4 Calculate the account balances after recording business transactions.

Real-World Business Connection

1154 LILL Studio

If custom accessories are your bag, you'll love the story of this unique sole proprietorship. It began at a Chicago street fair in 1999, where owner, Jennifer Velarde, an interior designer and entrepreneur, planned to launch her new line of handbags. She realized the night before, however, that she might not have enough bags to last through the two-day fair. So she decided to bring fabrics and have people custom-order their bags, using her sewn bags as "sample styles." It was a huge hit. Customers loved the experience of being in the designer's chair and having a creation that was one-of-a-kind. She began that weekend advertising "handbag parties" as a way to recreate the experience.

Connect to the Business

1154 LILL Studio now has multiple boutique locations, a network of home party reps, and a large handbag production staff. Its marketing strategies keep expanding, and now customers can go online to design a bag using virtual styles and swatches.

Analyze

How is a marketing strategy an important tool for growing the revenue of a new business?

Focus on the Photo

LILL Studio owner Jennifer Velarde has built a successful sole proprietorship by encouraging customers to be creative and design their own handbags. **What kinds of transactions would a business like this process?**

SECTION 4.1
Accounts and the Double-Entry Accounting System

You will now learn a concept that drives every accounting transaction in a company's financial reporting, the double-entry accounting system. You will specifically explore the rules for credit and debit, and how to employ these rules for business transaction analysis. When a large company records the dollar amount of a transaction in one account, it records an equal amount in another account. The same is true for a small company.

Section Vocabulary

- chart of accounts
- ledger
- double-entry accounting
- debit
- credit
- T account
- normal balance
- physical
- exceed

The Chart of Accounts

How Is the Chart of Accounts Organized?

A **chart of accounts** is a list of all accounts used by a business. A small business may require only 20 or 30 accounts, but a large one may have several thousand.

Recall that an account is a record of changes and balances of a specific asset, liability, or component of owner's equity. Accounts may have different **physical** forms, depending on the system. In a manual system, each account may be a separate page in a book or a separate card in a tray. In an electronic system, accounts are stored on disks or hard drives.

Whether a system is manual or electronic, accounts are grouped together in a **ledger**. The ledger is often referred to as the general ledger. "Keeping the books" refers to maintaining accounts in the ledger. Grouping accounts in a ledger makes information easy to find. Information is taken from the ledger and used to prepare financial statements.

A system for numbering accounts makes it easy to locate individual accounts in the ledger. Account numbers have two or more digits used for sorting information based on the kinds of reports the business needs. A small company may use a three-digit system. A very large corporation may use 35 or more digits. A typical numbering system used to prepare a chart of accounts is as follows:

- Asset accounts begin with 1.
- Liability accounts begin with 2.
- Owner's equity accounts begin with 3.
- Revenue accounts begin with 4.
- Expense accounts begin with 5.

Reading Check

Recall

What is the chart of accounts?

86 Chapter 4 • Transactions that Affect Assets, Liabilities, and Owner's Capital

CHART OF ACCOUNTS

ASSETS
101 Cash in Bank
105 Accounts Receivable—City News
110 Accounts Receivable—Green Company
115 Computer Equipment
120 Office Equipment
125 Delivery Equipment

LIABILITIES
201 Accounts Payable—Beacon Advertising
205 Accounts Payable—Coast to Coast Auto

OWNER'S EQUITY
301 Crista Vargas, Capital
302 Crista Vargas, Withdrawals
303 Income Summary

REVENUE
401 Delivery Revenue

EXPENSES
501 Advertising Expense
506 Maintenance Expense
507 Rent Expense
508 Utilities Expense

Double-Entry Accounting

How Does Double-Entry Accounting Work?

In Chapter 3 you used the accounting equation for analyzing and recording changes in account balances. Using the equation works well if a business has only a few accounts. It becomes awkward, however, if a business has many accounts and many transactions to analyze.

Accountants use the **double-entry accounting** system to analyze and record a transaction. Double-entry accounting records the different sides of business transactions as *debits* and *credits*. A **debit** is an entry on the left side of an account. A **credit** is an entry on the right side of an account. This system is more efficient than updating the accounting equation for each transaction.

T Accounts

An efficient tool for analyzing double-entry accounting transactions is a *T account*. The **T account**, so called because of its T shape, shows the dollar increase or decrease in an account that is caused by a transaction. T accounts help the accountant analyze the parts of a business transaction.

As in the illustration here, a T account has an account name, a left side, and a right side. The account name is at the top of the T. The left side of T accounts is *always* used for debit amounts. The right side of T accounts is *always* used for credit amounts. The words *debit* and *credit* are simply the accountant's terms for *left* and *right*. Accountants sometimes use **DR** for debit and **CR** for credit.

Account Name	
Left Side	Right Side
Debit Side	Credit Side
Debit	Credit

The Rules of Debit and Credit

Debits and credits are used to record the increase or decrease in each account affected by a business transaction. Under double-entry accounting, for each debit entry made in one account, a credit of an equal amount must be made in another account. Debit and credit rules vary according to whether an account is classified as an asset, a liability, or the owner's capital account. Regardless of the type of account, however, the left side of an account is always the debit side and the right side is always the credit side.

Each account classification has a specific side that is its normal balance side. An account's **normal balance** is always on the side used to record increases to the account. The word *normal* used here means *usual.* Throughout this book, note that the normal balance side of each account is shaded.

Rules for Asset Accounts.

Asset accounts follow three rules of debit and credit:

1. An asset account is *increased* (+) on the *debit* side (left side).
2. An asset account is *decreased* (−) on the *credit* side (right side).
3. The *normal balance* for an asset account is the *increase,* or the *debit* side. The normal balance side is shaded in the following T account.

Notice that assets appear on the left side of the accounting equation.

$$\text{ASSETS} = \text{LIABILITIES} + \text{OWNER'S EQUITY}$$

Asset Accounts	
Debit	Credit
+	−
(1) Increase Side	(2) Decrease Side
(3) Normal †Balance	

For asset accounts the *increase* side is the debit (left) side of the T account. The *decrease* side is the credit (right) side of the T account. Notice the (+) and (−) signs that are used to indicate the increase and decrease sides of the account. They do not mean the same thing as *debit* and *credit.*

Since the increase side of an asset account is always the debit side, asset accounts have a normal debit balance. For example, in the normal course of business, total increases to assets are larger than or **exceed** total decreases. You would expect an asset account, then, to have a normal debit balance.

Let's apply these rules to an actual asset account. Look at the entries in the T account for **Cash in Bank** shown here. The increases in the account are recorded on the left, or debit, side. The decreases in the account are recorded on the right, or credit, side. Total debits equal $350 ($200 + $150). Total credits equal $110 ($70 + $40). To find the balance, subtract total credits from total debits ($350 − $110). The debit balance is $240.

Cash in Bank	
Debit	Credit
+	−
200	70
150	40
350	110
Bal. 240	

88 Chapter 4 • Transactions that Affect Assets, Liabilities, and Owner's Capital

Rules for Liability and Owner's Capital Accounts.

The rules of debit and credit for liability and the owner's capital account are:

1. Liability and owner's capital accounts are *increased* on the *credit* (right) side.
2. Liability and owner's capital accounts are *decreased* on the *debit* (left) side.
3. The *normal balance* for liability and owner's capital accounts is the *increase*, or the *credit* side.

To illustrate these rules, let's look again at the accounting equation and the T accounts. Remember, the normal balance side is shaded.

Assets	=	Liabilities	+	Owner's Equity
Debit + Increase Side / Normal Balance Credit − Decrease Side		Debit − (2) Decrease Side Credit + (1) Increase Side / (3) Normal ††Balance		Debit − (2) Decrease Side Credit + (1) Increase Side / (3) Normal Balance

For all three types of accounts, the debit side is always the left side of the T account, and the credit side is always the right side. Notice, however, that the increase (+) and decrease (−) sides of the liability and owner's capital accounts are the opposite of those for assets. This difference exists because accounts classified as liabilities and owner's capital are on the opposite side of the accounting equation from accounts classified as assets. As a result, debit and credit rules on one side of the accounting equation—and the T accounts within it—are mirror images of those on the other side.

Let's apply these rules to actual accounts. First, look at the entries in the T account below for the liability account **Accounts Payable**. Increases are recorded on the right, or credit, side. The decreases are recorded on the left, or debit, side. Total credits equal $375 ($200 + $175); total debits equal $175 ($100 + $75). To find the balance, subtract the total debits from the total credits ($375 − $175). The credit balance is $200.

Accounts Payable	
Debit −	Credit +
100	200
75	175
175	375
	Bal. 200

Common Mistakes

Confusing Debits If you have a bank account, you may have a debit card. The term debit card can cause some confusion for someone taking the first accounting course. Throughout this course, we refer to an increase in cash as a debit to cash. However, using your debit card will result in a decrease in your cash account. It's the merchant's cash account that increases and is debited. Don't let this confuse you.

Now look at the entries in the T account for the owner's equity account **Crista Vargas, Capital**. Remember that the rules of debit and credit for the capital account are the same as those for a liability account.

Crista Vargas, Capital	
Debit	Credit
−	+
350	1,500
200	2,500
550	4,000
	Bal. 3,450

Increases to owner's capital are recorded on the right, or credit, side of the account. Decreases are recorded on the left, or debit, side. The capital account has a normal credit balance. If you subtract the total debits from the total credits ($4,000 − $550), you have a credit balance of $3,450.

SECTION 4.1
Assessment

After You Read

Reinforce the Main Idea

Create a chart like this one to summarize the rules of debit and credit. Fill in each blank box with the word *debit* or *credit*.

	Asset Accounts	Liability Accounts	Owner's Capital Account
Normal Balance			
Increase Side			
Decrease Side			

Problem 1 Applying the Rules of Debit and Credit

Speedy Appliance Repair, owned by R. Lewis, uses the following accounts:

> **General Ledger**
> Cash in Bank
> Accounts Receivable
> Office Equipment
> Accounts Payable
> R. Lewis, Capital

Instructions In the form provided in your working papers:

1. Classify each account as an asset, liability, or owner's capital account.
2. Indicate whether the increase side is a debit or a credit.
3. Indicate whether the decrease side is a debit or credit.
4. Indicate whether the normal balance for the account is a debit or credit balance.

The **Cash in Bank** account is completed as an example.

Account Name	Account Classification	Increase Side	Decrease Side	Normal Balance
Cash in Bank	Asset	Debit	Credit	Debit

Math for Accounting

On December 1, Poremba Pizza had a cash-in-bank balance of $7,000. During the month of December, Poremba Pizza wrote checks totaling $4,800. Two-thirds of this amount was used to purchase a computer for cash. The remaining amount was used to pay an outstanding invoice for kitchen equipment purchased from Restaurant City.

1. List the account(s) debited and the debit amount(s).
2. List the account(s) credited and the credit amount(s).
3. What was the cash-in-bank balance on December 31?

SECTION 4.2
Applying the Rules of Debit and Credit

Now that you are familiar with the rules of debit and credit for asset, liability, and owner's capital accounts, you can use those rules to analyze business transactions.

Section Vocabulary
- method
- demonstrate

Business Transaction Analysis

How Do You Analyze Transactions?

Whether a business is buying a new computer system, paying its utility bills, or receiving money for sales, the accountant must analyze how the transaction should be recorded. When analyzing business transactions, you should use the following step-by-step **method**:

Business Transaction

BUSINESS TRANSACTION ANALYSIS: Steps to Success

Analysis	Identify	1. Identify the accounts affected.
	Classify	2. Classify the accounts affected.
	+/−	3. Determine the amount of increase or decrease for each account affected.
Debit-Credit Rule		4. Which account is debited? For what amount?
		5. Which account is credited? For what amount?
T Accounts		6. What is the complete entry in T-account form?
		Account Name Account Name

Reading Check

Define

What do the terms debit and credit mean in accounting?

Assets and Equities Transactions

How Do You Use T Accounts for Assets and Equities?

The business transactions that follow are for Zip Delivery Service. Throughout the next several pages, you will learn how to analyze each Zip transaction, apply the rules of debit and credit, and complete the entry in T-account form. The T accounts **demonstrate** the effects of transactions on the accounting equation. A debit in one account is offset by a credit in another account. Refer to the Zip Delivery Service chart of accounts on page 84. These accounts will be used to analyze several business transactions.

Assets and Owner's Capital

Business Transactions 1 and 2 use T accounts to illustrate owner's investments in the business.

Business Transaction 1

On October 1 Crista Vargas took $25,000 from personal savings and deposited that amount to open a business checking account in the name of Zip Delivery Service.

Analysis		
	Identify	1. The accounts **Cash in Bank** and **Crista Vargas, Capital** are affected.
	Classify	2. **Cash in Bank** is an asset account. **Crista Vargas, Capital** is an owner's capital account.
	+/−	3. **Cash in Bank** is increased by $25,000. **Crista Vargas, Capital** is increased by $25,000.

Debit-Credit Rule	
	4. Increases in asset accounts are recorded as debits. Debit **Cash in Bank** for $25,000.
	5. Increases in the owner's capital account are recorded as credits. Credit **Crista Vargas, Capital** for $25,000.

T Accounts 6.

Cash in Bank		Crista Vargas, Capital
Debit + 25,000		Credit + 25,000

Business Transaction 2

On October 2 Crista Vargas took two telephones valued at $200 each from her home and transferred them to the business as office equipment.

Analysis	Identify	1. The accounts **Office Equipment** and **Crista Vargas, Capital** are affected.
	Classify	2. **Office Equipment** is an asset account. **Crista Vargas, Capital** is an owner's capital account.
	+/−	3. **Office Equipment** is increased by $400. **Crista Vargas, Capital** is increased by $400.
Debit-Credit Rule		4. Increases in asset accounts are recorded as debits. Debit **Office Equipment** for $400.
		5. Increases in owner's capital accounts are recorded as credits. Credit **Crista Vargas, Capital** for $400.
T Accounts		6.

```
         Office Equipment                    Crista Vargas, Capital
         Debit                                              Credit
           +                                                   +
          400                                                 400
```

Assets and Liabilities

The following examples show changes to assets and liabilities in T-account form:

- Business Transaction 3: increases an asset and decreases another asset
- Business Transaction 4: increases an asset and increases a liability
- Business Transaction 5: increases an asset and decreases another asset
- Business Transaction 6: decreases a liability and decreases an asset
- Business Transaction 7: increases an asset and decreases a liability

Business Transaction 3

On October 4 Zip issued Check 101 for $3,000 to buy a computer system.

Analysis	Identify	1. The accounts **Computer Equipment** and **Cash in Bank** are affected.
	Classify	2. **Computer Equipment** and **Cash in Bank** are asset accounts.
	+/−	3. **Computer Equipment** is increased by $3,000. **Cash in Bank** is decreased by $3,000.

94 Chapter 4 • Transactions that Affect Assets, Liabilities, and Owner's Capital

Debit-Credit Rule	4.	Increases in asset accounts are recorded as debits. Debit **Computer Equipment** for $3,000.
	5.	Decreases in asset accounts are recorded as credits. Credit **Cash in Bank** for $3,000.

T Accounts 6.

Computer Equipment	Cash in Bank
Debit + 3,000	Credit − 3,000

Business Transaction 4

On October 9 Zip bought a used truck on account from Coast to Coast Auto for $12,000.

Analysis	Identify	1.	The accounts **Delivery Equipment** and **Accounts Payable—Coast to Coast Auto** are affected.
	Classify	2.	**Delivery Equipment** is an asset account. **Accounts Payable—Coast to Coast Auto** is a liability account.
	+/−	3.	**Delivery Equipment** is increased by $12,000. **Accounts Payable—Coast to Coast Auto** is increased by $12,000.

Debit-Credit Rule	4.	Increases in asset accounts are recorded as debits. Debit **Delivery Equipment** for $12,000.
	5.	Increases in liability accounts are recorded as credits. Credit **Accounts Payable—Coast to Coast Auto** for $12,000.

T Accounts 6.

Delivery Equipment	Accounts Payable— Coast to Coast Auto
Debit + 12,000	Credit + 12,000

Business Transaction 5

On October 11 Zip sold one phone on account to Green Company for $200.

Analysis	Identify	1.	The accounts **Accounts Receivable—Green Company** and **Office Equipment** are affected.
	Classify	2.	**Accounts Receivable—Green Company** is an asset account. **Office Equipment** is also an asset account.
	+/−	3.	**Accounts Receivable—Green Company** is increased by $200. **Office Equipment** is decreased by $200.
Debit-Credit Rule		4.	Increases in asset accounts are recorded as debits. Debit **Accounts Receivable—Green Company** for $200.
		5.	Decreases in asset accounts are recorded as credits. Credit **Office Equipment** for $200.

6.

Accounts Receivable—Green Company		Office Equipment	
Debit + 200			Credit − 200

Business Transaction 6

On October 12 Zip mailed Check 102 for $350 as the first installment payment on the truck purchased from North Shore Auto on October 9.

Analysis	Identify	1.	The accounts **Accounts Payable—Coast to Coast Auto** and **Cash in Bank** are affected.
	Classify	2.	**Accounts Payable—Coast to Coast Auto** is a liability account. **Cash in Bank** is an asset account.
	+/−	3.	**Accounts Payable—Coast to Coast Auto** is decreased by $350. **Cash in Bank** is decreased by $350.
Debit-Credit Rule		4.	Decreases in liability accounts are recorded as debits. Debit **Accounts Payable—Coast to Coast Auto** for $350.
		5.	Decreases in asset accounts are recorded as credits. Credit **Cash in Bank** for $350.

T Accounts	6.	Accounts Payable—Coast to Coast Auto		Cash in Bank	
		Debit — 350			Credit — 350

Business Transaction 7

On October 14 Zip received and deposited a check for $200 from Green Company. The check is full payment for the telephone sold on account to Green Company on October 11.

Analysis Identify 1. The accounts **Cash in Bank** and **Accounts Receivable—Green Company** are affected.

Classify 2. **Cash in Bank** is an asset account. **Accounts Receivable—Green Company** is an asset account.

+/− 3. **Cash in Bank** is increased by $200. **Accounts Receivable—Green Company** is decreased by $200.

Debit-Credit Rule 4. Increases in asset accounts are recorded as debits. Debit **Cash in Bank** for $200.

5. Decreases in asset accounts are recorded as credits. Credit **Accounts Receivable—Green Company** for $200.

T Accounts	6.	Cash in Bank		Accounts Receivable—Green Company	
		Debit + 200			Credit — 200

Chapter 4 • Transactions that Affect Assets, Liabilities, and Owner's Capital

SECTION 4.2
Assessment

After You Read

Reinforce the Main Idea

Imagine that you have your own business. Write a description of a typical transaction your business would have. On a sheet of paper, express the same transaction in T-account form.

Problem 2 — Identifying Increases and Decreases in Accounts

Alice Roberts uses the following accounts in her business:

General Ledger

Cash in Bank	Office Equipment
Accounts Receivable	Accounts Payable
Office Furniture	Alice Roberts, Capital

Instructions Analyze each of the following transactions. In your working papers, explain the debit and the credit. Use the format shown in the example.

Example:

On June 2 Alice Roberts invested $5,000 of her own money in a business called Roberts Employment Agency.

a. The asset account **Cash in Bank** is increased. Increases in asset accounts are recorded as debits.

b. The owner's capital account **Alice Roberts, Capital** is increased. Increases in the owner's capital account are recorded as credits.

Date	Transactions
June 3	1. Purchased a computer on account from Computer Inc. for $2,500.
9	2. Transferred a desk (Office Furniture) to the business. The desk is worth $750.
15	3. Made a partial payment on account of $1,000 to Computer Inc.

Math for Accounting

Diane Hendricks always dreamed of owning a recording studio. On February 1 Diane withdrew $10,000 from personal savings and deposited it in a new business checking account called Hendricks Sound. On February 2 Hendricks Sound made a $2,000 down payment on equipment that cost $8,000. The remaining balance will be paid at a later date. What are the amounts shown in the accounting equation for Hendricks Sound after these transactions?

Chapter 4
Visual Summary

Concepts
Double Entry Accounting

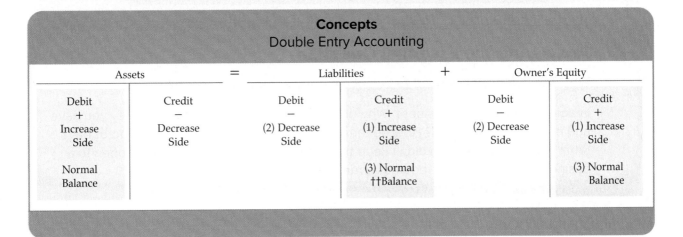

Analysis
Analyze a Transaction

Analysis	Identify Classify +/−	1. Identify the accounts affected. 2. Classify the accounts affected. 3. Determine the amount of increase or decrease for each account affected.
Debit-Credit Rule		4. Which account is debited? For what amount? 5. Which account is credited? For what amount?
T Accounts		6. What is the complete entry in T-account form? Account Name Account Name

Procedures
Normal Balance

Asset Rule:

Asset Accounts

| Debit
Increase
Side
+
Normal
Balance | Credit
Decrease
Side
− |

Liability and Owner's Capital Rule:

Liability and Owner's Capital Accounts

| Debit
Decrease
Side
− | Credit
Increase
Side
+
Normal
Balance |

Chapter 4
Review and Activities

Answering the Essential Question

Why would a business need to balance money earned with money spent?

Imagine that you are a business. Why do you need to know how much money you have and how much you spend? How do these things affect what you're able to do or buy? What would happen if you didn't keep track of your money? What would happen to a business if good financial records weren't kept?

Vocabulary Check

1. **Vocabulary** On a sheet of paper, write your own definition for each content and academic vocabulary term. Provide an illustration if necessary to make the definition clearer.

   - chart of accounts
   - ledger
   - physical
   - double-entry accounting
   - debit
   - credit
   - T account
   - exceed
   - normal balance
   - method
   - demonstrate

Concept Check

2. What is a chart of accounts and why are the accounts numbered? Prepare a sample chart of accounts to illustrate the typical numbering system.
3. Describe double-entry accounting.
4. What is the normal balance for asset accounts? Liability accounts? The owner's capital account?
5. Explain the three basic parts of a T account.
6. **Analyze** Why are the three rules for asset accounts different from the three rules for liability and owner's capital accounts?
7. **Predict** What are the six steps required to calculate the account balances after recording business transactions? Predict what might happen if step 2 is not done correctly.
8. **Evaluate** How would the charts of accounts for large corporations and small companies differ? How would they be alike?
9. **Math** The balance of a revenue account has a normal balance of $473.44. If the account is credited in the amount of $38.64, and the account is debited in the amount of $130.77, what is the new balance? What would be the new balance if it was an asset account instead of a revenue account?
10. **English Language Arts** Use the Internet and other sources to research the role of the double-entry accounting system. In a one-page essay, define the system, explain how it gets its name, and tell how it works to ensure the integrity of financial practice. Use at least two sources for your information and be sure to correctly cite them.

Chapter 4
Standardized Test Practice

Multiple Choice

1. The normal balance of an account is always on the _____
   a. side of the current account balance
   b. side used to record increases
   c. credit side of the account
   d. debit side of the account

2. At the beginning of the week, Cash in Bank has an account balance of $3,500.00. During the week, the following transactions occur: Office supplies in the amount of $236.12 are purchased; the owner takes a withdrawal of $450.00; a customer's payment of $780.00 is received; and the company's monthly electric bill in the amount of $178.43 is paid. What is the balance of the Cash in Bank account at the end of the week?
   a. Credit balance of $1,855.45
   b. Debit balance of $3,415.45
   c. Credit balance of $3,415.45
   d. Debit balance of $1,855.45

3. If the current assets of Random Electronics Inc. equal $789,541.00, and their current liabilities total $590,961.30, what is the current amount of owner's equity?
   a. $198,579.70
   b. $1,380,501.30
   c. $198,579.70
   d. $189,579.70

4. XYZ company shows the following account balances: Cash in Bank = $12,565.43; Owner's Capital = $40,875; Accounts Receivable = $22,578.44; Equipment = −$660.32. The following transactions occur: The owner takes a withdrawal in the amount of $500.00; a payment from customer Z in the amount of $3,000.00 is received; and equipment is purchased in the amount of $865.00. Which shows an incorrect account balance after all the transactions are recorded?
   a. Cash in Bank = $15,565.43
   b. Owner's Capital = $40,375.00
   c. Accounts Receivable = $19,578.44
   d. Equipment = $204.68

Chapter 4
Standardized Test Practice

Short Answer

5. In double-entry accounting, a debit is an entry on the _____ side of an account.

Extended Response

6. Explain how T accounts work and why they are useful for double-entry accounting.

7. Devise a study tool or plan that allows you to easily remember normal balances for asset, liability, and owner's equity accounts. Explain how your study tool or plan works.

Chapter 4
Computerized Accounting

Introduction to Computerized Accounting Systems

Making the Transition from a Manual to a Computerized System

MANUAL METHODS
• Transactions are recorded into journals by hand.
• The details of each transaction are then posted in the general ledger.
• The accountant computes account balances and prepares a trial balance to verify that the accounting equation is still in balance.
• Account names and balances are then transferred to the proper financial report and the report is summarized (totaled).

COMPUTERIZED METHODS
• Transactions are keyed to the appropriate screen in the accounting system.
• Posting to the general ledger accounts occurs automatically.
• The accounting system generates the trial balance.
• Financial reports pull the appropriate accounts and their current balances from the general ledger computer files. The reports automatically summarize and can be printed when the user chooses.

Chapter 4
Problems

Problem 3 Identifying Accounts Affected by Transactions

Ronald Hicks owns Wilderness Rentals and uses the following accounts in his business:

General Ledger

101 Cash in Bank
105 Accounts Receivable—
    Helen Katz
120 Office Equipment
125 Camping Equipment

201 Accounts Payable—
    Adventure Equipment Inc.
301 Ronald Hicks, Capital

Instructions For each of the following transactions:

1. Indicate the two accounts affected.
2. Indicate whether each account is debited or credited.

Date	Transactions
May 11	1. Sold on account to Helen Katz an unneeded fax machine.
19	2. Purchased camping equipment on credit from Adventure Equipment Inc. Payment is due within 30 days.
22	3. Ronald Hicks brought a filing cabinet from home and transferred it to the business **(Office Equipment)**.
23	4. Purchased tents and sleeping bags for cash.

Analyze

Identify the transactions that affect the Cash in Bank account.

Chapter 4 Problems

Problem 4 Using T Accounts to Analyze Transactions

Regina Delgado owns a business called Hot Suds Car Wash. She uses the following accounts:

General Ledger

101 Cash in Bank
110 Accounts Receivable—
    Valley Auto
125 Office Equipment
130 Office Furniture
135 Car Wash Equipment
201 Accounts Payable—
    Allen Vacuum Systems
301 Regina Delgado, Capital

Instructions For each transaction:

1. Determine which accounts are affected.
2. Prepare T accounts for the accounts affected.
3. Enter the debit and credit amounts in the T accounts.

Date	Transactions
May 5	1. Regina Delgado invested an additional $40,000 cash in her business.
12	2. Bought another car wash system on account for $27,000 from Allen Vacuum Systems.
17	3. Regina Delgado transferred some of her personal office furniture, valued at $3,750, to her business.
24	4. Hot Suds Car Wash purchased additional office equipment for $7,500. Payment was made by check.
29	5. Hot Suds Car Wash sold some surplus car washing equipment on account to Valley Auto for $1,200.

Analyze

Calculate the ending balance for the liability account **Accounts Payable—Allen Vacuum Systems.**

Chapter 4
Problems

Problem 5 Analyzing Transactions into Debit and Credit Parts

Abe Shultz owns Kits & Pups Grooming and uses the following accounts:

General Ledger

- 101 Cash in Bank
- 115 Accounts Receivable—Martha Giles
- 125 Office Equipment
- 130 Office Furniture
- 140 Grooming Equipment
- 205 Accounts Payable—Dogs & Cats Inc.
- 301 Abe Shultz, Capital

Instructions For each transaction:

1. In your working papers, prepare a T account for each account listed.
2. Using the appropriate T accounts, analyze and record each of the following business transactions. Identify each transaction by number.
3. After recording all transactions, write the word *Balance* on the normal balance side of each T account. Then compute and record the balance for each account.

Date		Transactions
May	1	1. Abe Shultz invested an additional $45,000 cash in his business.
	5	2. Bought grooming equipment on account from Dogs & Cats Inc. for $8,500.
	10	3. Purchased an office lamp for $85, Check 150.
	14	4. Abe Shultz transferred his personal fax machine, worth $200, to the business.
	19	5. Made a $3,000 payment on the grooming equipment bought on account, Check 151.
	22	6. Sold the typewriter on account to Martha Giles for $200.
	29	7. Bought a photocopier for $1,500, Check 152.
	31	8. Received a $100 payment for the typewriter sold on account.

Analyze

Calculate the total cash spent in the month of May.

Chapter 4
Problems

Problem 6 Analyzing Transactions into Debit and Credit Parts

Juanita Ortega runs Outback Guide Service. The accounts she uses to record and report business transactions are listed below.

General Ledger

101 Cash in Bank	150 Rafting Equipment
105 Accounts Receivable—Mary Johnson	205 Accounts Payable—Peak Equipment Inc.
130 Office Equipment	207 Accounts Payable—Premier Processors
140 Computer Equipment	301 Juanita Ortega, Capital
145 Hiking Equipment	

Instructions For each transaction:

1. In your working papers, prepare a T account for each account.

2. Analyze and record each of the following business transactions in the appropriate T accounts. Identify each transaction by number.

3. After recording all transactions, compute and record the account balance on the normal balance side of each T account.

4. Add the balances of those accounts with normal debit balances.

5. Add the balances of those accounts with normal credit balances.

6. Compare the two totals. Are they the same?

Chapter 4
Problems

Date		Transactions
May	2	1. Juanita Ortega transferred an additional $53,250 from her personal savings account into the business checking account.
	6	2. Bought hiking equipment for $550, Check 367.
	7	3. Bought rafting equipment on account from Peak Equipment Inc. for $2,675.
	11	4. Juanita Ortega transferred her own computer, valued at $850, to the business.
	16	5. Bought a cash register for the office on account from Premier Processors for $1,250.
	19	6. Sold the computer on credit for $850 to Mary Johnson.
	22	7. Paid $500 on account to Peak Equipment Inc., Check 368.
	24	8. Purchased shelves for the office for $650, Check 369.
	28	9. Paid $1,250 on account to Premier Processors, Check 370.
	31	10. Bought rafting oars for $175, Check 371.

Analyze

Calculate the ending balance in the **Computer Equipment** account.

Chapter 4
Problems

Problem 7 Analyzing Transactions Recorded in T Accounts

Richard Tang owns and operates a job place ment service, Job Connect. The T accounts below summarize several business transactions for May.

Instructions Use a form similar to the one presented below. For each of the 10 transactions:

1. Identify the account debited, and record the account name in the appropriate column.
2. Indicate whether the account debited is being increased or decreased.
3. Identify the account credited, and write the account name in the appropriate column.
4. Indicate whether the account credited is being increased or decreased.
5. Write a short description of the transaction.

	1.	2.	3.	4.	5.
Trans No.	Account Debited	Increase (I) or Decrease (D)	Account Credited	Increase (I) or Decrease (D)	Description
1	Cash in Bank	I	Richard Tang, Capital	I	Richard Tang invested $15,000 in the business.

```
           Cash in Bank                              Richard Tang, Capital
      Debit    |    Credit                          Debit    |    Credit
        +      |      −                               −      |      +
(1) 15,000     | (4) 1,225                                   | (1) 15,000
(9)    225     | (6)   900                                   | (2)    225
               | (7)   995
               | (8) 2,000


         Office Equipment                            Accounts Receivable
      Debit    |    Credit                          Debit    |    Credit
        +      |      −                               +      |      −
(2)    225     | (5)   225                      (5)   225    | (9)   225
(3)  8,000     |
(4)  1,225     |
```

Chapter 4 • Transactions that Affect Assets, Liabilities, and Owner's Capital 109

Chapter 4
Problems

Computer Equipment		Office Furniture	
Debit +	Credit −	Debit +	Credit −
(7) 995		(6) 900	
		(10) 145	

Accounts Payable	
Debit −	Credit +
(8) 2,000	(3) 8,000
	(10) 145

Analyze

Design a diagram that shows the accounting equation for Job Connect after all transactions have been completed.

Real-World Accounting Careers

Rachael Hamer
Protiviti Inc.

Q What do you do?

A I work with my clients, many times on-site, to ensure their compliance with the Sarbanes-Oxley Act of 2002. For example, I am currently working as an internal auditor for an oil and gas company, testing key controls as well as spreadsheets and reports.

Q What are your day-to-day responsibilities?

A I meet with the control owners, and we test whether the controls are operating the way they should. After making necessary changes, I document the results and update the project sponsor. The controls must operate correctly so that we know the information is accurate and can be used in the financial statements the company must publish.

Q What factors have been key to your success?

A I've been fortunate to work with an associate director at Protiviti who taught me how to manage a project and be very particular and thorough about the information we provide to the client.

Q What do you like most about your job?

A Depending on my workload, my schedule has flexibility, which helps with work/life balance.

Career Facts

Real-World Skills
Adaptability and flexibility; professionalism; project management skills; strong communication abilities

Training And Education
Bachelor's degree in accounting; CPA or CIA credential strongly encouraged; MBA not necessary but beneficial

Career Skills
Knowledge of accounting principles and government regulations; business writing and research skills; strong analytical and deductive skills

Career Path
Start out as a consultant and progress to senior consultant

Tips from Robert Half International

Prepare for the common job-interview request, "Tell me about yourself" by rehearsing a 15-second "sound bite" that describes your strongest skills and relevant experience in two or three sentences.

College and Career Readiness

Use the Internet to locate at least three resources detailing new accounting laws and regulations. Which sources do you think are the best and why?

Real-World Applications & Connections

CASE STUDY

Service Business: Landscaping

While in school Martin Hamilton gained experience working for a large landscaping company. Martin plans to start a business called Landscapes and Beyond. He has made a list of everything he owns, with the estimated value for each category.

Lawn mowers	$1,500
Shovels and lawn-care tools	180
Truck	8,700
Stereo equipment	1,000
50 books on landscaping	400
Desk, chair, and file cabinet	700

Martin also borrowed $5,000 from his family and will repay the debt in one year.

ACTIVITY Write the accounting equation for Martin's new business, listing each item in the appropriate part of the equation.

First Audit

Review the T accounts below for errors. On a separate sheet of paper, make any necessary corrections.

1. Purchased a delivery truck for $27,000 on account from A-1 Trucks, Inc.
2. Wrote a check for $1,500 to A-1 Trucks for payment on account.

Cash in Bank	Cash in Bank	Cash in Bank
Debit + 25,000	Debit + 25,000	Debit + 25,000

21st Century Skills

Teaching Others

Keeping up with technology and learning new skills are important in today's workplace. Learning from co-workers and supervisors is often the best way to increase your skills.

Activity

Assume that you work as the accountant for Westside Aquatics, a swim club in Florida. The club has 250 members. While some individuals pay cash on each visit, others are billed monthly for club use. A concession area featuring health foods and juices is a popular gathering place for members after swim lessons.

Instructions

List account names that would be used in a business like this. Describe to a classmate what kinds of transactions might affect each account. What is the normal balance side for each account?

Career Wise

Bookkeeper

While accounts design and maintain the overall system of record-keeping that runs a business, the bookkeeper is charged with the day-to-day maintenance of that system—managing the checkbooks, updating customer payments, disbursing cash to vendors and employees, and providing statements to the accountant. A bookkeeper must stay up to date on changes in bookkeeping, accounting, and tax.

Research And Share

a. Find the Web site of the American Institute of Professional Bookkeepers, and find the name of its certification program. Visit the Web site of the U.S. Department of Labor's Bureau of Labor Statistics, and locate the median salary for a bookkeeper.

b. What is the name of the AIPB's certification program? What is the median salary for a bookkeeper? What tax accounting functions are bookkeepers often responsible for?

Spotlight on Personal Finance

Your Spending Plan

Cash is an asset. If you spend it, the asset decreases. A good way to safeguard your cash is to create a spending and savings plan, and then follow it.

Activity

Suppose you want to save $20.00 per month. You have kept track of your personal finances for several months and you see the following pattern:

Money coming in each month: allowance, $20.00; part-time job, $112.00

Money going out each month: lunches, $25.00; snacks, $8.50; bus fare, $8.00; CDs, $28.00; contributions to charity, $10.00; movie theatres, $22.00; pizza, $18.00; school supplies, $14.50

Using the above information, create a spending plan that allows you to save $20.00 per month.

Chapter 5

Transactions That Affect Revenue, Expenses, and Withdrawals

Chapter Topics

5-1 Relationship of Revenue, Expenses, and Withdrawals to Owner's Equity

5-2 Applying the Rules of Debit and Credit to Revenue, Expense, and Withdrawals Transactions

Visual Summary

Review and Activities

Standardized Test Practice

Computerized Accounting

Problems

Real-World Accounting Careers

Real-World Applications & Connections

Essential Question

As you read this chapter, keep this question in mind:

Why is it important for businesses to monitor financial changes in the short-term?

Main Idea

Revenues, expenses, and withdrawals are temporary accounts. They start each new accounting period with zero balances. Double-entry accounting requires that total debit and total credits are always equal.

Chapter Objectives

Concepts	Analysis	Procedures
C1 Explain the difference between permanent accounts and temporary accounts.	**A1** List and apply the rules of debit and credit for revenue, expense, and withdrawals accounts.	**P1** Use the six-step method to analyze transactions affecting revenue, expense, and withdrawals accounts. **P2** Test a series of transactions for equality of debits and credits.

Real-World Business Connection

Mark

Mark Cosmetics, Avon's line of teen cosmetics and beauty products, takes an innovative approach to marketing beauty products specifically for the teen lifestyle.

The Mark business model relies on many independent representatives demonstrating and selling products to people they meet or know. A representative builds her own business, demonstrating the products, taking orders, and making a profit on each item she sells. It can be a great way to earn extra money for college or as an alternative to having a summer job.

Connect to the Business

Building your own business can be rewarding and profitable, but there is much effort and time involved. The most successful entrepreneurs are those that are passionate about the business and the product or service they are selling. Success also involves keeping careful records, analyzing profitability, and planning carefully to reach your goals.

English Language Arts/Writing

In a small business venture like the one featured here, explain how you could keep track of the sales you are making and the associated costs. Perform research, if necessary, and outline your plan in a one-page paper.

Focus on the Photo

The independent entrepreneurs of Mark are building their businesses, generating revenue and expenses. They earn money and spend it on items used to operate the business. **How might revenue and expense transactions affect the profits of a business?**

SECTION 5.1
Relationship of Revenue, Expenses, and Withdrawals to Owner's Equity

Do you ever wonder how a business is doing, or if there is a way to tell? You are about learn to record revenue, expense, and owner's withdrawals accounts transactions. These accounts provide that information about how the business is doing. A pilot for Southwest Airlines would never take off in a 737 **equipped** with only a speedometer and a gas gauge. These two instruments, although necessary, do not give a pilot all of the information needed to keep such a complex aircraft on course and operating smoothly. Operating a business is a bit like operating a 737. Owners need revenue and expense information to keep the business on course.

> **Section Vocabulary**
> - temporary accounts
> - permanent accounts
> - equipped
> - process

Temporary and Permanent Accounts

What Is the Difference Between Temporary Accounts and Permanent Accounts?

You learned earlier that the owner's capital account shows the amount of the owner's investment, or equity, in a business. Owner's equity is increased or decreased by transactions other than owner's investments. For example, the revenue, or income, earned by the business increases owner's equity. Both expenses and owner's withdrawals decrease owner's equity. (Remember that revenue is not the same as an owner's investment, and expense is not the same as an owner's withdrawal.)

Revenue, expenses, and withdrawals could be recorded as increases or decreases directly in the capital account. This method, however, makes classifying information about these transactions difficult. A more informative way to record transactions affecting revenue and expenses is to set up separate accounts for each type of revenue or expense. Such information helps the owner decide, for example, whether some expenses need to be reduced.

As you learned in Chapter 2, the life of a business is divided into periods of time called accounting periods. The activities for a given accounting period are summarized and then the period is closed. A new period starts, and transactions for the new period are entered into the accounting system. The **process** continues as long as the business exists.

Using Temporary Accounts.

- Revenue, expense, and withdrawals accounts are used to collect information for a single accounting period. These accounts are called **temporary accounts**. Temporary accounts start each new accounting period with zero balances. That is, the amounts in these accounts are not carried forward from one accounting

Figure 5–1 The Relationship of Temporary Accounts to the Owner's Capital Account

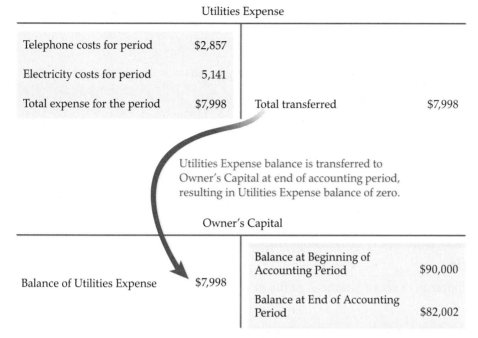

period to the next. Temporary accounts are not temporary in the sense that they are used for a short time and then discarded. They continue to be used in the accounting system, but the amounts recorded in them are totaled at the end of each accounting period. At the end of each period, the balances in the temporary accounts are transferred to the owner's capital account. (The procedure for transferring these balances to owner's capital is explained in Chapter 10.)

- Let's use **Utilities Expense,** a temporary account, as an example. During an accounting period, business transactions related to utilities such as electricity and telephones are recorded in **Utilities Expense.** By using this separate account, the owner can see at a glance how much expense is being added to the utilities account. The individual transaction amounts accumulate in the account as the accounting period progresses.

- At the end of the accounting period, the balance is transferred to the owner's capital account and subtracted from the beginning capital balance. Remember, expenses decrease owner's capital. In **Figure 5–1**, the account **Utilities Expense,** starts the new accounting period with a zero balance—ready for new expense transactions.

Using Permanent Accounts.

- In contrast to the temporary accounts, the owner's capital account is a permanent account. Asset and liability accounts are also permanent accounts. **Permanent accounts** are continuous from one accounting period to the next. In permanent accounts the dollar balances at the end of one accounting period become the dollar balances for the beginning of the next accounting period.

- For example, if a business has furniture totaling $2,875 at the end of one accounting period, the business will start with $2,875 in furniture at the beginning of the next accounting period. The ending balances in permanent accounts are carried forward to the next accounting period as the beginning balances.

End of Accounting Period	Beginning of Next Accounting Period
Furniture	Furniture
Debit | Credit	Debit | Credit
End Bal. 2,875	Beg. Bal. 2,875

- The permanent accounts show balances on hand or amounts owed at any time. They also show the day-to-day changes in assets, liabilities, and owner's capital.

Reading Check

What happens to temporary account balances at the end of each accounting period?

The Rules of Debit and Credit for Temporary Accounts

What Are the Normal Balances of Revenue, Expense, and Withdrawals Accounts?

In Chapter 4 you learned the rules of debit and credit for the asset, liability, and owner's capital accounts. In this chapter we will continue with the rules of debit and credit, this time for revenue, expense, and withdrawals accounts. Before looking at these rules, let's review quickly the T account showing the rules of debit and credit for the owner's capital account.

Owner's Capital Account	
Debit	Credit
−	+
Decrease Side	Increase Side Normal Balance

As you will see, the rules of debit and credit for accounts classified as revenue, expense, and withdrawals accounts are related to the rules for the owner's capital account.

Rules for Revenue Accounts

Accounts set up to record business income are classified as revenue accounts. The following rules of debit and credit apply to revenue accounts:

Rule 1: A revenue account is *increased* (+) on the *credit* side.

Common Mistakes

Permanent vs. Temporary Accounts
Permanent accounts carry their balances forward to the next accounting period. Their balances may change, but they exist over the life of the business. Temporary account balances relate only to one accounting period. Each new accounting period begins with zero balances, so that revenues, expenses, and withdrawals in one period are separated from the revenues, expenses, and withdrawals of the next period.

Rule 2: A revenue account is *decreased* (−) on the *debit* side.

Rule 3: The *normal balance* for a revenue account is the *increase* side, or the *credit* side. Revenue accounts normally have credit balances.

Revenue earned from selling goods or services increases owner's capital. The relationship of revenue accounts to the owner's capital account is shown by the T accounts in **Figure 5–2.** Can you explain why the T account for revenue is used to represent the credit (right) side of the capital account?

Figure 5–2 Rules of Debit and Credit for Revenue Accounts

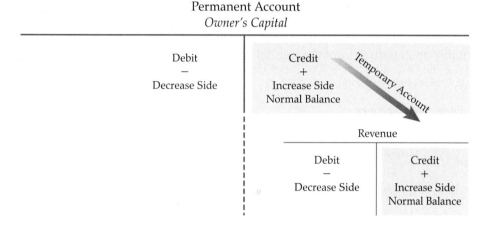

Increases in owner's capital are shown on the credit side of that account. Revenue increases owner's capital, so the revenue account is used to represent the credit side of the owner's capital account.

We can summarize the rules of debit and credit for revenue accounts with a T account illustration.

```
            Revenue Accounts
        Debit      |      Credit
          −        |        +
    (2) Decrease   |   (1) Increase
        Side       |        Side
                   |   (3) Normal
                   |      Balance
```

Let's apply the rules of debit and credit to an actual revenue account. Look at the entries in the T account for the revenue account called **Fees.** The increases to the revenue account are recorded on the right, or credit, side of the T account. The decreases are recorded on the left, or debit, side. To find the balance, subtract total debits ($200) from total credits ($500 + $1,000 + $2,000 = $3,500). You get a credit balance of $3,300 on the normal balance side for a revenue account.

```
              Fees
    Debit    |    Credit
      −      |      +
     200     |     500
             |   1,000
             |   2,000
             | Bal. 3,300
```

Rules for Expense Accounts

Accounts that record the costs of operating a business are expense accounts. These debit and credit rules apply to expense accounts:

Rule 1: An expense account is *increased* on the *debit* side.

Rule 2: An expense account is *decreased* on the *credit* side.

Rule 3: The *normal balance* for an expense account is the *increase* side, or the *debit* side. Expense accounts normally have debit balances.

Expenses are the costs of doing business. Expenses decrease owner's capital. Revenues have the opposite impact; revenues increase owner's capital.

Look at the T accounts in **Figure 5–3.** Can you explain why the T account for expenses is used to represent the debit (left) side of the capital account?

Figure 5–3 Rules of Debit and Credit for Expense Accounts

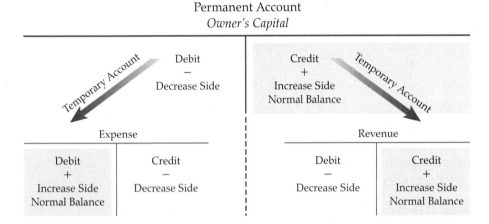

Decreases in owner's capital are shown on the debit side of that account. Since expenses decrease owner's capital, expense accounts are used to represent the debit side of the owner's capital account.

Let's use a T account to summarize the rules of debit and credit for expense accounts.

```
              Expense Accounts
        Debit           |    Credit
          +             |      −
    (1) Increase        |  (2) Decrease
        Side            |      Side
    (3) Normal          |
        Balance         |
```

Now look at the entries recorded in the T account called **Advertising Expense.** The increases to the expense account are recorded on the left, or debit, side of the T account. The decreases to the account are recorded on the right, or credit, side of the T account. When total credits ($125) are subtracted from total debits ($600), there is a balance of $475 on the debit side, which is the normal balance side for expense accounts.

```
         Advertising Expense
     Debit         |    Credit
       +           |      −
      400          |     125
      200          |
   Bal. 475        |
```

120   Chapter 5 • Transactions That Affect Revenue, Expenses, and Withdrawals

Rules for the Withdrawals Account

A withdrawal is an amount of money or an asset the owner takes out of the business. The withdrawals account is classified as a temporary owner's equity account. Recall that the permanent owner's equity account is the owner's capital account. Withdrawals, like expenses, decrease capital, so the rules of debit and credit are the same as for expense accounts.

Withdrawals Account	
Debit + (1) Increase Side (3) Normal Balance	Credit − (2) Decrease Side

Rule 1: The withdrawals account is *increased* on the *debit* side.

Rule 2: The withdrawals account is *decreased* on the *credit* side.

Rule 3: The *normal balance* for the withdrawals account is the *increase* side or *debit* side. The withdrawals account normally has a debit balance.

W. Smith, Withdrawals	
Debit + 500 1,500 Bal. 1,800	Credit − 200

Review the entries in the T account **W. Smith, Withdrawals**. The increases are recorded on the left, or debit side, of the T account. The decreases are recorded on the right, or credit, side of the T account. When total credits ($200) are subtracted from total debits ($2,000), there is a debit balance of $1,800 which is the normal balance side for the withdrawals account.

Summary of the Rules of Debit and Credit for Temporary Accounts

Figure 5–4 summarizes the rules of debit and credit for the temporary accounts and the basic accounting relationships of these accounts to the owner's capital account.

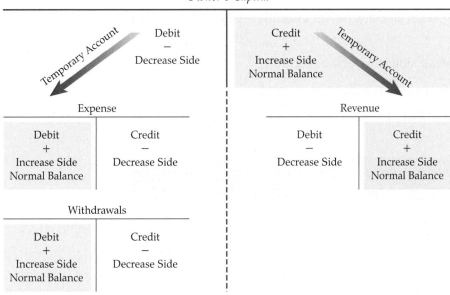

Figure 5–4 Rules of Debit and Credit for Temporary Accounts

SECTION 5.1
Assessment

After You Read

Reinforce the Main Idea

Create a chart like this one to illustrate the different types of accounts. Fill in each blank box with one or more of the following terms: *assets, liabilities, owner's capital, revenue, expenses, owner's withdrawals.*

	Normal Debit Balance	Normal Credit Balance
Permanent Accounts		
Temporary Accounts		

Problem 1 Applying the Rules of Debit and Credit

Caroline Palmer uses the following accounts in her San Francisco–Los Angeles commuter shuttle service.

General Ledger

Cash in Bank	Caroline Palmer, Capital
Advertising Expense	Accounts Receivable
Caroline Palmer, Withdrawals	Food Expense
Airplanes	Flying Fees
Fuel and Oil Expense	Accounts Payable
Repairs Expense	

Instructions In the form provided in your working papers, provide the following information for each account. The first account is completed as an example.

1. Classify the account as an asset, liability, owner's equity, revenue, or expense account.
2. Indicate whether the increase side is a debit or a credit.
3. Indicate whether the decrease side is a debit or a credit.
4. Indicate whether the account has a normal debit balance or a normal credit balance.

Account Name	Account Classification	Increase Side	Decrease Side	Normal Balance
Cash in Bank	Asset	Debit	Credit	Debit

Continued

122 Chapter 5 • Transactions That Affect Revenue, Expenses, and Withdrawals

SECTION 5.1
Assessment

Math for Accounting

A company had the following activity during the accounting period:

Made sales: $5,000

Incurred advertising expenses: 500

Incurred salaries expenses: 3,000

Deposited a check from the owner's personal savings account: 1,200

Additionally, the owner removed a computer from the business for personal use. The business had paid $650 for the computer. What was the total effect on owner's equity?

SECTION 5.2
Applying the Rules of Debit and Credit to Revenue, Expense, and Withdrawals Transactions

In Section 5.1 you learned the rules of debit and credit for revenue, expense, and withdrawals accounts. Learning to apply these rules to typical business transactions is our next **task.** In the course of a week, a business might receive money, pay rent, or pay utility bills. Let's look at more transactions for Zip Delivery Service.

Section Vocabulary
- revenue recognition
- task

Analyzing Transactions

How Do You Analyze Transactions That Involve Temporary Accounts?

In Chapter 4 Zip's transactions dealt with asset and liability accounts and with the permanent owner's equity account, **Crista Vargas, Capital.** Using the rules of debit and credit, let's analyze several business transactions that affect revenue, expense, and owner's withdrawals accounts. Use the same six-step method you learned in Chapter 4. Refer to Zip's chart of accounts on page 87 to analyze the following transactions.

Business Transaction 8

On October 15 Zip provided delivery services for Sims Corporation. A check for $1,200 was received in full payment.

Analysis	Identify	1. The accounts **Cash in Bank** and **Delivery Revenue** are affected.
	Classify	2. **Cash in Bank** is an asset account. **Delivery Revenue** is a revenue account.
	+/+	3. **Cash in Bank** is increased by $1,200. **Delivery Revenue** is increased by $1,200.

Debit-Credit Rule	4. Increases in asset accounts are recorded as debits. Debit **Cash in Bank** for $1,200.
	5. Increases in revenue accounts are recorded as credits. Credit **Delivery Revenue** for $1,200.

T Accounts 6.

```
         Cash in Bank                    Delivery Revenue
     Debit    |   Credit              Debit   |   Credit
       +      |     −                   −     |     +
     1,200    |                               |   1,200
```

124 Chapter 5 • Transactions That Affect Revenue, Expenses, and Withdrawals

Business Transaction 9

On October 16 Zip mailed Check 103 for $700 to pay the month's rent.

Analysis	Identify	1. The accounts **Rent Expense** and **Cash in Bank** are affected.
	Classify	2. **Rent Expense** is an expense account. **Cash in Bank** is an asset account.
	+/−	3. **Rent Expense** is increased by $700. **Cash in Bank** is decreased by $700.

Debit-Credit Rule	4. Increases in expense accounts are recorded as debits. Debit **Rent Expense** for $700.
	5. Decreases in asset accounts are recorded as credits. Credit **Cash in Bank** for $700.

T Accounts 6.

Rent Expense		Cash in Bank	
Debit + 700	Credit −	Debit +	Credit − 700

Business Transaction 10

On October 18 Rockport Advertising prepared an advertisement for Zip. Zip will pay Rockport's $75 fee later.

Analysis	Identify	1. The accounts **Advertising Expense** and **Accounts Payable—Rockport Advertising** are affected.
	Classify	2. **Advertising Expense** is an expense account. **Accounts Payable—Rockport Advertising** is a liability account.
	+/+	3. **Advertising Expense** is increased by $75. **Accounts Payable—Rockport Advertising** is increased by $75.

Debit-Credit Rule	4. Increases in expense accounts are recorded as debits. Debit **Advertising Expense** for $75.
	5. Increases in liability accounts are recorded as credits. Credit **Accounts Payable—Rockport Advertising** for $75.

Continued

T Accounts	6.	Advertising Expense		Accounts Payable—Rockport Advertising	
		Debit + 75	Credit −	Debit −	Credit + 75

Business Transaction 11

On October 20 Zip billed City News $1,450 for delivery services.

Analysis	Identify	1. The accounts **Accounts Receivable—City News** and **Delivery Revenue** are affected.
	Classify	2. **Accounts Receivable—City News** is an asset account. **Delivery Revenue** is a revenue account.
	+/+	3. **Accounts Receivable—City News** is increased by $1,450. **Delivery Revenue** is increased by $1,450.

Debit-Credit Rule		4. Increases in asset accounts are recorded as debits. Debit **Accounts Receivable—City News** for $1,450.
		5. Increases in revenue accounts are recorded as credits. Credit **Delivery Revenue** for $1,450.

T Accounts	6.	Accounts Receivable—City News		Delivery Revenue	
		Debit + 1,450	Credit −	Debit −	Credit + 1,450

In Transaction 11 Zip recorded revenue for services provided, even though the money had not been received. Following the GAAP principle of **revenue recognition**, revenue is recorded on the date earned, even if cash has not been received.

Business Transaction 12

On October 28 Zip paid a $125 telephone bill with Check 104.

Analysis Identify 1. The accounts **Utilities Expense** and **Cash in Bank** are affected.

Classify 2. **Utilities Expense** is an expense account. **Cash in Bank** is an asset account.

+/− 3. **Utilities Expense** is increased by $125. **Cash in Bank** is decreased by $125.

Debit-Credit Rule 4. Increases in expense accounts are recorded as debits. Debit **Utilities Expense** for $125.

5. Decreases in asset accounts are recorded as credits. Credit **Cash in Bank** for $125.

T Accounts 6.

Utilities Expense		Cash in Bank	
Debit + 125	Credit −	Debit +	Credit − 125

Business Transaction 13

On October 29 Zip wrote Check 105 for $600 to have the office repainted.

Analysis Identify 1. The accounts **Maintenance Expense** and **Cash in Bank** are affected.

Classify 2. **Maintenance Expense** is an expense account. **Cash in Bank** is an asset account.

+/− 3. **Maintenance Expense** is increased by $600. **Cash in Bank** is decreased by $600.

Debit-Credit Rule 4. Increases in expense accounts are recorded as debits. Debit **Maintenance Expense** for $600.

5. Decreases in asset accounts are recorded as credits. Credit **Cash in Bank** for $600.

T Accounts 6.

Maintenance Expense		Cash in Bank	
Debit + 600	Credit −	Debit +	Credit − 600

Business Transaction 14

On October 31 Crista Vargas wrote Check 106 to withdraw $500 cash for personal use.

Analysis	Identify	1. The accounts **Crista Vargas, Withdrawals** and **Cash in Bank** are affected.
	Classify	2. **Crista Vargas, Withdrawals** is an owner's equity account. **Cash in Bank** is an asset account.
	+/−	3. **Crista Vargas, Withdrawals** is increased by $500. **Cash in Bank** is decreased by $500.
Debit-Credit Rule		4. Increases in the owner's withdrawals account are recorded as debits. Debit **Crista Vargas, Withdrawals** for $500.
		5. Decreases in assets are recorded as credits. Credit **Cash in Bank** for $500.
T Accounts		6.

```
        Crista Vargas, Withdrawals              Cash in Bank
       Debit        |   Credit           Debit        |   Credit
         +          |     −                +          |     −
        500         |                                 |    500
```

Reading Check
According to the GAAP principle of revenue recognition, when is revenue recorded?

Testing for the Equality of Debits and Credits

Why Should You Make Sure the Ledger Is in Balance?

In a double-entry accounting system, correct analysis and recording of business transactions should result in total debits being equal to total credits. Testing for the equality of debits and credits is one way of finding out whether you have made any errors in recording transaction amounts. To test for the equality of debits and credits, follow these steps:

Step 1. Make a list of the account names used by the business.

Step 2. To the right of each account name, list the balance of the account. Use two columns, one for debit balances and the other for credit balances.

Step 3. Add the amounts in each column.

If you have recorded all the amounts correctly, the total of the debit column will equal the total of the credit column. The test for equality of debits and credits for the transactions in Chapters 4 and this chapter shows that total debits are equal to total credits, so the ledger is in balance.

	Account Name	Debit Balances	Credit Balances
101	Cash in Bank	$ 21,125	
105	Accounts Receivable—City News	1,450	
110	Accounts Receivable—Green Company		
115	Computer Equipment	3,000	
120	Office Equipment	200	
125	Delivery Equipment	12,000	
201	Accounts Payable—Rockport Advertising		$ 75
205	Accounts Payable—Coast to Coast Auto		11,650
301	Crista Vargas, Capital		25,400
302	Crista Vargas, Withdrawals	500	
303	Income Summary		
401	Delivery Revenue		2,650
501	Advertising Expense	75	
505	Maintenance Expense	600	
510	Rent Expense	700	
515	Utilities Expense	125	
		$ 39,775	$ 39,775

SECTION 5.2
Assessment

After You Read

Reinforce the Main Idea

Use a chart like this to organize the test for the equality of debits and credits. Write each of these terms in the appropriate box: *assets, liabilities, owner's capital, owner's withdrawals, revenue, expenses.* (Don't use the shaded boxes.)

Debit Balances	Credit Balances

Problem 2 Identifying Accounts Affected by Transactions

John Albers uses the following accounts in his business.

General Ledger

Cash in Bank	John Albers, Capital	Advertising Expense
Accounts Receivable	John Albers, Withdrawals	Rent Expense
Office Equipment	Service Fees	Utilities Expense
Accounts Payable		

Instructions In your working papers:

1. Identify the two accounts affected by each of the following transactions.
2. Indicate whether each account is debited or credited.

Date	Transactions
July 1	1. Issued Check 543 to pay the electric bill for the month.
3	2. Billed a customer for services provided on account.
10	3. John Albers took cash from the business for his personal use.
17	4. Issued Check 544 to pay for an advertisement.

Math for Accounting

This bar chart shows the first quarter's revenue and expense. In approximate amounts, what month had the most revenue and what was the amount? What month had the highest expense and what was the amount? Overall, did the business have a profit or loss? What was the amount?

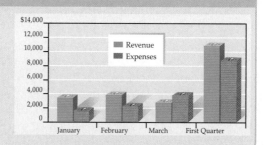

Chapter 5
Visual Summary

Concepts
Permanent accounts and temporary accounts.

Permanent Account	Temporary Account
Carries a balance forward from one accounting period to the next.	Accumulates dollar amounts for only one accounting period and then starts each new accounting period with a zero balance.
OWNER'S CAPITAL ASSET LIABILITY	OWNER'S WITHDRAWALS REVENUE EXPENSE

Analysis
The rules of debit and credit for revenue, expense, and withdrawals accounts.

Temporary Accounts

Revenue		Expense		Withdrawals	
Debit −	Credit +	Debit +	Credit −	Debit +	Credit −
Decrease Side	Increase Side Normal Balance	Increase Side Normal Balance	Decrease Side	Increase Side Normal Balance	Decrease Side

Procedures
The rules of debit and credit for revenue, expense, and withdrawals accounts.

1.	Identify the accounts affected.
2.	Classify the accounts affected.
3.	Determine the amount of increase or decrease for each account affected.
4.	Determine which account is debited and for what amount.
5.	Determine which account is credited and for what amount.
6.	Use T accounts to describe the entry.

Chapter 5
Review and Activities

Answering the Essential Question

Why is it important for businesses to monitor financial changes in the short-term?

You need to keep track of how much money you have on a daily basis in order to make decisions about what you're going to do or buy. The same is true of a business. Decisions hinge on finances as they are *today*, not how they will be at the end of the accounting period. What are some possible consequences of not monitoring finances in the short term?

Vocabulary Check

1. **Vocabulary** Write a paragraph using the content vocabulary terms. The paragraph should clearly explain how the terms relate to one another and to accounting. Underline each vocabulary term used.

   - equipped
   - process
   - permanent accounts
   - temporary accounts
   - task
   - revenue recognition

Concept Check

2. **Analyze** Describe the differences between temporary and permanent accounts.
3. State briefly the rules of debit and credit for increasing and decreasing each of the following types of accounts: • assets • owner's capital • liabilities
4. State briefly the rules of debit and credit for increasing and decreasing each of the following types of accounts: • revenue • expenses • owner's withdrawals
5. **Synthesize** If one side of a transaction involves a revenue account, can the other side involve the owner's capital account? Why or why not?
6. **Evaluate** Why is it important to test for equality of debits and credits? How would you test for equality of debits and credits?
7. **Math** Paul's boss told him that there is a 50% probability that his salary will increase next year, and a 25% probability that his bonus will increase. What is the probability that both will increase?
8. **English Language Arts** Explain using a graphic or visual aid how the accounting equation determines the normal balances for each of the main account categories: revenue, expense, and owner's equity. You may use your own illustrations or photos to communicate the concept.

Chapter 5
Standardized Test Practice

Multiple Choice

1. Which of these is a temporary account?

    a. Utilities expense

    b. Accounts payable

    c. Assets account

    d. Liability account

2. Which of these is true for temporary revenue accounts?

    a. They are increased on the debit (left) side.

    b. The normal balance is on the debit (left) side.

    c. They are decreased on the debit (left) side.

    d. They are set up to record the costs of operating the business.

3. If a business purchases a calculator on account, which accounts are affected by this transaction?

    a. Cash in Bank and Accounts Payable

    b. Office Equipment and Accounts Receivable

    c. Office Equipment and Cash in Bank

    d. Office Equipment and Accounts Payable

4. A skeleton form of an account showing only the debit and credit columns is called a(n)

    a. accounting equation.

    b. T account.

    c. account balance.

    d. none of these answers.

True or False

5. The normal balance for all three types of accounts is on the increase side.

Short Answer

6. Which type of account decreases the capital of a business and is classified as a temporary owner's equity account?

Extended Response

7. Explain how an accountant tests the equality of the ledger accounts.

Chapter 5
Computerized Accounting

The Chart of Accounts
Setting Up The Accounts in a General Ledger

MANUAL METHODS

- Accounts are set up in the general ledger using general ledger account forms.
- Accounts can be set up with or without beginning balances.

COMPUTERIZED METHODS

- Accounts are set up using defined account numbers and account types.
- Accounts can be set up with or without beginning balances.
- Accounting software programs offer hundreds of sample companies from which you can copy the chart of accounts instead of creating each account individually.

Chart of Accounts

Assets	Owner's Equity
Cash in Bank	Crista Vargas, Capital
Accounts Receivable	Crista Vargas, Withdrawals
Computer Equipment	
Office Equipment	**Revenue**
Delivery Equipment	Delivery Revenue
Liabilities	**Expenses**
Accounts Payable	Advertising Expense
	Maintenance Expense
	Rent Expense

Chapter 5
Problems

Problem 3 Identifying Increases and Decreases in Accounts

Ronald Hicks uses these accounts in his business, Wilderness Rentals.

General Ledger

101 Cash in Bank	301 Ronald Hicks, Capital
105 Accounts Receivable—Helen Katz	305 Ronald Hicks, Withdrawals
	401 Equipment Rental Revenue
120 Office Equipment	505 Maintenance Expense
125 Camping Equipment	525 Utilities Expense

Instructions Analyze each of the following transactions using the format shown in the example below. Record your answers in your working papers.

a. Explain the debit.
b. Explain the credit.

Example:
On Jan. 2 Ronald Hicks paid the bill for office cleaning, $100.

a. The expense account **505 Maintenance Expense** is increased. Increases in expenses are recorded as debits.
b. The asset account **101 Cash in Bank** is decreased. Decreases in assets are recorded as credits.

Date	Transactions
Jan. 3	1. Ronald Hicks withdrew $500 from his business for his own use, Check 225.
8	2. The business received $1,200 cash in rental fees from various customers.
12	3. The business paid a telephone bill of $85, Check 226.

Analyze
Calculate the amount of increase or decrease in the Cash in Bank account.

Chapter 5
Problems

Problem 4 Using T Accounts to Analyze Transactions

Regina Delgado, owner of Hot Suds Car Wash, uses these accounts:

General Ledger

- 101 Cash in Bank
- 125 Office Equipment
- 205 Accounts Payable—O'Brian's Office Supply
- 301 Regina Delgado, Capital
- 305 Regina Delgado, Withdrawals
- 401 Wash Revenue
- 510 Maintenance Expense
- 520 Rent Expense

Instructions In your working papers:

1. Determine which accounts are affected for each transaction.
2. Prepare two T accounts for each transaction.
3. Enter the amount of the debit and the amount of the credit in the T accounts.

Date		Transactions
Jan. 7	1.	Received a check for $1,675 for car wash services.
12	2.	Paid the monthly rent of $450 by writing Check 212.
15	3.	Regina Delgado withdrew $250 for her personal use, Check 213.
29	4.	Had the computer repaired at O'Brian's Office Supply for $245 and was given until next month to pay.

Analyze

Identify the transactions that affect expense accounts.

Problem 5 Analyzing Transactions into Debit and Credit Parts

Abe Shultz, owner of Kits & Pups Grooming, uses the following accounts to record transactions for the month.

Continued

Chapter 5
Problems

General Ledger

101	Cash in Bank	401	Boarding Revenue
105	Accounts Receivable—Juan Alvarez	405	Grooming Revenue
		501	Advertising Expense
140	Grooming Equipment	505	Equipment Repair Expense
205	Accounts Payable—Dogs & Cats Inc.	510	Maintenance Expense
		520	Rent Expense
301	Abe Shultz, Capital	530	Utilities Expense
305	Abe Shultz, Withdrawals		

Instructions For each transaction:

1. Prepare a T account for each account listed.

2. Enter a balance of $15,000 in the **Cash in Bank** account; also enter a balance of $15,000 in the **Abe Shultz, Capital** account.

3. Analyze and record each of the following business transactions, using the appropriate T accounts. Identify each transaction by number.

4. After all the business transactions have been recorded, write the word *Balance* on the normal balance side of each account.

5. Compute and record the balance for each account.

Date		Transactions
Jan. 1	1.	Purchased grooming equipment for $12,700, Check 283.
10	2.	Wrote Check 284 for advertising, $125.
12	3.	Received $1,850 cash for dog boarding services.
15	4.	Paid $150 for equipment repair, Check 285.
17	5.	Purchased a dog cage on account from Dogs & Cats Inc. for $75.
20	6.	Abe Shultz withdrew $150 for personal use, Check 286.
22	7.	Billed Juan Alvarez for $775 covering grooming services for all of the dogs boarded at the kennels he owns. Payment will be received later.
23	8.	Paid the first two weeks' rent by writing Check 287 for $325.
25	9.	Paid the electric bill at a cost of $115, Check 288.

Analyze

Identify the transactions that affect owner's equity.

Chapter 5
Problems

Problem 6 — Analyzing Transactions into Debit and Credit Parts

Juanita Ortega operates Outback Guide Service. She uses the following accounts to record and summarize her business transactions.

General Ledger

101	Cash in Bank	301	Juanita Ortega, Capital
105	Accounts Receivable—Mary Johnson	302	Juanita Ortega, Withdrawals
		401	Guide Service Revenue
150	Rafting Equipment	505	Maintenance Expense
205	Accounts Payable—Peak Equipment Inc.	515	Rent Expense
		520	Utilities Expense

Instructions For each transaction:

1. Prepare a T account for each account the business uses.
2. Analyze and record each of the following transactions using the appropriate T accounts. Identify each transaction by number.
3. After recording all transactions, compute and record the account balance on the normal balance side of each T account.
4. Test for the equality of debits and credits.

Date		Transactions
Jan. 2	1.	Juanita Ortega invested $12,000 cash in her business.
7	2.	Purchased two new whitewater rafts on account for $3,750 from Peak Equipment Inc.
10	3.	Billed, but did not collect, $750 for guide services provided to Mary Johnson.
12	4.	Repaired a raft at a cost of $123, Check 411.
14	5.	Wrote Check 412 to pay the electric bill of $95.
17	6.	Received $225 for guide service fees.
21	7.	Paid the $225 rent for the month, Check 413.
24	8.	Paid $1,750 toward the rafts bought on account, Check 414.
27	9.	Juanita Ortega withdrew $250 cash for personal use, Check 415.
29	10.	Received guide service fees of $250.

Analyze
Calculate the amount of revenue earned.

Chapter 5
Problems

Problem 7 Analyzing Transactions

Greg Failla owns Showbiz Video. He uses the following accounts to record business transactions.

General Ledger

101 Cash in Bank	205 Accounts Payable—
105 Accounts Receivable—	Computer Horizons
Gabriel Cohen	207 Accounts Payable—
110 Accounts Receivable—	New Media Suppliers
James Coletti	301 Greg Failla, Capital
140 Computer Equipment	305 Greg Failla, Withdrawals
145 Blu-ray discs	401 DVD Rental Revenue
	405 HD Projector Rental Revenue
	505 Equipment Repair Expense
	520 Rent Expense
	530 Utilities Expense

Instructions For each transaction:

1. Prepare a T account for each account listed above.
2. Analyze and record each of the following transactions using the appropriate T accounts. Identify each transaction by number.
3. After recording all transactions, compute a balance for each account.
4. Test for the equality of debits and credits.

Date	Transactions
Jan. 1	1. Greg Failla invested $17,500 cash in Showbiz Video.
3	2. Purchased computer equipment on account from Computer Horizons for $2,400.
8	3. Purchased videos on account from New Media Suppliers for $375.
10	4. Paid monthly rent of $750, Check 1183.
13	5. Wrote Check 1184 to pay for new videos, $265.
14	6. Sent a bill for $67 to Gabriel Cohen for a HD Projector rental.
16	7. Deposited the receipts from video rentals, $233.
19	8. Paid the gas and electric bill of $125, Check 1185.
21	9. Sent Check 1186 for $375 to New Media Suppliers as payment on account.

Continued

Chapter 5
Problems

Date	Transactions
22	10. Greg Failla withdrew $150 for his personal use, Check 1187.
25	11. Paid $45 for HD Projector repair, Check 1188.
30	12. Deposited HD Projector rental receipts of $264 in the bank.

Analyze
Identify the permanent accounts that have normal credit balances.

Problem 8 Completing the Accounting Equation

With the addition of temporary accounts, the basic accounting equation can be expressed as follows:

$$\text{Assets} = \text{Liabilities} + \underbrace{\text{Owner's Capital} - \text{Withdrawals} + \text{Revenue} - \text{Expenses}}_{\text{Owner's Equity}}$$

Instructions Using the expanded equation shown above, determine the missing amounts for the following accounting equations. Use the form in your working papers. The first equation is completed as an example.

	Assets	=	Liabilities	+	Owner's Capital	−	Withdrawals	+	Revenue	−	Expenses
1.	$64,400		$8,200		$56,300		$ 500		$10,000		$ 9,600
2.	$22,150		525		18,800		1,200		12,100		?
3.	17,500		75		21,650		?		4,115		3,250
4.	49,450		?		47,840		1,500		20,300		17,610
5.	21,900		1,150		20,005		950		?		16,570
6.	72,640		2,790		?		10,750		67,908		39,749
7.	?		1,988		41,194		6,196		52,210		42,597
8.	?		3,840		61,774		?		40,163		21,637

(Expenses plus withdrawals equal $27,749.)

9.	64,070		?		49,102		4,875		53,166		?

(Total owner's equity after adding revenue and subtracting expenses and withdrawals is $50,643.)

Analyze
For equation 7, calculate the sum of owner's equity at year end.

Real-World Accounting Careers

Christina Gerbe
Moody's Corp.

Q What do you do?

A I oversee a staff of 12 in general accounting (domestic expenses, cash and fixed assets) at Moody's Corp., an international company that provides independent credit ratings and financial information.

Q What are your day-to-day responsibilities?

A Each day is completely different, which is one of the reasons I like accounting. My duties include reviewing and analyzing accounts, supervising my team in monthly and quarterly closes, writing policy and procedure manuals, and meeting with various departments (legal, payroll, treasury, real estate and tax, for example) to discuss current transactions.

Q What factors have been key to your success?

A To be a successful accountant, you need sharp analytical and technical skills and attention to detail. Beyond that, you need to have good communication skills.

Q What do you like most about your job?

A The biggest challenge and most gratifying part of my job is leading and motivating my team.

Career Facts

Real-World Skills
Professional skepticism; critical thinking; technical competence

Training And Education
Bachelor's degree in accounting or finance; CPA credential

Career Skills
Familiarity with many corporate functions; knowledge of transactions and financial reporting

Career Path
Start in public accounting, then accept a position in a corporate environment to build knowledge of corporate accounting

---

Tips from Robert Half International

Being able to effectively communicate your ideas is a vital skill for accounting professionals. Especially during job interviews, speak clearly and use complete sentences free of slang and verbal crutches like "um" or "uh." Take a moment to organize your thoughts before you answer questions.

College and Career Readiness

What are some of the best ways to network with other accounting professionals? What questions would you ask to learn more about careers in accounting? Write a list of five to ten questions that would help you get more information from a professional in the field.

Real-World Accounting & Connections

Career Wise

International Accountant

International accounting can be a great career for those who are interested in other countries' financial practices. An international accountant gets to travel or work in another country.

Education Requirements

An international accountant must have at least a bachelor's degree in accounting or a related subject. Some also get a special certificate from the Association of International Accountants, but this is not required.

Roles

International accountants do accounting, financial analysis, auditing and taxes for companies that do business globally, rather than just in one country.

Responsibilities

International accountants must make sure the company's financial practices line up with accounting standards in the U.S.A. and in all foreign countries where they do business.

Activity

It is a good career strategy to plan ahead and figure out what skills and education you will need to get a certain job. Conduct research on international accounting careers and plan the skills you should learn and the classes you might take to prepare for an international accounting career.

Global Accounting

Time Zones

What time is it now where you live? What time is it in London or Hong Kong? Coordinating between countries around the globe requires knowledge of the world's time zones. Without first considering time zones, it would be difficult to contact a person at a distant location at the right place and the right time. There are 24 time zones in the world, each representing 15 degrees of longitude, or one hour intervals.

Activity

Imagine that your company has manufacturing locations in Dublin, Ireland and Denver, Colorado. Managers work from 8:00 a.m. to 3:30 p.m. at both locations. Dublin is 7 hours behind Denver. Schedule a telephone meeting so that both managers are present. What will you need to consider when scheduling the meeting?

A Matter of Ethics

Gossip in the Workplace

Assume that you are an accounting clerk for a large insurance company like Farmers Insurance. Your boss introduces you to the newest hire, and you recognize her as a former classmate from high school named Sally. You remember that Sally had been suspended from school for a series of locker thefts. During lunch you consider telling other co-workers about Sally's history. You also wonder if your boss knows.

Activity

Define the ethical issue involved. Then describe how you think ethics relates to workplace gossip. Who are the affected parties? What are the alternatives? How to the alternatives affect the parties? What would you do?

Spotlight on Personal Finance

Often Overlooked Bank Fees

There are three reasons listed below that indicate why you should use your banking privileges carefully.

Activity

Visit a local bank and inquire about the cost of each possible outcome and how you may protect yourself from the unnecessary costs of poor money management.

1. If you overdraw your checking account, will you be charged a "bank overdraft fee?"
2. If you make a late credit card payment, will you be charged a penalty interest rate?
3. If you go over your credit card limit, will you be charged a fee?

H.O.T. Audit

T Accounts

Review the T accounts below to verify that the following transactions have been recorded correctly. On a separate sheet of paper, make any necessary corrections.

1. Received $425 from Marty Prichard for printing services.
2. Provided printing services to Marty Prichard on account, $325.
3. Paid for repairs to the copier, $127.

Cash in Bank	
(1) 425	(3) 127
(2) 325	

Printing Revenue	
(3) 127	(2) 325
	(1) 425

Chapter 6

Recording Transactions in a General Journal

Chapter Topics

6-1 The Accounting Cycle

6-2 Recording Transactions in the General Journal

Visual Summary

Review and Activities

Standardized Test Practice

Computerized Accounting

Problems

Real-World Applications & Connections

Essential Question

As you read this chapter, keep this question in mind:

How do businesses keep permanent records of transactions?

Main Idea

The accounting cycle is a series of steps done in each accounting period to keep records in an orderly fashion. You can use the general journal to record all of the transactions of a business.

Chapter Objectives

Concepts	Analysis	Procedures
C1 Explain the first three steps in the accounting cycle.	**A3** Explain the purpose of journalizing.	**P1** Describe the steps to make a general journal entry.
C2 Give and describe several examples of source documents.	**A4** Apply information from source documents.	**P2** Make general journal entries.
		P3 Correct errors in general journal entries.

Real-World Business Connection

The Night Agency

Even advertising agencies need to advertise. New York City-based **The Night Agency,** founded by three Syracuse University graduates, enjoyed early success in 2004 on the streets and in the subways of New York with creative "guerilla-style" campaigns that earned plenty of media attention. Businesses called on **The Night Agency** for their digital advertising needs; the company has come through with innovative, award-winning campaigns.

Connect to the Business

Accountants for **The Night Agency** rely on the activities of the accounting cycle to keep accounting records accurate and in order. The records are kept for a certain period of time, usually a year. Just as most school years do not begin in January, the 12-month accounting cycle can begin in any month, depending on the business.

Analyze

If you were just starting an advertising agency, what types of financial information would you record?

Focus on the Photo

With numerous high-profile clients and projects, the daily financial transactions at a busy digital agency like The Night Agency can include everything from paying for advertising to hiring SEO specialists. **Why is it important for any business to keep accurate records of daily transactions?**

SECTION 6.1
The Accounting Cycle

Keeping accurate records is vital to a business. You will learn how to record business transactions in a journal using the first three steps of the accounting cycle.

Section Vocabulary

- accounting cycle
- source document
- invoice
- receipt
- memorandum
- check stub
- journal
- journalizing
- fiscal year
- calendar year
- occurs

The Steps of the Accounting Cycle

What Is the Accounting Cycle?

The accounting period of a business is separated into activities called the **accounting cycle**. These activities help the business keep its accounting records in an orderly fashion. Take a look at **Figure 6–1**, which illustrates accounting activities and their sequence.

In this chapter you will use Steps 1, 2, and 3 of the accounting cycle:

1. Collect and verify source documents.
2. Analyze each transaction.
3. Journalize each transaction.

After studying Chapters 3 through 10, you will have covered the entire accounting cycle for a service business organized as a sole proprietorship.

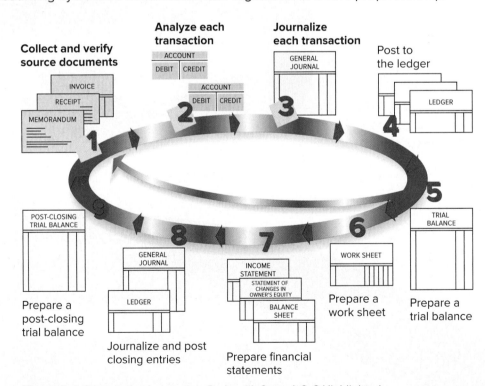

Figure 6–1 Steps in the Accounting Cycle with Steps 1, 2, 3 Highlighted

146 Chapter 6 • Recording Transactions in a General Journal

The First Step in the Accounting Cycle: Collecting and Verifying Source Documents

Most business transactions take place during the daily operations of a business. In the course of one day, a business may pay its rent, place an ad in a local newspaper, contract to have a Web site created, pay its employees, sell products, and purchase new equipment. When a business transaction **occurs**, a paper is prepared as evidence of that transaction. This paper is a **source document**.

There are several types of source documents that can be prepared by hand, by computer, or by a machine. The type of source document prepared depends on the nature of the transaction. **Figure 6–2** describes and illustrates commonly used source documents.

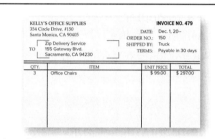

Invoice
The **invoice** lists specific information about a business transaction involving the buying or selling of an item on account. The invoice contains the date of the transaction; the quantity, description, and cost of each item; and the payment terms.

Receipt
A **receipt** is a record of cash received by a business. It indicates the date the payment was received, the name of the person or business from whom the payment was received, and the amount of the payment.

Memorandum
A **memorandum** is a brief written message that describes a transaction that takes place within a business. A memorandum is often used if no other source document exists for the business transaction.

Check Stub
The **check stub** lists the same information that appears on a check: the date written, the person or business to whom the check was written, and the amount of the check. The check stub also shows the balance in the checking account before and after each check is written.

Figure 6–2 Source Documents

The business owner, accountant, or accounting clerk (depending on the size of the business) uses source documents generated by business transactions to keep the records of the business. The accounting cycle starts by collecting and verifying the accuracy of source documents.

The Second Step in the Accounting Cycle: Analyzing Business Transactions

After collecting and verifying source documents, the second step in the accounting cycle can begin—analyzing information on the source documents to determine the debit and credit parts of each transaction.

You have already learned how to analyze business transactions using the rules of debit and credit. When you learned to analyze transactions, you were given a description of each transaction, such as: Zip Delivery Service bought a computer system from Info-Systems Inc. for $3,000 and issued Check 101 in payment. On the job you will not get a description of the transaction. Instead, you must examine a source document to determine what occurred during a business transaction.

The Third Step in the Accounting Cycle: Recording Business Transactions in a Journal

You are now ready to apply information from source documents. The third step in the accounting cycle is to record the debit and credit parts of each business transaction in a journal. A **journal** is a record of the transactions of a business. Journals are kept in chronological order, that is, the order in which the transactions occur. The process of recording business transactions in a journal is called **journalizing**. Keeping a journal can be compared to keeping a diary in which all important events are written. A journal is the only place where complete details of a transaction, including both the debit and credit parts, are recorded. The journal is sometimes called the *book of original entry* because it is where transactions are first entered in the accounting system.

Reading Check
Why are source documents necessary to a business?

The Accounting Period

What Are the Two Types of Accounting Periods?

As discussed in Chapter 2, accounting records are summarized for a certain period of time, called an accounting period. An accounting period may be for any designated length of time, such as a month, a quarter, or a year. Most businesses use a year as their accounting period. An accounting period of 12 months is called a **fiscal year**. If the fiscal year for a business begins on January 1 and ends on December 31, it is called a **calendar year** accounting period. Many businesses start their accounting periods in months other than January. For example, department stores often have fiscal years that begin on February 1 and end on January 31 of the following year. School districts usually have fiscal years that begin on July 1 and end on June 30.

SECTION 6.1
Assessment

After You Read

Reinforce the Main Idea

Create a table similar to this one to describe how the first three steps of the accounting cycle help organize the records of a business.

Step Number	Step Description	How This Step Helps Organize Business Records

Problem 1 Analyzing a Source Document

Instructions Analyze the invoice shown below and answer the following questions.

```
JAYMAX OFFICE SUPPLY                        INVOICE NO. 479
   554 Town Square
   Fort Myers, FL 33902              DATE:      Apr. 9, 20--
                                     ORDER NO.:
       Dario's Accounting Services   SHIPPED BY: Truck
TO     5821 Gulf Blvd.               TERMS:      Payable in 30 days
       Naples, FL 33940
```

QTY.	ITEM	UNIT PRICE	TOTAL
1	Fax Machine	$ 299.00	$ 299.00

1. What is the name of the company providing the service or merchandise?
2. What is the name of the business receiving the service or merchandise?
3. What is the date of the invoice?
4. What is the invoice number?
5. What item was sold?
6. What is the price for this item?
7. What are the payment terms?

continued

SECTION 6.1
Assessment

Math for Accounting

Glen's Catering received an invoice from Conover Restaurant Suppliers for the following supplies:

- 6 cartons of napkins at $4.88 per carton
- 3 boxes of salt packets at $3.19 per box
- 3 boxes of paper plates at $7.28 per box
- 4 boxes of medium paper cups at $8.24 per box

Calculate the total for each item on the invoice. Then calculate the total for all items.

SECTION 6.2
Recording Transactions in the General Journal

In Section 6.1 you learned about the first three steps in the accounting cycle. Let's apply these steps to business transactions for Zip Delivery Service.

Section Vocabulary
- general journal
- affect
- manual

Recording a General Journal Entry

How Do You Record a General Journal Entry?

Many kinds of accounting journals are used in business. One of the most common is the general journal. As its name suggests, the **general journal** is an all-purpose journal in which all of the transactions of a business may be recorded. **Figure 6–3** shows the general journal you will be using throughout the accounting cycle for Zip Delivery Service. The general journal has two amount columns. The first amount column, the amount column on the left, is used to record debit amounts. (Remember that debit means left.) The second amount column, the amount column on the right, is used to record credit amounts. (Remember that credit means right.) Look at **Figure 6–3** to find where each component of a general journal entry appears.

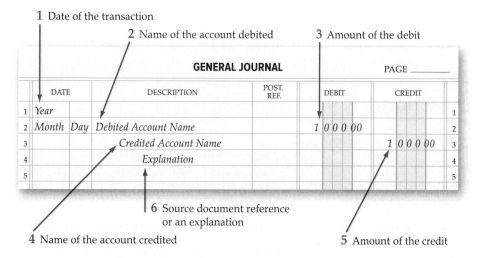

Figure 6–3 General Journal for Zip Delivery Service

In Chapters 4 and 5, you learned a step-by-step method for analyzing business transactions. In this chapter you will learn to complete the journal entry for a business transaction in the same manner. Review the following steps before you continue.

Business Transaction

BUSINESS TRANSACTION ANALYSIS: Steps to Success

Analysis	Identify	1. Identify the accounts **affected**.
	Classify	2. Classify the accounts affected.
	+/−	3. Determine the amount of increase or decrease for each account affected.

Debit-Credit Rule	4. Which account is debited? For what amount?
	5. Which account is credited? For what amount?

T Accounts	6. What is the complete entry in T-account form?

Journal Entry	7. What is the complete entry in general journal form?

Use these steps to determine the debit and credit parts of each journal entry. Remember, it is always helpful to use T accounts to analyze transactions. After analyzing many transactions, you will find that you need these tools less and less to determine the debit and credit parts of a journal entry. After the complete entry is recorded, verify that the total debits and total credits are equal.

Now, let's examine business transactions and their analysis for Zip Delivery Service.

Business Transaction 1

On October 1 Crista Vargas took $25,000 from personal savings and deposited that amount to open a business checking account in the name of Zip Delivery Service, Memorandum 1.

Zip Delivery Service
155 Gateway Blvd.
Sacramento, CA 94230

MEMORANDUM 1

TO: Accounting Clerk
FROM: Crista Vargas
DATE: October 1, 20--
SUBJECT: Contributed personal funds to the business

I have contributed $25,000 from my personal savings for a deposit to the business, Roadrunner Delivery Service.

Analysis	Identify	1. The accounts **Cash in Bank** and **Crista Vargas, Capital** are affected.
	Classify	2. **Cash in Bank** is an asset account. **Crista Vargas, Capital** is an owner's equity account.
	+/−	3. **Cash in Bank** is increased by $25,000. **Crista Vargas, Capital** is increased by $25,000.

Debit-Credit Rule	4. Increases in asset accounts are recorded as debits. Debit **Cash in Bank** for $25,000.
	5. 5. Increases in the owner's capital account are recorded as credits. **Credit Crista Vargas, Capital** for $25,000.

continued

T Accounts 6.

Cash in Bank		Crista Vargas, Capital	
Debit +	Credit −	Debit −	Credit +
25,000			25,000

Journal Entry 7.

GENERAL JOURNAL   PAGE 1

	DATE		DESCRIPTION	POST. REF.	DEBIT	CREDIT	
1	20--						1
2	Oct.	1	Cash in Bank		25 000 00		2
3			Crista Vargas, Owner's Equity			25 000 00	3
4			Memorandum 1				4

Look again at the general journal entry shown above. Notice that in the upper right-hand corner there is a line for the page number. Journal pages are numbered in consecutive order; that is, 1, 2, 3, and so on. When you fill one page with journal entries, go on to the next page. Be sure to properly number each new page.

Business Transaction 2

On October 2 Crista Vargas took two telephones valued at $200 each from her home and transferred them to the business as office equipment, Memorandum 2.

Zip Delivery Service
155 Gateway Blvd.
Sacramento, CA 94230

MEMORANDUM 2

TO: Accounting Clerk
FROM: Crista Vargas
DATE: October 2, 20--
SUBJECT: Contributed personal phones

I have contributed two telephones from my home to the business. The phones are valued at $200 each. Total contribution = $400.

Analysis

Identify
1. The accounts **Office Equipment** and **Crista Vargas, Capital** are affected.

Classify
2. **Office Equipment** is an asset account. **Crista Vargas, Capital** is an owner's equity account.

+/−
3. **Office Equipment** is increased by $400. **Crista Vargas, Capital** is increased by $400.

Debit-Credit Rule

4. Increases in asset accounts are recorded as debits. Debit **Office Equipment** for $400.

5. Increases in owner's capital accounts are recorded as credits. Credit **Crista Vargas, Capital** for $400.

continued

T Accounts	6.	Office Equipment		Crista Vargas, Capital	
		Debit + 400	Credit −	Debit −	Credit + 400

Journal Entry 7.

GENERAL JOURNAL PAGE 1

	DATE	DESCRIPTION	POST. REF.	DEBIT	CREDIT	
5	Oct. 2	Office Equipment		400 00		5
6		Crista Vargas, Capital			400 00	6
7		Memorandum 2				7

Business Transaction 3

On October 4 Zip issued Check 101 for $3,000 to buy a computer system.

$ 3,000.00   No. 101
Date October 4   20 --
To Info-Systems Inc.
For computer

	Dollars	Cents
Balance brought forward	25,000	00
Add deposits		
Total	25,000	00
Less this check	3,000	00
Balance carried forward	22,000	00

Analysis	Identify	1. The accounts **Computer Equipment** and **Cash in Bank** are affected.
	Classify	2. **Computer Equipment** and **Cash in Bank** are asset accounts.
	+/−	3. **Computer Equipment** is increased by $3,000. **Cash in Bank** is decreased by $3,000.

Debit-Credit Rule	4. Increases in asset accounts are recorded as debits. Debit **Computer Equipment** for $3,000.
	5. Decreases in asset accounts are recorded as credits. Credit **Cash in Bank** for $3,000.

T Accounts	6.	Computer Equipment		Cash in Bank	
		Debit + 3,000	Credit −	Debit +	Credit − 3,000

continued

Journal Entry 7.

	GENERAL JOURNAL			PAGE 1
DATE	DESCRIPTION	POST. REF.	DEBIT	CREDIT
8	4 Computer Equipment		3 000 00	
9	Cash in Bank			3 000 00
10	Check 101			
11				

Business Transaction 4

On October 9 Zip bought a used truck on account from Coast to Coast Auto for $12,000, Invoice 200.

Coast to Coast Auto
440 Lake Drive
Sacramento, CA 94230

TO: Zip Delivery Service
155 Gateway Blvd.
Sacramento, CA 94230

INVOICE NO. 200
DATE: Oct. 9, 20--
ORDER NO.: 99674
SHIPPED BY: n/a
TERMS: Installment

QTY.	ITEM	UNIT PRICE	TOTAL
1	Dodge Truck Used	$12,000.00	$12,000.00

Analysis

Identify 1. The accounts **Delivery Equipment** and **Accounts Payable—Coast to Coast Auto** are affected.

Classify 2. **Delivery Equipment** is an asset account. **Accounts Payable—Coast to Coast Auto** is a liability account.

+/− 3. **Delivery Equipment** is increased by $12,000. **Accounts Payable—Coast to Coast Auto** is increased by $12,000.

Debit-Credit Rule

4. Increases in asset accounts are recorded as debits. Debit **Delivery Equipment** for $12,000.

5. Increases in liability accounts are recorded as credits. Credit **Accounts Payable—Coast to Coast Auto** for $12,000.

T Accounts 6.

Delivery Equipment		Accounts Payable—Coast to Coast Auto	
Debit + 12,000	Credit −	Debit −	Credit + 12,000

Journal Entry 7.

	GENERAL JOURNAL			PAGE 1
DATE	DESCRIPTION	POST. REF.	DEBIT	CREDIT
11	9 Delivery Equipment		12 000 00	
12	Accts. Pay.—Coast to Coast Auto			12 000 00
13	Invoice 200			
14				

To separate the amounts to be paid to individual creditors, Zip uses a different account name for each creditor. The account name consists of **Accounts Payable** followed by the name of the creditor. You may have to abbreviate the name to fit it on one line of the journal. An acceptable abbreviation in the preceding journal entry is **Accts. Pay.—Coast to Coast Auto.**

Zip uses the same naming system for the amounts to be paid by individual customers. The account name consists of **Accounts Receivable** followed by the customer's name.

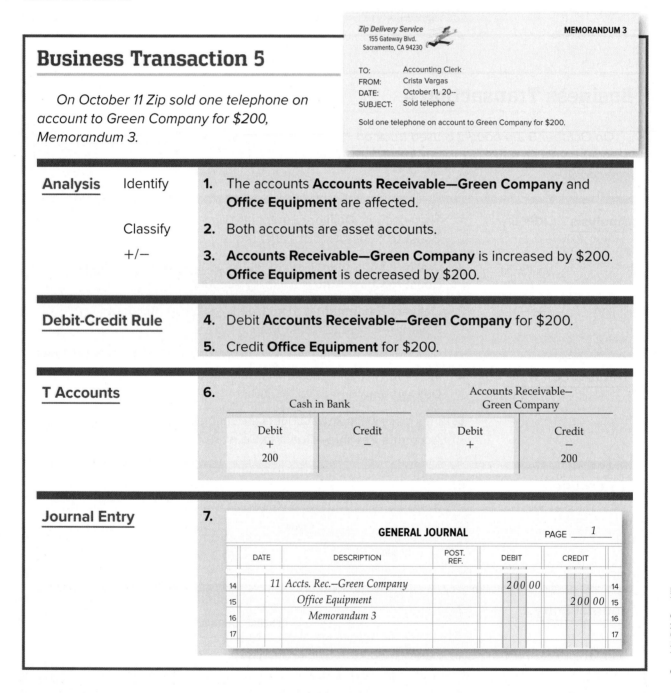

Business Transaction 5

On October 11 Zip sold one telephone on account to Green Company for $200, Memorandum 3.

Analysis Identify 1. The accounts **Accounts Receivable—Green Company** and **Office Equipment** are affected.

Classify 2. Both accounts are asset accounts.

+/− 3. **Accounts Receivable—Green Company** is increased by $200. **Office Equipment** is decreased by $200.

Debit-Credit Rule 4. Debit **Accounts Receivable—Green Company** for $200.
5. Credit **Office Equipment** for $200.

T Accounts 6.

Journal Entry 7.

156 Chapter 6 • Recording Transactions in a General Journal

Business Transaction 6

On October 12 Zip mailed Check 102 for $350 as the first installment on the truck purchased from Coast to Coast Auto on October 9.

$	350.00			No. 102
Date	October 12			20 --
To	Coast to Coast Auto			
For	installment on truck			

	Dollars	Cents
Balance brought forward	22,000	00
Add deposits		
Total	22,000	00
Less this check	350	00
Balance carried forward	21,650	00

Analysis

Identify 1. The accounts **Accounts Payable—Coast to Coast Auto** and **Cash in Bank** are affected.

Classify 2. **Accounts Payable—Coast to Coast Auto** is a liability account. **Cash in Bank** is an asset account.

+/− 3. Both accounts are decreased by $350.

Debit-Credit Rule

4. Debit **Accounts Payable—Coast to Coast Auto** for $350.
5. Credit **Cash in Bank** for $350.

T Accounts

6.

Accounts Payable—Coast to Coast Auto		Cash in Bank	
Debit − 350	Credit +	Debit +	Credit − 350

Journal Entry

7.

		GENERAL JOURNAL			PAGE 1	
	DATE	DESCRIPTION	POST. REF.	DEBIT	CREDIT	
17	12	Accts. Pay.—Coast to Coast Auto		350 00		17
18		Cash in Bank			350 00	18
19		Check 102				19
20						20

Business Transaction 7

On October 14 Zip received and deposited a check for $200 from Green Company, Receipt 1. The check is full payment for the telephone sold on account to Green on October 11.

Zip Delivery Service
155 Gateway Blvd.
Sacramento, CA 94230

RECEIPT No. 1

Oct. 14 20 --
RECEIVED FROM Green Company $ 200.00
Two hundred and no/100 DOLLARS
FOR Telephone
RECEIVED BY *Crista Vargas*

continued

Analysis	Identify	1. The accounts **Cash in Bank** and **Accounts Receivable—Green Company** are affected.
	Classify	2. **Cash in Bank** is an asset account. **Accounts Receivable—Green Company** is an asset account.
	+/−	3. **Cash in Bank** is increased by $200. **Accounts Receivable—Green Company** is decreased by $200.

Debit-Credit Rule	4. Debit **Cash in Bank** for $200.
	5. Credit **Accounts Receivable—Green Company** for $200.

T Accounts

6.
```
    Accounts Receivable—                      Office Equipment
       Green Company
    Debit      |  Credit                  Debit      |  Credit
      +        |    −                       +        |    −
     200       |                                     |   200
```

Journal Entry

7.

GENERAL JOURNAL PAGE 1

DATE	DESCRIPTION	POST. REF.	DEBIT	CREDIT
14	Cash in Bank		200 00	
	Accts. Rec.—Green Company			200 00
	Receipt 1			

Business Transaction 8

On October 15 Zip provided delivery services for Sims Corporation. A check for $1,200 was received in full payment, Receipt 2.

Receipt

Zip Delivery Service
155 Gateway Blvd.
Sacramento, CA 94230

RECEIPT No. 2

Oct. 15 20—

RECEIVED FROM Sims Corporation $ 1,200.00

One thousand two hundred and no/100 —————— DOLLARS

FOR Delivery services

RECEIVED BY Crista Vargas

Analysis	Identify	1. The accounts **Cash in Bank** and **Delivery Revenue** are affected.
	Classify	2. **Cash in Bank** is an asset account. **Delivery Revenue** is a revenue account.
	+/−	3. **Cash in Bank** is increased by $1,200. **Delivery Revenue** is increased by $1,200.

continued

Debit-Credit Rule	4. Increases in asset accounts are recorded as debits. Debit **Cash in Bank** for $1,200.
	5. Increases in revenue accounts are recorded as credits. Credit **Delivery Revenue** for $1,200.

T Accounts 6.

Cash in Bank		Delivery Revenue	
Debit + 1,200	Credit −	Debit −	Credit + 1,200

Journal Entry 7.

GENERAL JOURNAL PAGE 1

DATE	DESCRIPTION	POST. REF.	DEBIT	CREDIT
15	Cash in Bank		1 200 00	
	Delivery Revenue			1 200 00
	Receipt 2			

Business Transaction 9

On October 16 Zip mailed Check 103 for $700 to pay the month's rent.

Analysis	Identify	1. The accounts **Rent Expense** and **Cash in Bank** are affected.
	Classify	2. **Rent Expense** is an expense. **Cash in Bank** is an asset.
	+/−	3. **Rent Expense** is increased by $700. **Cash in Bank** is decreased by $700.

Check No. 103
$ 700.00
Date October 16 20 --
To Tooley & Co. Management
For rent

	Dollars	Cents
Balance brought forward	21,650	00
Add deposits 10/14	200	00
10/15	1,200	00
Total	23,050	00
Less this check	700	00
Balance carried forward	22,350	00

Debit-Credit Rule	4. Increases in expense accounts are recorded as debits. Debit **Rent Expense** for $700.
	5. Decreases in asset accounts are recorded as credits. Credit **Cash in Bank** for $700.

continued

T Accounts

6.

Rent Expense		Cash in Bank	
Debit + 700	Credit −	Debit +	Credit − 700

Journal Entry

7.

GENERAL JOURNAL PAGE 1

DATE	DESCRIPTION	POST. REF.	DEBIT	CREDIT	
16	Rent Expense		700 00		26
	Cash in Bank			700 00	27
	Check 103				28
					29

Business Transaction 10

On October 18 Rockport Advertising prepared an advertisement for Zip. Zip will pay Rockport's $75 fee later, Invoice 129.

INVOICE NO. 129

Rockport Advertising
10200 Prairie Parkway
Sacramento, CA 94206

DATE: Oct. 18, 20--
ORDER NO.: 699
SHIPPED BY: n/a
TERMS: Payable in 30 days

TO: Zip Delivery Service
155 Gateway Blvd.
Sacramento, CA 94230

QTY.	ITEM	UNIT PRICE	TOTAL
1	Print Ad	$75.00	$75.00

Analysis

Identify 1. The accounts **Advertising Expense** and **Accounts Payable—Rockport Advertising** are affected.

Classify 2. **Advertising Expense** is an expense account. **Accounts Payable—Rockport Advertising** is a liability account.

+/− 3. **Advertising Expense** is increased by $75. **Accounts Payable—Rockport Advertising** is increased by $75.

Debit-Credit Rule

4. Increases in expense accounts are recorded as debits. Debit **Advertising Expense** for $75.

5. Increases in liability accounts are recorded as credits. Credit **Accounts Payable—Rockport Advertising** for $75.

T Accounts

6.

Advertising Expense		Accounts Payable—Rockport Advertising	
Debit + 75	Credit −	Debit −	Credit + 75

continued

Journal Entry 7.

	GENERAL JOURNAL			PAGE 1
DATE	DESCRIPTION	POST. REF.	DEBIT	CREDIT
18	Advertising Expense		75 00	
	Accts. Pay.—Rockport Adv.			75 00
	Invoice 129			

Business Transaction 11

On October 20 Zip provided delivery services for a customer, City News. Zip billed City News $1,450, Sales Invoice 1.

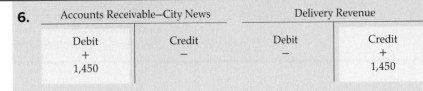

Analysis Identify
1. The accounts **Accounts Receivable—City News** and **Delivery Revenue** are affected.

Classify
2. **Accounts Receivable—City News** is an asset account. **Delivery Revenue** is a revenue account.

+/−
3. **Accounts Receivable—City News** is increased by $1,450. **Delivery Revenue** is increased by $1,450.

Debit-Credit Rule
4. Increases in asset accounts are recorded as debits. Debit **Accounts Receivable—City News** for $1,450.
5. Increases in revenue accounts are recorded as credits. Credit **Delivery Revenue** for $1,450.

T Accounts 6.

Accounts Receivable—City News		Delivery Revenue	
Debit + 1,450	Credit −	Debit −	Credit + 1,450

Journal Entry 7.

	GENERAL JOURNAL			PAGE 1
DATE	DESCRIPTION	POST. REF.	DEBIT	CREDIT
20	Accts. Rec.—City News		1 450 00	
	Delivery Revenue			1 450 00
	Sales Invoice 1			

Business Transaction 12

On October 28 Zip paid a $125 telephone bill with Check 104.

		No. 104
$	125.00	
Date	October 28	20 --
To	Pacific Bell Telephone	
For	telephone bill	

	Dollars	Cents
Balance brought forward	22,350	00
Add deposits		
Total	22,350	00
Less this check	125	00
Balance carried forward	22,225	00

Analysis

Identify
1. The accounts **Utilities Expense** and **Cash in Bank** are affected.

Classify
2. **Utilities Expense** is an expense account. **Cash in Bank** is an asset account.

+/−
3. **Utilities Expense** is increased by $125. **Cash in Bank** is decreased by $125.

Debit-Credit Rule

4. Increases in expense accounts are recorded as debits. Debit **Utilities Expense** for $125.

5. Decreases in asset accounts are recorded as credits. Credit **Cash in Bank** for $125.

T Accounts

6.

Utilities Expense		Cash in Bank	
Debit + 125	Credit −	Debit +	Credit − 125

Journal Entry

7.

		GENERAL JOURNAL			PAGE 1	
	DATE	DESCRIPTION	POST. REF.	DEBIT	CREDIT	
35	28	Utilities Expense		125 00		35
36		Cash in Bank			125 00	36
37		Check 104				37
38						38

162 Chapter 6 • Recording Transactions in a General Journal

Business Transaction 13

On October 29 Zip wrote Check 105 for $600 to have the office repainted.

			No. 105
$	600.00		
Date	October 29		20 --
To	Rainbow Painting		
For	office painted		
		Dollars	Cents
Balance brought forward		22,225	00
Add deposits			
Total		22,225	00
Less this check		600	00
Balance carried forward		21,625	00

Analysis

Identify
1. The accounts **Maintenance Expense** and **Cash in Bank** are affected.

Classify
2. **Maintenance Expense** is an expense account. **Cash in Bank** is an asset account.

+/−
3. **Maintenance Expense** is increased by $600. **Cash in Bank** is decreased by $600.

Debit-Credit Rule

4. Increases in expense accounts are recorded as debits. Debit **Maintenance Expense** for $600.

5. Decreases in asset accounts are recorded as credits. Credit **Cash in Bank** for $600.

T Accounts

6.

Maintenance Expense		Cash in Bank	
Debit + 600	Credit −	Debit +	Credit − 600

Journal Entry

7.

		GENERAL JOURNAL			PAGE 1	
	DATE	DESCRIPTION	POST. REF.	DEBIT	CREDIT	
38	29	Maintenance Expense		600 00		38
39		Cash in Bank			600 00	39
40		Check 105				40
41						41

Business Transaction 14

On October 31 Crista Vargas wrote Check 106 to withdraw $500 cash for personal use.

		No. 106
$ 500.00		
Date October 31		20 --
To Crista Vargas		
For withdrawal		

	Dollars	Cents
Balance brought forward	21,625	00
Add deposits		
Total	21,625	00
Less this check	500	00
Balance carried forward	21,125	00

Analysis

Identify 1. The accounts **Crista Vargas, Withdrawals** and **Cash in Bank** are affected.

Classify 2. **Crista Vargas, Withdrawals** is an owner's equity account. **Cash in Bank** is an asset account.

+/− 3. **Crista Vargas, Withdrawals** is increased by $500. **Cash in Bank** is decreased by $500.

Debit-Credit Rule

4. Increases in the owner's withdrawals account are recorded as debits. Debit **Crista Vargas, Withdrawals** for $500.

5. Decreases in asset accounts are recorded as credits. Credit **Cash in Bank** for $500.

T Accounts

6.
```
   Crista Vargas, Withdrawals          Cash in Bank
   Debit        |  Credit          Debit    |  Credit
     +          |    −               +      |    −
    500         |                           |   500
```

Journal Entry

7.

GENERAL JOURNAL PAGE 1

DATE	DESCRIPTION	POST. REF.	DEBIT	CREDIT	
41	31 Crista Vargas, Withdrawals		500 00		41
42	Cash in Bank			500 00	42
43	Check 106				43
44					44

Correcting the General Journal

How Do You Correct Errors in the General Journal?

Occasionally, errors occur when journalizing transactions. When an error is discovered, it must be corrected.

In a **manual** system, *an error should never be erased.* An erasure looks suspicious. It might be seen as an attempt to cover up a mistake or, worse, to change the accounting records illegally. To correct errors, use a pen and a ruler to draw a horizontal line through the entire incorrect item and write the correct information above the crossed-out error. A correction for an erroneous amount is shown in the general journal as follows:

GENERAL JOURNAL

DATE		DESCRIPTION	POST. REF.	DEBIT
20–				
Oct.	1	Cash in Bank		25 000 00 ~~52 000 00~~
		Crista Vargas, Capital		

To correct for an erroneous account name, cross out the incorrect information and write the correct account name above.

GENERAL JOURNAL PAGE 1

DATE		DESCRIPTION	POST. REF.	DEBIT	CREDIT
20–					
Oct.	1	Cash in Bank		25 000 00	
		Crista Vargas, Capital ~~Delivery Revenue~~			25 000 00
		Memorandum 1			

Common Mistakes

Writing too large When entering data in the general journal, be sure not to write too large. You will leave little space to write in correct data if you have to cross out incorrect information. Account titles and amounts should be written no larger the $\frac{3}{4}$ of the line space.

SECTION 6.2
Assessment

After You Read

Reinforce the Main Idea

Think of three different types of business transactions you might have in the next month. Use a table similar to this one to describe the general journal entry for each transaction.

Business Transaction	First Line of General Journal Entry	Second Line of General Journal Entry	Third Line of General Journal Entry

Problem 2 Recording Business Transactions

Instructions The six steps for recording a business transaction in the general journal are shown below, out of order. In your working papers or on a blank sheet of paper, indicate the proper order of these steps.

A. Amount of the credit

B. Name of the account credited

C. Source document reference

D. Date of the transaction

E. Amount of the debit

F. Name of the account debited

Problem 3 Analyzing Transactions

Glenda Hohn recently started a day-care center. She uses the following accounts.

General Ledger

Cash in Bank
Accts. Rec.—Tiny Tots Nursery
Office Furniture
Passenger Van
Accts. Pay.—Acme Bus Service

Glenda Hohn, Capital
Glenda Hohn, Withdrawals
Day-Care Fees
Utilities Expense
Van Expense

Instructions In your working papers or on a separate sheet of paper, for each transaction: Determine which accounts are affected. Classify each account. Determine whether the accounts are being increased or decreased. Use T-accounts or the table in the Working Papers to indicate which account is debited and which account is credited.

continued

166 Chapter 6 • Recording Transactions in a General Journal

SECTION 6.2
Assessment

Transactions

1. Bought a passenger van for cash.
2. Paid the telephone bill for the month.
3. Received cash from customers for day-care services.

Math for Accounting

Hania Dance Company bought a computer system on account from Tech World. The regular price for the system is $3,000, but Tech World reduced the price by 20 percent for a storewide sale. Answer the following questions about the journal entry for this transaction.

1. Which account is debited and for what amount?
2. Which account is credited and for what amount?

Chapter 6
Visual Summary

Concepts
Explain the first three steps in the accounting cycle.

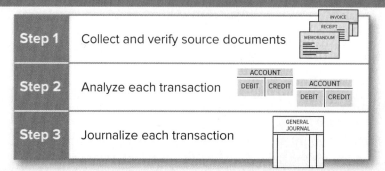

Step 1 — Collect and verify source documents
Step 2 — Analyze each transaction
Step 3 — Journalize each transaction

Analysis
Describe the steps to make a general journal entry.

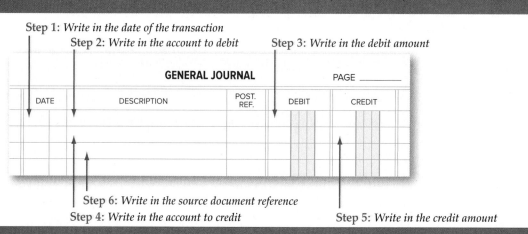

Step 1: Write in the date of the transaction
Step 2: Write in the account to debit
Step 3: Write in the debit amount
Step 4: Write in the account to credit
Step 5: Write in the credit amount
Step 6: Write in the source document reference

Procedures
Correct errors in general journal entries

DATE	DESCRIPTION	POST. REF.	DEBIT	CREDIT
20__				
Nov 2	Office Equipment		250 00	
	~~Maintenance Expense~~			
	Cash in Bank			250 00
	Check 110			

168 Chapter 6 • Recording Transactions in a General Journal

Chapter 6
Review and Activities

Answering the Essential Question

How do businesses keep permanent records of transactions?

Good record keeping in the form of journals make financial information readily available. Why might a business need to refer to a journal entry about a transaction?

Vocabulary Check

1. **Vocabulary** Arrange the vocabulary terms below into categories—groups of related words. Explain why you put the words together.

- accounting cycle
- check stub
- invoice
- memorandum
- Occur
- receipt
- source document
- calendar year
- fiscal year
- journal
- journalizing
- general journal
- Affect
- Manual

Concept Check

2. **Evaluate** Use a flowchart to illustrate the first three steps of the accounting cycle. What would happen if you skipped Step 2?
3. Describe four source documents. Be sure to include how and when each is used.
4. **Analyze** Why is a journal sometimes called the book of first entry?
5. How do you determine the debit and credit parts of a journal entry?
6. List—in the order they are recorded—the six types of information needed in a general journal entry.
7. **Analyze** Describe the relationship between the amounts entered in both columns of the general journal. Why must the amounts have this relationship?
8. What procedure is used to correct a general journal entry error in a manual system?
9. **Math** On Sunday, Dawson's flower shop sold 25 arrangements. The materials needed for 25 arrangements cost $250. If he sold each arrangement for $55, what was the net income?
10. **English Language Arts** You've been chosen as the local Business Owner of the Year. Write a one-page press release describing your business and why you've been successful. Be sure to include the skills you use in running your business, and how you plan to continue being successful in the future.

Chapter 6
Standardized Test Practice

Multiple Choice

1. Which is the purpose of a source document?

    a. To describe a transaction that takes place within a business

    b. To supply the information for journalizing a transaction

    c. To serve as evidence of a transaction

    d. All of these

2. A fiscal year is an accounting period

    a. that begins on January 1 and ends on December 31.

    b. of any length other than 12 months.

    c. of 12 months which begins in the month the business chooses.

    d. that may be for any designated length of time.

3. Which is the correct journal entry for the following transaction: A business buys $300 of office supplies on account from A1 Office Supply on March 13?

    a. 3/13   Office Supplies $300.00
            A1 Office Supply $300.00

    b. 3/13   A1 Office Supply $300.00
            Office Supplies $300.00

    c. 3/13   Cash $300.00
            Office Supplies $300.00

    d. 3/13   Office Supplies $300.00
            Accts. Pay. –A1 Office Supply $300.00

True or False

4. The general journal is a record of all the transactions that occur for a single accounting period.

Short Answer

5. Explain why it is important <u>not</u> to use an eraser when correcting entries in the general journal.

Chapter 6
Computerized Accounting

Recording General Journal Entries

Making the Transition from a Manual to a Computerized System

MANUAL METHODS
- Analyze the source document to determine which accounts are affected.
- Using a general journal form, enter the details of the transaction.
- Check for equality of debits and credits.

COMPUTERIZED METHODS
- Analyze the source document to determine which accounts are affected.
- Enter the transaction details in the general journal using the account numbers for each ledger account.
- The software will calculate the equality of debits and credits.

Chapter 6
Problems

Problem 4 Recording General Journal Transactions

Ronald Hicks owns and operates Wilderness Rentals. The following accounts are needed to journalize the month's transactions.

General Ledger

101	Cash in Bank	301	Ronald Hicks, Capital
105	Accts. Rec.—Helen Katz	305	Ronald Hicks, Withdrawals
110	Accts. Rec.—Polk and Co.	310	Income Summary
120	Office Equipment	401	Equipment Rental Revenue
125	Camping Equipment	501	Advertising Expense
201	Accts. Pay.—Adventure Equipment Inc.	505	Maintenance Expense
		515	Rent Expense
203	Accts. Pay.—Digital Tech Computers	520	Salaries Expense
		525	Utilities Expense
205	Accts. Pay.—Greg Mollaro		

Instructions Record the following transactions on page 1 of the general journal in your working papers. For each transaction:

1. Enter the date. Use the current year.
2. Enter the name of the account debited.
3. Enter the amount of the debit.
4. Enter the name of the account credited.
5. Enter the amount of the credit.
6. Enter a source document reference.

Date		Transactions
Jan.	1	Wrote Check 310 for the part-time secretary's salary, $270.
	3	Bought $2,000 of camping equipment on account from Adventure Equipment Inc., Invoice 320.
	5	Received $500 from a client for equipment rental, Receipt 150.
	7	Wrote Check 311 to pay the electricity bill of $110.
	11	Billed a client, Polk and Co., $1,700 for rental equipment, Sales Invoice 262.
	12	Ronald Hicks withdrew $800 for personal use, Check 312.

continued

Chapter 6
Problems

Date	Transactions
14	Bought a $300 scanner for the office computer from Digital Tech Computers, on account, Invoice 270.
16	Wrote Check 313 for $1,000 as an installment payment toward the amount owed to Adventure Equipment Inc.
25	Received $1,700 from Polk and Co. in payment on their account, Receipt 151.
30	Paid Digital Tech Computers $300 for the amount owed, Check 314.

Analyze
Calculate the amount of cash deducted from the Cash in Bank account in January.

Problem 5 Recording General Journal Transactions

Regina Delgado owns a business called Hot Suds Car Wash. She uses the following chart of accounts.

General Ledger

101 Cash in Bank	401 Wash Revenue
105 Accts. Rec.—Linda Brown	405 Wax Revenue
110 Accts. Rec.—Valley Auto	410 Interior Detailing Revenue
125 Office Equipment	501 Advertising Expense
130 Office Furniture	505 Equipment Rental Expense
135 Car Wash Equipment	510 Maintenance Expense
201 Accts. Pay.—Allen Vacuum Systems	520 Rent Expense
205 Accts. Pay.—O'Brian's Office Supply	525 Salaries Expense
301 Regina Delgado, Capital	530 Utilities Expense
305 Regina Delgado, Withdrawals	
310 Income Summary	

Instructions Record the following transactions on page 1 of the general journal in your working papers.

continued

Chapter 6
Problems

Date	Transactions
Jan. 1	Regina Delgado invested $12,000 in the business, Memorandum 41.
5	Purchased $5,000 in desks, chairs, and cabinets from O'Brian's Office Supply on account, Invoice 1632.
8	Deposited $1,600 for income received from car washes for the week, Receipt 101.
10	Paid the *Village Bulletin* $75 for running an ad, Check 301.
13	Regina Delgado withdrew $900 for personal use, Check 302.
17	Billed Valley Auto $400 for interior detailing, Sales Invoice 102.
18	Paid O'Brian's Office Supply $2,500 as an installment payment on account, Check 303.
20	Regina Delgado transferred to the business an electronic calculator valued at $350, Memorandum 42.
22	Wrote Check 304 for $600 to Shadyside Realty for the office rent.
24	Purchased $1,500 in car wash equipment from Allen Vacuum Systems on account, Invoice 312.
26	Received a $400 check from Valley Auto in full payment of its account, Receipt 102.
30	Issued Check 305 for $2,500 to O'Brian's Office Supply for the balance due on account.

Analyze
Identify the revenue account that was not used in the month of January.

Chapter 6
Problems

Problem 6 Recording General Journal Transactions

Abe Shultz owns and operates a pet grooming business called Kits & Pups Grooming. The following accounts are used to journalize transactions.

General Ledger

101	Cash in Bank	207	Accts. Pay.—Pet Gourmet
105	Accts. Rec.—Juan Alvarez	301	Abe Shultz, Capital
110	Accts. Rec.—N. Carlsbad	305	Abe Shultz, Withdrawals
115	Accts. Rec.—Martha Giles	310	Income Summary
125	Office Equipment	401	Boarding Revenue
130	Office Furniture	405	Grooming Revenue
135	Computer Equipment	501	Advertising Expense
140	Grooming Equipment	505	Equipment Repair Expense
145	Kennel Equipment	510	Maintenance Expense
201	Accts. Pay.—Able Store Equipment	520	Rent Expense
205	Accts. Pay.—Dogs & Cats Inc.	525	Salaries Expense
		530	Utilities Expense

Instructions Record the following transactions on page 7 of the general journal in your working papers.

continued

Chapter 6
Problems

Date		Transactions
Jan.	1	Received $125 for boarding a client's dog for one week, Receipt 300.
	3	Abe Shultz contributed to the business a computer valued at $2,500, Memorandum 33.
	5	Billed a client, Juan Alvarez, $80 for grooming his pets, Sales Invoice 212.
	9	Wrote Check 411 to Allegheny Power Co. for $150 in payment for the month's electricity bill.
	11	Abe Shultz withdrew $700 for personal use, Check 412.
	14	Purchased kennel equipment for $2,600 from Dogs & Cats Inc., on account, Invoice DC92.
	16	Paid the part-time receptionist's salary of $400 by issuing Check 413.
	18	Abe Shultz took from the business for his personal use a ten-key adding machine valued at $65, Memorandum 34.
	23	Juan Alvarez sent a check for $80 in full payment of his account, Receipt 301.
	28	Purchased on credit $250 in grooming equipment from the Pet Gourmet, Invoice PG333.
	31	Issued Check 414 for $1,300 as an installment payment for the amount owed to Dogs & Cats Inc.

Analyze

Calculate the total of the Accounts Receivable accounts as of January 31.

Chapter 6 Problems

Problem 7 Recording General Journal Transactions

Juanita Ortega is the owner of Outback Guide Service. The following accounts are used to record the transactions of her business.

General Ledger

101 Cash in Bank	205 Accts. Pay.—Peak Equipment Inc.
105 Accts. Rec.—M. Johnson	207 Accts. Pay.—Premier Processors
110 Accts. Rec.—Feldman, Jones & Ritter	301 Juanita Ortega, Capital
	302 Juanita Ortega, Withdrawals
115 Accts. Rec.—Podaski Systems Inc.	310 Income Summary
	401 Guide Service Revenue
130 Office Equipment	501 Advertising Expense
135 Office Furniture	505 Maintenance Expense
140 Computer Equipment	515 Rent Expense
145 Hiking Equipment	520 Salaries Expense
150 Rafting Equipment	525 Utilities Expense
201 Accts. Pay.—A-1 Adventure Warehouse	

Instructions Record the following transactions on page 1 of the general journal in your working papers.

Date		Transactions
Jan.	1	Juanita Ortega contributed the following assets to her business: cash, $1,500; hiking equipment, $2,000; rafting equipment, $2,500; and office furniture, $500; Memorandum 21.
	2	Issued Check 515 to *Town News* for a $75 ad.
	4	Purchased $3,000 in rafting equipment on account from A-1 Adventure Warehouse, Invoice AW45.
	6	A group from Feldman, Jones & Ritter went on a hiking trip. The group was billed $4,800 for guide services, Sales Invoice 300.
	10	Paid $300 to Dunn's Painting and Interior Co. for painting the office, Check 516.
	13	Made a $1,000 payment to A-1 Adventure Warehouse toward the amount owed, Check 517.
	15	Received a check for $4,800 from Feldman, Jones & Ritter in payment of their account, Receipt 252.
	18	Juanita Ortega paid herself $600 by issuing Check 518.

continued

Chapter 6
Problems

Date	Transactions
22	Billed a client, Mary Johnson, $1,200 for completing guide services on a hiking expedition, Sales Invoice 301.
25	Paid the monthly telephone bill for $175 by issuing Check 519.
30	Purchased a $3,600 computer system from Premier Processors. Made a down payment for $1,800 and agreed to pay the balance within 30 days, Check 520 and Invoice 749.

Analyze
Generalize about Outback's cash sales and credit sales.

Problem 8 Recording General Journal Transactions

Greg Failla operates Showbiz Video. The following accounts are used to record business transactions.

General Ledger

101 Cash in Bank	207 Accts. Pay.—New Media Suppliers
105 Accts. Rec.—G. Cohen	
110 Accts. Rec.—J. Coletti	209 Accts. Pay.—Palace Films
113 Accts. Rec.—S. Flannery	301 Greg Failla, Capital
115 Accts. Rec.—Spring Branch School District	305 Greg Failla, Withdrawals
	310 Income Summary
130 Office Equipment	401 DVD Rental Revenue
135 Office Furniture	405 HD projector Rental Revenue
140 Computer Equipment	501 Advertising Expense
145 Blu-ray discs	505 Equipment Repair Expense
150 Video Equipment	
201 Accts. Pay.—Broad Street Office Supply	510 Maintenance Expense
	520 Rent Expense
205 Accts. Pay.—Computer Horizons	525 Salaries Expense
	530 Utilities Expense

SOURCE DOCUMENT PROBLEM

Problem 6–8

Use the source documents in your working papers to complete this problem.

Instructions Record the following transactions on page 5 of the general journal in your working papers.

continued

Chapter 6
Problems

Date	Transactions
Jan. 1	Deposited $3,400 in receipts. Of that amount, $1,900 was HD Projector rentals and $1,500 was DVD rentals, Receipt 435.
3	Wrote Check 1250 for $325 of equipment repairs.
5	Purchased $400 in Blu-ray discs from Palace Films on account, Invoice PF32.
7	Bought from New Media Suppliers $2,600 in video equipment. Made a down payment of $600 and agreed to pay the balance in two installments, Check 1251 and Invoice NM101.
10	Rented DVDs to Spring Branch School District. The school district agreed to pay $1,800 at a later date, Sales Invoice 1650.
12	Issued Check 1252 for $750 to Computer Horizons for the amount owed to them.
15	Deposited $5,600 in receipts. HD Projector rentals amounted to $4,400 and DVD rentals were $1,200, Receipt 436.
18	Paid Clear Vue Window Cleaners $100 for monthly window cleaning, Check 1253.
25	Made a $1,000 installment payment toward the amount owed to New Media Suppliers by issuing Check 1254.

Analyze

Calculate the total expenses incurred in January.

Real-World Applications & Connections

CASE STUDY

Setting Up Accounting Records

Jasmine Lawrence recently purchased property (land and building) and opened Classic Auto Car Wash. The business provides a variety of services such as brushless automatic exterior car wash, interior shampooing, and detailing. Several workers have been hired to perform a variety of tasks. As Classic Auto's accountant, you have been asked to offer advice in setting up the accounting records for the business.

ACTIVITY Write Jasmine a business letter in which you suggest how she can set up and maintain accurate accounting records for her business.

INSTRUCTIONS Determine what kind of accounts you will suggest for Classic Auto's chart of accounts. What financial advice will you offer the owners?

21st Century Skills

Punctuality

People in different cultures have different ways of understanding and using time. This can be a challenge for businesses in the global workplace.

ACTIVITY Research cultural perceptions of time management and productivity in a country other than the United States. Summarize your findings in a one-page report.

Spotlight on Personal Finance

Your Personal Finance Records

Your day-to-day source documents are personal financial records. Personal financial records also include documents that are not related to everyday transactions. Vehicle titles, birth certificates, and tax returns are all personal financial documents. You can store your financial documents in home files, a home safe, or a safe-deposit box. You can also keep some financial records on a home computer.

Activity

Imagine a person your age who drives to a part-time job and has a credit card. Make a list of the types of records and documents such a person would probably have. Create a plan that describes which records and documents to store and where to store them.

H.O.T. Audit

Auditing General Journal Entries

In the following transactions, find any errors in the general journal entries presented.

1. Compare the transactions given below with the entries recorded in the general journal.

2. If any part of the transaction has been recorded incorrectly, make the necessary corrections to the journal entries on a sheet of paper.

Date		Transactions
Oct.	2	Bought $2,000 in office equipment, Check 601.
	4	Received $50 from a charge customer, Jack Lane, to apply to his account, Receipt 120.

GENERAL JOURNAL PAGE 1

	DATE		DESCRIPTION	POST. REF.	DEBIT	CREDIT	
1	20–						1
2	Oct.	2	Store Equipment		2 000 00		2
3			Cash in Bank			2 000 00	3
4			Check 601				4
5		4	Cash in Bank		50 00		5
6			Accts. Rec.–Jack Lane			50 00	6
7			Receipt 102				7
8							8

Chapter 7

Posting Journal Entries to General Ledger Accounts

Chapter Topics

7-1 The General Ledger

7-2 The Posting Process

7-3 Preparing A Trial Balance

Visual Summary

Review and Activities

Standardized Test Practice

Computerized Accounting

Problems

Real-World Accounting Careers

Real-World Applications & Connections

Essential Question

As you read this chapter, keep this question in mind:

Why is the general ledger important to the success of a business?

Main Idea

The general journal is a permanent record organized by account number. Posting is the process of transferring information from the journal to individual accounts in the ledger. The trial balance is a proof that total debits equal total credits in the ledger.

Chapter Objectives

Concepts	Analysis	Procedures
C1 Explain the purpose of the general ledger **C2** Describe the steps in the posting process.	**A1** Post general journal entries. **A2** Prepare a trial balance.	**P1** Locate and correct trial balance errors. **P2** Record correcting entries in the general journal.

Real-World Business Connection

The Solution People

Gerald Haman has a great job. Not only does he own a successful business, The Solution People, but he also gets to dress up as a superhero called Solutionman. Funny costumes and a gadget-and-toy-filled brainstorming room called the Thinkubator might sound silly, but the methods The Solution People use to help their clients generate innovative ideas are no joke. Over 160 *Fortune* 500 companies, 250 small businesses, and 120,000 people have benefited since the Chicago-based company began in 1989.

Connect to the Business

The accounting records kept by The Solution People include general ledger accounts. Journal entries to these ledger accounts create records of various business transactions that can be easily referenced. For example, an account might be created to track purchases of toys for use in the Thinkubator.

Analyze

What are some other general ledger accounts that might be used by The Solution People?

Focus on the Photo

The innovative ideas of The Solution People are generated in Chicago's "most creative meeting space," the Thinkubator, where you'll encounter everything from a super-sized abacus to thought-provoking furniture. **Why is brainstorming important to help generate innovative ideas?**

SECTION 7.1
The General Ledger

In Chapter 6 you learned to analyze business transactions and enter those transactions in a general journal.

In this chapter you will learn to post journal entries to the general ledger and to prepare a trial balance (Steps 4 and 5 in the accounting cycle illustrated in **Figure 7–1**). **Posting** is the process of transferring information from the journal to individual general ledger accounts.

Section Vocabulary
- posting
- general ledger
- ledger account forms
- vary

The Jeep dealer in your area records all business transactions in the journal and posts them to the general ledger. An up-to-date ledger allows the dealer's accountant to give management information such as sales of vehicles, service income, and salary and commission expense.

Figure 7–1 The Accounting Cycle with Steps 4 and 5 Highlighted

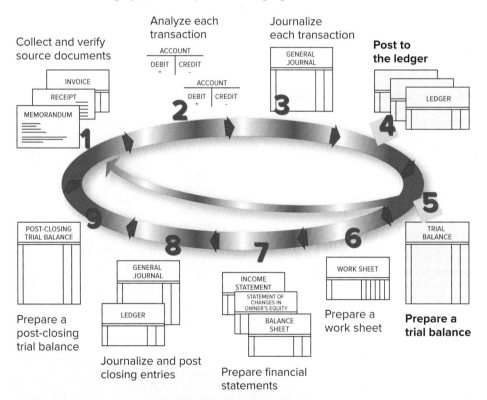

Setting Up the General Ledger

What Is a General Ledger?

Recall that the accounts used by a business are kept on separate pages or cards in a book or file called a ledger. In a computerized accounting system, the electronic files containing the accounts are still referred to as the ledger, or the ledger accounts. In either system the ledger is often called a general ledger. The **general ledger** is a permanent record organized by account number.

Posting journal entries to the ledger accounts creates a record of the impact of business transactions on each account used by a business. After journal entries have been posted, a business owner or manager can easily find the current balance of a specific account. If, for example, Crista Vargas wants to know how much money Zip Delivery Service has in its bank account, she can look at the balance of the **Cash in Bank** account.

The Four-Column Ledger Account Form

In a manual accounting system, information about specific accounts is recorded in **ledger account forms**. There are several common ledger account forms. These forms and other accounting stationery are usually described by the number of their amount columns. The number of columns refers only to those columns in which *dollar amounts* are recorded.

Zip uses the four-column ledger account form shown in **Figure 7–2**. The four-column ledger account form has spaces to enter the account name, the account number, the date, a description of the entry, and the posting reference. It also has four columns in which to record dollar amounts: Debit, Credit, Debit Balance, and Credit Balance.

Figure 7–2 Four-Column Ledger Account Form

Debit and credit amounts are posted from journal entries to the first two amount columns. The new account balance is entered in one of the last two amount columns. The type of account (expense, revenue, asset, etc.) determines which balance column to use. For example, accounts with a normal debit balance—such as asset or expense accounts—use the Debit Balance column. Accounts with a normal credit balance—such as liability or revenue accounts—use the Credit Balance column.

Reading Check

Recall

How is the general ledger organized?

Accounts in the Ledger

Before journal entries can be posted, a general ledger account is opened for each account listed on the chart of accounts.

Opening a General Ledger Account.

Two steps are required:

1. Write the account name at the top of the ledger account form.
2. Write the account number on the ledger account form.

These two steps are performed each time a ledger page is needed for a new account. The accounts opened for the first three asset accounts on Zip Delivery Service's chart of accounts (page 82) are shown in **Figure 7–3**.

In a computerized accounting system, an account is opened by entering its name and number from the chart of accounts. Computerized accounting systems **vary**, but all require entering information such as account numbers and names.

Figure 7–3 Opening General Ledger Accounts with Zero Balances

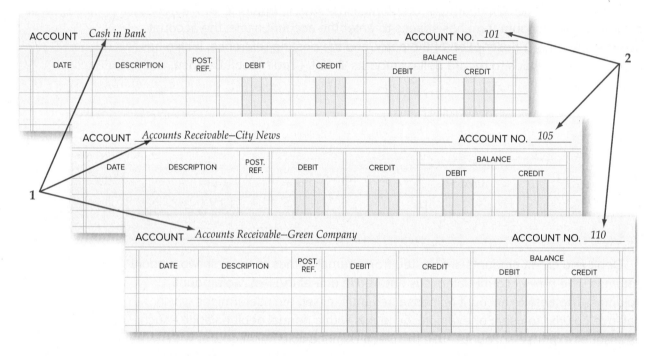

Starting a New Page for an Existing Account.

When a ledger account page is filled, continue posting on the next page.
Six steps are required to "open" a new page:

1. Write the account name at the top of the ledger account form.
2. Write the account number on the ledger account form.
3. Enter the complete date (year, month, and day) in the Date column.
4. Write the word *Balance* in the Description column.
5. Place a check mark (✓) in the Posting Reference column to show the amount entered on this line is not being posted from a journal.
6. Enter the balance in the appropriate Balance column. Usually asset, expense, and owner's withdrawals accounts have debit balances. Liability, owner's capital, and revenue accounts have credit balances.

Figure 7–4 shows an example of a new page opened for an account.

Figure 7–4 Starting a New Page for an Existing Account

ACCOUNT _Cash in Bank_ ①					ACCOUNT NO. _101_ ②	
DATE	DESCRIPTION	POST. REF.	DEBIT	CREDIT	BALANCE DEBIT	CREDIT
③ 20–						
Oct. 31	Balance ④	⑤ ✓			⑥ 21 1 2 5 00	

The Usefulness of Journals and Ledgers

How Are These Records Useful to Managers?

Managers continually use the information from accounting records. To find information about a specific business transaction, a manager can refer to the journal entry. To learn the current balance of important accounts like **Accounts Receivable** and **Accounts Payable**, managers look at the general ledger. Managers use ledgers to obtain summarized information.

SECTION 7.1
Assessment

After You Read

Reinforce the Main Idea

Use a chart like this one to compare and contrast a journal and a ledger. Add answer lines as needed.

Journal	Ledger
How are they alike?	
How are they different?	

Problem 1 Opening Ledger Accounts

Instructions Use the step-by-step processes presented in this section for starting new ledger pages for the following accounts. Use the accounting stationery provided in your working papers. January 1 of the current year is the date.

Account Name	Account Number	Balance
Cash in Bank	101	$10,000
Accounts Receivable—Mark Cohen	104	2,000
Accounts Payable—Jenco Industries	203	1,000
Tom Torrie, Capital	301	35,000
Admissions Revenue	401	- 0 -

Math for Accounting

The general ledger for Reese Delivery Service contains the following account balances:

Cash in Bank	$ 8,000
Supplies	$ 200
Delivery Equipment	?
Ed Reese, Capital	$16,200
Delivery Income	?

Using the following clues, determine the balances of the **Delivery Equipment** account and the **Delivery Income** account:

- Total debits equal $24,200.
- The balance of the **Delivery Income** account is one-half the balance of the **Delivery Equipment** account.

188 Chapter 7 · Posting Journal Entries to General Ledger Accounts

SECTION 7.2
The Posting Process

In the last section, you learned how to open accounts in the general ledger. In this section you will learn how to post general journal entries to the ledger. Recall that posting is the process of transferring information from the general journal to individual general ledger accounts. To provide current information to management, the accountant for the Jeep dealer in your area probably posts journal entries to the general ledger every day.

Section Vocabulary
- impact
- summarize

The Fourth Step in the Accounting Cycle: Posting

How Do You Post Transactions?

In Chapter 6 you learned that the general journal is a sort of business diary containing all of the trans actions of a business. It is not easy to see the effect of changes in an account by looking at journal entries. To provide a clear picture of how a business transaction changes an account's balance, the information in a journal entry is posted to the general ledger. The purpose of posting, therefore, is to show the **impact** of business transactions on the individual accounts. The ledger is sometimes called the *book of final entry.*

The size of the business, the number of transactions, and whether the accounting system is manual or computerized all affect how often posting occurs. Ideally, businesses post daily to keep their accounts up-to-date. Regardless of how often posting is performed, the process remains the same.

As in journalizing a transaction, posting to a ledger account is completed from left to right. Let's look at a journal entry for Zip that is ready to be posted to the ledger.

Posting to the Zip General Ledger

Zip's first transaction affects two accounts: **Cash in Bank** and **Crista Vargas, Capital**. The information in the journal entry is transferred item by item from the journal to each of the accounts affected. As you read about each step in the posting process, refer to **Figure 7–5**.

Locate the account to be debited in the ledger; in this example, **Cash in Bank** is to be debited.

1. Enter the date of the journal entry in the Date column of the account debited. Use the date of the journal entry, not the date on which the posting is done. Write the year and month in the left side of the Date column. It is not necessary to write the year and month for other postings to the same account on the same page unless the month or year changes. The day, however, is always entered.

Figure 7–5 Posting from the General Journal to Ledger Accounts

	GENERAL JOURNAL			PAGE 1	
DATE	DESCRIPTION	POST. REF.	DEBIT	CREDIT	
20—					1
Oct. 1	Cash in Bank	101	25 000 00		2
	Crista Vargas, Capital	301		25 000 00	3
	Memorandum 1				4
					5

1 Enter date of the journal entry

2 Description column is usually blank

3 Enter journal letter and page number in Post. Ref. column

4 Enter the debit amount

5 Compute the new account balance

6 Enter the account number in the general journal Post. Ref. column

ACCOUNT _Cash in Bank_ ACCOUNT NO. _101_

DATE	DESCRIPTION	POST. REF.	DEBIT	CREDIT	BALANCE	
					DEBIT	CREDIT
20—						
Oct. 1		G1	25 000 00		25 000 00	

7 Repeat steps 1–6 for the credit part of journal entry

ACCOUNT _Crista Vargas, Capital_ ACCOUNT NO. _301_

DATE	DESCRIPTION	POST. REF.	DEBIT	CREDIT	BALANCE	
					DEBIT	CREDIT
20—						
Oct. 1		G1		25 000 00		25 000 00

2. The Description column on the ledger account is usually left blank. Some businesses use this space to write in the source document number.

3. In the ledger account Posting Reference (Post. Ref.) column, identify where the journal entry is recorded. Enter a letter for the specific journal and the journal page number. In this example the letter "G" represents the general journal and the "1" indicates page 1 of the general journal.

4. Enter the debit amount in the Debit column of the ledger account.

5. Compute and record the new account balance in the appropriate balance column. Every amount posted will either increase or decrease the balance of that account.

6. Return to the journal and, in the Posting Reference column, enter the account number of the ledger account to which you just posted the debit part of the journal entry. Be sure it is entered on the same line as the debit

Common Mistakes

Posting Minimize posting errors by posting from left to right on the ledger account. First enter the date in the date column of the ledger account. Now write the journal and its page number in the Post Ref. column. Continue moving to the right by posting the amount in the debit or credit column, and finally compute the new account balance.

entry. In this example enter 101 in the Posting Reference column on the line for **Cash in Bank**.

This step in the posting process is very important. The notation in the Posting Reference column of the journal indicates that the journal entry has been posted. The posting reference also shows the account to which the entry was posted.

If the posting process is interrupted, perhaps by a telephone call, the posting reference signals the point at which posting stopped. *Never* write an account number in the Posting Reference column until *after* you have posted.

7. Repeat steps 1–6 for the credit part of the journal entry.
   - Locate the account to be credited. In this example **Crista Vargas, Capital**, is to be credited.
   - Enter the date.
   - Enter the posting reference on the ledger account form. In this example, G1 represents the first page of the general journal.
   - Enter the credit amount.
   - Compute the new account balance.
   - Enter the account number in the Posting Reference column of the general journal. In the example enter 301 to show that the credit was posted to **Crista Vargas, Capital.**

The journal entries made in Chapter 6 for Zip's transactions are shown in **Figure 7–6**.

Reading Check

Explain

Why should you only write an account number in the Posting Reference column of the ledger after you have posted all of the information from the journal entry?

The Importance of Posting

Posting organizes business transaction details into the proper accounts. As discussed earlier, transactions that are itemized in the general journal are helpful, but do not **summarize** similar transactions into the same location. Posting summarizes all business transactions so managers can see the cumulative effects on accounts like **Utilities Expense** or **Salaries Expense**.

The postings made to the general ledger accounts from these entries are shown in **Figure 7–7** on pages 193–194. Study these illustrations to check your understanding of the posting process.

GENERAL JOURNAL PAGE 1

DATE		DESCRIPTION	POST. REF.	DEBIT	CREDIT
20–					
Oct.	1	Cash in Bank	101	25 000 00	
		Crista Vargas, Capital	301		25 000 00
		Memorandum 1			
	2	Office Equipment	120	400 00	
		Crista Vargas, Capital	301		400 00
		Memorandum 2			
	4	Computer Equipment	115	3 000 00	
		Cash in Bank	101		3 000 00
		Check 101			
	9	Delivery Equipment	125	12 000 00	
		Accounts Payable–Coast to Coast Auto	205		12 000 00
		Invoice 200			
	11	Accounts Receivable–Green Co.	110	200 00	
		Office Equipment	120		200 00
		Memorandum 3			
	12	Accounts Payable–Coast to Coast Auto	205	350 00	
		Cash in Bank	101		350 00
		Check 102			
	14	Cash in Bank	101	200 00	
		Accounts Receivable–Green Co.	110		200 00
		Receipt 1			
	15	Cash in Bank	101	1 200 00	
		Delivery Revenue	401		1 200 00
		Receipt 2			
	16	Rent Expense	510	700 00	
		Cash in Bank	101		700 00
		Check 103			
	18	Advertising Expense	501	75 00	
		Accounts Payable–Rockport Advertising	201		75 00
		Invoice 129			
	20	Accounts Receivable–City News	105	1 450 00	
		Delivery Revenue	401		1 450 00
		Sales Invoice 1			
	28	Utilities Expense	515	125 00	
		Cash in Bank	101		125 00
		Check 104			
	29	Maintenance Expense	505	600 00	
		Cash in Bank	101		600 00
		Check 105			
	31	Crista Vargas, Withdrawals	302	500 00	
		Cash in Bank	101		500 00
		Check 106			

Figure 7–6 General Journal Entries for October Business Transactions

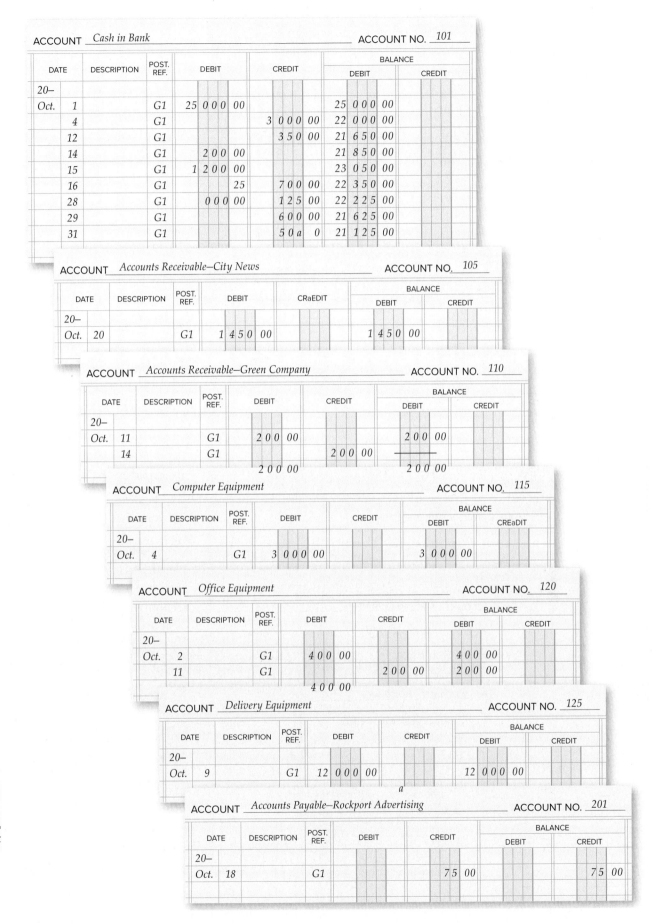

Figure 7–7 Postings to General Ledger Accounts for the Month of October

Chapter 7 • Posting Journal Entries to General Ledger Accounts

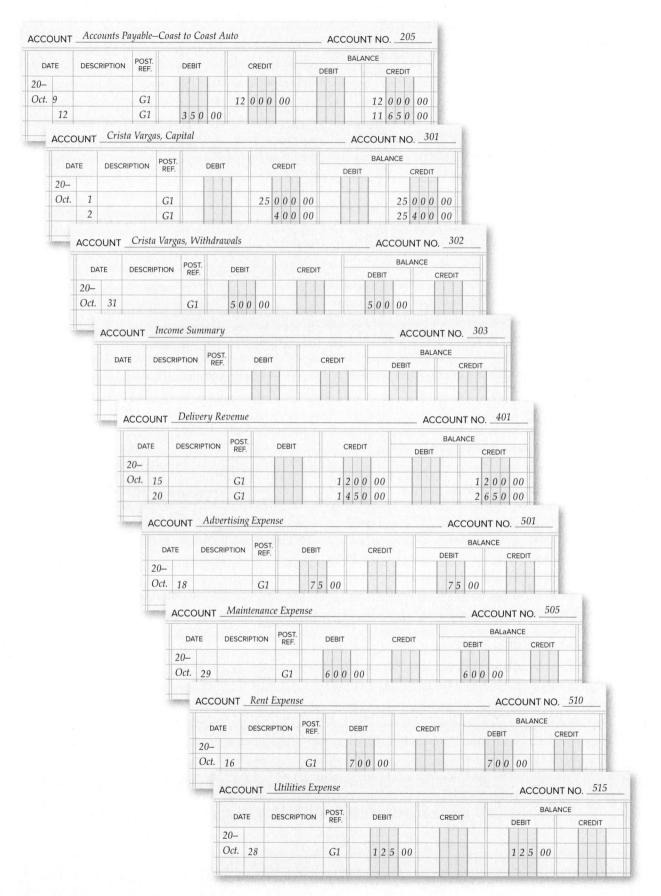

Figure 7–7 Postings to General Ledger Accounts for the Month of October (continued)

General Ledger Account Balances

How Do You Compute Account Balances?

On a four-column ledger account form, each time you post to an account, you also compute and show the new account balance.

Computing a New Account Balance

A rule of thumb for finding a new balance is that debits are added to debits, credits are added to credits, but debits and credits are subtracted. After you post to an account, compute the new account balance as follows:

When the existing account balance is a debit, and
- the amount posted is a debit, ADD the amounts.
- the amount posted is a credit, SUBTRACT the amounts.

When the existing account balance is a credit, and
- the amount posted is a debit, SUBTRACT the amounts.
- the amount posted is a credit, ADD the amounts.

A ledger account usually has space for several postings. Often, blank lines remain after the month's journal entries are posted. To save space the journal entries for more than one month are entered on the same ledger page. The new month and day are entered in the Date column, as in **Figure 7–8**.

Figure 7–8 A Ledger Account with Several Postings

ACCOUNT _Cash in Bank_   ACCOUNT NO. _101_

DATE		DESCRIPTION	POST. REF.	DEBIT	CREDIT	BALANCE DEBIT	BALANCE CREDIT
20–							
Oct.	1		G1	25 000 00		25 000 00	
	31		G2		500 00	24 500 00	
Nov.	1		G2		125 00	24 375 00	

Showing a Zero Balance in a Ledger Account

To show a zero balance after you post a transaction, draw a line across the center of the column—where the normal balance would appear. On October 11 Zip sold a phone for $200 on account to Green Company and received full payment on October 14. When the October 14 journal entry is posted, **Accounts Receivable—Green Company** has a zero balance. The line across the Debit Balance column in **Figure 7–9** means that the account has a zero balance. The line is drawn in the Debit column because the normal balance for this account is a debit.

ACCOUNT _Accounts Receivable–Green Company_   ACCOUNT NO. _110_

DATE		DESCRIPTION	POST. REF.	DEBIT	CREDIT	BALANCE DEBIT	BALANCE CREDIT
20–							
Oct.	11		G1	200 00		200 00	
	14		G1		200 00	——	

Figure 7–9 Showing a Ledger Account with a Zero Balance

SECTION 7.2 Assessment

After You Read

Reinforce the Main Idea

Create a diagram like this one to show how information is transferred between the journal and the ledger. For each line, draw an arrowhead to show the direction of the information transfer. The first line is provided as an example.

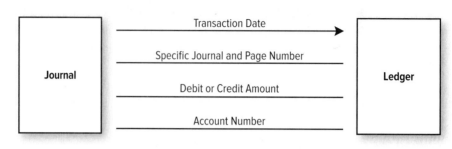

Problem 2 Posting from the General Journal to the Ledger

Instructions David Serlo made the following cash investment in his business. Use the six-step process to post the entry to the ledger accounts in your working papers.

GENERAL JOURNAL PAGE 1

	DATE	DESCRIPTION	POST. REF.	DEBIT	CREDIT
1	20–				
2	May 1	Cash in Bank		10 000 00	
3		David Serlo, Capital			10 000 00
4		Memorandum 101			
5					

Math for Accounting

As an employee of Always Fresh Bakery, you have been asked to analyze the impact that different sales levels have on the ultimate profit or loss of the business. After posting is completed, you prepare the following line graph to illustrate the sales figures for Always Fresh Bakery. Review the line graph and write a one-paragraph analysis of the impact of sales on the bakery's profit.

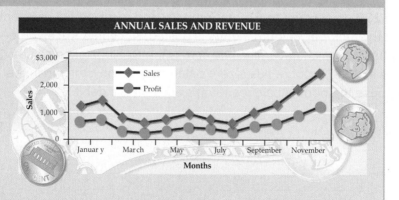

196 Chapter 7 • Posting Journal Entries to General Ledger Accounts

SECTION 7.3
Preparing A Trial Balance

In the last section, you learned how to post to the ledger. In this section you will learn how to prepare a trial balance. Accountants use a trial balance to prove that the accounting system is in balance. Preparing a trial balance is the fifth step in the accounting cycle. Every time the accountant for the Jeep dealer in your area posts to the ledger, he or she prepares a trial balance. The trial balance provides **assurance** that the journal entries are posted properly.

> **Section Vocabulary**
> - proving the ledger
> - trial balance
> - transposition error
> - slide error
> - correcting entry
> - assurance
> - error

The Fifth Step in the Accounting Cycle: The Trial Balance

What Is the Purpose of a Trial Balance?

After the journal entries have been posted to the accounts in the general ledger, the total of all of the debit balances should equal the total of all of the credit balances. Adding all the debit balances, then adding all the credit balances, and finally comparing the two totals to see whether they are equal is called **proving the ledger**.

A formal way to prove the ledger is to prepare a **trial balance**. A trial balance is a list of all the account names and their current balances. All of the debit balances are added. All of the credit balances are added. The totals are compared. If the totals are the same, the trial balance is in balance. If the totals are not equal, an **error** was made in journalizing, posting, or preparing the trial balance. You must find the error and correct it before continuing with the next step in the accounting cycle.

The equality of debits and credits does not, however, guarantee that the accounting records do not have errors. An amount might be posted to the wrong account. For example, suppose a credit sale for $500 was posted to the **Cash in Bank** account instead of **Accounts Receivable**. The trial balance remains in balance, but the company's cash is overstated by $500. What if a transaction did not get posted? Two accounts have wrong balances, but the total debits still equal the total credits.

The trial balance for Zip Delivery Service for the month of October is shown in **Figure 7–10** on page 198. The trial balance was prepared on two-column accounting stationery. The account numbers are listed in the far left column. The account names are listed in the next column. All of the debit balances are entered in the first amount column, and all of the credit balances are entered in the second amount column. Trial balances do not have to be prepared on accounting stationery, however. They can be handwritten on plain paper, typed, or prepared on a computer.

Figure 7–10 Trial Balance

Zip Delivery Service
Trial Balance
October 31, 20--

		Debit	Credit
101	Cash in Bank	21 1 2 5 00	
105	Accounts Receivable—City News	1 4 5 0 00	
110	Accounts Receivable—Green Company		
115	Computer Equipment	3 0 0 0 00	
120	Office Equipment	2 0 0 00	
125	Delivery Equipment	12 0 0 0 00	
201	Accounts Payable—Rockport Advertising		7 5 00
205	Accounts Payable—Coast to Coast Auto		11 6 5 0 00
301	Crista Vargas, Capital		25 4 0 0 00
302	Crista Vargas, Withdrawals	5 0 0 00	
303	Income Summary		
401	Delivery Revenue		2 6 5 0 00
501	Advertising Expense	7 5 00	
505	Maintenance Expense	6 0 0 00	
510	Rent Expense	7 0 0 00	
515	Utilities Expense	1 2 5 00	
	Totals	39 7 7 5	39 7 7 5 00

Finding and Correcting Errors

What Do You Do if You Are Out of Balance?

Anyone who works in accounting understands the saying, "To err is human . . ." If the debits do not equal the credits, you need to find the errors and correct them.

Finding Errors

Most trial balance errors can be located easily and quickly. When total debits do not equal total credits, follow these steps:

1. Add the debit and credit columns again. You may have added one or both of the columns incorrectly.

2. Find the difference between the debit and credit columns. If this amount is 10, 100, 1,000, and so on, you probably made an addition error. Suppose, for example, you have total debits of $35,245 and total credits of $35,345. The difference is $100, which indicates an addition error is likely. Add the columns again to find the error.

3. Check if the amount you are out of balance is evenly divisible by 9. For example, suppose the difference between the debits and credits is $27. That amount is evenly divisible by 9 (27 ÷ 9 = 3). If the difference is evenly divisible by 9, you may have a transposition error or a slide error. A **transposition error** occurs when two digits within an amount are accidentally reversed, or transposed. For example, the amount $325 may have been written as $352.

   A **slide error** occurs when a decimal point is moved by mistake. If you write $1,800 as either $180 or $18,000, you made a slide error.

198 Chapter 7 • Posting Journal Entries to General Ledger Accounts

To find a transposition error or a slide error, check the trial balance amounts against the general ledger account balances to make sure you copied the balances correctly.

4. Make sure that you included all general ledger accounts in the trial balance. Look in the general ledger for an account balance equal to the amount you are out of balance. For example, if the difference between total debits and credits is $725, look in the general ledger for an account with a balance of $725.

5. One of the account balances could have been recorded in the wrong column. That is, a debit was entered in the credit column or a credit was entered in the debit column. To find out if this happened, divide the out-of-balance amount by 2 and check whether the result matches the balance of an account. For example, suppose that the difference between the two columns is $300; $300 divided by 2 is $150. Look in the debit and credit columns for an account balance of $150. Then check to see if the $150 is entered in the wrong column.

6. If you still have not found the error, recompute the balance in each ledger account. You may have an addition or subtraction error on a ledger account form.

7. Finally, check the general ledger accounts to verify that the correct amounts are posted from the journal entries. Also, check to make sure that debit amounts are posted to the debit column and credit amounts are posted to the credit column.

Reading Check

Describe

Using the amount $456.00, give an example of a transposition error and a slide error.

Correcting Entries

When mistakes are made in accounting, one rule applies: *Never erase an error.* The method for correcting an error depends on when and where the error is found. There are three types of errors:

- Error in a journal entry that has not been posted.
- Error in posting to the ledger when the journal entry is correct.
- Error in a journal entry that has been posted.

In Chapter 6 you learned how to handle the first situation. When an error in a journal entry is discovered before posting, you draw a single line through the incorrect item in the journal and write the correction directly above it.

If the journal entry is correct but is posted incorrectly to the ledger, you draw a single line through the incorrect item in the ledger and write the correction directly above it.

When an error in a journal entry is discovered *after* posting, make a **correcting entry** to fix the error.

On November 15 the accountant for Zip found an error in a journal entry made on November 2. A $100 check to pay the electricity bill was journalized and posted to the **Maintenance Expense** account by mistake. The original journal entry is shown in the following T accounts.

Maintenance Expense		Cash in Bank	
Debit + 100	Credit −	Debit +	Credit − 100

The following T accounts show how the transaction *should* have been recorded.

Utilities Expense		Cash in Bank	
Debit + 100	Credit −	Debit +	Credit − 100

As you can see, the $100 credit to **Cash in Bank** is correct. The error is in the debit part of the November 2 transaction. **Maintenance Expense** is incorrectly debited for $100. To correct the error, **Maintenance Expense** is credited for $100 and **Utilities Expense** is debited for $100.

The accountant wrote Memorandum 70 to notify the accounting clerk of the mistake. The correcting entry, recorded in the general journal, is shown in **Figure 7–11**.

Figure 7–11 Correcting Entry

	DATE		DESCRIPTION	POST. REF.	DEBIT	CREDIT	
1	20–						1
2	Nov.	15	Utilities Expense		100 00		2
3			Maintenance Expense			100 00	3
4			Memorandum 70				4
5							5

GENERAL JOURNAL PAGE 3

Posting a correcting entry is similar to any other posting. In the Description column of the ledger accounts, however, the words *Correcting Entry* are written. **Figure 7–12** shows how the correcting entry is posted to the **Maintenance Expense** and **Utilities Expense** accounts.

ACCOUNT *Maintenance Expense*   ACCOUNT NO. 505

DATE		DESCRIPTION	POST. REF.	DEBIT	CREDIT	BALANCE DEBIT	BALANCE CREDIT
20–							
Oct.	29		G1			600 00	
Nov.	2		G2	100 00		700 00	
	15	Correcting Entry	G3		100 00	600 00	

ACCOUNT *Utilities Expense*   ACCOUNT NO. 515

DATE		DESCRIPTION	POST. REF.	DEBIT	CREDIT	BALANCE DEBIT	BALANCE CREDIT
20–							
Oct.	28		G1	125 00		125 00	
Nov.	15	Correcting Entry	G3	100 00		225 00	

Figure 7–12 Posting of Correcting Entry

SECTION 7.3
Assessment

After You Read

Reinforce the Main Idea

Create a chart like this one to show how to correct errors in three situations.

Where Error Was Found	When Error Was Found	How to Correct the Error
Journal	Before Posting	
Ledger	After Posting	
Journal	After Posting	

Problem 3 Analyzing a Source Document

Instructions Analyze the transaction that is described in Memorandum 47, and then record and post the required correcting entry in your working papers.

FUNTIME AMUSEMENT ARCADE MEMORANDUM 47

TO: Accounting Clerk
FROM: Dan Vonderhaar
DATE: May 20, 20--
SUBJECT: Correction of error

On May 10, we purchased an office copier for $1,500. I noticed in the general journal that the entry was recorded and posted to the Computer Equipment account. Please record the necessary entry to correct this error.

Problem 4 Recording and Posting a Correcting Entry

Instructions On July 7 Video Connection's accounting supervisor discovered that a July 3 transaction had been recorded incorrectly. The transaction, involving the purchase of advertising in the local newspaper with a $300 check, was incorrectly journalized and posted to the **Rent Expense** account. In your working papers, record and post the correcting entry using Memorandum 13 as the source document.

VIDEO Connection MEMORANDUM 13

TO: Accounting Clerk
FRaOM: Accounting Manager
DATE: July 7, 20--
SUBJECT: Correction of error

On July 3, we paid $300 for advertising in the *Daily News Record* that was incorrectly journalized and posted to the Rent Expense account. Please record the necessary entry to correct this error.

Math for Accounting

1. Compare the numbers in Column 1 to those in Column 2. Find any transposition, slide, or omission errors. Identify the type of error for each line.

2. Using a calculator or adding machine, total Column 1. Correct any errors in Column 2, and then total Column 2. Do the totals of Columns 1 and 2 match?

Column 1	Column 2
$18.00	$180.00
$15,000	$1,500
$222.52	$222.25
$187,235,499.05	$187,235,499.50
$47,988	$47,988
$578,334.99	$5,778,334.99

202 Chapter 7 • Posting Journal Entries to General Ledger Accounts

Chapter 7
Visual Summary

Concepts
Steps in the posting process.

1. Enter the date of the journal entry.
2. Description column is usually blank.
3. Enter journal letter and page number in Post. Ref. column.
4. Enter the debit amount.
5. Compute the new account balance.
6. Enter the account number in the general journal Post. Ref. column.
7. Repeat steps 1–6 for the credit part of journal entry.

Analysis
Prepare a trial balance.

For each ledger account, total the credits and debits

- If the debit total is larger, subtract the credit total from the debit total to get your ledger account total, which goes in the debit column of the trial balance.
- If the credit total is larger, subtract the debit total from the credit total to get your ledger account total, which goes in the credit column of the trial balance.

Put the ledger account total in the credit or debit column of your trial balance.

When you have debit or credit totals for each ledger account, add all of your credit totals to get a credit grand total.

Add all of your debit totals to get a debit grand total. This is your trial balance.

If the totals are not equal, an error was made somewhere.

Procedures
Record correcting entries in the general journal

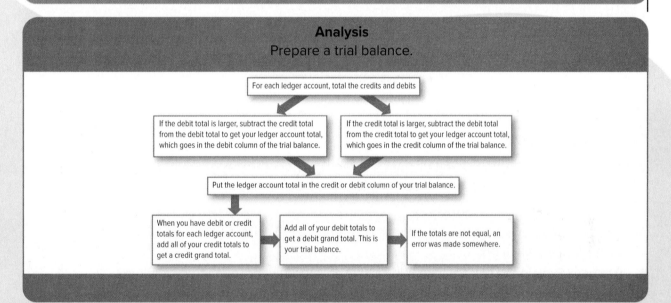

ORIGINAL ENTRY:

DATE		DESCRIPTION	POST. REF.	DEBIT	CREDIT
20–					
Apr	5	Office Supplies	140	250 00	
		Accounts Receivable	110		250 00
		Memorandum 3			

CORRECTING ENTRY:

DATE		DESCRIPTION	POST. REF.	DEBIT	CREDIT
20–					
May	2	Accounts Receivable	110	250 00	
		Accounts Payable	200		250 00
		Memorandum 3			
		Corrects April 5 Entry			

Chapter 7
Review and Activities

Answering the Essential Question

Why is the general ledger important to the success of a business?

Businesses need to keep track of their finances just as you and your family need to—money that comes in, expenses paid, money owed, and money saved or held on account. Why is the general ledger the best tool for tracking finances?

Vocabulary Check

1. **Vocabulary** Write a short essay about posting journal entries to general ledger accounts. Use as many of these content vocabulary terms as possible. Underline each vocabulary term used.

   - general ledger
   - posting
   - ledger account forms
   - vary
   - impact
   - summarize
   - assurance
   - error
   - proving the ledger
   - trial balance
   - slide error
   - transposition error
   - correcting entry

Concept Check

2. Illustrate the six steps required to open a new page in an existing general ledger account.
3. What information is entered in the Posting Reference (Post. Ref.) column of the ledger account? Give an example.
4. What information is entered in the Posting Reference (Post. Ref.) column of the general journal entry? Give an example.
5. **Evaluate** Explain the trial balance. Why is this an important accounting procedure?
6. What steps do you perform to locate trial balance errors?
7. **Analyze** Distinguish between a slide error and a transposition error and give an example of each.
8. **Analyze** What is the basic rule when correcting accounting errors? Why is this rule important?
9. Describe the three ways of correcting posting errors.
10. **Math** Dough House Bakery makes bread, cakes, cookies, and pies. At the end of the day, the prices of all the leftover products are reduced by 75%. If bread sells for $0.87 at the end of the day, what is its regular price?
11. **English Language Arts** Choose an accounting software package to investigate. In a report, compare and contrast the method of using the software program with the manual method for posting to journals and ledgers. Tell the benefits and drawbacks of each method, and whether or not you would use the software program you chose if you were setting up journals and ledgers for a company.

Chapter 7
Standardized Test Practice

Multiple Choice

1. When beginning a new page for a General Ledger account, placing a check mark (✓) in the Post. Ref. column indicates that
   a. you verified the balance is correct
   b. the amount entered is not being posted from a journal
   c. you verified all the lines from the previous page were filled
   d. an error has been corrected

2. What is the purpose of the trial balance?
   a. To prove the Cash in Bank account has the correct balance
   b. To correct any errors in the general journal
   c. To prove the ledger debit and credit balances are equal
   d. To prove that amounts were correctly posted from the journal

3. An error in posting may cause
   a. income to be overstated or understated on the income statement.
   b. a business to pay too much to a vendor.
   c. cash on hand to be less than the balance in the cash account.
   d. all of the above.

4. When preparing a trial balance, which of the following would not be considered a procedure?
   a. Check to see if both columns equal.
   b. Double rule both columns.
   c. Write the general ledger account names in the work sheet account name column.
   d. All of these are correct procedures.
   e. None of these answers.

Short Answer

5. Transferring information from the journal to individual general ledger accounts is known as _____.

Extended Response

6. Give two practical uses for the General Ledger in everyday business.

Chapter 7
Computerized Accounting

Posting to the General Ledger
Making the Transition from a Manual to a Computerized System

MANUAL METHODS	COMPUTERIZED METHODS
• Transfer the details of each journal entry to individual ledger accounts.	• After each journal entry is entered, amounts are posted automatically into the general ledger.
• Calculate the new account balance for each ledger account.	• New account balances are calculated for you.

Chapter 7
Problems

Problem 5 Posting General Journal Transactions

The accounts used by Wilderness Rentals have been opened and are included in the working papers accompanying this textbook. The general journal transactions for March of the current year are also included.

Instructions Post the transactions recorded on page 1 of the general journal to the accounts in the general ledger.

Analyze
Identify the account with the highest debit balance.

Problem 6 Preparing a Trial Balance

The ledger accounts for Hot Suds Car Wash are shown in your working papers.

Instructions Prepare a trial balance as of March 31 of the current year.

Analyze
Identify the account with the highest credit balance.

Problem 7 Journalizing and Posting Business Transactions

A partial chart of accounts for Kits & Pups Grooming follows.

General Ledger

101 Cash in Bank	301 Abe Shultz, Capital
105 Accts. Rec.—J. Alvarez	305 Abe Shultz, Withdrawals
120 Grooming Supplies	401 Boarding Revenue
130 Office Furniture	405 Grooming Revenue
140 Grooming Equipment	525 Salaries Expense
205 Accts. Pay.—Dogs & Cats Inc.	

Instructions

1. In your working papers, open an account in the general ledger for each of these accounts.
2. Record the March transactions on page 1 of the general journal.
3. Post each journal entry to the appropriate ledger account.
4. Prove the ledger by preparing a trial balance.

continued

Chapter 7
Problems

Date		Transactions
Mar.	1	Abe Shultz invested $5,000 in the business, Memorandum 51.
	3	Abe Shultz transferred a desk and chairs valued at $1,200 to the business, Memorandum 52.
	5	Issued Check 551 for $300 for grooming supplies.
	7	Bought grooming equipment on account for $1,800 from Dogs & Cats Inc., Invoice DC201.
	9	Groomed Juan Alvarez's show dogs on account, $400, Sales Invoice 350.
	12	Paid Dogs & Cats Inc. $900 on account, Check 552.
	15	Issued Check 553 for $500 to pay the office secretary's salary.
	18	Abe Shultz withdrew $1,000 cash from the business, Check 554.
	20	Deposited $1,400 received from clients for boarding their pets, Receipts 477–480.
	24	Wrote Check 555 for $900 to Dogs & Cats Inc. to apply on account.
	26	Received a check for $400 from Juan Alvarez on account, Receipt 481.

Analyze
Identify the temporary accounts.

Problem 8 Journalizing and Posting Business Transactions

The chart of accounts for Outback Guide Service follows.

General Ledger

- 101 Cash in Bank
- 115 Accts. Rec.—Podaski Systems Inc.
- 140 Computer Equipment
- 145 Hiking Equipment
- 150 Rafting Equipment
- 205 Accts. Pay.—Peak Equipment Inc.
- 207 Accts. Pay.—Premier Processors
- 301 Juanita Ortega, Capital
- 302 Juanita Ortega, Withdrawals
- 401 Guide Service Revenue
- 501 Advertising Expense

SOURCE DOCUMENT PROBLEM

Problem 7–8

Use the source documents in your working papers to complete this problem.

continued

Chapter 7
Problems

Instructions

1. In your working papers, open an account in the general ledger for each account in the chart of accounts.
2. Record the following transactions on page 1 of the general journal.
3. Post each journal entry to the appropriate accounts in the ledger.
4. Prove the ledger by preparing a trial balance.

Date		Transactions
Mar.	1	Invested $20,000 in cash and transferred rafting equipment valued at $5,000 to the business, Memorandum 35.
	3	Purchased $600 in hiking equipment on account from Peak Equipment Inc., Invoice 101.
	5	Bought a $2,800 computer from Premier Processors and agreed to pay for it within 60 days, Invoice 616.
	7	Deposited $700 cash received from clients, Receipts 310–315.
	9	Paid $400 to the Daily Courier for an ad, Check 652.
	12	Sent a bill for $900 to Podaski Systems Inc. for conducting a group rafting trip for them, Sales Invoice 352.
	15	Juanita Ortega wrote a check for $800 for personal use, Check 653.
	18	Paid Premier Processors $1,400 to apply on account, Check 654.
	22	Received $900 from Podaski Systems Inc. in full payment for the amount owed, Receipt 316.
	27	Bought a $500 television ad with LIVE TV, Check 655.
	28	Paid Peak Equipment Inc. $600 by writing Check 656.

Analyze

Calculate the total change in accounts payable.

Problem 9 Recording and Posting Correcting Entries

An auditor reviewed the accounting records of Showbiz Video. The auditor wrote a list of transactions, outlined below, describing the errors discovered in the March records. The general journal for March and a portion of the general ledger are included in your working papers.

continued

Chapter 7
Problems

Instructions

1. Record correcting entries on general journal page 22. Use *March 31* as the date and *Memorandum 50* as the source document for all correcting entries.
2. Some errors will not require correcting entries but will require a general ledger correction. Make the appropriate general ledger corrections.
3. Post all correcting entries to the general ledger accounts.

Date		Transactions
Mar.	3	The $125 purchase was for office supplies.
	7	A $200 payment to a creditor, Broad Street Office Supply, was not posted to the account.
	13	Greg Failla withdrew $1,200 from the business for personal use.
	17	Cash totaling $2,000 was received for video rentals.
	19	A $75 receipt from Shannon Flannery was posted as $57.
	27	Greg Failla invested an additional $3,000 in the business.
	29	The revenue of $1,000 was for video rentals.

Analyze

Compute the amount that Greg Failla, Capital was overstated or understated before the correcting entries were posted.

Real-World Accounting Careers

Kevin Maher
Designed Alloy Products

Q What do you do?

A I manage the financial operations, oversee the company's relationships with banks and insurance firms, and supervise all financial transactions. I also supervise a staff of 10 professionals.

Q What are your day-to-day responsibilities?

A There is no typical day in my role. For instance, on one recent day I was investigating ways to manage an increase in healthcare costs to minimize the impact on employees. Two days before that, I attended an export seminar to explore expanding our firm's presence overseas. The only function I perform every day is monitoring the company's cash position.

Q What factors have been key to your success?

A I have a background in information systems, and I understand the value of data. I can use data points to zero in on anomalies or trends and gain insight into the company's performance. Technology is a financial professional's most valuable tool, so it's important to know how to utilize it.

Q What do you like most about your job?

A I like the variety. We're a small company, so everyone here wears many different hats. That makes things exciting.

Career Facts

Real-World Skills
Strong communication, managerial and organizational skills; ability to collaborate effectively with internal and external stakeholders

Training And Education
A bachelor's degree in accounting or finance; MBA and CPA or CMA designation is preferred; at least seven years of experience.

Career Skills
Knowledge of U.S. GAAP; proficiency with accounting software

Career Path
Gain public accounting experience, then accept a position in a corporate environment, such as accounting manager, director of accounting or assistant controller

---

Tips from Robert Half International

Learn to manage your time by planning ahead. Each morning, draft a quick to-do list that includes all your short- and long-term projects. This document will help you determine the resources you need and allow you to prioritize your tasks.

College and Career Readiness

In what ways could you demonstrate to an employer that you are a valuable employee?

Real-World Applications & Connections

Career Wise

International Accountant

If you plan to be an accountant in the 21st century, chances are you're going to be an international accountant to some extent. In the modern business environment, transactions between countries are very likely, and mastering the international aspects of accounting can increase your value as an employee. International accountants may be responsible for translating financial statements prepared in foreign currencies into U.S. dollars, and analyzing financial statements prepared according to foreign or international standards. International accountants may also need to be familiar with foreign tax laws.

Research and Share

a. Use the Internet to locate three sources of information on international accounting.

b. What level of education is required to be an international accountant? In addition to those listed above, what are the roles and responsibilities of international accountants?

Global Accounting

Global E-Business

When you think about revenue and expenses for a company, consider how the Internet can make it faster and cheaper to do business. Business is conducted using Web sites, e-mail, and videoconferencing. Companies like Gap and K-Mart use Web sites to sell products at lower costs than store locations. Managers can communicate online with vendors or branch locations to avoid the high costs of travel.

Instructions

If you owned a small business and wished to expand your market to a global audience, discuss how you might use the Internet.

A Matter of Ethics

Meeting a Deadline

Imagine that you are an accounting clerk for Ace Hardware. The store manager has asked you to prepare the trial balance. The totals on the trial balance are not equal and you cannot find the error. You realize that the trial balance is due at the end of the day. You are frustrated and consider changing one of the account balances just to get the trial balance to balance.

Activity

Define the ethical issues involved. What are the alternatives? Who are the affected parties? How do the alternatives affect the parties? What would you do?

Spotlight on Personal Finance

Finding the Best Loan

Banks and lenders are more competitive than ever, and smart consumers can use this to their advantage when shopping for a loan. Many sites on the Internet let users easily compare interest rates and terms, and some actually have lenders bidding on the loan, trying to get the customer.

Research and Discuss

a. Using the Internet, search for "loan comparisons" and "loan finder." What are some of the popular services that make it easy to shop for loans?

b. As a class, make a list of the different services available. Discuss how loan comparison services can help consumers. What factors do you think are important to consider when shopping for a loan?

H.O.T. Audit

Posting Journal Entries to General Ledger Accounts

Sal Cerra, owner and operator of Sal's Silkscreen T-Shirts, has given you the final general ledger balances, and has provided the trial balance in your working papers. You are told that the Automotive Equipment account is incorrect.

Instructions

Make the necessary adjustment to Automotive Equipment and prepare a corrected trial balance on a separate sheet of paper as of September 30 of the current year.

Cash in Bank	$15,000
Accounts Receivable	11,800
Automotive Equipment	22,600
Automotive Tools	11,200
Office Equipment	300
Accounts Payable	3,000
Sal Cerra, Capital	42,250
Sal Cerra, Withdrawals	5,100
Repair Service	36,000
Advertising Expense	250
Auto Supplies Expense	5,600

Mini Practice Set 1

Main Task

Set up the accounting records and perform the daily activities for TechVision Web Design.

Summary of Steps

- Open general ledger accounts.
- Analyze, journalize, and post transactions.
- Prepare a trial balance.
- Create a clear and coherent oral presentation that analyzes the results of the accounting cycle.

Why It's Important

You could one day work with a sole proprietor as he or she starts a business from scratch. You might even open your own business!

Setting Up Accounting Records for a Sole Proprietorship

TechVision Web Design

Company Background Whether you are interested in a Web site for your new hair salon or a corporate intranet site, TechVision.com Web Sites can provide design, hosting, and maintenance services for the small mom-and-pop business and the multinational corporation alike.

Jack Hines, owner of TechVision, is known for his unique and dynamic presentations on the Web. His creativity and knowledge of Web culture makes him a success in electronic content delivery.

Organization TechVision is organized as a sole proprietorship. The business is fully owned and operated by Jack Hines.

Your Job Responsibilities As TechVision's accounting clerk, use the accounting stationery in your working papers for these tasks.

1. Open a general ledger account for each account in the chart of accounts.
2. Analyze each business transaction.
3. Enter each business transaction in the general journal, page 1.
4. Post each journal entry to the appropriate accounts in the general ledger.
5. Prove the general ledger by preparing a trial balance.

CHART OF ACCOUNTS
TechVision.com Web Sites

ASSETS
- 101 Cash in Bank
- 105 Accts. Rec.–Andrew Hospital
- 110 Accts. Rec.–Indiana Trucking
- 115 Accts Rec.–Sunshine Products
- 130 Office Supplies
- 135 Office Equipment
- 140 Office Furniture
- 145 Web Server

LIABILITIES
- 205 Accts. Pay.–Computer Specialists Inc.
- 210 Accts. Pay.–Office Systems
- 215 Accts. Pay.–Service Plus Software Inc.

OWNER'S EQUITY
- 301 Jack Hines, Capital
- 305 Jack Hines, Withdrawals

REVENUE
- 401 Web Service Fees

EXPENSES
- 505 Membership Expense
- 506 Telecommunications Expense
- 507 Rent Expense
- 508 Utilities Expense

Business Transactions Jack Hines, owner of TechVision.com Web Sites, began business operations on May 1 of this year. During the month of May, the business completed the transactions that follow.

Date		Transactions
May	1	Jack Hines invested $50,000 in the business, Memorandum 1.
	2	The owner, Jack Hines, invested a desktop computer and printer (Office Equipment) $3,500, Memorandum 2.
	2	Issued Check 101 for $125 for the purchase of office supplies.
	3	Bought office furniture for $2,700 on account from Office Systems, Invoice 457.
	7	Bought a Web server from Computer Specialists Inc. on account for $35,000, Invoice WS4658421.
	9	Received $1,000 from James Market for Web site services, Receipt 101.
	11	Completed Web site design services for Andrew Hospital to be paid later, Invoice 101, $3,000.
	12	Bought software for the Web server (Web Server) on account from Service Plus Software Inc., Invoice 876, $10,000.
	14	Wrote Check 102 for $118 to pay the electric bill.
	15	Jack Hines withdrew $2,500 for personal expenses, Check 103.
	17	Completed Web site design on account for Sunshine Products, Invoice 102, $5,000.
	18	Bought a filing cabinet (Office Furniture) for $275, Check 104.
	19	Received a check for $4,000 as payment for Web site maintenance for one year to a client, Receipt 102.
	20	Provided design services on account to Indiana Trucking, Invoice 103 for $2,000.
	21	Prepared Receipt 103, $2,500, received on account, Sunshine Products.
	22	Paid $4,900 for telecommunication services for the period May 1–May 31, Check 105.
	22	Wrote check 106 for $3,333 to Service Plus Software Inc. as payment on account.
	25	Sent Check 107 for $2,000 to Office Systems as payment on account.
	26	Received $1,000 for two months of Web services, Receipt 104.
	27	Paid the dues for membership in the All Inclusive Group for $7,000, Check 108.
	30	Wrote Check 109 for the monthly rent $750.
	30	Withdrew $2,500 for personal expenses, Check 110.
	30	Sent Check 111 for $25,000 to Computer Specialists Inc., as payment on account.

Analyze

Compute the total of all the checks written during May.

Chapter 8

The Six-column Work Sheet

Chapter Topics

8-1 Preparing the Work Sheet

8-2 Completing the Work Sheet

Visual Summary

Review and Activities

Standardized Test Practice

Computerized Accounting

Problems

Real-World Applications & Connections

Essential Question

As you read this chapter, keep this question in mind:

Why is the six-column work sheet an important accounting tool?

Main Idea

The work sheet organizes general ledger account information for the financial statements. After completing the work sheet, you will know the net income or net loss for the accounting period.

Chapter Objectives

Concepts	Analysis	Procedures
C1 Explain the purpose of the six-column work sheet. **C2** Describe the parts of a six-column work sheet.	**A1** Prepare a six-column work sheet.	**P1** Calculate net income and net loss.

Real-World Business Connection

Petkeepers Ltd.

The U.S. economy is driven by entrepreneurs, people who transform ideas for products or services into real-world businesses. Travis Mitchell, a pet owner with experience in animal sciences, started **Petkeepers Ltd.** after returning from a vacation to find his dog had a bad experience at a kennel. Mitchell's research indicated that St. Louis, Missouri, had a need for more in-home pet care services. What started as Mitchell's part-time job is now an established company with employees and a strong community presence.

Connect to the Business

A challenge for new entrepreneurs is developing a system for keeping accurate and organized accounting records. At the end of each accounting period, a business such as **Petkeepers Ltd.** uses tools such as the work sheet to summarize revenues and expenses.

Analyze

*Why do you think summarizing expenses and revenues might help manage a business like **Petkeepers Ltd.**?*

Focus on the Photo

Like any successful small business, Petkeepers collects and organizes its financial information with a work sheet. ***Why do you think it is important for financial information to be organized in a certain way?***

SECTION 8.1
Preparing the Work Sheet

Chapter 7 focused on steps four and five of the accounting cycle: posting journal entries to the general ledger and preparing a trial balance. The length of an accounting **period** can vary. The maximum period covered by the accounting cycle is one year. The first five steps of the accounting cycle are performed frequently during the cycle. The last four steps—preparing a work sheet, preparing financial statements, journalizing and posting closing entries, and preparing a post-closing trial balance—are performed at the end of the accounting period. Look at **Figure 8–1**. In this chapter you will learn how to prepare a work sheet, the sixth step of the accounting cycle. With this step, businesses like your local Midas Muffler shop or a Nike Outlet collect information from their ledger accounts and record this information on a single form.

Section Vocabulary
- work sheet
- ruling
- period
- implies

Figure 8–1 Steps in the Accounting Cycle with Step 6 Highlighted

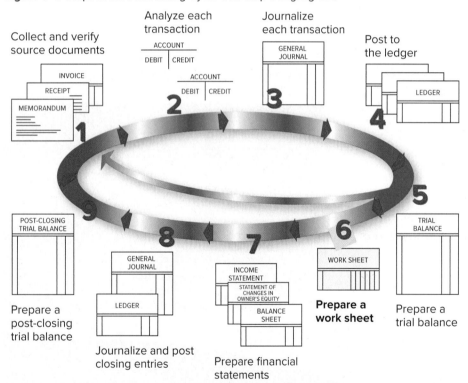

The Sixth Step of the Accounting Cycle: The Work Sheet

What Is the Purpose of a Work Sheet?

A **work sheet** is just what its name **implies**—a working paper used to collect information from the ledger accounts in one place. Like the T account, the work sheet is a tool that the accountant uses. With the work sheet, an accountant gathers all of the information needed to prepare the financial statements and to complete the other end-of-period activities.

218   Chapter 8 • The Six-Column Work Sheet

A work sheet may be prepared in pencil on standard multicolumn accounting paper. The paper comes in several sizes and is usually printed without column headings. Blank spaces for column headings allow the accountant to enter the headings needed by a particular business. The work sheet can also be prepared using accounting software.

Erasing an error is permitted on a work sheet because it is a working document. However, you should find and correct any errors in the journalizing or posting process before you prepare the work sheet. Otherwise, those errors will make it more difficult to prepare the work sheet, which must balance before it can be completed.

Reading Check

Recall

When are the last four steps of the accounting cycle completed?

The Work Sheet Sections

How Is the Work Sheet Organized?

The Zip Delivery Service work sheet, shown in **Figure 8–2**, has five sections:

1. the heading
2. the Account Name section
3. the Trial Balance section
4. the Income Statement section
5. the Balance Sheet section

Figure 8–2 Six-Column Work Sheet

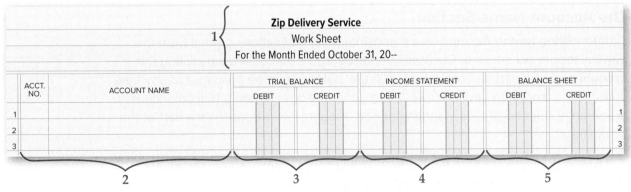

The Account Name section includes a column for the account number and a column for the account name. The Trial Balance, Income Statement, and Balance Sheet sections have Debit and Credit amount columns. The six amount columns give this work sheet its name: the six-column work sheet.

Chapter 8 • The Six-Column Work Sheet   **219**

The Work Sheet Heading

The work sheet heading contains three kinds of information:

1. The name of the business *(Who?)*
2. The name of the accounting form *(What?)*
3. The period covered by the work sheet *(When?)*

Notice how these elements are positioned on the work sheet above. Follow this format when preparing the heading of any work sheet.

Figure 8–3 Work Sheet with Account Names and Trial Balance Amounts

Zip Delivery Service
Work Sheet
For the Month Ended October 31, 20--

	ACCT. NO.	ACCOUNT NAME	TRIAL BALANCE DEBIT	TRIAL BALANCE CREDIT	INCOME STATEMENT DEBIT	INCOME STATEMENT CREDIT	BALANCE SHEET DEBIT	BALANCE SHEET CREDIT	
1	101	Cash in Bank	21 125 00						1
2	105	Accts. Rec.–City News	1 450 00						2
3	110	Accts. Rec.–Green Company	—						3
4	115	Computer Equipment	3 000 00						4
5	120	Office Equipment	200 00						5
6	125	Delivery Equipment	12 000 00						6
7	201	Accts. Pay.–Rockport Advertising		75 00					7
8	205	Accts. Pay.–Coast to Coast Auto		11 650 00					8
9	301	Crista Vargas, Capital		25 400 00					9
10	302	Crista Vargas, Withdrawals	500 00						10
11	303	Income Summary	—	—					11
12	401	Delivery Revenue		2 650 00					12
13	501	Advertising Expense	75 00						13
14	505	Maintenance Expense	600 00						14
15	510	Rent Expense	700 00						15
16	515	Utilities Expense	125 00						16
17									17

The Account Name Section

Information for the Account Name and Trial Balance sections comes from the general ledger accounts. In Chapter 7 you prepared a trial balance by listing the account names and their final balances. A trial balance can be prepared at any time during the accounting period to prove the general ledger. When a trial balance is prepared at the end of an accounting period, though, it is prepared as a part of the work sheet.

Look at the work sheet in **Figure 8–3**. The account numbers and names are from Zip's general ledger and are listed on the work sheet in the same order that they appear in the general ledger:

- Assets
- Liabilities
- Owner's Equity
- Revenue
- Expenses

All of the general ledger accounts are listed on the work sheet, even those that have a zero balance. Listing all of the accounts avoids accidentally omitting an account and ensures that the work sheet contains all accounts needed to prepare the financial reports.

Reading Check

Identify

What are the five sections that give the six-column work sheet its name?

The Trial Balance Section

The end-of-period balance in the general ledger for each account is entered in the appropriate amount column of the Trial Balance section. Accounts with debit balances are entered in the Trial Balance Debit column. Accounts with credit balances are entered in the Trial Balance Credit column.

If an account has a zero balance at the end of the period, a line is drawn in the normal balance column. Notice in **Figure 8–3** that a line was drawn in the Trial Balance Debit column for **Accounts Receivable—Green Company.** Lines were also recorded in the Trial Balance Debit and Credit columns for **Income Summary** since this account does not have a normal balance side. You will learn more about the **Income Summary** account in Chapter 10.

Ruling the Trial Balance Section.

- **Ruling** means "drawing a line." In accounting a single rule, or line, drawn under a column of amounts means that the entries above the rule are ready to be totaled.
- After all account names and balances have been entered on the work sheet, a single rule is drawn under the last entry and across both amount columns of the Trial Balance section as shown in **Figure 8– 4**.
- The Debit and Credit columns are now ready for totaling.

Totaling the Trial Balance Section.

- If the ledger balances, the total debits will equal the total credits. Look at **Figure 8–4**. The totals match, with each column totaling $39,775.
- Since total debits equal total credits, a double rule is drawn across both amount columns just beneath the totals.
- This double rule means that the amounts just above it are totals and that no other entries will be made in the Trial Balance columns.

If the total debits do not equal the total credits, there is an error. Find and correct the error before completing the work sheet. Procedures for locating errors were discussed in Chapter 7.

Figure 8–4 Work Sheet with Trial Balance Section Completed

Zip Delivery Service
Work Sheet
For the Month Ended October 31, 20--

	ACCT. NO.	ACCOUNT NAME	TRIAL BALANCE DEBIT	TRIAL BALANCE CREDIT	INCOME STATEMENT DEBIT	INCOME STATEMENT CREDIT	BALANCE SHEET DEBIT	BALANCE SHEET CREDIT	
1	101	Cash in Bank	21 125 00						1
2	105	Accts. Rec.–City News	1 450 00						2
3	110	Accts. Rec.–Green Company	—						3
4	115	Computer Equipment	3 000 00						4
5	120	Office Equipment	200 00						5
6	125	Delivery Equipment	12 000 00						6
7	201	Accts. Pay.–Rockport Advertising		75 00					7
8	205	Accts. Pay.–Coast to Coast Auto		11 650 00					8
9	301	Crista Vargas, Capital		25 400 00					9
10	302	Crista Vargas, Withdrawals	500 00						10
11	303	Income Summary	—	—					11
12	401	Delivery Revenue		2 650 00					12
13	501	Advertising Expense	75 00						13
14	505	Maintenance Expense	600 00						14
15	510	Rent Expense	700 00						15
16	515	Utilities Expense	125 00						16
17			39 775 00	39 775 00					17
18									18

SECTION 8.1
Assessment

After You Read

Reinforce the Main Idea

Create a diagram similar to this one. List the information that is transferred from the general ledger to the worksheet. Add answer lines as needed.

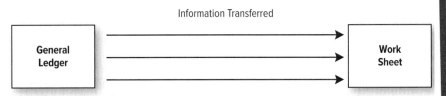

Problem 1 Entering Account Balances on the Work Sheet

The following accounts appear on the work sheet for Lee's Bike Shop.

Store Equipment	Scott Lee, Capital	Scott Lee, Withdrawals
Rent Expense	Advertising Expense	Maintenance Expense
Service Fees Revenue	Accts. Rec.—John Langer	Office Supplies
Accts. Pay.—Rubino Supply		

Instructions Use a form similar to the one below in your working papers. Classify each account and use an "X" to indicate whether the account balance is entered in the Debit or the Credit column of the Trial Balance section. The first account is completed as an example.

Account Name	Classification	Trial Balance	
		Debit	Credit
Store Equipment	Asset	X	

Problem 2 Analyzing a Source Document

Instructions Based on the invoice presented here, answer the following questions in your working papers.

1. Which company shipped the supplies?
2. Which company ordered the supplies?
3. On what date were the supplies received?
4. What does one box of file folders cost?
5. How many ring binders were ordered?
6. What is the invoice number?

Hailey Office Supply
2763 East Meadow Ave.
Richardson, TX 75080

INVOICE NO. 220
DATE: June 3, 20--
ORDER NO.: PO304
SHIPPED BY: Truck
TERMS: Net 30

TO: Garmot Electrical Co.
15638 Lone Star Highway
Plano, TX 75074

QTY.	ITEM	UNIT PRICE	TOTAL
2	Ring Binders, 37546	$ 2.94	$ 5.88
1	Bond Paper, 27361	5.83	5.83
2	Rotary Cards, 62744	1.75	3.50
4	Fine-Tip Markers, 28733	.84	3.36
4	File Folders, Box 36206	6.23	24.92
			$43.49
	Sales Tax		2.17
REC'D 6/15/20--	Total		$45.66

continued

SECTION 8.1
Assessment

Math for Accounting

The columns in the Trial Balance section of the work sheet have different totals:

Debit total = $34,800
Credit total = $35,600

The accountant discovered that the balance for **Advertising Expense** had been entered in the Credit column instead of the Debit column. Calculate the balance for the **Advertising Expense** account that should have been entered in the Debit column. What is the new total of the Debit column and Credit column?

SECTION 8.2
Completing the Work Sheet

In the previous section you learned how to set up the work sheet and prove the Trial Balance section. In this section you will learn how to extend the account balances and **compute** net income or net loss.

Section Vocabulary
- matching principle
- net income
- net loss
- compute
- accuracy

The Balance Sheet and Income Statement Sections

What Are the Balance Sheet and Income Statement?

The balance sheet and income statement are financial statements prepared at the end of the accounting period. The work sheet organizes the information for these reports.

The Balance Sheet Section

The Balance Sheet section of the work sheet contains the asset, liability, and owner's equity accounts. After proving the Trial Balance section, *extend,* or transfer, the appropriate amounts to the Balance Sheet section. To do this, copy the Trial Balance section amounts for the asset, liability, and owner's equity accounts to the appropriate Balance Sheet amount columns. Start with the first account and extend each account balance. Extend debit amounts to the Balance Sheet Debit column. Extend credit amounts to the Balance Sheet Credit column. **Figure 8–5** on page 226 shows the balances extended to the Balance Sheet section.

The Income Statement Section

The next step in completing the work sheet is to extend the appropriate account balances to the Income Statement section. The Income Statement section contains the revenue and expense accounts. After transferring the asset, liability, and owner's equity account balances to the Balance Sheet section, extend the revenue and expense account balances to the Income Statement section. Revenue accounts have a normal credit balance, so extend their balances to the Income Statement Credit column. Since expense accounts have a normal debit balance, extend expense account balances to the Debit column of the Income Statement section. **Figure 8–6** on page 226 shows the amounts in the Debit and Credit columns of the Income Statement section.

Totaling the Income Statement and Balance Sheet Sections

After all amounts have been extended to the Balance Sheet and Income Statement sections, these sections are totaled. A single rule drawn across the four Debit and Credit columns indicates that the columns are ready to be added.

Figure 8–5 Work Sheet with Trial Balance Amounts Extended to Balance Sheet Section

Zip Delivery Service
Work Sheet
For the Month Ended October 31, 20--

ACCT. NO.	ACCOUNT NAME	TRIAL BALANCE DEBIT	TRIAL BALANCE CREDIT	INCOME STATEMENT DEBIT	INCOME STATEMENT CREDIT	BALANCE SHEET DEBIT	BALANCE SHEET CREDIT
101	Cash in Bank	21 125 00				21 125 00	
105	Accts. Rec.–City News	1 450 00				1 450 00	
110	Accts. Rec.–Green Company	—				—	
115	Computer Equipment	3 000 00				3 000 00	
120	Office Equipment	200 00				200 00	
125	Delivery Equipment	12 000 00				12 000 00	
201	Accts. Pay.–Rockport Advertising		75 00				75 00
205	Accts. Pay.–Coast to Coast Auto		11 650 00				11 650 00
301	Crista Vargas, Capital		25 400 00				25 400 00
302	Crista Vargas, Withdrawals	500 00				500 00	
303	Income Summary	—	—				
401	Delivery Revenue		2 650 00				
501	Advertising Expense	75 00					
505	Maintenance Expense	600 00					
510	Rent Expense	700 00					
515	Utilities Expense	125 00					
		39 775 00	39 775 00				

Figure 8–6 Work Sheet with Trial Balance Amounts Extended to Income Statement Section

Zip Delivery Service
Work Sheet
For the Month Ended October 31, 20--

ACCT. NO.	ACCOUNT NAME	TRIAL BALANCE DEBIT	TRIAL BALANCE CREDIT	INCOME STATEMENT DEBIT	INCOME STATEMENT CREDIT	BALANCE SHEET DEBIT	BALANCE SHEET CREDIT
101	Cash in Bank	21 125 00				21 125 00	
105	Accts. Rec.–City News	1 450 00				1 450 00	
110	Accts. Rec.–Green Company	—				—	
115	Computer Equipment	3 000 00				3 000 00	
120	Office Equipment	200 00				200 00	
125	Delivery Equipment	12 000 00				12 000 00	
201	Accts. Pay.–Rockport Advertising		75 00				75 00
205	Accts. Pay.–Coast to Coast Auto		11 650 00				11 650 00
301	Crista Vargas, Capital		25 400 00				25 400 00
302	Crista Vargas, Withdrawals	500 00				500 00	
303	Income Summary	—	—	—	—		
401	Delivery Revenue		2 650 00		2 650 00		
501	Advertising Expense	75 00		75 00			
505	Maintenance Expense	600 00		600 00			
510	Rent Expense	700 00		700 00			
515	Utilities Expense	125 00		125 00			
		39 775 00	39 775 00				

Figure 8–7 Work Sheet with Income Statement and Balance Sheet Sections Totaled

Zip Delivery Service
Work Sheet
For the Month Ended October 31, 20--

	ACCT. NO.	ACCOUNT NAME	TRIAL BALANCE DEBIT	TRIAL BALANCE CREDIT	INCOME STATEMENT DEBIT	INCOME STATEMENT CREDIT	BALANCE SHEET DEBIT	BALANCE SHEET CREDIT	
1	101	Cash in Bank	21 125 00				21 125 00		1
2	105	Accts. Rec.–City News	1 450 00				1 450 00		2
3	110	Accts. Rec.–Green Company	—				—		3
4	115	Computer Equipment	3 000 00				3 000 00		4
5	120	Office Equipment	200 00				200 00		5
6	125	Delivery Equipment	12 000 00				12 000 00		6
7	201	Accts. Pay.–Rockport Advertising		75 00				75 00	7
8	205	Accts. Pay.–Coast to Coast Auto		11 650 00				11 650 00	8
9	301	Crista Vargas, Capital		25 400 00				25 400 00	9
10	302	Crista Vargas, Withdrawals	500 00				500 00		10
11	303	Income Summary							11
12	401	Delivery Revenue		2 650 00		2 650 00			12
13	501	Advertising Expense	75 00		75 00				13
14	505	Maintenance Expense	600 00		600 00				14
15	510	Rent Expense	700 00		700 00				15
16	515	Utilities Expense	125 00		125 00				16
17			39 775 00	39 775 00	1 500 00	2 650 00	38 275 00	37 125 00	17
18									18

Unlike the Trial Balance section debit and credit totals, the Debit and Credit column totals in these two sections will not be equal until the net income or net loss for the period is added. See **Figure 8–7**.

Showing Net Income or Net Loss on the Work Sheet

What Is Meant by Net Income or Net Loss?

The work sheet's Income Statement section shows the period's revenue and expenses. The GAAP **matching principle** requires matching expenses incurred in an accounting period with the revenue earned in the same period. Matching expenses with revenue gives a reliable measure of profit by showing the dollar value of resources used to produce the revenue. Owners and managers use this information to analyze results and make decisions.

> **Common Mistakes**
>
> **The Trial Balance Withdrawals Account**
> When extending amounts from the Trial Balance section to the Balance Sheet and Income Statement sections, you may be confused as to where the withdrawals account is entered. Although it is a temporary owner's equity account, it is not included in the calculation of net income or net loss. It is entered after the capital balance in the Balance Sheet Debit column.

Showing Net Income on the Work Sheet

After the Income Statement section columns have been totaled, total expenses (Debit column total) are subtracted from total revenue (Credit column total) to find net income. **Net income** is the amount of revenue that remains after expenses for the period have been subtracted. Net income is entered as a debit at the bottom of the Income Statement section of the work sheet. Look at **Figure 8–8** on page 228. Zip's net income is $1,150.

1. Skip a line after the last account and write the words *Net Income* in the Account Name column. **A**
2. On the same line, enter the net income amount in the Income Statement Debit column. **B**

Figure 8–8 Partial Work Sheet with Net Income

ACCT. NO.	ACCOUNT NAME	TRIAL BALANCE DEBIT	TRIAL BALANCE CREDIT	INCOME STATEMENT DEBIT	INCOME STATEMENT CREDIT	BALANCE SHEET DEBIT	BALANCE SHEET CREDIT
101	Cash in Bank	21 125 00				21 125 00	
105	Accts. Rec.–City News	1 450 00				1 450 00	
110	Accts. Rec.–Green Company	—				—	
115	Computer Equipment	3 000 00				3 000 00	
120	Office Equipment	200 00				200 00	
125	Delivery Equipment	12 000 00				12 000 00	
201	Accts. Pay.–Rockport Advertising		75 00				75 00
205	Accts. Pay.–Coast to Coast Auto		11 650 00				11 650 00
301	Crista Vargas, Capital		25 400 00				25 400 00
302	Crista Vargas, Withdrawals	500 00				500 00	
303	Income Summary	—		—			
401	Delivery Revenue		2 650 00		2 650 00		
501	Advertising Expense	75 00		75 00			
505	Maintenance Expense	600 00		600 00			
510	Rent Expense	700 00		700 00			
515	Utilities Expense	125 00		125 00			
		39 775 00	39 775 00	1 500 00	2 650 00	38 275 00	37 125 00
	Net Income			1 150 00			1 150 00

A   B   C

3. On the same line, enter the net income amount in the Balance Sheet Credit column. **C**

Why is the net income also shown in the Balance Sheet section of the work sheet? Remember, revenue and expense accounts are temporary accounts. As you can see in **Figure 8–9**, revenues increase capital, while expenses decrease capital. Net income, therefore, increases capital since revenues exceed expenses.

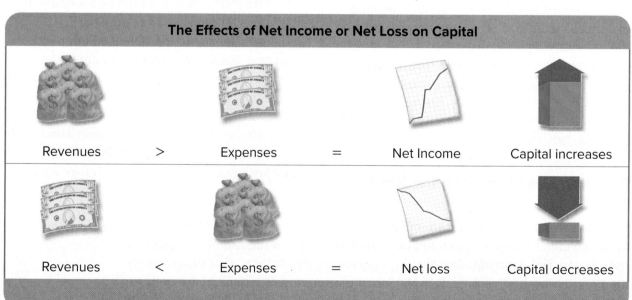

Figure 8–9 The Effects of Income or Loss on Capital

During the accounting period, revenue and expense amounts are recorded in the temporary accounts (like **Delivery Revenue** and **Utilities Expense**). At the end of the period, net income will be transferred to the owner's capital account. Since the capital account is increased by credits, the amount of the net income is entered in the Credit column of the Balance Sheet section of the work sheet.

To check the **accuracy** of the net income amount in the Balance Sheet section, subtract the total of the Credit column from the total of the Debit column. If the result does not equal net income, there is an error. Before continuing, you must find and correct the error. To find the error, check that the amounts from the Trial Balance section are extended correctly and that the totals of all columns are added correctly.

Reading Check

Explain

How is net income calculated on the Income Statement?

Figure 8–10 Completed Work Sheet

Zip Delivery Service
Work Sheet
For the Month Ended October 31, 20--

ACCT. NO.	ACCOUNT NAME	TRIAL BALANCE DEBIT	TRIAL BALANCE CREDIT	INCOME STATEMENT DEBIT	INCOME STATEMENT CREDIT	BALANCE SHEET DEBIT	BALANCE SHEET CREDIT
101	Cash in Bank	21 125 00				21 125 00	
105	Accts. Receivable–City News	1 450 00				1 450 00	
110	Accts. Receivable–Green Company	—				—	
115	Computer Equipment	3 000 00				3 000 00	
120	Office Equipment	200 00				200 00	
125	Delivery Equipment	12 000 00				12 000 00	
201	Accts. Payable–Rockport Advertising		75 00				75 00
205	Accounts Pay.–Coast to Coast Auto		11 650 00				11 650 00
301	Crista Vargas, Capital		25 400 00				25 400 00
302	Crista Vargas, Withdrawals	500 00				500 00	
303	Income Summary	—					
401	Delivery Revenue		2 650 00		2 650 00		
501	Advertising Expense	75 00		75 00			
505	Maintenance Expense	600 00		600 00			
510	Rent Expense	700 00		700 00			
515	Utilities Expense	125 00		125 00			
		39 775 00	39 775 00	1 500 00	2 650 00	38 275 00	37 125 00
	Net Income	21 125 00		1 150 00			1 150 00
		1 450 00		2 650 00	2 650 00	38 275 00	38 275 00

A   B D   E   F   G C

Completing the Work Sheet

The completed work sheet for Zip is shown in **Figure 8–10**. To complete the Income Statement and Balance Sheet sections, follow these steps:

1. On the line under the net income amount, draw a single rule across the four columns. **D**

2. In the Income Statement section Debit column, add the net income amount to the previous total and enter the new total. Bring down the total of the Income Statement section Credit column. Total debits should equal total credits. **E**

3. In the Balance Sheet section Credit column, add net income to the previous total and enter the new total. Bring down the total of the Balance Sheet section Debit column. The total debit amount should equal the total credit amount. **F**

4. In the Balance Sheet and Income Statement sections, draw a double rule under the four column totals. The double rule indicates that the Debit and Credit columns are equal and that no more amounts are to be entered in these columns. **G**

Showing a Net Loss on the Work Sheet

What if total expenses are more than total revenue? When that happens, a **net loss** occurs. A net loss decreases owner's equity. This decrease will eventually be shown as a debit to the capital account. When a net loss occurs, the steps to complete the work sheet are the same as described for Net Income except that the words *Net Loss* are written in the Account Name column. The net loss amount is entered in the Credit column of the Income Statement section and in the Debit column of the Balance Sheet section. The partial work sheet in **Figure 8–11** shows a net loss. Expenses exceed revenue by $746.

Figure 8–11 Partial Work Sheet Showing a Net Loss

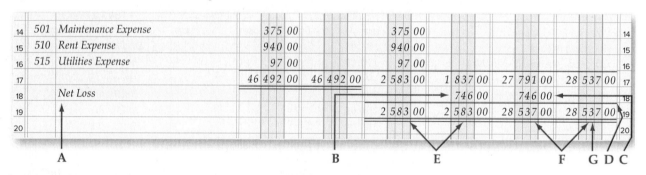

A Review of the Six-Column Work Sheet

How Do You Prepare a Work Sheet?

Follow these steps when preparing a six-column work sheet:

1. Enter the heading on the work sheet.
2. In the Account Name and Trial Balance sections, enter the account numbers, names, and balances for all general ledger accounts.
3. Prove the equality of the Trial Balance total debits and total credits.
4. Extend the amounts of the Trial Balance section to the appropriate columns in the Balance Sheet and Income Statement sections.
5. Total the columns in the Income Statement and Balance Sheet sections.
6. Determine the amount of the net income or net loss for the period.
7. Enter the amount of the net income or net loss in the appropriate columns in the Income Statement and Balance Sheet sections.
8. Total and rule the Income Statement and Balance Sheet sections.

SECTION 8.2
Assessment

After You Read

Reinforce the Main Idea

Create a chart like this one. Fill in the blanks to describe how net income and net loss are shown on the work sheet.

Situation	Is the result a net income or a net loss?	Extend to which column in the Income Statement section?	Extend to which column in the Balance Sheet section?
Revenue > Expenses			
Revenue < Expenses			

Problem 3 Extending Amounts Across the Work Sheet

The following accounts have balances and appear in the Account Name section of the work sheet for Lee's Bike Shop.

Store Equipment	Advertising Expense
Rent Expense	Accounts Receivable—John Langer
Service Fees Revenue	Scott Lee, Withdrawals
Accounts Payable—Rubino Supply	Maintenance Expense
Scott Lee, Capital	Office Supplies

Instructions Use a form similar to the one below in your working papers or on a separate sheet of paper. For the above accounts, enter an "X" in the column where the account balance is extended. The first account has been completed as an example.

Account Name	Income Statement		Balance Sheet	
	Debit	Credit	Debit	Credit
Store Equipment			X	

continued

SECTION 8.2
Assessment

Math for Accounting

For Art's Sake is a sole proprietorship that sells artwork. The first two quarterly financial statements for this year have been grim. Using the revenue and expense accounts for the third quarter, calculate the net income or loss. Compare the third-quarter figure with the figures from the first and second quarters. What do you predict the future of For Art's Sake will be?

Third Quarter

Rent Expense	$2,000
Utilities Expense	800
Advertising Expense	3,000
Maintenance Expense	1,200
Design Revenue	500
Picture Revenue	5,525
Frame Revenue	3,200

Second Quarter

Net Loss = $1,200

First Quarter

Net Loss = $2,000

Chapter 8
Visual Summary

Concepts
Describe the parts of a six-column work sheet.

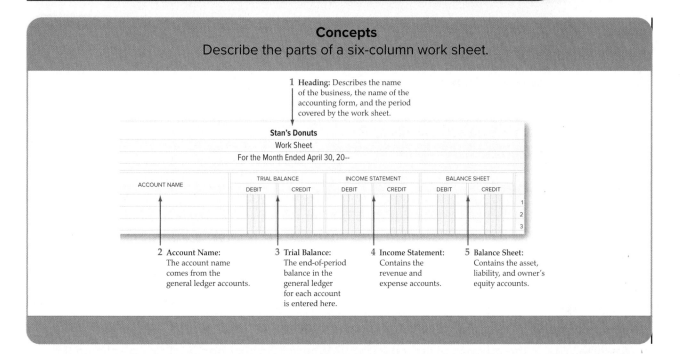

Analysis
Steps to prepare a six-column work sheet.

	Steps
1.	Create a heading
2.	In the Account Name and Trial Balance sections, enter account numbers, names, and balances.
3.	Complete the Trial Balance section, making sure that total debits equal total credits.
4.	Extend amounts from the Trial Balance section to the appropriate columns in the Balance Sheet and Income Statement sections
5.	Total the columns in the Income Statement and Balance Sheet columns
6.	Calculate the net income or net loss for the period.
7.	Enter net income or net loss on the work sheet.
8.	Total and rule the Income Statement and Balance Sheet sections.

Procedures
Calculate net income and net loss.

Net Income	Net Loss
If revenue > expenses, then: **Revenue − Expenses = Net Income**	If revenue < expenses, then: **Expenses − Revenue = Net Loss**

Chapter 8 • The Six-Column Work Sheet 233

Chapter 8
Review and Activities

Answering the Essential Question

Why is the six-column work sheet an important accounting tool?

A tool is used as an aid in accomplishing a specific task. How can a work sheet be a tool? What other tools do accountants use?

Vocabulary Check

1. **Vocabulary** Create multiple-choice test questions for each content vocabulary term.

   - imply
   - period
   - work sheet
   - ruling
   - compute
   - matching principle
   - net income
   - accurate
   - net loss

Concept Check

2. **Analyze** What is the purpose of the work sheet? Why is it an important tool for completing the accounting cycle?
3. Illustrate and briefly describe the five sections of a six-column work sheet.
4. List the items that are included in the work sheet heading.
5. **Predict** What might possibly be the result of not listing all accounts, including those with zero balances, on the work sheet?
6. Explain how to record net income or net loss on the work sheet.
7. **Evaluate** How does net income affect owner's equity? How does net loss affect owner's equity?
8. **Math** Ernesto owns an importing firm in Mexico. Last year, Ernesto's Imports imported 137 million dollars' worth of computers and transportation-related goods to Mexico from the United States. If these imports represent 0.1% of the total imports of U.S. goods to Mexico for the year, what is the value in dollars of all Mexican imports from the U.S. for that year?
9. **English Language Arts** Research the role of information technology in accounting. Discuss its impact on productivity and decision making, how it developed into what it is today, and how you think it might develop in the future. Tell what role IT plays in accounting and how it assists businesses in decision making. Report your findings in a two-page paper.

Chapter 8
Standardized Test Practice

Multiple Choice

1. Which information is not contained in the work sheet heading?

    a. The period covered

    b. The name of the accounting form

    c. The business name

    d. The accountant's name who prepared the work sheet

2. In what section of the work sheet is the end of period balance for each account contained?

    a. The Balance Sheet section

    b. The Trial Balance section

    c. The Income Statement section

    d. The Account Name section

3. The asset, liability, and owner's equity accounts are extended to the _____ of the work sheet.

    a. Balance Sheet section

    b. Income Statement section

    c. Retained Earnings statement

    d. Trial Balance section

4. Revenue and expense accounts are listed in the Trial Balance section of the work sheet and in the _____ of the work sheet.

    a. Retained Earnings statement

    b. Balance Sheet section

    c. Income Statement section

    d. Accounts Payable section

5. The amount of net income for the period is added to the Balance Sheet credit total because it increases the balance in the _____ account.

    a. accounts receivable

    b. accounts payable

    c. asset

    d. capital

Chapter 8
Standardized Test Practice

True or False

6. Matching expenses with revenue in the same period gives a reliable measure of profit by showing the value of resources used to produce the revenue.

Short Answer

7. What is the purpose of the work sheet?
8. When a net loss is shown on the work sheet, which account is affected?

Chapter 8
Computerized Accounting

Preparing the Trial Balance
Making the Transition from a Manual to a Computerized System

MANUAL METHODS

- After all journal entries have been posted and all general ledger balances have been updated, list account names and balances on two-column accounting stationery.
- List debit balances in one column and credit balances in the other.
- Add all debit balances.
- Add all credit balances.
- Compare the totals. If totals are not in balance, locate and correct any journalizing or posting errors.
- The trial balance is correct when total debits equal total credits.

COMPUTERIZED METHODS

- Although the Trial Balance is not necessary to check the equality of debits and credits since out-of-balance entries cannot be posted, it is one of the most commonly used accounting statements. It provides a clear list format for reviewing accounts. You can view and print the Trial Balance from the **Reports** menu.

Chapter 8
Problems

Problem 4 Preparing a Six-Column Work Sheet

The ending balances in the general ledger accounts of Wilderness Rentals for the period ended May 31 follow.

		Debit Balances	Credit Balances
101	Cash in Bank	$5,814	
105	Accts. Rec.—Helen Katz	717	
110	Accts. Rec.—Polk and Co.	590	
115	Office Supplies	847	
120	Office Equipment	4,360	
125	Camping Equipment	6,130	
201	Accts. Pay.—Adventure Equipment Inc.		$ 1,680
203	Accts. Pay.—Digital Tech Computers		3,554
205	Accts. Pay.—Greg Mollaro		635
301	Ronald Hicks, Capital		12,760
305	Ronald Hicks, Withdrawals	1,200	
310	Income Summary	—	—
401	Equipment Rental Revenue		9,716
501	Advertising Expense	1,940	
505	Maintenance Expense	1,083	
515	Rent Expense	3,500	
525	Utilities Expense	2,164	

Instructions Prepare a work sheet for the period ended May 31.

1. Enter the heading on the work sheet.
2. List all of the accounts in the Account Name and Trial Balance sections. For each account, include the account number, name, and balance.
3. Total and rule the Trial Balance section. Do total debits equal total credits? If not, find and correct the problem before continuing.
4. Extend the appropriate amounts to the Balance Sheet section.
5. Extend the appropriate amounts to the Income Statement section.
6. Total the amount columns in the Income Statement and Balance Sheet sections.
7. Enter the amount of net income or net loss in the appropriate columns in the Income Statement and Balance Sheet sections.
8. Total and rule the Income Statement and Balance Sheet sections.

Analyze
Identify the permanent accounts on the work sheet that have normal debit balances.

Chapter 8
Problems

Problem 5 Preparing a Six-Column Work Sheet

The general ledger for Hot Suds Car Wash shows the following account balances on May 31, the end of the period.

101	Cash in Bank	$9,400
105	Accts. Rec.—Linda Brown	429
110	Accts. Rec.—Valley Auto	372
115	Detailing Supplies	694
120	Detergent Supplies	418
125	Office Equipment	15,195
130	Office Furniture	2,029
135	Car Wash Equipment	7,486
201	Accts. Pay.—Allen Vacuum Systems	4,346
205	Accts. Pay.—O'Brian's Office Supply	2,730
301	Regina Delgado, Capital	26,530
305	Regina Delgado, Withdrawals	2,500
310	Income Summary	
401	Wash Revenue	7,957
405	Wax Revenue	5,329
410	Interior Detailing Revenue	2,970
501	Advertising Expense	1,940
505	Equipment Rental Expense	3,836
510	Maintenance Expense	1,424
520	Rent Expense	3,500
530	Utilities Expense	639

Instructions Prepare a work sheet for Hot Suds Car Wash in your working papers.

Analyze
Identify the temporary accounts on the work sheet that have normal credit balances.

Problem 6 Preparing a Six-Column Work Sheet

The account balances in the general ledger of Kits & Pups Grooming at the end of May are:

Continued

Chapter 8
Problems

101	Cash in Bank	$11,194
105	Accts. Rec.—Juan Alvarez	357
110	Accts. Rec.—Nathan Carlsbad	547
115	Accts. Rec.—Martha Giles	1,450
120	Grooming Supplies	842
125	Office Equipment	4,147
130	Office Furniture	935
135	Computer Equipment	2,200
140	Grooming Equipment	1,948
145	Kennel Equipment	7,305
201	Accts. Pay.—Able Store Equipment	7,945
205	Accts. Pay.—Dogs & Cats Inc.	1,205
207	Accts. Pay.—Pet Gourmet	2,846
301	Abe Shultz, Capital	23,048
305	Abe Shultz, Withdrawals	2,500
310	Income Summary	—
401	Boarding Revenue	11,596
405	Grooming Revenue	4,496
501	Advertising Expense	3,675
505	Equipment Repair Expense	932
510	Maintenance Expense	2,658
520	Rent Expense	7,500
530	Utilities Expense	2,946

Instructions Prepare a work sheet for the month ended May 31 for Kits & Pups Grooming in your working papers.

Analyze
Compute the total liabilities for Kits & Pups.

Chapter 8 Problems

Problem 7 Preparing a Six-Column Work Sheet

The general ledger account balances on May 31 for Outback Guide Service are:

101	Cash in Bank	$ 2,834
105	Accts. Rec.—Mary Johnson	384
125	Office Supplies	307
130	Office Equipment	5,902
135	Office Furniture	2,804
140	Computer Equipment	3,295
145	Hiking Equipment	922
150	Rafting Equipment	8,351
205	Accts. Pay.—Peak Equipment Inc.	1,204
301	Juanita Ortega, Capital	20,419
302	Juanita Ortega, Withdrawals	1,800
310	Income Summary	—
401	Guide Service Revenue	9,179
501	Advertising Expense	795
505	Maintenance Expense	125
515	Rent Expense	2,000
525	Utilities Expense	1,283

Instructions Prepare a work sheet for Outback Guide Service for the period ended May 31 in your working papers.

Analyze
Compute the total assets for Outback.

Problem 8 Completing the Work Sheet

The work sheet for Job Connect appears in your working papers. The amounts that have been entered are correct. Several amounts, however, are missing from various columns.

Instructions Calculate all missing amounts and complete the work sheet.

Analyze
Calculate the owner's equity. (Hint: Be sure to include the temporary accounts.)

Real-World Applications & Connections

CASE STUDY

The Balance Sheet and Net Income

You work for the Clinton Advertising Agency as an assistant to the company's accountant. You're preparing a work sheet for the month of April. The trial balance totals are $239,374 and you've extended the account columns in the Balance Sheet and Income Statement sections. After calculating the net income in the Income Statement section, you discover that the net income calculated in the Balance Sheet is not the same. The Balance Sheet section has a higher net income by $850. In examining your work sheet, you discover the balance entered in the Trial Balance section for the Withdrawals account is the same ($850).

ACTIVITY Provide a reason to explain why the net income amounts do not match.

Global Accounting

Time Zones

What time is it now on your watch? What time is it in London or New York? Without knowledge of the world's time zones, the international businessperson may have difficulty contacting a person at the right place and the right time. There are 24 time zones in the world, each representing 15 degrees of longitude, or one hour intervals.

Activity

Imagine that your company has manufacturing locations in Dublin, Ireland, and Denver, Colorado. Managers work from 8:00 a.m. to 3:30 p.m. at both locations. Dublin is 7 hours ahead of Denver. Schedule a telephone meeting so that both managers are present. What will you need to consider when scheduling the meeting?

21st Century Skills

Data Mining

Accounting involves a large amount of data, including a wide variety of business transactions, financial statements, and more. This data potentially represents a great deal of useful information for companies. Data mining is the process of extracting patterns from data. In cases where the volume of data available is far too large for individuals to analyze successfully, automated tools such as computer programs can analyze behavior and relationships in a large pool of data.

Activity

a. Use the Internet to research data mining. Make a list of potential uses for data mining.

b. As a class, discuss how data mining can be used for accounting purposes.

Career Wise

Bank Teller

As the person who handles money and deals directly with customers, a teller is one of the most important people in a bank. This job includes accepting cash for deposit, printing receipts, answering questions about balances, and sometimes helping customers open new accounts.

Research And Share

a. Visit the Web site of the U.S. Department of Labor's Bureau of Labor Statistics and obtain information about the career of bank teller.

Spotlight on Personal Finance

Your Savings

Businesses and individuals are faced with both expected and unexpected expenses. All expenses must be paid, and sound management of cash will usually provide sufficient cash in most situations.

Activity

List some typical cash expenditures a high school student might have in an normal week. Include the amount you think an average student might spend on each expense. Also list some expenses that are not common but might appear unexpectedly.

Chapter 9

Financial Statements for a Sole Proprietorship

Chapter Topics

9-1 The Income Statement

9-2 The Statement of Changes in Owner's Equity

9-3 The Balance Sheet and the Statement of Cash Flows

Visual Summary

Review and Activities

Standardized Test Practice

Computerized Accounting

Problems

Real-World Accounting Careers

Real-World Applications & Connections

Essential Question

As you read this chapter, keep this question in mind:
What is the basis for making good financial decisions?

Main Idea

The income statement reports the net income or net loss for an accounting period. The statement of changes in owner's equity shows how the owner's financial interest changes during the accounting period. The balance sheet reports the financial position at a specific point in time. The statement of cash flows reports the sources and uses of cash during the accounting period.

Chapter Objectives

Concepts	Analysis	Procedures
C1 Explain the purpose of the income statement.	**A1** Prepare a statement of changes in owner's equity.	**P1** Prepare a balance sheet.
C2 Prepare an income statement.	**A2** Explain the purpose of the balance sheet.	**P2** Explain the purpose of the statement of cash flows.
C3 Explain the purpose of the statement of changes in owner's equity.		**P3** Explain ratio analysis and compute ratios.

Real-World Business Connection

5boronyc

Skateboarding and New York City are in Steve Rodriguez's blood. The owner of 5boronyc founded his business in 1996, but his passion for skateboarding goes back more than 25 years to the long days he and his friends spent traveling through New York's five boroughs, seeking out the best spots for riding and practicing tricks. Rodriguez's passion for urban skateboarding culture is reflected in the creative decks, wheels, and apparel 5boronyc sells.

Connect to the Business

One step of the accounting cycle is to determine whether a business is making or losing money during a certain period of time. Financial statements gather information from different accounts and documents so that a sole proprietorship such as 5boronyc knows whether the owner's claims to business assets have increased or decreased.

Analyze

What types of accounts do you think 5boronyc used in 1996? What accounts do you think have been added?

Focus on the Photo

A sole proprietorship can be a small business like 5boronyc with few employees or a large business with many employees. Large and small businesses must prepare financial statements. **Would you prefer working as an accountant for a large or small company? Why?**

Chapter 9 • Financial Statements for a Sole Proprietorship

SECTION 9.1
The Income Statement

You will now learn how to prepare financial statements, the seventh step of the accounting cycle. We will discuss four different types of financial statements: the income statement, the statement of changes in owner's equity, the balance sheet, and the statement of cash flows. To operate a business profitably, the owner needs to have current financial information. Businesses ranging from an oil company to a dairy farm must organize financial information to **evaluate** profits or losses. **Financial statements** summarize the changes resulting from business transactions that occur during an accounting period. As you can see in **Figure 9–1**, preparing financial statements is the seventh step in the accounting cycle.

Section Vocabulary

- financial statements
- income statement
- evaluate
- illustrates

Figure 9–1 The Accounting Cycle with the Seventh Step Highlighted

1. Collect and verify source documents
2. Analyze each transaction
3. Journalize each transaction
4. Post to the ledger
5. Prepare a trial balance
6. Prepare a work sheet
7. Prepare financial statements
8. Journalize and post closing entries
9. Prepare a post-closing trial balance

Financial Statements

What Are the Four Financial Statements?

The primary financial statements prepared for a sole proprietorship are the income statement and the balance sheet. Two other statements, the statement of changes in owner's equity and the statement of cash flows, are also often prepared. The financial statements may be handwritten or typed but most often are prepared on a computer. With a computerized accounting system, the business owner can generate financial statements without first preparing a work sheet. Let's learn how to prepare financial statements for a service business such as Zip Delivery Service.

The Income Statement

What Is the Purpose of the Income Statement?

The **income statement** reports the net income or net loss for a specific period of time. As you recall from Chapter 8, net income or net loss is the difference between total revenue and total expenses. For this reason the income statement is sometimes called a *profit-and-loss statement* or an *earnings statement*.

Income Statement Sections

The income statement contains the following sections:

- the heading
- the revenue for the period
- the expenses for the period
- the net income or net loss for the period

Heading.

Like the work sheet heading, the heading of an income statement has three parts:

1. The name of the business *(Who?)*
2. The name of the report *(What?)*
3. The period covered *(When?)*

The heading for Zip's income statement is shown in **Figure 9–2**. Each line of the heading is centered on the width of the statement.

Figure 9–2 The Heading for an Income Statement

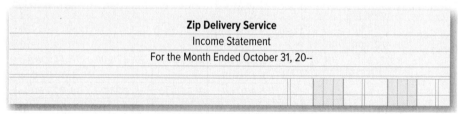

When preparing an income statement heading, be sure to follow the wording and capitalization shown in **Figure 9–2**. The date line is especially important because the reporting period varies from business to business.

Revenue Section.

- After the heading has been completed, enter the revenue earned for the period. Look at **Figure 9–3** on page 248.
- The information used to prepare the income statement comes from the Income Statement section of the work sheet.
- Zip's income statement is prepared on standard accounting stationery, which has a column for account names and two amount columns.
- The first amount column is used to enter the balances of the individual revenue and expense accounts.
- The second amount column is used to enter totals: total revenue, total expenses, and net income (or net loss).

Figure 9–3 shows the procedures to prepare an income statement:

1. Write *Revenue:* on the first line at the left side of the form. **A**

2. Enter the revenue account names beginning on the second line, indented about a half inch from the left edge of the form. **B**

3. Enter the balance of each revenue account. Since Zip has only one revenue account, **Delivery Revenue**, total revenue is the same as the balance of the one revenue account. The balance is thus written in the second, or totals, column. **C**

Figure 9–3 Preparing an Income Statement

Zip Delivery Service
Work Sheet
For the Month Ended October 31, 20--

	ACCT. NO.	ACCOUNT NAME	TRIAL BALANCE DEBIT	TRIAL BALANCE CREDIT	INCOME STATEMENT DEBIT	INCOME STATEMENT CREDIT	BALANCE SHEET DEBIT	BALANCE SHEET CREDIT	
1	101	Cash in Bank	21 125 00				21 125 00		1
2	105	Accts. Rec.–City News	1 450 00				1 450 00		2
3	110	Accts. Rec.–Green Company							3
4	115	Computer Equipment	3 000 00				3 000 00		4
5	120	Office Equipment	200 00				200 00		5
6	125	Delivery Equipment	12 000 00				12 000 00		6
7	201	Accts. Pay.–Rockport Advertising		75 00				75 00	7
8	205	Accts. Pay.–Coast to Coast Auto		11 650 00				11 650 00	8
9	301	Crista Vargas, Capital		25 400 00				25 400 00	9
10	302	Crista Vargas, Withdrawals	500 00				500 00		10
11	303	Income Summary							11
12	401	Delivery Revenue		2 650 00		2 650 00			12
13	501	Advertising Expense	75 00		75 00				13
14	505	Maintenance Expense	600 00		600 00				14
15	510	Rent Expense	700 00		700 00				15
16	515	Utilities Expense	125 00		125 00				16
17			39 775 00	39 775 00	1 500 00	2 650 00	38 275 00	37 125 00	17
18		Net Income			1 150 00			1 150 00	18
19					2 650 00	2 650 00	38 275 00	38 275 00	19
20									20

Zip Delivery Service
Income Statement
For the Month Ended October 31, 20--

Revenue: **A**				
Delivery Revenue **B**			2 650 00	**C**
Expenses: **D**				
Advertising Expense		75 00		
Maintenance Expense	**E**	600 00	**E**	
Rent Expense		700 00		
Utilities Expense		125 00		
Total Expenses **G**	**F**		**H** 1 500 00	**I**
Net Income **K**			**J** 1 150 00	**L**

248 Chapter 9 • Financial Statements for a Sole Proprietorship

Many businesses have more than one source of revenue and thus have a separate revenue account for each source. For example, a swim club might have accounts such as **Membership Fees** and **Pool Rental**. **Figure 9–4 illustrates** the Revenue section of the income statement for a business with more than one revenue account. Notice that the words *Total Revenue* are indented about one inch from the left edge of the form.

Figure 9–4 Income Statement with More Than One Revenue Account

Revenue:		
Rental Revenue	11 230 00	
Repair Revenue	4 850 00	
Total Revenue		16 080 00

Expenses Section.

- The expenses incurred during the period are reported next.
- The expense account names and the balances in the Income Statement section of the work sheet are used to prepare the Expenses section of the income statement.
- Refer to **Figure 9–3** as you read the following instructions:
  1. On the line following the revenue section, write *Expenses:* at the left side of the form. **D**
  2. On the following lines, write the names of the expense accounts, indented half an inch, in the order that they appear on the work sheet. Since there are several expense accounts, enter the individual balances in the first amount column. **E**
  3. Draw a single rule under the last expense account balance. **F**
  4. Write the words *Total Expenses* on the line following the last expense account name, indented about one inch. **G**
  5. Add the balances for all expense accounts. Write the total expenses amount in the second amount column, one line below the last expense account balance. **H**

Net Income Section.

- The next step is to enter net income. Net income, remember, occurs when total revenue is more than total expenses.
- Refer to **Figure 9–3** as you read the following instructions for the preparation of the Net Income section.
  1. Draw a single rule under the total expenses amount. **I**
  2. Subtract the total expenses from the total revenue to find net income. Enter the net income in the second amount column under the total expenses amount. **J**
  3. On the same line, write *Net Income* at the left side of the form. **K**
  4. If the amount of net income matches the amount on the work sheet, draw a double rule under the net income amount. **L**

> **Reading Check**
>
> **Identify**
>
> *What are the four sections of the Income Statement?*

Reporting a Net Loss

If total expenses are more than total revenue, a net loss exists. To determine the amount of net loss, subtract total revenue from total expenses. Enter the net loss in the second amount column under the total expenses amount. Write the words *Net Loss* on the same line at the left side of the form. **Figure 9–5** illustrates how to report a net loss.

Figure 9–5 Income Statement Showing a Net Loss

Revenue:		
Delivery Revenue		3 170 00
Expenses:		
Advertising Expense	275 00	
Maintenance Expense	750 00	
Miscellaneous Expense	270 00	
Rent Expense	1 900 00	
Utilities Expense	325 00	
Total Expenses		3 520 00
Net Loss		350 00

SECTION 9.1 Assessment

After You Read

Reinforce the Main Idea

Using a diagram like this one, summarize the steps for creating an income statement. Add answer boxes for steps as needed.

Creating an Income Statement

[ ]
↓
[ ]
↓
[ ]
↓
[ ]

Problem 1 Analyzing a Source Document

Instructions Based on the receipt shown here, answer the following questions in your working papers.

1. What is the name of the business that received the money?
2. How much money was received?
3. Who paid the money?
4. Why was the money paid?
5. When was the payment made?
6. Who received the money and made out the receipt?
7. Where is the business that received the money located?

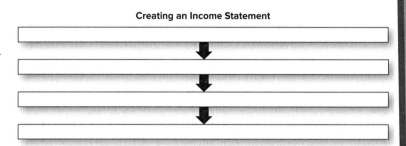

Math for Accounting

Earth-Friendly Cleaning Service prepares monthly financial statements. The two revenue accounts show ending balances of $8,740.00 and $2,080.00. The following expense account balances also appear in the Income Statement section of this month's work sheet.

Advertising Expense	$3,690.00
Maintenance Expense	2,745.00
Miscellaneous Expense	605.00
Rent Expense	3,400.00
Utility Expense	985.00

Calculate the amount of net income or net loss for the month.

SECTION 9.2
The Statement of Changes in Owner's Equity

An important concern of a business owner is whether the owner's equity has increased or decreased during the period. An increase in owner's equity means the owner's claims to the assets of the business have grown. A decrease means the owner's claims to the assets of the business have been reduced.

> **Section Vocabulary**
> - statement of changes in owner's equity

The Statement of Changes in Owner's Equity

Where Do You Find the Information Needed to Prepare This Financial Statement?

The **statement of changes in owner's equity** summarizes changes in the owner's capital account as a result of business transactions that occur during the period. Eventually, the balances of revenues, expenses, and the owner's withdrawal account will be transferred to the owner's capital account. This statement is prepared at the end of the accounting period.

The heading of the statement of changes in owner's equity is set up in the same manner as the heading for the income statement.

1. The first line consists of the name of the business. *(Who?)*
2. The second line indicates the name of the statement. *(What?)*
3. The third line indicates the period covered. *(When?)*

Because the statement of changes in owner's equity and the income statement cover the same period, the third line of the heading of both statements will include the same wording and date.

The information to prepare this statement is found in three places:

- the work sheet
- the income statement
- the owner's capital account in the general ledger

Completing the Statement of Changes in Owner's Equity

How Do You Prepare This Financial Statement?

Look at **Figure 9–6** on page 253. It is the statement of changes in owner's equity for Zip Delivery Service for the month ended October 31. The illustration shows the steps needed to complete this financial statement.

1. On the first line, write the words *Beginning Capital* followed by a comma and then by the first day of the period. For Zip that date is October 1, 20--. **A**

Figure 9–6 Statement of Changes in Owner's Equity

Zip Delivery Service
Statement of Changes in Owner's Equity
For the Month Ended October 31, 20--

Beginning Capital, October 1, 20– **A**				**B**
C Add: Investments by Owner	**D** 25 400 00			
Net Income	**E** 1 150 00			
Total Increase in Capital			26 550 00	**F**
Subtotal			26 550 00	**G**
Less: Withdrawals by Owner			500 00	**H**
Ending Capital, October 31, 20– **I**			26 050 00	**J**

2. In the second amount column, enter the balance of the capital account at the beginning of the period. The source of this information is the capital account in the general ledger. Since Zip was formed after the beginning of the period, there is no beginning capital balance. Place a line in the second amount column. **B**

3. Next, enter the *increases* to the capital account:
   - investments by the owner
   - net income

   Investments made by the owner during the period are recorded in the capital account. Crista Vargas, the owner of Zip, invested a total of $25,400 during October. This includes $25,000 cash and two phones valued at $400. Write *Add: Investments by Owner.* **C** Enter the total investment in the first amount column. **D**

   On the next line, write the words *Net Income*. Indent so that *Net Income* aligns on the left with *Investments by Owner* in the line above. In the first amount column, enter the net income amount from the income statement. Draw a single rule under the net income amount. **E**

4. Write the words *Total Increase in Capital* on the next line at the left side of the form. Add the investments by owner and net income amounts and enter the total in the second amount column. Draw a single rule under the amount. **F**

5. Write *Subtotal* on the next line, at the left side of the form. Add the amounts for beginning capital and total increase in capital. Enter the result in the second amount column. **G**

6. The next section of the statement lists the *decreases* to the capital account:
   - withdrawals
   - net loss

   Since Zip did not have a net loss for the period, write the words *Less: Withdrawals by Owner* at the left side of the form. Find the withdrawals amount on the work sheet. Enter the withdrawals amount in the second amount column. Draw a single rule under the withdrawals amount. **H**

7. On the next line, at the left side of the form, write the words *Ending Capital* followed by a comma and the last day of the period. **I**

8. Subtract the withdrawals amount from the subtotal to determine the ending balance of the capital account. Finally, draw a double rule below the ending capital amount. **J**

Reading Check

Compare and Contrast

In what ways are the income statement and statement of changes in owner's equity similar? How are they different?

Statement of Changes in Owner's Equity for an Ongoing Business

To prepare the statement of changes in owner's equity, you need to know the beginning balance of the owner's capital account. For an ongoing business, the balance entered on the work sheet for the capital account *may not* be the balance at the beginning of the period. If the owner made additional investments during the period, the investments would be recorded in the general journal, posted to the general ledger, and included in the amount shown on the work sheet.

For example, suppose the owner of Garo's Tree Service, James Garo, invested an additional $1,000 during the period. Look at **Figure 9–7**. The ledger account reflects the additional capital investment. The amount entered on the work sheet for **James Garo, Capital** includes the balance at the beginning of the period ($23,800) and the investment made during the period ($1,000).

ACCOUNT _James Garo, Capital_   ACCOUNT NO. _301_

DATE		DESCRIPTION	POST. REF.	DEBIT	CREDIT	BALANCE DEBIT	BALANCE CREDIT
20–							
Apr.	1	Balance	3				23 800 00
	15		G1		1 000 00		24 800 00

Garo's Tree Service
Work Sheet
For the Month Ended April 30, 20--

	ACCT. NO.	ACCOUNT NAME	TRIAL BALANCE DEBIT	TRIAL BALANCE CREDIT	INCOME STATEMENT DEBIT	INCOME STATEMENT CREDIT	BALANCE SHEET DEBIT	BALANCE SHEET CREDIT	
12	301	James Garo, Capital		24 800 00				24 800 00	12
13	302	James Garo, Withdrawals	700 00				700 00		13
20	525	Utilities Expense	620 00		620 00				20
21			21 924 00	21 924 00	6 033 00	9 309 00	29 877 00	26 601 00	21
22		Net Income			3 276 00			3 276 00	22
23					9 309 00	9 309 00	29 877 00	29 877 00	23
24									24

Figure 9–7 Statement of Changes in Owner's Equity for an Ongoing Business (continued)

Figure 9–7 Statement of Changes in Owner's Equity for an Ongoing Business

Garo's Tree Service			
Statement of Changes in Owner's Equity			
For the Month Ended April 30, 20--			
Beginning Capital, April 1, 20–			23 800 00
Add: Investments by Owner	1 000 00		
Net Income	3 276 00		
Total Increase in Capital		4 276 00	
Subtotal		28 076 00	
Less: Withdrawals by Owner		700 00	
Ending Capital, April 30, 20–		27 376 00	

There are two ways to determine the owner's capital account balance at the beginning of the period:

- Look at the ledger.
- Subtract the additional investments from the account balance shown on the work sheet.

For **James Garo, Capital**, the $1,000 additional investment in Garo's Tree Service is subtracted from the $24,800 account balance shown on the work sheet to arrive at the beginning account balance of $23,800.

Figure 9–7 shows the statement of changes in owner's equity for an ongoing business.

Statement of Changes in Owner's Equity Showing a Net Loss

Figure 9–8 shows a statement of changes in owner's equity for an ongoing business with a net loss. Notice that since there is only one item that increases capital—investments by owner—the amount of the individual item is entered in the second amount column. Since there are two items that decrease capital—withdrawals by owner and net loss—the amounts of the individual items are entered in the first amount column. The total decrease in capital is entered in the second amount column.

Figure 9–8 Statement of Changes in Owner's Equity Showing a Net Loss

Island Burgers			
Statement of Changes in Owner's Equity			
For the Month Ended April 30, 20--			
Beginning Capital, April 1, 20–			13 848 00
Add: Investments by Owner			1 500 00
Subtotal			15 348 00
Less: Withdrawals by Owner		900 00	
Net Loss		835 00	
Total Decrease in Capital			1 735 00
Ending Capital, April 30, 20–			13 613 00

SECTION 9.2
Assessment

After You Read

Reinforce the Main Idea

Using a diagram like this one, summarize the steps for creating a statement of changes in owner's equity. Add answer boxes for steps as needed.

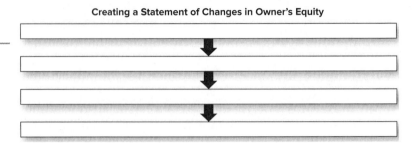

Creating a Statement of Changes in Owner's Equity

Problem 2 Determining Ending Capital Balances

The financial transactions affecting the capital accounts of several different businesses are summarized below.

Instructions Use the form in your working papers. Determine the ending capital balance for each business.

	Beginning Capital	Investments	Revenue	Expenses	Withdrawals
1.	$60,000	$ 500	$ 5,100	$2,400	$ 700
2.	24,075	0	13,880	7,240	800
3.	28,800	1,000	6,450	6,780	0
4.	0	10,500	5,320	4,990	200
5.	6,415	0	4,520	3,175	700
6.	20,870	1,300	13,980	9,440	1,700

continued

256 Chapter 9 • Financial Statements for a Sole Proprietorship

SECTION 9.2
Assessment

Math for Accounting

Tracy Murphy is a fashion designer who works from her home. She contributed her own funds and equipment to the business. She also continues to add new clients and increase her revenue. Tracy's investments in her business and the revenue she has earned over the past 10 months are shown below. Create a line graph, by date, comparing Tracy's investments to her revenue. What can you determine from this chart?

Investments

Date	Item	Amount
1/01/20--	Cash	$20,000
1/01/20--	Sewing machines	6,000
1/01/20--	Mannequins	3,500
1/01/20--	Material	25,000
1/01/20--	Computer equipment	5,600
1/01/20--	Business cards	500
2/15/20--	Cash	12,000
3/10/20--	Material	7,500
Total Investments		$80,100

Revenue

Date	Amount
2/01/20--	$ 5,000
3/01/20--	6,000
4/01/20--	12,000
5/01/20--	15,000
6/01/20--	17,500
7/01/20--	19,000
8/01/20--	20,000
9/01/20--	26,000
10/01/20--	20,000

SECTION 9.3
The Balance Sheet and the Statement of Cash Flows

The third and fourth financial statements prepared at the end of the period are the balance sheet and the statement of cash flows. The balance sheet reflects the accounting equation at the end of the period. The statement of cash flows shows how the business acquired and spent cash during the period.

The Balance Sheet

What Is the Purpose of the Balance Sheet?

The **balance sheet** is a report of the balances in the permanent accounts at the end of the period. The main purpose of the balance sheet is to report the assets of the business and the claims against those assets *on a specific date.* In other words, the balance sheet states the financial position of a business at a specific point in time. The balance sheet summarizes the following information:

- what a business owns
- what a business owes
- what a business is worth

For this reason the balance sheet is sometimes called a *statement of financial position.*

The balance sheet is prepared from information in the Balance Sheet section of the work sheet and from the statement of changes in owner's equity. The balance sheet may be handwritten, typed, or, as in most cases, prepared by computer.

Section Vocabulary

- balance sheet
- report form
- statement of cash flows
- ratio analysis
- profitability ratio
- return on sales
- current assets
- current liabilities
- working capital
- liquidity ratio
- current ratio
- quick ratio
- available
- converted

The Sections of the Balance Sheet

The balance sheet contains the following sections:

- the heading
- the assets section
- the liabilities and owner's equity sections

Heading.

Like the heading of the income statement, the heading of the balance sheet answers the questions *who? what? when?* The balance sheet heading includes:

1. The name of the business *(Who?)*
2. The name of the financial statement *(What?)*
3. The date of the balance sheet *(When?)*

Figure 9–9 Headings of Financial Statements

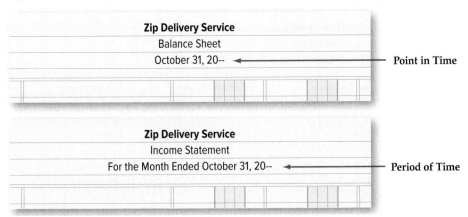

Unlike the income statement, which covers the entire period, the balance sheet relates to a specific point in time. The amounts shown on the balance sheet are the general ledger balances in the accounts on the last day of the period. Notice the difference between the date lines on the balance sheet and on the income statement in **Figure 9–9**.

Assets Section.

- Refer to Zip's balance sheet in **Figure 9–10** as you read how to prepare the balance sheet.
- Zip's work sheet and statement of changes in owner's equity are also included to show the information sources used to prepare the balance sheet.
- Zip's balance sheet is prepared in **report form**, listing the balance sheet sections one under the other.

The Assets section of the balance sheet is prepared as follows:

1. Write the word *Assets* on the first line in the center of the column containing the account names. **A**

2. On the following lines, list each asset account name and its balance in the same order as they appear in the Balance Sheet section of the work sheet. Enter the account balances in the first amount column. Draw a single rule under the last account balance. **B**

3. On the next line, write the words *Total Assets,* indented about half an inch. Add the individual asset balances and enter the total in the second amount column. **C**

Do *not* draw a double rule under the total yet. Enter it when the Liabilities and Owner's Equity sections are complete and equal to total assets.

Liabilities and Owner's Equity Sections.

- The information for the Liabilities and Owner's Equity sections is taken from the work sheet and from the statement of changes in owner's equity.
- Use the following steps to complete the Liabilities and Owner's Equity sections.

    1. On the line after *Total Assets,* write the heading *Liabilities* in the center of the column containing the account names. **D**

Figure 9–10 Preparing a Balance Sheet

Zip Delivery Service
Work Sheet
For the Month Ended October 31, 20--

ACCT. NO.	ACCOUNT NAME	TRIAL BALANCE DEBIT	TRIAL BALANCE CREDIT	INCOME STATEMENT DEBIT	INCOME STATEMENT CREDIT	BALANCE SHEET DEBIT	BALANCE SHEET CREDIT	
101	Cash in Bank	21 125 00				21 125 00		1
105	Accts. Rec.–City News	1 450 00				1 450 00		2
110	Accts. Rec.–Green Company	—				—		3
115	Computer Equipment	3 000 00				3 000 00		4
120	Office Equipment	200 00				200 00		5
125	Delivery Equipment	12 000 00				12 000 00		6
201	Accts. Pay.–Rockport Advertising		75 00				75 00	7
205	Accts. Pay.–Coast to Coast Auto		11 650 00				11 650 00	8
301	Crista Vargas, Capital		25 400 00				25 400 00	9
302	Crista Vargas, Withdrawals	500 00				500 00		10
303	Income Summary							11
401	Delivery Revenue		2 650 00		2 650 00			12
501	Advertising Expense	75 00		75 00				13
505	Maintenance Expense	600 00		600 00				14
510	Rent							15
515	Utili...							16
						37 125 00		17
	Net...					1 150 00		18
						38 275 00		19

Zip Delivery Service
Statement of Changes in Owner's Equity
For the Month Ended October 31, 20--

Beginning Capital, October 1, 20–			
Add: Investments by Owner	25 400 00		
Net Income	1 150 00		
Total Increase in Capital		26 550 00	
Subtotal		26 550 00	
Less: Withdrawals by Owner		500 00	
Ending Capital, October 31, 20–		26 050 00	

Zip Delivery Service
Balance Sheet
October 31, 20--

A Assets			
Cash in Bank	21 125 00		
Accounts Receivable–City News	1 450 00		
Accounts Receivable–Green Company	—		
Computer Equipment	3 000 00		
Office Equipment	200 00		
Delivery Equipment	12 000 00		
C Total Assets		37 775 00	**K**
D Liabilities			
Accounts Payable–Rockport Advertising	75 00		
Accounts Payable–Coast to Coast Auto	11 650 00		
F Total Liabilities		11 725 00	
G Owner's Equity			
H Crista Vargas, Capital		26 050 00	
I Total Liabilities and Owner's Equity	**J**	37 775 00	**K**

2. On the following lines, list the liability account names and their balances in the same order as on the Balance Sheet section of the work sheet. Enter the account balances in the first amount column. Draw a single rule under the last account balance. **E**

3. On the next line, write the words *Total Liabilities,* indented about half an inch. Add the individual liability balances and enter the total in the second amount column. **F**

4. On the next line, enter the heading *Owner's Equity* in the center of the column containing the account names. **G**

5. On the next line, write the name of the capital account. In the second amount column, enter the ending balance of the capital account as shown on the statement of changes in owner's equity. **H**

Proving the Equality of the Balance Sheet

Recall that the basic accounting equation must always be in balance. The balance sheet represents the basic accounting equation, so the Assets section total must equal the total of the Liabilities and Owner's Equity sections.

Assets = Liabilities + Owner's Equity

To prove the equality of the balance sheet, follow these steps:

1. Draw a single rule under the balance of the capital account. On the next line, write the words *Total Liabilities and Owner's Equity,* indented about half an inch. **I**

> **Common Mistakes**
>
> **The Balance Sheet** In preparing the balance sheet, you are taking the account balances from the Balance Sheet section of the work sheet. This is true of all accounts *except* the capital account. The account balance entered on the balance sheet must be the balance at the end of the period which is the ending capital reported on the statement of changes in owner's equity.

2. Add the total liabilities amount and the ending capital balance. Enter the total in the second amount column. **J** This total must equal the total assets amount. If the totals are not equal, there is an error. Most errors occur when transferring amounts from the work sheet or from the statement of changes in owner's equity. Verify that each account balance has been transferred properly. Find and correct the error and then complete the balance sheet.

3. When total assets equal total liabilities and owner's equity, draw a double rule under the total assets amount *and* under the total liabilities and owner's equity amount. **K** The balance sheet is now complete.

Refer again to the amounts and their placement in **Figure 9–10**. As you can see, completion of the work sheet is the basis for preparing the three financial statements studied so far.

The Statement of Cash Flows

What Is the Purpose of the Statement of Cash Flows?

Cash flowing through a business is like blood flowing through your body. The flow of cash keeps a business alive. It is essential to have cash **available** for the daily operations of the business and for unexpected expenses.

The **statement of cash flows** summarizes the following information:

- the amount of cash the business took in
- the sources of cash
- the amount of cash the business paid out
- the uses of cash

Like the income statement, the statement of cash flows covers a single accounting period.

This information is essential for sound management and investment decisions. You will learn more about the statement of cash flows in Chapter 19.

Reading Check

In Your Own Words

What four pieces of information does the statement of cash flows provide?

Ratio Analysis

What Is Ratio Analysis?

Ratio analysis is the process of evaluating the relationship between various amounts in the financial statements. Owners and managers use ratio analysis to determine the financial strength, activity, and debt-paying ability of a business.

Profitability Ratios

Profitability ratios are used to evaluate the earnings performance of the business during the accounting period. The earning power of a business is an important measure of its ability to grow and continue to earn revenue.

One commonly used profitability ratio is return on sales. Business owners use the **return on sales** ratio to examine the portion of each sales dollar that represents profit. To calculate this ratio, divide net income by sales. For example, the return on sales for Zip Delivery Service is calculated as follows:

$$\frac{\$1{,}150 \text{ net income}}{\$2{,}650 \text{ sales}} = 0.434 \text{ or } 43.4\%$$

This percentage indicates that each dollar of sales produced 43.4 cents of profit for Zip. It can be compared to other accounting periods, to determine whether it is increasing or decreasing.

For example, if net income next year is $2,750 and sales are $5,000, the return on sales would be computed as follows:

$$\frac{\$2{,}750 \text{ net income}}{\$5{,}000 \text{ sales}} = 0.550 \text{ or } 55.0\%$$

As you can see, profit per sales dollar would increase by 11.6 cents.

Liquidity Measures

Liquidity refers to the ease with which an asset can be converted to cash. **Current assets** are those used up or **converted** to cash during the normal operating cycle of the business. These might include **Accounts Receivable**, **Cash in Bank**, and **Supplies**. **Current liabilities** are debts of the business that must be paid within the next accounting period. **Accounts Payable** is an example of a current liability.

The amount by which current assets exceed current liabilities is known as **working capital.** Because current liabilities are usually paid out of current assets, working capital represents the excess assets available to continue operations.

The working capital for Zip is calculated as follows:

Current Assets   −   Current Liabilities   =   Working Capital
$22,575   − $11,725   = $10,850

A **liquidity ratio** is a measure of the ability of a business to pay its current debts as they become due and to provide for an unexpected need for cash. Two common ratios that are used to determine liquidity follow.

Current Ratio.

The **current ratio** reflects the relationship between current assets and current liabilities. The current ratio is calculated by dividing the dollar amount of current assets by the dollar amount of current liabilities. The current ratio for Zip based on the balance sheet in **Figure 9–10** is:

$$\frac{\text{Current Assets}}{\text{Current Liabilities}} = \text{Current Ratio} \qquad \frac{\$22,575}{\$11,725} = 1.92 \text{ or } 1.9{:}1$$

The current liabilities of a business must be paid within a year. These liabilities are paid from current assets.

A ratio of 2:1 or higher is considered favorable by creditors. It indicates that a business is able to pay its debts and that a business has twice as many current assets as current liabilities. A low ratio may indicate that a company could have trouble paying its debts.

Quick Ratio.

A **quick ratio** is a measure of the relationship between short-term assets and current liabilities. Short-term liquid assets—those that can be quickly converted to cash—are cash and net receivables. The quick ratio is computed by dividing the total cash and receivables by total current liabilities.

The quick ratio for Zip based on the Balance Sheet in **Figure 9–10** is:

$$\frac{\text{Cash and Receivables}}{\text{Current Liabilities}} = \text{Quick Ratio} \qquad \frac{\$22,575}{\$11,725} = 1.92{:}1$$

In some instances the current ratio and the quick ratio can be the same, as in the case in this example.

A quick ratio of 1:1 is considered adequate. This indicates that a business can pay its current debts with cash from incoming receivables. If a business has a quick ratio of 1:1 or higher, the business has $1.00 in liquid assets for each $1.00 of current liabilities.

Quick ratios may also be compared from one year to the next. For example, suppose that cash and receivables for the previous year were $48,653 and current liabilities were $53,245. That year's quick ratio would be computed as:

$$\frac{\$48{,}653}{\$53{,}245} = 0.91{:}1$$

As you can see, Zip has improved its liquidity position in the current year. The $1.92 in liquid assets per $1.00 in current liabilities (current year) is stronger than $0.91 in liquid assets per $1.00 in current liabilities (previous year).

SECTION 9.3
Assessment

After You Read

Reinforce the Main Idea

Using a diagram like this one, summarize the steps for creating a balance sheet. Add answer boxes for steps as needed.

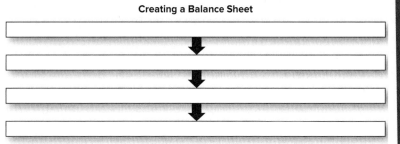

Creating a Balance Sheet

Problem 3 Calculating Return on Sales

The Gawle Company is a family-owned and operated appliance rental and repair business. The income statement for the month ended August 31 includes the following:

Rental Revenue	$3,256	Rent Expense	$2,100
Repair Revenue	2,140	Utilities Expense	483
Advertising Expense	575	Net Income	1,108
Maintenance Expense	1,130		

Instructions Calculate the return on sales for the month for The Gawle Company.

Math for Accounting

Compute the four amounts that are missing from the balance sheet below. On a separate sheet of paper, write the description and the dollar value of the four missing amounts.

Interactive Communication
Balance Sheet
December 31, 20--

Assets		
Cash in Bank	944 000 00	
Accounts Receivable—Chamber of Commerce	200 000 00	
Office Equipment	?	
Computer Equipment	1,000	
Total Assets	000 00	2,994 000 00
Liabilities		
Accounts Payable—Tip Top Advertising		
Interest Payable	275 500 00	
Salaries Payable	?	
Total Liabilities	2 100 00	?
Owner's Equity		
Chuck Thompson, Capital		2,715 100 00
Total Liabilities and Owner's Equity		?

Chapter 9
Visual Summary

Concepts
Prepare an income statement

Section	Information Presented
Heading	• Name of the business • Name of the report • Accounting period covered
Revenue	Balance of each revenue account for the period.
Expenses	Balance of each expense account for the period.
Net income	Total expenses subtracted from total revenue. The result is net income or net loss.

Analysis
Prepare a statement of changes in owner's equity.

Section	Information Presented
Heading	• Name of the business • Name of the report • Accounting period covered
Beginning Capital	This information comes from the owner's capital account in the general ledger.
Increases in Capital	• Investments by Owner • Net Income
Decreases in Capital	• Withdrawals by Owner • Net Loss
Ending Capital	Beginning Capital + Increases − Decreases

Procedures
Ratio Analysis

Return on Sales	Net Income/Sales = Return on Sales
Current Ratio	Current Assets/Current Liabilities = Current Ratio
Quick Ratio	Cash and Receivables/Current Liabilities = Quick Ratio

Chapter 9
Review and Activities

Answering the Essential Question

What is the basis for making good financial decisions?

Whether you're talking about a large corporation or yourself, certain information has to be considered. Using a table, illustrate the four financial statements that provide essential information for making sound decisions.

Vocabulary Check

1. **Vocabulary** Explain how each of the following terms relates to the chapter title, "Financial Statements for a Sole Proprietorship."

   - evaluate
   - financial statements
   - income statement
   - illustrates
   - statement of changes in owner's equity
   - balance sheet
   - report form
   - available
   - statement of cash flows
   - converted
   - current assets
   - current liabilities
   - profitability ratio
   - ratio analysis
   - return on sales
   - working capital
   - current ratio
   - liquidity ratio
   - quick ratio

Concept Check

2. What is the purpose of the income statement?
3. **Evaluate** Explain how the value of the income statement depends on the accuracy of the work sheet.
4. What is the purpose of the statement of changes in owner's equity?
5. In a statement of changes in owner's equity, what items are totaled in the "add" section?
6. What is the purpose of the balance sheet?
7. **Analyze** Why does the date in a balance sheet heading differ from the dates in other statements?
8. What information does the statement of cash flows provide?
9. What is the purpose of computing ratios from amounts on financial statement?
10. **Predict** To whom might the various ratios be useful and why?
11. **Math** Burnett just signed a contract at a dealership to buy a new car. The list price is $19,000. He made a down payment of $1,500. The monthly payments are $602.15 for 36 months. What is the total amount Burnett will pay for the car?
12. **English Language Arts** Perform research online to obtain the financial statements for a company. Analyze the statements using ratio analysis, and create a presentation on the financial position of the company. Tell whether or not you would advise extending credit to this company. Include at least one visual or graphic aid.

Chapter 9
Standardized Test Practice

Multiple Choice

1. The financial statement that reports net income or net loss is the
   a. Statement of changes in owner's equity
   b. Income statement
   c. Balance sheet
   d. Statement of cash flows

2. At the beginning of the accounting period, Frank's Supply Company had a capital balance of $210,050. During the period, the owner invested $10,500 additional capital, and did not take any withdrawals. The period showed a net loss of $24,850. What was the ending capital for the period?
   a. $195,700
   b. $196,150
   c. $224,400
   d. $220,550

3. The balance sheet shows how a business is doing
   a. for a period of 12 months.
   b. on a specific date during the year.
   c. regarding its profit or loss.
   d. All of the above

4. Financial statements are prepared in the following order:
   a. income statement, balance sheet, statement of changes in owner's equity.
   b. income statement, statement of changes in owner's equity, balance sheet.
   c. statement of changes in owner's equity, balance sheet, income statement.
   d. balance sheet, income statement, statement of changes in owner's equity.

5. The purpose of the Income Statement is to report
   a. all assets, liabilities, and owner's equity at a specified time.
   b. all the accounts used in journalizing a business's transactions.
   c. balances in the capital accounts in order to determine the net income or loss.
   d. the net income or loss for a fiscal period.

True or False

6. The information used to prepare the income statement comes from the Income Statement section of the work sheet.

Chapter 9
Computerized Accounting

Preparing Financial Statements
Making the Transition from a Manual to a Computerized System

MANUAL METHODS	COMPUTERIZED METHODS
Preparing an income statement • Transfer all revenue and expense accounts and their balances from the work sheet. • Subtract expenses from revenue to determine net income or net loss.	• From the **Reports** menu, select **Income Statement** or **Profit and Loss**, depending on the accounting software. • Click **Print**.
Preparing a statement of changes in owner's equity • Transfer the beginning balance of the capital account from the work sheet. • Add additional investments and net income or loss. • Subtract withdrawals. • Calculate the ending balance for the capital account.	• Since the software automatically computes the ending balance of the capital account for you, it is not necessary to prepare this statement.
Preparing a balance sheet • Transfer permanent accounts and their balances from the work sheet. • Transfer the ending capital account balance from the statement of changes in owner's equity. • Total all asset account balances. • Total all liabilities and owner's equity account balances. • Verify that assets equal liabilities plus owner's equity.	• From the **Reports** menu, select **Balance Sheet**. • Click **Print**.

Chapter 9
Problems

Problem 4 — Preparing an Income Statement

The work sheet for Wilderness Rentals for the month ended September 30, 20-- is in your working papers.

Instructions Using the work sheet, prepare an income statement for Wilderness Rentals.

Analyze

Compute the return on sales for the period.

Problem 5 — Preparing a Statement of Changes in Owner's Equity

Instructions Using the work sheet for Wilderness Rentals in your working papers and the income statement prepared in Problem 9–4, prepare a statement of changes in owner's equity and a balance sheet. Ronald Hicks made an additional investment in the business of $500 during the period.

Analyze

Compute the current ratio for Wilderness Rentals as of September 30.

Chapter 9
Problems

Problem 6 — Preparing Financial Statements

The trial balance for the Hot Suds Car Wash is listed below and in your working papers.

Instructions

1. Complete the work sheet in your working papers.
2. Prepare an income statement for the quarter ended September 30, 20--.
3. Prepare a statement of changes in owner's equity. Regina Delgado made no additional investments during the period.
4. Prepare a balance sheet in report form.

Hot Suds Car Wash
Trial Balance
For the Quarter Ended September 30, 20--

		Debit	Credit
101	Cash in Bank	8 4 5 7 00	
105	Accts. Rec.—Linda Brown		
110	Accts. Rec.—Valley Auto	5 8 4 00	
115	Detailing Supplies	6 1 9 00	
120	Detergent Supplies	8 1 0 00	
125	Office Equipment	4 6 0 00	15
130	Office Furniture	2 4 0 00	2
135	Car Wash Equipment	1 6 0 00	7
201	Accts. Pay.—Allen Vacuum Systems	5 2 2 00	3 5 2 8 00
205	Accts. Pay.—O'Brian's Office Supply		1 2 1 5 00
301	Regina Delgado, Capital		23 8 4 5 00
305	Regina Delgado, Withdrawals		1
310	Income Summary	5 0 0 00	
401	Wash Revenue		9 6 2 3 00
405	Wax Revenue		8 0 1 9 00
410	Interior Detailing Revenue		2 6 2 8 00
501	Advertising Expense		1
505	Equipment Rental Expense	9 6 3 00	4
510	Maintenance Expense	1 3 7 00	1
520	Rent Expense	1 8 6 00	3
530	Utilities Expense	5 0 0 00	
		7 2 0 00 48	48 8 5 8 00
		8 5 8 00	

Analyze
Calculate the return on sales for the period.

Chapter 9
Problems

Problem 7 Preparing Financial Statements

The general ledger accounts and balances for Kits & Pups Grooming follow.

General Ledger

101	Cash in Bank	$ 4,296
105	Accts. Rec.—Juan Alvarez	1,528
110	Accts. Rec.—Nathan Carlsbad	904
115	Accts. Rec.—Martha Giles	1,219
120	Grooming Supplies	1,368
125	Office Equipment	8,467
130	Office Furniture	3,396
135	Computer Equipment	2,730
140	Grooming Equipment	1,974
145	Kennel Equipment	7,412
201	Accts. Pay.—Able Store Equipment	3,876
205	Accts. Pay.—Dogs & Cats Inc.	2,746
207	Accts. Pay.—Pet Gourmet	1,281
301	Abe Shultz, Capital	30,928
305	Abe Shultz, Withdrawals	1,900
310	Income Summary	—
401	Boarding Revenue	11,989
405	Grooming Revenue	4,420
501	Advertising Expense	3,934
505	Equipment Repair Expense	943
510	Maintenance Expense	2,483
520	Rent Expense	8,850
530	Utilities Expense	3,836

continued

Chapter 9 Problems

Instructions

1. Prepare a work sheet for the month ended September 30, 20--.
2. Prepare an income statement for the period.
3. Prepare a statement of changes in owner's equity. Abe Shultz made an additional investment of $2,500 during the period.
4. Prepare a balance sheet in report form.

Analyze

Compute the quick ratio as of September 30.

Problem 8 Preparing a Statement of Changes in Owner's Equity

Instructions Use the balance sheet and income statement shown to prepare a statement of changes in owner's equity. (The owner made an additional investment of $4,000 and withdrew $1,500 during the period.)

Outback Guide Service		
Income Statement		
For the Month Ended September 30, 20--		
Revenue:		
Guide Service Revenue		8 913 00
Expenses:		
Advertising Expense	3 75 00	
Maintenance Expense	1 38 00	
Rent Expense	1 250 00	
Salaries Expense	1 500 00	
Utilities Expense	1 161 00	
Total Expenses		4 424 00
Net Income		4 489 00

continued

Chapter 9
Problems

Outback Guide Service
Balance Sheet
September 30, 20--

Assets			
Cash in Bank	3 1 1 7 00		
Accounts Receivable—Mary Johnson	4 2 3 00		
Accounts Receivable—Feldman, Jones & Ritter	4 4 3 00		
Accounts Receivable—Podaski Systems Inc.	1 0 0 8 00		
Hiking Supplies	1 5 3 00		
Office Supplies	3 3 8 00		
Office Equipment	6 4 9 2 00		
Office Furniture	3 0 8 4 00		
Computer Equipment	3 6 2 4 00		
Hiking Equipment	1 0 1 5 00		
Rafting Equipment	9 1 8 6 00		
Total Assets		28 8 8 3 00	
Liabilities			
Accounts Payable—A-1 Adventure Warehouse	6 5 4 5 00		
Accounts Payable—Peak Equipment Inc.	1 3 2 5 00		
Accounts Payable—Premier Processors	6 4 2 00		
Total Liabilities		8 5 1 2 00	
Owner's Equity			
Juanita Ortega, Capital		20 3 7 1 00	
Total Liabilities and Owner's Equity		28 8 8 3 00	

Analyze

Identify the largest expense for this business during the period.

Real-World Accounting Careers

Enrico Villanueva
Robert Half International

Q What do you do?

A I assist the international reporting manager in ensuring the timely and accurate consolidation of our company's international subsidiaries' financial results. I also provide support on international tax issues, system conversion activities and other international finance projects.

Q What are your day-to-day responsibilities?

A My duties include ensuring that the monthly financial reports for our international subsidiaries are properly consolidated into the U.S. parent company's books in accordance to U.S. GAAP. In addition, I facilitate the billing of charges between the parent company and its international subsidiaries, assist in the settlement of intercompany payables/receivables, and respond to external and internal auditors' inquiries relating to international entities.

Q What factors have been key to your success?

A My organizational and analytical skills. By being organized, I stay on top of projects. My analytical skills help me understand accounting issues and formulate corrective actions when necessary.

Q What do you like most about your job?

A My job is always challenging and never boring, and I get to put to use the skills I acquired in school and at previous jobs.

Career Facts

Real-World Skills
Excellent oral and written communication; research and analytical skills; proficiency with technology

Training And Education
Bachelor's degree in accounting or a related field; several years of experience in a single specialty

Career Skills
Knowledge of U.S. GAAP and foreign currency reporting requirements

Career Path
Start as a general ledger accountant or financial analyst, then positions involving international accounting and finance

---

Tips from Robert Half International

When job-hunting, consider your ideal work environment. Do you want to work for a corporation or government or in public accounting? Which is more important to you: a high salary or opportunities for advancement? You can learn more about prospective employers by conducting research online or at your library or school's career center.

College and Career Readiness

Use the Bureau of Labor Statistics web site to assess the employment outlook in the field of accounting. Present your findings to the class using a PowerPoint presentation or other multimedia application. Include details such as earnings, projection data, and the job outlook.

Real-World Applications & Connections

Career Wise

Management Accountant

Management accounting can be a great career for those who are more interested the executive side of a business than the accounting side. They use their accounting skills to help develop new products and determine the company's future plans.

Education Requirements

Management accountants do not have to be certified and the state boards do not have a minimum education requirement. Most employers will look for management accountants who have a college degree.

Roles

Management accountants analyze the company's financial information in order to help managers make decisions about the company's future.

Responsibilities

A management accountant is responsible for managing costs and assets, establishing work and production standards and then evaluating work performance. They also help executives plan and develop new products.

Activity

A career strategy that can give you an advantage is learning about useful skills for your chosen career and developing those skills early. Research management accounting careers and make a plan of things you could do in high school and college to develop those skills.

Global Accounting

The Euro

Financial statements are generally prepared in the currency of the country in which the business operates. Many European countries use a single currency, the *euro*. Nations in the European Union began using the euro in 2002. Individuals and businesses both benefit by having a single currency. Travelers can use one currency in multiple countries. Businesses have a more stable business environment due to the elimination of exchange rate fluctuations.

Instructions

Imagine that you own a business with locations in Italy and France. Describe how the adoption of the euro in both countries has changed your financial statement preparation.

Analyzing Financial Reports

Return on Sales

Business owners are especially interested in the *return on sales* (see page 237). This percentage shows how much of each revenue dollar becomes profit for the business.

Instructions

Use Zip's income statement on page 231 to calculate its return on sales. (Use $1,100 as net income instead of $1,150.) If total sales are about equal from month to month but return on sales decreases each month, what does this say about the business? What suggestions can you offer to improve the return on sales?

H.O.T. Audit

Statement of Changes in Owner's Equity

Using the amounts that are given below, calculate the missing amounts. Write the missing amounts on a separate sheet of paper.

SweepIt Carpet Cleaning Company
Statement of Changes in Owner's Equity
For the Month Ended July 31, 20--

Beginning Capital, July 1, 20--		?
Add. Investments by Owner		1 500 00
Subtotal		?
Less Withdrawals by Owner	1 500 00	
Net Loss	?	
Total Decrease in Capital		2 751 00
Ending Capital, July 31, 20--		19 729 00

Chapter 10

Completing the Accounting Cycle for a Sole Proprietorship

Chapter Topics

10-1 Preparing Closing Entries

10-2 Posting Closing Entries and Preparing a Post-closing Trial Balance

Visual Summary

Review and Activities

Standardized Test Practice

Computerized Accounting

Problems

Real-World Applications & Connections

Essential Question

As you read this chapter, keep this question in mind:

What activities have to be done at the end of an accounting period?

Main Idea

Closing entries transfer the temporary account balances to the owner's capital account. After the closing entries are posted, a post-closing trial balance is prepared to verify that debits equal credits.

Chapter Objectives

Concepts

C1 Explain why it is necessary to update accounts through closing entries.

C2 Explain the purpose of the Income Summary account.

C3 Explain the relationship between the Income Summary account and the capital account.

Analysis

A1 Analyze and journalize the closing entries.

Procedures

P1 Post the closing entries to the general ledger.

P2 Prepare a post-closing trial balance.

Real-World Business Connection

Trainz.com

The story of Scott Griggs and Trainz.com is a model example of dedication and hard work in business. Over 20 years ago, Griggs opened his first train store in Georgia while working a full-time job. Sales were strong, but the long hours and high business costs led Griggs to sell his company. A few years later, the Internet gave new life to his train dreams. Trainz.com is an auction Web site for model trains and accessories. Tens of thousands of trains move through the company's warehouses each year.

Connect to the Business

At the end of the accounting period, sole proprietorships such as Trainz.com close balances in temporary accounts, summarize revenue and expenses, and make sure that total debits equal total credits heading into the next accounting period.

Analyze

Why would **Trainz.com** employees want to compare sales of different types of model trains from one accounting period to another?

Focus on the Photo

Employees of a small business such as Trainz.com need to be well organized and prepared at the close of an accounting cycle when financial statements are completed. *As an employee, how would you handle the hectic close of an accounting cycle?*

SECTION 10.1
Preparing Closing Entries

Review these three types of financial statements:

- The income statement reports revenue, expenses, and net income or net loss for the accounting period.
- The statement of changes in owner's equity summarizes the impact of the business transactions on the owner's capital account.
- The balance sheet reports the financial position of the business at the end of the period.

Accountants for a company like Mattel prepare financial statements and then journalize and post the closing entries. They prepare a post-closing trial balance to verify that the accounting records still balance. These steps ensure that your final product is balanced and error-free!

> **Section Vocabulary**
> - closing entries
> - Income Summary account
> - compound entry
> - similarly
> - procedure
> - requires

Completing the Accounting Cycle

What Are the Last Two Steps of the Accounting Cycle?

During the accounting period, the accountant records transactions involving revenue, expenses, and withdrawals in temporary accounts. At the end of the period, the accountant transfers the balances in the temporary accounts to the owner's capital account to bring it up to date and to prepare the accounting records for the next period.

Closing entries are journal entries made to close, or reduce to zero, the balances in the temporary accounts and to transfer the net income or net loss for the period to the capital account.

After the closing entries have been journalized and posted, a trial balance is prepared to prove the equality of the general ledger after the closing process. The trial balance prepared after closing is called a *post-closing trial balance*. As you can see in **Figure 10–1** on page 281, the closing process and the post-closing trial balance complete the accounting cycle.

Starting the Eighth Step in the Accounting Cycle: Journalizing the Closing Entries

What Is the Purpose of Closing Entries?

Preparing financial records for the start of a new period is a little like keeping stats for a basketball team. For basketball stats, individual and team scores are recorded for every game, but each new game starts with a score of zero. **Similarly,** in keeping the stats or accounting for a business, entries are posted to the accounts during the accounting period (game), but the temporary accounts (**Rent Expense**, **Maintenance Expense**, **Revenue**, etc.) start each new accounting period (game) with zero balances.

Figure 10–1 The Accounting Cycle with Steps 8 and 9 Highlighted

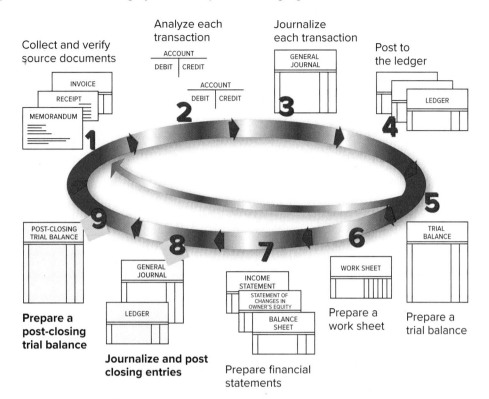

The income statement, you'll remember, reports the net income or net loss for *one accounting period*. The statement is prepared from information recorded and accumulated in the revenue and expense accounts. At the end of the period, the accountant records entries to close, or reduce to zero, the revenue and expense accounts because their balances also apply to only one accounting period. These closing entries also transfer the net income or net loss for the period to the capital account.

The closing process is shown in **Figure 10–2**.

- Prior to the closing process, you know that net income or net loss is calculated on the work sheet. **A**
- The net income or net loss amount then appears on the income statement. **B**
- On the statement of changes in owner's equity, the ending balance of the capital account includes net income or net loss. **C**
- The ending balance of the capital account then appears on the balance sheet. **D**
- At this point, however, the balance of the capital account in the general ledger does not equal the amount on the balance sheet because the closing entries need to be journalized and posted. **E**

For example, the balance on the work sheet for **Crista Vargas, Capital** is $25,400, but on the balance sheet, it is $26,050. These two amounts differ because the withdrawals and the net income have not been recorded in the capital account in the general ledger. The closing process updates accounts through closing entries and brings the balance of the general ledger capital account up to date.

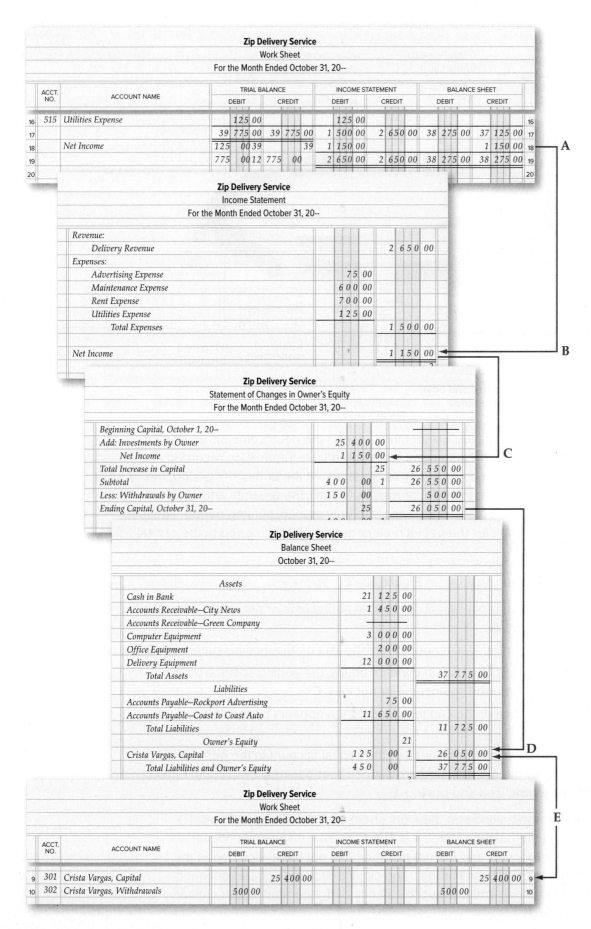

Figure 10–2 The Closing Process

The Income Summary Account

What Is the Purpose of the Income Summary Account?

Before the closing entries are journalized and posted, no single account in the general ledger shows all of the revenue and expenses for the period. This information is scattered among the individual revenue and expense accounts. There is, however, one general ledger account that, until this point, has not been used. That is the **Income Summary** account.

The **Income Summary account** is used to accumulate and summarize the revenue and expenses for the period. This account serves as a simple income statement in the ledger. Expenses, which have debit balances, are transferred as debits to **Income Summary**. Revenues, which have credit balances, are transferred as credits to **Income Summary**. The balance of the account equals the net income or net loss for the fiscal period.

Refer to the chart of accounts for Zip Delivery Service on page 82. Notice that **Income Summary** is in the Owner's Equity section of the general ledger. It is located there because of its relationship to the owner's capital account. Remember that the revenue and expenses transferred to the **Income Summary** account actually represent increases and decreases to owner's equity. The balance of **Income Summary** (the net income or net loss for the period) is transferred to the capital account at the end of the closing process.

Like the withdrawals account, **Income Summary** is a temporary account. However, it is quite different from the other temporary accounts.

- **Income Summary** is used only at the end of the accounting period to summarize the balances from the revenue and expense accounts.
- **Income Summary** does not have a normal balance, which means that it does not have an increase or a decrease side. As shown in the following T account, the debit and credit sides of the account are simply used to summarize the period's revenue and expenses.
- The balance of the **Income Summary** account before and after the closing process is zero.
- The **Income Summary** account does not appear on any financial statement.

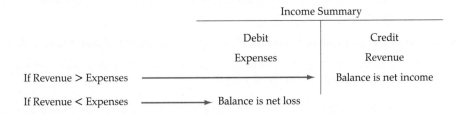

Reading Check

Interpret

How is the Income Summary account different from other temporary accounts?

Preparing Closing Entries

How Do You Journalize Closing Entries?

Four journal entries are prepared to close Zip's temporary accounts:

1. Transfer the balances of all revenue accounts to the credit side of the **Income Summary** account.

2. Transfer all expense account balances to the debit side of the **Income Summary** account.

3. Transfer the balance of the **Income Summary** account to the capital account (net income to the credit side; net loss to the debit side).

4. Transfer the balance of the withdrawals account to the debit side of the capital account.

Closing Revenue to Income Summary

The first step in the closing **procedure** is to transfer the balance of the revenue account to **Income Summary**. The balance for the revenue account is found in the Income Statement section of the work sheet. (Refer to the work sheet in **Figure 9–3** on page 248 when reading about closing entries.)

To record closing entries in the general journal, follow these steps:

1. Enter *Closing Entries* in the center of the Description column.

Closing Entry

First Closing Entry—Close Revenue to Income Summary

Analysis	Identify	1. Zip has only one revenue account, **Delivery Revenue**. The accounts affected are **Delivery Revenue** and **Income Summary**.
	Classify	2. **Delivery Revenue** is a revenue account. **Income Summary** is a temporary owner's equity account.
	+/−	3. The **Delivery Revenue** account balance is decreased by $2,650 to zero. That amount, $2,650, is transferred to the **Income Summary** account.
Debit-Credit Rule		4. Decreases in revenue accounts are recorded as debits. **Debit Delivery** Revenue for $2,650.
		5. To transfer the revenue to the **Income Summary** account, credit **Income Summary** for $2,650.
T Accounts		6.

	Delivery Revenue			Income Summary	
	Debit − Closing 2,650	Credit + Balance 2,650		Debit	Credit Closing 2,650

continued

Journal Entry

7.

		GENERAL JOURNAL			PAGE 3
	DATE	DESCRIPTION	POST. REF.	DEBIT	CREDIT
1	20—	Closing Entries			
2	Oct. 31	Delivery Revenue		2 650 00	
3		Income Summary			
4					2 650 00

2. Enter the date (the last day of the accounting period).
3. Enter the name(s) of the account(s) to be debited and the debit amount(s).
4. Enter the name of the account, **Income Summary**, to be credited and the amount to be credited.

Closing Expenses to Income Summary

The second closing entry transfers the balances of the expense accounts to **Income Summary**. The balances of the expense accounts are found in the Income Statement section of the work sheet.

It is not necessary to use a separate closing entry for each expense account. As you can see, Zip's entry has one debit and four credits.

Closing Entry

Second Closing Entry—Close Expenses to Income Summary

Analysis	Identify	1. The accounts affected by the second closing entry are **Advertising Expense**, **Maintenance Expense**, **Rent Expense**, **Utilities Expense**, and **Income Summary**.
	Classify	2. **Advertising Expense**, **Maintenance Expense**, **Rent Expense**, and **Utilities Expense** are expense accounts. **Income Summary** is a temporary owner's equity account.
	+/−	3. The balances of the four expense accounts are decreased to zero; the total decrease is $1,500. The total amount, $1,500, is transferred to the **Income Summary** account.
Debit-Credit Rule		4. To transfer the expenses to the **Income Summary** account, debit **Income Summary** for $1,500.
		5. Decreases in expense accounts are recorded as credits. Credit **Advertising Expense**, $75; **Maintenance Expense**, $600; **Rent Expense**, $700; **Utilities Expense**, $125.

continued

T Accounts

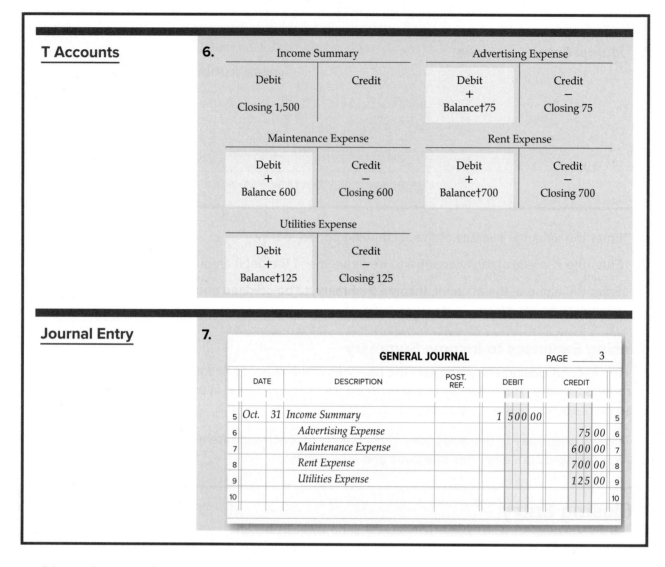

Journal Entry

A journal entry with two or more debits or two or more credits is called a **compound entry**. A compound entry saves both space and posting time. For example, each of Zip's expense accounts could be closed to **Income Summary** separately. That, however, **requires** four entries and postings to the **Income Summary** account instead of one entry and posting.

Closing Income Summary to Capital

The third closing entry transfers the balance of the **Income Summary** account to the capital account. As shown in the T account, after closing Zip's revenue and expense accounts, **Income Summary** has a credit balance of $1,150. A credit balance indicates net income for the period. It is the same amount that appears on the work sheet.

	Income Summary		
2 Closing entry for expenses	1,500	1 Closing entry for revenue	2,650
		Balance	1,150

Closing Entry

Third Closing Entry—Close Income Summary to Capital

Analysis

Identify
1. The accounts **Income Summary** and **Crista Vargas, Capital** are affected.

Classify
2. **Income Summary** is a temporary owner's equity account. **Crista Vargas, Capital** is an owner's capital account.

+/−
3. The **Income Summary** account balance is reduced to zero by transferring $1,150, the net income amount, to the capital account. **Crista Vargas, Capital** is increased by $1,150.

Debit-Credit Rule

4. To reduce the **Income Summary** balance to zero, debit **Income Summary** for $1,150.

5. Net income is recorded as a credit to the owner's capital account. Credit **Crista Vargas, Capital** for $1,150.

T Accounts

6.

Income Summary		Crista Vargas, Capital	
Debit	Credit	Debit	Credit
		−	+
Closing 1,150	Balance 1,150		Balance 25,400
			Closing 1,150

Journal Entry

7.

GENERAL JOURNAL   PAGE 3

DATE	DESCRIPTION	POST. REF.	DEBIT	CREDIT
Oct. 31	Income Summary		1,150 00	
	Crista Vargas, Capital			1,150 00

After the third closing entry has been posted, the **Income Summary** account appears in T-account form as follows.

Income Summary

2 Closing entry for expenses	1,500	1 Closing entry for revenue	2,650
3 Closing balance to owner's capital account	1,150		

If a business has a net loss, **Income Summary** has a debit balance. In that case, the third closing entry debits the capital account and credits **Income Summary** for the amount of the net loss. **Figure 10–3** shows this general journal entry.

Common Mistakes ⚠️

Closing the Withdrawals Account A mistake frequently made during the closing procedure is the closing of the Withdrawals account into the Income Summary account. The Income Summary account should be used to record the closing of the revenue and expense accounts, thus having a summary income statement in the Income Summary account. A withdrawal of cash or other assets by the owner for personal use is not an operating expense but a withdrawal of Owner's Equity (Capital). Remember that the Withdrawals account is closed directly into the Capital account.

Figure 10–3 Closing Income Summary for the Amount of Net Loss

	GENERAL JOURNAL			PAGE 3	
DATE	DESCRIPTION	POST. REF.	DEBIT	CREDIT	
20–	*Closing Entries*				1
Oct. 31	Crista Vargas, Capital		700 00		2
	Income Summary			700 00	3
					4

Closing Withdrawals to Capital

The fourth and last closing entry transfers the balance of the withdrawals account to the capital account. As you recall, withdrawals decrease owner's equity. The balance of the withdrawals account is transferred to the capital account to reflect the decrease in owner's equity. The balance of the withdrawals account is found in the Balance Sheet section of the work sheet.

Closing Entry

Fourth Closing Entry—Close Withdrawals to Capital

Analysis	Identify	1. The accounts affected by the fourth closing entry are **Crista Vargas, Withdrawals** and **Crista Vargas, Capital**.
	Classify	2. **Crista Vargas, Withdrawals** is a temporary owner's equity account. **Crista Vargas, Capital** is an owner's capital account.
	+/−	3. **Crista Vargas, Withdrawals** is decreased by $500. **Crista Vargas, Capital** is decreased by $500.
Debit-Credit Rule		4. Decreases in owner's capital accounts are recorded as debits. Debit **Crista Vargas, Capital** for $500.
		5. Decreases in owner's withdrawal accounts are recorded as credits. Credit **Crista Vargas, Withdrawals** for $500.

continued

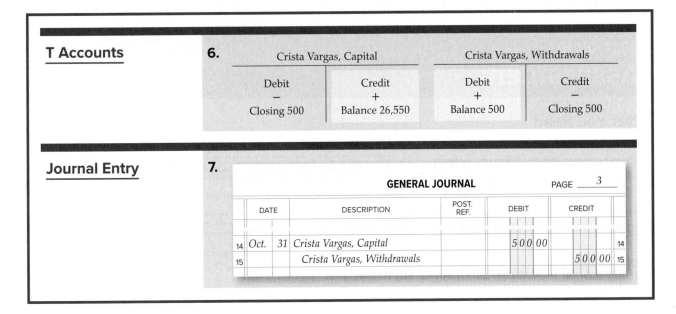

Figure 10–4 summarizes the closing entries for Zip.

DATE		DESCRIPTION	POST. REF.	DEBIT	CREDIT	
20--		Closing Entries				1
Oct.	31	Delivery Revenue		2650 00		2
		Income Summary			2650 00	3
						4
	31	Income Summary		1500 00		5
		Advertising Expense			75 00	6
		Maintenance Expense			600 00	7
		Rent Expense			700 00	8
		Utilities Expense			125 00	9
						10
	31	Income Summary		1150 00		11
		Crista Vargas, Capital			1150 00	12
						13
	31	Crista Vargas, Capital		500 00		14
		Crista Vargas, Withdrawals			500 00	15

GENERAL JOURNAL PAGE 3

Figure 10–4 Journalizing the Closing Entries

SECTION 10.1
Assessment

After You Read

Reinforce the Main Idea

Use a table like this one to describe the closing entries that are made at the end of each accounting period.

General Journal Entry Description	Account Debited	Account Credited

Problem 1 Preparing Closing Entries

Instructions Prepare closing entries for the following in your working papers.

1. A closing entry must be made for the account **Ticket Revenue**, which has a balance of $6,000.

2. A business has three expense accounts: **Gas and Oil Expense** (balance, $700), **Miscellaneous Expense** (balance, $600), and **Utilities Expense** (balance, $1,800). The end of the fiscal year is June 30.

Problem 2 Analyzing a Source Document

Instructions Using the source document:

1. Journalize the transaction in a general journal in your working papers.

2. Post the entry to the appropriate T accounts.

3. Assume it is the end of the accounting period. Record the closing entry for this account in the general journal.

4. Post the closing entry to the appropriate T accounts.

DEPARTMENT OF WATER & POWER
455 Main Street
Sarasota, FL 34230

ACCOUNT NO.: 00384843848-339
INVOICE NO.: 32004

TO Gulfview Tropical Fish and Supplies
4524 West Palm Bay Avenue
Sarasota, FL 34222

DATES		USAGE	RATE	AMOUNT DUE
FROM	TO			
March 15, 20--	April 14, 20--	1,000 kWh	$.129 per kWh	$129.00

Math for Accounting

Using these general ledger account balances, calculate the net income or net loss for the period. Then calculate the increase or decrease in owner's capital.

Anna Zarian, Capital	$25,000
Anna Zarian, Withdrawals	5,000
Accounting Fees Revenue	9,000
Advertising Expense	1,000
Miscellaneous Expense	2,000
Rent Expense	1,500
Utilities Expense	500

SECTION 10.2
Posting Closing Entries and Preparing a Post-closing Trial Balance

In Section 10.1 you learned how to journalize the closing entries. In this section you will complete the accounting cycle.

Section Vocabulary
- post-closing trial balance

Completing the Eighth Step in the Accounting Cycle: Posting the Closing Entries to the General Ledger

What Is Special About Posting the Closing Entries?

The next step in the closing process is to post the closing entries to the general ledger accounts. The posting procedure is the same as for any other general journal entry, with one exception. The words *Closing Entries* are written in the Description column of the general ledger account. The posting of the closing entries for Zip is shown in **Figure 10–5**. Note that *Closing Entries* can be abbreviated as *Clos. Ent*.

Figure 10–5 Closing Entries Posted to the General Ledger

	DATE		DESCRIPTION	POST. REF.	DEBIT	CREDIT	
1	20–		*Closing Entries*				1
2	Oct.	31	Delivery Revenue	401	2 6 5 0 00		2
3			Income Summary	303		2 6 5 0 00	3
4							4
5		31	Income Summary	303	1 5 0 0 00		5
6			Advertising Expense	501		7 5 00	6
7			Maintenance Expense	505		6 0 0 00	7
8			Rent Expense	510		7 0 0 00	8
9			Utilities Expense	515		1 2 5 00	9
10							10
11		31	Income Summary	303	1 1 5 0 00		11
12			Crista Vargas, Capital	301		1 1 5 0 00	12
13							13
14		31	Crista Vargas, Capital	301	5 0 0 00		14
15			Crista Vargas, Withdrawals	302		5 0 0 00	15
16							16

GENERAL JOURNAL PAGE 3

Reading Check

Recall

How is the posting procedure different from the procedure to record general journal entries?

ACCOUNT: Crista Vargas, Capital — ACCOUNT NO. 301

DATE	DESCRIPTION	POST. REF.	DEBIT	CREDIT	BALANCE DEBIT	BALANCE CREDIT
20–						
Oct. 1		G1		25 000 00		25 000 00
2		G1		400 00		25 400 00
31	Clos. Ent.	G3		1 150 00		26 550 00
31	Clos. Ent.	G3	500 00			26 050 00

ACCOUNT: Crista Vargas, Withdrawals — ACCOUNT NO. 302

DATE	DESCRIPTION	POST. REF.	DEBIT	CREDIT	BALANCE DEBIT	BALANCE CREDIT
20–						
Oct. 31		G1	500 00		500 00	
31	Clos. Ent.	G3		500 00	—	

ACCOUNT: Income Summary — ACCOUNT NO. 303

DATE	DESCRIPTION	POST. REF.	DEBIT	CREDIT	BALANCE DEBIT	BALANCE CREDIT
20–						
Oct. 31	Clos. Ent.	G3		2 650 00		2 650 00
31	Clos. Ent.	G3	1 500 00			1 150 00
31	Clos. Ent.	G3	1 150 00			—

ACCOUNT: Delivery Revenue — ACCOUNT NO. 401

DATE	DESCRIPTION	POST. REF.	DEBIT	CREDIT	BALANCE DEBIT	BALANCE CREDIT
20–						
Oct. 15		G1		1 200 00		1 200 00
20		G1		1 450 00		2 650 00
31	Clos. Ent.	G3	2 650 00			—

ACCOUNT: Advertising Expense — ACCOUNT NO. 501

DATE	DESCRIPTION	POST. REF.	DEBIT	CREDIT	BALANCE DEBIT	BALANCE CREDIT
20–						
Oct. 18		G1	75 00		75 00	
31	Clos. Ent.	G3		75 00	—	

ACCOUNT: Maintenance Expense — ACCOUNT NO. 505

DATE	DESCRIPTION	POST. REF.	DEBIT	CREDIT	BALANCE DEBIT	BALANCE CREDIT
20–						
Oct. 29		G1	600 00		600 00	
31	Clos. Ent.	G3		600 00	—	

Figure 10–5 Closing Entries Posted to the General Ledger (continued)

Figure 10–5 Closing Entries Posted to the General Ledger

ACCOUNT Rent Expense   ACCOUNT NO. 510

DATE	DESCRIPTION	POST. REF.	DEBIT	CREDIT	BALANCE DEBIT	BALANCE CREDIT
20—						
Oct. 16		G1	7 0 0 00		7 0 0 00	
31	Clos. Ent.	G3		7 0 0 00	—	
			7 0 0 00		7 0 0 00	

ACCOUNT Utilities Expense   ACCOUNT NO. 515

DATE	DESCRIPTION	POST. REF.	DEBIT	CREDIT	BALANCE DEBIT	BALANCE CREDIT
20—						
Oct. 28		G1	1 2 5 00		1 2 5 00	
31	Clos. Ent.	G3		1 2 5 00	—	
			1 2 5 00		1 2 5 00	

The Ninth Step in the Accounting Cycle: Preparing a Post-Closing Trial Balance

If We Already Have a Trial Balance, Why Do We Need a Post-Closing Trial Balance?

The last step in the accounting cycle is to prepare a post-closing trial balance. The **post-closing trial balance** is prepared to make sure total debits equal total credits after the closing entries are posted. This last step in the accounting cycle gives you the opportunity to correct any errors that might have been made after the trial balance was prepared. For example, a closing entry could have been posted to the wrong general ledger account, or an entire closing entry may not have been posted. In preparing the post-closing trial balance, all the balances in the ledger accounts are checked. The postclosing trial balance for Zip is shown in **Figure 10–6**.

Notice that only accounts with balances are listed on the post-closing trial balance. After the closing process, only permanent accounts have balances. Temporary accounts have zero balances, so there is no need to list those accounts on the post-closing trial balance.

Figure 10–6 Post-Closing Trial Balance

Zip Delivery Service		
Post-Closing Trial Balance		
October 31, 20--		
Cash in Bank	21 1 2 5 00	
Accounts Receivable–City News	1 4 5 0 00	
Computer Equipment	3 0 0 0 00	
Office Equipment	2 0 0 00	
Delivery Equipment	12 0 0 0 00	
Accounts Payable–Rockport Advertising		7 5 00
Accounts Payable–Coast to Coast Auto		11 6 5 0 00
Crista Vargas, Capital		26 0 5 0 00
Total	37 7 7 5 00	37 7 7 5 00

Reading Check

Explain

How is the post-closing trial balance similar to the trial balance? How is it different?

SECTION 10.2
Assessment

After You Read

Reinforce the Main Idea

Create a table like this one to list the general ledger account classifications. In the column titled "Appears on Post-Closing Trial Balance," place an X next to each account that carries a balance into the next accounting period.

Account Type	Appears on Post-Closing Trial Balance
Asset	
Liability	
Owner's Capital	
Owner's Withdrawals	
Revenue	
Expenses	

Problem 3 Determining Accounts Affected by Closing Entries

The following list contains some of the accounts used by Living Well Health Spa.

General Ledger

- Accts. Pay.—The Fitness Shop
- Accts. Rec.—Linda Brown
- Advertising Expense
- Cash in Bank
- Exercise Class Revenue
- Exercise Equipment
- Income Summary
- Laundry Equipment
- Maintenance Expense
- Membership Fees
- Miscellaneous Expense
- Office Furniture
- Rent Expense
- Repair Tools
- Ted Chapman, Capital
- Ted Chapman, Withdrawals
- Utilities Expense

Instructions Using the form in your working papers:

1. In the first column, indicate the financial statement where each account appears: balance sheet or income statement.
2. In the next column, indicate whether or not the account is affected by a closing entry.
3. In the last column, indicate whether or not the account appears on the post-closing trial balance.

The first account is shown as an example.

Account Name	Financial Statement	Is the account affected by a closing entry?	Does the account appear on the post-closing trial balance?
Accts. Pay.—The Fitness Shop	Balance Sheet	No	Yes

Do the Math

You work for a company with a large accounting department, and every accounting clerk has specific duties. Your co-worker responsible for preparing the postclosing trial balance is ill today. Your supervisor asked you to prepare the post-closing trial balance. Using your computer or lined paper, draft the post-closing trial balance using the following account balances: Cash, $27,800; Accounts Receivable, $33,000; Equipment, $81,000; Accounts Payable, $24,500; and Owner's Capital, $117,300.

Chapter 10
Visual Summary

Concepts
The Income Summary account has TWO purposes.

First Purpose: To *accumulate* the revenue and expenses for an accounting period.

Second Purpose: To *summarize* the revenue and expenses for an accounting period.

Analysis
Analyze and journalize the closing entries.

1. Close the revenue account(s) to **Income Summary**

	20–					
	April	30	Revenue Account		x x x xx	
			Income Summary			x x x xx

2. Close the expense accounts to **Income Summary**

			Expense Account			x x x xx

3. Close **Income Summary** to the owner's capital account.

4. Close the owner's withdrawals account to the owner's capital account.

Procedures
Prepare a post-closing trial balance.

Stan's Donuts
Post-Closing Trial Balance
April 30, 20___

Account	Debit	Credit
Cash in Bank	45 725 00	
Accts Rec.—Jerry's Catering	2 500 00	
Kitchen Equipment	4 500 00	
Office Equipment	500 00	
Tableware	250 00	
Accts. Pay.—Staunch Staffing		1 325 00
Accts. Pay.—USA Baking Supply		7 350 00
Joanne Chen, Capital		45 000 00
Totals	53 475 00	53 675 00

296 Chapter 10 • Completing the Accounting Cycle for a Sole Proprietorship

Chapter 10
Review and Activities

Answering the Essential Question

What activities have to be done at the end of an accounting period?

Just as you might have certain tasks that you need to do on a regular basis, there are tasks accountants need to do at the end of an accounting cycle. Why it is important that these tasks be done on a regular basis?

Vocabulary Check

1. **Vocabulary** Write a short essay about completing the accounting cycle using each of the content vocabulary words. You may use a word more than once. Underline each vocabulary word each time it is used.

   - closing entries
   - similarly
   - Income Summary account
   - procedure
   - compound entry
   - require
   - post-closing trial balance

Concept Check

2. **Analyze** Which accounts are considered temporary accounts? Why are they closed at the end of the fiscal year?
3. What is the purpose of the Income Summary account? How is it different from other temporary accounts?
4. How does the Income Summary account affect the capital account?
5. Illustrate and explain the steps for journalizing and posting the closing entries.
6. **Classify** What kind of account is the Income Summary account: asset, liability, owner's equity, revenue, or expense? Justify your answer.
7. **Compare and contrast** How is the post-closing trial balance similar to the trial balance? How is it different?
8. **Math** Jai is having a small warehouse built for his business. The warehouse measures 125 feet wide and 200 feet long. The height of the ceiling is 25 feet. Heating the building will cost $1.13 for every 1,500 cubic feet per day. What will it cost to heat the warehouse for a year?
9. **English Language Arts** Research the economic events that led up to the Great Depression of the 1930s. Tell what factors conspired to cause the collapse, what factors helped to rebuild the economy, and the government policy changes that resulted. Give your results in a two-page essay.

Chapter 10
Standardized Test Practice

Multiple Choice

1. The purpose of the Income Summary account is
   a. to close the temporary accounts.
   b. to accumulate and summarize the revenue and expenses for the period.
   c. to transfer the balance of the Income Summary account to the capital account.
   d. to report the net income or net loss for the accounting period.

2. Closing entries help ensure that
   a. balances of the temporary accounts are reduced to zero
   b. the net income or net loss for the period is transferred to the capital account
   c. accounts are prepared for the trial balance
   d. All of these

3. Transactions involving revenue, expenses, and withdrawals are recorded in the
   a. permanent accounts
   b. financial statements
   c. general journal
   d. temporary capital accounts

4. After adjusting and closing entries have been posted, a
   a. balance sheet is prepared.
   b. trial balance is prepared.
   c. post-closing balance sheet is prepared.
   d. post-closing trial balance is prepared.

5. The Capital account balance minus the net loss minus the withdrawals account balance equals
   a. net loss
   b. total liabilities
   c. the capital amount shown on a balance sheet.
   d. the capital amount shown on an income statement.

Short Answer

6. A journal entry with two or more debit or credit entries is called a _____.

Extended Response

7. Why is it necessary to prepare a post-closing trial balance?

Chapter 10
Computerized Accounting

Closing Entries
Making the Transition from a Manual to a Computerized System

MANUAL METHODS

- Using a general journal form, prepare closing entries for revenue, expense, income summary, and withdrawals accounts. Post the closing entries in the general ledger accounts.

COMPUTERIZED METHODS

- It is not necessary to journalize closing entries. Closing entries are performed by the computerized system.

Chapter 10 Problems

Problem 4 — Preparing Closing Entries

A portion of the work sheet for Wilderness Rentals for the period ended December 31 follows.

Instructions Using the information from the work sheet, prepare the journal entries to close the temporary accounts.

1. Record the closing entry for the revenue account.
2. Record the closing entry for the expense accounts.
3. Record the closing entry for the **Income Summary** account.
4. Record the closing entry for the withdrawals account.

Wilderness Rentals
Work Sheet
For the Period Ended December 31, 20--

	ACCT. NO.	ACCOUNT NAME	INCOME STATEMENT DEBIT	INCOME STATEMENT CREDIT	BALANCE SHEET DEBIT	BALANCE SHEET CREDIT	
1	101	Cash in Bank			7 000 00		1
2	105	Accts. Rec.—Helen Katz			2 000 00		2
3	110	Accts. Rec.—Polk and Co.			1 000 00		3
4	115	Office Supplies			900 00		4
5	120	Office Equipment			12 000 00		5
6	125	Camping Equipment			6 000 00		6
7	201	Accts. Pay.—Adventure Equip. Inc.				900 00	7
8	203	Accts. Pay.—Digital Tech Computers				400 00	8
9	205	Accts. Pay.—Greg Mollaro				500 00	9
10	301	Ronald Hicks, Capital				19 775 00	10
11	305	Ronald Hicks, Withdrawals			2 350 00		11
12	310	Income Summary					12
13	401	Equipment Rental Revenue		14 965 00			13
14	501	Advertising Expense	1 500 00				14
15	505	Maintenance Expense	1 560 00				15
16	515	Rent Expense	1 000 00				16
17	525	Utilities Expense	1 230 00				17
18			5 290 00	14 965 00	31 250 00	21 575 00	18
19		Net Income	9 675 00			9 675 00	19
20			14 965 00	14 965 00	31 250 00	31 250 00	20
21							21

Analyze

Predict the balance of the capital account after the closing entries are posted.

Chapter 10
Problems

Problem 5 — Preparing a Post-Closing Trial Balance

Instructions Use the accounts shown on the next two pages to prepare a Dec. 31 post-closing trial balance for Hot Suds Car Wash.

Cash in Bank
Debit +	Credit −
Balance 8,000	

Accounts Receivable—Linda Brown
Debit +	Credit −
Balance 875	

Accounts Receivable—Valley Auto
Debit +	Credit −
Balance 5,050	

Office Equipment
Debit +	Credit −
Balance 6,000	

Office Furniture
Debit +	Credit −
Balance 9,000	

Car Wash Equipment
Debit +	Credit −
Balance 65,000	

Accounts Payable—Allen Vacuum Systems
Debit −	Credit +
	Balance 41,000

Accounts Payable—O'Brian's Office Supply
Debit −	Credit +
	Balance 2,500

Regina Delgado, Capital
Debit −	Credit +
Closing 1,500	35,925
	Closing 16,000

Regina Delgado, Withdrawals
Debit +	Credit −
Balance 1,500	Closing 1,500

Income Summary
Debit	Credit
Closing 19,000	Closing 35,000
Closing 16,000	

Wash Revenue
Debit −	Credit +
Closing 15,000	Balance 15,000

Wax Revenue
Debit −	Credit +
Closing 8,000	Balance 8,000

Interior Detailing Revenue
Debit −	Credit +
Closing 12,000	Balance 12,000

Advertising Expense
Debit +	Credit −
Balance 2,500	Closing 2,500

Equipment Rental Expense
Debit +	Credit −
Balance 3,000	Closing 3,000

continued

Chapter 10
Problems

Maintenance Expense
Debit + Balance 5,000 | Credit − Closing 5,000

Rent Expense
Debit + Balance 5,000 | Credit − Closing 5,000

Utilities Expense
Debit + Balance 3,500 | Credit − Closing 3,500

Analyze
Predict the balance of the temporary accounts after the closing entries are posted.

Problem 6 Journalizing Closing Entries

The following account names and balances appear on the work sheet for Kits & Pups Grooming for the month ended December 31.

Kits & Pups Grooming
Work Sheet
For the Month Ended December 31, 20--

	ACCT. NO.	ACCOUNT NAME	INCOME STATEMENT DEBIT	INCOME STATEMENT CREDIT	BALANCE SHEET DEBIT	BALANCE SHEET CREDIT	
1		Cash in Bank			9 300 00		1
2		Accts. Rec.—Juan Alvarez			3 000 00		2
3		Accts. Rec.—Nathan Carlsbad			10 000 00		3
4		Accts. Rec.—Martha Giles			5 000 00		4
5		Office Equipment			8 000 00		5
6		Office Furniture			10 000 00		6
7		Computer Equipment			9 000 00		7
8		Grooming Equipment			15 000 00		8
9		Kennel Equipment			21 000 00		9
10		Accts. Pay.—Able Store Equip.				5 000 00	10
11		Accts. Pay.—Dogs & Cats Inc.				1 500 00	11
12		Accts. Pay.—Pet Gourmet				15 000 00	12
13		Abe Shultz, Capital				52 700 00	13
14		Abe Shultz, Withdrawals			7 000 00		14
15		Income Summary					15
16		Boarding Revenue		20 000 00			16
17		Grooming Revenue		8 000 00			17
18		Advertising Expense	700 00				18
19		Equipment Repair Expense	1 200 00				19
20		Maintenance Expense	500 00				20
21		Rent Expense	1 700 00				21
22		Utilities Expense	800 00				22

continued

Chapter 10 Problems

Instructions Using this information, record the closing entries for Kits & Pups Grooming. Use general journal page 11.

Analyze
Calculate the change in the capital account for the period.

Problem 7 Posting Closing Entries and Preparing a Post-Closing Trial Balance

Period ending December 31 closing entries for Outback Guide Service are:

	GENERAL JOURNAL			PAGE 1	
DATE	DESCRIPTION	POST. REF.	DEBIT	CREDIT	
20—	Closing Entries				1
Dec. 31	Guide Service Revenue		16 300 00		2
	Income Summary			16 300 00	3
31	Income Summary		10 000 00		4
	Advertising Expense			3 000 00	5
	Maintenance Expense			1 100 00	6
	Rent Expense			4 000 00	7
	Utilities Expense			1 900 00	8
31	Income Summary		6 300 00		9
	Juanita Ortega, Capital			6 300 00	10
31	Juanita Ortega, Capital		4 000 00		11
	Juanita Ortega, Withdrawals			4 000 00	12

Instructions Using your working papers, post the closing entries to the appropriate general ledger accounts and prepare a post-closing trial balance.

Analyze
Calculate the balance of the capital account after closing.

Chapter 10
Problems

Problem 8 Completing Period-End Activities

The general ledger for Show biz Video shows the following at December 31:

General Ledger

101	Cash in Bank	12,000	207 Accts. Pay.—	
105	Accts. Rec.—G. Cohen	3,000	New Media Suppliers	3,000
110	Accts. Rec.—J. Coletti	900	209 Accts. Pay.— Palace Films	14,000
113	Accts. Rec.—S. Flannery	1,800	301 Greg Failla, Capital	33,775
115	Accts. Rec.—Spring Branch School District	1,500	305 Greg Failla, Withdrawals	4,000
			310 Income Summary	—
130	Office Equipment	5,000	401 DVD Rental Revenue	9,600
135	Office Furniture	8,000	405 HD Projector Rental Revenue	3,500
140	Computer Equipment	10,000	501 Advertising Expense	1,600
145	Blu-ray discs	20,000	505 Equipment Repair Expense	1,200
150	Video Equipment	9,000	510 Maintenance Expense	400
201	Accts. Pay.—Broad Street Office Supply	400	520 Rent Expense	1,000
205	Accts. Pay.—Computer Horizons	15,500	530 Utilities Expense	375

Instructions Using the preceding account names and balances:

1. Prepare the six-column work sheet. The period covered is one month.
2. Prepare the financial statements. Greg Failla invested $10,000 during the month.
3. Record the closing entries on page 12 of the general journal.
4. Post the closing entries.
5. Prepare a post-closing trial balance.

Analyze
Calculate the total amount of all the accounts receivable accounts on December 31.

Chapter 10
Problems

Problem 9 Completing End-of-Period Activities

At the end of December, the general ledger for Job Connect showed the following account balances:

General Ledger

101	Cash in Bank	6,000	207	Accts. Pay.—Wildwood Furniture Sales	2,000
105	Accts. Rec.—CompuRite Systems	1,000	301	Richard Tang, Capital	23,600
110	Accts. Rec.—Marquez Manufact.	500	302	Richard Tang, Withdrawals	3,000
113	Accts. Rec.—Roaring Rivers Water Park	600	303	Income Summary	—
115	Accts. Rec.—M. Spencer	200	401	Placement Fees Revenue	6,900
130	Office Equipment	7,000	405	Technology Classes Revenue	2,400
135	Office Furniture	5,000	501	Advertising Expense	3,000
140	Computer Equipment	8,500	505	Maintenance Expense	800
201	Accts. Pay.—Micro Solutions Inc.	2,800	510	Miscellaneous Expense	800
205	Accts. Pay.—Vega Internet Services	1,600	520	Rent Expense	2,000
			530	Utilities Expense	900

Instructions Using the preceding account names and balances:

1. Prepare the six-column work sheet. The period covered is one month.
2. Prepare the financial statements.
3. Record the closing entries on page 28 of the general journal.
4. Post the closing entries.
5. Prepare a post-closing trial balance.

Analyze
Identify the largest expenditure for the period.

Real-World Applications & Connections

H.O.T. Audit

Auditing the Work Sheet

The worksheet and financial statements for Jones Company for December are complete. The balances of the revenue, expenses, and equity accounts on the financial statements are:

Fees Revenue	$ 3,000
Utilities Expense	100
Rent Expense	500
Gas Expense	700
Miscellaneous Expense	400
Bob Jones, Capital	16,400
Bob Jones, Withdrawals	600

Instructions

Answer the following questions about the closing entries that will be prepared:

1. What account was credited to close the Fees Revenue account?
2. What was the amount of the debit to Income Summary to close all expense accounts?
3. What was the amount of the net income?
4. What was the amount that withdrawals decreased Bob Jones, Capital account?

21st Century Skills

Job Stress

Many otherwise exciting jobs can also be quite stressful. For instance, police officers and air traffic controllers have very stressful jobs. Studies have found that a machine-controlled pace can also lead to job stress. Working alone at a monotonous job can be stressful. Over time, job stress can lead to serious health concerns, including high blood pressure and heart attacks. There is plenty of research into what a healthy career looks like, and stress researchers hope to offer solutions that will enhance both productivity and worker health.

Research and Write

a. Use the Internet to research what makes a healthy career. Look for ways researchers can gauge stress levels in people and the ways they determine the stress levels of different jobs.

b. Write a paragraph about what you find.

Career Wise

Auditor

Like accountants, auditors help make sure that organizations are run effectively and that their financial statements are accurate. Several different kinds of auditors exist. *Internal auditors* evaluate their company's performance, management, and monitor activities for potential fraud. *External auditors* are typically hired by an organization to verify whether or not an organization's financial statements are free of errors. *Government auditors* evaluate companies that are subject to government regulations or taxation to determine the accuracy of the companies' records. Auditors may have any of several specializations.

Research And Share

a. Use the Internet to locate three sources of information on auditors.

b. What level of education is required to be an auditor? What are the roles and responsibilities of auditors?

Spotlight on Personal Finance

Closing Activities You Perform

In your personal life, you perform certain activities for a period of time. You often bring closure to an activity and start another.

Activity

In your personal life, you have experienced "closing activities" without realizing it. Have you set a personal financial goal and achieved it through a financial planning process? When you achieved the goal, you actually brought closure to the financial activity. Have you ever saved money to buy something you wanted? When you bought it, that was a "closing activity". Did you make plans to attend an entertainment event with a friend and actually attended the event? Attending the event brought closure to your plans.

Create a table that lists a few of your personal goals, how you went about achieving the goals and in closing, what benefit(s) you received.

Analyzing Financial Reports

Calculating Return on Owner's Equity

One measure of business success is the **return on owner's equity (ROE):**

Net Income/Beginning Owner's Equity − Withdrawals = $3,600/$34,500 − $2,500 = 11.25%

The resulting percentage shows the amount earned during the period on each dollar invested by the owner.

Instructions

Use Zip's income statement (page 225) and statement of changes in owner's equity (page 229) to calculate October's ROE. How does this return compare with a bank savings account that pays 2 percent interest per year?

Chapter 11

Cash Controls and Banking Activities

Chapter Topics

11-1 Banking Procedures

11-2 Reconciling the Bank Statement

Visual Summary

Review and Activities

Standardized Test Practice

Computerized Accounting

Problems

Real-World Accounting Careers

Real-World Applications & Connections

Essential Question

As you read this chapter, keep this question in mind:

How do banks help us manage our money?

Main Idea

Internal controls are steps taken to protect assets and keep reliable records. The bank reconciliation is an important internal control.

Chapter Objectives

Concepts	Analysis	Procedures
C1 Describe the internal controls used to protect cash.	**A1** Prepare a check.	**P1** Journalize and post entries relating to bank service charges.
C2 Describe the forms needed to open and use a checking account.	**A2** Prepare bank deposits.	
	A3 Reconcile a bank statement.	**P2** Describe the uses of the electronic funds transfer system.
C3 Record information on check stubs.		

Real-World Business Connection

Donna Fenn

As an author and journalist, Donna Fenn has reported for over two decades about small business trends and entrepreneurship. At home, she is the mother of two Generation Y kids. Fenn combined her interests to write *Upstarts! How Gen Y Entrepreneurs Are Rocking The World of Business*, which profiles the success strategies of young entrepreneurs. Featuring more than 60 stories, Fenn's book demonstrates how young people are taking charge and changing the rules of business.

Connect to the Business

The entrepreneurs in *Upstarts!* all learned that effective cash control and banking activities are crucial to building a business and maintaining operations. You can prepare for an entrepreneurial future by being responsible for your own checking account.

Analyze

Imagine that you are starting a retail business that will handle several dozen cash transactions per day. List at least three ways you can ensure cash is protected and properly tracked.

Focus on the Photo

Donna Fenn's profiles of the young *Upstarts!* is designed to inspire entrepreneurs of any age by providing models of how new successful businesses grow through sound financial practices. **What do you think is the most common problem that retail businesses experience regarding cash?**

SECTION 11.1
Banking Procedures

In any business, cash (currency, coins, and checks) is used in daily transactions. An important part of accounting for a business, therefore, involves tracking the cash received and paid out. For example, Jamba Juice has established procedures for processing the cash received from the sale of drinks and other items. Jamba Juice also has a system of monitoring and controlling cash paid out for wages, utilities, supplies, and numerous other expenses.

Protecting Cash

How Does a Business Protect Cash?

It is important to protect cash from loss, waste, theft, forgery, and embezzlement. Cash is protected through internal controls and external controls. **Internal controls** refer to procedures within the business that are designed to protect cash and other assets and to keep **reliable** records. Internal controls for cash include:

1. Limiting the number of persons handling cash.
2. Separating accounting tasks involving cash. For example, one person handles cash receipts, another handles cash payments, and a different person keeps the accounting records showing the amounts received or paid.
3. Dual control of accounting tasks involving cash. For example, two employees handle and transport cash or cash records together.
4. Bonding (insuring) employees who handle cash or cash records.
5. Using a cash register and a safe to safeguard all cash, checks, check stock, credit cards, and other cash records.
6. Depositing cash receipts in the bank daily.
7. Making all cash payments by check.
8. Promptly reconciling the bank statement.

External controls are the measures and procedures provided outside the business to protect cash and other assets. For example, banks maintain controls to protect the funds their customers deposit. These controls include verifying the accuracy of signatures on checks and maintaining records of monetary transactions.

The Checking Account

How Do You Maintain a Checking Account?

A **checking account** allows a person or business to deposit cash in a bank and to write checks against the account balance. A **check** is a written order from a

Section Vocabulary

- internal controls
- external controls
- checking account
- check
- depositor
- signature card
- deposit slip
- endorsement
- blank endorsement
- special endorsement
- restrictive endorsement
- payee
- drawer
- drawee
- voiding a check
- reliable
- issued

depositor telling the bank to pay a stated amount of cash to the person or business named on the check. A **depositor** is a person or business that has cash on deposit in a bank.

Opening a Checking Account

A checking account helps protect cash and provides a record of cash transactions.

The Signature Card.

To open a checking account, a business owner fills out a **signature card** and deposits cash in the bank. A signature card contains the signature(s) of the person(s) authorized to write checks on the account. The bank keeps the signature card on file so that it can be matched against signed checks presented for payment. The use of a signature card protects both the account holder and the bank against checks with forged signatures. See **Figure 11–1** for the signature card used to open the checking account for Zip Delivery Service.

The Checkbook.

When a depositor opens a checking account, checks are printed for the depositor's use. Printed checks are packaged together in a *checkbook* like the one shown in **Figure 11–2**. Each page has several detachable checks attached to check stubs, and both are numbered in sequence. Using checks with preprinted numbers helps a business keep track of every check that it writes, an important internal control.

The ABA Number.

In addition to having a preprinted check number, each check is printed with the account number and an *American Bankers Association (ABA) number*. The ABA number is the number printed in the upper right corner of a check, just below the check number. The ABA number identifies the bank and speeds the hand sorting of checks.

Look at the ABA number on the check in **Figure 11–3**. The number above the line and to the left of the hyphen represents the city or state where the bank is located. The number to the right of the hyphen indicates the specific bank. The number below the line is the code for the Federal Reserve district where the bank is located.

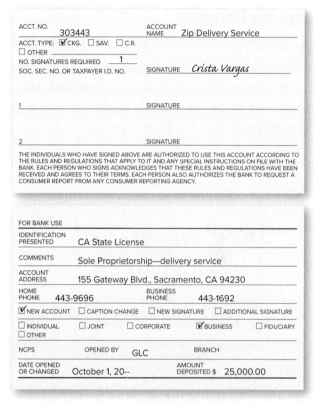

Figure 11–1 Checking Account Signature Card

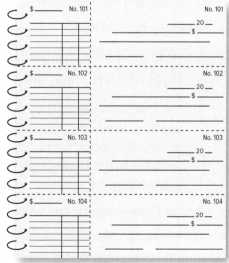

Figure 11–2 Checkbook

The ABA number was developed to speed the sorting of checks by hand. An updated version of the ABA number is also printed on the bottom of each check for electronic sorting. The ABA number, the account number, and the check number are printed at the bottom of the check in a special ink and typeface. These specially printed numbers are called *MICR (magnetic ink character recognition) numbers.* Can you identify the MICR numbers on the check in **Figure 11–3?**

Making Deposits

A business makes regular deposits to protect the currency, coins, and checks it receives. Most businesses make daily deposits.

Deposits are accompanied by a **deposit slip**, a bank form listing the cash and checks to be deposited. The deposit slip, also called a *deposit ticket,* gives both the depositor and the bank a detailed record of the deposit. Most banks provide printed deposit slips with the depositor's name, address, and account number. A deposit slip for Zip Delivery Service is shown in **Figure 11–4**.

Figure 11–3 Printed Check

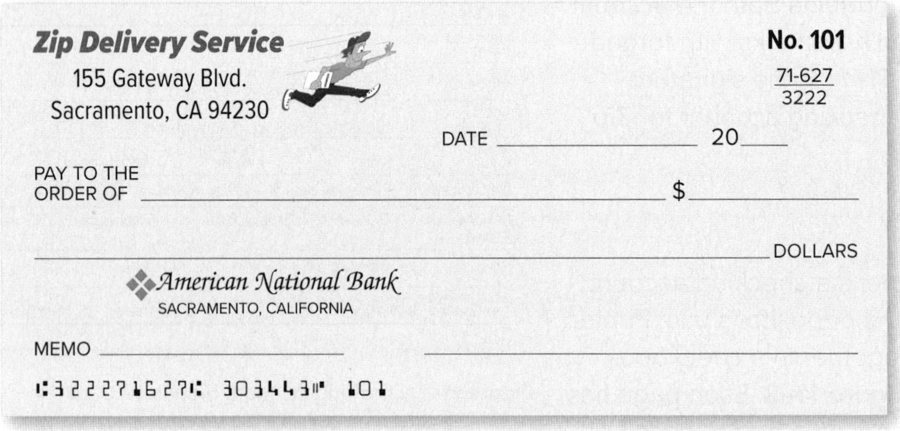
Zip Delivery Service
155 Gateway Blvd.
Sacramento, CA 94230

No. 101
71-627
3222

DATE ______________ 20____

PAY TO THE ORDER OF ______________________________ $__________

____ DOLLARS

American National Bank
SACRAMENTO, CALIFORNIA

MEMO ______________ ______________________

⑆322271627⑆ 303443⑈ 101

To complete a deposit slip, follow these steps:

1. Write the date on the Date line.
2. On the Cash line, indicate the total amount of currency and coins.
3. List checks separately by their ABA numbers. Write only the number that appears above the line in the ABA number, including the hyphen. If there are many checks, list the checks by amount on a calculator tape and attach the tape to the deposit slip. On the first Checks line, write "See tape listing," followed by the total amount of the checks.
4. Add the amounts, and write the total amount on the Total line.

The checks are arranged in the order listed on the deposit slip. The deposit slip and the cash and checks are handed to a bank teller. The teller verifies the deposit and gives the depositor a receipt. The deposit receipt is usually a machine-printed form, although it may be a copy of the deposit slip stamped and initialed by the teller.

Figure 11–4 Deposit Slip

Endorsing Checks.

A check is a form of property. When a business receives a check, it acquires the right to that check. To deposit the check in a checking account, the depositor endorses the check to transfer ownership to the bank. An **endorsement** is an authorized signature written or stamped on the back of a check that transfers ownership of the check.

A business can use three types of endorsements when transferring ownership of a check as shown in **Figure 11–5** on page 313:

- A **blank endorsement** includes only the signature or stamp of the depositor. It transfers ownership of the check, but it does not indicate who the new owner is. This is not a safe endorsement because the check can be cashed by anyone who presents it for payment.
- A **special endorsement** transfers ownership of the check to a specific individual or business.
- A **restrictive endorsement** transfers ownership to a specific owner and then limits, or restricts, how a check may be handled even after ownership is transferred. To protect checks from being cashed by anyone else, Zip uses a restrictive endorsement that reads "For Deposit Only." Zip stamps the endorsement on the back of each check as soon as it is received.

Recording Deposits.

The check stubs in the checkbook are a record of the **Cash in Bank** account. That is, the completed stubs reflect all checking account transactions: payments, deposits, and bank service charges.

To see how to record a deposit in the checkbook, refer to **Figure 11–6** and follow these steps:

1. Enter the date of the deposit on the check stub for the next unused check. Use the Add deposits line.
2. Enter the total amount of the deposit on the same line in the Dollars and Cents (amount) columns.
3. Add the deposit amounts to the amount on the Balance brought forward line. Enter the result on the Total line. This total is the new checkbook balance.

Figure 11–5
Endorsements

Reading Check

What are the three types of endorsements? Which type provides the safest handling?

Writing Checks

Writing a check is a simple procedure. You need to follow a few important rules, however, to ensure correct recordkeeping and proper handling of the money represented by the check.

- Write checks in ink, or prepare typewritten or computer-generated checks. Some businesses use a check-writing machine that perforates the amount of the check in words on the Dollars line. These perforations protect a check from alteration. Checks written in pencil are not acceptable because they can be easily altered.
- Complete the check stub *before* writing the check. This reduces the chance of forgetting to complete the stub.

Completing the Check Stub.

Because the stub serves as a permanent record of the check, it must be complete and accurate. A check stub has two parts. The upper part summarizes the details of the cash payment transaction. The lower part is a record of how the transaction affects the checking account. To see how to complete the check stub, refer to **Figure 11–7** and follow these steps:

1. In the upper part of the stub, enter the amount of the check, the date, the name of the payee (on the To line), and the purpose of the check (on the For line). A **payee** is the person or business to which a check is written.
2. If it has not already been done, enter the balance on the Total line on the lower part of the stub.
3. Enter the amount of the check on the Less this check line. This amount is the same as the amount shown on the first line in the upper part of the stub.
4. Subtract the check amount from the total. Enter the new balance on the Balance carried forward line.
5. Enter the new balance on the first line of the bottom part of the *next* check stub, the Balance brought forward line.

Figure 11–6 Recording a Deposit in the Checkbook

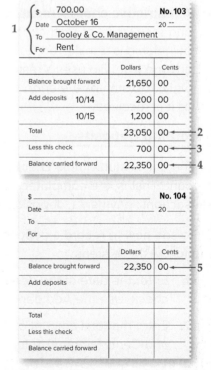

Figure 11–7 Filling Out the Check Stub

314 Chapter 11 • Cash Controls and Banking Activities

Filling Out the Check.

To see how to write a check, refer to **Figure 11–8** and follow these steps:

1. Write the date on which the check is being **issued**.

2. Write the payee's name on the Pay To The Order Of line. Start the payee's name as far left as possible.

3. Enter the amount of the check in numbers. Write clearly, and begin the first number as close to the printed dollar sign as possible.

4. On the next line, write the dollar amount of the check *in words*. Start at the left edge of the line. Write any cents as a fraction. For example, write 22 cents as 22/100. Draw a line from the cents fraction to the word *Dollars*.

5. Sign the check. Only an authorized person—one who has signed the signature card for the account—may sign the check. The person who signs a check is the **drawer**. The bank on which the check is written is the **drawee**.

The checkbook is now ready for the next transaction.

Voiding a Check.

If an error is made while writing a check, the check is marked "Void" in large letters across the front (in ink). This is known as **voiding a check**. A new check is then prepared. When a check is voided, the stub is also voided.

Since a business needs to account for each check used, a voided check is never destroyed. Instead it is filed with the business records. A voided check may be placed in a special file, or it may simply be folded and stapled to the check stub.

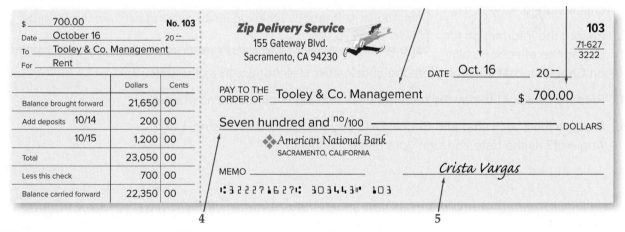

Figure 11–8 Completed Stub and Check

SECTION 11.1
Assessment

After You Read

Reinforce the Main Idea

Create a chart like this one to list the activities involved in maintaining a checking account. Identify how each activity helps protect cash. Add more rows as needed.

Activity	How This Activity Protects Cash

Problem 1 Preparing a Deposit Slip and Writing Checks

On August 14 Loretta Harper, owner of Peabody Cards and Gifts, deposited the following items in the checking account of the business.

Cash: $784.29

Checks: Charles Ling, drawn on American Bank of Commerce, ABA No. 32-7091; $39.44

Keith Lopez, drawn on People's Bank, ABA No. 84-268; $22.95

Marjorie Luke, drawn on Horizon Federal Savings and Loan, ABA No. 84-6249; $52.95

Mable Parker, drawn on Security National Bank, ABA No. 84-2242; $67.45

On August 15 Peabody received the July bill from Northeast Telephone for $214.80.

On August 17 Peabody received an advertising bill from the Bayside News for $275.00.

Instructions Using the forms provided in your working papers:

1. Complete a deposit slip.
2. Record the deposit on the check stub for Check 41.
3. Record the information for paying the telephone bill on Check Stub 41 and complete the check stub. Use August 16 as the date.
4. Prepare Check 41 to pay the telephone bill and sign your name as drawer.
5. Prepare the check stub and Check 42 to pay the bill for advertising. Use August 17 as the date and sign your name as drawer.

continued

SECTION 11.1
Assessment

Math for Accounting

Assume you work for a health food store that prepares two deposits each day. Using the following cash register summaries, determine how much cash should be deposited after the first shift ended at 3:00 p.m. and after the second shift ended at 9:00 p.m.

	First Shift (9:00 a.m. – 3:00 p.m.)			Second Shift (3:00 p.m. – 9:00 p.m.)		
	Cash Register			Cash Register		
	1	2	3	1	2	3
Coins	35.40	29.35	18.75	24.62	19.21	14.32
Currency	395.00	425.00	300.00	218.00	349.00	216.00
Checks	900.00	875.00	725.00	725.00	645.00	829.00

SECTION 11.2
Reconciling the Bank Statement

Have you ever been surprised to find that your bank account had less money than you thought because you forgot to record a withdrawal? If so, you realize the importance of reconciling (bringing into agreement) your checkbook with your bank statement. Business owners and their accountants also need to keep up-to-date records of cash.

Proving Cash

What Is Meant by Proving Cash?

The balance in the **Cash in Bank** account in the general ledger is regularly compared with the balance in the checkbook. If all cash receipts have been deposited, all cash payments have been made by check, and all transactions have been journalized and posted, the **Cash in Bank** account balance should agree with the checkbook balance. Comparing these two cash balances regularly is part of the internal control of cash. Some businesses prove cash daily or weekly, while others prove cash on a monthly basis.

If the **Cash in Bank** balance does not agree with the checkbook balance, and the trial balance has been proved, the error is probably in the checkbook. The following checkbook errors are the most common:

- faulty addition or subtraction
- failure to record a deposit or a check
- a mistake in copying the balance forward amount to the next check stub

If an error is made in the checkbook, the proper place to enter the correction is on the next unused check stub. For example, suppose Check 22 for $84.60 was recorded on the check stub as $48.60. The error is found when cash is proved. By this time several other checks have been written, so the next unused check stub is 31. In this case the amount of the error ($84.60 − $48.60 = $36.00) is subtracted from the balance brought forward on Check Stub 31 (see **Figure 11–9** on page 319). A note is made on Check Stub 22 to indicate that the error is corrected on Check Stub 31.

The Bank Statement

What Is a Bank Reconciliation?

A **bank statement** is an itemized record of all transactions in a depositor's account over a given period, usually a month. Typical bank statements include the following information:

1. the checking account balance at the beginning of the period
2. a list of all deposits made by the business during the period

Section Vocabulary

- bank statement
- canceled checks
- imaged checks
- reconciling the bank statement
- outstanding checks
- outstanding deposits
- bank service charge
- stop payment order
- NSF check
- Check 21
- postdated check
- electronic funds transfer system
- bankcard
- automated teller machine (ATM)
- funds

3. a list of all checks paid by the bank
4. a list of any other deductions from the depositor's account
5. the checking account balance at the end of the period

Find each of these items on the bank statement for Zip Delivery Service in **Figure 11–10**.

With the bank statement, a bank may include the **canceled checks** that it paid and deducted from the depositor's account. Instead of the actual checks, banks are increasingly sending images of the checks or simply a list of them. These **imaged checks** or *substitute checks* (copies of originals) or the list are used to verify the information on the bank statement. These checks and the bank statement should be kept in a file or storage box in case they are needed later as proof of payment or for other reasons.

Upon receipt, the bank statement is compared to the checkbook. The process of determining any differences between the bank statement and the checkbook is called **reconciling the bank statement**. It is also known as a *bank reconciliation*. The ending balance on the bank statement seldom agrees with the balance in the checkbook. There are three common reasons that the bank statement balance and the checkbook balance disagree:

- outstanding checks
- outstanding deposits
- bank charges

	Dollars	Cents
Balance brought forward	4,750	00
Add deposits		
Less: Error corr. Ck. 22	36	00
Total	4,714	00
Less this check		
Balance carried forward		

No. 31
Date _____ 20 __
To _____
For _____

Figure 11–9 Entering an Error Correction on the Check Stub

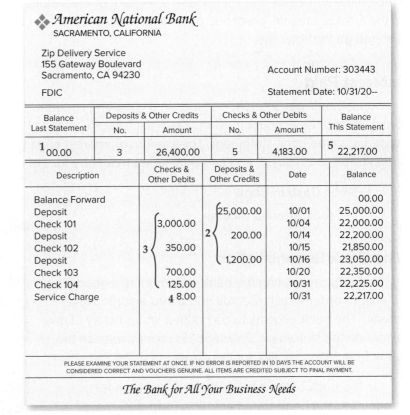

Figure 11–10 Bank Statement

Outstanding Checks and Deposits

In banking terms the word *outstanding* simply means "not yet received." **Outstanding checks**, therefore, are checks that have been written but have not yet been presented to the bank for payment. It is not unusual for checks written in one statement period to reach the bank in a later period. **Outstanding deposits** are deposits that have been made and recorded in the checkbook but do not appear on the bank statement. A deposit made the same day the bank statement is prepared may not appear on the statement.

Bank Service Charges

The bank statement balance also reflects any service charges made by the bank during the statement period. Banks impose a **bank service charge**, which is a fee for maintaining bank records and processing bank statement items for the depositor. This charge varies from bank to bank. It is frequently based on either the number of checks and deposits handled during the statement period or the balance in the depositor's account. The bank subtracts the service charge from the depositor's account. The depositor usually does not know about the service charge, or other bank charges handled in the same manner, until the statement is received.

	Dollars	Cents
Balance brought forward	21,125	00
Add deposits		
Less: Svc. Charge	−8	00
Total	21,117	00
Less this check		
Balance carried forward		

No. 107

Figure 11–11 Entering a Bank Service Charge on the Check Stub

Before the bank statement is reconciled, the checkbook balance is adjusted by the amount of the bank service charge. As shown in **Figure 11–11**, the words *Less: Service Charge* are written on the next unused check stub on the line above the Total line. The amount of the service charge is entered in the amount column, preceded by a minus sign. The balance is recalculated and entered on the Total line.

Interest Paid

Some banks pay interest on funds in a checking account. This is not a common practice, and the account must maintain a minimum balance to qualify. The interest appears on the bank statement. The amount of interest must be recorded in the checkbook, journalized, and posted. The journal entry for interest paid follows:

    Debit **Cash in Bank**
        Credit **Interest Income**

The Bank Reconciliation

Promptly reconciling the bank statement is a good way to ensure orderly cash records and guard against cash losses. The bank expects to be notified immediately of any errors on the statement. Failure to do so may release the bank from responsibility for the errors.

On the back of the bank statement is a form for reconciling the bank statement. This form documents the differences between the bank balance and the

checkbook balance. Refer to **Figure 11–12** and follow these steps to reconcile a bank statement:

1. Arrange the canceled checks in numerical order. Compare the canceled checks with those listed on the statement and with the stubs. When you match a check and a stub, place a check mark beside the check amount on the bank statement and on the check stub. The stubs without check marks represent outstanding checks. List the outstanding check numbers and amounts on the bank reconciliation form.

2. Enter the ending balance shown on the bank statement.

Figure 11–12 Bank Reconciliation Using a Bank Statement Form

3. Compare deposits listed on the bank statement to deposits listed in the checkbook. Enter the total of any outstanding deposits on the reconciliation form. Add this total to the bank statement balance and enter the result on the form.

4. Subtract the total of the outstanding checks from the amount calculated in Step 3. The result is the *adjusted bank balance*.

5. Compare the adjusted bank balance to the checkbook balance. When the balances match, the bank statement is reconciled.

Figure 11–13 illustrates a two-column account form.

If the adjusted bank balance does not match the checkbook balance, find and correct the error. Notify the bank immediately if it is a bank error. It is more likely, however, that the error is in the checkbook. Check the addition and subtraction on the check stubs and on the bank reconciliation form. Also look for any outstanding checks or deposits that have not been included in your calculations.

Figure 11–13 Bank Account Reconciliation Using a Two-Column Account Form

Journalizing Bank Service Charges

Like any other business, banks charge fees for their services. A bank service charge is an expense that is recorded in the accounting records.

The bank deducted the service charge from Zip's account, so it is not necessary to write a check for this expense. The bank statement is the source document for recording the bank service charge.

Business Transaction

On November 1 Zip received the bank statement (Figure 11–10 on page 319). A bank service charge of $8 appeared on the statement.

Analysis	Identify	1. Bank service charges are often recorded in the **Miscellaneous Expense** account. Therefore, **Miscellaneous Expense** and **Cash in Bank** are affected.
	Classify	2. **Miscellaneous Expense** is an expense account. **Cash in Bank** is an asset account.
	+/−	3. **Miscellaneous Expense** is increased by $8. **Cash in Bank** is decreased by $8.
Debit-Credit Rule		4. Increases in expense accounts are recorded as debits. Debit **Miscellaneous Expense** for $8.
		5. Decreases in asset accounts are recorded as credits. Credit **Cash in Bank** for $8.

T Accounts 6.

Miscellaneous Expense		Cash in Bank	
Debit + 8	Credit −	Debit +	Credit − 8

Journal Entry 7.

GENERAL JOURNAL PAGE 42

DATE		DESCRIPTION	POST. REF.	DEBIT	CREDIT
20–					
Nov.	1	Miscellaneous Expense		8 00	
		Cash in Bank			8 00
		Bank Statement			

Special Banking Procedures

Checks are usually written or received and deposited without any problems. However, three problems may occur:

- A business does not want the bank to pay a check that was issued.
- A business receives and deposits a check from a customer whose account does not have enough money to cover the check.
- A customer presents a check that has a date in the future.

Stopping Payment on a Check.

Occasionally, a drawer (Zip) orders the drawee (bank) not to honor, or pay, a check. A **stop payment order** is a demand by the drawer, usually in writing, that the bank not honor a specific check. A stop payment order is often used when a check is lost. The bank must receive the stop payment order *before* the check is presented for payment. Otherwise, it is too late.

To record a stop payment order, the accountant writes the words *Stopped Payment* on the check stub for the stopped check. The accountant then adds the amount of the stopped check on the next unused check stub, illustrated in **Figure 11–14**. If appropriate, the accountant then issues a replacement check.

Most banks charge a fee for a stop payment order. The fee appears on the bank statement. A journal entry, similar to that made for the bank service charge, is prepared. Most businesses record the fee in **Miscellaneous Expense**. The source document for this entry is the bank statement.

Recording NSF Checks.

An **NSF check** is a check returned to the depositor by the bank because the drawer's checking account does not have enough **funds** to cover the amount. *NSF* stands for *Not Sufficient Funds.* An NSF check is also known as a *dishonored* or *bounced* check.

Check Stub for Stopped Check

	No. 633
$ 500.00	
Date January 12	20 --
To Smith Engineering	
For Electrical Work	

	Dollars	Cents
Balance brought forward	12,723	00
Add deposits		
Total	STOPPED PAYMENT	
Less this check	500	00
Balance carried forward	12,223	00

Next Unused Check Stub

	No. 652
$	
Date	20
To	
For	

	Dollars	Cents
Balance brought forward	10,452	00
Add deposits Stopped Payment on Ck 633	500	00
Total	10,952	00
Less this check		
Balance carried forward		

Figure 11–14 Recording a Stopped Check on the Check Stub

Suppose that Burton Company (the drawer) wrote a check to Zip (the payee) to pay for delivery services. What if Burton's account does not have sufficient funds to cover the amount of the check? American National Bank (the drawee) has already shown this check as being deposited to Zip's account. When the bank finds out that Burton does not have enough funds to cover this check, it deducts the amount from Zip's account. The bank also sends the check back to Zip (not to Burton).

When the bank returns an NSF check, Zip subtracts the amount of the dishonored check from the checkbook balance. Zip also makes a journal entry to record the returned check. At this point Zip has not yet been paid for the delivery services and must go back to Burton Company to collect payment. Burton can then deposit enough money in its own bank to cover the check or find another way to pay.

The *Check Clearing for the 21st Century Act,* known as **Check 21**, went into effect in 2004. It allows the conversion of a paper check to an electronic image that can be processed quickly. A bank can now pay a check on the day it is written, instead of several days later. However, Check 21 does not require banks to process deposits more quickly. As a result, the speed of check processing could cause an increase in the number of NSF checks.

Postdated Checks.

A business might accept a check that has a future date instead of the actual date. This check is called a **postdated check**. It should not be deposited until the date on the check. Businesses sometimes accept postdated checks as a convenience to customers.

Reading Check

Summarize

What should you do if your bank sent you a dishonored check?

Business Transaction

On November 15 a check for $450, written by Burton Company for payment on account and deposited by Zip, was returned by the bank because of insufficient funds in Burton's account.

Analysis	Identify	1. **Accounts Receivable—Burton Company** and **Cash in Bank** are affected.
	Classify	2. **Accounts Receivable—Burton Company** and **Cash in Bank** are both asset accounts.
	+/−	3. **Accounts Receivable—Burton Company** is increased by the amount of the check returned by the bank, $450. **Cash in Bank** is decreased by $450.
Debit-Credit Rule		4. Increases in asset accounts are recorded as debits. Debit **Accounts Receivable—Burton Company** for $450.
		5. Decreases in asset accounts are recorded as credits. Credit **Cash in Bank** for $450.
T Accounts		6.

```
        Accounts Receivable—
          Burton Company                    Cash in Bank
        Debit      Credit              Debit         Credit
          +           −                  +              −
         450                                          450
```

continued

324   Chapter 11 • Cash Controls and Banking Activities

Journal Entry 7.

	GENERAL JOURNAL			PAGE 43	
DATE	DESCRIPTION	POST. REF.	DEBIT	CREDIT	
20—					1
Nov. 15	Accts. Rec.–Burton Company		450 00		2
	Cash in Bank			450 00	3
					4

Electronic Funds Transfer System

What Is EFTS?

Look at **Figure 11–15** on page 326. It illustrates the route a check follows from the time it is written until the time it is returned with the bank statement.

Since millions of checks are written each day, the transfer of checks and funds is routine. This transfer of funds among banks is a huge job, however. Banks use the **electronic funds transfer system (EFTS)** to handle such a large volume of transfers. The EFTS allows banks to transfer funds among accounts quickly and accurately without the exchange of checks.

The EFTS has a tremendous impact on banking activities:

- **Direct payroll deposit.** Employers can electronically transfer employees' pay to each employee's bank account.
- **Automated bill paying.** A depositor can authorize the bank to transfer funds from his or her checking account to the creditor's bank account.
- **Bankcards.** A **bankcard**, also known as an *ATM card,* is a bank-issued card that can be used at an **automated teller machine (ATM)** to conduct banking activities. An ATM is a computer terminal outside a bank or at a different location entirely. When a bankcard can be used for transactions at other businesses besides the bank, it is called a *debit card.*
- **Bank-by-phone service.** Account holders can complete transactions with their banks' computer systems by telephone.
- **Online banking.** Using the Internet, an account holder can access the bank's Web site to conduct banking transactions.

When using any electronic banking procedure, be sure to record all transactions to avoid errors in the checking account.

Routing of Checks

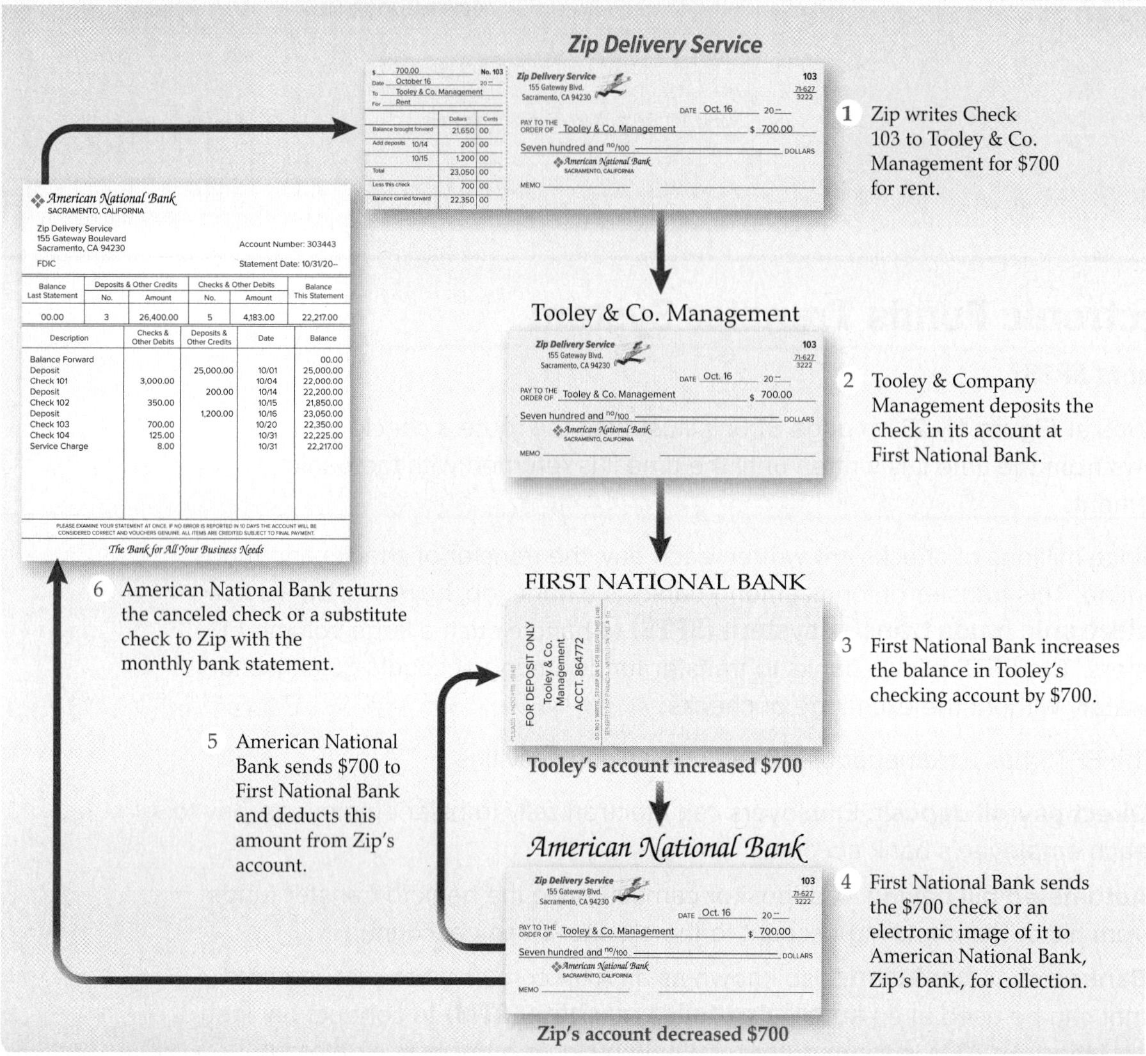
Zip Delivery Service
1 Zip writes Check 103 to Tooley & Co. Management for $700 for rent.
Tooley & Co. Management
2 Tooley & Company Management deposits the check in its account at First National Bank.
FIRST NATIONAL BANK
3 First National Bank increases the balance in Tooley's checking account by $700.
Tooley's account increased $700
American National Bank
4 First National Bank sends the $700 check or an electronic image of it to American National Bank, Zip's bank, for collection.
Zip's account decreased $700
5 American National Bank sends $700 to First National Bank and deducts this amount from Zip's account.
6 American National Bank returns the canceled check or a substitute check to Zip with the monthly bank statement.

Figure 11–15 Routing of Checks

SECTION 11.2
Assessment

After You Read

Reinforce the Main Idea

Create a chart like this one to list the adjustments that might be needed to reconcile the bank statement. Add answer rows as needed.

Reconciling the Bank Statement

Checkbook Balance Adjustments	Bank Statement Balance Adjustments

Problem 2 Analyzing a Source Document

Instructions Review the Global Travel Agency bank statement and answer the following questions in your working papers.

1. What is the amount of the returned check?
2. How much did the bank charge Global Travel Agency for the returned check?
3. What account will be debited for the $12 bank service charge?

Security National Bank
155 Flower Street, Cambridge, MA 02138

STATEMENT

Global Travel Agency
200 Brattle Street
Cambridge, MA 02138

Account Number: 2649-84
Statement Date: 2/28/20--

FDIC

Balance Last Statement	Deposits & Other Credits		Checks & Other Debits		Balance This Statement
	No.	Amount	No.	Amount	
2,714.00	3	2,395.00	7	4,036.00	1,073.00

Checks & Other Debits	Deposits & Other Credits	Date	Balance
	1,250.00	2/05	3,964.00
700.00		2/10	3,264.00
900.00		2/10	2,364.00
	845.00	2/15	3,209.00
600.00		2/15	2,609.00
(R) 100.00		2/20	2,509.00
(S) 25.00		2/20	2,484.00
725.00		2/25	1,759.00
600.00		2/25	1,159.00
(S) 12.00		2/26	1,147.00
214.00		2/26	933.00
	300.00	2/27	1,233.00
160.00		2/27	1,073.00

PLEASE EXAMINE YOUR STATEMENT AT ONCE. IF NO ERROR IS REPORTED IN 10 DAYS THE ACCOUNT WILL BE CONSIDERED CORRECT AND VOUCHERS GENUINE. ALL ITEMS ARE CREDITED SUBJECT TO FINAL PAYMENT.

C=CERTIFIED CHECK T=DEBIT OR CREDIT S=SERVICE CHARGE L=LIST CR=OVERDRAFT R=RETURNED CHECK

Math for Accounting

The balance in the checkbook of Valleyview Tennis Center on April 30 is $2,944.20. The balance shown on the April bank statement is $3,085.95. A deposit of $345.00 was made on April 29, and another deposit of $290.00 was made on April 30. Neither of these deposits appears on the bank statement. The service charge for the month was $5.25. Valleyview has four outstanding checks:

Check 344	$202.00
Check 346	55.00
Check 350	$ 25.00
Check 351	500.00

Instructions

1. Record the bank service charge in the checkbook.
2. Reconcile the bank statement.

Chapter 11
Visual Summary

Concepts
The following forms are used to maintain a checking account:

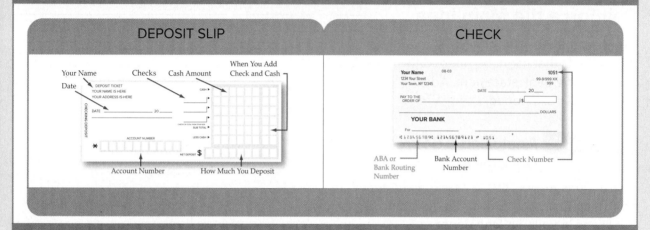

Analysis
Prepare a check.

- Write the payee's name on the Pay To The Order Of line starting as far left as possible.
- On the next line, write the dollar amount in words. Write any cents as a fraction.
- Write the issue date
- Write the amount of the check in numbers clearly, and begin the first number as close to the printed dollar sign as possible.
- Sign the check.

Procedures
Journalize the bank statement reconciliation.

DATE	DESCRIPTION	POST. REF.	DEBIT	CREDIT
	Miscellaneous Expense		x x xx	
	Cash in Bank			x x xx
	4/31/20—Bank Statement			

Chapter 11
Review and Activities

Answering the Essential Question

How do banks help us manage our money?

Whether you're in business or you're just managing your own money, you need to understand and utilize the services of a bank. What are some of the banking services you or your family use or have used?

Vocabulary Check

1. **Vocabulary** Prepare a banking glossary to help a younger student understand banking activities. Write a definition in your own words for each content vocabulary term. Add an illustration whenever it would make the definition clearer.

- check
- checking account
- depositor
- external controls
- internal controls
- reliable
- deposit slip
- signature card
- blank endorsement
- endorsement
- payee
- restrictive endorsement
- special endorsement
- drawer
- drawee
- issue
- voiding a check
- bank statement
- canceled checks
- imaged checks
- reconciling the bank statement
- bank service charge
- outstanding checks
- outstanding deposits
- Check 21
- funds
- NSF check
- postdated check
- stop payment order
- electronic funds transfer system (EFTS)
- automated teller machine (ATM)
- bankcard

Concept Check

2. **Synthesize** Describe at least four types of internal controls a business might use to protect funds. Include one that isn't discussed in this book.

3. **Analyze** Explain how a checking account can help you protect your money. How might high-security bank check stock increase security of a business's finances?

4. **Evaluate** Why is it important to keep track of checks written? Describe the process used for doing this.

5. Explain how to properly write a check. Why should the check be written in ink or by typewriter, computer, or check-writing machine?

6. How often should a business deposit cash as part of its internal controls?

7. Describe how to reconcile a bank statement. Explain why timely reconciliation is important.

Chapter 11
Review and Activities

8. What is a bank service charge? How does a business record these charges in its records?

9. How do banks use the electronic funds transfer system (EFTS)? What are the benefits of the EFTS?

10. **Math** You work at a grocery store and have rung up a customer's purchases. The total on the cash register tape reads $200, and the customer complains that there is obviously an error in the total. You find an overcharge caused by inputting the cost for fish fillets at $10 ten times instead of one time. How much must you subtract from the $200 total to get the correct total? What is the new total?

11. **English Language Arts** Explanatory paragraphs are often structured as a statement followed by examples. Imagine that you own and manage an import/export business. Write a paragraph explaining to your employees how to prepare for a meeting with clients from another country.

Chapter 11
Standardized Test Practice

Multiple Choice

1. Which of these is not an internal procedure designed to protect cash?

    a. Making any necessary payments using cash

    b. Limiting the number of people who handle cash

    c. Using a cash register and a safe

    d. Taking cash receipts to the bank frequently

2. The first step in balancing a checkbook is

    a. subtracting fees on the bank statement from the checkbook balance.

    b. comparing checks in the check record with those on the statement.

    c. adding interest earned to the checkbook balance.

    d. adding recent deposits to the bank statement balance.

3. If a check is cancelled, it means that it

    a. was written in error and has been voided

    b. has been paid to the payee

    c. has not yet been posted by the bank

    d. was returned by the bank for nonsufficient funds

4. The bank statement shows a checking account balance of $5,500. There are outstanding checks totaling $600, an outstanding deposit of $400, and a bank service charge of $15. The cash account balance should be

    a. $5,300.   b. $5,700.   c. $5,285   d. none of the above

True or False

5. Bank service charges should be journalized before reconciling the bank statement.

Short Answer

6. Name all the items needed to perform the bank reconciliation.

Chapter 11
Computerized Accounting

Reconciling the Bank Statement
Making the Transition from a Manual to a Computerized System

MANUAL METHODS

- Using the form on the back of the bank statement or accounting stationery, follow the steps on page 321 to reconcile the bank statement.

COMPUTERIZED METHODS

- Before attempting to reconcile the bank statement, make sure all transactions for the month have been posted.
- A computerized system will list all cash transactions and allow you to check off the appropriate cleared items.
- You may enter adjustments such as bank fees. These will be posted to the general ledger.

Chapter 11
Problems

Problem 3 Handling Deposits

On October 4 the owner of Wilderness Rentals deposited the following in the business checking account at First National Bank. The beginning balance in the account is $3,306.54 before these transactions.

Cash: Currency, $374.00; Coins, $7.42

Checks: Bob Warner, drawn on Consumers Bank, ABA No. 63-706; $64.98
Joan Walkman, drawn on Mountain Bank, ABA No. 63-699; $349.81
Ernesto Garcia, drawn on Progressive Savings and Loan, ABA No. 63-710; $29.44

Instructions In your working papers:

1. Place a restrictive endorsement on each check. Use "Wilderness Rentals."
2. Fill out a deposit slip. Use the ABA number to identify each check.
3. Record the deposit in the checkbook on Check Stub 651.

Analyze
Calculate the checkbook balance after the deposit is recorded on the check stub.

Problem 4 Maintaining the Checkbook

As the accounting clerk for Hot Suds Car Wash, you write checks and make deposits. The current checkbook balance is $3,486.29.

Instructions For each transaction:

1. Record the necessary information on the check stub. Determine the new balance and carry the balance forward.
2. Prepare the necessary checks and sign your name as drawer.

Date	Transactions
Oct. 3	1. Issued Check 504 for $868.45 to Custom Construction for construction supplies.
3	2. Deposited $601.35 in the checking account.
6	3. Issued Check 505 for $299.60 to CP Lumber for paint.

continued

Chapter 11
Problems

Date		Transactions
Oct. 7	4.	Issued Check 506 to Laverne Brothers for $1,000.00 for completing a painting job.
10	5.	Deposited $342.80 in the checking account.
10	6.	Issued Check 507 to Union Utilities for the September electricity bill of $175.50.

Analyze
Calculate the balance brought forward amount on Check Stub 508.

Problem 5 Reconciling the Bank Statement

On October 31 George Flaum, the accountant for Kits & Pups Grooming, received the bank statement dated October 30. After comparing the company's checkbook with the bank statement, George found the following:

1. The checkbook balance on October 31 is $960.
2. The ending bank statement balance is $1,380.
3. The bank statement shows a service charge of $10.
4. A deposit of $405 was made on October 30, but does not appear on the bank statement.
5. Check 768 for $529 and Check 772 for $306 are outstanding.

Instructions In your working papers:

1. Record the bank service charge in the checkbook.
2. Reconcile the bank statement.
3. Journalize the bank service charge in the general journal, page 4.
4. Post the bank service charge journal entry to the appropriate general ledger accounts.

Analyze
How many checks are outstanding? Identify the total amount that is outstanding.

Chapter 11
Problems

Problem 6 | Reconciling the Bank Statement

On October 31 Juanita Ortega, owner of Outback Guide Service, received a bank statement dated October 30. Juanita found the following:

1. The checkbook has a balance of $2,551.34.
2. The bank statement shows a balance of $2,272.36.
3. The statement shows a bank service charge of $20.00.
4. A check from Podaski Systems for $62.44, deposited on October 18, was returned by the bank. There is no fee for handling the NSF check.
5. A deposit of $672.48 made on October 30 does not appear on the bank statement.
6. These checks are outstanding:

   | Check 872 for | $126.84 | Check 883 for | $192.80 |
   | Check 881 for | 87.66 | Check 887 for | 68.64 |

Instructions Using the preceding information:

1. Record the service charge and the NSF check in the checkbook.
2. Reconcile the bank statement.
3. Record the service charge and NSF check on page 7 of the general journal.
4. Post the journal entries to the appropriate general ledger accounts.

Analyze
Review the Miscellaneous Expense account found in your working papers. Identify the Miscellaneous Expense account balance after posting the October 31 entry.

Problem 7 | Reconciling the Bank Statement

On October 31 Showbiz Video received the bank statement dated October 30. The accountant reviewed it and found the following:

1. The checkbook balance on October 31 is $13,462.96.
2. The ending bank statement balance is $13,883.80.
3. The bank statement shows a service charge of $17.50.
4. Deposits of $675.00 on October 28 and $925.00 on October 29 do not appear on the bank statement.
5. The following checks are outstanding:

   | Check 1766 | $125.00 | Check 1770 | $1,462.19 |
   | Check 1768 | 69.42 | Check 1771 | 381.73 |

continued

Chapter 11
Problems

Instructions In your working papers:

1. Record the bank service charge in the checkbook in your working papers.
2. Reconcile the bank statement.
3. Record the entry for the bank service charge on general journal page 13.
4. Post the bank service charge journal entry to the proper ledger accounts.

Analyze
Calculate the Cash in Bank balance after the October 31 posting. Does it agree with the adjusted checkbook balance?

Problem 8 Reconciling the Bank Statement Using the Account Form

SOURCE DOCUMENT PROBLEM

Problem 11–8

Use the source document in your working papers to complete this problem.

On October 20 Job Connect received its bank statement dated October 18.

1. The checkbook balance on October 20 is $880.84.
2. The ending bank statement balance is $344.58.
3. A $14.00 service charge appears on the bank statement.
4. The following checks are outstanding:

    | Check 864 | $ 88.41 | Check 871 | $129.88 |
    | Check 869 | 69.34 | Check 873 | 14.25 |

5. A $68.42 check from Tom McCrary deposited on October 13 was returned by the bank for insufficient funds. The bank charged Job Connect's account $7.00 for the NSF check. No journal entry was made for the NSF check.
6. A $938.72 deposit on October 19 is not on the bank statement.
7. A check for $200.00 to Fontenot Inc. was lost in the mail and has not been deposited. A stop payment order, which cost $10.00, was issued on October 15. No new check was issued.

Instructions Reconcile the bank statement using the account form in your working papers.

Analyze
Identify the account that will be debited for both the NSF check and the bank handling charge on the check.

Real-World Accounting Careers

Cliff Shepler
Brady Trane Services, Inc.

Q What do you do?

A My company provides commercial heating and cooling services, and I am responsible for the accounting and performance reporting for our New System Sales division. In addition to my accounting role, I serve as a lead on our risk management team.

Q What are your day-to-day responsibilities?

A I make sure the income statement, balance sheet and statement of cash flows produced each month are accurate and ready for upper management's review. In my risk management function, I work with the CFO to develop ways to save money on our company's insurance.

Q What factors have been key to your success?

A I have established solid working relationships with people throughout the organization. They have given me insight that I would not have received otherwise. I also have worked on projects outside of my comfort zone. That has helped me expand my responsibilities. And I'm not afraid to ask questions. By asking questions, I get a full understanding of what I'm working on and can offer viable solutions.

Q What do you like most about your job?

A I enjoy working with other people in the organization. It's interesting to get different opinions, and there is a sense of accomplishment when we solve issues together.

Career Facts

Real-World Skills
Strong communication, managerial and organizational skills; ability to collaborate effectively with internal and external stakeholders

Training and Education
A bachelor's degree in accounting or finance; MBA and CPA or CMA designation is preferred; at least seven years of experience.

Career Skills
Knowledge of U.S. GAAP; proficiency with accounting software

Career Path
Gain public accounting experience, then accept a position in a corporate environment, such as accounting manager, director of accounting or assistant controller

---

Tips from Robert Half International

In business, success is usually a team effort. Help the group get ahead by lending other coworkers a hand if you're able to. You'll not only show initiative but also gain allies who are likely to provide assistance when you need it.

College and Career Readiness

How can taking on additional responsibilities help advance your career?

Real-World Applications & Connections

CASE STUDY

Service Business: Entertainment

Dexter Shuman owns a bowling alley called Ten Pin Alley. Each night, Dexter counts the cash in the two cash registers and makes a night deposit at the local bank. For the month of May, Dexter made deposits totaling $6,400. During May, Dexter wrote checks totaling $2,900. The last three checks he wrote were Check 1408 for $180; Check 1409 for $560; and Check 1410 for $212. The beginning cash balance for the month was $13,840, which is the amount shown as the beginning balance on the May bank statement. That statement also includes a $12 service charge and an $18 charge for printing new checks.

Instructions

1. Determine the ending bank statement balance if all checks written have cleared.
2. Determine the ending bank statement balance if Checks 1408, 1409, and 1410 are outstanding.
3. Calculate the balance of the **Cash in Bank** account.

Global Accounting

The World Trade Organization

The World Trade Organization (WTO) negotiates most trade agreements between nations. The organization is dedicated to resolving disputes, stimulating economic growth, and lowering trade barriers. It handles issues including tariffs, customs processes, professional services, e-commerce, and import licensing.

Instructions

a. Locate the Web site for the WTO and navigate to the Events Calendar.
b. List the issues the WTO is reviewing this month.

21st century skills

Online Security

Criminals use identity theft to steal money, goods, and services. The Internet has led to new opportunities for criminals, and identity theft is now a serious concern. To guard against the theft of important "identity" factors online, be careful about the Web sites you use. Never click on links in e-mails that are sent to you by individuals who are "phishing" for identity information and passwords. Furthermore, it is wise to look for the SSL (secure socket layer) protection certificate symbol—signified by an "s" after the familiar "http" header on a URL—before you give out any personal data. This caution is particularly important when any aspect of your identity or financial history is at stake.

Research And Write

a. Use the Internet or your library to define "phishing."

b. Use the Internet to research technology protecting the security of people's identities. Look for information on how consumers can protect themselves as well as what businesses can do.

Spotlight on Personal Finance

Shopping with Savvy for a Checking Account

A checking account is not only a convenience but also a necessity for keeping your cash safe. Before opening a checking account, you should shop around at various banks and ask questions.

Instructions

Use the Internet or visit a bank and find answers to these questions:

1. How much is required for an initial deposit?
2. Does the account require a minimum balance? If so, how much?
3. If the account balance drops below the minimum required, will there be a charge? If so, how much?
4. Are checks free, or must I pay for them?
5. Are there any extra service charges with this account?
6. Does this account pay interest? If so, what is the current rate?
7. Do you offer special checking accounts to students at a minimal or no cost?

H.O.T. Audit

Reconciling the Bank Statement

The accountant for Landscaping Services received the October bank statement. While proving cash and reconciling the bank statement, the accountant discovered the following:

A check for $100, written by XYZ Pest Control for payment on account to Landscaping Services, was returned by the bank to Landscaping Services for not sufficient funds in XYZ Pest Control bank account. A fee of $15 was charged by Landscaping Services bank for the NSF check. This transaction was discovered by the accountant while reviewing the cash account during the bank reconciliation process.

Instructions

Record this transaction for $115 in the general journal of Landscaping Services on a separate sheet of paper.

Mini Practice Set 2

Main Task

Set up the accounting records and complete the accounting cycle for Scholastic Success Tutoring Service.

Summary of Steps

- Open general ledger accounts.
- Analyze, journalize, and post transactions.
- Reconcile the bank statement.
- Journalize and post the bank service charge.
- Prepare a trial balance, a work sheet, and the financial statements.
- Journalize and post the closing entries.
- Prepare a post-closing trial balance.
- Create a clear and coherent oral presentation that analyzes the results of the accounting cycle.

Why It's Important

This project pulls together all of the concepts and procedures you have learned.

Completing the Accounting Cycle for a Sole Proprietorship

Scholastic Success Tutoring Service

Company Background Scholastic Success Tutoring Service is owned and managed by Lisa Adams. It has been in business for one month. The business is organized as a sole proprietorship and provides tutoring services in a number of disciplines for students from pre-kindergarten through high school. The business earns revenue from tuition charged for one-on-one instruction and special classes.

Your Job Responsibilities As the accounting clerk for this business, use the accounting stationery in your working papers to complete the following activities.

1. Open a general ledger account for each account in the chart of accounts.
2. Analyze each business transaction.
3. Enter each business transaction in the general journal. Begin on journal page 1.
4. Post each journal entry to the appropriate accounts in the general ledger.
5. Reconcile the bank statement that was received on December 31. The statement is dated December 30. The checkbook has a current balance of $9,631. The bank statement shows a balance of $9,844. The bank service charge is $15. These checks are outstanding: Check 108, $183 and Check 109, $45. There are no outstanding deposits.
6. Make any necessary adjustments to the checkbook balance.
7. Journalize and post the entry for the bank service charge.
8. Prepare a trial balance and complete the work sheet.

9. Prepare an income statement, a statement of changes in owner's equity, and a balance sheet.

10. Journalize and post the closing entries.

11. Prepare a post-closing trial balance.

Business Transactions Scholastic Success Tutoring Service began business operations on December 1 of this year.

CHART OF ACCOUNTS
TechVision.com Web Sites

ASSETS
101 Cash in Bank
110 Accts. Rec.—Carla DiSario
120 Accts. Rec.—George McGarty
140 Office Supplies
150 Office Equipment
155 Instructional Equipment

LIABILITIES
210 Accts. Pay.—Educational Software
215 Accts. Pay.—T & N School Equip..

OWNER'S EQUITY
301 Lisa Adams, Capital
305 Lisa Adams, Withdrawals
310 Income Summary

REVENUE
401 Group Lessons Fees
405 Private Lessons Fees

EXPENSES
505 Maintenance Expense
510 Miscellaneous Expense
515 Rent Expense
525 Utilities Expense

Date		Transactions
Dec.	1	Lisa Adams invested $25,000 in the business, Memo 1.
	2	Bought a cash register (Office Equipment) for $525, Check 101.
	2	Purchased $73 in office supplies, Check 102.
	5	Purchased instructional computers for $13,924, Check 103.
	5	Received $950 for private instruction, Receipt 1.
	6	Bought $8,494 of instructional materials, Invoice 395, from Educational Software on account.
	8	Billed Carla DiSario for two group classes, $36, Invoice 101.
	9	Wrote Check 104 for $850 for the December rent.
	10	Billed George McGarty $275 for special group classes, Invoice 102.
	10	Received Invoice 5495 for a $2,375 microcomputer system, for office use, bought on account from T & N School Equipment.
	11	Prepared Receipt 2 for $695 for 20 private lessons given between December 1 and December 10.
	13	Received $36 from Carla DiSario on account, Receipt 3.
	14	Sent Check 105 for $200 to Educational Software on account.
	15	Wrote Check 106 for $750 to repaint two classrooms.
	18	Lisa Adams withdrew $500 for personal use, Check 107.
	20	Sent Check 108 for the electric bill of $183.
	24	Issued Check 109 for $45 for stamps (Miscellaneous Expense).

Analyze

Identify the creditor to which Scholastic Success Tutoring Service owed the most money on December 31.

UNIT 3

Accounting for a Payroll System

A Look Back

Unit 2 described the steps in the accounting cycle.

A Look Ahead

Unit 3 will describe the process of payroll accounting

Keys to Success

Q Why is developing teamwork skills important for an accounting career?

A Accounting careers involve more than just sitting alone at a desk, typing figures into a computer and calculating tax refunds. Accountants typically participate as members of a team, using interpersonal skills to effectively work with and for others. To complete tasks in an efficient manner, team leaders set out organizational goals and collaborate with individuals and teams.

Speaking and Listening

Teamwork Learning to work cooperatively within groups can help you accomplish goals. Working in a small group, have each member select a local business to contact to ask about how it pays employees. Organize your information and present it to the class. *How many employees does each business have? Do the businesses pay salary, hourly wage, commission, or a combination of these?*

Focus on the Photo

Most employees work to earn a living, but many people also enjoy their jobs. Regardless of employees' motivations, employers need to know how to pay appropriately. **Why should you learn about payroll?**

Chapter 12 • Payroll Accounting 343

Chapter 12
Payroll Accounting

Chapter Topics

12-1 Calculating Gross Earnings

12-2 Payroll Deductions

12-3 Payroll Records

Visual Summary

Review and Activities

Standardized Test Practice

Computerized Accounting

Problems

Real-World Applications & Connections

Essential Question

As you read this chapter, keep this question in mind:

What safeguards do employers use to ensure that employee payrolls are accurate and protected?

Main Idea

Gross earnings is the total amount an employee earns in a pay period. Payroll *deductions* are amounts withheld from an employee's gross earnings.

Chapter Objectives

Concepts

C1 Explain the importance of accurate payroll records.

C2 Compute gross pay using different methods.

Analysis

A1 Explain and compute employee-paid withholdings.

A2 Compute net pay.

Procedures

P1 Prepare payroll registers.

P2 Explain the methods of distributing payroll funds.

P3 Prepare an employee's earnings record.

Real-World Business Connection

Hot Studio

Maria Guidice, founder of the design firm Hot Studio, considers herself an "accidental entrepreneur." While self-employed, she began hiring others to help her complete a growing list of projects. As she gained more clients and gradually grew her staff, Hot Studio became a "real" company. Today, Guidice's business is known for its "human-centered" design solutions, from redesigning Gap Inc.'s Web site to help tie the company's Gap, Old Navy, and Banana Republic brands together, to making Salesforce.com more user-friendly.

Connect to the Business

Guidice says the best part about running Hot Studio is getting to pick the people she works with every day. Business owners know that high morale helps employees stay in jobs, building stability and improving work quality. Benefits—health-care plans, educational assistance, retirement plans, performance bonuses, and stock options—are just a few perks that keep employees happy.

Analyze

Describe the kind of work culture and benefits that would make you happy in a job.

Focus on the Photo

As the owner of Hot Studio, Maria Guidice is responsible for making sure her employees' payroll checks include certain deductions from their earnings. **What kinds of deductions do you think are taken from paychecks?**

SECTION 12.1
Calculating Gross Earnings

In a private enterprise economy, people are free to work for any business they choose—as long as they meet the requirements for employment. Employers such as Ford Motor Company, Pier 1, and Symantec Corporation rely on their employees to operate the business and pay their employees for the services they perform. In paying their employees, businesses follow certain guidelines. For example, both federal and state laws require businesses to keep accurate payroll records and to report employees' earnings.

Most companies set up a payroll system to ensure that their employees are paid on time and that payroll checks are accurate. Get ready to learn about the payroll system and how to calculate gross earnings and payroll deductions.

Section Vocabulary

- payroll
- pay period
- payroll clerk
- gross earnings
- salary
- wage
- time card
- electronic badge readers
- commission
- piece rate
- overtime rate
- percentage

Using a Payroll System

What Is a Payroll System?

A **payroll** is a list of the employees and the payments due to each employee for a specific pay period. A **pay period** is the amount of time over which an employee is paid. Most businesses use weekly, biweekly (every two weeks), semimonthly (twice a month), or monthly pay periods.

The payroll expense is a major expense for most companies. To compute salary expenses, most businesses set up a payroll system for recording and reporting employee earnings information. A well-designed payroll system achieves two goals:

1. The collection and processing of all information needed to prepare and issue payroll checks.
2. The generation of payroll records needed for accounting purposes and for reporting to government agencies, management, and others.

Businesses with many employees often hire a **payroll clerk**. The payroll clerk responsible for preparing the payroll

- makes sure employees are paid on time,
- makes sure each employee is paid the correct amount,
- completes payroll records,
- submits payroll reports, and
- pays payroll taxes.

All payroll systems have certain tasks in common, as shown in **Figure 12–1** on page 347.

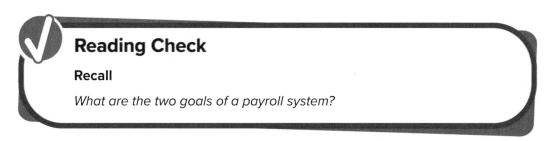

Reading Check

Recall

What are the two goals of a payroll system?

Figure 12–1 The Payroll System

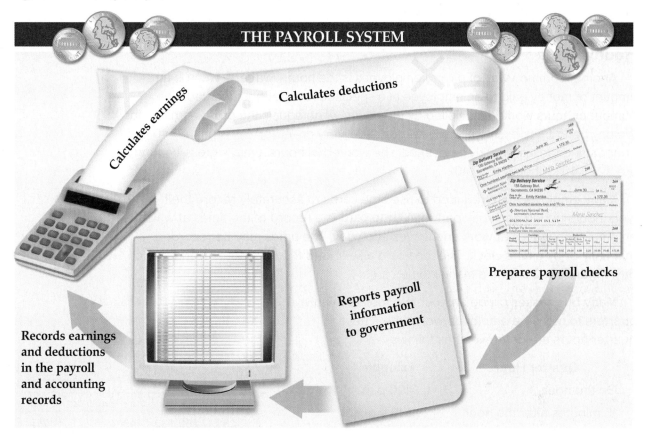

Computing Gross Pay

How Do You Calculate Gross Earnings?

Most employees are paid for the specific amount of time they work during a pay period. The total amount of money an employee earns in a pay period is the employee's **gross earnings**, or *gross pay*. The gross earnings expense is sometimes called *salary expense*. The method used to compute gross pay depends on the basis on which an employee is paid. Employees can be grouped into different pay categories:

- salary
- hourly wage
- commission
- salary plus commission or bonus
- piece rate

Some employees are entitled to *overtime pay*. Let's look at each method of paying employees.

Chapter 12 • Payroll Accounting

Salary

One common method of paying employees, especially those who are managers or supervisors, is by salary. A **salary** is a fixed amount of money paid to an employee for each pay period. In other words, an employee who is paid a salary earns the same amount regardless of the number of hours worked during the pay period. For example, Paula Ferguson, an administrative assistant, is paid a salary of $2,000 a month. Her gross earnings are $2,000 for each monthly pay period. Paula may work 160 hours in one month and 170 hours in the next, but her gross earnings for each of the two months are the same—$2,000.

Hourly Wage

Another common way of paying employees is the hourly wage. A **wage** is an amount of money paid to an employee at a specified rate per hour worked. The number of hours worked multiplied by the hourly wage equals the gross earnings for the pay period. For example, Emily Kardos, a delivery driver for Zip, is paid $7.40 per hour. During the last weekly pay period, she worked 31 hours. Emily's gross earnings are $229.40 (31 hours × $7.40).

Many employees are required to use time cards to accurately record their work hours during each pay period. A **time card** is a record of the times at which an employee arrives at work and leaves each day. The times may be recorded manually or by a time clock. The time card also shows the total hours worked each day. **Figure 12–2** shows a manual time card.

Many businesses divide an hour into four 15-minute quarters to measure employee work time. These quarter hours are determined as follows:

Quarter Hour	Example
On the hour	2:00 p.m.
15 minutes after the hour	2:15 p.m.
30 minutes after the hour	2:30 p.m.
45 minutes after the hour	2:45 p.m.

Employees seldom arrive and leave exactly on the quarter hour. As a result, some companies round arrival and departure times to the nearest quarter hour. Employees, therefore, are paid for working to the nearest quarter hour. Look at **Figure 12–2**. The times appearing on the time card for Emily Kardos for Monday will be rounded to the nearest quarter hour as follows:

Actual Time Recorded	Nearest Quarter Hour
7:58 a.m.	8:00 a.m.
12:25 p.m.	12:30 p.m.
1:32 p.m.	1:30 p.m.
4:18 p.m.	4:15 p.m.

NO.	11
NAME	Emily Kardos
SOC. SEC. NO.	201-XX-XXXX
WEEK ENDING	6/30/20--

DAY	IN	OUT	IN	OUT	IN	OUT	TOTAL
M	7:58	12:25	1:32	4:18			7¼
T	8:00	12:00	12:45	4:00			7¼
W	7:56	12:01	1:10	2:29			5¼
Th	8:01	11:55					4
F	7:45	12:02	1:05	3:58			7¼
S							
S							
					TOTAL HOURS		31

	HOURS	RATE	AMOUNT
REGULAR	31	$7.40	$240.25
OVERTIME	0		
		TOTAL EARNINGS	$229.40

SIGNATURE _Emily Kardos_ DATE _6/30/20--_

Figure 12–2 An Employee Time Card

Emily will be paid for working on Monday from 8:00 a.m. to 12:30 p.m. (4½ hours) and from 1:30 p.m. to 4:15 p.m. (2¾ hours) for a total of 7¼ hours.

Some businesses use computer technology to track employee arrival and departure times. One method uses **electronic badge readers**. The employee has an identification badge with a magnetic strip that contains employee information. The employee inserts the identification badge into a reader, which scans the magnetic strip and transfers the following information directly to the computer: the employee's name, the department or area where the employee works, and the arrival or departure time. This electronic equipment makes it fast and easy to prepare a daily printout of employee work hours.

Regardless of how employee work hours are recorded (manually, time clock, or electronic reader), business owners and supervisors check the accuracy of the hours reported and analyze the labor costs for every pay period.

Reading Check

Recall

What is the difference between a salary and a wage?

Commission

A **commission** is an amount paid to an employee based on a **percentage** of the employee's sales. Sales employees are often paid a commission to encourage them to increase their sales. For example, Joyce Torrez is paid a 5% commission on all her sales. Last week Joyce's total sales were $8,254. Joyce's gross earnings for the week are $412.70 ($8,254 × .05).

Salary Plus Commission or Bonus

Some salespeople earn a base salary plus a commission or a bonus on the amount of their sales. For example, Juan Espito, who works at a car stereo shop, is paid a salary of $200 per week plus a commission of 3% of his sales. Juan's sales were $4,810 last week. His gross earnings are $344.30 [$200 + ($4,810 × .03)].

Piece Rate

Some manufacturing companies pay employees a specific amount of money for each item the employee produces. This method of payment is called **piece rate**. Businesses often pay a low hourly rate in addition to the piece rate.

Overtime Pay

State and federal laws regulate the number of hours some employees may work in a week. Generally, employers are required to pay overtime when employees covered by these laws work more than 40 hours per week. The **overtime rate**, set by the Fair Labor Standards Act of 1938, is 1½ (1.5) times the employee's regular hourly pay rate. For example, Jesse Dubow, a photo-lab clerk at Fast Photo, worked 43 hours last week. Jesse's hourly rate of pay is $8.60. His hourly overtime rate is $12.90 ($8.60 × 1.5). His gross earnings for the week are $293.70 determined as follows:

	Hours		Rate		
Regular	40	×	$8.60	=	$344.00
Overtime	3	×	$12.90	=	38.70
Total					$382.70

Some employees who are paid a salary are also entitled to overtime pay. If a salaried employee is paid overtime, the salary is converted to an hourly rate using a standard number of hours for the period covered by the salary. Then the hourly overtime rate is calculated. For example, Jim Halley's salary is $600 per week. His hourly rate is $15 assuming a standard 40-hour work week ($600 ÷ 40). His hourly overtime rate is $22.50 ($15 × 1.5). Jim's gross earnings for a 44-hour week are $690, determined as follows:

	Hours		Rate		
Regular				=	$600.00
Overtime	4	×	$22.50	=	90.00
Total					$690.00

SECTION 12.1
Assessment

After You Read

Reinforce the Main Idea

Create a table similar to this one. For each category of employee, list the source of information for calculating gross earnings in a single pay period.

Method	Source of Information
Salary	
Hourly Wage	
Commission	
Salary plus Commission	

Problem 1 Calculating Gross Earnings

Cleary's Music Center has nine employees. The employees are paid weekly with overtime after 40 hours per week. The overtime rate is 1½ times the regular rate of pay. Payroll information interpreted from time cards for the week ending June 12 follows:

Employee	Pay Rate	Hours
David Clune	$7.75	33½
Richard Lang	$7.80	38
Jane Longas	$7.25	43
Betty Quinn	$8.30	44¼
John Sullivan	$8.30	39½
Kelly Talbert	$7.50	40
Gene Trimbell	$9.75	42½
Heidi Varney	$8.75	34¼
Kevin Wallace	$9.25	46

Instructions Prepare a form similar to the one that follows. Using the payroll information interpreted from time cards, calculate regular earnings, overtime earnings, and gross earnings for each employee.

Employee	Total Hours	Pay Rate	Regular Earnings	Overtime Earnings	Gross Earnings
Clune, David	33½	$7.75	$259.63	– 0 –	$259.63

Continued

SECTION 12.1
Assessment

Math for Accounting

You have been interviewing with several financial consulting firms for entry-level management positions, and you received two job offers. One company, Bryson Consulting, offers you a starting salary of $30,000 with a 2% bonus (on salary) if you can bring 10 new clients into the firm. The other company, The Patterson Group, offers a salary of $25,000 plus a 3% commission on all new client billings. If you bring 10 new clients to Bryson, what would the amount of your bonus be? How much in new business would you have to bring to Patterson to equal the bonus from Bryson? Which offer seems to have the higher earnings potential? Why?

SECTION 12.2
Payroll Deductions

Whether you work at Target, McDonald's, or Apple, your earnings and deductions are determined in the same way. In this section you will compute the amounts withheld from an employee's earnings.

The first time you received a payroll check, you were probably surprised that its amount was less than you expected. Various amounts had been taken out of your gross earnings. An amount subtracted from gross earnings is a **deduction**. Deductions include those required by law and those an employee wishes to have withheld from earnings.

Section Vocabulary
- deduction
- withholding allowance
- 401 (k) plan
- ensure

Deductions Required by Law

What Amounts Must Be Deducted from Earnings?

An employer is required by law to withhold payroll taxes. These taxes include the federal income tax, and the social security tax. In addition, employers often must withhold city and state income taxes.

Federal Income Tax

Most people pay the federal government a tax based on their annual income. To **ensure** that employees have the funds to pay their income taxes, employers are required to withhold a certain amount of money from each payroll check. The employer acts as a collection agent and sends the money withheld to the federal government.

The amount of income tax withheld is based on the estimated income tax the employee will actually owe. The exact income tax amount is determined when the employee prepares an income tax return. If too much money was withheld, the Internal Revenue Service (IRS) refunds the overpayment. If too little money was withheld, the employee pays the amount due when the income tax return is filed. To avoid penalties, an employee should have at least 90% of the actual tax liability deducted.

Form W-4.

The federal income tax amount withheld depends on three factors: (1) the employee's marital status, (2) the number of allowances the employee claims, and (3) the employee's gross earnings. The first two items are found on Form W-4, the Employee's Withholding Allowance Certificate. Each employee fills out a Form W-4 when starting a job and files a revised Form W-4 if the marital status or number of allowances changes. Employers keep a current Form W-4 on file for each employee.

Figure 12–3 shows the completed Form W-4 for Emily Kardos. The Form W-4 includes the employee's name, address, social security number, and marital status. The employee also lists the number of allowances claimed (refer to line 4). A **withholding allowance** reduces the amount of income tax to be withheld.

The more allowances a taxpayer claims, the lower the amount of income tax withheld from earnings. A taxpayer is usually allowed one personal allowance and one allowance for each person he or she supports (such as a child).

Some employees are *exempt* from federal income tax withholding. An employee is not required to have income tax withheld if he or she meets both of the following conditions.

- Last year I had a right to a refund of all federal income tax withheld because I had no tax liability.
- This year I expect a refund of all federal income tax withheld because I expect to have no tax liability.

Figure 12–3 Employee's Withholding Allowance Certificate—Form W-4

An exempt employee writes *Exempt* on Form W-4 so the employer will know not to withhold federal income tax.

Tax Tables.

An employee's gross earnings affect the amount withheld for federal income taxes. Most employers use IRS tables to determine the amount of federal tax to withhold. See **Figure 12–4** on page 355 for tables for single and married persons paid weekly. Other tax tables are available.

Let's use the tables to determine the tax withheld for Emily Kardos. She is single and claims no allowances. For the week ending June 30, she earned $229.40. This amount falls between $220 and $230 on the tax table for single persons. Reading across this line to the column for zero withholding allowances, you find that $26 is withheld from Emily's gross earnings.

Reading Check

List

What are the three factors that determine the amount of federal income tax withheld from employee earnings each pay period?

Social Security Tax

Employers also collect social security taxes for the federal government. Established by the Federal Insurance Contributions Act (FICA) in 1935, social security taxes are often called *FICA taxes*. FICA taxes have two components: social security and Medicare. Each is listed separately on payroll records. The social security tax funds programs that provide income to certain individuals:

1. The *old-age* and *disability* insurance programs provide income to people who are retired or disabled and their dependent children.

2. The *survivors'* benefits program provides income to the spouse and dependent children of a deceased worker.

The Medicare tax finances health insurance benefits for people who are elderly or disabled.

The FICA taxes are exact taxes in that, unlike the federal income tax, they do not involve estimation and are not affected by allowances or marital status. Congress

SINGLE Persons—WEEKLY Payroll Period

If the wages are –		And the number of withholding allowances claimed is –										
At least	But less than	0	1	2	3	4	5	6	7	8	9	10
		The amount of income tax to be withheld is –										
$0	$55	0	0	0	0	0	0	0	0	0	0	0
55	60	1	0	0	0	0	0	0	0	0	0	0
60	65	2	0	0	0	0	0	0	0	0	0	0
65	70	2	0	0	0	0	0	0	0	0	0	0
70	75	3	0	0	0	0	0	0	0	0	0	0
75	80	4	0	0	0	0	0	0	0	0	0	0
80	85	5	0	0	0	0	0	0	0	0	0	0
85	90	5	0	0	0	0	0	0	0	0	0	0
90	95	6	0	0	0	0	0	0	0	0	0	0
95	100	7	0	0	0	0	0	0	0	0	0	0
100	105	8	0	0	0	0	0	0	0	0	0	0
105	110	8	1	0	0	0	0	0	0	0	0	0
110	115	9	2	0	0	0	0	0	0	0	0	0
115	120	10	2	0	0	0	0	0	0	0	0	0
120	125	11	3	0	0	0	0	0	0	0	0	0
125	130	11	4	0	0	0	0	0	0	0	0	0
130	135	12	5	0	0	0	0	0	0	0	0	0
135	140	13	5	0	0	0	0	0	0	0	0	0
140	145	14	6	0	0	0	0	0	0	0	0	0
145	150	14	7	0	0	0	0	0	0	0	0	0
150	155	15	8	0	0	0	0	0	0	0	0	0
155	160	16	8	1	0	0	0	0	0	0	0	0
160	165	17	9	1	0	0	0	0	0	0	0	0
165	170	17	10	2	0	0	0	0	0	0	0	0
170	175	18	11	3	0	0	0	0	0	0	0	0
175	180	19	11	4	0	0	0	0	0	0	0	0
180	185	20	12	4	0	0	0	0	0	0	0	0
185	190	20	13	5	0	0	0	0	0	0	0	0
190	195	21	14	6	0	0	0	0	0	0	0	0
195	200	22	14	7	0	0	0	0	0	0	0	0
200	210	23	15	8	0	0	0	0	0	0	0	0
210	220	25	17	9	2	0	0	0	0	0	0	0
220	230	26	18	11	3	0	0	0	0	0	0	0
230	240	28	20	12	5	0	0	0	0	0	0	0
240	250	29	21	14	6	0	0	0	0	0	0	0
250	260	31	23	15	8	0	0	0	0	0	0	0
260	270	32	24	17	9	2	0	0	0	0	0	0
270	280	34	26	18	11	3	0	0	0	0	0	0
280	290	35	27	20	12	5	0	0	0	0	0	0
290	300	37	29	21	14	6	0	0	0	0	0	0
300	310	38	30	23	15	8	0	0	0	0	0	0
310	320	40	32	24	17	9	1	0	0	0	0	0
320	330	41	33	26	18	11	3	0	0	0	0	0
330	340	43	35	27	20	12	4	0	0	0	0	0
340	350	44	36	29	21	14	6	0	0	0	0	0

MARRIED Persons—WEEKLY Payroll Period

If the wages are –		And the number of withholding allowances claimed is –										
At least	But less than	0	1	2	3	4	5	6	7	8	9	10
		The amount of income tax to be withheld is –										
$0	$125	0	0	0	0	0	0	0	0	0	0	0
125	130	1	0	0	0	0	0	0	0	0	0	0
130	135	1	0	0	0	0	0	0	0	0	0	0
135	140	2	0	0	0	0	0	0	0	0	0	0
140	145	3	0	0	0	0	0	0	0	0	0	0
145	150	4	0	0	0	0	0	0	0	0	0	0
150	155	4	0	0	0	0	0	0	0	0	0	0
155	160	5	0	0	0	0	0	0	0	0	0	0
160	165	6	0	0	0	0	0	0	0	0	0	0
165	170	7	0	0	0	0	0	0	0	0	0	0
170	175	7	0	0	0	0	0	0	0	0	0	0
175	180	8	0	0	0	0	0	0	0	0	0	0
180	185	9	1	0	0	0	0	0	0	0	0	0
185	190	10	2	0	0	0	0	0	0	0	0	0
190	195	10	3	0	0	0	0	0	0	0	0	0
195	200	11	3	0	0	0	0	0	0	0	0	0
200	210	12	5	0	0	0	0	0	0	0	0	0
210	220	14	6	0	0	0	0	0	0	0	0	0
220	230	15	8	0	0	0	0	0	0	0	0	0
230	240	17	9	1	0	0	0	0	0	0	0	0
240	250	18	11	3	0	0	0	0	0	0	0	0
250	260	20	12	4	0	0	0	0	0	0	0	0
260	270	21	14	6	0	0	0	0	0	0	0	0
270	280	23	15	7	0	0	0	0	0	0	0	0
280	290	24	17	9	1	0	0	0	0	0	0	0
290	300	26	18	10	3	0	0	0	0	0	0	0
300	310	27	20	12	4	0	0	0	0	0	0	0
310	320	29	21	13	6	0	0	0	0	0	0	0
320	330	30	23	15	7	0	0	0	0	0	0	0
330	340	32	24	16	9	1	0	0	0	0	0	0

Figure 12–4 Internal Revenue Service Tax Tables

sets FICA tax rates and can change them at any time. Most employees are subject to FICA taxes, even those who are exempt from federal income taxes. The FICA tax rates are as follows:

Social security tax	6.20%
Medicare tax	1.45%
Total FICA taxes	7.65%

The social security tax is deducted from the employee's earnings until the maximum taxable earnings amount for the year is reached. This amount increases each year. For 2010, the maximum taxable earnings amount was $106,800. The maximum amount of social security tax that could be withheld from an employee in that year was $6,621.60 ($106,800 × .062). There is no maximum taxable earnings amount for the Medicare tax. For example, Lisa Gus earns $108,000 per year as a manager at a CPA firm. She has $8,187.60 in FICA taxes withheld from her earnings:

	Earnings Subject to Tax		Tax Rate		Tax
Social security tax	$106,800	×	.062	=	$6,621.60
Medicare tax	108,000	×	.0145	=	1,566.00
Total withheld					$8,187.60

State and Local Income Taxes

Most states and some cities tax the earnings of the people who live or work within their boundaries. Sometimes the withholdings are a percentage of gross earnings, like social security taxes. Amounts to be deducted also can be determined by tables similar to the ones for federal income tax.

Voluntary Deductions

What Deductions Can an Employee Choose?

Most employers agree to deduct other amounts from their employees' payroll check at the request of the employees. These deductions are withheld from each payroll check until the employee notifies the employer to stop. Voluntary employee deductions include union dues, health insurance payments, life insurance payments, pension and other retirement contributions, credit union deposits and payments, U.S. savings bonds, and charitable contributions.

The **401(k) plan** is a popular voluntary payroll deduction. The employee does not pay income tax on earnings contributed to the 401(k) plan until the money is withdrawn from the plan, usually after age 59½. In other words, taxable income is reduced by the amount of the contribution. Some employers make matching contributions to employees' 401(k) accounts.

Angelo Cappelli, a graphic artist, earns $644 each week and contributes $75 to his 401(k) account. Angelo will pay income tax on $569 ($644 − $75). He will not pay income tax on his $75 contribution until he withdraws it from his 401(k) account.

> **Common Mistakes**
>
> **Calculating Taxes** When calculating taxes, be sure to use the amount in the Total Earnings column for each employee. Do not use the amount in the Regular Earnings column because some employees may have overtime wages.

SECTION 12.2
Assessment

After You Read

Reinforce the Main Idea

Use an organizer like this one to list one major advantage and one major disadvantage, from the employee's point of view, for each type of deduction. Add as many answer rows as needed.

Legally Required			Voluntary		
Type	Advantage	Disadvantage	Type	Advantage	Disadvantage

Problem 2 Determining Taxes on Gross Earnings

Information related to the just-completed pay period of MegaCom Computer Upgrades is provided in the following chart. Determine the amounts to be withheld from each employee's gross earnings for FICA and income taxes. It is a weekly pay period, so use the tables on page 355 to determine the amount of federal income tax to be withheld. The state income tax is 2% of gross earnings. The social security tax rate is 6.2%, and the Medicare tax rate is 1.45%. Use the format shown in your working papers.

Employee	Marital Status	Allowances	Gross Earnings
Cleary, Kevin	S	0	$155.60
Halley, James	S	1	184.10
Hong, Kim	S	0	204.65
Jackson, Marvin	M	1	216.40
Sell, Richard	M	2	196.81
Total			$957.56

Problem 3 Analyzing a Source Document

Examine the following partially completed payroll check stub. The employee, Melanie Galvin, is single and claims one allowance. What amount should be deducted for:

1. Medicare tax?
2. Social Security tax?
3. Federal Income tax?

Continued

Chapter 12 • Payroll Accounting   357

SECTION 12.2
Assessment

Employee Pay Statement
Detach and retain this statement. 260

| Period Ending | Earnings ||| Deductions ||||||| Net Pay |
	Regular	Overtime	Total	Social Security Tax	Med. Tax	Federal Income Tax	State Income Tax	Hosp. Ins.	Other	Total	
1/15/20--	315.00		315.00								

Math for Accounting

As the president of Creative Craft Memory Books, you have decided to add five new sales consultants to your sales force. You offer these sales consultants a gross salary of $23,000 each, but the consultants will not take home $23,000. Calculate the FICA taxes to be withheld from each consultant's gross pay. What remains of their gross pay after you deduct FICA taxes?

SECTION 12.3
Payroll Records

In the previous sections you learned how to calculate gross earnings and deductions. Now you will learn how to compute net pay, prepare the payroll records for each period, and prepare payroll checks. Whether a business has many employees, like the Boeing Corporation, or a small staff, like a veterinarian's office, it is essential that the payroll be prepared on time and accurately.

Section Vocabulary
- payroll register
- net pay
- direct deposit
- employee's earnings record
- accumulated earnings
- previous
- assist

Preparing the Payroll Register

What Is the Purpose of a Payroll Register?

Federal and state laws require businesses to keep accurate payroll records. To help meet these requirements, businesses use a payroll register. The **payroll register** is a form that summarizes information about employees' earnings for each pay period. Let's learn how to prepare payroll registers.

Figure 12–5 shows the payroll register for Zip Delivery Service. As you can see, the payroll register lists each employee's I.D. number, name, marital status, and the number of allowances claimed. Refer to **Figure 12–5** as you read the descriptions of the payroll register.

Figure 12–5 Completed Payroll Register

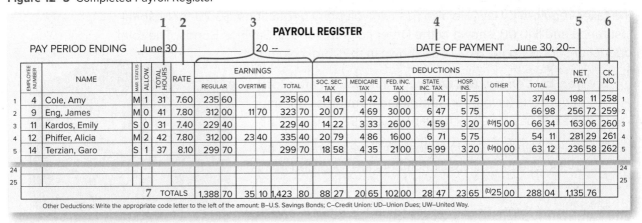

1. *Total Hours Column.* Regular and overtime hours from the employee's time card are added together, and the total is entered in this column.

2. *Rate Column.* This column shows the employee's current rate of pay, found on the employee's earnings record.

3. *Earnings Section.* The earnings section is divided into three columns:
   - regular
   - overtime
   - total earnings

   To complete these columns, the payroll clerk multiplies the hours worked by the employee's regular hourly rate or, when applicable, overtime hourly rate.

Employee 4, Amy Cole, worked 31 hours. Her regular hourly rate is $7.60. Amy earned $235.60 (31 × $7.60) for the week. Since there are no overtime hours, regular and total earnings are the same.

Alicia Phiffer, employee 12, worked 2 overtime hours in addition to 40 regular hours. Her regular hourly rate is $7.80. Her overtime hourly rate is $11.70 ($7.80 × 1.5). Her regular earnings are $312.00 (40 × $7.80) and her overtime earnings are $23.40 (2 × $11.70). Her total earnings are $309.60 ($312.00 + $23.40).

4. *Deductions Section.* The illustrated payroll register has seven deduction columns. The number of deduction columns, however, varies among businesses depending on the specific needs of each business. In the illustration, columns are provided for the deductions required by law:

    - social security tax
    - Medicare tax
    - federal income tax
    - state income tax

    Columns are also provided for voluntary deductions. Certain voluntary deductions taken by many employees on a regular basis will usually have their own columns, such as the column shown in the illustration for hospital insurance. Deductions taken less frequently are often placed in a column titled *Other*. In the illustration these deductions include credit union, union dues, savings bonds, and charitable contributions. Finally, a column is provided for the total deductions of each employee.

Look at the deductions for Garo Terzian, employee 14. Garo's deductions include the taxes required by law. He also has two voluntary deductions, $3.20 for hospital insurance and $10.00, shown in the Other column, for U.S. Savings Bonds. The total deductions for Garo are $63.12 as shown in the *Total* column.

5. *Net Pay Column.* **Net pay** is the amount left after total deductions have been subtracted from gross earnings. The net pay for Garo Terzian is $236.58.

6. *Check Number Column.* Most employees are paid by check. The payroll check numbers are recorded in this column.

7. *Column Totals.* Each amount column is totaled, and the totals are entered on the last line of the payroll register. To ensure that there are no mathematical errors, subtract the Total Deductions column total from Total Earnings column total. The result should equal the Net Pay column total.

Reading Check

Recall

How is net pay calculated?

Paying Employees

How Are Employees Paid?

Once the accuracy of the payroll register has been verified, a payroll check is prepared for each employee. Most businesses pay their employees by check as a means of cash control. When a company has only a few employees, payroll checks are written from the company's regular checking account. Companies with many employees have a separate checking account for payroll.

When a separate payroll checking account is used, funds are transferred to this account each pay period. A check for the total net pay amount is written from the company's regular checking account and deposited in the payroll checking account. Then individual payroll checks are issued to employees from the payroll account.

The payroll register is the source of information for preparing the payroll checks. Along with a payroll check, each employee is given a written or printed explanation showing how the employee's net pay is calculated. This explanation is provided on a stub attached to the payroll check. **Figure 12–6** shows a typical payroll check and stub. Notice that for Emily Kardos, the amounts on the stub are the same as the amounts in the payroll register. After each payroll check has been written, the check number is recorded in the payroll register.

Figure 12–6 Completed Payroll Check and Stub

Zip Delivery Service — 155 Gateway Blvd., Sacramento, CA 94230 — Check No. 260 — 63-947/670

Date: June 30, 20--
Pay to the Order of: Emily Kardos — $163.06
One hundred sixty-three and 06/100 Dollars

American National Bank
SACRAMENTO, CALIFORNIA

Crista Vargas

⑆067009471⑆ 3939 043 417⑈ 260

Employee Pay Statement — 260
Detach and retain this statement.

Period Ending	Earnings			Deductions							Net Pay
	Regular	Overtime	Total	Social Security Tax	Med. Tax	Federal Income Tax	State Income Tax	Hosp. Ins.	Other	Total	
6/30/20--	229.40		229.40	14.22	3.33	26.00	4.59	3.20	15.00	66.34	163.06

Employees can be paid by check or by direct deposit. The employer makes a **direct deposit** of the net pay electronically in the employee's personal bank account. No payroll check is prepared. The employee does, however, receive a printed record of the payroll calculation. Direct deposits are made through electronic funds transfer. With this system the employer informs the employee's bank of the amount to be deposited.

Reading Check

Explain

When paying employees, what is the difference between issuing a check and paying by direct deposit?

The Employee's Earnings Record

What Is the Purpose of an Employee's Earnings Record?

In addition to the payroll register, an employer must also keep an **employee's earnings record** for each employee. This record contains all of the payroll information related to an employee. **Figure 12–7** on page 363 shows an example of an employee's earnings record. The earnings record and payroll register have the same amount columns:

- three earnings columns
- four columns for deductions required by law
- additional columns for voluntary deductions (in this case, two columns)
- the Total column
- the Net Pay column

An additional column for the employee's accumulated earnings is also provided. **Accumulated earnings** are the employee's *year-to-date* gross earnings. That is the amount of the employee's gross earnings from the beginning of the year through the end of the pay period just completed.

The accumulated earnings for Emily Kardos as of June 30 are computed by adding her gross earnings for the pay period just completed to her accumulated earnings for the previous pay period as follows:

Gross Earnings for Pay Period Just Completed		Accumulated Earnings for Previous Pay Period		Accumulated Earnings for Pay Period Just Completed
$229.40	+	$4,178.60	=	$4,408.00

Businesses keep employees' earnings records on a quarterly basis. This makes it easier to complete government reports that are required each quarter. At the end of a quarter, the amount columns on each employee's earnings record are totaled. The final amount in the Accumulated Earnings column is carried forward to the top of the employee's earnings record for the next quarter. **Figure 12–7** illustrates how accumulated earnings amounts are carried forward.

As you can see from this chapter, preparing payroll is time consuming and detail oriented. A mistake that is not promptly detected and corrected can mean hours of rework. To reduce simple mathematical errors and to improve productivity, many businesses use computers and special software to prepare payroll.

Figure 12–7 Employee's Earnings Record

This amount was carried forward from the previous quarter's record.

EMPLOYEE'S EARNINGS RECORD FOR QUARTER ENDING June 30, 20--

Kardos (Last Name)   Emily (First)   M (Initial)

Address: 809 East Main Street

Sacramento, CA 94230

EMPLOYEE NO. 11   MARITAL STATUS S   ALLOWANCES 0

POSITION: Delivery Driver

RATE OF PAY: 6.75   SOC. SEC. NO. 201-XX-XXXX

PAY PERIOD		EARNINGS			DEDUCTIONS							NET PAY	ACCUMULATED EARNINGS
NO.	ENDED	REGULAR	OVERTIME	TOTAL	SOC. SEC. TAX	MEDICARE TAX	FED. INC. TAX	STATE INC. TAX	HOSP. INS.	OTHER	TOTAL		1,490 39
1	4/07/--	199 13		199 13	12 35	2 89	22 00	3 98	3 20	(B) 15 00	59 42	139 71	1,689 52
2	4/14/--	205 88		205 88	12 76	2 99	23 00	4 12	3 20		46 07	159 81	1,895 40
3	4/21/--	270 00	30 38	300 38	18 62	4 36	38 00	6 01	3 20		70 19	230 19	2,195 78
4	4/28/--	222 75		222 75	13 81	3 23	26 00	4 46	3 20		50 70	172 05	2,418 53
5	5/05/--	229 50		229 50	14 23	3 33	26 00	4 59	3 20	(B) 15 00	66 35	163 15	2,648 03
6	5/12/--	162 00		162 00	10 04	2 35	17 00	3 24	3 20		35 83	126 17	2,810 03
7	5/19/--	256 50		256 50	15 90	3 72	31 00	5 13	3 20		58 95	197 55	3,066 53
8	5/26/--	270 00	20 25	290 25	18 00	4 21	37 00	5 81	3 20		68 22	222 03	3,356 78
9	6/02/--	204 19		204 19	12 66	2 96	23 00	4 08	3 20	(B) 15 00	60 90	143 29	3,560 97
10	6/09/--	212 63		212 63	13 18	3 08	25 00	4 25	3 20		48 71	163 92	3,773 60
11	6/16/--	189 00		189 00	11 72	2 74	20 00	3 78	3 20		41 44	147 56	3,962 60
12	6/23/--	216 00		216 00	13 39	3 13	25 00	4 32	3 20		49 04	166 96	4,178 60
13	6/30/--	229 40		229 40	14 22	3 33	26 00	4 59	3 20	(B) 15 00	66 34	163 06	4,408 00
	QUARTERLY TOTALS	2,866 98	50 63	2,917 61	180 88	42 32	339 00	58 36	41 60	60 00	722 16	2,195 45	

Other Deductions: B—U.S. Savings Bonds; C—Credit Union; UD—Union Dues; UW—United Way.

This amount will be carried forward to the next quarter's record.

In a computerized system, the computer

- performs all payroll calculations,
- prepares and prints the payroll register,
- prints the payroll checks and stubs, and
- maintains the employees' earnings records.

Managerial Implications for Payroll Accounting

How Do Managers Use Payroll Information?

Wages and salaries form a large part of a company's expenses. Accurate and timely payroll records will **assist** management in planning and controlling these expenses. This information can be used in the following ways:

- To determine if overtime is justified or is a sign of possible inefficiency.
- To compare actual amounts to budgeted amounts to reveal any unplanned overtime and the reason for it.

This information helps managers investigate any amounts that were not expected or were unusual. After identifying any irregularities, managers can determine what caused them and how to resolve them.

Management should also put internal controls in place to prevent errors and fraud. For example, payroll records should be audited carefully and payroll procedures should be evaluated periodically.

SECTION 12.3
Assessment

After You Read

Reinforce the Main Idea

Use a diagram like this one to show the similarities and differences between the payroll register and the employee's earnings record.

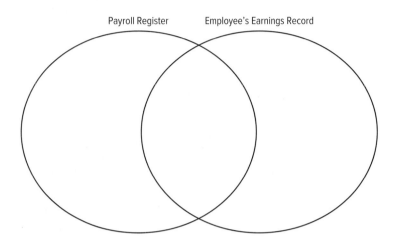

Problem 4 Preparing a Payroll Check

Use the information on the payroll register for Heather's Dance School and the form in the working papers to prepare a payroll check for Janice Burns for the week ending March 23, 20--.

PAYROLL REGISTER

PAY PERIOD ENDING March 23   20 --   DATE OF PAYMENT March 23, 20--

#	EMPLOYEE NUMBER	NAME	MAR. STATUS	ALLOW.	TOTAL HOURS	RATE	EARNINGS REGULAR	EARNINGS OVERTIME	EARNINGS TOTAL	SOC. SEC. TAX	MEDICARE TAX	FED. INC. TAX	STATE INC. TAX	HOSP. INS.	OTHER	TOTAL	NET PAY	CK. NO.
1	18	Burns, Janice	S	1	42	7.80	312 00	23 40	335 40	20 79	4 86	35 00	6 71	4 10	—	71 46	263 94	79
2																		

Math for Accounting

While reviewing the payroll records from the past two years for Sports Junction, an athletic supply store, you notice that the increasing payroll cost could be holding down overall company profits. The company's payroll costs and revenue for the last eight quarters are provided. Use spreadsheet, accounting, or graphics software to create a line chart to compare the costs and revenue graphically. Analyze the chart, and make a recommendation to Sports Junction's management about the salary levels. Do you recommend downsizing to increase profit?

	Year 1				Year 2			
	Qtr 1	Qtr 2	Qtr 3	Qtr 4	Qtr 1	Qtr 2	Qtr 3	Qtr 4
Payroll	24,000	24,500	28,670	35,280	34,000	42,000	53,000	68,000
Revenue	136,700	151,200	183,500	234,000	143,000	142,500	143,000	138,500

Chapter 12
Visual Summary

Concepts
Methods of computing gross pay:

SALARY: A fixed amount of money paid to an employee for each pay period.

HOURLY WAGE: An amount of money paid to an employee at a specified rate per hour worked.

COMMISSION: An amount paid to an employee based on a percentage of the employee's sales.

SALARY + COMMISSION/BONUS: Some sales employees get a salary as well as a commission or bonus for the sales they make.

PIECE RATE: A specific amount of money paid for each item an employee produces.

COMBINATION: Some employees are paid using more than one of these methods at the same time.

Analysis
Net pay is the amount of money actually received by the employee:

GROSS PAY − TOTAL PAYROLL DEDUCTIONS = NET PAY

Procedures
A payroll register summarizes information for all employees in a single pay period. A pay period is the amount of time over which employees are paid. The payroll register includes:

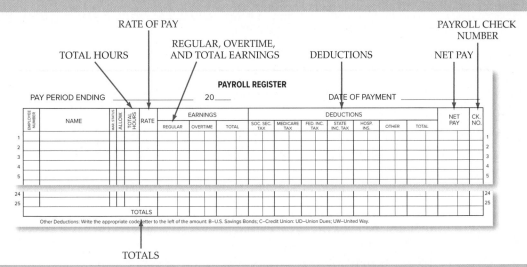

- TOTAL HOURS
- RATE OF PAY
- REGULAR, OVERTIME, AND TOTAL EARNINGS
- DEDUCTIONS
- NET PAY
- PAYROLL CHECK NUMBER
- TOTALS

Chapter 12 • Payroll Accounting 365

Chapter 12
Review and Activities

Answering the Essential Question

What safeguards do employers use to ensure that employee payrolls are accurate and protected?

If you work, your pay check is very important to you. You want to be assured there are no errors, as do your employers. What should be done to guarantee that your pay checks are accurate?

Vocabulary Check

1. **Vocabulary** Create a fill-in-the-blank test question for each content vocabulary term.

   - payroll
   - payroll clerk
   - pay period
   - gross earnings
   - salary
   - electronic badge readers
   - time card
   - wage
   - A *percentage*
   - commission
   - overtime rate
   - piece rate
   - A synonym for *ensure*
   - deduction
   - withholding allowance
   - 401(k) plan
   - payroll register
   - net pay
   - accumulated earnings
   - direct deposit
   - employee's earnings record

Concept Check

2. List five tasks that an effective payroll system performs. What are the benefits of direct deposit?
3. List and describe five employee bases or categories used to compute gross pay.
4. List the three federal taxes businesses are required to withhold from employees' wages.
5. **Analyze** How do payroll deductions affect net pay?
6. **Compare and contrast** How are legally required deductions and voluntary deductions similar? How are they different?
7. What information does the payroll register contain?
8. What is the purpose of the employee's earnings record?
9. **Evaluate** Describe some advantages of a computer payroll accounting system?
10. **Math** Alice McGuiness sells printers at Office Express. She receives a graduated commission of 5% on her first $4,000 and 8% on all sales over $4,000. Her sales total was $6,980 this past week. What is her commission for the week?
11. **English Language Arts** Conduct research to find out what kinds of public services are paid for with personal income tax. In a one page report, outline these services, tell the ways they benefit society, and tell how paying personal income tax relates to a citizen's responsibility.

Chapter 12
Standardized Test Practice

Multiple Choice

1. A certain employee is paid every week on an hourly pay basis. She worked 38.5 hours for the week, and her hourly rate of pay is $12.00. The federal and state taxes combined equal $114.02. If there are no additional deductions, what is the employee's net pay for the week?

    a. $462.00

    b. $348.00

    c. $576.02

    d. $347.98

2. Which method of paying employees involves a fixed amount paid for each period?

    a. Piece rate

    b. Commission

    c. Salary

    d. Hourly wage

3. Which form is considered the Withholding Allowances Certificate?

    a. W-2

    b. W-4

    c. 1040

    d. 1099

4. Julie earns $18.50 per hour. She is paid overtime (time and one-half) for hours worked on Sunday. The first Sunday in June she worked 6 hours, giving her 46 hours for the week. What were her gross wages if total taxes withheld equaled 18%?

    a. $851.00

    b. $1,069.67

    c. $743.33

    d. $906.50

Short Answer

5. To summarize the payroll information for the employees of a business, the accountant prepares a _____.

Extended Response

6. List and explain the four types of deductions discussed in the text.

Chapter 12
Computerized Accounting

Preparing the Payroll
Enter and Maintain Employee Information; Process Payroll

MANUAL METHODS
- Calculate employee gross earnings based on time cards or salary information.
- Calculate required and voluntary deductions for each employee.
- Complete the payroll register.
- Prepare payroll checks.
- Update employee earnings records.
- Prepare journal entries.
- Post journal entries to the general ledger.

COMPUTERIZED METHODS
- Set up information related to employee earnings and deductions in employees' records.
- Based on rate, deduction, and withholding information for each employee, the software automatically calculates gross earnings and deductions.
- The payroll checks, journal entries, and ledger postings are generated automatically.

Chapter 12
Problems

Problem 5 Calculating Gross Pay

Wilderness Rentals pays employees either an hourly wage or a salary plus commission based on rental revenue. Hourly wage employees can earn overtime. The overtime rate is 1½ times the regular hourly rate of pay for hours worked over 40 in a week.

Instructions For each of the following employees, determine the total gross pay for the pay period.

John Gilmartin

- Earns an hourly wage of $7.60.
- Worked 43 hours this week.

Arlene Stone

- Receives a salary of $250 per week plus a 3% commission on rental revenue.
- Had rental revenue of $760 this week.

Tom Driscoll

- Earns an hourly wage of $7.35.
- Worked 39 hours this week.

Ann Ryan

- Receives a salary of $185 per week plus 3% commission.
- Had rental revenue of $1,235 this week.

Analyze
Identify the employee who had the lowest gross pay for the week.

Problem 6 Preparing a Payroll Register

Hot Suds Car Wash has four employees. They are paid on a weekly basis with overtime paid for all hours worked over 40 in a week. The overtime rate is 1½ times the regular rate of pay. The payroll information follows.

Employee	Employee Number	Rate per Hour	Marital Status	Allowances	Union Member
James Dumser	108	$7.95	Single	0	No
Gail Job	112	$8.00	Married	1	Yes
James Liptak	102	$8.00	Married	2	Yes
Bruce Stern	109	$7.80	Single	1	Yes

Continued

Chapter 12
Problems

During the week ending October 9, Dumser worked 39 hours, Job worked 41 hours, and Liptak and Stern each worked 36 hours.

Instructions On the forms provided in your working papers:

1. Prepare a payroll register for the week ending October 9. The date of payment is also October 9. List employees in alphabetical order by *last* name. Use the tables on page 355 to determine the federal income tax withholding. The rate for the state income tax is 2%. Compute social security tax at 6.2% and Medicare tax at 1.45%. Union members pay weekly dues of $4.50. Both Gail Job and Bruce Stern had $6.75 deducted for health and hospital insurance.

2. Total the amount columns. Subtract total deductions from total earnings. Does the result equal the sum of the Net Pay column? If not, find and correct any error(s) on the payroll register.

Analyze
Identify the employee who had the highest amount withheld for federal income tax for the week.

Problem 7 Preparing Payroll Checks and Employee's Earnings Records

The payroll register for Kits & Pups Grooming is presented here and also appears in your working papers.

PAYROLL REGISTER

PAY PERIOD ENDING October 17 20 --   DATE OF PAYMENT October 17, 20--

EMPLOYEE NUMBER	NAME	MAR. STATUS	ALLOW.	TOTAL HOURS	RATE	EARNINGS			DEDUCTIONS							NET PAY	CK. NO.
						REGULAR	OVERTIME	TOTAL	SOC. SEC. TAX	MEDICARE TAX	FED. INC. TAX	STATE INC. TAX	HOSP. INS.	OTHER	TOTAL		
162	Hurd, Mildred	S	0	38	7.60	288 80		288 80	17 91	4 19	35 00	7 22		(B) 5 00	69 32	219 48	1
157	Montego, José	S	1	39	7.90	308 10		308 10	19 10	4 47	30 00	7 70	5 10		66 37	241 73	2
151	Pilly, Amanda	M	2	36	8.10	291 60		291 60	18 08	4 23	10 00	7 29	7 60	(B) 5 00	52 20	239 40	3
163	Steams, Margaret	S	0	41	7.60	304 00	11 40	315 40	19 55	4 57	40 00	7 89			72 01	243 39	4
				TOTALS		1,192 50	11 40	1,203 90	74 64	17 46	115 00	30 10	12 70	(B)10 00	259 90	944 00	

Other Deductions: Write the appropriate code letter to the left of the amount: B–U.S. Savings Bonds; C–Credit Union; UD–Union Dues; UW–United Way.

370 Chapter 12 • Payroll Accounting

Chapter 12
Problems

Instructions On the forms in your working papers:

1. Prepare a payroll check and stub for each employee. Where would you record the check numbers in the payroll register?

2. Record the payroll information in the employee's earnings records for José Montego and Amanda Pilly.

Analyze
Identify the net pay for José Montego for this week.

Problem 8 Preparing the Payroll

Outback Guide Service has six employees and pays each week. Hourly employees are paid overtime for all hours worked over 40 in a week. The overtime rate is 1½ times the regular rate of pay. Outback pays its employees by hourly rate, salary, or salary plus a 5% commission on total amount of sales.

Method of Computing Earnings

Employee	Hourly Wage	Salary	Salary Plus Commission
Cummings, Carol		$270.00	
Dame, Ted	$7.60		
Lengyel, Tom			$160.00 plus 5%
Robinson, Jean			$140.00 plus 5%
Usdavin, James	$7.40		
Wong, Kim			$140.00 plus 5%

Employee deductions include federal income taxes (use the tables on page 355), FICA taxes at 6.2% for social security and 1.45% for Medicare, state income taxes of 1.5%, and hospital insurance premiums of $5.43 for single employees and $9.37 for married employees. Also, Kim Wong and Tom Lengyel have $10.00 withheld each week to purchase U.S. savings bonds.

Continued

Chapter 12
Problems

NO. 73
NAME Ted Dame
SOC. SEC. NO. 093-XX-XXXX
WEEK ENDING 10/17/20--

DAY	IN	OUT	IN	OUT	IN	OUT
M	8:58	12:03	12:55	5:09		
T	8:55	11:55	1:00	4:00		
W	9:30	12:10	1:04	3:30		
Th	8:57	12:03	12:59	6:00		
F	8:58	12:00	1:00	6:05		
S	9:00	12:00				
S						

TOTAL HOURS _____

	HOURS	RATE	
REGULAR			
OVERTIME			
		TOTAL EARNINGS	

SIGNATURE _____ DATE _____

NO. 92
NAME James Usdavin
SOC. SEC. NO. 087-XX-XXXX
WEEK ENDING 10/17/20--

DAY	IN	OUT	IN	OUT	IN	OUT	TOTAL
M	8:55	12:06	1:01	5:40			
T	7:58	11:01	12:03	6:38			
W	9:03	1:10	2:00	6:00			
Th	7:59	11:55	1:10	4:51			
F	9:01	12:06	1:05	3:47			
S	9:00	12:03					
S							

TOTAL HOURS _____

	HOURS	RATE	AMOUNT
REGULAR			
OVERTIME			
		TOTAL EARNINGS	

SIGNATURE _____ DATE _____

During the pay period ending October 17, the sales were: Tom Lengyel $1,204.76, Jean Robinson $1,925.80, and Kim Wong $2,135.65.

The hourly employees filled in the following time cards.

Instructions On the forms in your working papers:

1. Complete the time cards to the nearest quarter hour.
2. Prepare a payroll register for the week ending October 17. The date of payment is also October 17. Each employee's number, marital status, and number of allowances claimed are listed on her or his employee's earnings record.
3. Prepare a payroll check and stub for each employee.
4. Record the payroll information for each employee on her or his employee's earnings record.

Continued

Chapter 12
Problems

Analyze
Calculate the total amount deducted from employees' gross pay for FICA taxes (social security and Medicare).

Problem 9 Preparing the Payroll Register

Showbiz Video has six employees who are paid weekly. The hourly employees are paid overtime for hours worked over 40 in a week, at a rate 1½ times their regular rate of pay. Employee information follows:

Employee	Employee Number	Marital Status	Allowances
Mary Arcompora	105	Married	2
Barbara Fox	137	Married	1
John French	135	Single	1
Chris German	141	Married	4
David Izbecki	139	Single	0
Susan Tilbert	129	Married	1

Mary Arcompora, the store manager, is paid a salary of $300.00 per week plus 1% of all rental sales. Barbara Fox and John French, salespeople, are paid a salary of $200.00 per week plus a 6% commission on all rentals from the "Oldies but Goodies" section. Chris German and David Izbecki, office workers, are paid an hourly wage of $7.25. Susan Tilbert, a stock person, is paid $7.40 per hour.

The payroll deductions include federal income tax, social security tax of 6.2%, Medicare tax of 1.45%, and state income tax of 1.8%. Chris German and Susan Tilbert have $12.50 deducted each week for hospital insurance.

Continued

Chapter 12
Problems

NO. 141
NAME Chris German
SOC. SEC. NO. 449-XX-XXXX
WEEK ENDING October 24, 20--

DAY	IN	OUT	IN	OUT	IN	OUT	TOTAL
M	8:05	12:03	1:00	5:05			
T	8:00	12:05	1:10	5:00			
W	9:05	12:05	1:15	6:05			
Th	8:30	11:55	1:15	6:00			
F	8:00	12:00	1:00	6:15			
S							
S							

TOTAL HOURS

	HOURS	RATE	AMOUNT
REGULAR			
OVERTIME			
TOTAL EARNINGS			

During the week ending October 24, "Oldies but Goodies" rentals were $484.90 for Barbara Fox and $641.70 for John French. Total rental revenue for the week was $3,917.30.

Instructions Prepare a payroll register in your working papers for the week ending October 24. Complete time cards to the nearest quarter of an hour. Use the federal income tax tables provided in the chapter.

NO. 139
NAME David Izbecki
SOC. SEC. NO. 461-XX-XXXX
WEEK ENDING October 24, 20--

DAY	IN	OUT	IN	OUT	IN	OUT
M	8:00	12:05	1:00	5:00		
T	8:10	12:00	1:00	5:05		
W	8:05	12:30	1:30	5:00		
Th	8:00	12:00	1:00	5:10		
F	8:05	12:05	1:05	3:30		
S						
S						

TOTAL HOURS

	HOURS	RATE
REGULAR		
OVERTIME		
TOTAL EARNINGS		

SIGNATURE _____ DATE _____

NO. 129
NAME Susan Tilbert
SOC. SEC. NO. 401-XX-XXXX
WEEK ENDING October 24, 20--

DAY	IN	OUT	IN	OUT	IN	OUT	TOTAL
M	8:05	12:00	1:00	5:00			
T	8:10	12:00	1:15	5:00			
W	8:30	12:00	1:00	6:00			
Th	9:00	12:00	12:45	5:00			
F	9:00	11:50					
S							
S							

TOTAL HOURS

	HOURS	RATE	AMOUNT
REGULAR			
OVERTIME			
TOTAL EARNINGS			

SIGNATURE _____ DATE _____

Chapter 12
Problems

Analyze
Compute the total net pay of all employees for the pay period.

Problem 10 Calculating Gross Earnings

Job Connect has seven employees, all of whom are paid weekly. For hourly wage employees, overtime is paid at 1½ times the regular rate of pay for hours worked over 40 in a week.

Barbara Miller, the office manager, is paid a salary of $375.00 per week plus a bonus of 3% of all revenue over $6,000 per week. Lynn Austin, an office assistant, is paid a salary of $250.00 per week plus 5% of all telephone sales made in the office. Charlene Womack, the office secretary, is paid a salary of $230.00 per week. Susan Dilloway and Doris Franco, placement workers, are paid an hourly wage of $8.95. Pam Darrah is also a placement worker but is paid a commission of $35.00 for every job placement that she completes. David Facini, a part-time maintenance worker, is paid $7.60 per hour.

SOURCE DOCUMENT PROBLEM

Problem 12–10

Use the source documents in your working papers to complete this problem.

For the week ending October 24, the office recorded the following payroll information.

- Total office sales for the week were $8,420.00.
- Susan Dilloway worked a total of 38½ hours.
- Doris Franco worked a total of 41¼ hours.
- Phone sales for the week were $1,375.00.
- Pam Darrah made seven job placements.
- David Facini worked a total of 23 hours.

Instructions Using the form provided in your working papers, calculate the gross earnings for the workers at Job Connect for the week ending October 24.

Analyze
Identify the employee who had the highest gross earnings.

Real-World Applications & Connections

CASE STUDY

Payroll: Number of Employees

You are employed by Delzell Financial Services, which is a financial advisory company for small businesses. Lucey Shoes has hired Delzell to advise them on their employees. Lucey presently has two full-time employees that work 40 hours per week and make $9.90 per hours. Lucey wants to hire a part-time employee who will earn $8.40 per hour. Lucey presently has weekly sales of about $5,000. They want their payroll to be no more than 20% of the weekly sales.

Instructions Draft a report for Delzell explaining how many hours the new part-time employee can work per week and stay within their payroll guidelines. Check all calculations and check for correct spelling, grammar, and punctuation. Prepare the final report.

21st Century Skills

Ergonomics

Ergonomics is the understanding of how the interaction of humans and other elements in a system or setting affects people's health. Ergonomics was first used during World War II in the design of products to reduce workplace medical issues. Since then, the study of ergonomics has grown a great deal. With so many people spending much of their time at a computer every day, new health challenges have surfaced. It is the task of designers to create technological tools that not only accomplish their purpose but also help us stay healthy.

Research And Write

a. Use the Internet to research the different ways that ergonomics can be applied in the design of products. Be sure to look at products in various industries.

b. Write a report about how items are designed ergonomically and the health concerns they are intended to address.

Career Wise

Certified Payroll Professional

Certified Payroll Professional (CPP) is one of two certifications offered by the American Payroll Association. A CPP must have at least three years of experience in payroll administration, payroll production, or payroll accounting. A CPP also must pass an exam demonstrating skills in dealing with employee issues such as taxes, benefits, regulatory requirements, and auditing.

Research And Write

a. Locate the Web site of the American Payroll Association (APA). Download the "CPP Knowledge, Skills, and Abilities" guidelines.

b. What other certification does the APA offer, and how does it differ from the CPP? Put your results in a table.

H.O.T. Audit

Ken Boehm is paid a hourly wage of $7.60. Last week Ken worked a total of 41½ hours. He is paid overtime for all work over 40 hours. The social security tax rate is 6.2%, Medicare tax is 1.45% and the state income tax is 3.5% of total earnings. Ken is single and claims one allowance.

Instructions

Examine the following payroll check stub for Ken. Check all calculations and create a new payroll check stub using a separate sheet of paper.

Audit with Alex

Employee Pay Statement
Detach and retain this statement. 260

Period Ending	Earnings			Deductions						Net Pay
	Regular	Overtime	Total	Social Security Tax	Med. Tax	Federal Income Tax	State Income Tax	Hosp. Ins.	Other	Total	
3/15/20--	304.00	17.10	321.10	19.26	4.66	23.00	9.63	9.10	---	65.65	255.45

Chapter 13: Payroll Liabilities and Tax Records

Chapter Topics

13-1 Journalizing and Posting the Payroll

13-2 Employer's Payroll Taxes

13-3 Tax Liability Payments and Tax Reports

Visual Summary

Review and Activities

Standardized Test Practice

Computerized Accounting

Problems

Real-World Accounting Careers

Real-World Applications & Connections

Essential Question

As you read this chapter, keep this question in mind:

What responsibilities does an employer accept in the payroll process?

Main Idea

Employers are legally required to make tax deposits on time and to report the earnings of each employee. Employers must pay taxes in addition to the amounts withheld from their employees. This is an expense for the business.

Chapter Objectives

Concepts	Analysis	Procedures
C1 Record payroll transactions in the general journal. **C2** Describe the employer's payroll taxes.	**A1** Compute and complete payroll tax expense forms.	**P1** Record the payment of tax liabilities in the general journal. **P2** Complete payroll tax reports.

Real-World Business Connection

Accountemps

Bookkeepers, financial analysts, certified public accountants—these are some of the professionals employed by Accountemps, the largest temporary staffing company for accounting-related labor. Founded in 1948, Accountemps is the first company of its kind and today has more than 360 offices; its client companies have access to more than 3 million temporary professionals to serve their accounting, finance, and bookkeeping needs.

Connect to the Business

Temporary staffing agencies such as Accountemps make a profit by charging client companies a percentage over what the temporary worker's agreed-upon wage. Temporary or contract accountants receive their checks from Accountemps, not the company that provided the work. Part of the appeal of using a staffing service is that the agency takes on the paperwork and payroll expenses associated with temporary employees.

Analyze

What are some other reasons companies that work with Accountemps choose to hire temporary or contract employees rather than full-time employees?

Focus on the Photo

Accountemps handles payroll for its staff who are temporary professionals. It must send employees' payroll taxes to the appropriate government agencies within a certain amount of time. **What might happen to the company if the accounting department failed to pay the employees' taxes to the government?**

SECTION 13.1
Journalizing and Posting the Payroll

Employees, **ranging** from design engineers at Ford Motor Company to the waitresses at a neighborhood coffee shop, expect their payroll checks to arrive on time and to be accurate. You have learned that various amounts are withheld from employees' earnings for taxes and voluntary deductions. When the payroll register is complete, the payroll entry is journalized and the amounts are posted to the general ledger.

> **Section Vocabulary**
> - Salaries Expense
> - ranging
> - appropriate

After the payroll has been prepared, a check is written to transfer the *total net pay* amount from the regular checking account of a business to its payroll checking account. The check is deposited in the payroll account, and all payroll checks for the period are written on the payroll account. The next step is to record the payment of the payroll in the accounting records.

Analyzing and Journalizing the Payroll

How Do You Journalize Payroll?

Let's analyze the effect of payroll on the employer's accounting system. Each pay period, the business pays out a certain amount of money to its employees in the form of wages and salaries. Employee earnings are a normal operating expense of a business. The expense account often used to record employees' earnings is called **Salaries Expense.** To increase the amount in **Salaries Expense**, the account is debited for the gross earnings for the pay period.

The business withholds various deductions, such as income and FICA taxes, from gross earnings each pay period. Employees also request voluntary withholdings such as premiums for insurance coverage. The employer retains the amounts withheld until it is time to pay the appropriate government agencies and businesses. The amounts withheld but not yet paid are liabilities of the business. Remember, a liability is an amount *owed* by a business.

Each type of payroll liability is recorded in a separate account.

Type of Deduction	Ledger Account
Federal income tax	**Employees' Federal Income Tax Payable**
Social security tax	**Social Security Tax Payable**
Medicare tax	**Medicare Tax Payable**
Hospital insurance premiums	**Hospital Insurance Premiums Payable**

Depending on the business, it is possible that several different types of deductions are recorded in the Other Deductions column of the payroll register. If so, the total for each type of deduction is credited to the **appropriate** liability account. Refer to the Zip Delivery Service payroll register shown in **Figure 12–5** on page 359. The deductions that may appear in the Other Deductions column each have an account in Zip's general ledger:

380 Chapter 13 • Payroll Liabilities and Tax Records

- **U.S. Savings Bonds Payable**
- **Credit Union Payable**
- **Union Dues Payable**
- **United Way Payable**

The credit part of the payroll journal entry is made up of several items. The largest item is for net pay. Net pay is the amount actually paid out in cash by the employer to the employees. **Cash in Bank** is credited for total net pay.

The difference between gross earnings and net pay equals the employer's payroll liabilities. Each payroll liability account is separately credited for the total amount shown on the payroll register.

Reading Check

Critical Thinking

"Employee earnings are a normal operating expense of the business." What does this mean?

Business Transaction

Zip's payroll register in **Figure 12–5** on page 359 is the source document for the payroll journal entry.

Analysis		
Identify	1.	The accounts **Salaries Expense, Employees' Federal Income Tax Payable, Employees' State Income Tax Payable, Social Security Tax Payable, Medicare Tax Payable, Hospital Insurance Premiums Payable, U.S. Savings Bonds Payable,** and **Cash in Bank** are affected.
Classify	2.	**Salaries Expense** is an expense account. **Employees' Federal Income Tax Payable, Employees' State Income Tax Payable, Social Security Tax Payable, Medicare Tax Payable, Hospital Insurance Premiums Payable,** and **U.S. Savings Bonds Payable** are liability accounts. **Cash in Bank** is an asset account.
+/−	3.	**Salaries Expense** is increased by $1,423.80; **Employees' Federal Income Tax Payable** is increased by $102.00; **Employees' State Income Tax Payable** is increased by $28.47; **Social Security Tax Payable** is increased by $88.27; **Medicare Tax Payable** is increased by $20.65; **Hospital Insurance Premiums Payable** is increased by $23.65; **U.S. Savings Bonds Payable** is increased by $25.00; **Cash in Bank** is decreased by $1,135.76

Continued

Debit-Credit Rule

4. Increases in expense accounts are recorded as debits. Debit **Salaries Expense** for $1,423.80.

5. Decreases in asset accounts are recorded as credits. Credit **Cash in Bank** for $1,135.76. Increases in liability accounts are recorded as credits. Credit **Employees' Federal Income Tax Payable** for $102.00; **Employees' State Income Tax Payable** for $28.47; **Social Security Tax Payable** for $88.27; **Medicare Tax Payable** for $20.65; **Hospital Insurance Premiums Payable** for $23.65; **U.S. Savings Bonds Payable** for $25.00.

T Accounts

6.

Salaries Expense		Employees' Federal Income Tax Payable	
Debit + 1,423.80	Credit −	Debit −	Credit + 102.00

Employees' State Income Tax Payable		Social Security Tax Payable	
Debit −	Credit + 28.47	Debit −	Credit + 88.27

Medicare Tax Payable		Hospital Insurance Premiums Payable	
Debit −	Credit + 20.65	Debit −	Credit + 23.65

U.S. Savings Bonds Payable		Cash in Bank	
Debit −	Credit + 25.00	Debit +	Credit − 1,135.76

Journal Entry

7.

GENERAL JOURNAL PAGE 29

DATE		DESCRIPTION	POST. REF.	DEBIT	CREDIT
20–					
June	30	Salaries Expense		1 423 80	
		Emplys' Fed. Inc. Tax Pay.			102 00
		Emplys' State Inc. Tax Pay.			28 47
		Social Sec. Tax Pay.			88 27
		Medicare Tax Pay.			20 65
		Hosp. Ins. Premiums Pay.			23 65
		U.S. Savings Bonds Pay.			25 00
		Cash in Bank			1 135 76
		Pay. Reg. 6/30–Ck 186			

The payroll expense is $1,423.80. The employees receive $1,135.76 in cash (net pay). Later the business will pay the federal government $210.92 ($102.00 federal income tax, $88.27 social security tax, and $20.65 Medicare tax). The business will also pay the state $28.47 for state income tax. A check for $23.65 will be written to the insurance company for hospital insurance premiums. Finally, a check for $25 will be sent to the federal government to purchase savings bonds.

These liabilities are the result of deductions that were taken from employees' earnings. In the next section, you will learn about the payroll tax liabilities of the employer.

Reading Check

Explain

Why are the amounts withheld from employees' payroll checks liabilities of the employer?

Posting the Payroll Entry

How Do You Post the Payroll?

Figure 13–1 shows the general journal entry and the individual ledger accounts after posting.

GENERAL JOURNAL PAGE 29

	DATE	DESCRIPTION	POST. REF.	DEBIT	CREDIT
1	20–				
2	June 30	Salaries Expense	514	1 4 2 3 80	
3		Employees' Fed. Inc. Tax Pay.	210		1 0 2 00
4		Employees' State Inc. Tax Pay.	215		28 47
5		Social Security Tax Pay.	220		88 27
6		Medicare Tax Pay.	225		20 65
7		Hosp. Ins. Premiums Pay.	230		23 65
8		U.S. Savings Bonds Pay.	235		25 00
9		Cash in Bank	101		1 1 3 5 76
10		Payroll Reg. 6/30–Ck. 186			

ACCOUNT **Salaries Expense** ACCOUNT NO. **514**

DATE	DESCRIPTION	POST. REF.	DEBIT	CREDIT	BALANCE DEBIT	BALANCE CREDIT
20–						
June 23	Balance	✓			34 6 8 3 10	
30		G29	1 4 2 3 80		36 1 0 6 90	

ACCOUNT **Employees' Federal Income Tax Payable** ACCOUNT NO. **210**

DATE	DESCRIPTION	POST. REF.	DEBIT	CREDIT	BALANCE DEBIT	BALANCE CREDIT
20–						
June 23	Balance	✓				2 9 3 18
30		G29		1 0 2 00		3 9 5 18

Figure 13–1 Posting the Payroll Entry to the General Ledger (continued)

ACCOUNT Employees' State Income Tax Payable   ACCOUNT NO. 215

DATE		DESCRIPTION	POST. REF.	DEBIT	CREDIT	BALANCE DEBIT	BALANCE CREDIT
20–							
June	23	Balance	✓				3 2 1 12
	30		G29		2 8 47		3 4 9 59

ACCOUNT Social Security Tax Payable   ACCOUNT NO. 220

DATE		DESCRIPTION	POST. REF.	DEBIT	CREDIT	BALANCE DEBIT	BALANCE CREDIT
20–							
June	23	Balance	✓				2 5 0 35
	30		G29		8 8 27		3 3 8 62

ACCOUNT Medicare Tax Payable   ACCOUNT NO. 225

DATE		DESCRIPTION	POST. REF.	DEBIT	CREDIT	BALANCE DEBIT	BALANCE CREDIT
20–							
June	23	Balance	✓				5 8 63
	30		G29		2 0 65		7 9 28

ACCOUNT Hospital Insurance Premiums Payable   ACCOUNT NO. 230

DATE		DESCRIPTION	POST. REF.	DEBIT	CREDIT	BALANCE DEBIT	BALANCE CREDIT
20–							
June	23	Balance	✓				4 73
	30		G29		2 3 65		2 8 38

ACCOUNT U.S. Savings Bonds Payable   ACCOUNT NO. 235

DATE		DESCRIPTION	POST. REF.	DEBIT	CREDIT	BALANCE DEBIT	BALANCE CREDIT
20–							
June	23	Balance	✓				7 5 00
	30		G29		2 5 00		1 0 0 00

ACCOUNT Cash in Bank   ACCOUNT NO. 101

DATE		DESCRIPTION	POST. REF.	DEBIT	CREDIT	BALANCE DEBIT	BALANCE CREDIT
20–							
June	23	Balance	✓			19 6 1 0 30	
	30		G29		1 1 3 5 76	18 4 7 4 54	

Figure 13–1 Posting the Payroll Entry to the General Ledger

SECTION 13.1 Assessment

After You Read

Reinforce the Main Idea

Use an organizer like this one to express the payroll journal entry as an equation. Use broad account categories. (For example, use *liabilities* instead of listing all possible accounts.) Draw arrows (↓↑) in each box to show whether the account category increases or decreases.

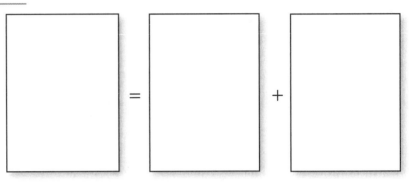

Problem 1 Determining Payroll Amounts

SweepIt Cleaning Service reported the following amounts for the week ending November 4. The total amount earned by all employees is $2,193.40. The amount withheld for federal income tax is $263.00. Social security tax is $136.00, and Medicare tax is $31.80. Three of the employees each have $11.25 deducted for hospital insurance. The amount withheld for state income tax is $38.70.

Instructions Answer the following questions concerning the November 4 payroll for SweepIt Cleaning Service.

1. What amount is recorded in the **Salaries Expense** account?
2. What amount is recorded in the **Medicare Tax Payable** account?
3. What is the total amount of liabilities for the weekly payroll?
4. What amount is entered in the **Hospital Insurance Premiums Payable** account?
5. What amount is recorded as a credit for the **Cash in Bank** account?

Math for Accounting

You are the payroll clerk for Queen City Motors. As you review the payroll records, you notice that two employees are nearing the $90,000 limit for social security tax. As commission-only sales employees, Marcie Laliberte and Kevin Hogan have earned $87,200 and $88,700, respectively.

1. How much more must Marcie and Kevin earn in commission to reach the social security tax limit?
2. If Marcie and Kevin are paid 7% commission on each sale, how much more in sales must each make to reach the social security tax limit?

Chapter 13 • Payroll Liabilities and Tax Records 385

SECTION 13.2
Employer's Payroll Taxes

In Section 13.1 you learned how to journalize and post the payroll entry. This entry, in part, records taxes that employees are required to pay on their earnings. Your local florist, employing designers, delivery workers, and sales clerks, must also pay taxes on these workers' earnings. These amounts need to be calculated, journalized, and posted.

Computing Payroll Tax Expenses

Which Payroll Taxes Are Paid by the Employer?

In addition to withholding taxes from employees' wages, the employer *pays* taxes on these wages. The employer's taxes, considered operating expenses of the business, consist of the employer's FICA taxes, the federal unemployment tax, and the state unemployment tax.

> **Section Vocabulary**
> - Federal Unemployment Tax Act (FUTA)
> - State Unemployment Tax Act (SUTA)
> - unemployment taxes
> - Payroll tax expense
> - federal
> - accumulated
> - temporarily

The Employer's FICA Taxes

Under the **Federal** Insurance Contributions Act, both the employee and the employer pay FICA taxes. As you recall, the employer withholds a percentage of gross earnings for social security and Medicare taxes. In addition, the employer pays FICA taxes using the same percentage of gross earnings. Recall that the current rates are 6.2% for social security tax and 1.45% for Medicare tax.

The employer and the employee pay social security tax on gross earnings up to the maximum taxable limit per employee ($106,800 in 2010). The employer and the employee pay Medicare tax on all gross earnings; there is no maximum taxable limit. The payroll clerk checks the **accumulated** earnings on each employee's earnings record to determine whether that employee has reached the maximum taxable amount. When an employee reaches the limit, the social security tax is no longer computed.

In determining social security tax and Medicare tax for both employee and employer, it makes no difference whether an employee is full-time, part-time, temporary, or permanent. A full-time adult employee and a student employed part-time only for the summer are subject to the same taxes.

At Zip Delivery Service for the week ending June 30, the employees' total social security taxes are $88.27 and total Medicare taxes are $20.65. The employer's taxes on the total gross earnings are $88.28 (6.2% of $1,423.80) and $20.65 (1.45% of $1,423.80), respectively. Notice that the social security tax for the employees ($88.27) and the employer ($88.28) do not match. The same situation could also exist for the Medicare tax. This is because the employer's tax is calculated on the total gross earnings ($1,423.80). The employees' taxes are calculated on each employee's gross earnings and the individual tax amounts are totaled. This may result in small differences between the employees' and employer's taxes.

Federal and State Unemployment Taxes

Two unemployment laws, the **Federal Unemployment Tax Act (FUTA)** and the **State Unemployment Tax Act (SUTA),** require employers to pay unemployment taxes. **Unemployment taxes** are collected to provide funds for workers who are **temporarily** unemployed. Unemployment taxes are based on a percentage of the employees' gross earnings.

> **Common Mistakes**
>
> **Unemployment Taxes** The unemployment taxes are paid by the employer based on the amount of total gross earnings for the pay period. The amount of total earnings is used to calculate the taxes, but the amounts are entered as part of the payroll taxes entry on the general journal.

The employer pays both federal and state unemployment taxes. The maximum federal unemployment tax is 6.2% on the first $7,000 of an employee's annual wages. State unemployment tax rates and maximum taxable amounts vary among states. Employers may deduct up to 5.4% of the state unemployment taxes from federal unemployment taxes. Most employers, therefore, pay a federal tax of .8% (6.2% − 5.4%) of taxable gross earnings.

In a few states, employees are also required to pay unemployment taxes. The percentage amount varies among these states.

For Zip, since none of the employees has reached the maximum taxable amount, the federal unemployment tax for the week ended June 30 is $11.39 ($1,423.80 × .008, or .8%). The state unemployment tax is $76.89 ($1,423.80 × .054, or 5.4%).

Reading Check

Recall

State and federal unemployment taxes are percentages of what amount?

Journalizing the Employer Payroll Taxes

How Do You Journalize Payroll Tax Expense?

The employer's payroll taxes are business expenses recorded in the **Payroll Tax Expense** account. Until paid, the employer's payroll taxes are liabilities of the business.

Use the **Social Security Tax Payable** and the **Medicare Tax Payable** accounts to record both the employees' and the employer's FICA taxes. Record the employer's unemployment taxes in the **Federal Unemployment Tax Payable** and **State Unemployment Tax Payable** accounts.

In the next business transaction, we will analyze the accounts affected when an employer pays its payroll taxes. This entry takes place in each payroll period.

Business Transaction

*Zip's payroll register in **Figure 12–5** on page 359 is the source document for the payroll tax journal entry.*

Analysis	Identify	1. The accounts **Payroll Tax Expense, Social Security Tax Payable, Medicare Tax Payable, State Unemployment Tax Payable,** and **Federal Unemployment Tax Payable** are affected.
	Classify	2. **Payroll Tax Expense** is an expense account. **Social Security Tax Payable, Medicare Tax Payable, State Unemployment Tax Payable,** and **Federal Unemployment Tax Payable** are liability accounts.
	+/−	3. **Payroll Tax Expense** is increased by $197.21; **Social Security Tax Payable** is increased by $88.28; **Medicare Tax Payable** is increased by $20.65; **State Unemployment Tax Payable** is increased by $76.89; **Federal Unemployment Tax Payable** is increased by $11.39.
Debit-Credit Rule		4. Increases in expense accounts are recorded as debits. Debit **Payroll Tax Expense** for $197.21.
		5. Increases in liability accounts are recorded as credits. Credit **Social Security Tax Payable** for $88.28; **Medicare Tax Payable** for $20.65; **State Unemployment Tax Payable** for $76.89; **Federal Unemployment Tax Payable** for $11.39.
T Accounts		6.

Payroll Tax Expense		Social Security Tax Payable	
Debit + 197.21	Credit −	Debit −	Credit + 88.28

Medicare Tax Payable		State Unemployment Tax Payable	
Debit −	Credit + 20.65	Debit −	Credit + 76.89

Federal Unemployment Tax Payable	
Debit −	Credit + 11.39

Continued

Journal Entry 7.

	GENERAL JOURNAL			PAGE 30	
DATE	DESCRIPTION	POST. REF.	DEBIT	CREDIT	
20–					1
June 30	Payroll Tax Expense		197 21		2
	Social Security Tax Pay.			88 28	3
	Medicare Tax Pay.			20 65	4
	State Unemplymnt. Tax Pay.			76 89	5
	Fed. Unemplymnt. Tax Pay.			11 39	6
	Payroll Reg. 6/30				7

Posting Payroll Taxes to the General Ledger

How Do You Post Payroll Tax Expense?

Figure 13–2 shows the individual ledger accounts after posting the payroll taxes entry.

Notice that the **Social Security Tax Payable** and the **Medicare Tax Payable** accounts have two entries for the June 30 payroll. The first entry is the amount of taxes withheld from the *employees'* earnings. The second entry is the amount of taxes paid by the *employer*.

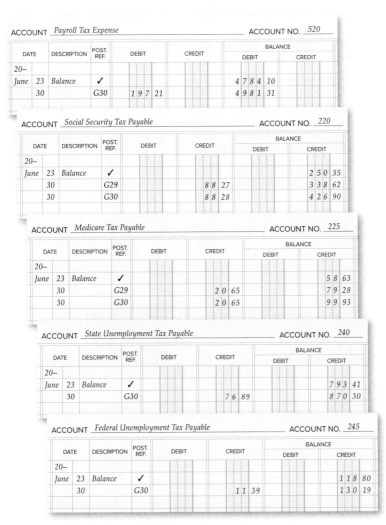

Figure 13–2 General Ledger Accounts after Posting the Payroll Taxes Entry

SECTION 13.2
Assessment

After You Read

Reinforce the Main Idea

Employer payroll taxes are paid at different rates. Use a chart like this one to show the rates of the employer payroll taxes.

	Employer Payroll Tax Rates	
Tax	Percent of Total Earnings	Earnings Limit
Social security		
Federal unemployment		
Medicare		
State unemployment		

Problem 2 Calculating Employer's Payroll Taxes

For the week ending June 30, EZ Copy Center's payroll has total gross earnings of $4,836.60. Calculate the employer's payroll taxes. Use the following percentages:

Social security tax	6.2%	Federal unemployment tax	0.8%
Medicare tax	1.45%	State unemployment tax	5.4%

Problem 3 Identifying Entries for Payroll Liabilities

The following list includes several payroll-related items used in preparing the weekly payroll for Outdoor Adventures. These items are included in either the entry to record the payroll or the entry to record the employer's payroll taxes.

Employees' federal income tax	Employees' state income tax
Employer's social security tax	Union dues
U.S. savings bonds	Employees' social security tax
Employer's Medicare tax	State unemployment tax
Federal unemployment tax	Employees' Medicare tax

Instructions Use the form provided in your working papers. Place a check mark in the column that describes the entry in which the item is recorded:

- entry to record the payroll
- entry to record the employer's payroll taxes.

Math for Accounting

Calculate the employer's federal unemployment tax rate for each of the following states:

State	State Unemployment Tax Rate
State A	1.5%
State B	3.0%
State C	4.5%
State D	5.4%
State E	6.0%

SECTION 13.3
Tax Liability Payments and Tax Reports

After journalizing and posting the payroll entries, a business pays the amounts owed to government agencies and other institutions.

Paying the Payroll Tax Liabilities

How and When Do Employers Pay Their Liabilities?

Payroll liabilities are paid at regular intervals.

FICA and Federal Income Taxes

A business makes one payment combining (1) social security and Medicare taxes (for both employees and employer), and (2) employees' federal income taxes withheld. It makes the payment at an authorized financial institution or Federal Reserve Bank. Most small businesses, like Zip, make this payment monthly. It is due by the 15th day of the month following the payroll month. Payment for the month ending June 30 is due by July 15. Larger businesses make the payment every two weeks.

Many small businesses prepare and send a **Form 8109** with the check. The Form 8109, or *Federal Tax Deposit Coupon,* identifies the type of tax and the tax period. Notice the ovals on the right side of Form 8109 in **Figure 13–3.** The 941 oval indicates FICA and federal income taxes. The 2nd Quarter oval indicates the period ending June 30.

> **Section Vocabulary**
> - Form 8109
> - Electronic Federal Tax Payment System (EFTPS)
> - Form W-2
> - Form W-3
> - Form 941
> - Form 940
> - intervals
> - annual

Figure 13–3 Federal Tax Deposit Coupon (Form 8109) for FICA and Federal Income Taxes

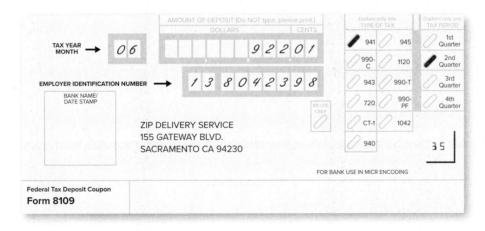

The **Electronic Federal Tax Payment System (EFTPS)** is used by larger businesses to make deposits. Eventually even small businesses will use EFTPS.

Business Transaction

Zip pays $922.01 payroll tax liabilities July 15 including $395.18 employees' federal income taxes, $426.90 social security taxes, and $99.93 Medicare taxes (refer to ledger accounts in **Figures 13–1** and **13–2**).

Analysis

Identify
1. The accounts **Employees' Federal Income Tax Payable, Social Security Tax Payable, Medicare Tax Payable,** and **Cash in Bank** are affected.

Classify
2. **Employees' Federal Income Tax Payable, Social Security Tax Payable,** and **Medicare Tax Payable** are liability accounts. **Cash in Bank** is an asset account.

+/−
3. **Employees' Federal Income Tax Payable** is decreased by $395.18; **Social Security Tax Payable** is decreased by $426.90; **Medicare Tax Payable** is decreased by $99.93; **Cash in Bank** is decreased by $922.01.

Debit-Credit Rule

4. Decreases in liability accounts are recorded as debits. Debit **Employees' Federal Income Tax Payable** for $395.18; **Social Security Tax Payable** for $426.90; **Medicare Tax Payable** for $99.93.

5. Decreases in asset accounts are recorded as credits. Credit **Cash in Bank** for $922.01.

T Accounts

6.

Employees' Federal Income Tax Payable		Social Security Tax Payable	
Debit − 395.18	Credit +	Debit − 426.90	Credit +

Medicare Tax Payable		Cash in Bank	
Debit − 99.93	Credit +	Debit +	Credit − 922.01

Journal Entry

7.

GENERAL JOURNAL   PAGE 31

DATE	DESCRIPTION	POST. REF.	DEBIT	CREDIT
20–				
July 15	Emplys' Fed. Inc. Tax Pay.		395 18	
	Social Security Tax Pay.		426 90	
	Medicare Tax Pay.		99 93	
	Cash in Bank			922 01
	Check 208			

State and Local Income Taxes

At regular **intervals** businesses pay the amounts withheld for state and local income taxes. Each state and local government determines how and when the payments are made and what reports are filed.

Business Transaction

*Zip pays $349.59 to the state. This is the amount of state income tax withheld from employees' earnings, as indicated in the **Employees' State Income Tax Payable** account shown in **Figure 13–1**.*

Analysis

Identify
1. The accounts **Employees' State Income Tax Payable** and **Cash in Bank** are affected.

Classify
2. **Employees' State Income Tax Payable** is a liability account. **Cash in Bank** is an asset account.

+/−
3. **Employees' State Income Tax Payable** is decreased by $349.59. **Cash in Bank** is decreased by $349.59.

Debit-Credit Rule

4. Decreases in liabilities are debits. Debit **Employees' State Income Tax Payable** for $349.59.

5. Decreases in assets are credits. Credit **Cash in Bank** for $349.59.

T Accounts

6.
```
       Employees' State
       Income Tax Payable              Cash in Bank
       Debit    Credit              Debit    Credit
        −          +                  +         −
       349.59                                 349.59
```

Journal Entry

7.

	GENERAL JOURNAL			PAGE 32	
DATE	DESCRIPTION	POST. REF.	DEBIT	CREDIT	
20–					1
July 15	Emplys' State Inc. Tax Pay.		349 59		2
	Cash in Bank			349 59	3
	Check 209				4

Federal Unemployment Taxes

Most businesses pay the FUTA tax quarterly. If a business has accumulated FUTA taxes of less than $100, only one **annual** payment is necessary.

Business Transaction

Zip pays $130.19 for FUTA taxes. This is the balance of the **Federal Unemployment Tax Payable** account shown in **Figure 13–2**.

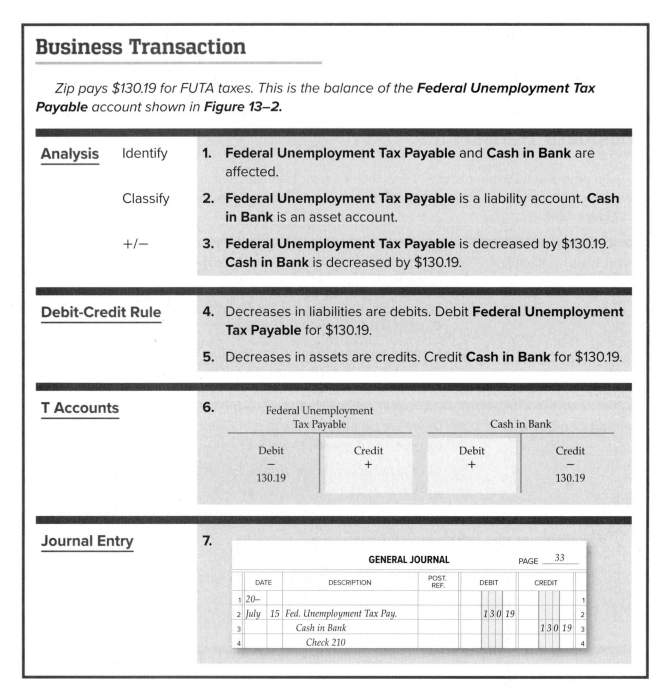

A Form 8109 is prepared and sent with the check for the FUTA tax. To indicate that FUTA taxes are being paid, the 940 oval is filled in. **Figure 13–4** shows the Form 8109 that Zip sends with the FUTA payment.

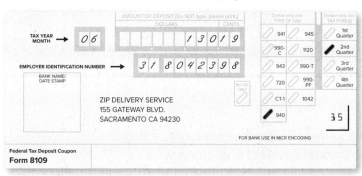

Figure 13–4 Federal Tax Deposit Coupon (Form 8109) for Federal Unemployment Taxes

State Unemployment Taxes

State unemployment taxes are usually paid on a quarterly basis.

Business Transaction

Zip pays $870.30 to the state, as shown in **Figure 13–2.**

Analysis Identify 1. **State Unemployment Tax Payable** and **Cash in Bank** are affected.

Classify 2. **State Unemployment Tax Payable** is a liability account. **Cash in Bank** is an asset account.

+/− 3. **State Unemployment Tax Payable** is decreased by $870.30. **Cash in Bank** is decreased by $870.30.

Debit-Credit Rule 4. Decreases in liability accounts are recorded as debits. Debit **State Unemployment Tax Payable** for $870.30.

5. Decreases in asset accounts are recorded as credits. Credit **Cash in Bank** for $870.30.

T Accounts 6.

State Unemployment Tax Payable		Cash in Bank	
Debit −	Credit +	Debit +	Credit −
870.30			870.30

Journal Entry 7.

GENERAL JOURNAL   PAGE 34

	DATE		DESCRIPTION	POST. REF.	DEBIT	CREDIT	
1	20–						1
2	July	15	State Unemployment Tax Pay.		870 30		2
3			Cash in Bank			870 30	3
4			Check 211				4

Other Payroll Liabilities

Employers also make payment for employees' voluntary deductions.

Business Transaction

On July 15 Zip pays $28.38 for hospital insurance premiums. (Refer to **Figure 13–1**.)

Analysis

Identify
1. The accounts **Hospital Insurance Premiums Payable** and **Cash in Bank** are affected.

Classify
2. **Hospital Insurance Premiums Payable** is a liability account. **Cash in Bank** is an asset account.

+/−
3. **Hospital Insurance Premiums Payable** is decreased by $28.38. **Cash in Bank** is decreased by $28.38.

Debit-Credit Rule
4. Decreases in liabilities are debits. Debit **Hospital Insurance Premiums Payable** for $28.38.
5. Decreases in assets are credits. Credit **Cash in Bank** for $28.38.

T Accounts
6.

Hospital Insurance Premiums Payable		Cash in Bank	
Debit − 28.38	Credit +	Debit +	Credit − 28.38

Journal Entry
7.

GENERAL JOURNAL   PAGE 35

DATE	DESCRIPTION	POST. REF.	DEBIT	CREDIT
20–				
July 15	Hospital Ins. Premiums Pay.		28 38	
	Cash in Bank			28 38
	Check 212			

Posting the Payment of Payroll Liabilities

After the payments for the employer's payroll liabilities have been journalized, the entries are posted to the appropriate general ledger accounts. **Figure 13–5** shows Zip's general ledger accounts after posting.

Figure 13–5 General Ledger Accounts after Posting of Payroll Liabilities Payments

ACCOUNT _Employee's Federal Income Tax Payable_ ACCOUNT NO. _210_

DATE	DESCRIPTION	POST. REF.	DEBIT	CREDIT	BALANCE DEBIT	BALANCE CREDIT
20–						
June 23	Balance	✓				2 9 3 18
30		G29		1 0 2 00		3 9 5 18
July 15		G31	3 9 5 18			

ACCOUNT _Employee's State Income Tax Payable_ ACCOUNT NO. _215_

DATE	DESCRIPTION	POST. REF.	DEBIT	CREDIT	BALANCE DEBIT	BALANCE CREDIT
20–						
June 23	Balance	✓				3 2 1 12
30		G29		2 8 47		3 4 9 59
July 15		G32	3 4 9 59			

ACCOUNT _Social Security Tax Payable_ ACCOUNT NO. _220_

DATE	DESCRIPTION	POST. REF.	DEBIT	CREDIT	BALANCE DEBIT	BALANCE CREDIT
20–						
June 23	Balance	✓				2 5 0 35
30		G29		8 8 27		3 3 8 62
30		G30		8 8 28		4 2 6 90
July 15		G31	4 2 6 90			

ACCOUNT _Medicare Tax Payable_ ACCOUNT NO. _225_

DATE	DESCRIPTION	POST. REF.	DEBIT	CREDIT	BALANCE DEBIT	BALANCE CREDIT
20–						
June 23	Balance	✓				5 8 63
30		G29		2 0 65		7 9 28
30		G30		2 0 65		9 9 93
July 15		G31	9 9 93			

ACCOUNT _Hospital Insurance Premium Payable_ ACCOUNT NO. _230_

DATE	DESCRIPTION	POST. REF.	DEBIT	CREDIT	BALANCE DEBIT	BALANCE CREDIT
20–						
June 23	Balance	✓				4 73
30		G29		2 3 65		2 8 38
July 15		G35	2 8 38			

ACCOUNT	State Unemployment Tax Payable				ACCOUNT NO.	240	
DATE	DESCRIPTION	POST. REF.	DEBIT	CREDIT	BALANCE		
					DEBIT		CREDIT
20—							
June 23	Balance	✓					793 41
30		G30		76 89			870 30
July 15		G34	870 30				

ACCOUNT	Federal Unemployment Tax Payable				ACCOUNT NO.	245	
DATE	DESCRIPTION	POST. REF.	DEBIT	CREDIT	BALANCE		
					DEBIT		CREDIT
20—							
June 23	Balance	✓					118 80
30		G30		11 39			130 19
July 15		G33	130 19				

Preparing Payroll Tax Forms

How Does the Government Receive Payroll Information?

Employers must complete a variety of payroll-related tax forms.

Form	Name of Form	When Filed
W-2	Wage and Tax Statement	Annually
W-3	Transmittal of Wage and Tax Statements	Annually
941	Employer's Quarterly Federal Tax Return	Quarterly
940	Employer's Annual Federal Unemployment (FUTA) Tax Return	Annually

Reading Check

Recall

What payroll taxes are deposited monthly in a Federal Reserve Bank or other authorized financial institution?

Forms W-2 and W-3

The Wage and Tax Statement, **Form W-2,** summarizes an employee's earnings and withholdings for the calendar year. Form W-2 reports (1) gross earnings, (2) federal income tax withheld, (3) FICA taxes withheld, and (4) state and local income taxes withheld.

Each employee receives a Form W-2 by January 31 of the following year. Employees use Form W-2 to prepare their individual income tax returns. **Figure 13–6** shows the Form W-2 for Emily Kardos.

Employers prepare many copies of Form W-2. The employer sends Copy A to the IRS and gives Copies B and C to the employee. The employer keeps Copy D. Additional copies are sent to city or state government if necessary.

The employer files **Form W-3,** Transmittal of Wage and Tax Statements, with the federal government along with the Forms W-2. Form W-3 summarizes the information on the Forms W-2. Forms W-2 and W-3 are due by February 28. The federal government uses the Form W-2 information to check individual income tax returns. **Figure 13–7** shows Zip's Form W-3.

Forms 941 and 940

Form 941, illustrated in **Figure 13–8** on page 400, is the employer's quarterly federal tax return that reports accumulated amounts of FICA and federal income taxes withheld from employees' earnings, as well as FICA tax owed by the employer. The employer's federal unemployment tax is reported annually on **Form 940.**

Figure 13–6 Form W-2 Wage and Tax Statement

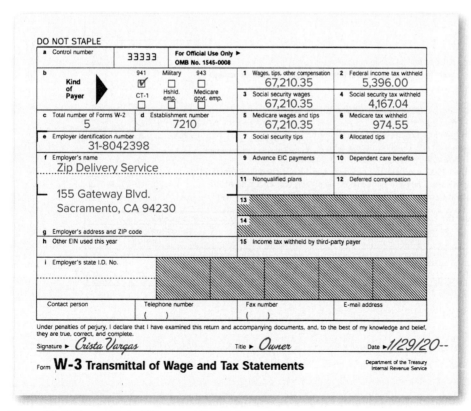

Figure 13–7 Transmittal of Wage and Tax Statements

Chapter 13 • Payroll Liabilities and Tax Records

Form **941**
Department of the Treasury
Internal Revenue Service

Employer's Quarterly Federal Tax Return
▶ See separate instructions for information on completing this return.
Please type or print.

Enter state code for state in which deposits were made ONLY if different from state in address to the right ▶ ☐ (see page 3 of instructions).

Name (as distinguished from trade name): Zip Delivery Service
Trade name, if any:
Address (number and street): 155 Gateway Blvd.

Date quarter ended: June 30, 20--
Employer identification number: 31-8042398
City, state, and ZIP code: Sacramento, CA 94230

OMB No. 1545-0029

T	
FF	
FD	
FP	
I	
T	

If address is different from prior return, check here ▶ ☐

IRS Use

If you do not have to file returns in the future, check here ▶ ☐ and enter date final wages paid ▶
If you are a seasonal employer, see **Seasonal employers** on page 1 of the instructions and check here ▶ ☐

1	Number of employees in the pay period that includes March 12th . ▶	1	
2	Total wages and tips, plus other compensation	**2**	9,252 96
3	Total income tax withheld from wages, tips, and sick pay	**3**	747 64
4	Adjustment of withheld income tax for preceding quarters of calendar year	**4**	0 00
5	Adjusted total of income tax withheld (line 3 as adjusted by line 4—see instructions) . .	**5**	773 55
6	Taxable social security wages . . . . . **6a** 9,252 96 × 12.4% (.124) =	**6b**	1,147 37
	Taxable social security tips . . . . . **6c** × 12.4% (.124) =	**6d**	0 00
7	Taxable Medicare wages and tips . . . **7a** 9,252 96 × 2.9% (.029) =	**7b**	268 33
8	Total social security and Medicare taxes (add lines 6b, 6d, and 7b). Check here if wages are not subject to social security and/or Medicare tax . . . . ▶ ☐	**8**	1,415 70
9	Adjustment of social security and Medicare taxes (see instructions for required explanation) Sick Pay $_____ ± Fractions of Cents $_____ ± Other $_____ =	**9**	0 00
10	Adjusted total of social security and Medicare taxes (line 8 as adjusted by line 9—see instructions)	**10**	1,415 70
11	**Total taxes** (add lines 5 and 10)	**11**	2,163 34
12	Advance earned income credit (EIC) payments made to employees	**12**	0 00
13	Net taxes (subtract line 12 from line 11). **This should equal line 17, column (d) below (or line D of Schedule B (Form 941))**	**13**	2,163 34
14	Total deposits for quarter, including overpayment applied from a prior quarter . . .	**14**	2,163 34
15	**Balance due** (subtract line 14 from line 13). See instructions	**15**	0 00
16	**Overpayment,** if line 14 is more than line 13, enter excess here ▶ $_____ and check if to be: ☐ Applied to next return OR ☐ Refunded.		

- **All filers:** If line 13 is less than $500, you need not complete line 17 or Schedule B (Form 941).
- **Semiweekly schedule depositors:** Complete Schedule B (Form 941) and check here . . . . ▶ ☐
- **Monthly schedule depositors:** Complete line 17, columns (a) through (d), and check here . . . . ▶ ☐

17	**Monthly Summary of Federal Tax Liability.** Do not complete if you were a semiweekly schedule depositor.		
(a) First month liability	(b) Second month liability	(c) Third month liability	(d) Total liability for quarter
627.29	627.29	908.76	2,163.34

Sign Here
Under penalties of perjury, I declare that I have examined this return, including accompanying schedules and statements, and to the best of my knowledge and belief, it is true, correct, and complete.
Signature ▶ *Crista Vargas*
Print Your Name and Title ▶ Crista Vargas, Owner
Date ▶ 6/30/20--

Figure 13–8 Form 941

SECTION 13.3
Assessment

After You Read

Reinforce the Main Idea

Different payroll tax forms are used for different purposes. Using a chart like this one, indicate the purpose of each form.

FORM	PURPOSE
W-2	
W-3	
941	
940	

Problem 4 Payment of Payroll Liabilities

The following account balances appear in the general ledger for Soap Box Laundry on April 30.

General Ledger

Social Security Tax Payable	$318.55
State Income Tax Payable	205.60
Medicare Tax Payable	113.28
Employees' Federal Income Tax Payable	286.00

Instructions Record the payment of the four liability accounts on page 34 of the general journal. (Hint: Two checks are written.)

Problem 5 Analyzing a Source Document

PAYROLL REGISTER

PAY PERIOD ENDING May 19   20--   DATE OF PAYMENT May 19, 20--

TOTALS: Regular 1,218.93 | Overtime 109.14 | Total 1,328.07 | Soc. Sec. Tax 82.34 | Medicare Tax 19.26 | Fed. Inc. Tax 184.00 | State Inc. Tax 26.56 | Hosp. Ins. 20.00 | Other — | Total 332.16 | Net Pay 995.91

Other Deductions: Write the appropriate code letter to the left of the amount: B–U.S. Savings Bonds; C–Credit Union; UD–Union Dues; UW–United Way.

Instructions Based on this payroll register, record the appropriate journal entry in your working papers. Use page 14 in the general journal.

Math for Accounting

Small businesses (those that have revenues of less than $200,000 per year) pay their federal taxes by the 15th day of the month following the payroll month. Large businesses pay their taxes every two weeks. Review the graph, and answer the following questions.

1. Which businesses pay their taxes every two weeks?
2. What percent of businesses represented in the circle graph are large businesses?

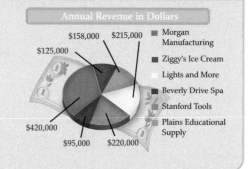

Annual Revenue in Dollars
- Morgan Manufacturing $215,000
- Ziggy's Ice Cream $158,000
- Lights and More $125,000
- Beverly Drive Spa $420,000
- Stanford Tools $95,000
- Plains Educational Supply $220,000

Chapter 13
Visual Summary

Concepts
Record payroll transactions in the general journal:

		GENERAL JOURNAL			PAGE 10
DATE		DESCRIPTION	POST. REF.	DEBIT	CREDIT
20—					
Date		Salaries Expense		xxxxx	
		Emplys' Fed. Inc. Tax Pay.			xxxxx
		Emplys' State Inc. Tax Pay.			xxxxx
		Social Sec. Tax Pay.			xxxxx
		Medicare Tax Pay.			xxxxx
		Hosp. Ins. Premiums Pay.			xxxxx
		U.S. Savings Bond Pay.			xxxxx
		Cash in Bank			xxxxx

Analysis
Compute and complete payroll tax expense forms.

Tax Type	Tax Rate	Paid By — Employee (Withholding)	Paid By — Employer (Expense)	How Paid
Federal Income Tax	Based on Form W-4	✓		EFTPS or Form 8109 (941 oval)
Social Security	6.2% of earnings up to $106,800 (2 × 6.2% = 12.4%)	✓	✓	EFTPS or Form 8109 (941 oval)
Medicare	1.45% of total earnings (2 × 1.45% = 2.9%)	✓	✓	EFTPS or Form 8109 (941 oval)
FUTA	6.2% minus SUTA rate up to 5.4% (usually 6.2% − 5.4% = .8%) of earnings up to $7,000		✓	

Procedures
Record the payment of tax liabilities in the general journal.

	GENERAL JOURNAL			PAGE 15
DATE	DESCRIPTION	POST. REF.	DEBIT	CREDIT
20—				
Date	Emplys' Fed. Inc. Tax Pay.		xxxxx	
	Social Security Tax Pay.		xxxxx	
	Medicare Tax Pay.		xxxxx	
	Cash in Bank			xxxxx

Chapter 13
Review and Activities

Answering the Essential Question

What responsibilities does an employer accept in the payroll process?

If you receive a pay check, you might not know everything the employer does "behind the scenes" to make sure that your pay is accurate and that all laws are followed. What laws does an employer have to follow regarding paying its employees?

Vocabulary Check

1. **Vocabulary** Explain how each of the following terms is related to payroll. You might want to draw a diagram to show some or all of the relationships.

   - Give three examples of *range* used as a noun.
   - Salaries Expense
   - Write a sentence using *range* as a verb.
   - Write a synonym for *appropriate*. An antonym.
   - Explain what *federal* means.
   - Write a synonym for *accumulate*.
   - Federal Unemployment Tax Act (FUTA)
   - Payroll Tax Expense
   - State Unemployment Tax Act (SUTA)
   - unemployment taxes
   - Write an antonym for *temporary*.
   - Electronic Federal Tax Payment System (EFTPS)
   - Form 8109
   - Explain what *interval* means.
   - Write a synonym for *annual*.
   - Form W-2
   - Form W-3
   - Form 940
   - Form 941

Concept Check

2. In what expense account are employees' earnings often recorded?
3. Which payroll taxes are expenses to the business?
4. **Compare** Explain how the employer's and the employee's share of FICA are calculated.
5. What is the journal entry to record the FUTA tax expense?
6. **Analyze** The amount of the FICA tax deposit is twice as much as the employer's FICA tax expense. Why?
7. What information is reported on Form 940? 941?
8. **Predict** Will an employer file more Forms 940 or 941? Why?
9. **English Language Arts** Research the history of personal income tax in the United States. Where did the idea come from and when did it begin? What were the initial conditions of the tax and how did it develop? What were the initial laws regarding penalty for nonpayment? Write a one-page report that summarizes your findings and give your views about whether or not taxing an individual's personal income tax should be a right of the government.

Chapter 13
Standardized Test Practice

Multiple Choice

1. The expense account used to journalize employees' earnings is called
   a. Employment expense.
   b. Salaries expense.
   c. Payroll expense.
   d. Employee earnings expense.

2. When posting payroll taxes to the General Ledger, which account is credited?
   a. State unemployment tax payable
   b. Social Security tax payable
   c. Payroll tax expense
   d. Payroll expense

3. When a semimonthly payroll is paid, the credit to Cash is equal to the
   a. total earning of all employees.
   b. total deductions for income tax and social security tax.
   c. total deductions.
   d. net pay for all employees.

4. The form that accompanies the payment of taxes to the federal government is a
   a. Form W-2.
   b. Form 8109.
   c. Form 940.
   d. Form 941.

True or False

5. Until paid, an employer's payroll taxes are liabilities of the business.

Short Answer

6. _____ summarizes an employee's earnings and withholdings for the calendar year.

Extended Response

7. List the tax payments an employer must make for each payroll period.

Chapter 13
Computerized Accounting

Recording and Paying Payroll Tax Liabilities
Record and Post Payroll Entries, Pay Payroll Tax Liabilities

MANUAL METHODS
- Prepare the payroll journal entry based on the payroll register totals.
- Post the payroll entries to the general ledger.
- Prepare tax liability checks to pay federal, state, and local taxing authorities.
- Prepare journal entries to record the payment of the liabilities.

COMPUTERIZED METHODS
- When the payroll checks are generated, the payroll tax journal entries are automatically prepared and posted.

Chapter 13
Problems

Problem 6 Calculating Employer's Payroll Taxes

Instructions For each of the total gross earnings amounts recorded in the past five pay weeks for Hot Suds Car Wash, determine these taxes:

- employer's FICA taxes (social security 6.2%, Medicare 1.45%)
- federal unemployment tax (.8%)
- state unemployment tax (5.4%)

Use the form provided in the working papers. None of the employees has reached the taxable earnings limit.

Total gross earnings:

1. $914.80
2. $1,113.73
3. 2,201.38
4. $791.02
5. $1,245.75

Analyze
Explain the relationship between federal and state unemployment tax rates.

Problem 7 Recording the Payment of the Payroll

The totals of the payroll register for Kits & Pups Grooming are shown in your working papers. On December 31 the owner, Abe Shultz, wrote Check 1400 to pay the payroll.

Instructions In your working papers:

1. Record the payroll entry in the general journal.
2. Post the entry to the general ledger accounts.

Analyze
Identify the payroll liability account that has the largest credit entry.

Problem 8 Journalizing Payroll Transactions

The Outback Guide Service payroll register for the week ending Dec. 31 follows:

Chapter 13
Problems

PAYROLL REGISTER

PAY PERIOD ENDING Dec. 31   20--   DATE OF PAYMENT Dec. 31, 20--

	EMP. NO.	NAME	MAR. STATUS	ALLOW	TOTAL HOURS	RATE	EARNINGS			DEDUCTIONS							NET PAY	CK. NO.
							REGULAR	OVERTIME	TOTAL	SOC. SEC. TAX	MEDICARE TAX	FED. INC. TAX	STATE INC. TAX	HOSP. INS.	OTHER	TOTAL		
1	31	Coleman, Clarence	M	1	41	7.60	304 00	11 40	315 40	19 55	4 57	21 00	6 31			51 43	263 97	1
2	28	Lorman, Victoria	S	1	30	8.00	240 00		240 00	14 88	3 48	21 00	4 80		(UD)5 40	49 56	190 44	2
3	33	Peterson, Peg	S	1	38	7.25	275 50		275 50	17 08	3 99	26 00	5 51			52 58	222 92	3
4	35	Torrez, Joyce	M	2	36	7.25	261 00		261 00	16 18	3 78	6 00	5 22		(UD)5 40	36 58	224 42	4
					TOTALS		1,080 50	11 40	1,091 90	67 69	15 82	74 00	21 84		10 80	190 15	901 75	

Other Deductions: Write the appropriate code letter to the left of the amount: B–U.S. Savings Bonds; C–Credit Union; UD–Union Dues; UW–United Way.

Instructions In your working papers:

1. Record the entry for the payment of the payroll on page 15 of the general journal. (Check 1201, dated Dec. 31).

2. Use the information in the payroll register to compute the employer's payroll taxes:
   - FICA taxes (6.2% for social security, 1.45% for Medicare)
   - federal unemployment tax (0.8%)
   - state unemployment tax (5.4%)

   None of the employees has reached the taxable earnings limit.

3. Record the entry for the employer's payroll taxes on page 15 of the general journal.

Analyze
Examine the entry recording the payment of payroll. Which account, Social Security Tax Payable or Employees' Federal Income Tax Payable, had the larger credit?

Problem 9 Recording and Posting Payroll Transactions

Showbiz Video completed the following payroll transactions during the first two weeks of December. Showbiz Video pays its employees on a biweekly basis (every two weeks).

Instructions In your working papers:

1. Record the December 13 transactions on page 38 in the general journal.
2. Post both payroll entries to the appropriate general ledger accounts.
3. Journalize and post the December 16 transactions.

Chapter 13 • Payroll Liabilities and Tax Records **407**

Chapter 13
Problems

Date	Transactions
Dec. 13	Wrote Check 2206 to pay the payroll of $3,840.58 (gross earnings) for the pay period ending December 13. The following amounts were withheld: FICA taxes, $238.12 for social security and $55.69 for Medicare; employees' federal income taxes, $639.00; employees' state income taxes, $96.02; insurance premium, $21.00; U.S. savings bonds, $20.00.
13	Recorded the employer's payroll taxes (FICA tax rates, 6.2% for social security and 1.45% for Medicare; federal unemployment tax rate, 0.8%; state unemployment tax rate, 5.4%). No employee has reached the maximum taxable amount.
16	Paid the amounts owed to the federal government for employees' federal income taxes and FICA taxes, Check 2215.
16	Purchased U.S. savings bonds for employees for $100.00, Check 2216.
16	Paid $148.00 to American Insurance Company for employees' insurance, Check 2217.

SOURCE DOCUMENT PROBLEM

Problem 13–9

Use the source documents in your working papers to complete this problem.

Analyze

Identify the payroll accounts that have a balance after entries have been posted.

Problem 10 Recording and Posting Payroll Transactions

Job Connect pays its employees twice a month. Employee earnings and tax amounts for the pay period ending December 31 are:

Gross earnings	$12,543.40
Social security tax	777.69
Medicare tax	181.88
Employees' federal income tax	662.00
Employees' state income tax	250.87

Instructions In your working papers:

1. Prepare Check 1602 (payable to "Job Connect Payroll Account") to transfer the net pay amount to the payroll checking account.

Chapter 13
Problems

2. On page 19 of the general journal, record the payment of the payroll.

3. Post the payroll transaction to the general ledger.

4. Compute payroll tax expense forms and journalize the entry to record employer's payroll taxes using these rates:
   - social security, 6.2%
   - Medicare, 1.45%
   - state unemployment, 5.4%
   - federal unemployment, 0.8%

   No employee has reached the taxable earnings limit.

5. Post the entry to the general journal.

6. Prepare checks dated December 31 to pay the following payroll liabilities:
   a. Federal unemployment taxes, payable to First City Bank (Check 1603).
   b. State unemployment taxes, payable to the State of North Carolina (Check 1604).
   c. Employees' federal income taxes and FICA taxes, payable to First City Bank (Check 1605).

7. Journalize and post the entries for the payment of the payroll liabilities.

8. Complete payroll tax expense forms. Prepare a Form 8109 for each of the two federal tax deposits paid in Instruction 6, parts (a) and (c). The oval for FICA and federal income tax is 941. The oval for the federal unemployment tax is 940.

Analyze
Calculate the employer's total payroll-related expense for the pay period.

Real-World Accounting Careers

Sarah Yates
Center Theatre Group

Q What do you do?

A I work in the Finance Department for a non-profit theatre company.

Q What are your day-to-day responsibilities?

A Some of my day-to-day responsibilities include conducting internal audits, keeping all system data up-to-date, continually updating the payroll process with new, more efficient procedures, reconciling payroll accounts, and analyzing, auditing, and approving all payroll data prior to processing.

Q What factors have been key to your success?

A Some of the key factors to my success as a payroll professional include constantly staying on top of new laws and regulations, and paying close attention to detail while still being able to see the big picture. It is very important to understand the consequences of any actions I might take. Another key to success is staying on the lookout for more efficient ways to handle the many procedures that payroll has — no matter how small.

Q What do you like most about your job?

A I love working in such a creative environment and knowing that I am truly an integral part in producing these amazing shows. If I ever start to feel bogged down by spreadsheets or tax forms, I can always go catch a show, see the crowd stand and cheer and applaud, and remember why I love my job so much.

Career Facts

Real-World Skills
Critical thinking, communication, technology

Training and Education
Bachelor's degree in accounting or finance; a CPP or FPC title

Career Skills
Advanced knowledge of payroll laws and tax regulations; basic knowledge of GAAP

Career Path
Start as an entry-level payroll specialist or generalist, then assume more responsibility as you gain experience

Tips from Robert Half International

If you'd like to learn more about a potential career path, arrange an informational interview with someone already working in that field. You can gain real-world insight and, if you make a good impression, you may even be considered for an entry-level position when one becomes available.

College and Career Readiness

Use the Internet to locate at least three resources detailing new accounting laws and regulations. Which sources do you think are the best and why?

Real-World Accounting & Connections

Career Wise

Forensic Accountant

Forensic accounting can be a great career for those who are interested in criminal justice. Forensic accountants use their accounting skills to analyze financial documents related to lawsuits or criminal proceedings and provide information about fraud or help locate assets that were gained by criminal activity.

Education Requirements

Forensic accountants must have a degree in accounting. They are also typically CPAs or Certified Fraud Examiners (CFE). It is best to have auditing or accounting work experience before becoming a forensic accountant.

Roles

Forensic accounting is a special field of accounting used in criminal investigations and lawsuits. Forensic accountants examine financial documents involved in legal proceedings and deliver their findings to the court or law enforcement officials.

Responsibilities

A forensic accountant helps recover the proceeds of crime and provides information to court and law enforcement officials to help them resolve trials and investigations. They may also testify in court as an expert witness.

Activity

Conduct research online about forensic accounting. Then, create a list of pros and cons of choosing forensic accounting as a career. When you are finished write a paragraph explaining whether you are interested in this type of accounting, and the reasons why or why not.

Global Accounting

Offshoring

Offshoring is the practice of U.S. companies employing overseas providers to perform certain business processes, such as manufacturing or even accounting. For example, technology companies like IBM and Microsoft might use programmers in India, mainly because overseas salaries are lower than U.S. salaries. Results have been mixed. Some companies report problems in employee turnover and communication. Others find that investors like the lower costs.

Activity

Define *offshoring* in your own words and discuss its potential advantages and disadvantages for a business.

21st Century Skills

Public Speaking

Imagine you are the senior payroll clerk for the Fashion Square Gift Shop. It is your duty to explain the deductions from employees' payroll checks at a new employee orientation.

Activity

1. Prepare an outline and visuals for the items you need to discuss.

2. With a classmate, take turns delivering your presentations. Focus on both verbal and nonverbal communication skills, including eye contact, gestures, and facial expressions. Give each other feedback and suggestions for improvements.

A Matter of Ethics

Money Shuffling

Suppose that you are an accounting clerk at Cybercafé. The business owner wants you to use employees' 401(k) withholdings to pay off a pressing debt. She believes she can replace the funds within two months and that the employees will not lose very much by not having their money invested during that time.

Activity

Define the ethical issue involved. What are the alternatives? Who are the affected parties? How do the alternatives affect the parties? What would you do?

Analyzing Financial Reports

Employee Costs

Employee benefits might include paid vacations and health insurance. Many employers calculate the cost of benefits as a percentage of total salaries.

Activity

Use the payroll register on page 323 and payroll tax information in this chapter to answer these questions.

1. If the employee benefits cost 24% of employee regular earnings, what is the benefits cost for this salary period?

2. If a new employee is to be hired for a 40-hour work week at a wage of $8 an hour, what would be the total weekly cost for the new employee? (The company estimates its employee benefits cost at 22%.)

Mini Practice Set 3

Main Task
Prepare the weekly payroll for Happy Trees Landscaping Service.

Summary of Steps
- Calculate gross earnings, deductions, and net pay.
- Prepare the payroll register.
- Write paychecks for the employees.
- Enter information on employee's earnings records.
- Journalize and post the payroll transaction.
- Calculate, journalize, and post the employer's payroll tax liability.
- Make the payroll tax deposit.
- Journalize and post the payroll tax deposit.
- Pay the insurance premium, and journalize and post the entry.
- Create a clear and coherent oral presentation that analyzes the results of the accounting cycle.

Why It's Important
Payroll is a basic cost of doing business.

Payroll Accounting

Happy Trees Landscaping Service

Company Background Happy Trees Landscaping Service is a plant maintenance company that is located in Kingsbury, Michigan. It is a service business that is organized as a sole proprietorship. It is owned and operated by Joanna Ecke. The company places and maintains plants, flowers, small trees outside offices and corporate buildings.

The business has been in operation for almost five years. During that time its revenue has increased each year, additional employees have been hired, and the business is now showing a good profit.

Payroll Information The business presently employs eight people. A Form W-4 is on file for each employee. The list that follows summarizes the data on those documents.

Michael Alter	Single	claims 1 exemption
Christine Cuddy	Single	claims 0 exemptions
Jesse Dubow	Single	claims 0 exemptions
Joclyn Filley	Single	claims 1 exemption
Greg Millette	Married	claims 2 exemptions
Heather Repicky	Married	claims 3 exemptions
Daniel Ripp	Married	claims 2 exemptions
Yourself	Single	claims 1 exemption

The business pays its employees on a weekly basis. Overtime is paid at the rate of 1½ times the regular rate of pay for all hours worked over 40. The weekly pay period runs from Monday through Saturday, with employees being paid on Saturday for the week's work. Because most office buildings are closed on Sunday, the business is also closed then.

The employees are paid by one of three methods: hourly rate, salary, or salary plus 10% commission on any new accounts they acquire for the company. The following table lists the employees, the method by which their wages are computed, and other pertinent information. Next, you will find the time cards for the employees who are paid on an hourly basis.

continued

Employee	Empl. no.	Position	Employee Status	Rate of Pay
Heather Repicky	010	Manager	Full-time	$725.00/ weekly salary
Greg Millette	011	Salesperson	Full-time	$450.00/ week plus 10% Commission
Jesse Dubow	012	Bookkeeper	Part-time	$250.00/ week
Joclyn Filley	013	Supply clerk	Full-time	$7.40/ hour
Daniel Ripp	016	Service	Full-time	$8.30/ hour
Christine Cuddy	018	Service	Full-time	$8.30/ hour
Michael Alter	019	Supply clerk	Part-time	$7.10/ hour
Yourself	022	Accounting clerk	Part-time	$175.00/ week

The time cards for the hourly-rate employees are shown here.

NO. 019
NAME Michael Alter
SOC. SEC. NO. 049-XX-XXXX
WEEK ENDING 7/25/20--

DAY	IN	OUT	IN	OUT	IN	OUT	TOTAL
M			2:00	5:00			
T			2:00	6:00			
W			3:00	5:00			
Th			2:00	6:00			
F			2:00	6:00			
S			9:00	2:00			
S							

NO. 018
NAME Christine Cuddy
SOC. SEC. NO. 223-XX-XXXX
WEEK ENDING 7/25/20--

DAY	IN	OUT	IN	OUT	IN	OUT	TOTAL
M	9:00	12:00	12:30	5:00			
T	9:00	11:30	12:00	5:00			
W	9:00	1:00					
Th	9:00	12:00	12:30	4:00			
F	8:30	1:00	1:30	3:00			
S	9:00	1:30					
S							

NO. 013
NAME Joclyn Filley
SOC. SEC. NO. 042-XX-XXXX
WEEK ENDING 7/25/20--

DAY	IN	OUT	IN	OUT	IN	OUT	TOTAL
M	9:00	12:00	1:00	3:00			
T	9:00	12:00	1:00	5:00			
W	8:00	12:00	1:00	5:00			
Th	9:00	12:00	1:00	3:30			
F	9:00	12:00	1:00	4:00			
S	9:00	12:00					
S							

	HOURS	RATE	AMOUNT
REGULAR			
OVERTIME			
TOTAL EARNINGS			

SIGNATURE _____ DATE _____

NO. 016
NAME Daniel Ripp
SOC. SEC. NO. 011-XX-XXXX
WEEK ENDING 7/25/20--

DAY	IN	OUT	IN	OUT	IN	OUT	TOTAL
M	9:00	12:00	12:30	5:00			
T	9:00	12:30	1:00	6:00			
W	9:00	12:00	1:00	4:30			
Th	8:30	12:30	1:00	5:00			
F	9:00	11:30	12:00	5:00			
S	9:00	1:00					
S							

	HOURS	RATE	AMOUNT
REGULAR			
OVERTIME			
TOTAL EARNINGS			

SIGNATURE _____ DATE _____

Use the federal tax tables in the working papers to determine federal income tax withholding. These are the rates for other taxes: state income tax, 2%; FICA (employee and employer contributions) social security tax, 6.2% and Medicare tax, 1.45%; state unemployment tax, 5.4%; and federal unemployment tax, 0.8%.

Happy Trees Landscaping Service

YOUR JOB RESPONSIBILITIES

The business entered the third quarter of its fiscal year at the beginning of July. It is presently the last week of July. Jesse Dubow, the bookkeeper, is leaving on vacation. In his absence you are to prepare this week's payroll.

1. Complete the timecards for the four hourly employees. Enter the total hours worked at the bottom of each card. Complete each timecard to the nearest quarter hour.

2. Greg Millette recorded additional clients that brought in $925.00 in new business. Calculate his commission and add it to his salary to determine his gross earnings.

3. Enter the payroll information for all employees in the payroll register. Each employee was recently assigned an employee number because the business is soon converting to an automated payroll system. Since this payroll is being prepared manually, list the employees in the payroll register in alphabetical order by last name.

4. Use the following information to complete the payroll register.

   a. Use the federal tax chart to determine income tax amounts to be withheld.

   b. Michael Alter, Christine Cuddy, Joclyn Filley, and Greg Millette each have $5.00 deducted for the purchase of U.S. savings bonds.

   c. Jesse Dubow, Heather Repicky, and Daniel Ripp each have $4.00 deducted for donations to the United Way.

   d. All employees pay an insurance premium each week. Married employees pay $9.00 and single employees pay $6.00.

   e. None of Happy Trees' employees has reached the maximum taxable amount for the social security tax.

5. Calculate the net pay for each employee.

6. Total all amount columns in the payroll register. Prove the accuracy of the totals.

7. Write Check 972 for the amount of the total net pay on the regular checking account of the business. Make the check payable to the Happy Trees Landscaping Service—Payroll Account. In Jesse's absence Joanna Ecke will sign the check for you. Complete the deposit slip for the payroll account.

8. Record the payroll transaction in the general journal, page 21. Use the information contained in the payroll register and Check 972 as the source documents. Post the transaction to the general ledger accounts.

9. Write the paychecks for the employees. Use the information in the payroll register to complete the check stubs. After a check has been written for an employee, enter the check number in the payroll register.

10. Enter this week's payroll information on the employee's earnings records for Michael Alter and Greg Millette only. Be sure to add the current gross earnings amount to the accumulated total.

continued

11. Calculate and record the employer's taxes for this pay period in the general journal. The source of information is the payroll register. Post the journal entry to the general ledger.

12. Make a deposit for the taxes owed to the federal government. The total includes the amounts withheld for employees' federal income tax, social security tax, and Medicare tax. Complete Form 8109 by entering the amount owed. Write Check 973, payable to the First Federal Bank for the taxes.

13. Enter the transaction in the general journal and post the entry to the general ledger.

14. Pay the monthly insurance premium by writing Check 974 to the American Insurance Company for $228.00. Record the payment in the general journal and post the entry to the general ledger.

Analyze

Identify the pay method that had the highest employee gross earnings this week.

UNIT 4

The Accounting Cycle for a Merchandising Corporation

A LOOK BACK
Unit 3 described the payroll system, including calculating and journalizing payroll for a sole proprietorship.

A LOOK AHEAD
In Unit 4, you will switch from sole proprietorships to learn how to complete the accounting cycle for a merchandising corporation.

Keys to Success

Q Why are good speaking and listening skills essential to career advancement?

A A common trait among managers and leaders is the ability to effectively communicate with others. People in leadership positions often speak to small and large groups. Whether to inform or to inspire, strong public speakers present their remarks in a style appropriate to the topic, purpose, and audience. They demonstrate command of English and get across their information clearly and concisely, enhancing their presentation with visual displays of data.

English Language Arts

Speaking and Listening Using spoken language effectively will help you lead others in the workplace. Assume you are a representative of the merchandising corporation of your choice. Give a short presentation about the company, including such information as product line, current sales figures and number of stores. *What image of the company are you trying to project? How might visual data help get across your message?*

Focus on the Photo

Because you purchase most products, such as clothing, from merchandising corporations, learning about these companies can help you be a wise consumer. *What are some large merchandising corporations and what popular items do they sell?*

Chapter 14
Accounting for Sales and Cash Receipts

Chapter Topics

14-1 Accounting for a Merchandising Business

14-2 Analyzing Sales Transactions

14-3 Analyzing Cash Receipt Transactions

Visual Summary

Review and Activities

Standardized Test Practice

Computerized Accounting

Problems

Real-World Applications & Connections

Essential Question

As you read this chapter, keep this question in mind:

How do merchandising businesses keep track of what is sold and how much money is collected?

Main Idea

A *wholesaler* sells to retailers, and a *retailer* sells to the final users. In addition to using the general ledger, a business keeps a subsidiary ledger of individual customer accounts. Merchandising businesses receive cash from cash sales, payments on account, bankcard sales, and occasionally from other types of transactions.

Chapter Objectives

Concepts

C1 Explain the difference between a service business and a merchandising business.

C2 Explain the difference between a retailer and a wholesaler.

Analysis

A1 Analyze transactions relating to the sale of merchandise.

Procedures

P1 Record sales and cash receipt transactions in a general journal.

Real-World Business Connection

Dutch Valley Foods

Nuts from supermarket bulk bins, jars of jam from a gift basket, maple syrup on a restaurant table. There is a chance some or even all of them moved through the warehouses of Pennsylvania-based distributor Dutch Valley Foods before reaching you. The company buys food and nonfood products from vendors that include such well-known brands as Hershey, Keebler and Smucker, and then sells them in bulk to the retail businesses that sell directly to you. Dutch Valley Foods is a wholesaler, a business that sells to retailers.

Connect to the Business

Retailers buy merchandise from Dutch Valley Foods for resale to the public. Depending on sales policy, the retailer will accept payment in a variety of forms. Each form of payment affects a different general ledger account.

Analyze

When Dutch Valley Foods sells foods to retailers, what general ledger accounts do you think are affected?

Focus on the Photo

As a wholesale merchandising business, Dutch Valley Foods depends on sales of its bulk items, such as nuts and dried fruit. Accurate tracking of such sales transactions helps this company decide what products to offer. **What information from sales do you think should be recorded in the accounting records?**

SECTION 14.1
Accounting for a Merchandising Business

A service business is one that provides a service to the public for a fee. In contrast a merchandising business buys goods (such as computers, clothing, and furniture) and then sells those goods to customers for a profit. You're probably familiar with merchandising businesses like Amazon.com or Wal-Mart. Most merchandising businesses operate either as retailers or as wholesalers. Some merchandising businesses are both retailers and wholesalers. A **retailer** is a business that sells to the final user, that is, to you—the consumer. A **wholesaler** is a business that sells to retailers. In this chapter we will analyze transactions relating to the sale of merchandise for The Starting Line Sports Gear, a retailer. You will learn how to carry out the accounting cycle steps for a merchandising corporation and the processes merchandising corporations use to conduct financial transactions. Refer to the chart of accounts for The Starting Line Sports Gear on page 426.

Section Vocabulary
- retailer
- wholesaler
- merchandise
- inventory
- sales
- series
- enables
- sequence
- indicated

The Operating Cycle of a Merchandising Business

How Is the Operating Cycle Different from the Accounting Cycle?

Recall that the accounting cycle is a **series** of tasks performed in a single period to maintain records. The merchandising business *operating cycle* is a series of transactions, as illustrated in **Figure 14–1**.

The collection of cash from sales **enables** the business to purchase more items to sell, pay expenses, and make a profit. As long as the company is in business, this is a continuous, repeating **sequence**.

Knowledge Extension

To organize a corporation, an application is submitted to the appropriate official (usually the corporation commissioner or secretary of state) of the state in which the company is to be incorporated. The application is called the *articles of incorporation*.

Generally, the application includes the business's name, address, the name of the person(s) organizing the corporation, stock information, and structure of the business.

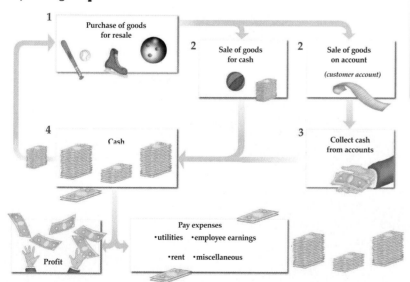

Figure 14–1 The Operating Cycle for a Merchandising Business

CHART OF ACCOUNTS

ASSETS

101	Cash in Bank	130	Supplies
105	Change Fund	135	Prepaid Insurance
110	Petty Cash Fund	140	Delivery Equipment
115	Accounts Receivable	142	Accumulated Depreciation—Delivery Equipment
117	Allowance for Uncollectible Accounts	145	Office Equipment
118	Notes Receivable	147	Accumulated Depreciation—Office Equipment
120	Interest Receivable	150	Store Equipment
125	Merchandise Inventory	152	Accumulated Depreciation—Store Equipment

LIABILITIES

201	Accounts Payable	212	Social Security Tax Payable
202	Notes Payable	213	Medicare Tax Payable
203	Discount on Notes Payable	214	Federal Unemployment Tax Payable
204	Federal Corporate Income Tax Payable	215	State Unemployment Tax Payable
205	Employees' Federal Income Tax Payable	220	Sales Tax Payable
211	Employees' State Income Tax Payable		

STOCKHOLDERS' EQUITY

301	Capital Stock	310	Income Summary
305	Retained Earnings		

REVENUE

401	Sales	410	Sales Returns and Allowances
405	Sales Discounts	415	Interest Income

COST OF MERCHANDISE

501	Purchases	510	Purchases Discounts
505	Transportation In	515	Purchases Returns and Allowances

EXPENSES

601	Advertising Expense	645	Loss/Gain on Disposal of Plant Assets
605	Bankcard Fees Expense	650	Maintenance Expense
610	Cash Short and Over	655	Miscellaneous Expense
612	Delivery Expense	657	Payroll Tax Expense
615	Depreciation Expense—Delivery Equipment	660	Rent Expense
620	Depreciation Expense—Office Equipment	665	Salaries Expense
625	Depreciation Expense—Store Equipment	670	Supplies Expense
630	Federal Corporate Income Tax Expense	675	Uncollectible Accounts Expense
635	Insurance Expense	680	Utilities Expense
640	Interest Expense		

Accounts Receivable Subsidiary Ledger

BRE	Break Point Sports Club
DIM	Dimaio, Joe
GAL	Galvin, Robert
KLE	Klein, Casey
MON	Montero, Anita

Accounts Payable Subsidiary Ledger

CHA	Champion Store Supply
COM	Computer Solutions
DAR	Dara's Delivery Service
FAS	FastLane Athletics
GEA	Geary Office Supply

Continued

Accounts Receivable Subsidiary Ledger		Accounts Payable Subsidiary Ledger	
RAH	Rahim, Shashi	PRO	Pro Runner Warehouse
RAM	Ramos, Gabriel	SLF	Sports Link Footwear
SOU	South Branch High School Athletics	SNS	Sports Nutrition Supply
SUL	Sullivan, Megan		
TAM	Tammy's Fitness Club		
WON	Wong, Kim		
YOU	Young, Lara		

Accounts Used by a Merchandising Business

What Accounts Does a Merchandising Business Use?

A merchandising business buys goods from a wholesaler or a manufacturing business and then sells these goods to its customers. Goods bought for resale are called **merchandise**. The items of merchandise the business has in stock are referred to as **inventory.**

Merchandise Inventory Account

The inventory of a business is represented in the general ledger by the asset account **Merchandise Inventory.** Increases to **Merchandise Inventory** are recorded as debits, and decreases are recorded as credits. The normal balance of the **Merchandise Inventory** account is a debit. At the beginning of each period, the dollar amount of merchandise in stock is **indicated** by the debit balance in **Merchandise Inventory.**

During the operating cycle, the business sells merchandise that is in stock and purchases new items to replace the inventory sold. The sale of merchandise and the purchase of new merchandise are recorded in separate accounts.

Merchandise Inventory	
Debit + Increase Side Normal Balance	Credit − Decrease Side

Sales	
Debit − Decrease Side	Credit + Increase Side Normal Balance

Sales Account

When a retail merchandising business sells goods to a customer, the amount of the merchandise sold is recorded in the **Sales** account. **Sales** is a revenue account. Increases to the **Sales** account are recorded as credits, and decreases are recorded as debits. The normal balance of the **Sales** account is a credit. Both cash sales and sales on account are recorded as credits to the **Sales** account.

Sales on account affect the **Accounts Receivable** account, and cash sales affect the **Cash in Bank** account.

Reading Check

Define

In your own words, define the content vocabulary terms retailer, wholesaler, merchandise, inventory, and sales.your lifestyle?

International Sales

What Challenges Face a Company That Has International Sales?

When companies have sales transactions on an international level, many complexities arise. The obligations and rights of each party to the sale extend across borders and into different sets of legal requirements.

The *United Nations Convention on Contracts for the International Sales of Goods (CISG)* was created to provide guidelines and laws governing the international sale of goods. While "The Convention" does not cover sales of all goods, it governs most business-to-business transactions.

International sales also introduce the challenge of multiple currencies. Which currency will be used for the transaction? How will currency exchange rates affect revenue? These are just a few considerations that must be examined when conducting international sales.

SECTION 14.1
Assessment

After You Read

Reinforce the Main Idea

Create a table similar to this one to describe service businesses and merchandising businesses.

Business Type	What Is Sold?	Who Is the Customer?
Service		
Retailer		
Wholesaler		

Problem 1 Recording Merchandising Transactions

Instructions Record the following transactions in T-account form in your working papers for Sharp Shot Camera Shop. A partial chart of accounts follows:

General Ledger
- Cash in Bank
- Accounts Receivable
- Merchandise Inventory
- Accounts Payable
- Sales

Date		Transactions
Apr.	4	Sold 10 Canon cameras on account for $3,000, Sales Slip 224.
	10	Sold 2 dozen photo albums for $150, cash, Sales Receipt 302.
	20	Sold 4 rolls of 35mm film for $24 cash, Sales Receipt 303.
	25	Sold a Canon camera to a customer for $380 cash, Sales Receipt 304.

Math for Accounting

Alpine Outfitters estimates the annual cost of maintaining merchandise inventory to be 10% of the inventory value. Alpine's accountants are preparing a budget for the coming year, and they plan to maintain an inventory valued at $1.5 million. Answer the following questions:

1. What is the estimated cost of maintaining the inventory?
2. If the inventory was valued at $2 million, and the estimated rate of maintenance was 11%, what would be the estimated annual maintenance cost?

SECTION 14.2
Analyzing Sales Transactions

In a merchandising business, the most frequent transaction is the sale of merchandise. Some businesses sell on a cash-only basis. Others sell only on credit. Most businesses handle both cash and credit sales.

Sales on Account

What Does a Sale on Account Involve?

The sale of merchandise that will be paid for at a later date is called a **sale on account,** a *charge sale,* or a *credit sale.* The sale on account is made to a **charge customer**; this credit option is also called a *charge account.*

Store Credit Card Sales

Charge customers use **credit cards** issued by a business such as Target to make their purchases. A store credit card, imprinted with the customer's name and account number, facilitates the sale on account.

Nonbank Credit Card Sales

In the next section, you will learn about bank credit cards. We consider nonbank credit cards here because they are similar to a store credit card. A *nonbank credit card* is a credit card issued by corporations such as American Express and Discover. Nonbank credit card sales are considered a form of credit sales because payment is collected at a later date.

Items Related to Sales on Account

A charge sale involves a sales slip, which shows the amount of tax charged and the credit terms.

The Sales Slip.

A **sales slip** is a form that lists these details: date of the sale; customer account identification; and description, quantity, and price of the item(s) sold.

The description may include the physical details (such as "white athletic socks"), a stock number, or both. A sales slip is usually prepared in multiple copies. The customer receives the original as a receipt and as proof of purchase. The number of copies kept by the business varies with its needs. A copy is always used for accounting purposes as the source document for recording the journal entry.

Prenumbered sales slips help businesses keep track of all sales made on account. The Starting Line uses prenumbered sales slips printed with its name and address. The Starting Line's sales slip is shown in **Figure 14–2** on page 426.

Section Vocabulary

- sale on account
- charge customer
- credit cards
- sales slip
- sales tax
- credit terms
- accounts receivable subsidiary ledger
- subsidiary ledger
- controlling account
- sales return
- sales allowance
- credit memorandum
- contra account
- proportion
- detect

Notice that the total amount on the sales slip includes cost of the items sold and *sales tax.*

Sales Tax.

Most states and some cities tax the retail sale of goods and services with a **sales tax.** Items subject to sales tax and sales tax rates vary from state to state. The sales tax rate is usually stated as a percentage of the sale, such as 5%. Sales tax rates are determined by the proper taxing authority.

The sales tax is paid by the customer and collected by the business. The business acts as the collection agent for the state or city government. (In the future we will refer only to the state government.) At the time of the sale, the business adds the sales tax to the total selling price of the goods. Periodically, the business sends the collected sales tax to the state. Until the state is paid, however, the sales tax collected from customers represents a liability of the business. The business keeps a record of the sales tax owed to the state in a liability account called **Sales Tax Payable.** For **Sales Tax Payable,** the increase and balance side is a credit and the decrease side is a debit.

To calculate the sales tax, multiply the merchandise subtotal by the sales tax rate (see **Figure 14–2**). Casey Klein bought $200 worth of merchandise. The sales tax rate is 6%. The sales clerk multiplied $200 by 6% (.06) to compute the $12 sales tax. The total transaction amount is $212.

Figure 14–2 The Starting Line Sports Gear Sales Slip

```
STARTING LINE
SPORTS GEAR ▶▶▶
595 Leslie Street ▶ Dallas, TX 75207
```

DATE: December 1, 20--			NO. 50	
SOLD TO	Casey Klein 3345 Spring Creek Parkway Plano, Texas 75074			
CLERK B.E.	CASH	CHARGE ✓	TERMS n/30	
QTY.	DESCRIPTION		UNIT PRICE	AMOUNT
1	Pair Running Shoes		$ 100.00	$ 100 00
6	Pair Athletic Socks		10.00	60 00
1	Vinyl Jacket/PantsPair		40.00$	40 00
			SUBTOTAL	$ 200 00
			SALES TAX	12 00
	Thank You!		TOTAL	$ 212 00

Not all sales of retail merchandise are taxed. In most states, sales to tax-exempt organizations, such as schools, are not taxed. For example, South Branch High School purchased $1,500 worth of merchandise on account. Schools are tax exempt, so no sales tax is added to the amount of the sale.

Sales Tax Payable	
Debit	Credit
–	+
Decrease Side	Increase Side Normal Balance

Credit Terms.

The sales slip in **Figure 14–2** has space to indicate the credit terms of the sale. **Credit terms** state the time allowed for payment. The credit terms for the sale to Casey Klein are n/30. The "n" stands for the *net,* or total, amount of the sale. The "30" stands for the number of days the customer has to pay for the merchandise. Casey Klein owes The Starting Line $212 (the net amount) by December 31 (30 days after December 1).

The Accounts Receivable Subsidiary Ledger

What Is a Subsidiary Ledger?

Businesses with few charge customers usually include an Accounts Receivable account for each customer in the general ledger. A large business, however, with many charge customers sets up a separate ledger that contains an account for each charge customer. This ledger is called the **accounts receivable subsidiary ledger.** A **subsidiary ledger** is a ledger, or book, that contains detailed data summarized to a controlling account in the general ledger. For example, the accounts receivable subsidiary ledger contains details of all the individuals and businesses that owe money to a company. Summary information about accounts receivable appears in the **Accounts Receivable** account in the general ledger. **Accounts Receivable** is a **controlling account** because its balance equals the total of all account balances in the subsidiary ledger. The balance of **Accounts Receivable** thus serves as a control on the accuracy of the balances in the accounts receivable subsidiary ledger after all posting is complete.

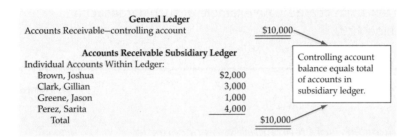

Figure 14–3 shows the accounts receivable subsidiary ledger form used by The Starting Line. The subsidiary ledger account form has lines at the top for the name and address of the customer. In a manual accounting system, subsidiary ledger accounts are arranged in alphabetical order. They are not usually numbered. In a computerized system, however, each charge customer is assigned a specific account number.

Notice that the subsidiary ledger account form has only three amount columns. The Debit and Credit columns are used to record increases and decreases to the customer's account. There is only one Balance column. Since **Accounts Receivable** is an asset account, the normal balance is a debit, so one balance amount is sufficient.

> **Knowledge Extension**
>
> Customer files are frequently maintained by businesses to keep track of relevant information on customers to maintain ongoing sales and delivery activities. Customer files may include contracts, quotes, purchase orders, warranty information, and other business documents.

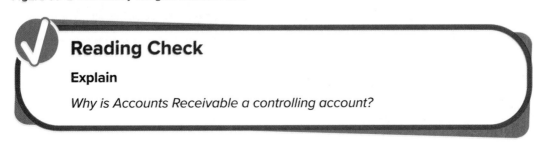

Figure 14–3 Subsidiary Ledger Account Form

✓ Reading Check

Explain

Why is Accounts Receivable a controlling account?

Recording Sales on Account

How Are Sales on Account Recorded?

According to the revenue recognition principle, revenue for a sale on account is recognized and recorded at the time of the sale, when it is earned. Revenue must also be *realizable,* which means that it is expected to be converted to cash. Look at The Starting Line's sale on account to Casey Klein in the next business transaction.

Notice that the debit in the general journal entry is to "Accounts Receivable/Casey Klein." The slash indicates that two accounts are debited: **Accounts Receivable** (controlling) and **Accounts Receivable—Casey Klein** (subsidiary).

As mentioned earlier, when merchandise is sold to tax-exempt organizations, such as school districts, sales tax is not charged. An example of such a transaction and sales receipt follows on page 426.

Business Transaction

On December 1 The Starting Line sold merchandise on account to Casey Klein for $200 plus sales tax of $12, Sales Slip 50.

STARTING LINE SPORTS GEAR
595 Leslie Street • Dallas, TX 75207

DATE: December 1, 20--   NO. 50
SOLD TO: Casey Klein, 3345 Spring Creek Parkway, Plano, Texas 75074
CLERK: B.E.   CASH: CHARGE: ✓   TERMS: n/30

QTY.	DESCRIPTION	UNIT PRICE	AMOUNT
1	Pair Running Shoes	$100.00	$100 00
6	Pair Athletic Socks	10.00	60 00
1	Vinyl Jacket/Pants	40.00	40 00
	SUBTOTAL		$200 00
	SALES TAX		12 00
	TOTAL		$212 00

Thank You!

Analysis

Identify   1. The accounts affected are **Accounts Receivable** (controlling), **Accounts Receivable—Casey Klein** (subsidiary), **Sales,** and **Sales Tax Payable.**

Classify   2. **Accounts Receivable** (controlling) and **Accounts Receivable—Casey Klein** (subsidiary) are asset accounts. **Sales** is a revenue account. **Sales Tax Payable** is a liability account.

+/−   3. **Accounts Receivable** (controlling) and **Accounts Receivable—Casey Klein** (subsidiary) are increased by the total amount, $212 (the dollar amount of merchandise sold plus sales tax). **Sales** is increased by the dollar amount of merchandise sold, $200. **Sales Tax Payable** is increased by the amount of sales tax charged, $12.

Debit-Credit Rule

4. Increases to asset accounts are recorded as debits. Debit **Accounts Receivable** (controlling) for $212. Also debit **Accounts Receivable—Casey Klein** (subsidiary) for $212.

5. Increases to revenue and liability accounts are recorded as credits. Credit **Sales** for $200 and **Sales Tax Payable** for $12.

Continued

T Accounts

6.

Accounts Receivable		Sales	
Debit + 212	Credit −	Debit −	Credit + 200

Accounts Receivable Subsidiary Ledger Casey Klein		Sales Tax Payable	
Debit + 212	Credit −	Debit −	Credit + 12

Journal Entry

7.

GENERAL JOURNAL   PAGE 20

DATE	DESCRIPTION	POST. REF.	DEBIT	CREDIT
20–				
Dec. 1	Accts. Rec./Casey Klein	✓	212 00	
	Sales			200 00
	Sales Tax Payable			12 00
	Sales Slip 50			

Business Transaction

On December 3 The Starting Line sold merchandise on account to South Branch High School Athletics for $1,500, Sales Slip 51.

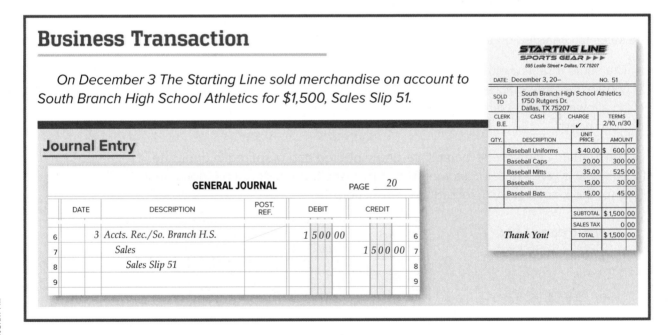

Journal Entry

GENERAL JOURNAL   PAGE 20

DATE	DESCRIPTION	POST. REF.	DEBIT	CREDIT
3	Accts. Rec./So. Branch H.S.		1 500 00	
	Sales			1 500 00
	Sales Slip 51			

This transaction is analyzed and recorded in the same manner as the December 1 entry for Casey Klein *except* there is no sales tax. The Starting Line's accountant debits **Accounts Receivable/South Branch High School Athletics** for $1,500 and credits the **Sales** account for $1,500.

Sales Returns and Allowances

All merchandising businesses expect that some customers will be dissatisfied with their purchases. The reasons for dissatisfaction vary. An item may be damaged or defective. The color or size may be incorrect. Whatever the reason, merchants usually allow dissatisfied customers to return merchandise. Any merchandise returned for credit or a cash refund is called a **sales return.**

Sometimes a customer discovers that merchandise is damaged or defective but still usable. When this happens, the merchant may reduce the sales price for the damaged merchandise. A price reduction granted for damaged goods kept by the customer is called a **sales allowance.**

The Credit Memorandum.

If the sales return or allowance occurs on a charge sale, the business usually prepares a credit memorandum. A **credit memorandum** lists the details of a sales return or allowance. The charge customer's account is credited (decreased) for the amount of the return or allowance.

Figure 14–4 on page 430 shows a credit memorandum, or credit memo, used by The Starting Line. The credit memo was prepared when Gabriel Ramos returned merchandise that he bought on account on November 29. Note that the credit memo includes a description of the returned item, the reasons for the return, and the amount to be credited to Gabriel Ramos' account.

The Starting Line's credit memo also includes spaces for the date and sales slip number of the original sale. Notice too that the total on the credit memo includes the sales tax charged on the original sale.

The same form is used if Gabriel Ramos is instead given a sales allowance. Of course, the amount credited to his account would be less. The credit granted for an allowance is the difference between the original sales price and the reduced price.

The Starting Line's credit memos are prenumbered and prepared in duplicate. The original is given to the customer. The copy is the business's source document used for the journal entry to record the transaction.

Figure 14–4 Credit Memorandum

The Sales Returns and Allowances Account.

Sales returns and allowances decrease the total revenue earned by a business. This decrease, however, is not recorded in the **Sales** account. Instead, a separate account called **Sales Returns and Allowances** is used. **Sales Returns and Allowances** summarizes the total returns and allowances for damaged, defective, or otherwise unsatisfactory merchandise. If the **Sales Returns and Allowances** account balance is large in **proportion** to the **Sales** account balance, there may be merchandising problems. The **Sales Returns and Allowances** account is carefully analyzed to **detect** any trouble.

The **Sales Returns and Allowances** account is a contra account. As a **contra account,** its balance decreases the balance of its related account. **Sales Returns and Allowances** is more specifically classified as a *contra revenue* account because it is related to a revenue account, **Sales.** Since the normal balance side of **Sales** is a credit, the normal balance side of **Sales Returns and Allowances** is a debit. This relationship is shown here:

Sales		Sales Returns and Allowances	
Debit	Credit	Debit	Credit
−	+	+	−
Decrease Side	Increase Side Normal Balance	Increase Side Normal Balance	Decrease Side

Cash Refunds

Sometimes a merchant will give a customer a cash refund instead of a credit. The Starting Line's store policy is to give a cash refund only if the original sale was a cash sale. For cash refunds the **Cash in Bank** account is credited instead of **Accounts Receivable.**

Posting to the Accounts Receivable Subsidiary Ledger

How Do You Post to the Accounts Receivable Subsidiary Ledger?

Refer to **Figure 14–5**. Look at the general journal entry. The credit is to **Accts Rec./Gabriel Ramos.** The slash indicates that both **Accounts Receivable** (controlling) and **Accounts Receivable—Gabriel Ramos** (subsidiary) are credited. Notice that a diagonal line is entered in the Post. Ref. column. This diagonal line indicates that the amount, $159, is posted in *two* places: *first* to the **Account Receivable** controlling account in the general ledger and *then* to the **Gabriel Ramos** account in the accounts receivable subsidiary ledger.

> **Common Mistakes**
>
> **Contra Accounts** Since contra revenue accounts decrease the balance of the sales account, their balance and increase sides will be on the side opposite that of the sales account. To illustrate, draw a T account, label it Sales, and write in the balance, increase, and decrease sides. Draw another T account and label it Sales Returns and Allowances or Sales Discounts. Now write in the balance, increase, and decrease sides. Compare the two.

Business Transaction

On December 4 The Starting Line issued Credit Memorandum 124 to Gabriel Ramos for the return of merchandise purchased on account, $150 plus $9 sales tax.

CREDIT MEMORANDUM NO. 124
STARTING LINE SPORTS GEAR
595 Leslie Street • Dallas, TX 75207
ORIGINAL SALES DATE: Nov. 29, 20--
ORIGINAL SALES SLIP: No. 35
APPROVAL: J.R.
☒ MDSE RET
DATE: December 4, 20--
NAME: Gabriel Ramos
ADDRESS: 278 Summit Avenue, Dallas, TX 75206

QTY	DESCRIPTION	AMOUNT
1	Athletic Suit	$150.00

REASON FOR RETURN: wrong color
SUB TOTAL: $150.00
SALES TAX: 9.00
TOTAL: $159.00

Gabriel Ramos — CUSTOMER SIGNATURE

Analysis

Identify
1. The accounts affected are **Accounts Receivable** (controlling), **Accounts Receivable—Gabriel Ramos** (subsidiary), **Sales Returns and Allowances,** and **Sales Tax Payable.**

Classify
2. **Accounts Receivable** (controlling) and **Accounts Receivable—Gabriel Ramos** (subsidiary) are asset accounts. **Sales Returns and Allowances** is a contra revenue account. **Sales Tax Payable** is a liability account.

+/−
3. **Sales Returns and Allowances** is increased by $150. **Sales Tax Payable** is decreased by $9. **Accounts Receivable** (controlling) and **Accounts Receivable—Gabriel Ramos** (subsidiary) are decreased by $159.

Debit-Credit Rule

4. Increases to a contra revenue account are recorded as debits. Debit **Sales Returns and Allowances** for $150. Decreases to liability accounts are recorded as debits. Debit **Sales Tax Payable** for $9.

5. Decreases to asset accounts are recorded as credits. Credit **Accounts Receivable** (controlling) for $159. Also credit **Accounts Receivable—Gabriel Ramos** (subsidiary) for $159.

T Accounts

6.

Accounts Receivable		Sales Returns and Allowances	
Debit +	Credit − 159	Debit + 150	Credit −

Accounts Receivable Subsidiary Ledger Gabriel Ramos		Sales Tax Payable	
Debit +	Credit − 159	Debit − 9	Credit +

Journal Entry

7.

GENERAL JOURNAL PAGE 20

DATE	DESCRIPTION	POST. REF.	DEBIT	CREDIT
4	Sales Returns and Allowances		150 00	
	Sales Tax Payable		9 00	
	Accts. Rec./Gabriel Ramos	✓		
	Credit Memorandum 124			159 00

After the amount is posted to the Accounts Receivable controlling account, the account number (115) is entered to the *left* of the diagonal line in the Posting Reference column. After the amount is posted to the subsidiary ledger account, Gabriel Ramos, a check mark (✓) is entered to the *right* of the diagonal line.

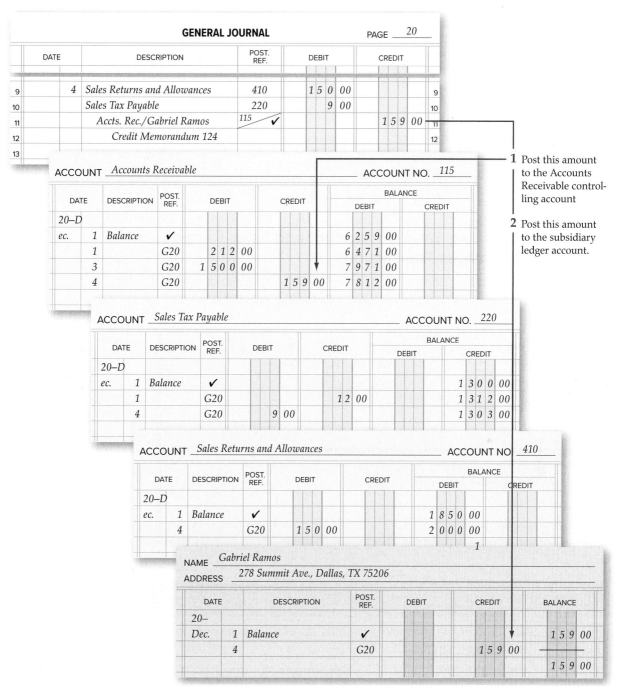

Figure 14–5 Posting to the Accounts Receivable Subsidiary Ledger

SECTION 14.2
Assessment

After You Read

Reinforce the Main Idea

Create a flowchart like this one. Enter labels in the boxes and next to the arrows. Use these terms to create the labels: general ledger, journal, posted to, recorded in, sales slip, subsidiary ledger. Terms can be used more than once.

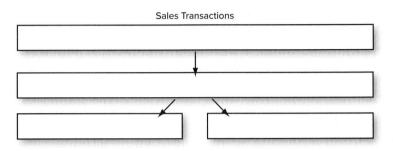

Problem 2 Recording Sales on Account and Sales Returns and Allowances Transactions

Instructions In your working papers, record the following transactions of Alpine Ski Shop on page 2 of the general journal. Use the following accounts:

General Ledger
- Cash in Bank
- Accounts Receivable
- Merchandise Inventory
- Sales Tax Payable
- Accounts Payable
- Sales
- Sales Returns and Allowances

Accounts Receivable Subsidiary Ledger
- Palmer, James
- Rodriguez, Anna

Date		Transactions
Sept.	1	Sold $300 in merchandise plus sales tax of $18 on account to James Palmer, Sales Slip 101.
	4	Sold $600 in merchandise plus $36 sales tax to Anna Rodriguez on account, Sales Slip 102.
	7	Issued Credit Memorandum 15 to James Palmer for the return of $300 in merchandise plus sales tax of $18.
	19	Anna Rodriguez telephoned the manager of Alpine Ski Shop and said that the zipper on her ski jacket is broken. The manager agreed to give her a $40 credit on her purchase, plus a $2.40 sales tax credit, Credit Memorandum 16.

Continued

SECTION 14.2
Assessment

Math for Accounting

Assume the lighting fixture industry has $.065 in sales returns and allowances for every $1.00 in sales (in other words, an industry average of 6.5%). Last year Light House Gallery had sales of $900,000 and returns and allowances of $46,800. Answer the following questions:

1. What was Light House Gallery's percentage of returns and allowances to sales?

2. Is the percentage favorable or unfavorable compared to the industry average?

SECTION 14.3
Analyzing Cash Receipt Transactions

Each business must account for the cash it receives. In this section you will explore cash sales, charge sales, bankcard sales, and cash discounts.

Section Vocabulary
- cash receipt
- cash sale
- cash discount
- sales discount

Cash Transactions

How Does Cash Come into a Business?

A transaction in which money is received by a business is called a **cash receipt.** The three most common sources of cash for a merchandising business are payments for cash sales, charge sales, and bankcard sales. Cash is also received, though much less frequently, from other types of transactions. Let's learn how to handle these four kinds of cash receipts.

Cash Sales

In a **cash sale** transaction, the business receives full payment for the merchandise sold *at the time of the sale.* The proof of sale and the source document generated by a cash sale transaction differ from those for a sale on account.

Most retailers use a cash register to record cash sales. Instead of using preprinted sales slips, cash sales are recorded on two rolls of paper tape inside the cash register. The details of a cash sale are printed on the two tapes at the same time. The portion of one tape that contains a record of the sale is torn off and given to the customer as a receipt. The other tape remains in the register.

A business totals and clears its cash register daily. The cash register tape lists the total cash sales and the total sales tax collected on these sales. The tape also shows the day's total charge sales. A proof is usually prepared to show that the amount of cash in the cash register equals the amount of cash sales and sales tax recorded on the cash register tape. The proof and the tape are sent to the accounting clerk, who uses the tape like the one in **Figure 14–6** as the source document for the journal entry to record the day's cash sales.

Figure 14–6 Cash Register Tape

Charge Customer Payments

Businesses record cash received on account from charge customers by preparing receipts. A receipt, shown in **Figure 14–7**, is a form that serves as a record of cash received. Receipts are prenumbered and may be prepared in multiple copies. The receipt is a source document for the journal entry.

Reading Check

Recall

What are the three most common sources of cash for a merchandising business?

Bankcard Sales

Many businesses accept bankcards. Unlike a store credit card, which is issued by a business and is used only at that business, a bankcard is issued by a bank and honored by many businesses. The most widely used bank *credit cards* in North America are VISA, MasterCard, and Discover.

Figure 14–7 Receipt for Cash Received from a Charge Customer

A debit card requires the entry of a personal identification number (PIN) on a keypad. The advantage of both cards to a store is that it does not have to wait to receive payment until the bank collects from the cardholder.

Both bank credit card and debit card transactions are usually recorded as though they are cash sales. However, some companies use a separate account for credit card sales.

Bankcard sales can be processed manually using multicopy bankcard slips or electronically. Either way, the total bankcard sales and related sales taxes are totaled and listed on the end-of-day cash register tape. **Figure 14–8** shows a cash register tape indicating the day's bankcard sales and related sales tax. The cash register tape is the source document to record bankcard sales. Bankcard sales are included on the daily cash proof.

In a manual system, the business uses a special deposit slip to deposit the bankcard and credit card slips in its checking account. There is often a three- or four-day delay before the amount is credited to the checking account. This is due to the time it takes the store's bank to collect the funds from the various banks that issued the customers' bankcards. In an electronic system, bankcard and credit cards are usually transmitted in daily batches, and the amount may be credited to the checking account of the business on the same or the next business day. Deposits of bankcard sales slips or electronic batch transmittal records are treated the same as cash deposits.

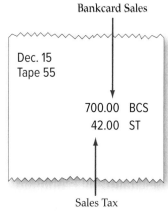

Figure 14–8 Bankcard Sales Tape

Other Cash Receipts

Merchants may also receive cash from infrequent transactions, such as a bank loan or the sale of assets other than merchandise. A receipt is prepared to indicate the source of the cash received.

Cash Discounts

Why Do Businesses Give Cash Discounts?

To encourage charge customers to pay promptly, some merchandisers offer a cash discount. A **cash discount,** or **sales discount,** is the amount a customer can deduct from the amount owed for purchased merchandise if payment is made within a certain time. A cash discount is an advantage to both the buyer, who receives merchandise at a reduced cost, and the seller, who receives cash quickly.

Businesses do not offer cash discounts to all customers. Some offer them only to business customers. The Starting Line offers a cash discount to charge customers who buy merchandise in large quantities. Its credit terms are *2/10, n/30.* These terms mean that the customer can deduct 2% of the merchandise cost if it pays within 10 days of the sale date. Otherwise, the full (net) amount is due within 30 days. A cash discount decreases the amount the business actually receives from the sale. Let's look at an example.

On December 3 The Starting Line sold $1,500 worth of merchandise on account to South Branch High School Athletics. It records the transaction as a credit to **Sales** and a debit to **Accounts Receivable** for $1,500. If South Branch pays within 10 days (by December 13), it will receive a cash discount. Starting Line will receive $1,470, or the original price less the cash discount of $30.

> **Knowledge Extension**
>
> **Additional Sales Tax Considerations** In some states, such as Texas, the sales tax amount owed by the customer is reduced in addition to the cash discount when a customer pays within the specified early payment terms. For example, if a business sells $1,000 worth of merchandise with a state general tax rate of 6.25% and 2/10, n/30 credit terms, the customer may pay $1,041.25 if paid within 10 days of the sale date. The 2% cash discount decreases the sales tax amount owed in addition to the merchandise discount amount. Cash in Bank would be debited for $1,041.25, Sales Discount debited for $20.00, Sales Tax Payable debited for $1.25, and Accounts Receivable credited for $1,062.50. The full (net) amount of $1,062.50 is otherwise due within 30 days.

Merchandise Sold	×	Discount Rate	=	Discount
$1,500	×	.02	=	$30

Sales Slip Amount	−	Discount Amount	=	Amount Paid Within Discount Period
$1,500	−	$30	=	$1,470

Cash in Bank is debited for $1,470, the amount of cash actually received. **Accounts Receivable** is credited for the full $1,500 because the customer paid for the merchandise and does not owe any more on the purchase. The difference between $1,500 and $1,470, $30, is the discount amount. A cash discount is recorded only when the customer pays for he merchandise within the time stated. The discount is on the price of the merchandise *before* taxes.

A separate account is used to record cash discounts taken by customers. The $30 discount is entered in the contra revenue account **Sales Discounts,** which reduces the revenue earned from sales. The normal balance of the **Sales** account is a credit. The normal balance of the **Sales Discounts** account is a debit.

Sales		Sales Discounts	
Debit − Decrease Side	Credit + Increase Side Normal Balance	Debit + Increase Side Normal Balance	Credit − Decrease Side

✓ Reading Check

Explain

How is a purchase made with a credit card different from a purchase made with a bankcard?

Recording Cash Receipts

How Do Businesses Record the Receipt of Cash?

This section discusses recording cash from the four sources.

Charge Customer Payments

Let's look at a payment from a charge customer.

Business Transaction

On December 5 The Starting Line received $212 from Casey Klein to apply to her account, Receipt 301.

STARTING LINE SPORTS GEAR
595 Leslie Street ▸ Dallas, TX 75207

RECEIPT No. 301
Dec. 5 20—
RECEIVED FROM Casey Klein $ 212.00
Two hundred twelve and no/100 ———————— DOLLARS
FOR On account
RECEIVED BY *Michael Smith*

Analysis	Identify	1. The accounts affected are **Cash in Bank**, **Accounts Receivable** (controlling), **Accounts Receivable—Casey Klein** (subsidiary).
	Classify	2. **Cash in Bank**, **Accounts Receivable** (controlling), and **Accounts Receivable—Casey Klein** (subsidiary) are asset accounts.
	+/−	3. **Cash in Bank** increases by $212. **Accounts Receivable** (controlling) and **Accounts Receivable—Casey Klein** (subsidiary) decrease by $212.
Debit-Credit Rule		4. Increases to asset accounts are recorded as debits. Debit **Cash in Bank** for $212.
		5. Decreases to asset accounts are recorded as credits. Credit **Accounts Receivable** (controlling) for $212. Also credit **Accounts Receivable—Casey Klein** (subsidiary) for $212.

Continued

T Accounts

6.

```
        Cash in Bank                    Accounts Receivable
    Debit    |   Credit              Debit    |   Credit
      +      |     −                   +      |     −
     212     |                                |    212
```

```
                                   Accounts Receivable
                                   Subsidiary Ledger
                                       Casey Klein
                                   Debit    |   Credit
                                     +      |     −
                                            |    212
```

Journal Entry

7.

GENERAL JOURNAL   PAGE 20

DATE	DESCRIPTION	POST. REF.	DEBIT	CREDIT
5	Cash in Bank		212 00	
	Accts. Rec./Casey Klein			212 00
	Receipt 301			

Cash Discount Payments

When a customer pays for a purchase on account within the discount period, the amount paid equals the invoice amount less the cash discount.

Business Transaction

On December 12 The Starting Line received $1,470 from South Branch High School Athletics in payment of Sales Slip 51 for $1,500 less the discount of $30, Receipt 302.

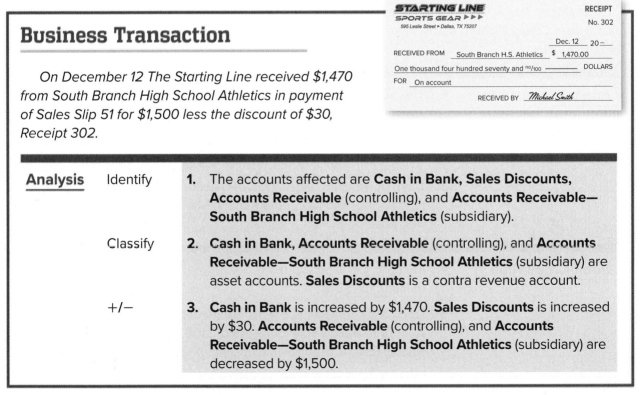

STARTING LINE SPORTS GEAR
595 Leslie Street • Dallas, TX 75207
RECEIPT No. 302
Dec. 12 20—
RECEIVED FROM South Branch H.S. Athletics $ 1,470.00
One thousand four hundred seventy and no/100 DOLLARS
FOR On account
RECEIVED BY Michael Smith

Analysis

Identify 1. The accounts affected are **Cash in Bank**, **Sales Discounts**, **Accounts Receivable** (controlling), and **Accounts Receivable— South Branch High School Athletics** (subsidiary).

Classify 2. **Cash in Bank**, **Accounts Receivable** (controlling), and **Accounts Receivable—South Branch High School Athletics** (subsidiary) are asset accounts. **Sales Discounts** is a contra revenue account.

+/− 3. **Cash in Bank** is increased by $1,470. **Sales Discounts** is increased by $30. **Accounts Receivable** (controlling), and **Accounts Receivable—South Branch High School Athletics** (subsidiary) are decreased by $1,500.

Continued

Debit-Credit Rule	4. Increases to asset accounts are recorded as debits. Debit **Cash in Bank** for $1,470. Increases to contra revenue accounts are recorded as debits. Debit **Sales Discounts** for $30.
	5. Decreases to asset accounts are recorded as credits. Credit **Accounts Receivable** (controlling) for $1,500. Also credit **Accounts Receivable—South Branch High School Athletics** (subsidiary) for $1,500.

T Accounts

6.

Cash in Bank		Accounts Receivable	
Debit + 1,470	Credit −	Debit +	Credit − 1,500

Sales Discounts		Accounts Receivable Subsidiary Ledger South Branch High School Athletics	
Debit + 30	Credit −	Debit +	Credit − 1,500

Journal Entry

7.

	GENERAL JOURNAL			PAGE 20
DATE	DESCRIPTION	POST. REF.	DEBIT	CREDIT
12	Cash in Bank		1 470 00	
	Sales Discounts		30 00	
	Accts. Rec./South Br. H.S.			1 500 00
	Receipt 302			

Cash Sales

As a rule, businesses journalize cash sales and make cash deposits daily. Let's analyze transactions relating to sale of merchandise for cash on December 15.

Business Transaction

The Starting Line had bankcard sales of $700 and collected $42 in relate sales taxes on December 15, Tape 55.

Dec. 15
Tape 55

3000.00 CA
180.00 ST

Analysis	Identify	1. The accounts affected are **Cash in Bank, Sales,** and **Sales Taxes Payable.**
	Classify	2. **Cash in Bank** is an asset account. **Sales** is a revenue account. **Sales Tax Payable** is a liability account.
	+/−	3. **Cash in Bank** is increased by $3,180. **Sales** is increased by $3,000. **Sales Tax Payable** is increased by $180.

Continued

Debit-Credit Rule	**4.** Increases in asset accounts are recorded as debits. Debit **Cash in Bank** for $3,180.
	5. Increases in revenue and liability accounts are recorded as credits. Credit **Sales** for $3,000, and **Sales Tax Payable** for $180.

T Accounts

6.

```
       Cash in Bank                           Sales
   Debit    |    Credit              Debit    |    Credit
     +      |      −                   −      |      +
   3,180    |                                 |    3,000

                                     Sales Tax Payable
                                      Debit    |    Credit
                                        −      |      +
                                               |     180
```

Journal Entry

7.

	GENERAL JOURNAL			PAGE 20
DATE	DESCRIPTION	POST. REF.	DEBIT	CREDIT
15	Cash in Bank		3 1 8 0 00	
	Sales			3 0 0 0 00
	Sales Tax Payable			1 8 0 00
	Tape 55			

Bankcard Sales

Let's record sales paid by bankcard. Note the similarity to cash sales.

Business Transaction

The Starting Line had bankcard sales of $700 and collected $42 in relate sales taxes on December 15, Tape 55.

```
Dec. 15
Tape 55

   700.00  BCS
    42.00  ST
```

Analysis	Identify	**1.** The accounts affected are **Cash in Bank, Sales,** and **Sales Tax Payable.**
	Classify	**2. Cash in Bank** is an asset account. **Sales** is a revenue account. **Sales Tax Payable** is a liability account.
	+/−	**3. Cash in Bank** is increased by $742. **Sales** is increased by $700. **Sales Tax Payable** is increased by $42.

Continued

Debit-Credit Rule	4. Increases in asset accounts are recorded as debits. Debit **Cash in Bank** for $3,180. 5. Increases in revenue and liability accounts are recorded as credits. Credit **Sales** for $3,000, and **Sales Tax Payable** for $180.

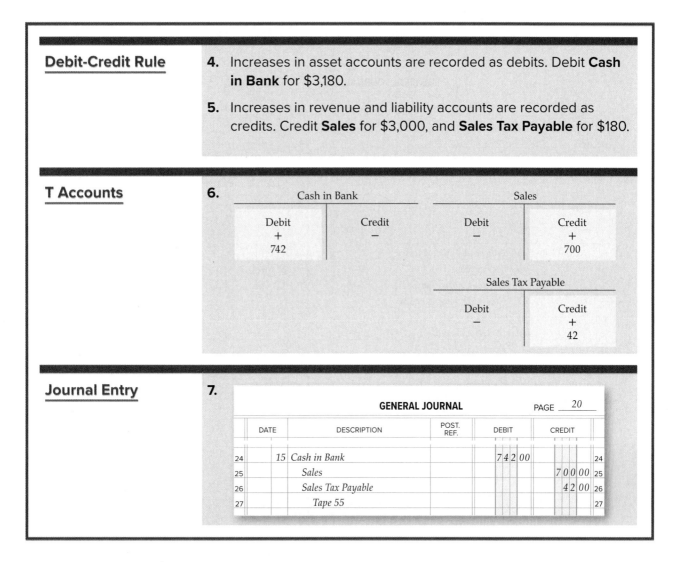

Other Cash Receipts

Occasionally a business receives cash from a transaction that does not involve the sale of merchandise. The **Sales** account is not used because the item is not a *merchandise* item.

Business Transaction

On December 16 The Starting Line received $30 from Mandy Harris, an office employee. She purchased a calculator that the business was no longer using, Receipt 303.

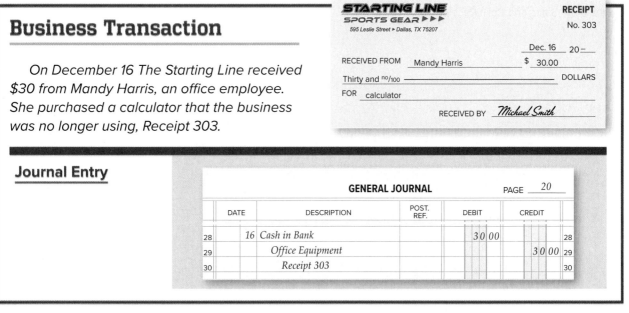

Chapter 14 • Accounting for Sales and Cash Receipts **443**

The transactions discussed in this chapter are shown in **Figure 14–9**.

	DATE		DESCRIPTION	POST. REF.	DEBIT	CREDIT	
	\multicolumn{2}{l}{GENERAL JOURNAL}				PAGE 20	
1	20–						1
2	Dec.	1	Accts. Rec./Casey Klein	/	2 1 2 00		2
3			Sales			2 0 0 00	3
4			Sales Tax Payable			1 2 00	4
5			Sales Slip 50				5
6		3	Accts. Rec./South Branch H.S.	/	1 5 0 0 00		6
7			Sales			1 5 0 0 00	7
8			Sales Slip 51				8
9		4	Sales Returns and Allowances		1 5 0 00		9
10			Sales Tax Payable		9 00		10
11			Accts. Rec./Gabriel Ramos	/		1 5 9 00	11
12			Credit Memorandum 124				12
13		5	Cash in Bank		2 1 2 00		13
14			Accts. Rec./Casey Klein	/		2 1 2 00	14
15			Receipt 301				15
16		12	Cash in Bank		1 4 7 0 00		16
17			Sales Discounts		3 0 00		17
18			Accts. Rec./South Branch H.S.	/		1 5 0 0 00	18
19			Receipt 302				19
20		15	Cash in Bank		3 1 8 0 00		20
21			Sales			3 0 0 0 00	21
22			Sales Tax Payable			1 8 0 00	22
23			Tape 55				23
24		15	Cash in Bank		7 4 2 00		24
25			Sales			7 0 0 00	25
26			Sales Tax Payable			4 2 00	26
27			Tape 55				27
28		16	Cash in Bank		3 0 00		28
29			Office Equipment			3 0 00	29
30			Receipt 303				30
31							31

Figure 14–9 Sales and Cash Receipt Transactions

SECTION 14.3
Assessment

After You Read

Reinforce the Main Idea

Create a table similar to this one to analyze four different types of cash receipt transactions, the debit and credit parts of each type, and the source document for each.

1. Cash Transaction	2. Account(s) Debited	3. Account(s) Credited	4. Source Document

Problem 3 Analyzing a Source Document

As the accounting clerk for Super Cycle Shop, you record the business transactions. The store's manager hands you the source document shown here.

Instructions Analyze the source document and record the necessary entries on page 17 of the general journal.

```
May 15
Tape 40

        1000.00  CA
          60.00  ST
         800.00  BCS
          48.00  ST
```

Problem 4 Recording Cash Receipts

Commerce Technology, a computer equipment retailer, had the following selected transactions in March.

Instructions Record each transaction on page 4 of the general journal in your working papers.

Date	Transactions
Mar. 1	Sold one modem for $130 plus $10.40 sales tax, Sales Slip 49.
5	Sold one computer monitor to Kelly Wilson on account for $300 plus $24 sales tax, Sales Slip 55.
17	Bankcard sales totaled $750 plus $60 sales tax, Tape 65.

Math for Accounting

This graph illustrates the sales of flowers throughout the year for Randy's Florist.

1. The sale of tulips was highest in which month?
2. Which type of flower sells at a steady rate, regardless of the month?

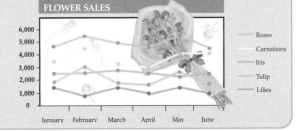

Chapter 14
Visual Summary

Concepts

Service Business	Merchandising Business
Provides a service to the public for a fee.	Buys goods and then sells those goods to customers for a profit.

Analysis
Graphic of receipt with the following information:

Chuck's Furniture

DATE: April 1, 20--   NO. 121

SOLD TO: Abner Little, 8220 Nasturtium Ln., Altus, OK 73521

CLERK: B.E.   CASH   CHARGE: ✓   TERMS: n/30

QTY	DESCRIPTION	UNIT PRICE	AMOUNT
1	Recliner	$599.00	$599.00
2	End Tables	$40.00	80.00
4	Pillows	$25.00	100.00
		SUBTOTAL	$779.00
		SALES TAX	68.16
	Thank You!	TOTAL	847.16

Identify
The accounts affected are: **Accounts Receivable** (controlling), **Accounts Receivable—Abner Little** (subsidiary), **Sales**, and **Sales Tax Payable**.

Classify
Accounts Receivable (controlling), and **Accounts Receivable—Abner Little** (subsidiary) are asset accounts. **Sales** is a revenue account. **Sales Tax Payable** is a liability account.

+/−
Accounts Receivable (controlling) and **Accounts Receivable—Abner Little** (subsidiary) are increased by the total amount, $847.16 (the dollar amount of merchandise sold plus tax). **Sales** is increased by the dollar amount of the merchandise sold ($779.00). **Sales Tax Payable** is increased by the amount of sales tax charged ($68.16).

Procedures
General Journal entries with the following information:

GENERAL JOURNAL   PAGE 25

	DATE	DESCRIPTION	POST. REF.	DEBIT	CREDIT	
1	20--					1
2	Apr 1	Accts. Rec./Abner Little		847.16		2
3		Sales			779.00	3
4		Sales Tax Payable			68.16	4
5		Sales Slip 121				5
6						6
7	Apr 5	Cash in Bank		635.24		7
8		Accts. Rec./John Eckhardt			635.24	8
9		Sales Slip 122				9
10						10

Chapter 14
Review and Activities

Answering the Essential Question

How do merchandising businesses keep track of what is sold and how much money is collected?

How does this benefit you as a consumer?

Vocabulary Check

1. **Vocabulary** Find each content vocabulary term below in the text and write your own definition of the term based on what you find.

   - enable
   - retailer
   - sequence
   - series
   - wholesaler
   - indicate
   - inventory
   - merchandise
   - Sales
   - charge customer
   - credit cards
   - sale on account
   - sales slip
   - accounts receivable subsidiary ledger
   - controlling account
   - credit terms
   - sales tax
   - subsidiary ledger
   - credit memorandum
   - sales allowance
   - sales return
   - contra account
   - detect
   - proportion
   - cash receipt
   - cash sale
   - cash discount
   - sales discount

Concept Check

2. **Compare and Contrast** How are a service business and a merchandising business alike? How are they different?

3. Name the two types of merchandising businesses. Can they both collect sales tax? If not, which one can?

4. What accounts are posted for a sale on account?

5. **Apply** True or False: State laws prevent your local department store from acting as a tax collector.

6. **Classify** Classify the following accounts: **Merchandise Inventory, Sales, Sales Returns and Allowances, Sales Discounts,** and **Sales Tax Payable.**

7. Which accounts are affected when a business receives a payment from a charge customer who has taken a cash discount?

8. **Math** Johnson Auto Repair Shop is purchasing a cruise-control module for a 6-year old sports coupe. The list price in the wholesaler's catalog is $845. Johnson receives a 40% trade discount. What is the net price?

9. **English Language Arts** Research the history of the modern corporation. Describe how it was impacted by industrialization in Britain and the United States, and comment on how it has, impacted the standard of living in the U.S. Present your findings in a two-page essay.

Chapter 14
Standardized Test Practice

Multiple Choice

1. Which of the following actions does not impact the Sales account?

    a. cash sales

    b. sales on account

    c. projected sales

    d. increases to accounts receivable

2. Which type of account has a balance that equals the total of all account balances in the subsidiary ledger?

    a. subsidiary account

    b. managing account

    c. cash account

    d. controlling account

3. What type of account is Sales Returns and Allowances?

    a. subsidiary account

    b. contra account

    c. managing account

    d. controlling account

4. If the merchandise is purchased for $1,000 on August 1, with terms of 2/10 n/30, the amount due to the vendor on August 11 is

    a. $1,000.

    b. $990.

    c. $980.

    d. $20.

5. The journal entry for a cash receipt on account is

    a. debit Cash; credit Accounts Receivable.

    b. debit Cash; credit Accounts Payable.

    c. debit Accounts Payable; credit Cash.

    d. debit Accounts Payable; credit Accounts Receivable

Extended Response

6. Name one thing that cash sales, sales on account, and bankcard sales have in common.

7. Explain the differences between cash sales and sales on account.

Chapter 14
Computerized Accounting

Sales and Cash Receipts
Recording Sales Transactions and Cash Receipts Transactions

MANUAL METHODS	COMPUTERIZED METHODS
• Prepare general journal entries based on a sales slip or an invoice.	• Invoices can be created with the software and posted to the general ledger accounts at the same time.
• Post journal entries to the appropriate general ledger accounts.	• New account balances are calculated for you.
• Calculate new account balances.	
• Prepare journal entries based on deposit slips, receipts, or cash register tapes.	• Deposits are recorded and posted to the general ledger using the cash receipts task iteam.
• Post journal entries to the appropriate general ledger accounts.	• General ledger accounts are updated automatically.
• Calculate new account balances.	

Chapter 14
Problems

Problem 5 Recording Sales and Cash Receipts

Sunset Surfwear, a California-based merchandising store, had the following sales and cash receipt transactions for January. The partial chart of accounts for Sunset Surfwear follows.

General Ledger

101	Cash in Bank	401	Sales
115	Accounts Receivable	405	Sales Discounts
215	Sales Tax Payable	410	Sales Returns and Allowances

Accounts Receivable Subsidiary Ledger

ADA	Adams, Martha	MOU	Moulder, Nate
HAM	Hamilton, Alex	WES	Westwood High School Athletics
JUN1	Jun, Helen		

Instructions Record the following transactions on page 20 of the general journal.

Date	Transactions
Jan. 1	Sold $300 in merchandise plus a sales tax of $18 on account to Martha Adams, Sales Slip 777.
5	Sold $1,500 in merchandise on account to Westwood High School Athletics, Sales Slip 778.
7	Received $400 from Alex Hamilton on account, Receipt 345.
10	Issued Credit Memorandum 102 to Martha Adams for $318 covering $300 in returned merchandise plus $18 sales tax.
15	Recorded cash sales of $800 plus $48 in sales tax, Tape 39.
15	Recorded bankcard sales of $900 plus $54 in sales tax, Tape 39.
20	Received $1,500 from Westwood High School Athletics in payment of Sales Slip 778, Receipt 346.
25	Sold $1,200 in merchandise plus sales tax of $72 on account to Helen Jun, Sales Slip 779.
28	Granted a $106 sales allowance to Helen Jun, which includes $100 for damaged merchandise she kept and sales tax of $6, Credit Memorandum 103.
30	Received $500 from a charge customer, Nate Moulder, in payment of his $500 account, Receipt 347.

Continued

Chapter 14 Problems

Analyze
Calculate the sum of all the debits to the Sales Returns and Allowances account during January.

Problem 6 — Posting Sales and Cash Receipts

The January transactions for InBeat CD Shop are recorded on page 15 of the general journal.

Instructions Post the transactions to the general ledger and subsidiary ledger in your working papers.

GENERAL JOURNAL PAGE 15

DATE		DESCRIPTION	POST. REF.	DEBIT	CREDIT
20–					
Jan.	1	Cash in Bank		388 00	
		Sales Discounts		12 00	
		Accts. Rec./Alicia Alvarez			400 00
		Receipt 92			
	4	Accts. Rec./Dena Greenburg		735 00	
		Sales			700 00
		Sales Tax Payable			35 00
		Sales Slip 60			
	6	Cash in Bank		2 100 00	
		Sales			2 000 00
		Sales Tax Payable			100 00
		Tape 32			
	6	Cash in Bank		3 150 00	
		Sales			3 000 00
		Sales Tax Payable			150 00
		Tape 32			
	8	Sales Returns and Allowances		700 00	
		Sales Tax Payable		35 00	
		Accts. Rec./ Dena Greenburg			735 00
		Credit Memorandum 15			
	10	Cash in Bank		1 358 00	
		Sales Discounts		42 00	
		Accts. Rec./Joe Montoya			1 400 00
		Receipt 93			
	15	Accts. Rec./Alicia Alvarez		420 00	
		Sales			400 00
		Sales Tax Payable			20 00
		Sales Slip 61			
	27	Cash in Bank		1 365 00	
		Accts. Rec./Chelsea Wright			1 365 00
		Receipt 94			

Analyze
Identify the customer with the highest balance at the end of January.

Chapter 14 Problems

Problem 7 Recording Sales and Cash Receipts

Shutterbug Cameras had the following transactions during January. The partial chart of accounts for Shutterbug is shown here.

General Ledger

101	Cash in Bank	401	Sales
115	Accounts Receivable	405	Sales Discounts
130	Supplies	410	Sales Returns and Allowances
215	Sales Tax Payable		

Accounts Receivable Subsidiary Ledger

DIA	Diaz, Arturo	NAK	Nakata, Yoko
FAS	FastForward Productions	SUL	Sullivan, Heather

Instructions Record the transactions on page 5 of the general journal in your working papers.

Date	Transactions
Jan. 1	Sold merchandise on account to Yoko Nakata for $250 plus a 4% sales tax of $10, Sales Slip 90.
3	Received $50 in cash from the sale of supplies to Betty's Boutique, Receipt 201.
7	Sold $300 in merchandise plus a sales tax of $12 to Arturo Diaz on account, Sales Slip 91.
12	Sold on account $1,500 in merchandise plus a sales tax of $60 to FastForward Productions, credit terms 2/10, n/30, Sales Slip 92.
13	Issued Credit Memorandum 20 for $312 to Arturo Diaz, which includes $300 in merchandise returned by him plus sales tax of $12.
14	Received a check for $260 from Yoko Nakata in full payment of his account, Receipt 202.
15	Cash sales amounted to $2,500 plus $100 in sales tax, Tape 75.
15	Bankcard sales were $3,000 plus $120 in sales tax, Tape 75.
21	Received a check for $1,530 from FastForward Productions in payment of their $1,560 account balance less a cash discount of $30, Receipt 203.
28	Granted Heather Sullivan a $104 allowance for damaged merchandise of $100 plus a 4% sales tax of $4, Credit Memorandum 21.

Continued

Chapter 14
Problems

Analyze
Compute the net amount of sales tax for the month based on these transactions.

Problem 8 Recording Sales and Cash Receipt Transactions

River's Edge Canoe & Kayak is a merchandising business in Wyoming. The partial chart of accounts follows:

General Ledger
- 101 Cash in Bank
- 115 Accounts Receivable
- 135 Supplies
- 215 Sales Tax Payable
- 401 Sales
- 405 Sales Discounts
- 410 Sales Returns and Allowances

Accounts Receivable Subsidiary Ledger
- ADV Adventure River Tours
- DRA Drake, Paul
- WILD Wildwood Resorts
- WU Wu, Kim

Instructions Record January transactions on page 10 of the general journal.

Date	Transactions
Jan. 1	Sold $2,000 in merchandise on account to Wildwood Resorts, a tax-exempt agency, credit terms 3/15, n/30, Sales Slip 103.
5	Granted Wildwood Resorts a $150 credit allowance for defective merchandise, Credit Memorandum 33.
8	Received $485 from Adventure River Tours for $500 in merchandise sold to it on Dec. 27 less a 3% cash discount of $15, Receipt 96.
10	Sold $500 in merchandise plus a 5% sales tax of $25 to Paul Drake on account, credit terms 3/15, n/30, Sales Slip 104.
12	Received a check for $1,794.50 from Wildwood Resorts on account ($1,850 less a 3% cash discount of $55.50), Receipt 97.
15	Cash sales were $3,500 plus sales tax of $175, Tape 22.
15	Bankcard sales amounted to $4,000 plus sales tax of $200, Tape 22.
20	Sold to Adventure River Tours $75 in supplies. Cash received recorded on Receipt 98.

Continued

Chapter 14
Problems

Date	Transactions
Jan. 22	Granted Kim Wu $63 credit for $60 in damaged merchandise sold to her last month and 5% sales tax of $3 on the merchandise, Credit Memorandum 34.
25	Paul Drake sent a check for $510 in payment of his account. The account balance was $525 ($500 in merchandise and $25 sales tax). He took a 3% cash discount of $15 on the merchandise, Receipt 99.

Analyze
Compute the total amount of cash that would have been collected if customers had not taken any cash discounts.

Problem 9 Recording and Posting Sales and Cash Receipts

Buzz Newsstand had the following transactions for the month of January.

General Ledger
- 101 Cash in Bank
- 115 Accounts Receivable
- 135 Supplies
- 215 Sales Tax Payable
- 401 Sales
- 405 Sales Discounts
- 410 Sales Returns and Allowances

SOURCE DOCUMENT PROBLEM

Problem 14–9

Use the source documents in your working papers to complete this problem.

Accounts Receivable Subsidiary Ledger
- ADK Adkins, Lee
- JAV Java Shops In
- NAD Nadal, Saba
- ROL Rolling Hills Pharmacies

Instructions

1. Record the transactions on page 9 of the general journal.
2. Post each transaction to the appropriate general ledger and accounts receivable subsidiary ledger accounts. A partial general ledger and accounts receivable subsidiary ledger are included in the working papers. The current account balances are recorded in the accounts.

Chapter 14
Problems

Date	Transactions
Jan. 1	Lee Adkins returned $200 in damaged merchandise purchased on account last month, issued Credit Memorandum 10 for $212 ($200 in merchandise plus 6% sales tax of $12).
3	Received a check from Rolling Hills Pharmacies for $2,254 in payment of its account of $2,300 less a 2% cash discount of $46, Receipt 75.
7	Gave credit to Saba Nadal for the return of $300 in merchandise sold to him on account, plus sales tax of $18. Issued Credit Memorandum 11 for $318.
10	Java Shops Inc. sent a check for $1,470 in payment of its account of $1,500 less a 2% cash discount of $30, Receipt 76.
15	Cash sales were $2,500 plus $150 in sales tax, Tape 25.
15	Bankcard sales were $2,000 plus $120 in sales tax, Tape 25.
20	Janson Lee, a neighboring store, needed supplies urgently. Sold it $40 in supplies and received cash from the sale, Receipt 77.
25	Received a check for $636 from Lee Adkins on account, Receipt 78.
31	Sold $3,000 in merchandise plus sales tax of $180 on account to Rolling Hills Pharmacies, Sales Slip 114.

Analyze

Calculate the net sales, which is Sales less Sales Discounts and Sales Returns and Allowances.

Real-World Applications & Connections

CASE STUDY

Merchandising Business

Lisa Marie is the owner of a merchandising business, Work It Out, which sells fitness and athletic equipment to gyms and personal trainers. Lisa's line has always been purchased from an upscale manufacturing business. In an effort to increase her profit margin, Lisa has changed to a lower-priced manufacturer. Prior to this change, sales returns had been 2%, which is an industry average. Recently, after reviewing her financial reports, her accountant tells her that her sales returns have doubled.

Instructions Write a short memo that explains to Lisa why her sales returns may have increased.

21st century Skills

Articles of Incorporation

Governments have basic rules regarding the management of corporations. In the United States and Canada, this primary set of rules is called the Articles of Incorporation. Specific articles of incorporation vary from one corporation to another and one jurisdiction to another, including from one state to another, but generally provide similar information. Incorporating a business protects the business owners from personal liability should the business ever be sued. Additionally, corporations have greater potential to attract investors and raise capital.

Research And Discuss

a. Use the Internet to locate the Articles of Incorporation for your state. Compile the various items into a list.

b. As a class, discuss your findings, and analyze why you think each item is required in order to form a corporation.

Career Wise

Retail Sales Professional

People love to shop—and since more products are available than ever before, helpful retail salespeople are always in demand. Whether it is in a mall clothing store, a local book shop or a luxury auto dealership, friendly retail sales workers who understand their products are beneficial to both the retailer and the customer.

Research and Share

a. Visit the Web site of the Bureau of Labor Statistics and locate the page on retail salespersons. How many retail salespersons does the BLS say worked in the U.S. at last count? What are the prospects for growth of this job?

b. Make a list of traits that make a person a good retail sales professional. Share your list with the class.

Spotlight on Personal Finance

Your Budget

Businesses plan ahead, estimating their revenue and expenses. It is important for you to plan ahead and spend your money wisely. This is accomplished by developing a budget, which is a plan for spending money.

Instructions

Develop a weekly budget for a person your age. Create two columns on a sheet of paper. Label one *Income* and the other *Expenses*. List all sources of income for the week in the first column and the planned spending in the second.

H.O.T Audit

Sales Tax

Country Farms Inc. has hired you as an auditor to examine the financial records of the business. You discover that the September 15 cash register tape shows:

```
Sept. 15
Tape 30

          5000.00  CA
           300.00  ST
```

The entry made in the general journal indicates the accounting clerk neglected to record the amount of the sales tax.

Instructions

On a separate sheet of paper, record the entry that will update the accounting records of the business. If the updated entry is not made, what effect will this error have on the business's books?

Chapter 15

Accounting for Purchases and Cash Payments

Chapter Topics

15-1 Purchasing Items Needed by a Business

15-2 Analyzing and Recording Purchases on Account

15-3 Analyzing and Recording Cash Payments

Visual Summary

Review and Activities

Standardized Test Practice

Computerized Accounting

Problems

Real-World Accounting Careers

Real-World Applications & Connections

Essential Question

As you read this chapter, keep this question in mind:

Why is it important for businesses to carefully track all money spent?

Main Idea

Merchandising businesses follow an orderly process for purchasing merchandise, supplies, and equipment. The accounting department is responsible for making the cash payments for the business. Individual accounts payable transactions are posted to a subsidiary ledger.

Chapter Objectives

Concepts	Analysis	Procedures
C1 Explain the procedures for processing a purchase on account. **C2** Describe the accounts used in the purchasing process.	**A1** Analyze transactions relating to the purchase of merchandise. **A2** Record a variety of purchases and cash payment transactions.	**P1** Post to the accounts payable subsidiary ledger. **P2** Identify controls over cash.

Real-World Business Connection

GEICO

The GEICO gecko has been promoting GEICO insurance for more than 10 years as the company's official "spokes-creature." An invention of The Martin Agency in Richmond, Virginia, the gecko has captivated audiences, generated revenue for the company, and, in 2005, was voted America's favorite advertising icon.

Connect to the Business

Advertising costs money, but the revenue it can produce is essential to a company's success. Every dollar spent by a business should generate two, or more, in revenue. GEICO has also developed a business structure that focuses on phone and Internet sales which has maximized their cost efficiency.

Analyze

Think of some business purchases that are important for generating revenue. Write down each purchase and tell how it can contribute to a company's bottom line. Can you think of any costs that do little or nothing to generate revenue?

Focus on the Photo

Commercials have made GEICO famous for insuring automobiles, but this company also sells insurance for motorcycles, ATVs, RVs, homes, apartments, condos, mobile homes, boats, and even ID theft. **What kinds of purchases and cash payments might an insurance company need to make?**

SECTION 15.1
Purchasing Items Needed by a Business

The **primary** source of income for a merchandising business is from the sale of its merchandise. However, to sell merchandise, a business must first buy the items. The items you find for sale at Wal-Mart—housewares, clothing, sporting goods, jewelry—are all purchased from wholesalers and suppliers. In this chapter you will learn how to analyze transactions relating to the purchase of merchandise.

Section Vocabulary
- purchase requisition
- purchase order
- packing slip
- processing stamp
- purchases discount
- discount period
- Purchases account
- cost of merchandise
- primary
- obtained

The Purchasing Process
What Are the Four Stages of Purchasing?

All businesses, from the corner grocery store to a giant international corporation, are involved in the purchasing process. Retail businesses need shopping bags for customers, sales slips, and cash register tapes. They also need to purchase supplies, equipment, and merchandise.

The purchase of supplies, equipment, and merchandise is divided into four stages:

- requesting needed items
- ordering from a supplier
- verifying items received
- processing the supplier's invoice

Let's take a look at each of these stages.

Requesting Needed Items

In a small business, the owner does all the buying. In a large business, a separate purchasing department buys items for the entire company. When the company needs to buy equipment or supplies, or when the inventory of merchandise on hand is low, a purchase requisition is prepared.

A **purchase requisition** is a written request to order a specified item or items. Usually a purchase requisition is a prenumbered, multicopy form. The manager of the department requesting the item(s) approves the purchase requisition. Then the original copy of the purchase requisition goes to the purchasing department or the purchasing agent. The person making the request keeps a copy. **Figure 15–1** on page 461 shows the purchase requisition form used by The Starting Line Sports Gear.

Ordering from a Supplier

A **purchase order** is a written *offer* to a supplier to buy specified items. Much of the information on the purchase order comes directly from the purchase requisition. Other information may be **obtained** from the supplier's catalog.

460 Chapter 15 • Accounting for Purchases and Cash Payments

Figure 15–1 Purchase Requisition

	STARTING LINE SPORTS GEAR	
	PURCHASE REQUISITION	
		NO. 9421
FOR DEPARTMENT: Shoes		DATE: Nov. 12, 20--
NOTIFY: Vic Ventura ON DELIVERY		DATE WANTED: Dec. 15, 20--
QUANTITY	DESCRIPTION	STOCK NO.
20 Pair	Soft Cushion, White	94682
10 Pair	Soft Cushion, Black	94788
10 Pair	Low Cut, White	94281
10 Pair	Low Cut, Black	94666
ORDER FROM: Pro Runner Warehouse	APPROVED BY: *Jennifer Mack*	SHOE DEPARTMENT MANAGER

Look at the purchase order prepared by The Starting Line in **Figure 15–2**. The purchase order contains:

1. quantity
2. description
3. unit price
4. total cost
5. supplier's name and address
6. date needed
7. shipping method (optional)

The purchase order is a prenumbered multicopy form. The original of the purchase order goes to the supplier. One copy goes to the department requesting the items. The purchasing department keeps another copy.

Figure 15–2 Purchase Order

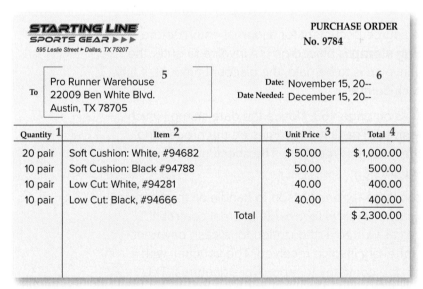

Reading Check

Compare and Contrast

How are a purchase requisition and a purchase order similar? How are they different?

Verifying Items Received

Note that a purchase order is only an *offer* to buy items. Until the items are actually received, the buyer does not know whether or not the supplier has accepted the offer. A supplier may not be able to fill the purchase order because an item is out of stock or has been discontinued. The mailing of a purchase order, therefore, does not require a journal entry. A supplier accepts a purchase order by shipping the items requested and billing the buyer for these items.

When a supplier ships to a buyer, the shipment includes a **packing slip**, which is a form that lists the items included in the shipment. When a shipment arrives, the buyer immediately unpacks and checks the contents against the quantities and items listed on the packing slip. If the shipment contents do not agree with those listed on the packing slip, a note about the differences is made on the packing slip. The packing slip is then sent to the accounting department to be checked against both the purchase order and the supplier's bill. A buyer does not have to pay for items that it did not receive, that were damaged, or that were not ordered and will be returned to the supplier.

Processing the Supplier's Invoice

When it ships to a buyer, the supplier prepares a bill called an *invoice*. It lists the credit terms; the quantity, description, unit price, and total cost of the items shipped; the buyer's purchase order number; and the method of shipment.

The supplier sends the invoice directly to the buyer's accounting department, where it is date stamped to indicate when the invoice was received. The accounting clerk checks each detail (item, quantity, and price) on the invoice against the packing slip and the purchase order. This procedure verifies that the buyer is billed for the quantities and items actually ordered and received and that the prices are correct.

Once verified, the invoice is the source document for a journal entry. Before the invoice is recorded, a **processing stamp** is placed on the invoice to enter the following information: the date the invoice is to be paid, the discount amount, if any, the amount to be paid, and the check number.

Look at the invoice in **Figure 15–3** on page 463. Notice the date stamp, which indicates when the invoice was received. The first three lines on the processing stamp are completed at the time the invoice is received. The check number is entered later, when the check is issued.

When a small business grows too large for one person to handle all the financial responsibilities, it may adopt the *voucher system* to provide internal control. A *voucher* is a document that serves as written authorization for a cash payment. The business prepares a voucher for every invoice received. The voucher, with invoice attached, is circulated within the company for approval signatures. The approved voucher is authorization to issue a check.

Purchases Discounts

Suppliers frequently offer charge customers a cash discount for early payment. For the buyer this discount is called a **purchases discount**. A purchases discount and cash discount are calculated in the same way.

For example, The Starting Line Sports Gear purchased $2,300 of merchandise on account from Pro Runner Warehouse. **Figure 15–3** shows the invoice dated December 14. The credit terms are 2/10, n/30. If The Starting Line Sports Gear pays for the merchandise on or before December 24, it may deduct 2% of the value of the merchandise. The 10 days, called the **discount period**, is the time within which an invoice must be paid if the discount is taken. If The Starting Line Sports Gear does not pay for the merchandise within the discount period, it pays the net, or total amount, within 30 days of the invoice date.

Figure 15–3 Invoice

PRO RUNNER WAREHOUSE
22009 Ben White Blvd.
Austin, TX 78705

REC'D DEC. 14, 20--

INVOICE NO. 7894
DATE: Dec. 14, 20--
ORDER NO.: 9784
SHIPPED BY: Federal Trucking
TERMS: 2/10, n/30

TO Starting Line Sports Gear
    595 Leslie Street
    Dallas, TX 75207

QTY.	ITEM	UNIT PRICE	TOTAL
20 pair	Soft Cushion: White, #94682	$ 50.00	$ 1,000.00
10 pair	Soft Cushion: Black, #94788	50.00	500.00
10 pair	Low Cut: White, #94281	40.00	400.00
10 pair	Low Cut: Black, #94666	40.00	400.00
	Total		$ 2,300.00

Due Date: 12/24
Discount: $ 46.00
Net Amount: $ 2,254.00
Check No.: _____

The Starting Line can save $46 if it pays the invoice within the 10-day discount period ending December 24. The end of the discount period can be determined by adding 10 days to the date of the invoice (December 14 + 10 days = December 24). The amount to be paid within the discount period is calculated as follows:

> **Common Mistakes**
>
> **Discount Periods** The discount period is calculated from the date of the invoice, not the date the invoice is received.

1. $\dfrac{\text{Merchandise Purchased}}{\$2{,}300} \times \dfrac{\text{Discount Rate}}{.02} = \dfrac{\text{Discount}}{\$46}$

2. $\dfrac{\text{Invoice Amount}}{\$2{,}300} - \dfrac{\text{Discount Amount}}{\$46} = \dfrac{\text{Amount Paid Within Discount Period}}{\$2{,}254}$

The Purchases Account

What Is the Purpose of the Purchases Account?

When a business buys merchandise to sell to customers, the cost of the merchandise is recorded in the **Purchases account**. The **Purchases** account is a temporary account, classified as a **cost of merchandise** account. Cost of merchandise accounts contain the actual cost to the business of the merchandise sold to customers.

Purchases	
Debit	Credit
+	−
Increase Side	Decrease Side
Normal Balance	

Merchandise that is purchased for resale is a cost of doing business. Therefore, the **Purchases** account follows the rules of debit and credit for expense accounts. The **Purchases** account is increased by debits and decreased by credits. The normal balance of the **Purchases** account is a debit.

Chapter 15 • Accounting for Purchases and Cash Payments

SECTION 15.1
Assessment

After You Read

Reinforce the Main Idea

Create a table similar to this one to identify the four stages in the purchasing process, the activity involved in each stage, and the document used in that stage.

Stage	Stage Title	Activity	Document
1			
2			
3			
4			

Problem 1 Analyzing a Purchase Order

Instructions Analyze the purchase order shown here and answer the following questions in your working papers.

1. What company ordered the merchandise?
2. What company was asked to supply the merchandise?
3. What is the purchase order number?
4. When was the purchase order prepared?
5. When is the merchandise needed?
6. How many gallons of paint were ordered?
7. How many different colors of paint were ordered?
8. What colors were ordered?
9. How much does each gallon of paint cost?
10. What is the total cost of the order?

CASE CONSTRUCTION COMPANY
601 Mt. Lebanon Road, Shaker Heights, OH 44120

PURCHASE ORDER
No. 7894

To: Westmoreland Paint and Supply Co.
1714 Peak Road
Cleveland, OH 44109

Date: November 15, 20--
Date Needed: December 1, 20--

Quantity	Item	Unit Price	Total
4 gal.	Exterior paint, white, #682	$ 20.00	$ 80.00
4 gal.	Exterior paint, gray, #788	20.00	80.00
6 gal.	Exterior paint, brown, #281	20.00	120.00
6 gal.	Exterior paint, beige, #66	20.00	120.00
5 gal.	Exterior paint, peach, #711	20.00	100.00
	Total		$ 500.00

Math for Accounting

As a new accountant for the South City School District, one of your primary duties involves preparing purchase requisitions. For each item ordered, you compute the extensions (quantity ordered multiplied by the cost per unit). On a separate sheet of paper, calculate the extensions for each of the following items ordered.

Quantity	Item Description	Unit Price
1	Box of copy paper	$34.00/box
5	Reams of art paper	$12/ream
2	Globes	$55/ea.
100	No. 2 pencils	$.12/ea.
4 doz.	Transparency markers	$6.50/doz.

SECTION 15.2
Analyzing and Recording Purchases on Account

Small retailers may rely on a bookkeeper or part-time accountant to record purchase transactions. Large companies such as Kohl's Department Store can have thousands of suppliers. They need an entire accounts payable department to verify invoices and correct any discrepancies before recording purchases in the accounting records.

Section Vocabulary

- accounts payable subsidiary ledger
- tickler file
- due date
- purchases return
- purchases allowance
- debit memorandum

Purchases of Assets on Account

How Are the Purchases of Assets on Account Recorded?

Regardless of its type, size, or purpose, a retail business needs to buy supplies, equipment, and other assets. Most importantly, it buys merchandise to resell. Merchandise can be bought on a cash basis or on account. The business posts purchases on account in the accounts payable subsidiary ledger.

The Accounts Payable Subsidiary Ledger

In Chapter 14 you learned that when a business sells to many customers on credit, using an accounts receivable subsidiary ledger is efficient. Likewise, a business that purchases from many suppliers on account finds it efficient to set up an **accounts payable subsidiary ledger** with an account for each supplier or creditor. These individual accounts are summarized in the general ledger controlling account, **Accounts Payable.** The balance of the **Accounts Payable** controlling account and the total of all account balances in the accounts payable subsidiary ledger must agree after posting.

The Accounts Payable Subsidiary Ledger Form

The ledger account form in **Figure 15-4** is used for both the accounts payable subsidiary ledger and the accounts receivable subsidiary ledger. The ledger account form has lines for the creditor's name and address. A manual accounting system arranges the accounts payable subsidiary ledger in alphabetical order with no account numbers. A computerized accounting system assigns each creditor an account number.

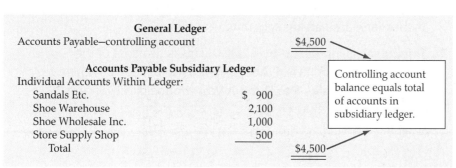

The ledger account form has three amount columns. The normal balance of an Accounts Payable subsidiary account is a credit. The Debit column for payments to creditors records decreases to the account. The Credit column for purchases on account is for increases to it. The Balance column shows the amount owed to the creditor.

Figure 15–4 Subsidiary Ledger Account Form

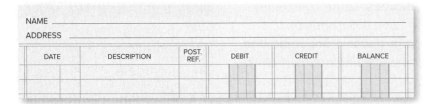

Merchandise Purchases on Account

The Starting Line's first purchase transaction in December involves a purchase of merchandise on account. When this entry is recorded in the journal, a diagonal line in the Posting Reference column indicates that the credit amount is posted in *two* places.

- *First* post to the **Accounts Payable** controlling account in the general ledger.
- *Then* post to the **Pro Runner Warehouse** account in the accounts payable subsidiary ledger.

After the invoice from Pro Runner Warehouse is journalized, it is put in a **tickler file**, which has a folder for each day of the month. Each invoice is placed in a folder according to its **due date**, the date it is to be paid. For example, an invoice due December 24 is placed in the "24" folder.

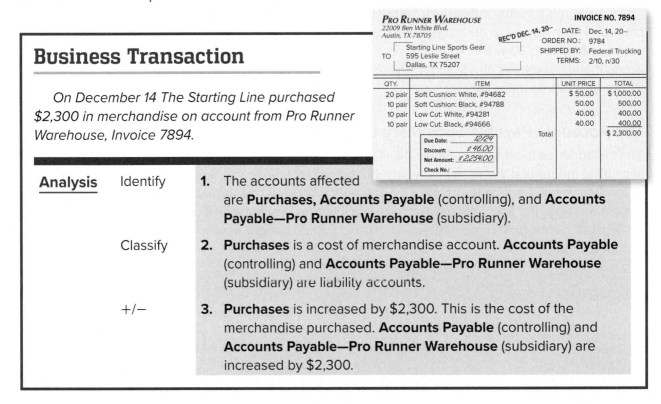

Business Transaction

On December 14 The Starting Line purchased $2,300 in merchandise on account from Pro Runner Warehouse, Invoice 7894.

Analysis

Identify
1. The accounts affected are **Purchases, Accounts Payable** (controlling), and **Accounts Payable—Pro Runner Warehouse** (subsidiary).

Classify
2. **Purchases** is a cost of merchandise account. **Accounts Payable** (controlling) and **Accounts Payable—Pro Runner Warehouse** (subsidiary) are liability accounts.

+/−
3. **Purchases** is increased by $2,300. This is the cost of the merchandise purchased. **Accounts Payable** (controlling) and **Accounts Payable—Pro Runner Warehouse** (subsidiary) are increased by $2,300.

continued

Debit-Credit Rule	4. Increases to cost of merchandise accounts are recorded as debits. Debit **Purchases** for $2,300.
	5. Increases to liability accounts are recorded as credits. Credit **Accounts Payable** (controlling) for $2,300. Also credit **Accounts Payable—Pro Runner Warehouse** (subsidiary) for $2,300.

T Accounts	6.

Purchases
| Debit + 2,300 | Credit − |

Accounts Payable
| Debit − | Credit + 2,300 |

Accounts Payable Subsidiary Ledger
Pro Runner Warehouse
| Debit − | Credit + 2,300 |

Journal Entry	7.

GENERAL JOURNAL PAGE 21

DATE	DESCRIPTION	POST. REF.	DEBIT	CREDIT
20—				
Dec. 14	Purchases		2 3 0 0 00	
	Accts. Pay./Pro Runner Whs.			2 3 0 0 00
	Invoice 7894			

Other Purchases on Account

The Starting Line purchases assets other than merchandise, such as supplies, computer equipment, and store equipment. The following example illustrates the purchase of store equipment on account.

Business Transaction

On December 15 The Starting Line received Invoice 3417, dated December 13, from Champion Store Supply for store equipment bought on account for $1,200, terms n/30.

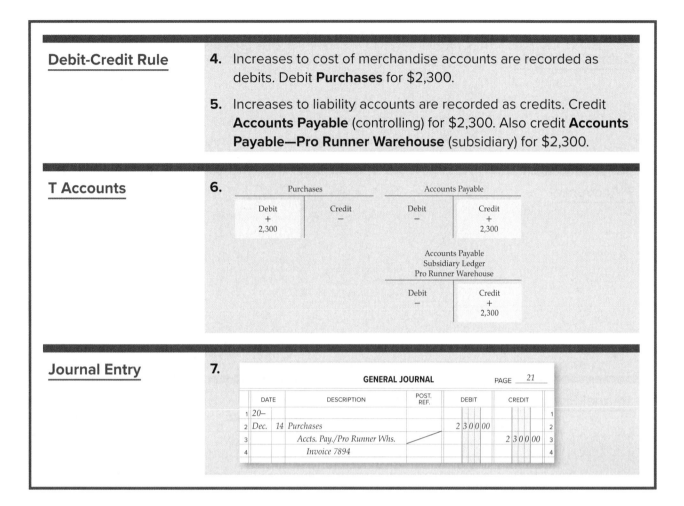

Analysis	Identify	1. The accounts affected are **Store Equipment, Accounts Payable** (controlling), and **Accounts Payable—Champion Store Supply** (subsidiary).
	Classify	2. **Store Equipment** is an asset account. **Accounts Payable** (controlling) and **Accounts Payable—Champion Store Supply** (subsidiary) are liability accounts.
	+/−	3. **Store Equipment** is increased by $1,200. **Accounts Payable** (controlling) and **Accounts Payable—Champion Store Supply** (subsidiary) are increased by $1,200.

continued

Debit-Credit Rule	**4.** Increases to asset accounts are recorded as debits. Debit **Store Equipment** for $1,200. **5.** Increases to liability accounts are recorded as credits. Credit **Accounts Payable** (controlling) for $1,200. Also credit **Accounts Payable—Champion Store Supply** (subsidiary) for $1,200.
T Accounts	**6.** 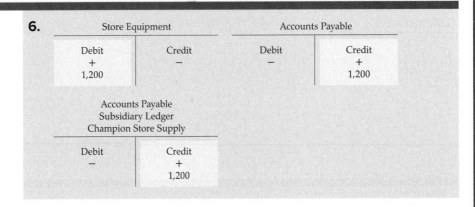
Journal Entry	**7.** 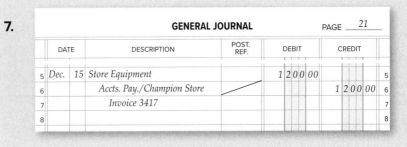

Reading Check

Define

What is a tickler file? How are invoices filed in a tickler file?

Purchases Returns and Allowances

How Are Purchases Returns and Allowances Recorded?

Occasionally, a business buys merchandise that, upon inspection, is unacceptable. A **purchases return** occurs when a business returns merchandise to the supplier for full credit. A **purchases allowance** occurs when a business keeps less than satisfactory merchandise and pays a reduced price.

A **debit memorandum**, or debit memo, is used to notify suppliers (creditors) of a return or to request an allowance. The "debit" in debit memorandum indicates that the creditor's account will be debited, or decreased.

Figure 15–5 Debit Memorandum

```
                DEBIT MEMORANDUM       No. 51
                                  Date: December 16, 20--
                                  Invoice No.: FL610
STARTING LINE
   SPORTS GEAR ▶▶▶
   595 Leslie Street ▶ Dallas, TX 75207

To: FastLane Athletics              This day we have
    35992 Fletcher Blvd. #334       debited your
    Boston, MA 02206                account as follows:
```

Quantity	Item	Unit Price	Total
5 pair	All-Star Athletic Shoes	$ 40.00	$ 200.00

Figure 15–5 shows a debit memorandum prepared by The Starting Line. As you can see, the debit memorandum is prenumbered and has spaces for the creditor's name and address and the invoice number. The original is sent to the creditor. The copy is the source document for the journal entry.

A debit memorandum always results in a debit (decrease) to the **Accounts Payable** controlling account in the general ledger and to the creditor's account in the subsidiary ledger. The account credited depends on whether the debit memorandum is for merchandise or another asset.

The **Purchases Returns and Allowances** account is used to record the return of merchandise to a supplier or to record an allowance. It is classified as a *contra cost of merchandise* account to the **Purchases** account. The normal balance of **Purchases** is a debit, so the normal balance of **Purchases Returns and Allowances** is a credit. This relationship is shown in these T accounts.

Purchases	
Debit	Credit
+	−
Increase Side	Decrease Side
Normal Balance	

Purchases Returns and Allowances	
Debit	Credit
−	+
Decrease Side	Increase Side
	Normal Balance

Recording a Purchases Returns and Allowances Transaction

The Starting Line Sports Gear prepared and sent the debit memorandum shown in **Figure 15–5**.

Business Transaction

On December 16 The Starting Line issued Debit Memorandum 51 for the return of $200 in merchandise purchased on account from FastLane Athletics.

continued

Analysis	Identify	1. The accounts affected are **Accounts Payable** (controlling), **Accounts Payable—FastLane Athletics** (subsidiary), and **Purchases Returns and Allowances**.
	Classify	2. **Accounts Payable** (controlling) and **Accounts Payable—FastLane Athletics** (subsidiary) are liability accounts. **Purchases Returns and Allowances** is a contra cost of merchandise account.
	+/−	3. **Accounts Payable** (controlling) and **Accounts Payable—FastLane Athletics** (subsidiary) are decreased by $200. **Purchases Returns and Allowances** is increased by $200.
Debit-Credit Rule		4. Decreases to liability accounts are recorded as debits. Debit **Accounts Payable** (controlling) for $200. Also debit **Accounts Payable—FastLane Athletics** (subsidiary) for $200.
		5. Increases to contra cost of merchandise accounts are recorded as credits. Credit **Purchases Returns and Allowances** for $200.
T Accounts		6.

Accounts Payable
Debit	Credit
−	+
200	

Purchases Returns and Allowances
Debit	Credit
−	+
	200

Accounts Payable Subsidiary Ledger
FastLane Athletics
Debit	Credit
−	+
200	

Journal Entry 7.

GENERAL JOURNAL PAGE 21

	DATE	DESCRIPTION	POST. REF.	DEBIT	CREDIT	
8	Dec. 16	Accts. Pay./FastLane Athletics		200 00		8
9		Purchases Returns and Allow.			200 00	9
10		Debit Memorandum 51				10
11						11

Posting to the Accounts Payable Subsidiary Ledger

In **Figure 15–6** note how the transaction is posted. The $200 debit is posted to two accounts—the **Accounts Payable** controlling account and the accounts payable subsidiary ledger account, **FastLane Athletics.** After it is posted to **Accounts Payable**, the account number *201* is entered to the *left* of the diagonal in the Posting Reference column. After the amount is posted to **FastLane Athletics**, a check mark (✓) is entered to the *right* of the diagonal line.

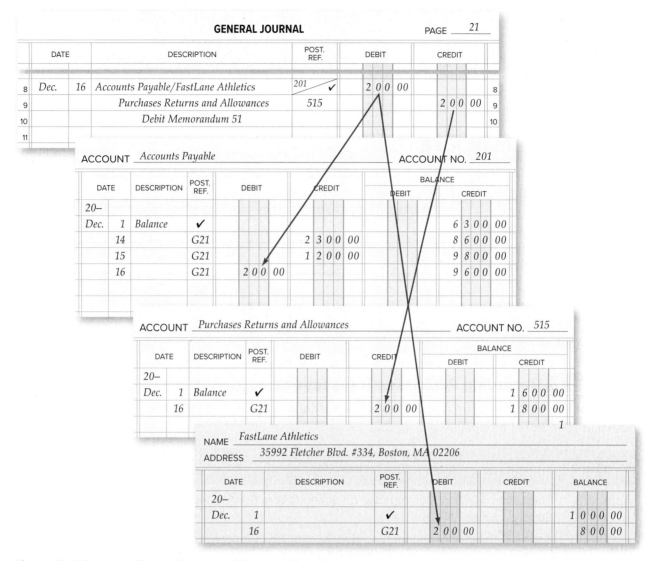

Figure 15–6 Posting to General Ledger and Accounts Payable Ledger

SECTION 15.2
Assessment

After You Read

Reinforce the Main Idea

Create a flowchart like this one and write the correct labels in the boxes and next to the arrows. Create the labels from these terms: *general ledger, invoice, journal, posted to, recorded in, subsidiary ledger*. Terms can be used more than once.

Problem 2 — Recording Purchases Transactions

Instructions Record the following purchases transactions on page 7 of the general journal in your working papers.

Date		Transactions
Sept.	2	Purchased $900 in merchandise on account from Sunrise Novelty Supply, Invoice SN110.
	7	Issued Debit Memorandum 18 to Sunrise Novelty Supply for a $50 allowance granted on damaged merchandise.

Problem 3 — Analyzing a Source Document

As an accounting intern for Kaleidoscope Comics, you perform a variety of accounting tasks. The accountant hands you this debit memo.

Instructions

1. Analyze the source document. Determine which accounts are to be debited or credited.

2. Record the entry on page 15 of the general journal in your working papers.

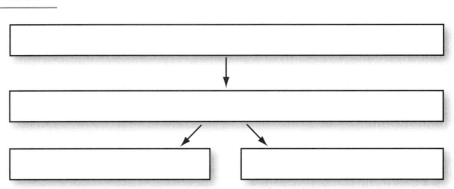

Continued

472   Chapter 15 • Accounting for Purchases and Cash Payments

SECTION 15.2
Assessment

Math for Accounting

Alpha Enterprises received an invoice dated May 2 for the purchase of $3,000 of merchandise. Terms of sale are 2/10, n/30. Answer the following questions:

1. What is the due date of the invoice?
2. What is the amount of the cash discount?
3. What is the net amount to be paid?
4. If the discount period is missed, what is the last day the invoice is to be paid? How much would be paid?

SECTION 15.3
Analyzing and Recording Cash Payments

If cash is the lifeblood of a business, the accounting department is its heart. All cash entering or leaving a business is "pumped" through the accounting department at some time. If it is not, cash losses can occur.

Section Vocabulary
- premium
- FOB destination
- FOB shipping point
- bankcard fee
- benefit

Controls over Cash

How Does a Business Manage Cash Payments?

Earlier chapters explained ways to guard against losses of cash receipts. For example, businesses should deposit all cash receipts in a bank account. Businesses must also properly manage cash payments so that losses do not occur. The following are procedures to manage cash payments:

- Require proper authorization of all cash payments. Support each payment with an approved source document, such as an invoice.
- Write checks for all payments. Allow only authorized persons to sign checks.
- Use prenumbered checks.
- Retain and account for spoiled checks. Mark these checks "Void," and file them in sequence.

Cash Payment Transactions

How Does a Merchandising Business Make Cash Payments?

Businesses buy merchandise and other assets on account or by paying cash. The Starting Line makes all cash payments by check. When it makes a cash payment, the clerk records the details on the check stub. The check stub is the source document for the journal entry. Then the clerk prepares a check, which an authorized person signs. Let's look at how to record several types of cash payment transactions that occur frequently.

Recording Cash Purchase of Insurance

Businesses buy insurance to protect against losses from hazards such as theft, fire, and flood. Insurance policies cover varying time periods, such as six months or one year. The cost of insurance protection is called the **premium**. A premium is paid in advance at the beginning of the covered period. Insurance paid in advance is an asset because until the insurance protection expires, it represents a **benefit** to the company. The insurance premium is recorded in the asset account, **Prepaid Insurance**.

Business Transaction

On December 17 The Starting Line paid $1,500 to Keystone Insurance Company for the premium on a six-month insurance policy, Check 1001.

Analysis

Identify
1. The accounts affected are **Prepaid Insurance** and **Cash in Bank**.

Classify
2. Both **Prepaid Insurance** and **Cash in Bank** are asset accounts.

+/−
3. **Prepaid Insurance** is increased by $1,500. **Cash in Bank** is decreased by $1,500.

Debit-Credit Rule

4. Increases to asset accounts are recorded as debits. Debit **Prepaid Insurance** for $1,500.

5. Decreases to asset accounts are recorded as credits. Credit **Cash in Bank** for $1,500.

T Accounts

6.

Prepaid Insurance		Cash in Bank	
Debit +	Credit −	Debit +	Credit −
1,500			1,500

Journal Entry

7.

GENERAL JOURNAL PAGE 21

DATE	DESCRIPTION	POST. REF.	DEBIT	CREDIT
Dec. 17	Prepaid Insurance		1 500 00	
	Cash in Bank			1 500 00
	Check 1001			

Recording Cash Purchases of Merchandise

Usually businesses purchase merchandise on account. Sometimes a business buys merchandise for cash. Let's look at an example of a cash purchase of merchandise.

Business Transaction

On December 19 The Starting Line purchased merchandise from FastLane Athletics for $1,300, Check 1002.

continued

Chapter 15 • Accounting for Purchases and Cash Payments   475

Analysis	Identify	1. The accounts affected are **Purchases** and **Cash in Bank.**
	Classify	2. **Purchases** is a cost of merchandise account. **Cash in Bank** is an asset account.
	+/−	3. **Purchases** is increased by $1,300. **Cash in Bank** is decreased by $1,300.
Debit-Credit Rule		4. Increases to cost of merchandise accounts are recorded as debits. Debit **Purchases** for $1,300.
		5. Decreases to asset accounts are recorded as credits. Credit **Cash in Bank** for $1,300.
T Accounts		6. 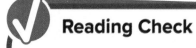

Purchases — Debit + 1,300 | Credit −

Cash in Bank — Debit + | Credit − 1,300

Journal Entry 7.

GENERAL JOURNAL PAGE 21

DATE	DESCRIPTION	POST. REF.	DEBIT	CREDIT
Dec. 19	Purchases		1 300 00	
	Cash in Bank			1 300 00
	Check 1002			

Reading Check

Recall

What are four procedures that a business should use to control its cash payments?

Recording Cash Payments for Items Purchased on Account

An invoice received by a business is verified for items, quantities, and prices, recorded in the journal, and filed by due date in a tickler file.

Each day, the invoices due for payment are removed from the tickler file. Checks are prepared, signed by an authorized person, and mailed to creditors.

The amount of the check in payment of an invoice depends on the credit terms and the payment date. For example, The Starting Line purchased $2,300 of merchandise on account from Pro Runner Warehouse. The invoice, dated December 14, listed credit terms of 2/10, n/30. If The Starting Line pays for the merchandise on or before December 24, it can take a discount of $46.

When it pays the invoice on December 24, Starting Line debits **Accounts Payable** (controlling) **and Accounts Payable—Pro Runner Warehouse** for the full amount of the invoice, $2,300. The Starting Line is paying for all of the merchandise and doesn't owe any more money on this purchase. **Cash in Bank** is credited for $2,254, the actual amount of the check. The difference between $2,300 and $2,254 is the cash discount, which is credited to the **Purchases Discounts** account.

The **Purchases Discounts** account tracks the cash discounts a business takes. **Purchases Discounts** is a contra cost of merchandise account. Its balance reduces the balance of the **Purchases** account. Its normal balance is therefore a credit.

Purchases	
Debit	Credit
+	−
Increase Side	Decrease Side
Normal Balance	

Purchases Discounts	
Debit	Credit
−	+
Decrease Side	Increase Side
	Normal Balance

Business Transaction

On December 24 The Starting Line paid $2,254 to Pro Runner Warehouse for merchandise purchased on account, $2,300 less a discount of $46, Check 1003.

STARTING LINE SPORTS GEAR ▶▶▶ 1003
595 Leslie Street ▶ Dallas, TX 75207
22-523 / 4210
DATE Dec. 24 20—
PAY TO THE ORDER OF Pro Runner Warehouse $ 2,254.00
Two thousand two hundred fifty-four and no/100 —————— DOLLARS
Security National Bank
DALLAS, TEXAS
MEMO _____ *Michael Brown*
⑆4210 225 23⑆ 727596⑈ 1003

Analysis	Identify	1. The accounts affected are **Accounts Payable** (controlling), **Accounts Payable—Pro Runner Warehouse** (subsidiary), **Cash in Bank,** and **Purchases Discounts.**
	Classify	2. **Accounts Payable** (controlling) and **Accounts Payable—Pro Runner Warehouse** (subsidiary) are liability accounts. **Cash in Bank** is an asset account. **Purchases Discounts** is a contra cost of merchandise account.
	+/−	3. **Accounts Payable** (controlling) and **Accounts Payable—Pro Runner Warehouse** (subsidiary) are decreased by $2,300. **Cash in Bank** is decreased by $2,254. **Purchases Discounts** is increased by $46.
Debit-Credit Rule		4. Decreases to liability accounts are recorded as debits. Debit **Accounts Payable** (controlling) for $2,300. Also debit **Accounts Payable—Pro Runner Warehouse** (subsidiary) for $2,300.
		5. Decreases to asset accounts are recorded as credits. Credit **Cash in Bank** for $2,254. Increases to contra cost of merchandise accounts are recorded as credits. Credit **Purchases Discounts** for $46.

continued

T Accounts

6.

Accounts Payable		Cash in Bank	
Debit −	Credit +	Debit +	Credit −
2,300			2,254

Accounts Payable Subsidiary Ledger Pro Runner Warehouse		Purchases Discounts	
Debit −	Credit +	Debit −	Credit +
2,300			46

Journal Entry

7.

GENERAL JOURNAL   PAGE 21

DATE	DESCRIPTION	POST. REF.	DEBIT	CREDIT
Dec. 24	Accts. Pay./Pro Runner Ware.		2 300 00	
	Cash in Bank			2 254 00
	Purchases Discounts			46 00
	Check 1003			

Other Cash Payments

When a company buys merchandise from a supplier, there is often a charge for shipping the goods. The shipping terms determine whether the buyer or shipper will pay the shipping charges.

Shipping terms are stated as either FOB destination or FOB shipping point. "FOB" stands for "free on board." **FOB destination** means that the *supplier* pays the shipping cost to the buyer's destination or location. When merchandise is shipped **FOB shipping point**, the *buyer* pays the shipping charge from the supplier's shipping point.

Terms	Shipping Cost Paid By
FOB destination	Supplier
FOB shipping point	Buyer

A shipping charge is an additional cost of the merchandise. The account set up to handle shipping charges is **Transportation In**, which is a cost of merchandise account. **Transportation In** follows the rules of debit and credit for expense accounts. **Transportation In** is increased by debits and decreased by credits. The normal balance of **Transportation In** is a debit.

Here is an example of the payment of shipping charges.

Business Transaction

On December 24 The Starting Line issued Check 1004 for $275 to Dara's Delivery Service for shipping charges on merchandise purchased from Sports Link Footwear.

Journal Entry

	DATE	DESCRIPTION	POST. REF.	DEBIT	CREDIT	
21	Dec. 24	Transportation In		275 00		21
22		Cash in Bank			275 00	22
23		Check 1004				23

GENERAL JOURNAL   PAGE 21

Recording Bankcard Fees

The Starting Line makes the following entry to record the bankcard fee deducted from the business checking account. Banks charge a fee for handling bankcard sales slips. This **bankcard fee** is usually a percentage of the total of the amounts recorded on the bankcard sales slips processed.

Business Transaction

On December 31 The Starting Line records the bankcard fee of $75, December bank statement.

Journal Entry

	DATE	DESCRIPTION	POST. REF.	DEBIT	CREDIT	
24	Dec. 31	Bankcard Fees Expense		75 00		24
25		Cash in Bank			75 00	25
26		December Bank Statement				26

GENERAL JOURNAL   PAGE 21

The journal entries for all of the transactions discussed and illustrated in this chapter are shown in **Figure 15–7**.

GENERAL JOURNAL
PAGE 21

DATE		DESCRIPTION	POST. REF.	DEBIT	CREDIT
20—					
Dec.	14	Purchases		2 300 00	
		Accounts Payable/Pro Runner Warehouse			2 300 00
		Invoice 7894			
	15	Store Equipment		1 200 00	
		Accounts Payable/Champion Store Supply			1 200 00
		Invoice 3417			
	16	Accounts Payable/FastLane Athletics		200 00	
		Purchases Returns & Allowances			200 00
		Debit Memorandum 51			
	17	Prepaid Insurance		1 500 00	
		Cash in Bank			1 500 00
		Check 1001			
	19	Purchases		1 300 00	
		Cash in Bank			1 300 00
		Check 1002			
	24	Accounts Payable/Pro Runner Warehouse		2 300 00	
		Cash in Bank			2 254 00
		Purchases Discounts			46 00
		Check 1003			
	24	Transportation In		275 00	
		Cash in Bank			275 00
		Check 1004			
	31	Bankcard Fees Expense		75 00	
		Cash in Bank			75 00
		December Bank Statement			

Figure 15–7 Journal Entries for December Business Transactions

SECTION 15.3
Assessment

After You Read

Reinforce the Main Idea

Create a chart similar to the one here to analyze each cash payment transaction for a merchandising business discussed in the chapter. Identify the accounts affected, the account(s) credited, and the account(s) debited.

Action	Accounts Affected	Account to Credit	Account to Debit

Problem 4 Recording Cash Payment Transactions

Meadow Link Golf Club had the following cash payment transactions.

Instructions Record the following transactions on page 6 of the general journal in your working papers.

Date		Transactions
May.	1	Purchased $10,500 in golf equipment (merchandise) from TopMax Golf Manufacturers, Check 1150.
	5	Issued Check 1151 for $325 to Franco's Trucking for delivery charges on merchandise bought from TopMax Golf Manufacturers.
	7	Paid Lone Star Insurance Company $2,500 for the annual premium on an insurance policy, Check 1152.

Math for Accounting

Clara's Designs is a crafts store with a large inventory of seasonal crafts items. As the inventory clerk, you are to create a chart or graph to compare the cost of the items purchased and the related shipping charges. Using the data provided, design a chart or graph that depicts both sets of data. What conclusions can you draw from your chart or graph?

	Cost of Item	Shipping Charges
Holiday decorator ribbon	$2,000	$200
Door wreaths and hangers	3,200	320
Potpourri	1,200	120
Styrofoam trees	4,350	435

Chapter 15
Visual Summary

Concepts
Describe the accounts used in the purchasing process.

PURCHASES	PURCHASES DISCOUNTS
Cost of merchandise	Contra cost of merchandise
PURCHASES RETURNS AND ALLOWANCES	**TRANSPORTATION IN**
Contra cost of merchandise	Cost of merchandise

Analysis
Graphic of Invoice

CRAZY WAYNE'S FURNITURE OUTLET
55234 W. Jane St.
Sacramento, CA 94203

INVOICE NO. 3852
DATE: April 15, 20--
ORDER NO.: 1292
SHIPPED BY: USA LOGISTICS
TERMS: 2/10, n/30

TO: Chuck's Furniture
425 Main St.
Altus, OK 73521

QTY.	ITEM	UNIT PRICE	TOTAL
10	Halogen Lamps	$100.00	$1,000.00
5	Ultrasuede Futon Cvr.	60.00	300.00
15	Cherry Bookshelf	130.00	1,950.00
		Total	$3,250.00

Due Date: 4/25
Discount: $75.00
Net Amount: $3,175.00
Check No.:

Procedures
Accounts Payable Ledger

ACCOUNT: Accounts Payable   ACCOUNT NO. 201

DATE	DESCRIPTION	POST. REF.	DEBIT	CREDIT	BALANCE DEBIT	BALANCE CREDIT
20--						
April 1	Balance	✓				7 500 00
15		G5		2 500 00		10 000 00
16		G5		1 500 00		11 500 00
17		G5	500 00			11 000 00

Chapter 15 • Accounting for Purchases and Cash Payments

Chapter 15
Review and Activities

Answering the Essential Question

Why is it important for businesses to carefully track all money spent?

The main goal of a business is to make a profit, which is basically the amount of money taken in minus the amount of money spent. For this reason, careful records need to be kept so the amount of money spent is exact. How does a business keep track of the money it spends?

Vocabulary Check

1. **Vocabulary** Work with a partner to pair these terms. Write an explanation for why the terms are paired.

   - obtain
   - purchase requisition
   - purchase order
   - primary
   - packing slip
   - processing stamp
   - purchases discount
   - cost of merchandise
   - discount period
   - purchases account
   - accounts payable subsidiary ledger
   - due date
   - tickler file
   - debit memorandum
   - purchases allowance
   - purchases return
   - benefit
   - bankcard fee
   - FOB destination
   - FOB shipping point
   - premium

Concept Check

2. **Synthesize** Write a brief description of the steps involved in making a purchase on account.
3. **Analyze** Why is an invoice checked against both the purchase order and the packing slip?
4. **Apply** What type of account is **Transportation In**?
5. List the four procedures that a business should use to control its cash payments.
6. Which accounts are debited and credited when merchandise is purchased for cash? On account?
7. What account is debited when insurance is purchased?
8. What is the purpose of the accounts payable subsidiary ledger?
9. Why do you post a purchase on account to two different ledgers?
10. **Math** An invoice from the Computer Superstore shows a net price of $15,988.32. The terms are 4/15, net/30. The invoice is dated July 18. The invoice is paid August 3. What is the last day to take the discount? What is the last day to pay the invoice? How much is paid on August 3?
11. **English Language Arts** Imagine you purchased an item online that did not meet your expectations? How would you handle the situation? Write a letter to the company explaining why the product did not meet your expectations. Be sure to include all information the company would need to process your refund or product replacement.

Chapter 15 • Accounting for Purchases and Cash Payments   **483**

Chapter 15
Standardized Test Practice

Multiple Choice

1. What is the name of the source document used for journalizing a purchase?
    a. purchase order
    b. purchase requisition
    c. receipt
    d. invoice

2. Which document is needed to purchase items from a specified supplier?
    a. purchase order
    b. purchase requisition
    c. receipt
    d. invoice

3. The terms "2/10 n/30" on an invoice indicate that the purchaser
    a. must pay the invoice within 30 days or pay 2% interest.
    b. forfeits a 2% discount if he pays after 10 days.
    c. must pay at least two-tenths of the invoice within 30 days.
    d. can take a 2% discount if he pays within 30 days.

4. A business transaction that involves a purchase on account is considered to be a(n)
    a. cash transaction.
    b. credit transaction.
    c. investment of the owner.
    d. expense transaction.

5. A liability resulting from the purchase of goods or services on credit is usually
    a. an account receivable.
    b. an account payable.
    c. a revenue.
    d. a reduction of equity.
    e. a net loss.

Chapter 15
Standardized Test Practice

True or False

6. When a business makes a purchase on account, the purchase is recorded in the accounts payable subsidiary ledger.

Extended Response

7. List the steps for making a purchase of merchandise on account.

Chapter 15
Computerized Accounting

Recording Purchases and Cash Payments
Recording Purchase and Cash Payment Transactions

MANUAL METHODS	COMPUTERIZED METHODS
• Prepare general journal entries and post them based on an invoice for merchandise purchased on account or for cash.	• The system automatically generates journal entries from the invoice information and posts them to the general ledger.
• Write checks in payment of an invoice or memorandum.	• The system automatically generates journal entries as it prepares the check for payment.
• Prepare journal entries to record the checks and post them to the general ledger.	• The system posts journal entries and updates account balances.
• Calculate new account balances.	

Chapter 15
Problems

Problem 5 Determining Due Dates and Discount Amounts

Sunset Surfwear frequently purchases merchandise on account. When it receives invoices, a clerk puts a processing stamp on the invoice indicating the due date, the amount of any discount, and the amount to be paid. The following invoices were received during March.

	Invoice Number	Invoice Date	Credit Terms	Invoice Amount
1.	24574	March 5	2/10, n/30	$3,000.00
2.	530992	March 7	3/10, n/30	5,550.00
3.	211145	March 12	2/15, n/60	729.95
4.	45679	March 16	n/45	345.67
5.	34120	March 23	2/10, n/30	1,526.50
6.	00985	March 27	n/30	700.00

Instructions Prepare a form similar to the one that follows. The first invoice has been completed as an example. For each invoice do the following:

1. Determine the due date. Assume that Sunset Surfwear always pays invoices within the discount period.
2. Compute the discount amount, if any.
3. Compute the amount to be paid.

Invoice Number	Invoice Date	Credit Terms	Invoice Amount	Due Date	Discount Amount	Amout To Be Paid
24574	Mar. 5	2/10, n/30	$3,000.00	Mar. 15	$60.00	$2,940.00

Analyze
Compute how much money Sunset Surfwear will save by taking all discounts.

Problem 6 Analyzing Purchases and Cash Payments

InBeat CD Shop had the following transactions in March.

Instructions Use the T accounts in your working papers. For each transaction:

1. Determine which accounts are affected.
2. Enter the debit and credit amounts in the T accounts.

Continued

Chapter 15
Problems

Date		Transactions
Mar.	2	Purchased merchandise on account from NightVision and Company, $2,000, Invoice NV-20, terms 2/10, n/30.
	6	Issued Check 250 for $85 to Penn Trucking Company for delivering merchandise from NightVision and Company.
	7	Purchased $300 in supplies on account from Temple Store Supply, Invoice 6011, terms n/30.
	12	Issued Check 251 for $1,960 to NightVision and Company in payment of Invoice NV-20 for $2,000 less a cash discount of $40.
	15	Paid Keystone Insurance Company $2,500 for the annual premium on business insurance, Check 252.
	16	Purchased $3,000 in merchandise on account from NightVision and Company, Invoice NV-45, terms 2/10, n/30.
	18	Issued Debit Memorandum 25 for $100 to NightVision and Company for the return of merchandise.
	20	Purchased $900 in merchandise from Dandelion Records, Check 253.
	22	Issued Check 254 to Temple Store Supplies for $300 for Invoice 6011.

Analyze
Determine the total decrease to the checking account for the month.

Problem 7 Recording Purchases Transactions

Shutterbug Cameras, a retail merchandising store, had the following purchases transactions in March.

Instructions In your working papers, record the transactions on page 31 of the general journal.

Date		Transactions
Mar.	3	Purchased $4,500 in merchandise on account from Photo Emporium, Invoice 1221, terms 2/10, n/30.
	5	Bought $750 in supplies on account from State Street Office Supply, Invoice 873, terms n/30.
	9	Returned $225 in merchandise to Photo Emporium, issued Debit Memorandum 72.

Continued

Chapter 15
Problems

Date	Transactions
Mar. 10	Purchased $3,600 in merchandise on account from Video Optics Inc., Invoice VO94, terms 3/15, n/45.
15	State Street Office Supply granted a $60 credit for damaged supplies purchased on August 5, Debit Memorandum 73.
20	Bought $4,800 in store equipment from Digital Precision Equipment on account, Invoice 1288, terms n/30.
25	Issued Debit Memorandum 74 to Video Optics Inc. for the return of $120 in merchandise.
28	Purchased $1,800 in merchandise on account from U-Tech Products, Invoice UT66, terms n/30.
29	Bought $270 in supplies on account from ProStudio Supply, Invoice 4574, terms n/30.
30	Returned $150 in merchandise to U-Tech Products, Debit Memorandum 75.
31	Issued Debit Memorandum 76 to ProStudio Supply for the return of $35 in supplies bought on March 29.

Analyze
Identify the total credit amount made to the Purchases Returns and Allowances account for the month.

Problem 8 Recording Cash Payment Transactions

Cycle Tech Bicycles had the following cash payment transactions in March.

Instructions In your working papers, record the transactions on page 19 of the general journal.

Date	Transactions
Mar. 1	Purchased $1,800 in merchandise from Summit Bicycles, Check 2111.
3	Issued Check 2112 for $2,450 to Spaulding Inc. in payment of the $2,500 account balance less a 2% cash discount of $50.
7	Issued Check 2113 for $3,100 to Desert Palms Insurance Company for the annual business insurance premium.
12	Paid $175 to Viking Express for delivery of merchandise purchased from Schwinn Inc., FOB shipping point, Check 2114.

Continued

Chapter 15
Problems

Date	Transactions
Mar. 17	Issued Check 2115 to Suspension Specialists for $3,880 in payment of Invoice 1492 for $4,000 less a 3% cash discount of $120.
20	Received the March bank statement and recorded bankcard fees of $275.
24	Paid All-Star News $130 to run an advertisement promoting the store, Check 2116.
28	Bought $100 in supplies and $700 in store equipment from Superior Store Equipment Inc., Check 2117 for $800.
31	Issued Check 2118 for $2,185.50 in payment of monthly wages of $3,000 less deductions for the following taxes: Employees' Federal Income Tax Payable, $480; Employees' State Income Tax Payable, $105; Social Security Tax Payable, $186; and Medicare Tax Payable, $43.50.

Analyze
Determine the total decrease to the checking account for the month.

Problem 9 Recording Purchases and Cash Payment Transactions

River's Edge Canoe & Kayak had the following purchases and cash payment transactions for the month of March.

Instructions In your working papers, record the following transactions on page 16 of the general journal.

Date	Transactions
Mar. 1	Purchased $8,200 in merchandise on account from Trailhead Canoes, Invoice TC202, terms 2/10, n/30.
5	Issued Check 887 for $98 to Santini Trucking Company for delivery of merchandise purchased from Trailhead Canoes.
7	Bought $230 in supplies on account from StoreMart Supply, Invoice SM101, terms n/30.
9	Issued Check 888 to World-Wide Insurance Company for $2,500 in payment of business insurance premium.
11	Issued Check 889 to Trailhead Canoes for $8,036 in payment of Invoice TC202 for $8,200 less a 2% cash discount of $164.

Continued

Chapter 15
Problems

Date	Transactions
Mar. 15	Purchased $6,200 in merchandise on account from Mohican Falls Kayak Wholesalers, Invoice 45332, terms 3/15, n/45.
18	Issued Debit Memorandum 67 for $25 to StoreMart Supply for damaged supplies purchased on March 7.
21	Paid StoreMart Supply the balance due on their account, $205 ($230 less $25 credit), Check 890.
25	Returned $200 in merchandise to Mohican Falls Kayak Wholesalers, Debit Memorandum 68.
28	Rollins Plumbing Service completed $120 in repair work (**Maintenance Expense**) on account, Invoice RP432.

Analyze
State the total debit amount to the Purchases account for the month.

Problem 10 Recording and Posting Purchases and Cash Payment Transactions

Buzz Newsstand's purchases and cash payment transactions for the month of March are described below.

Instructions In your working papers:
1. Record the transactions in the general journal, page 7.
2. Post the transactions to the ledger accounts.

SOURCE DOCUMENT PROBLEM

Problem 15–10

Use the source documents in your working papers to complete this problem.

Date	Transactions
Mar. 1	Purchased $5,600 in merchandise on account from ADC Publishing, Invoice 785, terms 3/15, n/45.
3	Issued Check 1400 for $735 to Pine Forest Publications in payment of Invoice PFP98 for $750 less a 2% cash discount of $15.
5	Issued Check 1401 for $275 to Rizzo's Trucking Company for transportation charges.
7	Issued Check 1402 for $588 to Delta Press in payment of Invoice DP166 for $600 less a 2% cash discount of $12.

Continued

Chapter 15
Problems

Date	Transactions
Mar. 9	American Trend Publishers granted a $100 allowance for damaged merchandise, Debit Memorandum 33.
11	Issued Check 1403 for $3,200 to Keystone Insurance Company for the annual premium for business insurance.
15	Issued Check 1404 for $5,432 to ADC Publishing in payment of Invoice 785 for $5,600 less a 3% cash discount of $168.
18	Bought $2,800 in merchandise on account from Delta Press, Invoice DP204, terms 2/10, n/30.
22	Issued Debit Memorandum 34 to Delta Press for the return of $300 in merchandise.
28	Paid American Trend Publishers the balance due on their account, $800, no discount, Check 1405.
30	Bought $120 in merchandise for cash from ADC Publishing, Check 1406.

Analyze

Calculate how much Buzz Newsstand saved in March by taking cash discounts.

Real-World Accounting Careers

Roger Hussey

Odette School of Business, University of Windsor

Q What do you do?

A As a professor, I have three areas of responsibility: teaching, research, and service to the university and the wider community.

Q What are your day-to-day responsibilities?

A On teaching days, I may present a lecture to as many as 60 undergraduate students or 10 to 12 graduate students. I teach for three or four hours, then have individual discussions with students who are having difficulties. On non-teaching days, I may mark essays, prepare lessons or do my own research, or I may have a meeting with my colleagues to discuss future accounting courses we might offer.

Q What factors have been key to your success?

A A tremendous amount of weight is put on research and publications in academic journals. That is key to success; you must publish in academic journals.

Q What do you like most about your job?

A I have a tremendous amount of discretion over how I schedule my time outside of classes. That is the greatest advantage, particularly if you want to research and write.

Career Facts

Real-World Skills
Communication and public speaking skills; creativity; intellectual curiosity; empathy

Training And Education
The education needed to work as a university professor differs depending on the institution. Many positions require a master's degree, and some require a Ph.D.

Career Skills
Knowledge of advanced accounting concepts

Career Path
Obtain undergraduate and master's degrees, then enter a doctoral program; it takes several years to earn tenure at a university

Tips from Robert Half International
Ninety-three percent of CFOs surveyed by Robert Half said it's important for accounting and finance majors to work in the field while in college. Summer internships and temporary assignments are good ways to gain the necessary real-world, hands-on experience.

College and Career Readiness

How might you acquire the basic accounting experience you need for an entry-level position? Conduct research and list five ways to acquire the necessary experience. Create a PowerPoint presentation to share the information you find and present it to the class.

Career Wise

Forensic Accountant

The word "forensic" means "relating to or dealing with the application of scientific knowledge to legal problems." Much like forensic scientists apply a broad range of sciences to answer questions of interest to a legal system, forensic accountants specialize in analyzing financial information in a way that is suitable for use in a court of law. Forensic accountants often serve as expert witnesses in legal trials for white collar crimes such as embezzlement, money laundering, tax evasion, and tax fraud. Most large accounting firms have a specialized forensics department. Forensic accountants need a thorough knowledge of business practices, financial reporting systems, accounting and auditing standards and procedures, and litigation processes.

Research And Share

a. Use the Internet to locate three sources of information on forensic accounting.

b. What level of education is required to be a forensic accountant? What are the roles and responsibilities of a forensic accountant?

Global Accounting

Utilizing E-Procurement

E-procurement is the sale and purchase of supplies, work, and services through the Internet and other information and networking systems. When KingCo Oilfield Services needs supplies for its global operations, it uses an e-procurement system to connect buyers and suppliers via the Internet. Bidders around the world can compete online for contracts in different currencies and languages. The easy-to-use system saves KingCo nearly $400,000 a month in transaction costs and has decreased its order fulfillment time.

Activity

Imagine that your company has manufacturing plants in both the United States and Japan. Describe how an e-procurement system might be helpful.

A Matter of Ethics

Showing Favoritism

Manufacturers make purchases to conduct their business. Imagine that you work in the purchasing department of a large toy manufacturer like Mattel. Your job is to place orders for the parts used to make toy cars and trucks. One supplier, whose prices are only slightly higher than others, has indicated that it would send free tickets to a sports event if you order most of the parts from it.

Activity

a. Determine the ethical issues. What are the alternatives? Who are the affected parties? How do the alternatives affect the parties? What would you do?

Analyzing Financial Reports

Identifying Corporate Goals

The information in an annual report is directed to many audiences: existing and potential stockholders; financial analysts and advisors; government regulators, such as the Securities and Exchange Commission (SEC); employees; and creditors.

Activity

Use the Internet to locate and save or print a copy of Amazon.com's annual report to complete the following.

1. List Amazon.com's measurable financial successes for the current year of the report.

2. Why is the accuracy of the information presented in an annual report essential to both management and investors?

H.O.T. Audit

Correcting Entries

Nuclear Med Industries hired Alex to audit the financial records of the business. He discovered on June 10 a check for $275 was paid for shipping charges. The recorded entry was a debit to Purchases and a credit to Cash in Bank.

Activity

On a separate sheet of paper, answer the following questions.

1. What entry should have been made?

2. What should the correcting entry be?

3. If a correcting entry is not made, what effect would this error have on the records of the business?

Chapter 16

Special Journals: Sales and Cash Receipts

Chapter Topics

16-1 The Sales Journal

16-2 The Cash Receipts Journal

    Visual Summary

    Review and Activities

    Standardized Test Practice

    Computerized Accounting

    Problems

    Real-World Applications & Connections

Essential Question

As you read this chapter, keep this question in mind:

How do special journals save time and increase accuracy?

Main Idea

The sales journal is used to record credit sales of merchandise. The cash receipts journal is used to record the cash a business receives.

Chapter Objectives

Concepts

C1 Identify the special journals and explain how they are used in a merchandising business.

C2 Record transactions in sales and cash receipts journals.

Analysis

A1 Post from the sales and cash receipts journals to customer accounts in the accounts receivable subsidiary ledger.

A2 Foot, prove, total, and rule the sales and cash receipts journals.

Procedures

P1 Post column totals from the sales and cash receipts journals to general ledger accounts.

P2 Prepare a schedule of accounts receivable.

Real-World Business Connection

JCPenney

During its 100 plus years in the retail industry, JCPenney has adapted to many changes in the way retail companies operate. For a merchandising company, the prime objective is finding ways to increase sales and cash receipts. James Cash Penney began his business in 1902 as a single store, and by 1917 had increased to 175 stores over 22 states. By 1941, the company operated 1,600 stores in 48 states. To further augment sales, they began a catalog business in 1963 to reach customers in remote areas. Now, online sales represent much of their total business.

Connect to the Business

The ability to adapt to change is what keeps a company like JCPenney in business year after year. Monitoring the way customers live and shop tells a merchandising company what direction to take in what it sells, how it advertises, and how it grows.

Analyze

What changes can you think of that have taken place over the last decade that affect the way consumers buy goods and services in the United States?

Focus on the Photo

JCPenney's ability to follow shopping trends and stay one step ahead has allowed this retail company to connect with consumers and stay profitable. **What types of retail sales transactions do you think are recorded in accounting records?**

SECTION 16.1
The Sales Journal

You have learned how to record a variety of business transactions in a general journal. A merchandising business can record all of its transactions in a general journal. However, each transaction requires at least three journal lines—one line for the debit, one line for the credit, and one line for the explanation. Each debit and credit is posted separately to the general ledger. For merchandising businesses with many sales transactions, such as Crate & Barrel, this would be very time consuming. To improve efficiency many merchandising businesses use special journals. In this chapter and the next chapter, you will learn how to record transactions in special journals.

> **Section Vocabulary**
> - special journals
> - sales journal
> - footing
> - impose

Using Special Journals

Why Are Special Journals Used?

Special journals have amount columns that are used to record debits and credits to specific general ledger accounts. Most transactions are recorded on one line. Special journals thus simplify the journalizing and posting process. The four most commonly used special journals and the type of transaction recorded in each journal are:

Journal	Transaction
Sales journal	sale of merchandise on account
Cash receipts journal	receipt of cash
Purchases journal	purchase of any asset on account
Cash payments journal	payment of cash, including payment by check

Businesses that use special journals still need the general journal to record transactions that cannot be entered in the special journals. Let's first look at the sales journal.

Journalizing and Posting to the Sales Journal

How Do You Use the Sales Journal?

The **sales journal** is a special journal used to record sales of merchandise on account. **Figure 16–1** on page 497 shows a page from the sales journal used by The Starting Line Sports Gear.

Like the general journal, the sales journal has a space for the page number and columns for the date and the posting reference. There is a separate column in which to record the sales slip number and a column in which to record the name of the charge customer. There are also three special amount columns. **Figure 16–1** illustrates what is to be recorded in each column.

Figure 16–1 Sales Journal

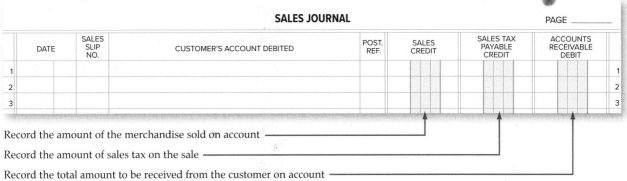

Record the amount of the merchandise sold on account
Record the amount of sales tax on the sale
Record the total amount to be received from the customer on account

Recording Sales of Merchandise on Account

Are you ready to record a sale on account in the sales journal? It's simple. Let's look at the same transactions you analyzed in Chapter 14 for The Starting Line Sports Gear. This time we will record transactions using the sales journal.

Business Transaction

On December 1 The Starting Line sold merchandise on account to Casey Klein for $200 plus $12 sales tax, Sales Slip 50.

Journal Entry

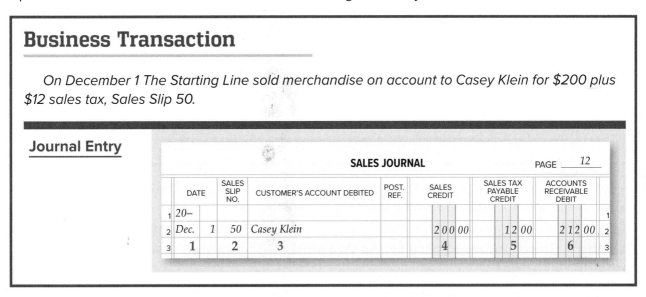

To record the transaction in the sales journal, journalize from left to right by following these steps.

1. Enter the date of the sales slip in the Date column.
2. Enter the sales slip number in the Sales Slip No. column.
3. Enter the name of the customer in the Customer's Account Debited column.
4. Enter the total of the merchandise sold in the Sales Credit column. (This is the amount shown in the subtotal box of the sales slip.)
5. Enter the amount of the sales tax in the Sales Tax Payable Credit column.
6. Enter the total amount to be received from the customer in the Accounts Receivable Debit column. (This is the amount shown in the total box of the sales slip.)

Chapter 16 • Special Journals: Sales and Cash Receipts 499

You learned in Chapter 14 that businesses do not **impose** sales tax on the sale of merchandise to tax-exempt organizations. The next business transaction involves a sale that The Starting Line makes to an organization that has tax-exempt status.

This transaction is analyzed in the same manner as the previous transaction—except there is no sales tax. Journalizing this transaction involves only five steps. The first four steps are the same. There is no sales tax so the fifth step is to enter the total amount to be received from the customer in the Accounts Receivable Debit column.

STARTING LINE SPORTS GEAR
595 Leslie Street ▸ Dallas, TX 75207

DATE: December 3, 20-- NO. 51

SOLD TO: South Branch High School Athletics
1750 Rutgers Dr.
Dallas, TX 75207

CLERK: B.E. CASH: CHARGE: ✓ TERMS: 2/10, n/30

QTY.	DESCRIPTION	UNIT PRICE	AMOUNT
15	Baseball Uniforms	$40.00	$600.00
15	Baseball Caps	20.00	300.00
15	Baseball Mitts	35.00	525.00
2	Baseballs	15.00	30.00
3	Baseball Bats	15.00	45.00

SUBTOTAL $1,500.00
SALES TAX 0.00
Thank You! TOTAL $1,500.00

Business Transaction

On December 3 The Starting Line sold merchandise on account to South Branch High School Athletics for $1,500, Sales Slip 51.

Journal Entry

SALES JOURNAL PAGE 12

DATE	SALES SLIP NO.	CUSTOMER'S ACCOUNT DEBITED	POST. REF.	SALES CREDIT	SALES TAX PAYABLE CREDIT	ACCOUNTS RECEIVABLE DEBIT
3	51	South Branch H.S. Ath.		1 500 00		1 500 00
1	2	3		4		5

For sales that do not include a sales tax, amounts entered in the Sales Credit column and in the Accounts Receivable Debit column are the same.

Posting a Sales Journal Entry to the Accounts Receivable Subsidiary Ledger

In Chapter 14 you learned about posting to the accounts receivable subsidiary ledger. To keep the balances of the customer accounts current, sales journal transactions are posted daily to the accounts receivable subsidiary ledger. Whether you use the general journal or the sales journal when posting to the accounts receivable subsidiary ledger, the process is similar. Refer to **Figure 16–2** on page 501 and follow these steps.

1. Enter the date of the transaction in the Date column of the subsidiary ledger account. Use the same date as the journal entry.

2. In the Posting Reference column of the subsidiary ledger account, enter the journal letter and the journal page number. Use the letter *S* for the sales journal.

3. In the Debit column of the subsidiary ledger account, enter the total amount to be received from the customer.

4. Compute the new balance and enter it in the Balance column. To find the new balance, add the amount in the Debit column to the previous balance amount.

Figure 16–2 Posting a Sales Journal Entry to the Accounts Receivable Subsidiary Ledger

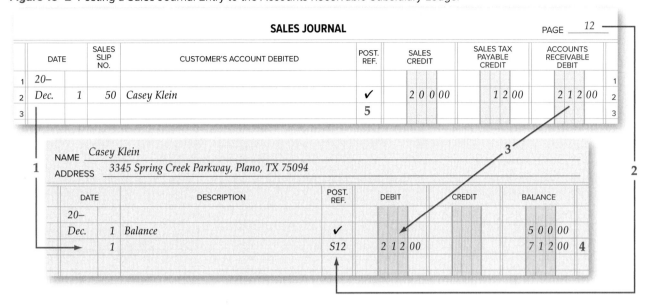

5. Return to the sales journal and enter a check mark (✓) in the Posting Reference column. The check mark indicates that the transaction has been posted to the accounts receivable subsidiary ledger. In manual accounting systems, customer accounts are not numbered, so a check mark is used in the Posting Reference column.

Completing the Sales Journal

All special journals have amount columns used to record debits and credits to specific general ledger accounts. These amount columns simplify posting. Instead of posting each transaction separately to the general ledger, only the amount column totals are posted. For the sales journal, the column totals posted are the Sales Credit, the Sales Tax Payable Credit, and the Accounts Receivable Debit. Therefore, only three postings are made to the general ledger from the sales journal.

Footing, Proving, Totaling, and Ruling the Sales Journal

Before posting amounts to the general ledger, calculate and verify the column totals. Refer to **Figure 16–3** for each step.

1. Draw a single rule across the three amount columns, just below the last transaction.

2. Foot the amount columns. A **footing** is a column total written in small penciled figures. A footing must be verified. It is written in pencil so that it can be erased if a mistake is discovered.

3. On a separate sheet of paper, test for the equality of debits and credits. The total of the debit column should equal the total of the two credit columns.

Debit Column		**Credit Columns**	
Accounts Receivable	$11,305	Sales Tax Payable	$ 555
		Sales	10,750
Total	$11,305	Total	$11,305

Chapter 16 • Special Journals: Sales and Cash Receipts

Figure 16–3 Totaled and Ruled Sales Journal

SALES JOURNAL PAGE 12

DATE		SALES SLIP NO.	CUSTOMER'S ACCOUNT DEBITED	POST. REF.	SALES CREDIT	SALES TAX PAYABLE CREDIT	ACCOUNTS RECEIVABLE DEBIT	
20–								1
Dec.	1	50	Casey Klein	✓	2 0 0 00	1 2 00	2 1 2 00	2
	3	51	South Branch High School Athletics	✓	1 5 0 0 00		1 5 0 0 00	3
	5	52	Break Point Sports Club	✓	3 5 0 0 00	2 1 0 00	3 7 1 0 00	4
	9	53	Gabriel Ramos	✓	3 0 0 00	1 8 00	3 1 8 00	5
	10	54	Kim Wong	✓	1 0 0 00	6 00	1 0 6 00	6
	12	55	Robert Galvin	✓	2 5 0 00	1 5 00	2 6 5 00	7
	18	56	Joe Dimaio	✓	4 0 0 00	2 4 00	4 2 4 00	8
	20	57	Megan Sullivan	✓	5 0 0 00	3 0 00	5 3 0 00	9
	24	58	Tammy's Fitness Club	✓	1 0 0 0 00	6 0 00	1 0 6 0 00	10
	26	59	Anita Montero	✓	3 5 0 00	2 1 00	3 7 1 00	11
	28	60	Shashi Rahim	✓	2 0 0 00	1 2 00	2 1 2 00	12
	30	61	Lara Young	✓	2 4 5 0 00	1 4 7 00	2 5 9 7 00	13
	31		Totals		10 7 5 0 00	5 5 5 00	11 3 0 5 00	14
								15

4. In the Date column, on the line below the single rule, enter the date the journal is being totaled.

5. On the same line, in the Customer's Account Debited column, enter the word *Totals*.

6. Enter the column totals, in ink, just below the footings.

7. Double-rule the three amount columns. A double rule, as you know, indicates that the totals have been verified.

After the sales journal has been footed, proved, totaled, and ruled, the column totals are posted to the general ledger.

Posting the Total of the Sales Credit Column

Refer to **Figure 16–4** on page 503 as you read the procedure for posting the total of the Sales Credit column to the **Sales** account in the general ledger.

1. In the Date column of the **Sales** account in the general ledger, enter the date from the Totals line of the sales journal.

2. Enter the sales journal letter and page number in the Posting Reference column. Remember that *S* is the letter for the sales journal.

3. In the Credit column, enter the total from the Sales Credit column of the sales journal.

4. Compute the new balance and enter it in the Credit Balance column. To determine the new balance, add the amount entered in the Credit column to the previous balance.

5. Return to the sales journal and enter the **Sales** account number, in parentheses, below the double rule in the Sales Credit column. The number written in parentheses indicates that the column total has been posted to the general ledger account.

Figure 16–4 Posting the Sales Credit Total to the General Ledger Account

Posting the Total of the Sales Tax Payable Credit Column

The next amount to be posted is the Sales Tax Payable Credit column total. Refer to **Figure 16–5** as you read the following procedure.

1. In the Date column of the **Sales Tax Payable** account, enter the date from the Totals line of the sales journal.

2. Enter the sales journal letter and page number in the Posting Reference column.

3. In the Credit column, enter the total from the Sales Tax Payable Credit column of the sales journal.

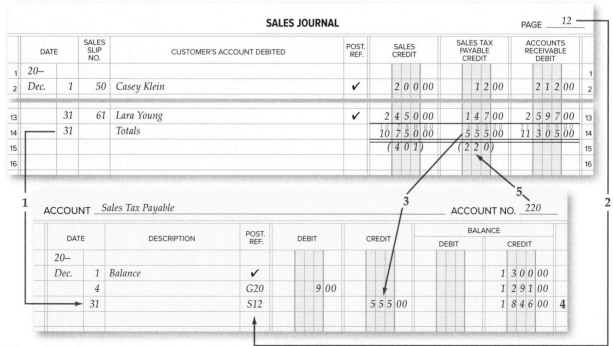

Figure 16–5 Posting the Sales Tax Payable Credit Total to the General Ledger

Chapter 16 • Special Journals: Sales and Cash Receipts **503**

4. Compute the new balance and enter it in the Credit Balance column.

5. Return to the sales journal and enter the Sales Tax Payable account number, in parentheses, below the double rule in the Sales Tax Payable Credit column.

>
>
> **Reading Check**
>
> **Explain**
>
> *How do you know when the column total has been posted to the general ledger?*

Posting the Total of the Accounts Receivable Debit Column

The last column of the sales journal to be posted is the Accounts Receivable Debit column. Refer to **Figure 16–6** as you read these steps.

1. In the Date column of the **Accounts Receivable** account, enter the date from the Totals line of the sales journal.

2. Enter the sales journal letter and page number in the Posting Reference column.

3. In the Debit column, enter the total from the Accounts Receivable Debit column of the sales journal.

4. Compute the new balance and enter it in the Debit Balance column.

5. Return to the sales journal and enter the **Accounts Receivable** account number, in parentheses, below the double rule in the Accounts Receivable Debit column.

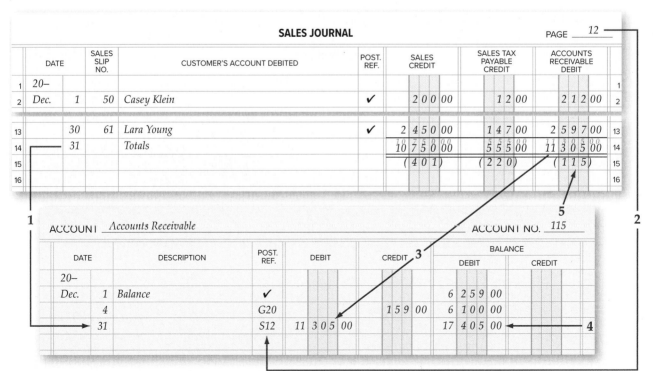

Figure 16–6 Posting Accounts Receivable Debit Total to the General Ledger

504 Chapter 16 • Special Journals: Sales and Cash Receipts

Figure 16–7 Starting a New Journal Page

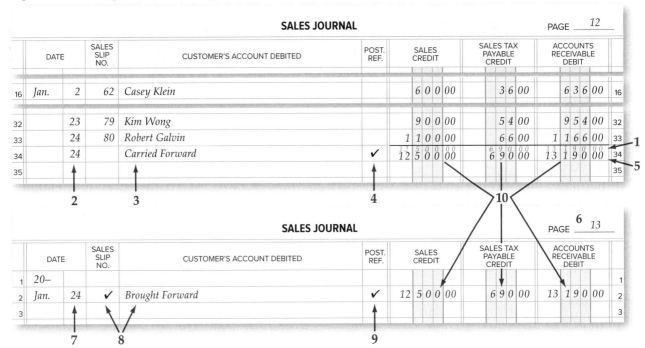

Proving the Sales Journal at the End of a Page

All special journals are totaled and proved at the end of every month. Sometimes, however, a business has so many transactions in one month that it is impossible to fit them all on one journal page. When this occurs, the journal page is totaled and ruled before a new page is started.

Refer to **Figure 16–7** and follow these steps to record the totals and carry them forward to the next page.

1. Draw a single rule across the three amount columns below the last transaction. Foot the columns and prove the equality of debits and credits.
2. On the line following the last transaction, enter the date of the last transaction in the Date column.
3. In the Customer's Account Debited column, write *Carried Forward*.
4. Place a check mark (✓) in the Posting Reference column. This check mark indicates that these totals are not to be posted.
5. Enter the column totals in ink. *Do not* place a double rule under the columns.
6. On the next journal page, enter the new page number.
7. On lines 1 and 2, enter the complete date (year, month, and day) in the Date column. Use the same date as on the last line of the previous page.
8. Place a check mark (✓) in the Sales Slip No. column and write *Brought Forward* in the Customer's Account Debited column.
9. Place a check mark (✓) in the Posting Reference column.
10. Enter the column totals from the previous page on line 2.

The journal page is now ready for the recording of the next transaction.

SECTION 16.1
Assessment

After You Read

Reinforce the Main Idea

Create a table similar to this one to identify the four special journals and the kind of transaction recorded in each

Special Journal	Kind of Transaction

Problem 1 Posting Column Totals from the Sales Journal

Instructions Here are the column totals of the sales journal for the month of April. In your working papers, post these totals to the appropriate general ledger accounts.

SALES JOURNAL   PAGE 4

	DATE	SALES SLIP NO.	CUSTOMER'S ACCOUNT DEBITED	POST. REF.	SALES CREDIT	SALES TAX PAYABLE CREDIT	ACCOUNTS RECEIVABLE DEBIT	
1	20--							1
2	Apr. 1	47	Amy Anderson	✓	800 00	48 00	848 00	2
32	30		Totals		12 000 00	720 00	12 720 00	32
33								33

Problem 2 Analyzing a Source Document

Metro Sports Distributors had this transaction that occurred on June 15.

Instructions

1. Analyze the sales slip shown. In your working papers, record the required entry on page 3 of the sales journal.

2. Post to the customer's account in the accounts receivable subsidiary ledger.

METRO SPORTS DISTRIBUTORS
Kingston Mall, Williamsburg, Virginia 23185

DATE: June 15, 20--   NO. 113

SOLD TO: M&M Consultants
2816 Mt. Odin Drive
Williamsburg, VA 23185

CLERK	CASH	CHARGE	TERMS
RDP		3	3/15, n/45

QTY.	DESCRIPTION	UNIT PRICE	AMOUNT
10	Golf Bags	$ 50.00	$ 500 00
2 doz.	Ladies Golf Shirts	100.00/doz	200 00
20	Golf CapsGolf Bags	6.00	120 00
		SUBTOTAL	$ 820 00
		SALES TAX	41 00
	Thank You!	TOTAL	$ 861 00

506 Chapter 16 • Special Journals: Sales and Cash Receipts

SECTION 16.1
Assessment

Math for Accounting

Vijay Products had the following amounts entered in the Sales Credit Column of its October Sales Journal.

Oct.	2	$1,500
	10	3,400
	17	8,200
	30	7,600

Assume a 3% sales tax was imposed on each sale. Answer the following questions.

1. What amount should be entered in the Sales Tax Payable Credit Column for each sale?

2. What amount should be entered in the Accounts Receivable Debit Column for each sale?

SECTION 16.2
The Cash Receipts Journal

In Chapter 14 you learned that the three most common sources of cash for a merchandising business are payments from charge customers, cash sales, and bankcard sales. These businesses also receive cash, but less frequently, from the sale of other business assets.

Can you imagine how many checks local cable television, electric, and telephone companies receive daily? They must have a streamlined cash receipts process to record checks efficiently. In this section you will learn to use a cash receipts journal as an efficient way to record cash receipts.

Section Vocabulary

- cash receipts journal
- schedule of accounts receivable
- sum

Journalizing and Posting to the Cash Receipts Journal

How Do You Use the Cash Receipts Journal?

The **cash receipts journal** is a special journal used to record all cash receipt transactions. It always has a Cash in Bank Debit column since every transaction in it debits the **Cash in Bank** account. The number of credit columns varies, depending on the company's needs. The Starting Line's cash receipts journal in **Figure 16-8** has six amount columns plus the date, source document, account name, and posting reference columns.

To keep the customer account balances current, the entries in the Accounts Receivable Credit column are posted daily to the accounts receivable subsidiary ledger. The entries in the General Credit column are also posted daily to the individual general ledger accounts. At month-end all special amount column totals are posted to the general ledger accounts named in the column headings.

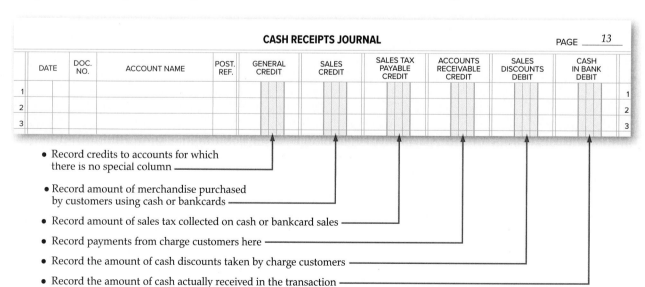

- Record credits to accounts for which there is no special column
- Record amount of merchandise purchased by customers using cash or bankcards
- Record amount of sales tax collected on cash or bankcard sales
- Record payments from charge customers here
- Record the amount of cash discounts taken by charge customers
- Record the amount of cash actually received in the transaction

Figure 16-8 Cash Receipts Journal

508 Chapter 16 • Special Journals: Sales and Cash Receipts

Recording Cash from Charge Customers

Follow these steps when journalizing cash receipts from charge customers. The source document for this transaction is the receipt.

1. Enter the date of the transaction in the Date column.
2. Enter the receipt number in the Document Number column. Write the letter *R* (for receipt) before the receipt number.
3. Enter the name of the customer in the Account Name column.
4. Enter the decrease in the amount owed by the customer in the Accounts Receivable Credit column.
5. Enter the amount of cash received in the Cash in Bank Debit column.

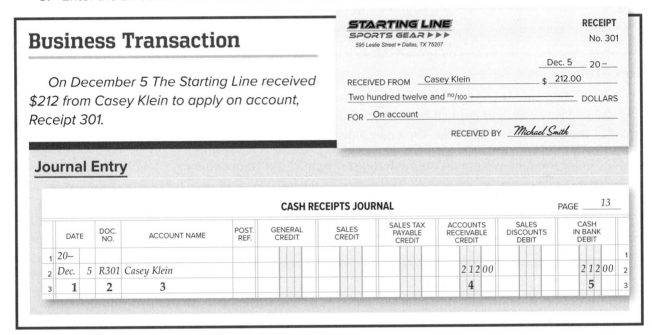

Business Transaction

On December 5 The Starting Line received $212 from Casey Klein to apply on account, Receipt 301.

When this entry has been posted to the customer's account in the accounts receivable subsidiary ledger, enter a check mark (✓) in the Posting Reference column of the cash receipts journal.

Recording Cash Received on Account, Less a Cash Discount

Let's use these steps to record a cash receipt transaction (shown on page 493) with a cash discount.

1. Enter the date of the receipt in the Date column.
2. Enter the receipt number in the Document Number column. Remember to write the letter *R* before the receipt number.
3. Enter the name of the customer in the Account Name column.
4. In the Accounts Receivable Credit column, enter the amount of the original sales transaction (the amount that was debited in the sales journal) less any related sales returns or allowances.
5. Enter the cash discount amount in the Sales Discounts Debit column.
6. Enter the amount of cash received in the Cash in Bank Debit column.

Chapter 16 • Special Journals: Sales and Cash Receipts

Business Transaction

On December 12 The Starting Line received $1,470 from South Branch High School Athletics in payment of Sales Slip 51 for $1,500 less the discount of $30, Receipt 302.

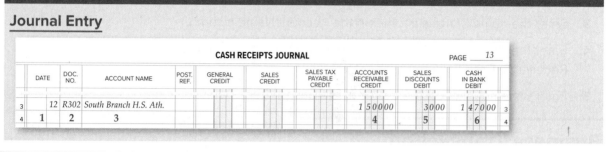

Journal Entry

	DATE	DOC. NO.	ACCOUNT NAME	POST. REF.	GENERAL CREDIT	SALES CREDIT	SALES TAX PAYABLE CREDIT	ACCOUNTS RECEIVABLE CREDIT	SALES DISCOUNTS DEBIT	CASH IN BANK DEBIT	
3	12	R302	South Branch H.S. Ath.					1 500 00	30 00	1 470 00	3
4	1	2	3					4	5	6	4

CASH RECEIPTS JOURNAL   PAGE 13

Recording Cash Sales

The Starting Line records cash sales every two weeks, but many businesses journalize cash sales and deposit cash daily. Follow these steps to enter cash sales in the cash receipts journal. The source document is the cash register tape.

1. Enter the date written on the cash register tape in the Date column.

2. Enter the number of the tape in the Document Number column. Write the letter *T* (for tape) before the tape number.

3. Enter the words *Cash Sales* in the Account Name column.

4. Enter a dash in the Posting Reference column to indicate that no entry has been posted individually to the general ledger accounts. Cash sales amounts are posted to the general ledger as part of the column totals at month-end.

5. Enter the amount of merchandise sold in the Sales Credit column.

6. Enter the amount of the sales taxes collected in the Sales Tax Payable Credit column.

7. Enter the total cash received in the Cash in Bank Debit column.

Business Transaction

On December 15 The Starting Line records the cash sales for the first two weeks of December, $3,000, and $180 in related sales taxes, Tape 55.

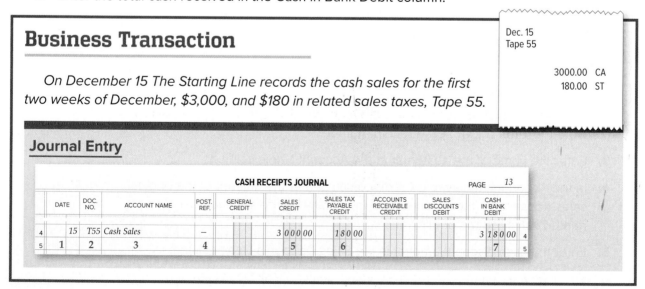

Journal Entry

CASH RECEIPTS JOURNAL   PAGE 13

	DATE	DOC. NO.	ACCOUNT NAME	POST. REF.	GENERAL CREDIT	SALES CREDIT	SALES TAX PAYABLE CREDIT	ACCOUNTS RECEIVABLE CREDIT	SALES DISCOUNTS DEBIT	CASH IN BANK DEBIT	
4	15	T55	Cash Sales	—		3 000 00	180 00			3 180 00	4
5	1	2	3	4		5	6			7	5

510   Chapter 16 • Special Journals: Sales and Cash Receipts

Recording Bankcard Sales

A business can record and deposit bankcard sales slips at any interval. The Starting Line processes them every two weeks. Follow these steps to enter bankcard sales in the cash receipts journal. Note the similarity to recording a cash sale. The entry's source document is the cash register tape.

1. Enter the date of the cash register tape in the Date column.
2. Enter the number of the tape in the Document Number column. Remember to write the letter *T* before the tape number.
3. Enter the words *Bankcard Sales* in the Account Name column.
4. Enter a dash in the Posting Reference column. The amounts recorded in the bankcard sales entry will be posted to the general ledger as part of the column totals at the end of the month.
5. Enter the amount of merchandise sold in the Sales Credit column.
6. Enter the sales taxes collected in the Sales Tax Payable Credit column.
7. Enter the total cash received in the Cash in Bank Debit column.

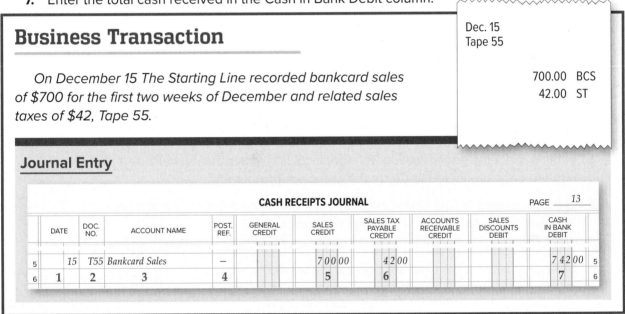

Recording Other Cash Receipts

Occasionally, a retail business receives cash from a transaction (shown on page 495) that does not involve the sale of merchandise. Since the business is receiving cash, it enters the transaction in the cash receipts journal. These are the steps to enter the next business transaction in the cash receipts journal:

1. Enter the date of the receipt in the Date column.
2. Enter an *R* and the receipt number in the Document Number column.
3. Enter *Office Equipment* in the Account Name column.
4. Enter the amount of the credit in the General Credit column. Use the General Credit column when the credit part of the entry is to an account that does not have a special amount column.
5. Enter the amount of cash received in the Cash in Bank Debit column.

Chapter 16 • Special Journals: Sales and Cash Receipts

Business Transaction

On December 16 The Starting Line received $30 from Mandy Harris, an office employee. She purchased a calculator that the business was no longer using, Receipt 303.

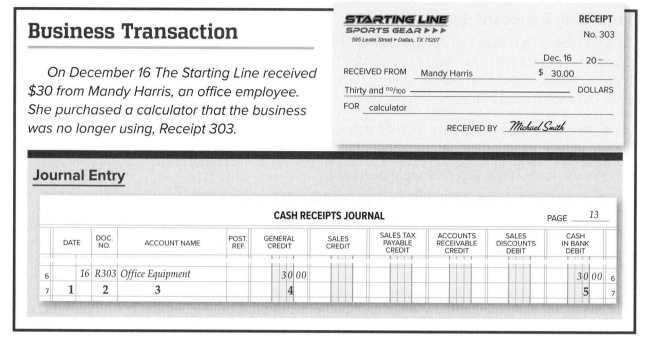

Journal Entry

Posting to the Accounts Receivable Subsidiary Ledger

A business posts daily from the Accounts Receivable Credit column to the accounts receivable subsidiary ledger so that customer accounts are always current. To post a cash receipt transaction to an account in the accounts receivable subsidiary ledger, follow these steps. Refer to **Figure 16–9**.

1. Enter the date of the transaction in the Date column of the subsidiary ledger account.

2. In the subsidiary ledger Posting Reference column, enter the journal letters and the page number. Use *CR* for the cash receipts journal.

3. In the subsidiary ledger Credit column, enter the amount from the Accounts Receivable Credit column of the cash receipts journal.

4. Compute and enter the new balance in the Balance column. If the account balance is zero, draw a line through the Balance column.

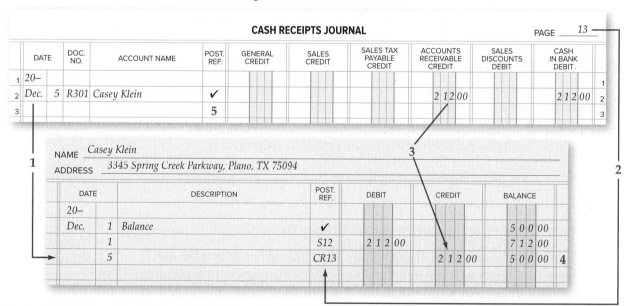

Figure 16–9 Posting from the Cash Receipts Journal to the Accounts Receivable Subsidiary Ledger

5. Return to the cash receipts journal and enter a check mark (✓) in the Posting Reference column.

Posting the General Credit Column

A business posts daily from the General Credit column of the cash receipts journal to the appropriate accounts in the general ledger. Refer to **Figure 16–10** as you read the steps for posting amounts from this column.

1. Enter the transaction date in the general ledger account Date column.

2. Enter the journal letters and page number in the Posting Reference column of the ledger account. Use *CR* for cash receipts journal.

3. In the Credit column of the ledger account, enter the amount from the General Credit column of the cash receipts journal.

4. Compute and enter the new account balance in the proper Balance column. For the **Office Equipment** account shown in **Figure 16–10**, use the Debit Balance column. If the account balance is zero, draw a line through the appropriate Balance column.

5. Return to the cash receipts journal and enter the general ledger account number in the Posting Reference column.

The cash receipts journal entry for the December 16 transaction is now posted. All transactions in the General Credit column are posted this way.

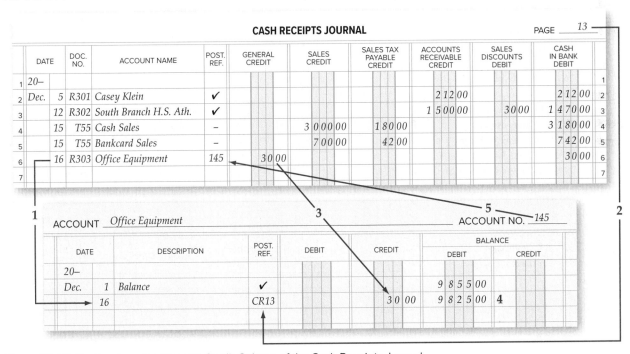

Figure 16–10 Posting from the General Credit Column of the Cash Receipts Journal

Footing, Proving, Totaling, and Ruling the Cash Receipts Journal

Follow these steps to complete the cash receipts journal. Refer to **Figure 16–11** on page 514.

1. Draw a single rule across the six amount columns, just below the last transaction.

2. Foot the columns.

Chapter 16 • Special Journals: Sales and Cash Receipts   513

3. Test for the equality of debits and credits as follows.

Debit Column			**Credit Columns**		
Sales Discounts	$	30	General	$	30
Cash in Bank		17,283	Sales		9,700
			Sales Tax Payable		582
			Accounts Receivable		7,001
Total	$	17,313	Total	$	17,313

4. In the Date column, on the line below the single rule, enter the date the journal is being totaled.

5. On the same line in the Account Name column, enter the word *Totals*.

6. Enter the column totals, in ink, just below the footings.

7. Double-rule the amount columns.

Review the completed cash receipts journal. Consider time saved by using this special journal to record cash receipts in one location instead of on many general journal pages. After the cash receipts journal is footed, proved, totaled, and ruled, the column totals are posted to the general ledger.

CASH RECEIPTS JOURNAL PAGE 13

	DATE	DOC. NO.	ACCOUNT NAME	POST. REF.	GENERAL CREDIT	SALES CREDIT	SALES TAX PAYABLE CREDIT	ACCOUNTS RECEIVABLE CREDIT	SALES DISCOUNTS DEBIT	CASH IN BANK DEBIT	
1	20–										1
2	Dec. 5	R301	Casey Klein	✓				212 00		212 00	2
3	12	R302	South Branch H.S. Ath.	✓				1 500 00	30 00	1 470 00	3
4	15	T55	Cash Sales	–		3 000 00	180 00			3 180 00	4
5	15	T55	Bankcard Sales	–		700 00	42 00			742 00	5
6	16	R303	Office Equipment	145	30 00					30 00	6
7	18	R304	Break Point Sports Club	✓				1 000 00		1 000 00	7
8	20	R305	Anita Montero	✓				100 00		100 00	8
9	22	R306	Gabriel Ramos	✓				500 00		500 00	9
10	23	R307	South Branch H.S. Ath.	✓				1 000 00		1 000 00	10
11	24	R308	Tammy's Fitness Club	✓				1 300 00		1 300 00	11
12	26	R309	Kim Wong	✓				300 00		300 00	12
13	27	R310	Lara Young	✓				200 00		200 00	13
14	27	R311	Robert Galvin	✓				465 00		465 00	14
15	28	R312	Joe Dimaio	✓				424 00		424 00	15
16	31	T56	Cash Sales	–		5 000 00	300 00			5 300 00	16
17	31	T56	Bankcard Sales	–		1 000 00	60 00			1 060 00	17
18	31		Totals		30 00	9 700 00	582 00	7 001 00	30 00	17 283 00	18
19					(✓)						19

Figure 16–11 Totaled and Ruled Cash Receipts Journal

Posting Column Totals to the General Ledger

Starting Line uses a cash receipts journal with six amount columns. Only five of them are posted to the general ledger. The General Credit column total is *not* posted. The entries in this column have already been posted individually to the general ledger accounts. (See **Figure 16–10**.)

514 Chapter 16 • Special Journals: Sales and Cash Receipts

The total in each amount column is posted to the general ledger account named in the column heading. See **Figure 16–12**. The column heading indicates whether to post the amount to the ledger account's Debit or Credit column.

Refer to **Figure 16–12** on page 515 and follow these steps to post column totals to the general ledger.

1. Place a check mark in parentheses under the General Credit column total to indicate that this total is not posted.

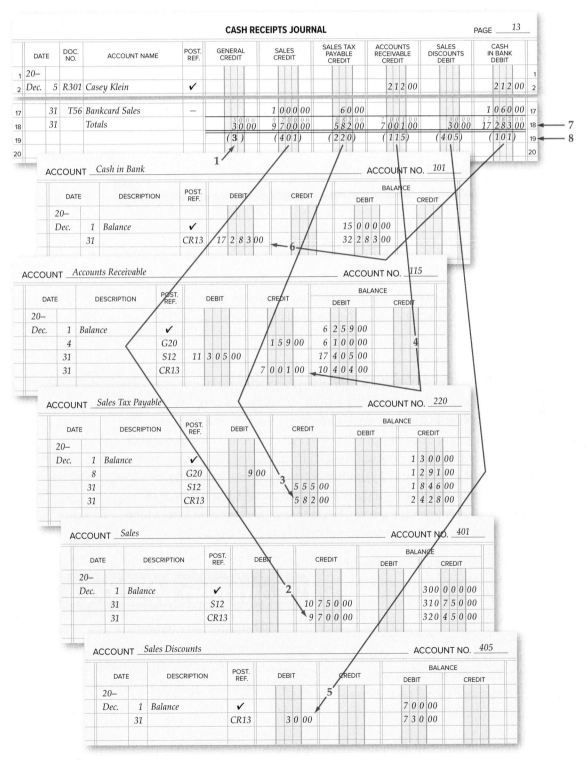

Figure 16–12 Posting Column Totals from the Cash Receipts Journal

Chapter 16 • Special Journals: Sales and Cash Receipts 515

2. Post the **Sales** total to the **Sales** account Credit column.
3. Post the **Sales Tax Payable** total to the **Sales Tax Payable** account Credit column.
4. Post the **Accounts Receivable** total to the **Accounts Receivable** controlling account Credit column.
5. Post the **Sales Discounts** total to the **Sales Discounts** account Debit column.
6. Post the **Cash in Bank** total to the **Cash in Bank** account Debit column.
7. Compute new balances for each general ledger account.
8. Write each account number in parentheses below the double rule in the cash receipts journal.

Reading Check

Define

What does it mean to "foot" the journal?

Using the Schedule of Accounts Receivable

What Is a Schedule of Accounts Receivable?

Accountants prepare a schedule for each subsidiary ledger to determine whether the subsidiary ledger's **sum** equals the controlling account's ending balance.

Proving the Accounts Receivable Subsidiary Ledger

To prove the accounts receivable subsidiary ledger, accountants prepare a **schedule of accounts receivable** that lists each charge customer, the balance in the customer's account, and the total amount due from all customers. The **Accounts Receivable** account is the controlling account for the accounts receivable subsidiary ledger. That is, the controlling account is a summary of all customer accounts in the subsidiary ledger. At month-end after all posting is completed, the balance of **Accounts Receivable** (controlling) in the general ledger should equal the total of the balances of the individual accounts receivable subsidiary ledger accounts.

The schedule of accounts receivable may be prepared on plain paper, accounting stationery, or on a computer. The heading identifies the schedule type and the date. All customer accounts appear in alphabetical order, as they do in the accounts receivable subsidiary ledger. All accounts are included, even those with zero balances. This prevents omitting a customer by mistake. **Figure 16–13** shows The Starting Line's schedule of

Common Mistakes

Schedule You might think of a schedule as a timetable. In accounting, a schedule is a detailed list of the items that make up a general ledger account balance.

516 Chapter 16 • Special Journals: Sales and Cash Receipts

Figure 16–13 Schedule of Accounts Receivable

ACCOUNT: Accounts Receivable   ACCOUNT NO. 115

DATE	DESCRIPTION	POST. REF.	DEBIT	CREDIT	BALANCE DEBIT	BALANCE CREDIT
20-- Dec. 1	Balance	✓			6 259 00	
4		G20		159 00	6 100 00	
31		S12	11 305 00		17 405 00	
31		CR13		7 001 00	10 404 00	

Starting Line Sports Gear
Schedule of Accounts Receivable
December 31, 20--

Break Point Sports Club	3 710 00	
Joe Dimaio	—	
Robert Galvin	—	
Casey Klein	500 00	
Anita Montero	371 00	
Shashi Rahim	212 00	
Gabriel Ramos	318 00	
South Branch High School Athletics	1 000 00	
Megan Sullivan	530 00	
Tammy's Fitness Club	1 060 00	
Kim Wong	106 00	
Lara Young	2 597 00	
Total Accounts Receivable		10 404 00

The balance of the **Accounts Receivable** account should equal the total of the schedule of accounts receivable.

accounts receivable for the end of December. Its total amount to be received from all customers is $10,404, which matches the balance of the **Accounts Receivable** account in the general ledger. This proves the accounts receivable subsidiary ledger.

Detecting Errors in the Subsidiary Ledger

Proving the accounts receivable subsidiary ledger with the controlling account verifies that the sum of the subsidiary ledger equals the controlling account's ending balance. This internal control procedure uncovers certain errors such as failing to post a transaction, or miscalculating an account balance. It does *not* ensure that transactions were posted to the correct customer account. The subsidiary ledger and the controlling account can balance even if an amount was posted to the wrong account. Often this type of error is discovered when a customer finds the error on a bill and reports it.

Internet Sales

How Do Internet Companies Receive Cash?

Companies can sell products with little expense and effort over the Internet, but they must have procedures and systems for processing online sales transactions. Companies can use an *Internet merchant account* for credit and debit card payments or an online payment service such as PayPal. Most Internet sales systems batch transactions daily to provide updated sales and inventory entries for the company's computer systems.

Online sales transactions must be secure. Businesses can use a card verification service to authenticate bankcard security codes and personal identification data to protect against online fraud.

SECTION 16.2
Assessment

After You Read

Reinforce the Main Idea

Create a table similar to this one to identify the journal to be used and the accounts to be debited and credited for each transaction.

Transaction	Journal	Account(s) Debited	Account(s) Credited
Bankcard Sales			
Sale of Merchandise on Account			
Sale of Asset for Cash			
Cash Sales			

Problem 3 Completing the Cash Receipts Journal

The cash receipts journal for the month of January is provided in your working papers and illustrated here.

Instructions In your working papers, foot, prove, total, and rule the cash receipts journal.

CASH RECEIPTS JOURNAL PAGE 10

DATE	DOC. NO.	ACCOUNT NAME	POST. REF.	GENERAL CREDIT	SALES CREDIT	SALES TAX PAYABLE CREDIT	ACCOUNTS RECEIVABLE CREDIT	SALES DISCOUNTS DEBIT	CASH IN BANK DEBIT
20– Jan. 3	R502	Jennifer Smith	✓				80 00		80 00
5	R503	Wilton High School	✓				3 100 00	62 00	3 038 00
8	R504	Store Equipment	155	75 00					75 00
15	T42	Cash Sales	—		5 000 00	300 00			5 300 00
15	T42	Bankcard Sales	—		1 200 00	72 00			1 272 00
20	R505	Norwin High School	✓				2 400 00	48 00	2 352 00
30	R506	Supplies	115	30 00					30 00

Math for Accounting

You are hired to audit the books for a computer software retailer, Software Biz. As you review the books, you realize that the sales tax was not calculated on a sales slip. You also want to estimate cash receipts from a sale. On a separate sheet of paper, make the following calculations:

1. Accounts Receivable—Business World; Sales Slip 47: software for $550; sales tax at 6%. What is the total due?

2. Accounts Receivable—Cindy Caskey; Sales Slip 48:
   Software $46.53
   Sales tax   2.79
   Total $49.32

   Terms: 1/10, n/30. How much will be received if it is paid within the discount period?

Chapter 16
Visual Summary

Concepts
Record transactions in sales and cash receipts journals

SALES JOURNAL PAGE 1

	DATE	SALES SLIP NO.	CUSTOMER'S ACCOUNT DEBITED	POST. REF.	SALES CREDIT	SALES TAX PAYABLE CREDIT	ACCOUNTS RECEIVABLE DEBIT	
1	Date	Sales	Customer's Account Name		xxx xx	xx xx	xxx xx	1
2		Slip						2
3		No.						3

CASH RECEIPTS JOURNAL PAGE 1

	DATE	DOC. NO.	ACCOUNT NAME	POST. REF.	GENERAL CREDIT	SALES CREDIT	SALES TAX PAYABLE CREDIT	ACCOUNTS RECEIVABLE CREDIT	SALES DISCOUNTS DEBIT	CASH IN BANK DEBIT	
1	Date	Receipt No.	Customer's Name					xxx xx		xxx xx	1
2	Date	Receipt No.	Customer's Name					xxx xx	xxx xx	xxx xx	2
3	Date	Tape No.	Cash Sales			xxx xx	xxx xx			xxx xx	3
4	Date	Tape No.	Bankcard Sales			xxx xx	xxx xx			xxx xx	4
5	Date	Receipt No.	Account Affected		xxx xx					xxx xx	5

Analysis
Post from the sales journal to customer accounts in the accounts receivable subsidiary ledger.

Journal	Individual Column Amounts	Are Posted To:
Sales Journal	Accounts Receivable Debit Column	➡ Subsidiary Ledger Customer Account

Procedures
Post column totals from the sales journal to general ledger accounts.

Journal	Column Total	Is Posted To:
Sales Journal	Sales Credit Column	➡ **Sales** Account
	Sales Tax Payable Credit Column	➡ **Sales Tax Payable** Account
	Accounts Receivable Debit Column	➡ **Accounts Receivable** Account

Chapter 16
Review and Activities

Answering the Essential Question

How do special journals save time and increase accuracy?

The details of transactions can be reflected in the general ledger. However, this requires many journal lines. Special journals are used to divide the record keeping, and the totals are transferred to the general ledger. How does this increase accuracy?

Vocabulary Check

1. **Vocabulary** Define each of the terms below in your own words. Then explain why each is important to the accounting process
   - sales journal
   - special journals
   - impose
   - footing
   - cash receipts journal
   - sum
   - schedule of accounts receivable

Concept Check

2. **Evaluate** True or False: The larger the company, the more likely special journals are used. Defend your response.
3. Name four commonly used special journals and describe the transactions entered in each.
4. List the steps required to record transactions in the sales journal.
5. **Analyze** When are transactions in the cash receipts journal posted to the accounts receivable subsidiary ledger? Why?
6. Which columns in the sales journal are double-ruled? What does the double rule indicate?
7. Which sales journal column total is posted to the general ledger Sales account?
8. **Synthesize** When posting column totals from the cash receipts journal to the general ledger, which total is *not* posted? What would be the result of posting this total to the general ledger?
9. What is the purpose of a schedule of accounts receivable? Explain how to prepare a schedule of accounts receivable.
10. **English Language Arts** Partner with another student in your class to role play a sales interaction. One student acts as a buyer, and the other as the seller of an item you choose. The interaction should include discussing the benefits of the item and why the buyer should want to purchase it, demonstrating how the item works, questions from the buyer to the seller, and negotiation of the price. At the end of the purchase, the buyer and seller work together to process the sales order according to the instructions in the text.

Chapter 16
Standardized Test Practice

Multiple Choice

1. A check mark next to a transaction entered in the sales journal indicates that the
   a. transaction has been posted to the accounts receivable subsidiary ledger.
   b. transaction was recorded using a verified source document.
   c. balance has been carried over to the accounts receivable subsidiary ledger.
   d. payment for the sale has been received.

2. Special journals are used
   a. in place of the general journal.
   b. to simplify the journalizing and posting process.
   c. to record transactions that cannot be recorded in the general journal.
   d. for types of transactions that are made infrequently.

3. A journal amount column that is not headed with an account name is called a
   a. special amount column.
   b. general amount column.
   c. date column.
   d. posting reference column.

4. The Accounts Receivable controlling account in the general ledger
   a. shows the total amount owed by all customers when posting is complete.
   b. helps keep the general ledger a balancing ledger in which debits equal credits.
   c. offers a means of testing the accuracy of the customer account balances.
   d. does all of the above.
   e. does none of the above.

Short Answer

5. What is done after the sales journal has been footed, proved, totaled, and ruled?

True or False

6. The cash receipts journal is used only for cash receipts resulting from sales.

Extended Response

7. Explain how the accounts receivable subsidiary ledger is proven.

Chapter 16
Computerized Accounting

Mastering Sales and Cash Receipts
Making the Transition from a Manual to a Computerized System

MANUAL METHODS

Setting up Customer Records
- Ledger sheets or cards are prepared with customer account details such as customer name, address, and contact information.
- Account activities are posted to the customer's subsidiary ledger account.

Issuing and Recording a Credit Memorandum
- A credit memorandum is prepared.
- A journal entry is prepared to record the credit memorandum.
- The journal entry is posted to the customer's account and to the general ledger.

COMPUTERIZED METHODS

Setting up Customer Records
- Assign each customer an ID code.
- Record details of customer accounts into customer records.
- The customer's sales and payment history is maintained automatically.

Issuing and Recording a Credit Memorandum
- The accounting software will create a credit memo and apply it to the appropriate outstanding customer invoice.
- The credit memo is posted automatically to the customer account and to the general ledger.

Chapter 16
Problems

Problem 4 | Recording and Posting Sales Transactions

Shutterbug Cameras records its transactions in special journals. The accounts receivable subsidiary ledger and certain general ledger accounts are included in your working papers.

Instructions In your working papers:

1. Record each transaction on page 12 of the sales journal or on page 13 of the cash receipts journal.
2. Post to the customer accounts in the accounts receivable subsidiary ledger on a daily basis.
3. Foot, prove, total, and rule the sales journal.
4. Post the column totals from the sales journal and the cash receipts journal to the general ledger accounts named in the column headings.
5. Prepare a schedule of accounts receivable.

Date		Transactions
May	1	Sold merchandise on account to Yoko Nakata, $500 plus 4% sales tax of $20, Sales Slip 220.
	3	Cash sales totaled $1,500 plus $60 in sales tax, Tape 10.
	7	Cash sales totaled $200 plus $8 in sales tax, Tape 11.
	9	Sold merchandise on account to Heather Sullivan for $100 plus sales tax of $4, Sales Slip 223.
	12	Cash sales totaled $400 plus $16 in sales tax, Tape 12.
	15	Sold $3,000 in merchandise on account to FastForward Productions plus sales tax of $120, Sales Slip 225.
	18	Sold merchandise on account to Yoko Nakata, $200 plus sales tax of $8, Sales Slip 226.

Analyze

Based on these transactions, calculate how much cash was collected for the month.

Problem 5 | Recording and Posting Cash Receipts

River's Edge Canoe & Kayak uses special journals and an accounts receivable subsidiary ledger for recording business transactions. The accounts receivable subsidiary ledger accounts and certain general ledger accounts are included in your working papers. The current account balances are recorded in the accounts.

continue

Chapter 16
Problems

Instructions In your working papers:

1. Record the transactions on page 7 in the cash receipts journal.
2. Post the amounts from the Accounts Receivable Credit column to the customers' accounts in the accounts receivable subsidiary ledger. Post the individual amounts in the General Credit column to the general ledger accounts.
3. Foot, prove, total, and rule the cash receipts journal.
4. Post the column totals to the general ledger accounts named in the column headings.
5. Prepare a schedule of accounts receivable.

Date		Transactions
May	2	Received $400 from Paul Drake to apply on his account, Receipt 505.
	5	Received $2,940 from Adventure River Tours in payment of its $3,000 account less a 2% cash discount of $60, Receipt 506.
	7	Sold old shelving (**Store Equipment**) for $50, Receipt 507.
	10	Wildwood Resorts sent us a check for $2,425 in payment of its account, $2,500 less a 3% cash discount of $75, Receipt 508.
	15	Cash sales totaled $2,000 plus $100 in sales tax, Tape 20.
	15	Bankcard sales were $3,000 plus $150 in sales tax, Tape 20.
	18	Celeste Everett sent us a check for $150 to apply on her account, Receipt 509.
	20	Isabel Rodriquez sent us $100 to apply on her account, Receipt 510.
	30	Cash sales were $4,500 plus $225 sales tax, Tape 21. Bankcard sales totaled $3,800 plus sales tax of $190, Tape 21.

Analyze
Identify the source document that would be used to record bankcard sales and cash sales.

Problem 6 **Recording and Posting Sales and Cash Receipts**

Buzz Newsstand had the following sales and cash receipt transactions for May. In your working papers, the accounts receivable subsidiary ledger and general ledger accounts have been opened with current balances.

Chapter 16
Problems

Instructions

1. Record the sales and cash receipts for May on page 11 of the sales journal and page 11 of the cash receipts journal.
2. Post to the customer accounts in the accounts receivable subsidiary ledger on a daily basis.
3. Post from the General Credit column of the cash receipts journal on a daily basis.
4. Foot, prove, total, and rule both journals.
5. Post the column totals of the sales journal to the general ledger accounts named in the column headings.
6. Post the column totals from the cash receipts journal to the general ledger accounts named in the column headings.
7. Prepare a schedule of accounts receivable.

> **SOURCE DOCUMENT PROBLEM**
>
> **Problem 16–8**
>
> Use the source documents in your working papers to complete this problem.

Date		Transactions
May	1	Sold $300 in merchandise plus 6% sales tax of $18 on account to Ilya Bodonski, Sales Slip 170.
	3	Received $490 from Katz Properties in payment of its $500 account balance less a 2% discount of $10, Receipt 145.
	5	As a favor, sold $60 in supplies to Straka Stores, received cash, Receipt 146.
	7	Rothwell Management Inc. sent us a check for $294 in payment of its $300 account balance less a $6 cash discount, Receipt 147.
	9	Sold $100 in merchandise plus $6 sales tax on account to Saba Nadal, Sales Slip 171.
	10	Sold $600 in merchandise plus $36 sales tax on account to Java Shops Inc., terms 2/10, n/30, Sales Slip 172.
	12	Sold $50 in merchandise plus $3 sales tax on account to Lee Adkins, Sales Slip 173.
	15	Received a check for $196 from Rolling Hills Pharmacies in payment of its $200 account less a 2% cash discount of $4, Receipt 148.
	15	Cash sales were $2,400 plus $144 sales tax, Tape 33.
	15	Bankcard sales totaled $2,000 plus sales tax of $120, Tape 33.
	18	Sold $1,000 in merchandise plus sales tax of $60 on account to Katz Properties, terms 2/10, n/30, Sales Slip 174.
	20	Received $100 from Ilya Bodonski to apply on her account, Receipt 149.
	22	Sold $800 in merchandise plus $48 sales tax on account to Rothwell Management Inc., terms 2/10, n/30, Sales Slip 175.

continue

Chapter 16
Problems

Date		Transactions
May	23	Lee Adkins sent us a check for $53 in payment of his account, Receipt 150.
	24	Received a $106 check from Saba Nadal to apply on account, Receipt 151.
	26	Java Shops Inc. sent us a check for $200 to apply on account, Receipt 152.
	27	Sold $200 in merchandise plus sales tax of $12 on account to Lee Adkins, Sales Slip 176.
	28	Received $75 from the sale of excess store equipment, Receipt 153.
	30	Cash sales totaled $2,600 plus $156 sales tax, Tape 34.
	30	Bankcard sales were $2,200 plus sales tax of $132, Tape 34.

Analyze
Calculate the percent of sales discounts given in May to total May sales.

Real-World Applications & Connections

CASE STUDY

Merchandising Business: Movie Theater

You work in the accounting department for Springdale Movie Theater. The owners are considering enlarging the concession area one and one-half times its current size to offer a wider variety of foods. No theater seats would be lost. The cost is estimated at $200,000. Sales figures for the past three years follow.

Year	Ticket Sales	Concession Sales
Year 1	$4,800,000	$1,800,000
Year 2	5,100,000	1,600,000
Year 3	5,400,000	1,480,000

Instructions

1. Use a spreadsheet program to analyze the sales and planned expansion. What percent of total sales are ticket sales and concession sales each year?

2. Assuming concession sales remain at $1,480,000, what percentage increase in concession revenue is needed to pay for the renovation costs in one year?

21st Century Skills

Allocating Resources

When you operate a business, you must constantly evaluate the effective use of resources, including time, money, materials, space, and staff. Imagine you own Retro Café, a late-night spot for young professionals and college students, open 11 a.m. until midnight. Your review of the month's financial reports reveals that the café is showing a net loss for the third month in a row.

Instructions Design a form to gather opinions on the menu, staffing, and operating hours. Think about what it is important for you, as the business owner, to know. What are the staffing implications if most sales occur after 3 p.m.?

Career Wise

Government Accountant

Government accountants and auditors maintain the financial records of government agencies and audit private companies and individuals subject to government regulations and taxation. Federal government employees may work for the Internal Revenue Service (IRS), or in financial management, financial institution examination, or budget analysis and administration. Government accountants may also keep track of how the government spends taxpayer money. They help create budgets, analyze publicly funded programs, and make sure money is collected and spent as required by law. Government accountants play a crucial role in ensuring a country operates smoothly.

Research and Share

a. Use the Internet to locate three sources of information on the job of a government accountant.

b. What level of education is required to be a government accountant? What are the roles and responsibilities of a government accountant?

Spotlight on Personal Finance

Examining Your Currency

You have learned that businesses have cash receipts in various forms including transactions with charge cards and bankcards; checks; and currency.

Instructions

What do you know about the cash you use? Write a brief report about U.S. paper currency. Address what currency is made of, the seven denominations in circulation today and the portrait appearing on each, the purpose of a security thread, and whether torn bills are acceptable.

H.O.T. Audit

Bank Fees

As the controller for Tri-State Furniture and Carpet Company, you review all source documents before approving bills for payment. After a bill is approved, the bill is sent to the company treasurer for payment. You also review the monthly bank statements for fees and charges made against Tri-State's checking account. For all bankcard deposits in excess of $35,000 for the month, the bank charges a 4% fee. If less than $30,000, the bank charges a 5% fee. During the month Tri-State made the following bank card deposits:

April 9: $8,800

April 16: $9,200

April 23: $7,700

April 30: $10,200

Upon reviewing the bank statement, the bank deducted $1795 for bank card fees.

Instructions

For what amount should the bank have charged Tri-State for bankcard fees? If Tri-State recorded the bank card fees for the month based on what the bank charged, what correcting entry would have to be made?

Chapter 17

Special Journals: Purchases and Cash Payments

Chapter Topics

17-1 The Purchases Journal

17-2 The Cash Payments Journal

Visual Summary

Review and Activities

Standardized Test Practice

Computerized Accounting

Problems

Real-World Accounting Careers

Real-World Applications & Connections

Essential Question

As you read this chapter, keep this question in mind:

How does keeping special journals help businesses organize information about what a company purchases on account and what it pays out in cash?

Main Idea

The purchases journal is used to record credit purchases. The cash payments journal is used to record the cash a business pays out.

Chapter Objectives

Concepts	Analysis	Procedures
C1 Explain the purpose of the purchases and cash payments journals. **C2** Record transactions in the purchases and cash payments journals.	**A1** Record payroll transactions in the cash payments journal. **A2** Post from the purchases and cash payments journals to the general ledger and the accounts payable subsidiary ledger.	**P1** Total, prove, and rule the purchases and cash payments journals. **P2** Prepare a schedule of accounts payable. **P3** Prove cash.

Real-World Business Connection

Education.com

Believing that family involvement helps improve student success at school, Education.com set out to equip parents with online resources for finding the best information fast. A one-stop source of information on education, development, and parenting, this Web site provides expert advice features and articles that tackle a variety of issues affecting pre-school to high school students. Besides providing fun worksheets and other printable resources, the site also offers School Finder, a tool for learning about schools or researching new ones. As a new online service, Education.com found a large community. In its first year, nearly a million visitors used its site each month.

Connect to the Business

Web-based companies like Education.com make many purchases, including state-of-the-art computers, software, servers, office equipment, and services of writers and Web developers. The company's accounting system documents each purchase.

Analyze

When Education.com purchases new computers for processing and storing data, what general ledger accounts are affected?

Focus on the Photo

With the variety of educational resources that Education.com offers to parents and students, the company must buy many items and supplies on credit to keep its Web site operating. **What kinds of items do you think companies buy on credit?**

SECTION 17.1
The Purchases Journal

Businesses use special journals to record transactions that are similar and occur frequently. If the accounting system were like a transit system, special journals would be like traffic police in helping avoid accounting traffic jams by channeling incoming data into appropriate "lanes." They also provide a shortcut for much of the data headed for the general ledger.

Sales and cash receipts journals record the sale of merchandise and other assets. In this chapter you will learn how accountants use purchases and cash payments journals to record the purchase of merchandise and other assets. You will also learn about the accounts payable subsidiary ledger.

Businesses like Target purchase merchandise for resale from hundreds of suppliers. Target and other companies use purchases and cash payments journals to simplify the purchase and payment processes.

> **Section Vocabulary**
> - purchases journal

> **Knowledge Extension**
> Vendor files are frequently maintained by businesses to keep track of relevant information on suppliers to maintain ongoing activities for purchases of merchandise. Vendor files may include contracts, quotes, purchase orders, warranty information, and other business documents.

Using the Purchases Journal

What Is the Purpose of the Purchases Journal?

Accountants use the **purchases journal** as a special journal to record all purchases on account. **Figure 17–1** shows the purchases journal that The Starting Line Sports Gear uses.

The purchases journal includes space for the page number, a column for the date, a column for the invoice number, a column for the name of the creditor, and a column for the posting reference. It also has three amount columns. Refer to **Figure 17–1** for a description of what amounts should be recorded in each column.

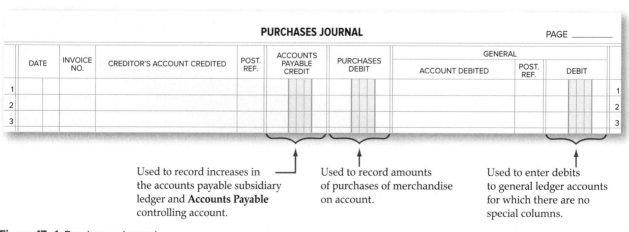

Figure 17–1 Purchases Journal

532 Chapter 17 • Special Journals: Purchases and Cash Payments

Recording the Purchase of Merchandise on Account

After verifying an invoice, the accounting clerk records the purchase in the purchases journal. Refer to the purchases journal and follow these steps:

1. Enter the date in the Date column. Use the date the invoice was *received,* not the date the invoice was prepared.
2. Enter the invoice number in the Invoice Number column.
3. Enter the creditor's name in the Creditor's Account Credited column.
4. Enter the *total* of the invoice in the Accounts Payable Credit column.
5. For purchases of merchandise on account, enter the total amount of the invoice in the Purchases Debit column.

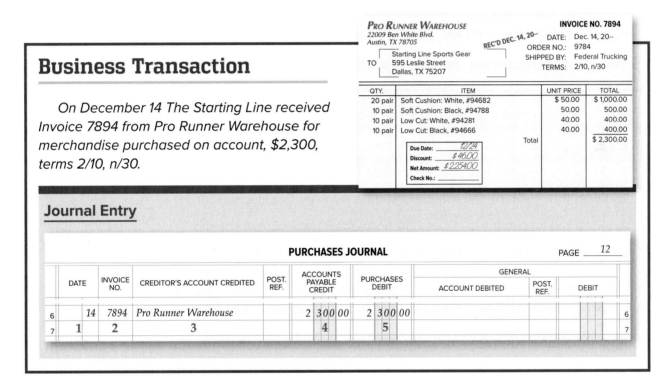

After journalizing the invoice from Pro Runner Warehouse, the accounting clerk places it in a tickler file by due date. In this case it is filed in the December 24 folder. Starting Line plans to take the discount, and December 24 is ten days after the invoice date.

Journalizing Other Purchases on Account

Starting Line also purchases supplies and other assets on account. These purchases do not occur often enough to set up special columns in the purchases journal. When these purchases do occur, they are recorded in the General Debit column of the purchases journal.

Business Transaction

On December 15 The Starting Line received Invoice 3417, dated December 13, from Champion Store Supply for store equipment bought on account, $1,200, terms n/30.

CHAMPION STORE SUPPLY
47249 Randall Parkway
Dallas, TX 75207

REC'D DEC. 15, 20--

INVOICE NO. 3417
DATE: Dec. 13, 20--
ORDER NO.: 9795
SHIPPED BY: Federal Trucking
TERMS: n/30

TO: Starting Line Sports Gear
595 Leslie Street
Dallas, TX 75207

QTY.	ITEM	UNIT PRICE	TOTAL
3	Corner Shelf Units	$	$ 900.00
1	Shirt Rack	300.00	300.00
	Total	300.00	$1,200.00

Due Date: 1/12
Discount: $0.00
Net Amount: $1,200.00
Check No.:

Journal Entry

PURCHASES JOURNAL PAGE 12

DATE	INVOICE NO.	CREDITOR'S ACCOUNT CREDITED	POST. REF.	ACCOUNTS PAYABLE CREDIT	PURCHASES DEBIT	GENERAL ACCOUNT DEBITED	POST. REF.	DEBIT
15	3417	Champion Store Supply		1 200 00		Store Equipment		1 200 00
1	2	3		4		5		6

Refer to the purchases journal and follow these steps:

1. Enter the date the invoice was *received* in the Date column.

2. Enter the invoice number in the Invoice Number column.

3. Enter the creditor's name in the Creditor's Account Credited column.

4. Enter the *total* of the invoice in the Accounts Payable Credit column.

5. Store equipment is not merchandise purchased for resale, so this transaction is not recorded in the Purchases Debit column. Instead, it is recorded in the General column. Write the name of the general ledger account being debited in the General Account Debited column.

6. Enter the total amount of the invoice in the General Debit column.

After journalizing the invoice, the accounting clerk places it in the tickler file. Because Champion Store Supply does not offer its credit customers a cash discount, the due date is January 12, 30 days after the date of the invoice. The invoice is filed in folder "12" to indicate this date.

Posting from the Purchases Journal

How Do You Post Entries from the Purchases Journal?

As you have learned, special journals save time in recording and posting business transactions. Each transaction in the purchases journal is a purchase on account. Therefore, each transaction is separately posted *daily* to the accounts payable subsidiary ledger to keep creditor accounts current.

Refer to **Figure 17–2** on page 535 as you read about posting to the accounts payable subsidiary ledger.

1. In the Date column of the subsidiary ledger account, enter the date of the transaction.

Figure 17–2 Posting from the Purchases Journal to the Accounts Payable Subsidiary Ledger

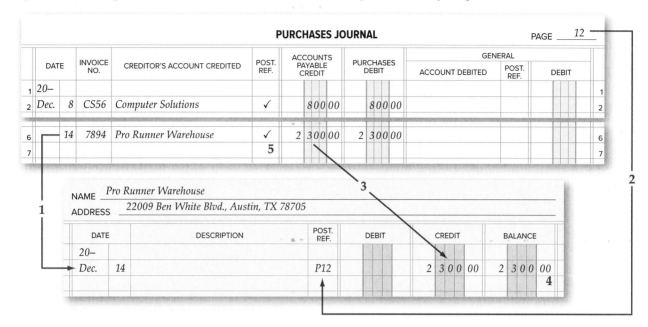

2. In the Posting Reference column of the subsidiary ledger account, record the journal letter and the page number. *P* is the letter used for the purchases journal.

3. In the Credit column of the subsidiary ledger account, enter the amount owed to the creditor.

4. Compute the new account balance by adding the amount in the Credit column to the previous balance amount. Since there was no previous balance in Pro Runner's account, enter $2,300 in the Balance column.

5. Return to the purchases journal and place a check mark (✓) in the *first* Posting Reference column (next to the Creditor's Account Credited column).

Posting from the General Debit Column

The clerk makes daily postings from the General Debit column of the purchases journal to the appropriate accounts in the general ledger. Refer to **Figure 17–3** as you read the following steps:

1. Enter the date of the transaction in the Date column of the general ledger account.

2. In the Posting Reference column of the general ledger account, record the journal letter (*P* for purchases journal) and page number.

3. In the Debit column of the general ledger account, enter the amount recorded in the General Debit column of the purchases journal.

4. Compute and record the new balance in the Debit Balance column.

5. Return to the purchases journal and place the general ledger account number in the General Posting Reference column (following the General Account Debited column).

Figure 17–3 Posting from the Purchases Journal to the General Ledger

		PURCHASES JOURNAL						PAGE 12	
DATE	INVOICE NO.	CREDITOR'S ACCOUNT CREDITED	POST. REF.	ACCOUNTS PAYABLE CREDIT	PURCHASES DEBIT	GENERAL ACCOUNT DEBITED	POST. REF.	DEBIT	
20–									1
Dec. 8	CS56	Computer Solutions	✓	800 00	800 00				2
15	3417	Champion Store Supply	✓	1 200 00		Store Equipment	150	1 200 00	7
							5		8

ACCOUNT Store Equipment   ACCOUNT NO. 150

DATE	DESCRIPTION	POST. REF.	DEBIT	CREDIT	BALANCE DEBIT	BALANCE CREDIT
20–						
Dec. 1	Balance	✓			4 000 00	
15		P12	1 200 00		5 200 00	

Totaling, Proving, and Ruling the Purchases Journal

To complete the purchases journal, refer to **Figure 17–4** and follow these steps:

1. Draw a single rule across the three amount columns: Accounts Payable Credit, Purchases Debit, and General Debit.
2. Foot each amount column.
3. Test for the equality of debits and credits.

Debit Column		**Credit Columns**	
Purchases	$15,400	Accounts Payable	$16,850
General	1,450		
	$16,850		$16,850

		PURCHASES JOURNAL						PAGE 12	
DATE	INVOICE NO.	CREDITOR'S ACCOUNT CREDITED	POST. REF.	ACCOUNTS PAYABLE CREDIT	PURCHASES DEBIT	GENERAL ACCOUNT DEBITED	POST. REF.	DEBIT	
20–									1
Dec. 8	CS56	Computer Solutions	✓	800 00	800 00				2
10	4692	Sports Link Footwear	✓	4 000 00	4 000 00				3
12	SN63	Sports Nutrition Supply	✓	600 00	600 00				4
14	2560	FastLane Athletics	✓	3 000 00	3 000 00				5
14	7894	Pro Runner Warehouse	✓	2 300 00	2 300 00				6
15	3417	Champion Store Supply	✓	1 200 00		Store Equipment	150	1 200 00	7
15	9881	Geary Office Supply	✓	250 00		Supplies	130	250 00	8
18	8560	FastLane Athletics	✓	2 000 00	2 000 00				9
20	6593	Computer Solutions	✓	1 200 00	1 200 00				10
27	5200	Sports Link Footwear	✓	1 500 00	1 500 00				11
31		Totals		16 850 00	15 400 00			1 450 00	12

Figure 17–4 The Completed Purchases Journal

4. In the Date column, on the line below the single rule, enter the date the journal is being totaled.

5. On the same line, write the word *Totals* in the Creditor's Account Credited column.

6. Enter the three column totals, in ink, just below the footings.

7. Draw a double rule across the three amount columns.

The completed purchases journal is a quick reference tool for the accountant to review current transactions that affect the **Purchases** account and the **Accounts Payable** account. Miscellaneous purchases are also reflected in the purchases journal. If a general journal had been used, these transactions would have been mixed with other transactions (cash receipts, payments, and more) of the period.

What else can the accountant verify from the completed purchases journal? A quick review of the Post. Ref. column tells the accountant that amounts have been posted to the accounts payable subsidiary ledger.

Reading Check

In Your Own Words

What does "proving" the purchases journal mean?

Posting the Special Column Totals to the General Ledger

After totaling and ruling the purchases journal, the clerk posts the totals of the Accounts Payable Credit column and the Purchases Debit column to the general ledger accounts. Then the clerk calculates the new balance for each account, and enters the new balance in the appropriate Balance column. **Figure 17–5** shows the posting of these column totals.

After posting each column total, write the general ledger account number, in parentheses, in the column below the double rule as shown in **Figure 17–5.** The total of the General Debit column is not posted because the individual amounts were posted during the month. Place a check mark (✓) in parentheses below the double rule in the General Debit column to indicate that the total is not posted.

Figure 17–5 Posting Column Totals from the Purchases Journal to the General Ledger

PURCHASES JOURNAL PAGE 12

	DATE	INVOICE NO.	CREDITOR'S ACCOUNT CREDITED	POST. REF.	ACCOUNTS PAYABLE CREDIT	PURCHASES DEBIT	GENERAL ACCOUNT DEBITED	POST. REF.	DEBIT	
1	20–									1
2	Dec. 8	CS56	Computer Solutions	✓	800 00	800 00				2
11	27	5200	Sports Link Footwear	✓	1 500 00	1 500 00				11
12	31		Totals		16 850 00	15 400 00			1 450 00	12
13					(201)	(501)			(✓)	13
14										14

ACCOUNT __Accounts Payable__   ACCOUNT NO. __201__

DATE	DESCRIPTION	POST. REF.	DEBIT	CREDIT	BALANCE DEBIT	BALANCE CREDIT
20–						
Dec. 1	Balance	✓				6 300 00
16		G21	200 00			6 100 00
31		P12		16 850 00		22 950 00

ACCOUNT __Purchases__   ACCOUNT NO. __501__

DATE	DESCRIPTION	POST. REF.	DEBIT	CREDIT	BALANCE DEBIT	BALANCE CREDIT
20–						
Dec. 1	Balance	✓			190 000 00	
19		CP14	1 300 00		191 300 00	
31		P12	15 400 00		206 700 00	

SECTION 17.1
Assessment

After You Read

Reinforce the Main Idea

Create a table similar to this one to describe how these three separate transactions would appear in the purchases journal.

Transaction	Account Debited	Account Credited	Amount Column Debited	Amount Column Credited
Purchased Merchandise on Account				
Purchased Store Equipment on Account				
Purchased Supplies on Account				

Problem 1 Recording Transactions in the Purchases Journal

The Design Den, a retail merchandising business, uses special journals. On page 3 of the purchases journal in the working papers, record the following purchases:

Date	Transactions
Feb. 1	Purchased $1,400 in merchandise on account from Woodstock Furnishings, terms 3/15, n/30, Invoice WF39.
2	Bought $900 in store equipment on account from Holmes Equipment Company, terms n/30, Invoice 98.
4	Purchased $700 in merchandise from Fuller Fabrics, terms 3/10, n/30, Invoice 72.
7	Purchased computer speakers on account for $50 from Digital Solutions, terms 2/10, n/30, Invoice AB220.
10	Purchased fabric from Valley Upholstery for $1,500, terms 2/10, n/30, Invoice 947.

Math for Accounting

Dynamo Industries received an invoice from Santos Suppliers for merchandise purchased on July 5 for $12,000 with terms of 3/15, n/30. Answer the following questions:

1. What is the due date of the invoice?
2. What is the amount of cash discount?
3. What is the net amount to be paid?
4. What account is debited and for what amount?
5. What account is credited and for what amount?

SECTION 17.2
The Cash Payments Journal

You have learned about three special journals: the sales journal, the cash receipts journal, and the purchases journal. Now you will study the *cash payments journal*. For many businesses like grocers who must pay for the purchases of a wide variety of merchandise, frequent payments make the use of the *cash payments journal* necessary.

Section Vocabulary
- cash payments journal
- schedule of accounts payable
- proving cash
- constantly
- automatically
- adjust

Using the Cash Payments Journal

How Do You Record Cash Payments?

The **cash payments journal** is used to record all transactions in which cash is paid out or decreased. These transactions include: payments to creditors for items bought on account, cash purchases of merchandise and other assets, payments for various expenses, payments for wages and salaries, and cash decreases for bank service charges and bankcard fees. The source documents for the journal entries are check stubs and the bank statement. The cash payments journal is also called the *cash disbursements journal*.

Figure 17–6 shows the cash payments journal that The Starting Line uses. Notice that it has five amount columns.

The seven transactions that follow are typical of those recorded in the cash payments journal. Note that each transaction in the cash payments journal results in a credit to the **Cash in Bank** account.

	DATE	DOC NO.	ACCOUNT NAME	POST REF.	GENERAL DEBIT	GENERAL CREDIT	ACCOUNTS PAYABLE DEBIT	PURCHASES DISCOUNTS CREDIT	CASH IN BANK CREDIT	
1										1
2										2
3										3

CASH PAYMENTS JOURNAL PAGE _____

Used to enter debits and credits to general ledger accounts for which there are no special columns.

Used to record decreases to accounts in the accounts payable subsidiary ledger and to the **Accounts Payable** controlling account.

Used to enter the amount of purchases discounts taken.

Every transaction recorded here decreases **Cash in Bank.**

Figure 17–6 Cash Payments Journal

Recording the Cash Purchase of an Asset

Cash purchases of various assets are recorded in the cash payments journal. One asset commonly purchased for cash is insurance. Let's record a transaction involving a cash purchase of insurance.

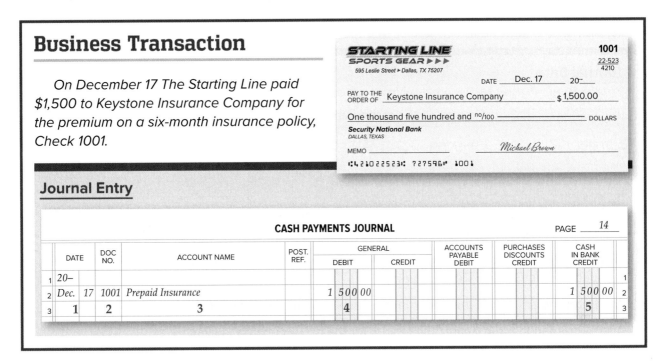

Refer to the cash payments journal above and follow these steps:

1. Enter the date of the transaction in the Date column.
2. Enter the check number in the Document Number column.
3. Enter the name of the account debited in the Account Name column.
4. Because there is no special column for **Prepaid Insurance,** enter the amount of the debit in the General Debit column.
5. Enter the amount of the credit in the Cash in Bank Credit column.

After payment, the invoice for the insurance is filed.

Recording a Cash Purchase of Merchandise

Retail businesses are **constantly** purchasing merchandise for resale. While they make most purchases on account, many are for cash. Let's record a cash purchase of merchandise for resale.

Refer to The Starting Line's purchase from FastLane Athletics and the cash payments journal as you follow these steps:

1. Enter the date of the transaction in the Date column.
2. Enter the check number in the Document Number column.
3. Enter the name of the account debited in the Account Name column.

Chapter 17 • Special Journals: Purchases and Cash Payments **541**

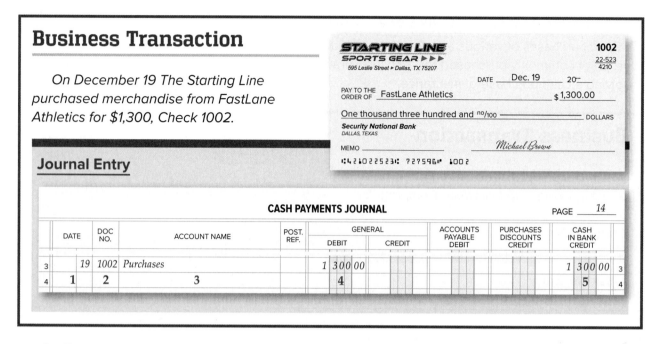

4. Because there is no special column for **Purchases,** enter the amount of the debit in the General Debit column.

5. Enter the amount of the check in the Cash in Bank Credit column.

After recording the transaction, the receipt for the cash purchase is filed.

Recording a Payment on Account

Now let's learn how to make a payment on account and take a purchase discount.

Business Transaction

On December 24 The Starting Line paid $2,254 to Pro Runner Warehouse for merchandise purchased on account, $2,300 less a discount of $46, Check 1003.

Journal Entry

CASH PAYMENTS JOURNAL   PAGE 14

	DATE	DOC NO.	ACCOUNT NAME	POST. REF.	GENERAL DEBIT	GENERAL CREDIT	ACCOUNTS PAYABLE DEBIT	PURCHASES DISCOUNTS CREDIT	CASH IN BANK CREDIT	
4	24	1003	Pro Runner Warehouse				2 300 00	46 00	2 254 00	4
5	1	2	3				4	5	6	5

542   Chapter 17 • Special Journals: Purchases and Cash Payments

Refer to the purchases journal above and follow these steps:

1. Enter the date of the transaction in the Date column.
2. Enter the check number in the Document Number column.
3. Enter the creditor's name in the Account Name column.
4. Enter the amount of the original purchase in the Accounts Payable Debit column.
5. Enter the amount of the purchase discount in the Purchases Discounts Credit column.
6. Enter the amount of the check in the Cash in Bank Credit column.

Remember that a processing stamp is placed on each invoice when it is verified. After the cash payment has been journalized, the accounting clerk records the check number on the "Check No." line of the processing stamp. The paid invoice is then filed.

Companies offer no discount for some purchases. For others they offer a discount, but the business cannot pay within the discount period. In these cases the check is written for the full amount of the purchase.

Recording Other Cash Payments

Let's record a check written to pay for shipping charges when merchandise is sent FOB shipping point.

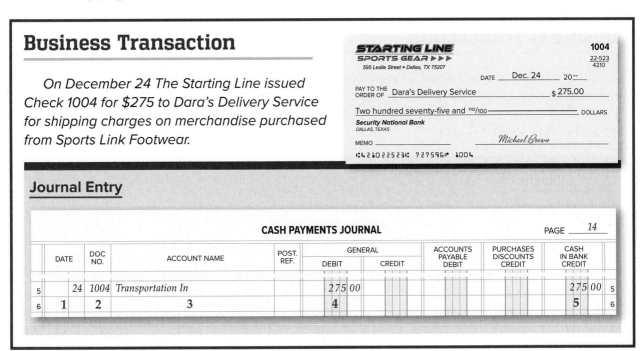

Refer to the cash payments journal above and follow these steps:

1. Enter the date of the transaction in the Date column.
2. Enter the check number in the Document Number column.
3. Enter the name of the account debited in the Account Name column.

4. Because there is no special column for **Transportation In,** enter the amount in the General Debit column.

5. Enter the amount of the check in the Cash in Bank Credit column.

Recording Payment of Payroll

In an earlier chapter, you learned how to record the payroll entry in the general journal. When a business uses special journals, the entry for payment of the payroll is recorded in the cash payments journal. The information to record payroll transactions in journals is taken from the payroll register.

To record the payroll transaction, refer to the cash payments journal on the following page and follow these steps:

1. On the first line of the entry, enter the date of the transaction in the Date column.

2. Enter the check number in the Document Number column.

3. Enter the name of the account debited in the Account Name column.

4. Enter the amount of the payroll (gross pay) in the General Debit column.

5. Enter the net pay in the Cash in Bank Credit column.

6. On the next four lines, enter the names of the accounts credited in the Account Name column. Enter the amount of each liability in the General Credit column.

Business Transaction

On December 31 The Starting Line wrote Check 1012 for $2,974 to pay the payroll of $4,000 (gross earnings) for the pay period ended December 31. The following amounts were withheld: Employees' Federal Income Tax, $640; Employees' State Income Tax, $80; Social Security Tax, $248; and Medicare Tax, $58.

Journal Entry

CASH PAYMENTS JOURNAL PAGE 14

DATE	DOC NO.	ACCOUNT NAME	POST. REF.	GENERAL DEBIT	GENERAL CREDIT	ACCOUNTS PAYABLE DEBIT	PURCHASES DISCOUNTS CREDIT	CASH IN BANK CREDIT
31	1012	Salaries Expense **3**		4 000 00				5 2 974 00
1	**2**	Employees' Fed. Inc. Tax Pay.		**4**	640 00			
		Employees' State Inc. Tax Pay.	**6**		80 00			
		Social Security Tax Pay.			248 00			
		Medicare Tax Pay.			58 00			

Recording Bank Service Charges

Bank service charges are **automatically** deducted from the checking account. Although no check is written to pay these charges, the transactions are recorded in the cash payments journal because the charges decrease the **Cash in Bank** account.

Business Transaction

On December 31 The Starting Line recorded a bank service charge for $20 indicated on the bank statement.

Journal Entry

					POST.	GENERAL		ACCOUNTS PAYABLE DEBIT	PURCHASES DISCOUNTS CREDIT	CASH IN BANK CREDIT	
	DATE		DOC NO.	ACCOUNT NAME	REF.	DEBIT	CREDIT				
18		31		Miscellaneous Expense		20 00				20 00	18
19											19

CASH PAYMENTS JOURNAL — PAGE 14

Recording Bankcard Fees

Most banks charge a fee for handling bankcard sales. This fee is automatically deducted from the business's checking account. For example, Starting Line's bank deducted a bankcard fee of $75. The fee appeared on the bank statement as a deduction from the checking account balance. The accounting clerk recorded this decrease in cash in the cash payments journal.

✓ Reading Check

List

What are some examples of transactions in which cash is paid out or decreased?

The clerk also enters the bank charges in the checkbook records. **Figure 17–7** illustrates one way to **adjust** the balance on the check stub. The deposit heading is crossed out, and the words *Bankcard Fees* are written in its place. A deduction of $75 is entered on the stub. On the next line, the words *Less Bank Service Charge* are written and an entry is made for the $20 deduction. Both amounts are subtracted from the balance brought forward.

	Dollars	Cents
Balance brought forward	15,274	00
~~Add deposits~~ Bankcard Fees	−75	00
Less Bank Svc. Chg.	−20	00
Total	15,179	00
Less this check		
Balance carried forward		

No. 1013

Figure 17–7 Recording Bankcard Fees and Service Charges in the Checkbook

Business Transaction

On December 31 The Starting Line recorded the bankcard fee of $75 that appeared on the bank statement.

Journal Entry

				POST.	GENERAL		ACCOUNTS	PURCHASES	CASH
	DATE	DOC NO.	ACCOUNT NAME	REF.	DEBIT	CREDIT	PAYABLE DEBIT	DISCOUNTS CREDIT	IN BANK CREDIT
19		31	Bankcard Fees Expense		75 00				75 00

CASH PAYMENTS JOURNAL — PAGE 14

Posting from the Cash Payments Journal

How Do You Post from the Cash Payments Journal?

Individual amounts in the Accounts Payable Debit column and the General Debit column are posted daily. Column totals are posted at the end of the month.

To keep creditors' accounts current, clerks make daily postings from the Accounts Payable Debit column to the accounts payable subsidiary ledger. Refer to **Figure 17–8** on page 547 and follow these steps:

1. Enter the date of the transaction in the Date column of the subsidiary ledger account.

2. In the subsidiary ledger account's Posting Reference column, enter the journal letters (*CP* for the cash payments journal) and the page number.

3. In the Debit column of the subsidiary ledger account, enter the amount recorded in the Accounts Payable Debit column of the journal.

4. Compute the new account balance and enter it in the Balance column. If the account has a zero balance, draw a line through the Balance column.

5. Return to the cash payments journal and enter a check mark (✓) in the Posting Reference column.

Reading Check

Recall

How often are individual cash amounts posted to the cash payments journal? When are column totals posted?

Figure 17–8 Posting from the Cash Payments Journal to the Accounts Payable Subsidiary Ledger

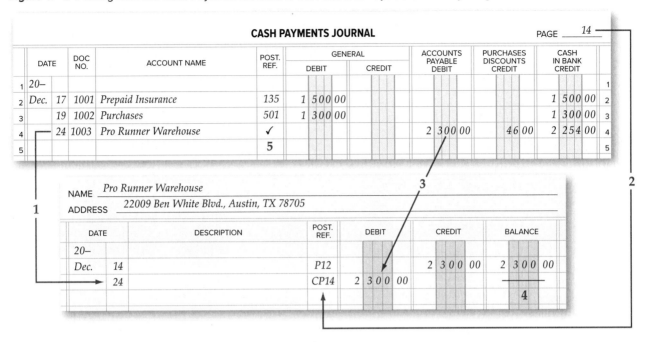

Figure 17–9 shows The Starting Line's accounts payable subsidiary ledger after all postings have been made. Notice that the accounts contain entries from the purchases, cash payments, and general journals.

Figure 17–9 The Completed Accounts Payable Subsidiary Ledger

Chapter 17 • Special Journals: Purchases and Cash Payments **547**

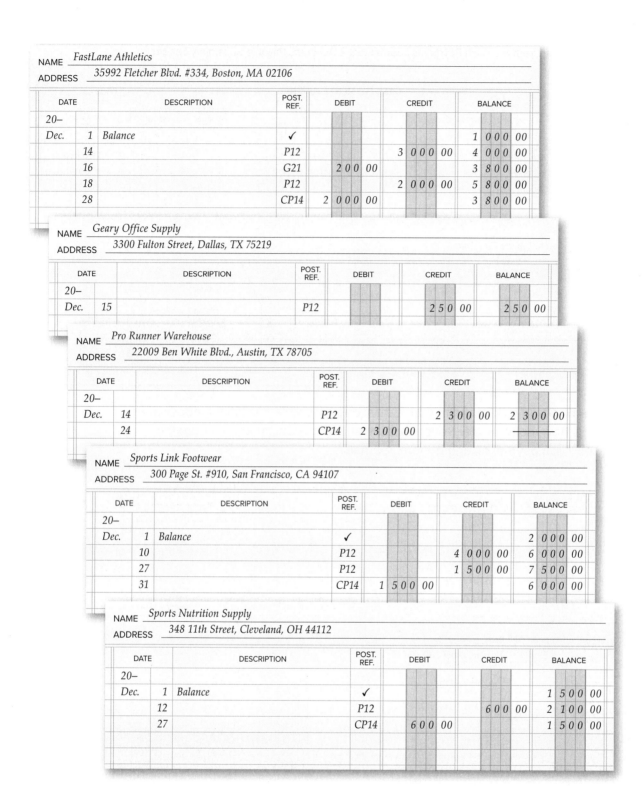

Figure 17–9 The Completed Accounts Payable Subsidiary Ledger (continued)

548 Chapter 17 • Special Journals: Purchases and Cash Payments

Figure 17–10 Posting from the General Debit Column of the Cash Payments Journal

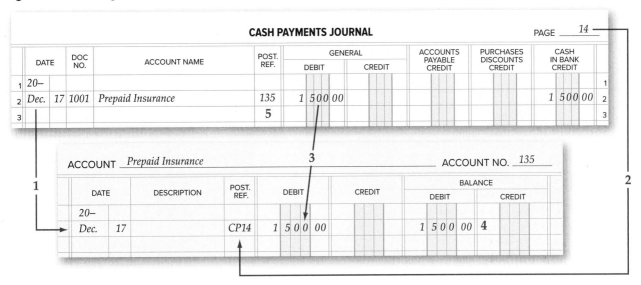

Posting from the General Debit Column

Accountants post daily from the General Debit column to the appropriate general ledger accounts. Refer to **Figure 17–10** as you read the following steps.

1. Enter the date of the transaction in the Date column of the general ledger account.

2. Enter the journal letters (*CP* for the cash payments journal) and the page number in the Posting Reference column of the general ledger account.

3. In the Debit column, enter the amount from the General Debit column of the cash payments journal.

4. Compute the new balance and enter it in the appropriate Balance column. (Because the example used here has no previous balance, the amount recorded in the Debit column is also entered in the Debit Balance column.)

5. Return to the cash payments journal and enter the account number in the Posting Reference column.

All the transactions in the General Debit column are posted to the general ledger accounts in the same way.

Accountants also post entries daily from the General Credit column to the appropriate general ledger accounts. This is done in the same way as shown for General Debit Column entries.

Totaling, Proving, and Ruling the Cash Payments Journal

Accountants total the cash payments journal following the same steps that they use for other special journals. Before they rule the journal, they prove the equality of debits and credits.

Chapter 17 • Special Journals: Purchases and Cash Payments **549**

Figure 17–11 The Completed Cash Payments Journal

CASH PAYMENTS JOURNAL   PAGE 14

DATE	DOC NO.	ACCOUNT NAME	POST. REF.	GENERAL DEBIT	GENERAL CREDIT	ACCOUNTS PAYABLE DEBIT	PURCHASES DISCOUNTS CREDIT	CASH IN BANK CREDIT
20–								
Dec. 17	1001	Prepaid Insurance	135	1 500 00				1 500 00
19	1002	Purchases	501	1 300 00				1 300 00
24	1003	Pro Runner Warehouse	✓			2 300 00	46 00	2 254 00
24	1004	Transportation In	505	275 00				275 00
26	1005	Champion Store Supply	✓			1 200 00		1 200 00
27	1006	Sports Nutrition Supply	✓			600 00		600 00
28	1007	FastLane Athletics	✓			2 000 00	40 00	1 960 00
30	1008	Dara's Delivery Service	✓			300 00		300 00
31	1009	Rent Expense	660	2 000 00				2 000 00
31	1010	Computer Solutions	✓			1 200 00	24 00	1 176 00
31	1011	Sports Link Footwear	✓			1 500 00	30 00	1 470 00
31	1012	Salaries Expense	665	4 000 00				2 974 00
		Employee's Federal Inc. Tax Pay.	205		640 00			
		Employee's State Inc. Tax Pay.	211		80 00			
		Social Security Tax Payable	212		248 00			
		Medicare Tax Payable	213		58 00			
31	—	Miscellaneous Expense	655	20 00				20 00
31	—	Bankcard Fees Expense	605	75 00				75 00
31		Totals		9 170 00	1 026 00	9 100 00	140 00	17 104 00

Debit Column		**Credit Columns**	
General	$ 9,170	General	$ 1,026
Accounts Payable	9,100	Purchases Discounts	140
		Cash in Bank	17,104
	$18,270		$18,270

Since debits equal credits, the cash payments journal can be double-ruled, as shown in **Figure 17–11.**

Posting Column Totals to the General Ledger

At the end of the month, the accountant posts the total of each special amount column to the general ledger account named in the column heading. For the cash payments journal, column totals are posted to **Accounts Payable, Purchases Discounts,** and **Cash in Bank. Figure 17–12** on page 551 shows the posting of the three special column totals to the general ledger accounts. Note that the account numbers for the three general ledger accounts are written in parentheses below the double rule in the appropriate columns of the cash payments journal.

The totals of the General Debit and Credit columns are not posted. Each entry in those columns was posted individually to the general ledger accounts. A check mark (✓) is entered below the double rule in the General Debit and Credit columns.

Reading Check

In Your Own Words

List the steps to post from the General Debit column to general ledger accounts.

Figure 17–12 Posting Column Totals from the Cash Payments Journal

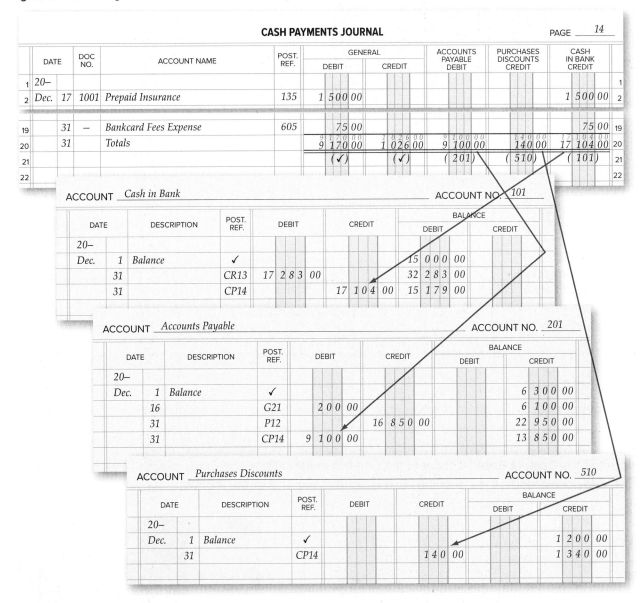

Proving the Accounts Payable Subsidiary Ledger

The accountant prepares a **schedule of accounts payable** after posting the column totals. This schedule lists all creditors in the accounts payable subsidiary ledger, the balance in each account, and the total amount owed to all creditors. The clerk proves the accounts payable subsidiary ledger when the total of the schedule of accounts payable agrees with the balance of the **Accounts Payable** (controlling) account in the general ledger.

Figure 17–13 shows The Starting Line's schedule of accounts payable for December. The accounts are listed in alphabetical order. All creditors are listed, even those with zero balances. Notice that the total listed on the schedule ($13,850) agrees with the balance of the **Accounts Payable** (controlling) account.

Figure 17–13 Schedule of Accounts Payable

ACCOUNT _Accounts Payable_   ACCOUNT NO. _201_

DATE	DESCRIPTION	POST. REF.	DEBIT	CREDIT	BALANCE DEBIT	BALANCE CREDIT
20--						
Dec. 1	Balance	✓				6 300 00
16		G21	200 00			6 100 00
31		P12		16 850 00		22 950 00
31		CP14	9 100 00			13 850 00

The Starting Line Sports Gear
Schedule of Accounts Payable
December 31, 20--

Champion Store Supply	1 000 00	
Computer Solutions	1 300 00	
Dara's Delivery Service	—	
FastLane Athletics	3 800 00	
Geary Office Supply	250 00	
Pro Runner Warehouse	—	
Sports Link Footwear	6 000 00	
Sports Nutrition Supply	1 500 00	
Total Accounts Payable		13 850 00

The balance of the Accounts Payable account should equal the total of the schedule of accounts payable.

Proving Cash

Proving cash is the process of verifying that cash recorded in the accounting records agrees with the amount entered in the checkbook. Ideally businesses should prove cash each day. When a business uses special journals, however, it updates the **Cash in Bank** account in the general ledger at the end of the month. For many businesses, then, proving cash is done at the end of the month.

The cash proof may be prepared on plain paper, on accounting stationery, on a special cash proof form, or on a computer. The cash proof for The Starting Line shown in **Figure 17–14,** is prepared on two-column accounting stationery. To prove cash, follow these steps:

1. On the first line, record the beginning balance of **Cash in Bank** according to the general ledger account.

2. On the next line, enter the total cash received during the month. This is the total of the Cash in Bank Debit column from the cash receipts journal.

3. Add the first and second lines.

4. From this subtotal subtract the cash payments for the month. This is the total of the Cash in Bank Credit column from the cash payments journal.

5. Compare this figure to the balance shown on the last check stub in the checkbook. If the ending balance of **Cash in Bank** and the balance on the check stub match, you have proved cash. In this example the ending balance of **Cash in Bank** is $15,179. The balance shown on the last check stub is also $15,179; therefore, cash is proved.

Common Mistakes

Schedule of Accounts Payable You may be surprised to see that liability account balances are listed in the debit column of the Schedule of Accounts Payable (Figure 17–13). This schedule uses a format that disregards debit and credit balances. Therefore, individual amounts are listed in the first column and the column total in the second column so that the total can be easily read.

If the balances are not equal, you should look for errors. Recording a bank service charge or a bankcard fee in the checkbook but not in the general ledger can cause the cash proof to be out of balance. Next, you should verify that all disbursements and deposits were recorded in the accounting records. If cash is being proved at month-end, the accountant can then continue making month-end entries.

Figure 17–14 Cash Proof

The Starting Line Sports Gear		
Cash Proof		
December 31, 20--		
Beginning Cash in Bank Balance		15 000 00
Plus: Cash Receipts for the Month		17 283 00
Subtotal		32 283 00
Less: Cash Payments for the Month		17 104 00
Ending Cash in Bank Balance		15 179 00
Check Stub Balance		15 179 00

Note that proving cash is different from reconciling a bank statement, which is taught in Chapter 11 (pages 318–319). Proving cash verifies that amounts recorded in the general ledger, cash receipts journal, and cash payments journal agree with the checking account balance. It does *not* confirm that the bank has processed all the deposits from the business or that all outstanding checks have cleared.

SECTION 17.2
Assessment

After You Read

Reinforce the Main Idea

Create a table similar to the one here to identify the journal in which each of the following transactions should be recorded. Identify debit and credit parts for each transaction.

Transaction	Journal	Account(s) Debited	Account(s) Credited
1. Sold merchandise on account			
2. Purchased merchandise on account			
3. Paid a creditor			
4. Sold merchandise for cash			
5. Bankcard sales			
6. Paid the payroll			
7. Discovered that purchase of supplies was incorrectly debited to **Purchases**			

Problem 2 Preparing a Cash Proof

Apple Tree Boutique uses special journals. On September 30 the total of the Cash in Bank Debit column of the cash receipts journal is $18,750.12. The total of the Cash in Bank Credit column of the cash payments journal is $16,890.43. The checkbook balance on September 30 is $5,610.59.

Instructions Prepare a cash proof for September in your working papers. The balance of **Cash in Bank** on September 1 is $3,750.90.

Problem 3 Analyzing a Source Document

The Country Peddler, which is a retail merchandising business, had the following transaction that occurred on November 2.

Instructions Analyze Check Stub 104 that is shown here. In your working papers, make the necessary entry to record the transaction on page 11 of the cash payments journal.

		No. 104
$ 873.00		
Date November 2		20—
To Colonial Products Inc.		
For Inv. 323 $900 less 3% disc. $27.00		

	Dollars	Cents
Balance brought forward	3,468	29
Add deposits		
Total	3,468	29
Less this check	873	00
Balance carried forward	2,595	29

Math for Accounting

At Car Wash Palace, hourly wage earners are paid weekly. These employees earned $8,000 in total gross earnings this week.

1. Calculate each withholding amount.
2. What is the total net pay for this week?

Tax	Rate
Social Security	6.2%
Medicare	1.45%
Federal Unemployment	0.8%
State Unemployment	5.4%
State Income	4.0%

Chapter 17
Visual Summary

Concepts
The purpose of the purchases and cash payments journals

Special Journals

- The purchases journal records all purchases on account.
- The cash payments Journal records all transactions in which cash is paid out or decreased.

Analysis
Record transactions in the purchases and cash payments journals.

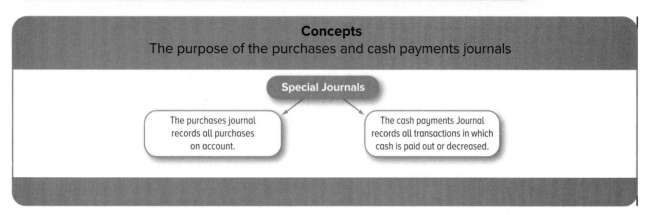

Procedures
Total, prove, and rule the purchases and cash payments journals.

STEP	PROCEDURE
1.	Rule the amount columns
2.	Foot the amount columns
3.	Prove the journal
4.	Under the last transaction, enter the date the journal is totaled in the Date column.
5.	On the same line, enter the word *Totals:* (1) In the purchases journal Creditors' Account Credited Column; and (2) in the cash payments journal Account Name column
6.	Enter the column totals, in ink, just below the footings.
7.	Double-rule the amount columns.

Chapter 17
Review and Activities

Answering the Essential Question

How does keeping special journals help businesses organize information about what a company purchases on account and what it pays out in cash? These journals help avoid confusion by keeping data up-to-date and in the right place. How do you keep track of your purchases?

Vocabulary Check

1. **Vocabulary** Write a brief description for each of the terms below.

   - purchases journal
   - cash payments journal
   - constant
   - automatic
   - adjust
   - schedule of accounts payable
   - proving cash

Concept Check

2. What are the source documents for the purchases journal? The cash payments journal?
3. **Analyze** Can a transaction amount be recorded in three columns? Explain why or why not.
4. In recording payroll, what four liabilities are entered in the General Credit column of the cash payments journal?
5. **Analyze** Compare the time frames for posting purchases transactions and cash payments transactions to the accounts payable subsidiary ledger. Explain why these transactions are posted at these intervals.
6. When totaling, proving, and ruling the purchases journal, where are the double rules placed?
7. **Evaluate** What information does the schedule of accounts payable include? Describe the purpose of the schedule of accounts payable. Why is this an important tool?
8. How does an accountant prove cash?
9. **Math** Imagine that you have two job offers. The first pays $2,000 a month and requires a 48-mile round-trip commute. The second is closer to home and you could easily walk or ride to a bike to work, but it pays only $1,920 a month. You will be working five-days a week. Taking into account the current price of gas and a car that gets 20 miles per gallon, how much will the commute to the first job cost? Which job will benefit more?
10. **English Language Arts** In your own words, define what it is to be a corporation. Research the requirements a company must meet to become a corporation. Tell what benefits and drawbacks incorporation represents to a company, and what protections it provides. Then give your opinion about whether or not a small company should seek to incorporate. Summarize your findings and views in a one-page essay.

Chapter 17
Standardized Test Practice

Multiple Choice

1. The purchases journal is used to journalize

    a. all purchases.

    b. charge purchases only.

    c. cash purchases only.

    d. purchases of merchandise only.

2. Placing a check mark in the first posting reference column of the purchases journal indicates the balance has been

    a. posted to the subsidiary ledger account.

    b. paid in full.

    c. totaled, verified, and ruled.

    d. posted to the general ledger.

3. A listing of vendor accounts, account balances, and total amount due all vendors is a

    a. schedule of accounts payable.

    b. schedule of accounts receivable.

    c. schedule of vendors.

    d. schedule of creditors.

4. Just for You Shoes purchased shoes for its two retail locations. The mall store ordered 20 pairs at $29.90/pair and 40 pairs at $18.50/pair. The downtown store ordered 35 pairs at $29.90/pair and 24 pairs at $18.50/pair. What was the total cost of the shoes for the two stores?

    a. $96.50

    b. $193.50

    c. $2,828.50

    d. $2,703.10

Short Answer

5. The journal used to journalize a cash payment on account is _____.

Extended Response

6. Describe the purpose of the purchases and the cash payments journals.

Chapter 17
Computerized Accounting

Mastering Purchases and Cash Payments
Making the Transition from a Manual to a Computerized System

MANUAL METHODS	COMPUTERIZED METHODS
Setting up Vendor Records • Ledger sheet or card is prepared with vendor details such as name, address, and contact information. • Account activities are posted to the vendor subsidiary ledger accounts.	**Setting up Vendor Records** • Record details of vendor accounts into vendor records. • The software will access this information automatically each time the vendor ID code is used.
Recording a Debit Memorandum • A journal entry is prepared to record the debit memorandum. • The journal entry is posted to the subsidiary account and to the general ledger.	**Recording a Debit Memorandum** • The accounting software creates a credit memo and applies it to the appropriate outstanding vendor invoice. • The accounting software automatically posts the credit memo to the vendor account and to the general ledger.

Chapter 17
Problems

Problem 4 Recording Payment of the Payroll

Denardo's Country Store pays its employees on a biweekly basis. This week the payroll is $2,000. You issued Check 949 for $1,487 in payment of the payroll less the following amounts: Employees' Federal Income Tax, $320; Employees' State Income Tax, $40; Social Security Tax, $124; and Medicare Tax, $29.

Instructions In your working papers, record the July 15 payroll on page 4 of the cash payments journal.

Analyze

Calculate the total deductions from employees' wages for federal, state, social security, and Medicare taxes. What percentage of gross payroll is this amount?

Problem 5 Recording Transactions in the Purchases Journal

Sunset Surfwear had the following purchases transactions for the month of July.

Instructions Use the purchases journal in your working papers.

1. Record each of the following transactions on page 4 of the purchases journal.
2. Foot, prove, total, and rule the purchases journal.

Date		Transactions
July	1	Purchased merchandise on account from Waverunner Designs for $1,200, Invoice WD121.
	3	Received Invoice CA552 from Capital Accessories for the purchase of $1,600 in merchandise on account.
	5	Purchased $2,000 in store equipment on account from Neilson Store Equipment, Invoice NS444.
	9	Purchased $870 in merchandise on account from Kelley Apparel Inc., Invoice KA772.
	12	Purchased $250 in supplies from Moore Paper & Office Supply Co., Invoice MPS266.
	15	Purchased $1,800 in merchandise on account from AcaTan Products, Invoice ATP99.
	18	Purchased $500 in office equipment on account from Moore Paper & Office Supply Co., Invoice MPS275.
	22	Received Invoice WD156 from Waverunner Designs for the purchase of $900 in merchandise on account.
	25	Purchased $475 in merchandise on account from Capital Accessories, Invoice CA560.

continued

Chapter 17
Problems

Date	Transactions
July 28	Purchased from Kelley Apparel Inc. $390 in merchandise on account, Invoice KA800.
30	Received Invoice NS460 from Neilson Store Equipment for the purchase of $1,200 in store equipment on account.

Analyze
Calculate the total purchases on account for the month.

Problem 6 Recording and Posting Purchases

Shutterbug Cameras, a retail merchandising business, had the following purchases on account for the month of July. In the working papers, the beginning balances in the accounts are opened for you.

Instructions In your working papers:

1. Record July's transactions on page 18 of the purchases journal.
2. Post to the accounts payable subsidiary ledger accounts daily.
3. Post amounts entered in the General Debit column daily.
4. Foot, prove, total, and rule the purchases journal at the end of the month.
5. Post the column totals at the end of the month to the account named in the column heading.
6. Prepare a schedule of accounts payable.

Date	Transactions
July 1	Purchased $1,200 in merchandise on account from U-Tech Products, Invoice UT220.
3	Purchased $140 in supplies on account from State Street Office Supply, Invoice 983, n/30.
6	Received Invoice 1338 from Photo Emporium for the purchase of $150 in merchandise on account, 2/10, n/30.
9	Invoice 445 for $90 was sent by Allen's Repair for plumbing repairs completed at the store (**Maintenance Expense**).
12	Purchased $800 in merchandise on account from Video Optics Inc., Invoice VO167, 3/15, n/45.

continued

Chapter 17
Problems

Date		Transactions
July	15	Received Invoice 1322 from Digital Precision Equipment for the purchase of $2,500 in store equipment on account, n/30.
	18	Purchased $1,300 in merchandise on account from U-Tech Products, Invoice UT257.
	20	Purchased $400 in supplies on account from ProStudio Supply, Invoice 4677.
	23	Photo Emporium sent Invoice 1359 for the purchase of $600 of merchandise on account.
	26	Received Invoice 478 for $120 from Allen's Repair for additional plumbing work completed at the store.
	28	Purchased $200 in merchandise on account from Video Optics Inc., Invoice VO183.

Analyze

Explain what problems would occur if the transaction of July 23 (Photo Emporium) was posted to the accounts payable subsidiary ledger as $60.

Problem 7 Recording and Posting Cash Payments

River's Edge Canoe & Kayak is a retail merchandising business located in Jackson Hole, Wyoming. The beginning balances of the accounts needed to complete this problem are opened in your working papers.

Instructions In your working papers:

1. Record the transactions on page 19 of the cash payments journal.
2. Post the individual amounts in the Accounts Payable Credit column on a daily basis to the creditors' accounts.
3. Post the individual amounts in the General Debit and Credit columns on the date the transaction occurred.
4. Foot, prove, total, and rule the cash payments journal.
5. Post the column totals at the end of the month.
6. Prepare a schedule of accounts payable.
7. Prove cash at the end of the month. The beginning **Cash in Bank** balance was $8,000; cash receipts were $7,000; and the ending check stub balance is $8,402.

continued

Chapter 17
Problems

Date	Transactions
July 1	Issued Check 1405 for $1,372 to North American Waterways Suppliers in payment on account of $1,400 invoice less a 2% discount.
3	Received insurance premium statement from Rocky Mountain Insurance Company. Issued Check 1406 for $1,800.
5	Issued Check 1407 to Pacific Wholesalers for $500 to apply on account.
8	Received an invoice for $300 from Jackson News for advertising. Issued Check 1408.
12	Issued Check 1409 for $686 to Trailhead Canoes in payment on account of $700 invoice less a 2% discount.
15	Issued Check 1410 for $200 to Rollins Plumbing Service in full payment of the amount owed on account.
18	Issued Check 1411 for $75 to Ben Jacobs for completing odd jobs in the store (**Miscellaneous Expense**).
20	Purchased $90 in supplies from StoreMart Supply by issuing Check 1412.
22	Paid StoreMart Supply $400 to apply on account, Check 1413.
24	Issued Check 1414, $700 to Office Max, store equipment.
25	Paid Mohican Falls Kayak Wholesalers $200 to apply on account, Check 1415.
26	Issued Check 1416 to Office Max, $150 to apply on account.
28	Paid transportation charges of $125 to Stein's Trucking, Check 1417.

Analyze
Identify the amount that River's Edge owes its creditors at month end.

Problem 8 Recording and Posting Purchases and Cash Payments

Buzz Newsstand had the following purchases and cash payment transactions for July. The balance in the accounts payable subsidiary ledger and general ledger accounts are opened in the working papers.

Instructions In your working papers:

1. Record the purchases and cash payment transactions on page 12 of the purchases journal and page 12 of the cash payments journal.
2. Post to the creditors' accounts in the accounts payable subsidiary ledger daily.

continued

Chapter 17
Problems

3. Post from the General Debit and Credit columns of the journals on the date the transaction occurred.

4. Foot, prove, total, and rule both journals.

5. Post the column totals of the purchases journal to the general ledger accounts named in the column headings.

6. Post the column totals of the cash payments journal to the general ledger accounts named in the column headings.

7. Prepare a schedule of accounts payable.

8. Prepare a cash proof at the end of the month. The beginning **Cash in Bank** balance was $9,000. Cash receipts were $10,500. The check stub balance on July 31 is $10,178.

SOURCE DOCUMENT PROBLEM

Problem 17–8

Use the source documents in your working papers to complete this problem.

Date		Transactions
July	1	Issued Check 2455 for $1,552 to ADC Publishing in payment on account of $1,600 invoice less a 3% discount.
	2	Purchased $400 in merchandise on account from Pine Forest Publications, Invoice PFP144, terms 2/10, n/30.
	2	Paid Candlelight Software $1,358 in payment of $1,400 account less a 3% discount, Check 2456.
	4	Issued Check 2457 to Nomad Computer Sales for $350 to apply on account.
	5	Purchased $2,000 in store equipment on account from CorpTech Office Supply, Invoice CT67.
	7	Issued Check 2458 for $125 to Wolfe Trucking for transportation charges.
	9	Purchased $900 in merchandise on account from American Trend Publishers, Invoice ATP98.
	12	Purchased $300 in supplies on account from CorpTech Office Supply, Invoice CT72.
	14	Issued Check 2459 to Delta Press for $750 to apply on account.
	15	Received insurance premium statement from SeaTac Insurance Co. for $1,600. Issued Check 2460.
	16	Check 2461 was issued for $882 to American Trend Publishers in payment of $900 account less a 2% discount.
	18	Purchased $500 in merchandise on account from Candlelight Software, Invoice CS101, terms n/30.
	20	Purchased $200 in merchandise on account from Nomad Computer Sales, Invoice NC56, terms 2/10, n/30.
	22	Paid Pine Forest Publications $100 to apply on account, Check 2462.

continued

Chapter 17
Problems

Date		Transactions
July	23	Issued Check 2463 for $2,000 to CorpTech Office Supply to apply on account.
	25	Purchased $600 in merchandise on account from ADC Publishing, Invoice ADC70.
	28	Issued Check 2464 for $450 to Nomad Computer Sales to apply on account.
	30	Recorded the bank service charge of $25. Recorded the bankcard fees, $130. July bank statement.

Analyze
Examine the purchases made during July and determine the dollar amount by which merchandise inventory increased.

Real-World Accounting Careers

Larry Brandon

Brandon, Rackley and Dukes, PC

Q What do you do?

A My primary focus is on retirement plans, including the drafting of those retirement plans. I also prepare tax returns for partnerships, corporations and individuals and do some small business consulting.

Q What are your day-to-day responsibilities?

A I manage staff and handle employee issues—we're a small firm, so I'm the "human resources department." I also oversee the firm's administration and operational issues.

Q What factors have been key to your success?

A In public accounting, you have to have a strong work ethic, good interpersonal skills to build relationships with clients and the desire to keep learning.

Q What do you like most about your job?

A I'm a people person, so I love the client contact. When I prepare a client's personal income tax return, I may have contact with that particular person only once a year. But I enjoy those relationships because the clients seek my advice and opinions.

Career Facts

Real-World Skills
Interpersonal skills; the ability to stay organized; proficiency with technology

Training And Education
Bachelor's degree in accounting or finance; CPA credential

Career Skills
Knowledge of U.S. GAAP; regulatory knowledge is increasingly essential

Career Path
Join a public accounting firm and earn your CPA license. Several years of experience are needed to become a partner

Tips from Robert Half International
When accepting your first job offer, you may not know what to expect in terms of starting salary. Determine the going rate by conducting research on the Internet, reading industry publications and consulting resources like Robert Half's annual Salary Guide.

College and Career Readiness

What steps would you need to take to obtain a CPA license? Develop a PowerPoint presentation that shows each of the steps and how you would achieve them. Be sure to include why a strong work ethic, good interpersonal skills, and the desire to keep learning are important parts of the process.

Real-World Applications & Connections

Career Wise

General Ledger Accountant

The general ledger is the primary source of information from which companies generate their financial reports. The general ledger accountant is responsible for making sure that all information in the ledger, including all business transactions, is accurate and up to date, and properly accounted for. General ledger accountants may themselves be responsible for recording the day-to-day bookkeeping activities of an organization, or may work in a supervisory role, directing other accountants or clerks and reviewing their work. A general ledger accountant may also be referred to as a corporate accountant or a management accountant.

Research and Share

a. Use the Internet to locate three sources of information on general ledger accounting.

b. What level of education is required to be a general ledger accountant? What are the roles and responsibilities of a general ledger accountant?

Global Accounting

International Competitive Advantage

If your company has resources, products, or skills that are valuable, unique, or difficult to imitate globally, it has an *international competitive advantage*. For example, Wal-Mart has developed its inventory management and purchasing system into a powerful resource that has given it a sustained competitive advantage in the international retailing industry. This system supports vendor management to maintain the timely flow of merchandise to stores to meet customer demand.

Instructions

Brainstorm other resources or skills a company might use to gain international competitive advantage. Comment on how maintaining vendor and customer files may be used as a tool to maintain competitive advantages.

A Matter of Ethics

Insufficient Funds

As the owner of a sporting goods store, you manage the accounting records. Your delivery van has been in the shop for repairs and you need it back as soon as possible. The mechanic calls with the repair cost, which is much higher than you expected. Your checking account cannot cover the bill, but you authorize the work anyway. If you write a check for the repairs, you believe the account will have enough money when the mechanic cashes the check.

Instructions

Determine the ethical issues? What are the alternatives? Who are the affected parties? How do the alternatives affect the parties?

Analyzing Financial Reports

Cost of Sales

To increase profits, corporations must increase sales or decrease costs. They share information about both in their annual reports. Amazon.com's annual report has a section labeled *Management's Discussion and Analysis of Financial Condition and Results of Operations Overview*. It includes a comparison of net sales and gross profit for the current year and previous year.

Instructions

Use the Internet to locate and save or print a copy of Amazon.com's annual report to complete the following. Refer to the second paragraph of this Overview. How has Amazon.com increased operating income? How has the company increased product sales?

H.O.T. Audit

Cash Payments Journal

Alpha Enterprises had the following business transactions in May:

- Alpha issued a check to Metro Ace Realty for rent, $800.
- Alpha issued a check to Cray Hardware for $1,940 in payment for a $2,000 invoice due, less cash discount of $60.
- Alpha paid $300 to Ron's Paint Shop for painting supplies.

The accounting clerk for Alpha made the following postings:

$800 was posted as a credit to Rent Expense; $1,940 was posted as a debit to Accounts Payable; and $300 was posted as a credit to Paint Supplies.

Instructions

Determine whether the transactions were posted correctly. If not, what error(s) were made?

Chapter 18

Adjustments and the Ten-Column Work Sheet

Chapter Topics

18-1 Identifying Accounts to be Adjusted and Adjusting Merchandise Inventory

18-2 Adjusting Supplies, Prepaid Insurance, and Federal Corporate Income Tax

18-3 Completing the Work Sheet and Journalizing and Posting the Adjusting Entries

Visual Summary

Review and Activities

Standardized Test Practice

Computerized Accounting

Problems

Real-World Applications & Connections

Essential Question

As you read this chapter, keep this question in mind:

What factors other than transactions can cause financial changes within a business?

Main Idea

Adjustments transfer the cost of "used up" assets to expense accounts. Adjustments for changes in merchandise inventory are made directly to the Income Summary account.

Chapter Objectives

Concepts	Analysis	Procedures
C1 Describe the parts of a ten-column work sheet. **C2** Generate trial balances and end-of-period adjustments.	**A1** Determine which general ledger accounts to adjust. **A2** Calculate the adjustments.	**P1** Prepare a ten-column work sheet. **P2** Journalize the adjustments.

Real-World Business Connection

Foot Locker, Inc.

Foot Locker, Inc. is the world's leading retailer of athletically inspired footwear and apparel. Headquartered in New York City, it operates approximately 3,500 retail stores worldwide under the brand names Foot Locker, Lady Foot Locker, Kids Foot Locker, Footaction, Champs Sports, and CCS. Foot Locker, Inc. also operates a direct-to-customers business offering athletic footwear, apparel, and equipment through its Internet and catalog channels.

Connect to the Business

Foot Locker, Inc. takes a physical count of its store inventory two to three times each year. Based on the results of these counts, the company adjusts its accounting records accordingly to ensure the ongoing accuracy of its Merchandise Inventory. Because of ongoing shifts in consumer tastes and fashion trends, these adjustments are essential for retailers to effectively manage inventory.

Analyze

Think of how customers might change their shoe-buying habits during an economic downturn. Make a list of factors that might influence trends and ways Foot Locker, Inc. might adjust its inventory accordingly.

Focus on the Photo

Foot Locker built its business around the concept of offering the right shoe for the right sport. Customers can choose from a vast inventory of styles for different athletic activities. **What are ten items that might be on Foot Locker's Merchandise Inventory account?**

SECTION 18.1
Identifying Accounts to be Adjusted and Adjusting Merchandise Inventory

Recall that the work sheet is the basis for preparing end-of-period journal entries and financial statements. The purpose of all period-end reports is to provide essential information about a company's financial position.

Section Vocabulary
- adjustment
- beginning inventory
- ending inventory
- physical inventory

Completing End-of-Period Work

What Is the Purpose of the Ten-Column Work Sheet?

The general ledger summarizes the effects of business transactions on individual accounts for an accounting period. Managers, stockholders, and creditors need more than account totals, however, to evaluate performance. They need to know net income and the value of stockholders' equity, which for a corporation is like owner's equity for a sole proprietorship. They need this information to make sound business decisions.

The accounting worksheet is the basis for preparing end-of-period journal entries and financial statements. You will expand on this idea by learning how to use the ten-column worksheet to calculate and journalize end-of-period adjustments.

The Ten-Column Work Sheet

How Is the Ten-Column Work Sheet Different from the Six-Column Work Sheet?

The work sheet in Chapter 8 had six amount columns. The work sheet in this chapter, however, has ten amount columns. The additional columns are for the *Adjustments* and *Adjusted Trial Balance* sections.

Prepared in the same way as the six-column work sheet, the ten-column work sheet has five amount sections instead of three:

- Trial Balance
- Adjustments
- Adjusted Trial Balance
- Income Statement
- Balance Sheet

Completing the Trial Balance Section

A trial balance is prepared to prove the equality of debits and credits in the general ledger. **Figure 18–1** on page 571 shows the end-of-period trial balance for The Starting Line Sports Gear.

To prepare the trial balance, the number and name of each account in the general ledger are entered on the work sheet in the Account Number and Account Name columns. The accounts are listed in the order that they appear in the general ledger (Asset, Liability, Stockholders' Equity, Revenue, Cost of Merchandise, Expense). The balance of each account is entered in the appropriate Debit or

Credit column of the Trial Balance section. Notice that every general ledger account is listed, even those with zero balances. After all balances are entered, the Trial Balance Debit and Credit columns are ruled, totaled, and proved. Then a double rule is drawn across both columns.

Calculating Adjustments

Not all changes in account balances result from daily business transactions. Some result from internal business operations or the passage of time. For example, supplies such as paper, pens, shopping bags, and sales slips are bought for use by the business. They are recorded in an asset account called **Supplies.** These

	ACCT. NO.	ACCOUNT NAME	TRIAL BALANCE DEBIT	TRIAL BALANCE CREDIT
1	101	Cash in Bank	15 179 00	
2	115	Accounts Receivable	10 404 00	
3	125	Merchandise Inventory	84 921 00	
4	130	Supplies	5 549 00	
5	135	Prepaid Insurance	1 500 00	
6	140	Delivery Equipment	19 831 00	
7	145	Office Equipment	9 825 00	
8	150	Store Equipment	5 200 00	
9	201	Accounts Payable		13 850 00
10	204	Fed. Corp. Inc. Tax Payable		
11	205	Employees' Fed. Inc. Tax Pay.		640 00
12	211	Employees' State Inc. Tax Pay.		80 00
13	212	Social Security Tax Payable		248 00
14	213	Medicare Tax Payable		58 00
15	214	Fed. Unemployment Tax Payable		18 36
16	215	State Unemployment Tax Payable		114 73
17	220	Sales Tax Payable		2 428 00
18	301	Capital Stock		75 000 00
19	305	Retained Earnings		19 771 19
20	310	Income Summary		
21	401	Sales		320 450 00
22	405	Sales Discounts	730 00	
23	410	Sales Returns & Allowances	2 000 00	
24	501	Purchases	206 700 00	
25	505	Transportation In	4 036 18	
26	510	Purchases Discounts		1 340 00
27	515	Purchases Returns & Allowances		1 800 00
28	601	Advertising Expense	2 450 00	
29	605	Bankcard Fees Expense	4 199 27	
30	630	Fed. Corp. Income Tax Expense	9 840 00	
31	635	Insurance Expense		
32	650	Maintenance Expense	3 519 25	
33	655	Miscellaneous Expense	348 28	
34	657	Payroll Tax Expense	3 826 83	
35	660	Rent Expense	14 000 00	
36	665	Salaries Expense	29 374 60	
37	670	Supplies Expense		
38	680	Utilities Expense	2 364 87	
39			435 798 28	435 798 28
40				

Figure 18–1 The Trial Balance Section of the Work Sheet

supplies are used gradually during the accounting period. Another example is insurance premiums, which cover a certain period of time. The premiums are recorded in an asset account called **Prepaid Insurance.** During the period some insurance is used up, or expires. At the end of the period, the balances in accounts such as **Supplies** and **Prepaid Insurance** are brought up to date.

To illustrate, suppose that supplies costing $500 were purchased during an accounting period and recorded in the **Supplies** account. At the end of the period, the **Supplies** account balance is $500; however, only $200 of supplies are still on hand. The account balance for **Supplies** is adjusted downward to show that the business used $300 of supplies.

Permanent Accounts and Temporary Accounts.

Up to this point, the general ledger account balances have been changed by journal entries made to record transactions that are supported by source documents. There are no source documents, however, for the changes in account balances caused by the internal operations of a business or the passage of time. Such changes are recorded through adjustments made at the end of the period to the account balance. An **adjustment** is an amount that is added to or subtracted from an account balance to bring that balance up to date. Every adjustment affects at least one permanent account and one temporary account.

At the end of the period, adjustments are made to transfer the costs of the assets consumed from the asset accounts (permanent accounts) to the appropriate expense accounts (temporary accounts). Accountants say that these assets are "expensed" because the costs of consumed assets are expenses of doing business. Thus, when an adjustment is recorded, the expenses for a given period are matched with the revenue for that period.

The work sheet's Adjustments section is used to record the adjustments made at the period-end to bring various account balances up to date.

Determining the Adjustments Needed.

How do you generate end-of-period adjustments? Review each account balance in the work sheet's Trial Balance section. If the balance for an account is not up to date *as of the last day of the fiscal period,* that account balance must be adjusted.

Refer to **Figure 18–1.** The first account listed is **Cash in Bank.** All cash received or paid out during the period was journalized and posted to the **Cash in Bank** account. This balance is up to date. The next account, **Accounts Receivable** (controlling), is also up to date since all amounts owed or paid by charge customers were journalized and posted.

The third account is **Merchandise Inventory,** used to report the cost of merchandise on hand. The balance reported in the Trial Balance section ($84,921) is the merchandise on hand at the *beginning* of the period.

The amount of merchandise on hand is constantly changing during the period. The changes are not recorded in the **Merchandise Inventory** account. During the period the cost of merchandise purchased is recorded in the **Purchases** account. As merchandise sales reduce inventory, the amount of each sale is recorded in the **Sales** account, not **Merchandise Inventory.** At the end of the period, the balance in the **Merchandise Inventory** account does not reflect the amount of merchandise on hand. So the **Merchandise Inventory** account balance must be adjusted.

The other account balances on the work sheet are reviewed in the same manner. Additional adjustments are made as described later in this chapter.

Reading Check

Recall

What two types of accounts are affected by adjustments?

Adjusting the Merchandise Inventory Account

Merchandise Inventory is an asset account used by merchandising businesses. **Beginning inventory** is the merchandise a business has on hand and available for sale at the *beginning* of a period. **Ending inventory** is the merchandise on hand at the *end* of a period. The ending inventory for one period becomes the beginning inventory for the next period.

The account balance of **Merchandise Inventory** does not change during the period. It is changed *only* when a physical inventory is taken. A **physical inventory** is an actual count of all merchandise on hand and available for sale. A physical inventory can be taken at any time. One is always taken at the end of a period.

For example, if a toy store counts inventory at the end of a period, it can calculate the cost of inventory. This is done by multiplying the quantity of each item by its unit cost.

At the end of each period, after the physical inventory has been taken and the cost of ending inventory has been calculated, the ending inventory amount replaces the beginning inventory amount recorded in **Merchandise Inventory.** This is accomplished by an adjustment to **Merchandise Inventory.**

Calculating the Adjustment for Merchandise Inventory.

When calculating the adjustment for **Merchandise Inventory,** you need to know (1) the **Merchandise Inventory** account balance and (2) the physical inventory amount. For The Starting Line, this is:

Merchandise Inventory account balance	$84,921
Physical inventory	81,385

Adjustment

*To adjust the **Merchandise Inventory** account to reflect the physical inventory amount ($81,385), the following transaction is recorded.*

Analysis	Identify	1. The accounts **Merchandise Inventory** and **Income Summary** are affected.
	Classify	2. **Merchandise Inventory** is an asset account (permanent). **Income Summary** is a stockholder's equity account.
	+/−	3. **Merchandise Inventory** is decreased by $3,536. This amount is transferred to **Income Summary.**
Debit-Credit Rule		4. To transfer the decrease in **Merchandise Inventory,** debit **Income Summary** for $3,536.
		5. Decreases to asset accounts are recorded as credits. Credit **Merchandise Inventory** for $3,536.
T Accounts		6.

Income Summary		Merchandise Inventory	
Debit	Credit	Debit +	Credit −
3,536			3,536

The effect of all the purchases and sales during the period is a decrease to **Merchandise Inventory** of $3,536 ($84,921 − $81,385). This reduction in inventory needs to be recorded as an adjustment in the accounting records. The two accounts affected by the inventory adjustment are **Merchandise Inventory** and **Income Summary.**

If the ending inventory amount is higher than beginning inventory, **Merchandise Inventory** is debited and **Income Summary** is credited. For example, suppose that the beginning inventory was $84,921 and the ending inventory is $88,226. Inventory increased by $3,305 ($88,226 − $84,921). The **Merchandise Inventory** account is debited for $3,305, and **Income Summary** is credited for $3,305.

Merchandise Inventory	
Debit + 3,305	Credit −

Income Summary	
Debit	Credit 3,305

Entering the Adjustment for Merchandise Inventory on the Work Sheet.

Adjustments are entered in the Adjustments columns of the work sheet. The debit and credit parts of each adjustment are given a unique label. The label consists of a small letter in parentheses and is placed just above and to the left of the adjustment amounts. The adjustments are labeled as follows:

First adjustment	(a)
Second adjustment	(b)
Third adjustment	(c)

The number of adjustments varies depending on the business. Once the adjustments have been entered, the work sheet provides the information needed to make the adjusting journal entries.

Use the T accounts in step 6 of the preceding example as a guide to entering the inventory adjustment on the work sheet. Refer to **Figure 18–2.** To record the adjustment for **Merchandise Inventory:**

1. In the Adjustments Debit column, enter the debit amount of the adjustment on the Income Summary line. Label this amount *(a)*.

2. In the Adjustments Credit column, enter the credit amount of the adjustment on the Merchandise Inventory line. Label it *(a)* also.

Figure 18–2 Recording the Adjustment for Merchandise Inventory on the Work Sheet

	ACCT. NO.	ACCOUNT NAME	TRIAL BALANCE DEBIT	TRIAL BALANCE CREDIT	ADJUSTMENTS DEBIT	ADJUSTMENTS CREDIT
1	101	Cash in Bank	15 179 00			
2	115	Accounts Receivable	10 404 00			
3	125	Merchandise Inventory	84 921 00			(a) 3 536 00
20	310	Income Summary			(a) 3 536 00	
21						

SECTION 18.1
Assessment

After You Read

Reinforce the Main Idea

Using a diagram like this one, summarize the steps for determining the **Merchandise Inventory** adjustment and entering it on the work sheet. Add answer boxes for steps as needed.

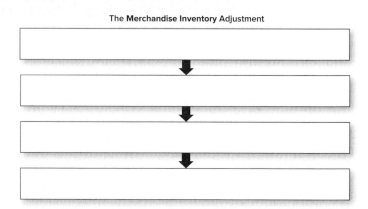

The **Merchandise Inventory Adjustment**

Problem 1 Analyzing the Adjustment for Merchandise Inventory

Ely Corporation, a custom furniture manufacturer, has a general ledger account balance of $73,395 for **Merchandise Inventory** as of July 1. On the following June 30, the end of the fiscal period, Ely took a physical inventory and determined it had $74,928 in merchandise on hand. In your working papers, answer the following questions regarding the adjustment for **Merchandise Inventory**:

1. Is the value of the ending inventory more or less than the value of the beginning inventory?
2. What is the amount of the inventory adjustment?
3. Which account is debited?
4. Which account is credited?

Math for Accounting

Your company, Photo Shots, is looking to expand its merchandise, but the accounting manager wants to analyze inventory data for the last three years before recommending expansion. Using the data given, create a bar graph to depict the total amount of merchandise sold by year. Use the formula (Beginning Inventory + Purchases) − Ending Inventory = Cost of Merchandise Sold. Write a short paragraph summarizing your analysis of the results. Explain how Photo Shots might benefit from determining the most efficient levels of inventory.

Merchandise Inventory in U.S.$

Year	Beginning	Ending	Purchases
Year 1	900	1,000	5,000
Year 2	1,000	2,000	8,000
Year 3	2,000	1,500	7,500

576 Chapter 18 • Adjustments and the Ten-Column Work Sheet

SECTION 18.2
Adjusting Supplies, Prepaid Insurance, and Federal Corporate Income Tax

In Section 18.1 you learned that at the end of the period, the accountant must update the **Merchandise Inventory** account and adjust other accounts in the general ledger. For example insurance is a costly expense for high-risk businesses, such as building construction and demolition. Thus, the financial statements must reflect the proper insurance expenses each period. Supplies and federal corporate income taxes also represent **significant** expenses that must be reported accurately.

Section Vocabulary
- prepaid expense
- significant
- estimates

Adjusting the Supplies Account

Why Do You Make an Adjustment for Supplies?

A merchandising business buys various supplies for employees to use in the everyday operations of the business. Pencils, pens, computer paper, shopping bags, sales slips, price tags, and cash register tapes are purchased, and the cost is debited to the **Supplies** account.

Supplies are used daily. As they are consumed, they become expenses of the business. Keeping daily records of each item as it is used is inefficient, so the **Supplies** account is updated at the end of the period.

In the Trial Balance section of the work sheet in **Figure 18–1** on page 571, the balance of the **Supplies** account is $5,549. This amount is the cost of the supplies on hand on January 1 *plus* the cost of the additional supplies purchased during the period.

At the end of December, The Starting Line took a physical inventory. It found that $1,839 of supplies were on hand, meaning that it used $3,710 of supplies during the period ($5,549 − $1,839 = $3,710). The amount of supplies on hand decreased by $3,710. Therefore, **Supplies** (a permanent asset account) is credited for $3,710. **Supplies Expense** (a temporary account) is debited (increased) to record the cost of supplies used in the period.

Adjustment

Record the adjustment for supplies.

Analysis		
Identify	1.	The accounts affected are **Supplies** and **Supplies Expense.**
Classify	2.	**Supplies** is an asset account (permanent). **Supplies Expense** is an expense account (temporary).
+/−	3.	**Supplies** is decreased by $3,710. **Supplies Expense** is increased by $3,710.

Continued

Debit-Credit Rule	4. Increases to expense accounts are recorded as debits. Debit **Supplies Expense** for $3,710.
	5. Decreases to asset accounts are recorded as credits. Credit **Supplies** for $3,710.
T Accounts	6.

Supplies Expense		Supplies	
Debit + 3,710	Credit −	Debit +	Credit − 3,710

The adjustment for supplies is shown in **Figure 18–3**. To enter the adjustment on the work sheet, follow these steps:

1. In the Adjustments Debit column, enter the debit amount of the adjustment on the Supplies Expense line. Since this is the second adjustment, label it *(b)*.

2. In the Adjustments Credit column, enter the credit amount of the adjustment on the Supplies line. Label it *(b)* also.

Adjusting the Prepaid Insurance Account

Why Do You Make an Adjustment for Insurance?

On December 17 The Starting Line purchased an insurance policy for six months, mid-December through mid-May. The accounting clerk debited the premium of $1,500 ($250 per month) to the **Prepaid Insurance** account. This is an example of a **prepaid expense,** an expense paid in advance.

At the end of December, the **Prepaid Insurance** balance is $1,500 (see the work sheet's Trial Balance section in **Figure 18–1**). However, the coverage for half of a month, costing $125, has expired. The value of the unexpired portion of the coverage has decreased to $1,375 ($1,500 − $125), so the clerk adjusts **Prepaid Insurance** with a $125 credit to update its balance. The adjustment records the expired portion as a business expense.

Adjustment

Record the adjustment for the expiration of one-half month's insurance coverage.

Analysis	Identify	1. The accounts affected are **Insurance Expense** and **Prepaid Insurance.**
	Classify	2. **Insurance Expense** is an expense account (temporary). **Prepaid Insurance** is an asset account (permanent).
	+/−	3. **Insurance Expense** is increased by $125. **Prepaid Insurance** is decreased by $125.

Continued

Debit-Credit Rule	4. Increases to expense accounts are recorded as debits. Debit **Insurance Expense** for $125.
	5. Decreases to asset accounts are recorded as credits. Credit **Prepaid Insurance** for $125.
T Accounts	6. Insurance Expense — Debit + 125, Credit − Prepaid Insurance — Debit +, Credit − 125

ACCT. NO.	ACCOUNT NAME	TRIAL BALANCE DEBIT	TRIAL BALANCE CREDIT	ADJUSTMENTS DEBIT	ADJUSTMENTS CREDIT
1	101 Cash in Bank	15 179 00			
2	115 Accounts Receivable	10 404 00			
3	125 Merchandise Inventory	84 921 00			(a) 3 536 00
4	130 Supplies	5 549 00			(b) 3 710 00
5	135 Prepaid Insurance	1 500 00			(c) 125 00
6	140 Delivery Equipment	19 831 00			
7	145 Office Equipment	9 825 00			
8	150 Store Equipment	5 200 00			
9	201 Accounts Payable		13 850 00		
10	204 Fed. Corp. Inc. Tax Payable				(d) 155 00
11	205 Employees' Fed. Inc. Tax Pay.		640 00		
12	211 Employees' State Inc. Tax Pay.		80 00		
13	212 Social Security Tax Payable		248 00		
14	213 Medicare Tax Payable		58 00		
15	214 Fed. Unemployment Tax Payable		18 36		
16	215 State Unemployment Tax Payable		114 73		
17	220 Sales Tax Payable		2 428 00		
18	301 Capital Stock		75 000 00		
19	305 Retained Earnings		19 771 19		
20	310 Income Summary			(a) 3 536 00	
21	401 Sales		320 450 00		
22	405 Sales Discounts	730 00			
23	410 Sales Returns and Allowances	2 000 00			
24	501 Purchases	206 700 00			
25	505 Transportation In	4 036 18			
26	510 Purchases Discounts		1 340 00		
27	515 Purchases Returns and Allowances		1 800 00		
28	601 Advertising Expense	2 450 00			
29	605 Bankcard Fees Expense	4 199 27			
30	630 Fed. Corp. Income Tax Expense	9 840 00		(d) 155 00	
31	635 Insurance Expense			(c) 125 00	
32	650 Maintenance Expense	3 519 25			
33	655 Miscellaneous Expense	348 28			
34	657 Payroll Tax Expense	3 826 83			
35	660 Rent Expense	14 000 00			
36	665 Salaries Expense	29 374 60			
37	670 Supplies Expense			(b) 3 710 00	
38	680 Utilities Expense	2 364 87			
39		435 798 28	435 798 28	7 526 00	7 526 00
40					

Figure 18–3 Recording the Adjustments on the Work SheetSheet

The adjustment for **Prepaid Insurance** is shown in **Figure 18–3**. To enter the adjustment on the work sheet, follow these steps:

1. In the Adjustments Debit column, enter the debit amount of the adjustment on the Insurance Expense line. Since this is the third adjustment, label it *(c)*.

2. In the Adjustments Credit column, enter the credit amount of the adjustment on the Prepaid Insurance line. Label it *(c)* also.

Reading Check

Identify

What type of product or service can be prepaid?

Adjusting the Federal Corporate Income Tax Accounts

Why Does This Company Make an Adjustment for Income Tax?

The Starting Line is organized as a corporation. A corporation is considered to be a legal entity separate from its owners. The Starting Line owns assets, pays its own debts, and enters into legal contracts.

A corporation pays federal corporate income taxes on its net income. Many states and cities also tax corporate income. For now, we will discuss only federal corporate income taxes. A corporation's accountant **estimates** its federal corporate income taxes for the coming year and pays that amount to the federal government in quarterly installments. At the end of the year, the exact net income and the tax on that income are determined. If the corporation owes additional taxes, it pays them when it files its corporate income tax return.

At the beginning of the year, The Starting Line's accountant estimated that its federal corporate income taxes would be $9,840. The business made quarterly payments of $2,460 in March, June, September, and December. These payments were journalized as debits to **Federal Corporate Income Tax Expense** and credits to **Cash in Bank.**

At the end of the year, The Starting Line's accountant determined that the federal corporate income tax for the year is $9,995. The Starting Line has already paid $9,840 (2,460 × 4). Therefore the business owes an additional $155 ($9,995 − $9,840).

To bring the accounting records up to date, **Federal Corporate Income Tax Expense** and **Federal Corporate Income Tax Payable** must both be increased by $155. The following T accounts illustrate this adjustment.

Federal Corporate Income Tax Expense		Federal Corporate Income Tax Payable	
Debit + 155	Credit −	Debit −	Credit + 155

The adjustment for **Federal Corporate Income Tax Expense** is shown in **Figure 18–3.** Since this is the fourth adjustment, it is labeled *(d)*.

After all adjustments have been entered, the Adjustments section of the work sheet is totaled and ruled. Each adjustment has an equal debit and credit, so the totals of the Adjustments Debit and Credit columns should be the same. When the Adjustments section has been proved, a double rule is drawn under the totals and across both columns, as shown in **Figure 18–3.**

SECTION 18.2
Assessment

After You Read

Reinforce the Main Idea

Use a chart like this one to describe three adjustments discussed in Section 2. List the accounts to be debited and credited for each adjustment.

Description of Adjustment	Account Debited	Account Credited

Problem 2 Analyzing Adjustments

Information related to accounts requiring adjustment on the work sheet for the year ended December 31 for Star City Resorts Corporation follows. Indicate in your working papers the amount of each adjustment, which account is debited, and which account is credited.

1. The Trial Balance section shows a balance of $3,347.45 for **Supplies.** The amount of supplies actually on hand is $892.75.
2. Star City paid an annual insurance premium of $4,440 on November 1.
3. Star City made quarterly federal corporate income tax payments of $945 each. Its actual tax, calculated at the end of the year, is $3,885.

Math for Accounting

You own a personal-training franchise called Your Body. You want to provide health insurance for your 10 employees. As you review the alternative plans, you have to make a decision based not only on affordability, but also on total benefits. Determine the total annual premium of each plan and decide which plan best fits your needs.

Type of plan	Plan #1	Plan #2	Plan #3
Monthly cost per employee	$600	$500	$800
Co-pay for each employee	$10	$20	None

SECTION 18.3
Completing the Work Sheet and Journalizing and Posting the Adjusting Entries

After the adjustments have been entered in the Adjustments section of the work sheet, proving that the accounts are still in balance is important. This is done by completing an *adjusted trial balance*. After proving the Adjusted Trial Balance section, the accountant can complete the work sheet.

Section Vocabulary
- adjusting entries

Extending Work Sheet Balances

What Does Extending Balances Involve?

At this point the amounts for each account must be *extended to,* or carried over to, the Adjusted Trial Balance, the Income Statement, and the Balance Sheet sections.

Completing the Adjusted Trial Balance Section

The next step after entering all adjustments is to finish the Adjusted Trial Balance section. This work sheet section shows the updated balances of all general ledger accounts. To complete this section, the accountant combines the balance of each account in the Trial Balance section with the adjustment, if any, in the Adjustments section. The new balance is then entered in the appropriate Adjusted Trial Balance column.

Note the way new balances are computed in **Figure 18–4.** If there is no adjustment, the account balance shown in the Trial Balance section is simply extended to the same column (Debit or Credit) in the Adjusted Trial Balance section. The first two accounts, **Cash in Bank** and **Accounts Receivable,** have no adjustments, so those balances are extended to the Adjusted Trial Balance Debit column.

If the account balance in the Trial Balance section has an adjustment, the accountant calculates a new balance. The amount of the adjustment (from the Adjustments section) is added to or subtracted from the amount in the Trial Balance section. Add debits to debits; add credits to credits; subtract debits and credits.

The first account in the Trial Balance section to have an adjustment is **Merchandise Inventory.** As you know from Section 18.1, **Merchandise Inventory** has an unadjusted debit balance of $84,921. Adjustment (a) is a credit of $3,536. To calculate the new balance, the accountant subtracts the credit adjustment from the debit balance. The adjusted balance of $81,385 ($84,921 − $3,536) is extended to the Adjusted Trial Balance Debit column.

The adjusted balances for **Supplies, Prepaid Insurance,** and **Federal Corporate Income Tax Expense** are calculated in the same way.

ACCT. NO.	ACCOUNT NAME	TRIAL BALANCE DEBIT	TRIAL BALANCE CREDIT	ADJUSTMENTS DEBIT	ADJUSTMENTS CREDIT	ADJUSTED TRIAL BALANCE DEBIT	ADJUSTED TRIAL BALANCE CREDIT
1	101 Cash in Bank	15 179 00				15 179 00	
2	115 Accounts Receivable	10 404 00				10 404 00	
3	125 Merchandise Inventory	84 921 00			(a) 3 536 00	81 385 00	
4	130 Supplies	5 549 00			(b) 3 710 00	1 839 00	
5	135 Prepaid Insurance	1 500 00			(c) 125 00	1 375 00	
6	140 Delivery Equipment	19 831 00				19 831 00	
7	145 Office Equipment	9 825 00				9 825 00	
8	150 Store Equipment	5 200 00				5 200 00	
9	201 Accounts Payable		13 850 00				13 850 00
10	204 Fed. Corp. Inc. Tax Payable				(d) 155 00		155 00
11	205 Employees' Fed. Inc. Tax Pay.		640 00				640 00
12	211 Employees' State Inc. Tax Pay.		80 00				80 00
13	212 Social Security Tax Payable		248 00				248 00
14	213 Medicare Tax Payable		58 00				58 00
15	214 Fed. Unemployment Tax Payable		18 36				18 36
16	215 State Unemployment Tax Payable		114 73				114 73
17	220 Sales Tax Payable		2 428 00				2 428 00
18	301 Capital Stock		75 000 00				75 000 00
19	305 Retained Earnings		19 771 19				19 771 19
20	310 Income Summary			(a) 3 536 00		3 536 00	
21	401 Sales		320 450 00				320 450 00
22	405 Sales Discounts	730 00				730 00	
23	410 Sales Returns and Allowances	2 000 00				2 000 00	
24	501 Purchases	206 700 00				206 700 00	
25	505 Transportation In	4 036 18				4 036 18	
26	510 Purchases Discounts		1 340 00				1 340 00
27	515 Purchases Returns and Allowances		1 800 00				1 800 00
28	601 Advertising Expense	2 450 00				2 450 00	
29	605 Bankcard Fees Expense	4 199 27				4 199 27	
30	630 Fed. Corp. Income Tax Expense	9 840 00		(d) 155 00		9 995 00	
31	635 Insurance Expense			(c) 125 00		125 00	
32	650 Maintenance Expense	3 519 25				3 519 25	
33	655 Miscellaneous Expense	348 28				348 28	
34	657 Payroll Tax Expense	3 826 83				3 826 83	
35	660 Rent Expense	14 000 00				14 000 00	
36	665 Salaries Expense	29 374 60				29 374 60	
37	670 Supplies Expense			(b) 3 710 00		3 710 00	
38	680 Utilities Expense	2 364 87				2 364 87	
39		435 798 28	435 798 28	7 526 00	7 526 00	435 953 28	435 953 28
40							

Figure 18–4 Extending Balances to the Adjusted Trial Balance Section of the Work Sheet

If an account has a zero balance in the Trial Balance section, the amount listed in the Adjustments section is extended to the Adjusted Trial Balance section. **Federal Corporate Income Tax Payable,** for example, has a zero balance in the Trial Balance section. Adjustment (d) is a credit of $155. This amount is extended to the Adjusted Trial Balance Credit column.

After extending all account balances to the Adjusted Trial Balance section, the accountant totals both columns. If total debits equal total credits, this section has been proved. The accountant then draws a double rule under the totals and across both columns. If total debits do not equal total credits, an error exists. To find an error, re-add each column. If the error still exists, ensure that the Trial Balance and Adjustment amounts were extended properly to the Adjusted Trial Balance section.

The Starting Line
Work
For Year Ended

Line	ACCT. NO.	ACCOUNT NAME	TRIAL BALANCE DEBIT	TRIAL BALANCE CREDIT	ADJUSTMENTS DEBIT	ADJUSTMENTS CREDIT
1	101	Cash in Bank	15 179 00			
2	115	Accounts Receivable	10 404 00			
3	125	Merchandise Inventory	84 921 00			(a) 3 536 00
4	130	Supplies	5 549 00			(b) 3 710 00
5	135	Prepaid Insurance	1 500 00			(c) 125 00
6	140	Delivery Equipment	19 831 00			
7	145	Office Equipment	9 825 00			
8	150	Store Equipment	5 200 00			
9	201	Accounts Payable		13 850 00		
10	204	Fed. Corp. Inc. Tax Payable				(d) 155 00
11	205	Employees' Fed. Inc. Tax Pay.		640 00		
12	211	Employees' State Inc. Tax Pay.		80 00		
13	212	Social Security Tax Payable		248 00		
14	213	Medicare Tax Payable		58 00		
15	214	Fed. Unemployment Tax Payable		18 36		
16	215	State Unemployment Tax Payable		114 73		
17	220	Sales Tax Payable		2 428 00		
18	301	Capital Stock		75 000 00		
19	305	Retained Earnings		19 771 19		
20	310	Income Summary			(a) 3 536 00	
21	401	Sales		320 450 00		
22	405	Sales Discounts	730 00			
23	410	Sales Returns and Allowances	2 000 00			
24	501	Purchases	206 700 00			
25	505	Transportation In	4 036 18			
26	510	Purchases Discounts		1 340 00		
27	515	Purchases Returns and Allowances		1 800 00		
28	601	Advertising Expense	2 450 00			
29	605	Bankcard Fees Expense	4 199 27			
30	630	Fed. Corp. Income Tax Expense	9 840 00		(d) 155 00	
31	635	Insurance Expense			(c) 125 00	
32	650	Maintenance Expense	3 519 25			
33	655	Miscellaneous Expense	348 28			
34	657	Payroll Tax Expense	3 826 83			
35	660	Rent Expense	14 000 00			
36	665	Salaries Expense	29 374 60			
37	670	Supplies Expense			(b) 3 710 00	
38	680	Utilities Expense	2 364 87			
39			435 798 28	435 798 28	7 526 00	7 526 00
40		Net Income				

Figure 18–5 The Starting Line Sports Gear Completed Work Sheet (continued)

Sports Gear
Sheet
December 31, 20--

ADJUSTED TRIAL BALANCE		INCOME STATEMENT		BALANCE SHEET		
DEBIT	CREDIT	DEBIT	CREDIT	DEBIT	CREDIT	
15 179 00				15 179 00		1
10 404 00				10 404 00		2
81 385 00				81 385 00		3
1 839 00				1 839 00		4
1 375 00				1 375 00		5
19 831 00				19 831 00		6
9 825 00				9 825 00		7
5 200 00				5 200 00		8
	13 850 00				13 850 00	9
	155 00				155 00	10
	640 00				640 00	11
	80 00				80 00	12
	248 00				248 00	13
	58 00				58 00	14
	18 36				18 36	15
	114 73				114 73	16
	2 428 00				2 428 00	17
	75 000 00				75 000 00	18
	19 771 19				19 771 19	19
3 536 00		3 536 00				20
	320 450 00		320 450 00			21
730 00		730 00				22
2 000 00		2 000 00				23
206 700 00		206 700 00				24
4 036 18		4 036 18				25
	1 340 00		1 340 00			26
	1 800 00		1 800 00			27
2 450 00		2 450 00				28
4 199 27		4 199 27				29
9 995 00		9 995 00				30
125 00		125 00				31
3 519 25		3 519 25				32
348 28		348 28				33
3 826 83		3 826 83				34
14 000 00		14 000 00				35
29 374 60		29 374 60				36
3 710 00		3 710 00				37
2 364 87		2 364 87				38
435 953 28	435 953 28	290 915 28	323 590 00	145 038 00	112 363 28	39
		32 674 72			32 674 72	40
		323 590 00	323 590 00	145 038 00	145 038 00	41

Figure 18–5 The Starting Line Sports Gear Completed Work Sheet

Extending Amounts to the Balance Sheet and Income Statement Sections

Beginning with line 1, each account balance in the Adjusted Trial Balance section is extended to the appropriate column of either the Balance Sheet or the Income Statement section. See these extensions in **Figure 18–5** on pages 585 and 586.

The Income Statement section contains the balances of all temporary accounts. You will find the **Income Summary** and all revenue, cost of merchandise, and expense accounts in this section.

The Balance Sheet section contains the balances of all permanent accounts. In that section you will find all asset, liability, and capital accounts (**Capital Stock** and **Retained Earnings**).

> ### Common Mistakes
>
> **Extended Amounts** When extending amounts on the ten-column work sheet, it is often confusing as to where the Income Summary account is entered. Although the account is classified as a temporary Stockholders' Equity account, it is entered in the Income Statement section of the work sheet. It is not entered in the Balance Sheet section with the permanent Stockholders' Equity accounts.

Completing the Work Sheet

After all amounts have been extended to the Balance Sheet and Income Statement sections, the accountant draws a *single* rule across the columns in these sections and totals all four columns (see **Figure 18–5**). Notice that the words *Net Income* have been written in the Account Name column on the same line as the net income amount.

As you learned in Chapter 8, the totals of the debit and credit columns within the Balance Sheet and Income Statement sections are not equal at this point. The difference between the two column totals in each section is the amount of net income (or net loss) for the period. After the net income (or net loss) has been recorded, the columns in the Balance Sheet and Income Statement sections are ruled and totaled as shown in **Figure 18–5.**

If the totals of the two Income Statement columns are equal, and the totals of the two Balance Sheet columns are equal, a double rule is drawn across all four columns. The double rule indicates that these sections of the work sheet have been proved.

Journalizing and Posting Adjusting Entries

How Do You Journalize and Post Adjusting Entries?

The journal entries that update the general ledger accounts at the end of a period are called **adjusting entries**. The Adjustments section of the work sheet is the source of information for journalizing the adjusting entries. The accounts debited and credited in the Adjustments section are entered in the general journal.

Journalizing Adjustments

Before recording the first adjusting entry, the accountant writes the words *Adjusting Entries* in the Description column of the general journal. Writing this heading eliminates the need for an explanation to be written after each adjusting entry.

Figure 18–6 Recording Adjusting Entries in the General Journal

	DATE		DESCRIPTION	POST. REF.	DEBIT	CREDIT	
1	20–		*Adjusting Entries*				1
2	Dec.	31	Income Summary		3 5 3 6 00		2
3			Merchandise Inventory			3 5 3 6 00	3
4		31	Supplies Expense		3 7 1 0 00		4
5			Supplies			3 7 1 0 00	5
6		31	Insurance Expense		1 2 5 00		6
7			Prepaid Insurance			1 2 5 00	7
8		31	Fed. Corporate Income Tax Exp.		1 5 5 00		8
9			Fed. Corp. Income Tax Pay.			1 5 5 00	9
10							10

GENERAL JOURNAL PAGE 22

The following entries are recorded in the Adjustments columns of the work sheet:

(a) adjusting merchandise inventory

(b) adjusting supplies

(c) adjusting insurance

(d) adjusting income tax

The first adjustment, which was labeled *(a)* on the work sheet, is recorded in the general journal in **Figure 18–6** as a debit to **Income Summary** for $3,536 and a credit to **Merchandise Inventory** for $3,536. The label *(a)* is not recorded in the general journal.

Remaining adjustments are entered in the general journal in the same manner, with the debit part of the entry recorded first. The date for each adjusting entry is the last day of the period.

Reading Check

Explain

List the steps to journalize adjustments

Posting Adjusting Entries to the General Ledger

After the adjusting entries have been recorded in the general journal, the accountant posts them to the general ledger accounts. Once the adjusting entries have been posted, the general ledger accounts are up to date. The balances in the general ledger accounts all agree with the amounts entered on the Income Statement and Balance Sheet sections of the work sheet.

Posting these entries accomplishes the following:

- The **Supplies Expense** account has been "charged" with the value of the supplies used in the period.

- The **Supplies** account reflects only the amount of items still remaining in inventory.

588 Chapter 18 • Adjustments and the Ten-Column Work Sheet

- The **Merchandise Inventory** account reflects the correct inventory value.
- The **Income Summary** account has been "charged" with the cost of goods sold for the period.
- The **Insurance Expense** account reflects the appropriate insurance expense for the period.
- The **Prepaid Insurance** account has been reduced by the amount of insurance expired.
- The **Federal Corporate Income Tax Payable** account has been increased to reflect the appropriate payable amount.
- The **Federal Corporate Income Tax Expense** account reflects the tax expense for the period.

Figure 18–7 shows the general journal and the general ledger accounts after the posting of the adjusting entries has been completed. Notice that the words *Adjusting Entry* have been written in the Description column of the general ledger accounts.

GENERAL JOURNAL PAGE 22

	DATE		DESCRIPTION	POST. REF.	DEBIT	CREDIT
1	20–		*Adjusting Entries*			
2	Dec.	31	Income Summary	310	3 5 3 6 00	
3			Merchandise Inventory	125		3 5 3 6 00
4			Supplies Expense	670	3 7 1 0 00	
5			Supplies	130		3 7 1 0 00
6			Insurance Expense	635	1 2 5 00	
7			Prepaid Insurance	135		1 2 5 00
8			Fed. Corporate Income Tax Exp.	630	1 5 5 00	
9			Fed. Corp. Income Tax Pay.	204		1 5 5 00

ACCOUNT **Merchandise Inventory** ACCOUNT NO. 125

DATE		DESCRIPTION	POST. REF.	DEBIT	CREDIT	BALANCE DEBIT	BALANCE CREDIT
20–							
Jan.	1	Balance	✓			84 9 2 1 00	
Dec.	31	Adjusting Entry	G22		3 5 3 6 00	81 3 8 5 00	

ACCOUNT **Supplies** ACCOUNT NO. 130

DATE		DESCRIPTION	POST. REF.	DEBIT	CREDIT	BALANCE DEBIT	BALANCE CREDIT
20–							
Dec.	1	Balance	✓			5 2 9 9 00	
	15		P12	2 5 0 00		5 5 4 9 00	
	31	Adjusting Entry	G22		3 7 1 0 00	1 8 3 9 00	

ACCOUNT **Prepaid Insurance** ACCOUNT NO. 135

DATE		DESCRIPTION	POST. REF.	DEBIT	CREDIT	BALANCE DEBIT	BALANCE CREDIT
20–							
Dec.	17		CP14	1 5 0 0 00		1 5 0 0 00	
	31	Adjusting Entry	G22		1 2 5 00	1 3 7 5 00	

Figure 18–7 Adjusting Entries Posted to the General Ledger (continued)

ACCOUNT	Federal Corporate Income Tax Payable					ACCOUNT NO. 204	
DATE	DESCRIPTION	POST. REF.	DEBIT	CREDIT		BALANCE	
						DEBIT	CREDIT
20– Dec. 31	Adjusting Entry	G22		1 5 5 00			1 5 5 00

ACCOUNT	Income Summary					ACCOUNT NO. 310	
DATE	DESCRIPTION	POST. REF.	DEBIT	CREDIT		BALANCE	
						DEBIT	CREDIT
20– Dec. 31	Adjusting Entry	G22	3 5 3 6 00			3 5 3 6 00	

ACCOUNT	Federal Corporate Income Tax Expense					ACCOUNT NO. 630	
DATE	DESCRIPTION	POST. REF.	DEBIT	CREDIT		BALANCE	
						DEBIT	CREDIT
20– Dec. 1	Balance	✓				9 8 4 0 00	
31	Adjusting Entry	G22	1 5 5 00			9 9 9 5 00	

ACCOUNT	Insurance Expense					ACCOUNT NO. 635	
DATE	DESCRIPTION	POST. REF.	DEBIT	CREDIT		BALANCE	
						DEBIT	CREDIT
20– Dec. 31	Adjusting Entry	G22	1 2 5 00			1 2 5 00	

ACCOUNT	Supplies Expense					ACCOUNT NO. 670	
DATE	DESCRIPTION	POST. REF.	DEBIT	CREDIT		BALANCE	
						DEBIT	CREDIT
20– Dec. 31	Adjusting Entry	G22	3 7 1 0 00			3 7 1 0 00	

Figure 18–7 Adjusting Entries Posted to the General Ledger

SECTION 18.3
Assessment

After You Read

Reinforce the Main Idea

Use a diagram like this one to describe the process of updating the general ledger accounts.

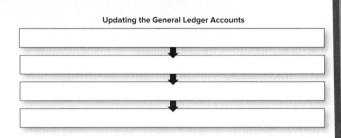

Updating the General Ledger Accounts

Problem 3 Analyzing the Work Sheet

Refer to **Figure 18–5** and answer the following questions in your working papers:

1. What amount is extended to the Income Statement section for **Federal Corporate Income Tax Expense?**
2. To which section of the work sheet is the balance of **Prepaid Insurance** extended?
3. What is the total amount of supplies consumed during the period?
4. What is the total amount still owed to the federal government for corporate income tax?

Problem 4 Analyzing a Source Document

Answer these questions, based on the debit memorandum shown here.

1. Which company is returning the merchandise?
2. How many items are being returned?
3. What amount is entered in the journal entry?
4. Which account is debited?
5. Which account is credited?

DEBIT MEMORANDUM No. 284
Date: May 4, 20--
Invoice No.: 378456

ROTARY SUPPLY CORPORATION
634 West Washington Avenue, Lincoln, SC 30856

To: K & L Electrical
845 Morgan Street
Plantsville, SC 30455

This day we have debited your account as follows:

Quantity	Item	Unit Price	Total
4	Locking Fixture #27304-78	$ 36.95	$ 147.80
2	Mirror/Reflectors #8935-231	11.24	22.48
	Subtotal		$ 170.28
	Tax		6.81
	Total		$ 177.09

Math for Accounting

The following column totals appear in the Balance Sheet section of the work sheet for Tonya's Toys on December 31.

Debit column $317,290
Credit column $323,730

Calculate the net income or net loss for the period.

Chapter 18
Visual Summary

Concepts
The Ten-Column Worksheet contains debit and credit columns for these five sections:

Trial Balance	Adjustments	Adjusted Trial Balance	Income Statement	Balance Sheet
Includes all general ledger accounts, even those with zero balances.	Records the adjustments made at the end of the period to bring various account balances up to date.	Shows the updated balances of all general ledger accounts.	Contains the balances of all temporary accounts, including income, summary and revenue, cost of merchandise and expense accounts.	Contains the balances of all permanent accounts, including assets, liabilities, and stockholders' equity.

Analysis
Common adjusting entries and the information needed to calculate them:

Adjustment	Information Needed	
Merchandise Inventory	• Merchandise Inventory account balance • Physical inventory amount	
Supplies	• Supplies account balance	• Cost of supplies on hand
Prepaid insurance	• Expired portion of insurance coverage	• Cost of insurance
Federal Corporate Income Tax	• Federal corporate income tax expense account balance (estimated quarterly payments) • Federal corporate income tax based on exact income.	

Procedures
To prepare a ten-column worksheet:

STEP	PROCEDURE
1.	Complete the Trial Balance section. Enter the account numbers, names and balances for all general ledger accounts. Total, prove, and rule the section.
2.	Calculate the adjustments needed and enter them in the Adjustments section.
3.	Complete the Adjusted Trial Balance section. For each account, combine the Trial Balance section amount with the Adjustment section amount and enter the total in the Adjusted Trial Balance section. (Add debits to debits, add credits to credits, subtract debits and credits).
4.	Total, prove, and rule the Adjustments and Adjusted Trial Balance sections.
5.	Extend the amounts in the Adjusted Trial Balance section to the appropriate columns in the Balance Sheet and Income Statement Sections.
6.	Complete the work sheet. Calculate the net income (or net loss) and enter it into the appropriate columns in the Income Statement and Balance Sheet sections. Total, prove, and rule the Income Statement and Balance Sheet sections.

Chapter 18
Review and Activities

Answering the Essential Question

What factors other than transactions can cause financial changes within a business? Sometimes it isn't only about buying and selling. Other things can influence the finances of a business. How do businesses check finances at the end of a period to make sure all the numbers are accurate? How do businesses benefit from analyzing inventory data?

Vocabulary Check

1. **Vocabulary** Use the content vocabulary terms to create a crossword puzzle on graph paper. Use the definitions as clues.

   - adjustment
   - beginning inventory
   - ending inventory
   - physical inventory
   - significant
   - prepaid expense
   - estimate
   - adjusting entries

Concept Check

2. **Analyze** What is the purpose of the ten-column work sheet? Explain exactly how the ten-column worksheet differs from the six-column work sheet.

3. **Evaluate** True or False: The trial balance section of the work sheet is not essential to the adjusting process. Explain your answer in detail.

4. Why might a general ledger account be updated at the end of the period?

5. How is the amount of merchandise on hand determined at the end of a period? What is the accounting step that follows this?

6. Which accounts are included on the Balance Sheet section of the work sheet? On the Income Statement section of the work sheet?

7. What is an adjusting entry? How are adjusting entries journalized?

8. **Analyze** Explain why the date of adjusting entries is the last day of the period.

9. **Math** Three accounting supervisors in a large corporation earn different salaries based on their experience and time on the job. The median base salary is $76,948, with a range of $64,494 to $91,617. Write an inequality comparing the mean and median.

10. **English Language Arts** List the qualities that an entrepreneur should possess in order to be able to grow a successful business. Explain how organization plays a major role in day to day operations. Tell why you think a person would decide to take the responsibilities and risks involved in beginning a new business and what benefits such risks can have.

Chapter 18
Standardized Test Practice

Multiple Choice

1. Which of the following statements is incorrect concerning an adjusted trial balance?

    a. An adjusted trial balance lists account balances and their locations in the ledger.

    b. An adjusted trial balance shows proper balance sheet and income statement amounts.

    c. An adjusted trial balance is prepared before the adjusting entries have been journalized and posted.

    d. An adjusted trial balance can be used to prepare the financial statements.

    e. All of the above are true statements.

2. A store purchased a one-year insurance policy for $1,800 on September 1. Its fiscal period ended December 31. What is the amount of the adjustment and what accounts are debited and credited on December 31?

    a. $1,800; insurance expense and prepaid insurance

    b. $600; insurance expense and prepaid insurance

    c. $1,200; insurance expense and prepaid insurance

    d. $600; prepaid insurance and insurance expense

3. The Income Summary amount in a work sheet's Adjustments Debit column represents the

    a. decrease in Merchandise Inventory.

    b. increase in Merchandise Inventory.

    c. beginning Merchandise Inventory.

    d. ending Merchandise Inventory.

4. The ending balance of the Supplies account appears

    a. in the Trial Balance columns of the work sheet.

    b. in the Balance Sheet columns of the work sheet.

    c. in the Income Statement columns of the work sheet.

    d. on the statement of changes in owner's equity.

5. What entries are recorded at the end of the fiscal period to update general ledger accounts?

   a. Closing

   b. Adjusting

   c. Post-closing

   d. Reversing

Short Answer

6. What are the five sections of the ten-column work sheet?

Extended Response

7. Why are general ledger accounts with zero balances entered in the Trial Balance section of the work sheet?

Chapter 18
Computerized Accounting

Recording Adjusting Entries
Making the Transition from a Manual to a Computerized System

MANUAL METHODS

- From the general ledger, prepare a trial balance in the first two columns of a work sheet.
- Record adjustments on the work sheet.
- Calculate adjusted account balances in the Adjusted Trial Balance columns.
- Extend account balances to Income Statement or Balance Sheet columns in the work sheet.
- Calculate net income (loss) for the accounting period.
- Record adjusting entries in the general journal.
- Post adjusting entries to general ledger accounts.

COMPUTERIZED METHODS

- Print a Trial Balance.
- Record the adjusting entries in the general journal.
- General ledger accounts are automatically updated.

Chapter 18
Problems

Problem 5 Completing a Ten-Column Work Sheet

The August 31 trial balance for InBeat CD Shop is entered on the work sheet in your working papers. Listed below is the data needed to make the adjustments.

Instructions In your working papers, complete the ten-column work sheet for InBeat CD Shop for the month ended August 31.

Data for Adjustments

Merchandise Inventory, August 31	$ 77,872
Supplies consumed during the period	2,171
Insurance premium expired during the period	489
Additional federal corporate income taxes owed	118

Analyze

After the adjustments are entered on the work sheet, assess which expense account has the highest balance.

Problem 6 Completing a Ten-Column Work Sheet

The August 31 trial balance for Shutterbug Cameras is listed on the next page. Also listed is the data needed for the adjustments.

Instructions In your working papers, complete the ten-column work sheet for Shutterbug Cameras for the month ended August 31.

		Trial Balance	
		Debit	**Credit**
101	Cash in Bank	$13,873	
115	Accounts Receivable	5,382	
125	Merchandise Inventory	82,981	
130	Supplies	2,397	
135	Prepaid Insurance	1,350	
140	Store Equipment	30,769	
201	Accounts Payable		$ 8,481
207	Federal Corporate Income Tax Payable		—
210	Employees' Federal Income Tax Payable		194
211	Employees' State Income Tax Payable		48
212	Social Security Tax Payable		119
213	Medicare Tax Payable		25
215	Sales Tax Payable		381
216	Federal Unemployment Tax Payable		19
217	State Unemployment Tax Payable		92
301	Capital Stock		80,000
305	Retained Earnings		28,568

Continued

Chapter 18
Problems

		Trial Balance	
		Debit	Credit
310	Income Summary	—	—
401	Sales		95,487
405	Sales Discounts	37	
410	Sales Returns and Allowances	945	
501	Purchases	39,491	
505	Transportation In	2,039	
510	Purchases Discounts		656
515	Purchases Returns and Allowances		219
601	Advertising Expense	128	
605	Bankcard Fees Expense	219	
620	Federal Corporate Income Tax Expense	1,580	
630	Insurance Expense	—	
640	Maintenance Expense	2,513	
645	Miscellaneous Expense	652	
647	Payroll Tax Expense	1,953	
650	Rent Expense	9,000	
655	Salaries Expense	18,631	
660	Supplies Expense	—	
670	Utilities Expense	349	

Data for Adjustments

Merchandise Inventory, August 31	$78,672
Supplies on hand, August 31	389
Insurance premium expired during the period	490
Additional federal income taxes owed	252

Analyze
Identify which account in the Balance Sheet section, Accounts Payable or Accounts Receivable, has the higher balance.

Problem 7 Completing a Ten-Column Work Sheet

The balances of the general ledger accounts of Cycle Tech Bicycles, as of August 31, are listed below. Also listed is the data needed for the adjustments.

Instructions In your working papers:

Prepare a ten-column work sheet for Cycle Tech Bicycles for the month ended August 31.

101	Cash in Bank	$ 22,323
115	Accounts Receivable	1,737
125	Merchandise Inventory	23,654

Continued

Chapter 18
Problems

130	Supplies	3,971
135	Prepaid Insurance	1,800
140	Store Equipment	25,395
145	Office Equipment	15,239
201	Accounts Payable	11,051
210	Federal Corporate Income Tax Payable	—
211	Employees' Federal Income Tax Payable	519
212	Employees' State Income Tax Payable	142
213	Social Security Tax Payable	408
214	Medicare Tax Payable	137
215	Sales Tax Payable	1,871
216	Federal Unemployment Tax Payable	51
217	State Unemployment Tax Payable	263
301	Capital Stock	40,000
305	Retained Earnings	14,908
310	Income Summary	—
401	Sales	128,231
405	Sales Discounts	214
410	Sales Returns and Allowances	1,289
501	Purchases	67,118
505	Transportation In	1,172
510	Purchases Discounts	810
515	Purchases Returns and Allowances	322
601	Advertising Expense	2,938
605	Bankcard Fees Expense	185
625	Federal Corporate Income Tax Expense	3,650
630	Insurance Expense	—
645	Maintenance Expense	2,450
650	Miscellaneous Expense	3,929
655	Payroll Tax Expense	834
657	Rent Expense	10,750
660	Salaries Expense	4,670
665	Supplies Expense	—
675	Utilities Expense	5,395

Data for Adjustments

Ending merchandise inventory	$ 24,188
Ending supplies inventory	1,049
Insurance premium expired	675
Federal income tax expense for the month	3,827

Continued

Chapter 18
Problems

Analyze
Determine which section of the work sheet, the Trial Balance section or the Balance Sheet section, had the higher balance for Merchandise Inventory.

Problem 8 Completing a Ten-Column Work Sheet

The balances of the general ledger accounts of River's Edge Canoe & Kayak, as of August 31, are listed below.

Instructions In your working papers:

1. Prepare a ten-column work sheet for River's Edge Canoe & Kayak for the year ended August 31. The account names are entered on the work sheet. The data for the adjustments follows:

   (a) The cost of the ending merchandise inventory is $45,669.

   (b) The cost of the supplies on hand on August 31 is $619.

   (c) The one-year insurance premium of $1,680 was paid on April 1.

   (d) The total federal income taxes owed for the year are $2,635.

2. Enter the journal entries for the adjustments on page 31 of the general journal.

3. Post the journal entries to the general ledger accounts.

101	Cash in Bank	$ 15,387
115	Accounts Receivable	2,852
130	Merchandise Inventory	49,205
135	Supplies	3,027
140	Prepaid Insurance	1,680
145	Delivery Equipment	19,437
150	Store Equipment	29,504
201	Accounts Payable	13,339
204	Federal Corporate Income Tax Payable	—
210	Employees' Federal Income Tax Payable21	632
211	Employees' State Income Tax Payable	117
212	Social Security Tax Payable	472
213	Medicare Tax Payable	108
215	Sales Tax Payable	2,931
216	Federal Unemployment Tax Payable	77
217	State Unemployment Tax Payable	315
219	U.S. Savings Bonds Payable	150
301	Capital Stock	50,000
305	Retained Earnings	25,425
310	Income Summary	—
401	Sales	144,945
405	Sales Discounts	203
410	Sales Returns and Allowances	1,381

Continued

600 Chapter 18 • Adjustments and the Ten-Column Work Sheet

Chapter 18
Problems

501	Purchases	79,310
505	Transportation In	1,192
510	Purchases Discounts	1,292
515	Purchases Returns and Allowances	576
601	Advertising Expense	3,151
605	Bankcard Fees Expense	288
625	Federal Corporate Income Tax Expense	2,480
635	Insurance Expense	—
650	Maintenance Expense	1,381
655	Miscellaneous Expense	3,772
658	Payroll Tax Expense	1,219
660	Rent Expense	10,350
665	Salaries Expense	11,965
670	Supplies Expense	—
680	Utilities Expense	2,595

Analyze
Calculate River's Edge net income (net loss) if it had *not* taken any purchases discounts for paying creditors within the discount period.

Problem 9 Locating Errors on the Work Sheet

The Trial Balance and adjustment sections for Buzz Newsstand have been prepared in your working papers. It is apparent from the totals on the work sheet that errors have been made in preparing these portions of the work sheet.

The accounting records show:

1. The merchandise on hand at the end of the month is valued at $12,950.
2. The supplies on hand on August 31 are valued at $529.
3. The insurance premium was paid on August 1. The premium was $980 and covers the period from August 1 to November 30.
4. The total federal income tax owed for the period is $249.

Instructions In your working papers:

1. Find and correct the error(s) in the Trial Balance section.
2. On the line provided on the work sheet, write in the corrected totals for the Trial Balance section.
3. Find and correct the error(s) in the Adjustments section.
4. Write in the corrected totals for the Adjustments columns.

Analyze
Explain your calculations for the Merchandise Inventory adjustment.

Real-World Applications & Connections

CASE STUDY

Merchandising Business: Training Videos

You are an accountant for EZ Training Systems. The company CEO has asked you to prepare financial statements without the appropriate adjusting entries.

INSTRUCTIONS Analyze this information for adjustments and complete the tasks.

	Beginning	Ending
Supplies	$5,000	$1,200
Prepaid Insurance	6,400	800

1. Net income before adjustments is $22,400. Calculate net income after the adjustments have been made.

2. Explain why the financial statements do not present an accurate picture of the company without considering the adjusting entries.

21st Century Skills

The Sarbanes-Oxley Act

Irresponsible financial reporting or manipulation of financial information can have devastating results. Government officials created the Sarbanes-Oxley Act of 2002 as a reaction to a number of major corporate and accounting scandals that shook the public's confidence in the nation's financial system. Scandals involving companies including Enron, WorldCom, Tyco International, and Adelphia Communications emphasized the need for new accounting standards for U.S. public companies, management, and accounting firms.

Research and Write

a. Use the Internet to locate articles describing the accounting scandals at one of the following companies: Enron, WorldCom, Tyco International, Adelphia Communications, Arthur Andersen.

b. Describe what happened in the incident you chose to analyze. How were regulation and internal controls used to address the problem? What role does the Securities and Exchange Commission play in regulating the accounting industry?

Spotlight on Personal Finance

Monitoring Supplies

Your family, like most families, purchases many different paper supplies for your home. You notice when certain supplies are running low and purchase more. This is very similar to how a business handles its supplies.

Personal Finance Activity

List some common paper goods that a family might purchase and consume within a week or month. List some other paper supplies that might be purchased just once a year.

Analyzing Financial Reports

Classifying the Balance Sheet: Assets and Liabilities

Accounts on the balance sheet are *classified* or grouped into related categories. These classifications provide subtotals which can be used to compute ratios and do comparisons. For example any portion of a long-term liability that is due within the next year is classified as a current liability. This affects the company's liquidity ratios and working capital. (See page 262 in Chapter 9.)

Instructions

Use the Internet to locate and save or print a copy of Amazon.com's annual report to complete the following. Use the balance sheet to answer these questions.

1. One of the line items in the assets section is Goodwill. What do you think Goodwill means in this context? Use the Internet to find the definition of Goodwill as it is used on financial statements. How is the definition similar or different from what you thought?

2. How does this balance sheet prove the accounting equation?

H.O.T Audit

Adjusting Entries

The ten-column work sheet for Alpha Training Corporation was completed and a net income of $37,458 was reported. Upon examination, you discovered that the adjustment for Supplies was never entered in the Adjustments section of the work sheet. The Supplies amount entered in the Trial Balance section was $4,172. You estimate that there is only $2,181 in supplies still on hand at the end of the period.

Instructions

Answer the following questions regarding the omission of the adjusting entry on the work sheet.

1. What is the amount that should have been entered in the Adjustment section of the work sheet? What general ledger accounts are used for the adjustment?

2. Would this adjustment cause the reported net income to be higher or lower? What would be the correct net income for the period?

3. What amount is entered in the Balance Sheet section for Supplies?

Chapter 19

Financial Statements for a Corporation

Chapter Topics

19-1 The Ownership of a Corporation

19-2 The Income Statement

19-3 The Statement of Retained Earnings, Balance Sheet, and Statement of Cash Flows

Visual Summary

Review and Activities

Standardized Test Practice

Computerized Accounting

Problems

Real-World Applications & Connections

Essential Question

As you read this chapter, keep this question in mind:

What are some of the unique accounting requirements of a merchandising corporation?

Main Idea

Owner's equity in a corporation is called stockholders' equity. A merchandiser's income statement has a Cost of Merchandise Sold section, and a corporation's income statement shows income tax expense. Additionally, a corporation prepares the statement of retained earnings, the balance sheet, and the statement of cash flows.

Chapter Objectives

Concepts	Analysis	Procedures
C1 Explain how to record ownership of a corporation. **C2** Explain the relationship between the work sheet and the financial statements for a merchandising corporation.	**A1** Explain how a corporation's financial statements differ from a sole proprietorship. **A2** Analyze the financial data contained on the statements.	**P1** Prepare an income statement, statement of retained earnings, and balance sheet. **P2** Describe the statement of cash flows for a merchandising corporation.

Real-World Business Connection

San Antonio Spurs

In a city that has only one major professional sports franchise, the San Antonio Spurs command major attention and excitement. A franchise of the National Basketball Association since 1976, the Spurs' history includes four NBA championships and great players such as George Gervin, David Robinson, Tim Duncan, and Manu Ginobili. South Texas residents also appreciate the Spurs' generous involvement in educational programs and other community initiatives.

Connect to the Business

With all the focus on basketball, it can be easy to forget the San Antonio Spurs are first and foremost a for-profit business. The franchise is owned by 20 individuals and groups; a seven-member board of directors makes business decisions. The investors include corporations such as ARAMARK, Clear Channel Communications, and AT&T.

Analyze

In purchasing stock in a corporation like Clear Channel, on which financial statements might you find the San Antonio Spurs?

Focus on the Photo

The San Antonio Spurs is a popular basketball franchise financed by corporations that sell stock in their own companies. As public corporations, they must disclose their financial information. ***Where might you find reports that disclose such financial information?***

SECTION 19.1
The Ownership of a Corporation

Get ready to learn how to utilize the worksheet to prepare financial statements for a merchandising corporation. You will specifically explore how a corporation's financial statements differ from a sole proprietorship's financial statements. This is an important task for all businesses.

Accounting for a Corporation

Who Owns a Corporation?

One person owns a sole proprietorship. A corporation may be owned by one person or by thousands of people. The ownership of a corporation is represented by shares of stock.

Recording the Ownership of a Corporation

As you recall, investments by the owner of a sole proprietorship are recorded in the owner's capital account. A $25,000 owner's investment in a sole proprietorship is recorded as shown in the T accounts. **Cash in Bank** is debited for $25,000, and **Crista Vargas, Capital** is credited for $25,000.

Cash in Bank		Crista Vargas, Capital	
Debit + 25,000	Credit −	Debit −	Credit + 25,000

> **Section Vocabulary**
> - Capital Stock
> - stockholders' equity
> - retained earnings
> - comparability
> - reliability
> - relevance
> - full disclosure
> - materiality
> - contributed
> - retained
> - potential

Businesses organized as corporations have a **Capital Stock** account instead of the owner's capital account in a sole proprietorship. **Capital Stock** represents investments in the corporation by its stockholders (owners).

Capital Stock is classified as a stockholders' equity account. **Stockholders' equity** is the value of the stockholders' claims to the corporation. Like the owner's capital account in a sole proprietorship, increases to **Capital Stock** are recorded as credits and decreases are recorded as debits.

Reporting Stockholders' Equity in a Corporation

The form of business organization does not affect the amount of equity in the business. That is, one person may have an ownership interest in a sole proprietorship worth $80,000, or 10 people may have shares of stock in a corporation worth a total of $80,000. The difference is in the way these two amounts are reported on the balance sheet.

A sole proprietorship reports the balance of the owner's capital account in the owner's equity section of the balance sheet. For a corporation the owner's equity section of the balance sheet is called stockholders' equity. The law requires that stockholders' equity be reported in two parts: (1) equity **contributed** by stockholders and (2) equity earned through business profits.

606 Chapter 19 • Financial Statements for a Corporation

Business Transaction

On January 1 stockholders invested $25,000 in exchange for shares of stock of the corporation, Receipt 997.

Analysis	Identify	1. The accounts affected are **Cash in Bank** and **Capital Stock.**
	Classify	2. **Cash in Bank** is an asset account. **Capital Stock** is a stockholders' equity account.
	+/−	3. **Cash in Bank** is increased by $25,000. **Capital Stock** is increased by $25,000.
Debit-Credit Rule		4. Increases to asset accounts are recorded as debits. Debit **Cash in Bank** for $25,000.
		5. Increases to stockholders' equity accounts are recorded as credits. Credit **Capital Stock** for $25,000.

T Accounts 6.

Cash in Bank		Capital Stock	
Debit + 25,000	Credit −	Debit −	Credit + 25,000

Journal Entry 7.

	GENERAL JOURNAL			PAGE 1	
DATE	DESCRIPTION	POST. REF.	DEBIT	CREDIT	
20-- Jan. 1	Cash in Bank		25 000 00		
	Capital Stock			25 000 00	
	Receipt 997				

Equity Contributed by Stockholders

The first part of stockholders' equity is the amount of money invested by stockholders. This amount is comparable to the investments made by the owner in a sole proprietorship. In a corporation stockholders contribute to equity by buying shares of stock issued by the corporation. Stockholders' investments are recorded in the **Capital Stock** account.

Equity Earned Through Business Profits

The second part of stockholders' equity is the amount of accumulated net income earned and **retained** by the corporation. This amount is comparable to the amount of net income less any withdrawals by the owner in a sole proprietorship. In a corporation this amount, called **retained earnings**, represents the increase in stockholders' equity from the portion of net income not distributed to the stockholders.

Earnings retained by a corporation are recorded in the **Retained Earnings** account. **Retained Earnings** is classified as a stockholders' equity account. Like the **Capital Stock** account, it is increased by credits and decreased by debits. **Retained Earnings** has a normal credit balance.

Retained Earnings	
Debit	Credit
—	+
Decrease Side	Increase Side
	Normal Balance

In a sole proprietorship, net income increases owner's capital. This increase in owner's capital represents an increase in the assets of the business. In a corporation net income increases retained earnings, which represents the growth, or increase, in the assets of the corporation.

Balance Sheet Presentation

A comparison of the capital section of the balance sheet for a sole proprietorship and for a corporation follows:

Sole Proprietorship
Owner's Equity:
  Owner's Capital

Corporation
Stockholders' Equity:
  Capital Stock
  Retained Earnings

Reading Check

Recall

How do stockholders contribute equity to a corporation?

Characteristics of Financial Information

What Qualities Are Required in Financial Statements?

At the end of a period, a business prepares various financial statements. These statements summarize the changes that have taken place during the period and report the financial condition of the business at the end of the period. Financial statements are used by many groups:

- Managers analyze financial statements to evaluate past performance and to make informed decisions and predictions for future operations.
- Stockholders are interested in the performance, **potential** future growth, and success of the business.
- Creditors want to know a company's ability to pay its debts in a timely manner and the amount of credit that should be extended.
- Government agencies, employees, consumers, and the general public are also interested in the financial position of the business.

608 Chapter 19 • Financial Statements for a Corporation

Comparability

For accounting information to be useful, it must be understandable and comparable. The data must be presented in a way that lets users recognize similarities, differences, and trends from one period to another. **Comparability** allows accounting information to be compared from one fiscal period to another. The same types of statements, therefore, are prepared at the end of each period for the same length of time (for example, one month or one year). By comparing financial statements in different periods of equal length, financial patterns and relationships can be identified and analyzed, and the information from the analysis can be used to make decisions regarding business operations. Comparability also allows the comparison of financial information between businesses.

Reliability

The users of accounting data assume that the data is reliable. **Reliability** refers to the confidence users have that the financial information is reasonably free from bias and error.

Relevance

Relevance is the requirement that all information that would affect the decisions of financial statement users be disclosed in the financial reports.

Full Disclosure

"To disclose" means "to uncover or to make known." **Full disclosure** means that financial reports include enough information to be complete.

Materiality

If something is "material," it is important. In accounting, **materiality** means that relevant information should be included in financial reports.

Reading Check

Explain

Why is comparability important when analyzing financial statements?

A Corporation's Financial Statements

What Financial Statements Does a Merchandising Corporation Prepare?

The Starting Line Sports Gear, a merchandising corporation, prepares four financial statements: the income statement, the statement of retained earnings, the balance sheet, and the statement of cash flows. Three statements report the changes that have taken place over the period. One statement, the balance sheet, shows the financial position of the business on a specific date—the last day of the period. The work sheet provides most of the information needed to complete all four statements.

Today most businesses rely on automated equipment or computers to maintain the general and subsidiary ledgers and to prepare the end-of-period financial statements. Computers offer the advantages of speed and accuracy. The impact of a change in an estimate, operating procedure, or accounting method can be seen instantaneously. For example, electronic spreadsheets allow *what-if analysis,* which is when one or more variables are changed to see how the final outcome would be affected.

SECTION 19.1
Assessment

After You Read

Reinforce the Main Idea

Use a diagram like this one to describe the basic accounting equation for a corporation. Use three key terms from Section 19.1.

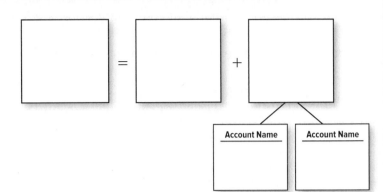

Problem 1 Analyzing Stockholders' Equity Accounts

1. An investment of $60,000 by Kevin Cleary in his sole proprietorship is recorded as a credit to which account?

2. The sale of 100 shares of stock for $8,500 by the Sims Corporation is recorded as a credit to which account?

3. Stockholders' equity consists of which two accounts?

Problem 2 Analyzing a Source Document

A sales slip for Cindy's Curtains is presented at right. The accountant noticed errors in the calculations.

Instructions Check all calculations and recalculate the sales tax using a rate of 4 percent.

Math for Accounting

Stock in Middlewood Corporation sells for $20 per share. The balance in the **Capital Stock** account at the beginning of the period was $72,400. The amount entered in the Balance Sheet section of the end-of-period work sheet is $83,100. How many shares of stock were sold during the fiscal period?

SECTION 19.2
The Income Statement

In Section 19.1 you learned that corporations prepare four financial statements at the end of each period. In this section you will learn how to prepare and analyze the income statement.

The Income Statement

How Is a Merchandising Business Income Statement Different from a Service Business Income Statement?

As you know, the income statement reports the net income or loss earned by a business. In Chapter 9 you prepared an income statement for Zip Delivery Service, a service business organized as a sole proprietorship. You subtracted total expenses from revenue to find the period's net income or loss. When preparing the income statement, whether for a service or merchandising business, the *revenue realization* principle is applied. Revenue for a credit sale is recorded at the time of the sale because the account receivable is expected to be converted to cash. The matching principle is also applied when preparing the income statement. Expenses are matched with revenue earned during the same period.

Section Vocabulary
- net sales
- net purchases
- gross profit on sales
- operating expenses
- selling expenses
- administrative expenses
- operating income
- vertical analysis
- expanded
- assigned
- interpret

Merchandising businesses have an additional cost—the cost of the merchandise that is purchased and then resold to customers. The income statement for a merchandising business is thus **expanded** to include the cost of merchandise sold.

An income statement for a merchandising business has five sections:

- Revenue
- Cost of Merchandise Sold
- Gross Profit on Sales
- Operating Expenses
- Net Income (or Loss)

A comparison of the income statements for a service business and for a merchandising business follows:

Service Business	Merchandising Business
Zip Delivery Service	*The Starting Line Sports Gear*
Revenue	Revenue
− Expenses	− Cost of Merchandise Sold
Net Income (Loss)	Gross Profit on Sales
	− Operating Expenses
	Net Income (Loss)

The Starting Line's income statement has four amount columns (see **Figure 19–1**). Totals are entered in the far right column. Balances that are added or subtracted are entered in the other columns. The format of a computer-generated income statement can vary from the format of an income statement prepared manually on accounting stationery. Regardless of how an income statement is prepared, the formats are very similar.

Figure 19–1 The Heading and Revenue Section of the Income Statement

The Starting Line Sports Gear				
Income Statement				
For the Year Ended December 31, 20--				
Revenue:				
Sales			320 450 00	
Less: Sales Discounts		730 00		
Sales Returns and Allowances		2 000 00	2 730 00	
Net Sales				317 720 00

All information needed to prepare the income statement comes from the work sheet, particularly the Income Statement section. As with all other financial statements, the income statement begins with a three-line heading. The income statement for The Starting Line is prepared for the year ended December 31.

The Revenue Section

The first section on the income statement is the revenue section. This section reports the net sales for the period. The balances of the **Sales** revenue account and the **Sales Discounts** and **Sales Returns and Allowances** contra revenue accounts are reported in this section. Remember that contra revenue accounts *decrease* the revenue account. Therefore, **net sales** is the amount of sales for the period less any sales discounts, returns, and allowances. Refer to **Figure 19–1** as you learn how to complete the revenue section.

1. On the first line, enter the classification *Revenue:* at the left edge of the accounting stationery.

2. On the second line, enter the name of the revenue account *Sales,* indented about half an inch. Enter the balance of the account in the *third* amount column.

3. On the next lines, enter the deductions from **Sales.** Write the word *Less:* followed by the names and balances of the two contra revenue accounts. (You may have to abbreviate the account names.) Enter the balances of the accounts in the *second* amount column.

4. Add the balances of the two contra revenue accounts. Write the total below the **Sales** balance on the fourth line, in the *third* amount column.

5. On the next line, enter the words *Net Sales,* indented about an inch. Subtract the total of the two contra accounts from the balance of the **Sales** account. Enter the amount in the *fourth* amount column. The Starting Line's net sales for the year are $317,720.

The Cost of Merchandise Sold Section

The cost of merchandise sold section follows the revenue section. As the words indicate, the *cost of merchandise sold* is the actual cost to the business of the merchandise it sold to customers during the period.

The cost of merchandise sold is calculated as follows:

>   Beginning Merchandise Inventory
> + Net Purchases During the Period
> ───────────────────────────────
>   Cost of Merchandise Available for Sale
> − Ending Merchandise Inventory
> ───────────────────────────────
>   Cost of Merchandise Sold

Computing the cost of merchandise sold requires two steps:

1. Determine the cost of all merchandise available for sale.
2. Calculate the cost of merchandise sold.

Calculating Cost of Merchandise Available for Sale.

To calculate the cost of merchandise available for sale, add net purchases to the beginning inventory amount. **Net purchases** represents all costs related to merchandise purchased during the period. To calculate net purchases, add the transportation charges for the period (**Transportation In**) to the **Purchases** balance and then subtract the balances of **Purchases Discounts** and **Purchases Returns and Allowances**.

> **Common Mistakes**
>
> **Transportation In** It is easy to forget that the cost of getting the merchandise to your place of business adds to the cost of the purchases. When calculating the cost of merchandise available for sales, be sure to add the balance in the Transportation In account to the balance in the Purchases account. This gives you the actual cost of the delivered merchandise.

>   Purchases
> + Transportation In
> ───────────────────────────────
>   Cost of Delivered Merchandise
> − Purchases Discounts
> − Purchases Returns and Allowances
> ───────────────────────────────
>   Net Purchases

Calculating Cost of Merchandise Sold.

To calculate the cost of merchandise sold, subtract the ending merchandise inventory amount from the cost of merchandise available for sale. Refer to **Figure 19–2** as you learn how to complete the cost of merchandise sold section.

1. On the line below net sales, enter the words *Cost of Merchandise Sold:* at the left edge.
2. Next, enter *Merchandise Inventory, January 1, 20—* indented about half an inch. Enter the amount of the beginning inventory in the *third* amount column. (The beginning inventory is found on the work sheet, in the Trial Balance section, on the Merchandise Inventory line.)
3. Next, enter *Purchases,* indented about half an inch, and place the **Purchases** account balance in the *first* amount column.
4. On the next line, enter *Plus: Transportation In* indented about half an inch. Enter the balance of **Transportation In** in the *first* amount column, below the

Purchases amount. Draw a line across the *first* amount column under the **Transportation In** amount.

5. On the next line, write *Cost of Delivered Merchandise* indented about half an inch. Add the balances of **Purchases** and **Transportation In.** Enter the result in the *second* amount column.

6. On the next line, write *Less: Purchases Discounts,* indented about half an inch, and place the **Purchases Discounts** account balance in the *first* amount column.

7. On the next line, write *Purchases Returns and Allowances* so that it lines up with **Purchases Discounts** in the line above and place the **Purchases Returns and Allowances** account balance in the *first* amount column.

The Starting Line Sports Gear
Income Statement
For the Year Ended December 31, 20--

Revenue:					
Sales				320 4 5 0 00	
Less: Sales Discounts		7 3 0 00			
Sales Returns and Allowances		2 0 0 0 00	2 7 3 0 00		
Net Sales					317 7 2 0 00
Cost of Merchandise Sold:					
Merchandise Inventory, January 1, 20--				84 9 2 1 00	
Purchases	206 7 0 0 00				
Plus: Transportation In	4 0 3 6 18				
Cost of Delivered Merchandise		210 7 3 6 18			
Less: Purchases Discounts	1 3 4 0 00				
Purchases Returns and Allowances	1 8 0 0 00	3 1 4 0 00			
Net Purchases				207 5 9 6 18	
Cost of Merchandise Available				292 5 1 7 18	
Merchandise Inventory, December 31, 20--				81 3 8 5 00	
Cost of Merchandise Sold					211 1 3 2 18
Gross Profit on Sales					106 5 8 7 82

Figure 19-2 Income Statement Through Gross Profit on Sales

8. To find the total deduction from **Purchases,** add the balances of the **Purchases Discounts** and **Purchases Returns and Allowances** accounts. Enter the total on the Purchases Returns and Allowances line, in the *second* amount column. Draw a line across the *first* and *second* amount columns under this total.

9. On the next line, write *Net Purchases* indented about half an inch. Subtract the total of the **Purchases Discounts** and **Purchases Returns and Allowances** accounts from the cost of delivered merchandise. The difference is the amount of net purchases for the period. Enter the amount in the *third* amount column. Draw a line across the *third* amount column under this amount.

10. On the next line, write *Cost of Merchandise Available* indented about half an inch. Add the net purchases amount to the beginning inventory amount. The total is the cost of merchandise available for sale. Enter the total in the *third* amount column.

11. On the next line, write *Merchandise Inventory, December 31, 20—* indented about half an inch. Enter the amount of the ending inventory in the *third* amount column. (The ending inventory is found on the work sheet, in the Balance Sheet section, on the Merchandise Inventory line.) Draw a line across the *third* amount column under this amount.

12. On the next line, write *Cost of Merchandise Sold* indented about one inch. Subtract the ending inventory amount from the cost of merchandise available for sale. The difference is the cost of merchandise sold during the period. Enter the amount in the *fourth* amount column. Draw a line across the *fourth* amount column under this amount.

The Gross Profit on Sales Section

After the cost of merchandise sold has been calculated, the gross profit on sales can be determined. The **gross profit on sales** during the period is the profit made before operating expenses are deducted. Gross profit on sales is found by subtracting the cost of merchandise sold from net sales. *Gross Profit on Sales* is entered at the left edge, and the amount is placed in the *fourth* amount column. In **Figure 19–2** you can see that The Starting Line's gross profit on sales is $106,587.82.

The Operating Expenses Section

The next section of the income statement shows the operating expenses for the period. **Operating expenses** are the costs of the goods and services used in the process of earning revenue for the business. Some businesses choose to further classify operating expenses into **selling expenses** (incurred to sell or market the merchandise sold) and **administrative expenses** (related to the management of the business). Look at **Figure 19–3**. *Operating Expenses* is entered at the left edge on the line following the gross profit on sales. On the following lines, the names and balances of all expense accounts except **Federal Corporate Income Tax Expense** are listed in the same order as on the work sheet. Federal corporate income tax is a normal expense for a corporation, but it is not considered to be an operating expense. Rather than a cost related to earning revenue, income tax represents a cost *resulting* from the revenue earned.

Notice that the balances of the expense accounts are entered in the *third* amount column. The balances are totaled. The total, $63,918.10, is entered in the *fourth* amount column.

The Net Income Section

The final section of the income statement reports the net income (or net loss) for the period, both before and after federal corporate income taxes. It is customary to present the federal corporate income tax amount *separately* on the income statement. This is done so that the income statement shows the amount of operating income. **Operating income** is the excess of gross profit over operating expenses. It is the amount of income earned before deducting federal corporate income taxes.

Look at **Figure 19–3** again. To calculate operating income, subtract the total operating expenses from the gross profit on sales. The Starting Line's operating income for the period is $42,669.72.

To calculate net income, follow these steps.

1. Enter *Less: Federal Corporate Income Tax Expense* on the next line, indented about half an inch.
2. Enter the amount of income taxes, $9,995, in the *fourth* amount column. Federal corporate income taxes appear in the Income Statement section of the work sheet. Draw a line across the *fourth* column under this amount.
3. Enter *Net Income* (or *Net Loss*) on the next line at the left edge.
4. Subtract the amount of federal corporate income taxes from the operating income. The result is net income or net loss.

Figure 19–3 The Starting Line's Completed Income Statement

The Starting Line Sports Gear
Income Statement
For the Year Ended December 31, 20--

Revenue:				
Sales			320 450 00	
Less: Sales Discounts		730 00		
Sales Returns and Allowances		2 000 00	2 730 00	
Net Sales				317 720 00
Cost of Merchandise Sold:				
Merchandise Inventory, January 1, 20--			84 921 00	
Purchases	206 700 00			
Plus: Transportation In	4 036 18			
Cost of Delivered Merchandise		210 736 18		
Less: Purchases Discounts	1 340 00			
Purchases Returns and Allowances	1 800 00	3 140 00		
Net Purchases			207 596 18	
Cost of Merchandise Available			292 517 18	
Merchandise Inventory, December 31, 20--			81 385 00	
Cost of Merchandise Sold				211 132 18
Gross Profit on Sales				106 587 82
Operating Expenses:				
Advertising Expense			2 450 00	
Bank Card Fees Expense			4 199 27	
Insurance Expense			125 00	
Maintenance Expense			3 519 25	
Miscellaneous Expense			348 28	
Payroll Tax Expense			3 826 83	
Rent Expense			14 000 00	
Salaries Expense			29 374 60	
Supplies Expense			3 710 00	
Utilities Expense			2 364 87	
Total Operating Expenses				63 918 10
Operating Income				42 669 72
Less: Federal Corporate Income Tax Expense				9 995 00
Net Income				32 674 72

5. Enter the difference, $32,674.72, in the *fourth* amount column. Net income on the income statement must agree with the net income shown on the work sheet. If it does, draw a double rule under the amount to show that the income statement is proved and complete. If it does not, check the addition and subtraction on the income statement. Also check that all the accounts and balances in the Income Statement section of the work sheet appear correctly on the income statement.

Reading Check

Compare and Contrast

Explain the difference between operating income and net income.

Analyzing Amounts on the Income Statement

Why Do People Look at Percentages?

Managers use financial analysis to evaluate the company's financial performance. The information reported on the income statement and other financial statements is expressed in dollars. Dollar amounts are useful, but the analysis can be expanded and made more meaningful by expressing the dollar amounts as percentages. These percentages more clearly indicate the relationships among the items on the financial statements. They also enable financial statement users to compare the relationships within an accounting period and changes in these relationships between accounting periods.

One type of analysis is called **vertical analysis.** With vertical analysis, each dollar amount reported on a financial statement is also reported as a percentage of another amount, called a base amount, appearing on that same statement. For example, on the income statement, each amount is reported as a percentage of net sales. Current-period percentages can be compared with percentages from past periods or with percentages from other companies within the same industry.

Figure 19–4 shows a comparative income statement. As you can see, the net sales amount for each year is **assigned** a percentage of 100. Every other amount on the income statement is stated as a percentage of the net sales amount. Notice that net income was 14.68 percent of net sales in the previous year and only 10.28 percent of net sales for the current year. Managers would want to find the cause to **interpret** the decrease. Analysis like this helps managers make informed decisions about future operations.

The Starting Line Sports Gear
Comparative Income Statement
For the Current and Previous Years Ended December 31

	Current Year		Previous Year	
	Dollars	Percent	Dollars	Percent
Revenue:				
Sales	$ 320,450.00	100.86 %	$ 296,350.00	100.79 %
Less: Sales Discounts	730.00	0.23	625.00	0.21
Sales Ret. and Allow.	2,000.00	0.63	1,700.00	0.58
Net Sales	$ 317,720.00	100.00 %	$ 294,025.00	100.00 %
Cost of Merchandise Sold:				
Merch. Inventory, Jan. 1	$ 84,921.00	26.73 %	$ 82,100.00	27.92 %
Net Purchases	207,596.18	65.34	186,836.56	63.54
Merch. Available for Sale	292,517.18	92.07 %	$ 268,936.56	91.47 %
Merch. Inventory, Dec. 31	81,385.00	25.62	84,921.00	28.88
Cost of Merchandise Sold	$ 211,132.18	66.45 %	$ 184,015.56	62.59 %
Gross Profit on Sales	$ 106,587.82	33.55 %	$ 110,009.44	37.41 %
Operating Expenses:				
Advertising Expense	$ 2,450.00	0.77 %	$ 1,779.00	0.61 %
Bankcard Fees Expense	4,199.27	1.32	3,569.37	1.21
Insurance Expense	125.00	0.04	0.00	0.00
Maintenance Expense	3,519.25	1.11	3,308.10	1.13
Miscellaneous Expense	348.28	0.11	742.00	0.25
Payroll Tax Expense	3,826.83	1.20	3,444.15	1.17
Rent Expense	14,000.00	4.41	13,200.00	4.49
Salaries Expense	29,374.60	9.25	26,437.14	8.99
Supplies Expense	3,710.00	1.17	2,968.00	1.01
Utilities Expense	2,364.87	0.74	2,305.75	0.78
Total Operating Expenses	$ 63,918.10	20.12 %	$ 57,753.51	19.64 %
Operating Income	$ 42,669.72	13.43 %	$ 52,255.93	17.77 %
Fed. Corp. Inc. Tax Exp.	9,995.00	3.15	9,085.00	3.09
Net Income	$ 32,674.72	10.28 %	$ 43,170.93	14.68 %

Figure 19–4 Comparative Income Statement Showing Vertical Analysis

SECTION 19.2
Assessment

After You Read

Reinforce the Main Idea

Use a diagram like this one to describe the steps for preparing an income statement for a merchandising corporation.

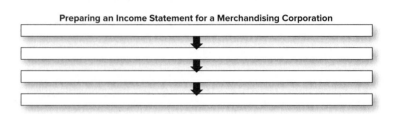

Problem 3 Calculating Amounts on the Income Statement

Instructions For each group of figures that follows, determine the missing amount.

1. Beginning merchandise inventory — $81,367
   Net purchases — 15,139
   Cost of merchandise available for sale — ?

2. Net sales — $52,935
   Cost of merchandise sold — 36,232
   Gross profit on sales — ?

3. Purchases — $26,472
   Transportation in — 1,311
   Cost of delivered merchandise — ?

4. Cost of merchandise available for sale — $49,769
   Ending merchandise inventory — 32,621
   Cost of merchandise sold — ?

Comparative Income Statement
For the Current and Previous Years Ended December 31

	Current Year Dollars	Current Year Percent	Previous Year Dollars	Previous Year Percent
Revenue:				
Sales	$ 1,500,000	100.13 %	$ 850,000	100.26 %
Less: Sales Ret. and Allow.	2,000	0.13	2,200	0.26
Net Sales	$ 1,498,000	100.00 %	$ 847,800	100.00 %
Cost of Merchandise Sold:				
Merch. Inventory, Jan. 1	$ 250,000	16.69 %	$ 100,000	11.80 %
Net Purchases	800,000	53.40	650,000	76.67
Merch. Available for Sale	$ 1,050,000	70.09 %	$ 750,000	88.46 %
Merch. Inventory, Dec. 31	60,000	4.01	250,000	29.49
Cost of Merchandise Sold	$ 990,000	66.09 %	$ 500,000	58.98 %
Gross Profit on Sales	$ 508,000	33.91 %	$ 347,800	41.02 %
Operating Expenses:				
Advertising Expense	$ 50,000	3.34 %	$ 25,000	2.95 %
Bankcard Fees Expense	10,000	0.67	8,000	0.94
Insurance Expense	5,300	0.35	5,300	0.63
Maintenance Expense	9,000	0.60	7,500	0.88
Payroll Tax Expense	6,200	0.41	3,200	0.38
Rent Expense	20,000	1.34	20,000	2.36
Salaries Expense	58,000	3.87	48,000	5.66
Supplies Expense	3,000	0.20	2,200	0.26
Utilities Expense	2,400	0.16	1,850	0.22
Total Operating Expenses	$ 163,900	10.94 %	$ 121,050	14.28 %
Operating Income	$ 344,100	22.97 %	$ 226,750	26.75 %
Fed. Corp. Inc. Tax Exp.	44,940	3.00	25,434	3.00
Net Income	$ 299,160	19.97 %	$ 201,316	23.75 %

continued

Math for Accounting

Look at the comparative income statement provided here and answer the following questions using the vertical analysis of this financial statement.

1. What is the trend in sales?
2. How is that trend affecting the net income?
3. What expenses have decreased in the past year (as a percent of sales)?
4. Are there any significant changes in inventory?
5. What is the largest expense?
6. Would you invest your money in this business? Why or why not?

Comparative Income Statement
For the Current and Previous Years Ended December 31

	Current Year		Previous Year	
	Dollars	Percent	Dollars	Percent
Revenue:				
Sales	$ 1,500,000	100.13 %	$ 850,000	100.26 %
Less: Sales Ret. and Allow.	2,000	0.13	2,200	0.26
Net Sales	$ 1,498,000	100.00 %	$ 847,800	100.00 %
Cost of Merchandise Sold:				
Merch. Inventory, Jan. 1	$ 250,000	16.69 %	$ 100,000	11.80 %
Net Purchases	800,000	53.40	650,000	76.67
Merch. Available for Sale	$ 1,050,000	70.09 %	$ 750,000	88.46 %
Merch. Inventory, Dec. 31	60,000	4.01	250,000	29.49
Cost of Merchandise Sold	$ 990,000	66.09 %	$ 500,000	58.98 %
Gross Profit on Sales	$ 508,000	33.91 %	$ 347,800	41.02 %
Operating Expenses:				
Advertising Expense	$ 50,000	3.34 %	$ 25,000	2.95 %
Bankcard Fees Expense	10,000	0.67	8,000	0.94
Insurance Expense	5,300	0.35	5,300	0.63
Maintenance Expense	9,000	0.60	7,500	0.88
Payroll Tax Expense	6,200	0.41	3,200	0.38
Rent Expense	20,000	1.34	20,000	2.36
Salaries Expense	58,000	3.87	48,000	5.66
Supplies Expense	3,000	0.20	2,200	0.26
Utilities Expense	2,400	0.16	1,850	0.22
Total Operating Expenses	$ 163,900	10.94 %	$ 121,050	14.28 %
Operating Income	$ 344,100	22.97 %	$ 226,750	26.75 %
Fed. Corp. Inc. Tax Exp.	44,940	3.00	25,434	3.00
Net Income	$ 299,160	19.97 %	$ 201,316	23.75 %

SECTION 19.3
The Statement of Retained Earnings, Balance Sheet, and Statement of Cash Flows

In the previous section, you learned how to prepare and analyze the income statement, which reports the net income or loss for the period. In this section you will learn about a corporation's statement of retained earnings, balance sheet, and statement of cash flows.

Section Vocabulary

- statement of retained earnings
- horizontal analysis
- base period
- cash inflows
- cash outflows
- operating activities
- investing activities
- financing activities
- consists
- adequate

The Statement of Retained Earnings

What Does This Statement Report?

A corporation has two stockholders' equity accounts, **Capital Stock** and **Retained Earnings.** The **Capital Stock** account represents the stockholders' investment in the corporation. Its balance changes only when the corporation issues additional shares of stock. The **Retained Earnings** account summarizes the accumulated profits of a corporation minus any amounts paid to stockholders as returns on their investments.

In Chapter 9 you learned about the statement of changes in owner's equity for a sole proprietorship. It shows the changes in the owner's capital account during the period. A corporation prepares a similar statement, the **statement of retained earnings**, that reports the changes in the **Retained Earnings** account during the period. These changes result from business operations and *dividends,* which are the distributions of earnings to stockholders.

The changes to **Retained Earnings** are summarized as follows:

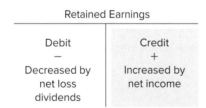

The statement of retained earnings is prepared from information found on the work sheet. The statement of retained earnings is a supporting document for the balance sheet. The final balance of the **Retained Earnings** account, as calculated on the statement of retained earnings, is used when preparing the balance sheet.

Figure 19–5 shows the statement of retained earnings for The Starting Line. The first line shows the balance of the **Retained Earnings** account at the beginning of the period. This balance comes from the Balance Sheet section of the work sheet. The second line is the net income for the period. This is from the Income Statement columns of the work sheet. Add net income to the beginning balance of the **Retained Earnings** account.

Since The Starting Line did not distribute any of its net income to stockholders during the period, there are no deductions from **Retained Earnings**. The new balance of the **Retained Earnings** account is $52,445.91.

Figure 19–5 Statement of Retained Earnings

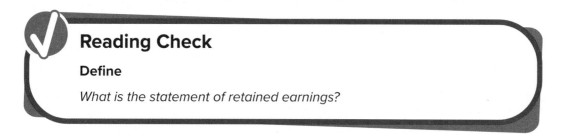

Reading Check

Define

What is the statement of retained earnings?

The Balance Sheet

How Is the Balance Sheet for a Corporation Different from the Balance Sheet for a Sole Proprietorship?

The balance sheet reports the balances of all asset, liability, and stockholders' equity accounts for a specific date. The balance sheet is prepared from the information in the Balance Sheet section of the work sheet and from the statement of retained earnings.

Figure 19–6 on page 624 shows The Starting Line's balance sheet. This balance sheet is prepared in report form. In the report form, classifications (assets, liabilities, and stockholders' equity) are shown one under the other.

The assets are listed first. The classification *Assets* is centered on the first line. The account names are listed at the left edge in the same order as they appear on the work sheet. The individual balances are entered in the first amount column. *Total Assets* is entered on the line below the last account name, indented about half an inch. The total assets amount is entered in the second amount column. The double rule, however, is not drawn until the Liabilities and Stockholders' Equity sections are complete, and the total of these two sections equals the total of the Assets section.

The Liabilities section begins with *Liabilities* centered on the second line below total assets. As in the Assets section, the account names are listed at the left edge in the same order as they appear on the work sheet. The individual balances are entered in the first amount column. *Total Liabilities* follows on the line below the last account name, indented about half an inch. The total liabilities amount is then entered in the second amount column.

Next, the Stockholders' Equity section begins on the second line below total liabilities. *Stockholders' Equity,* which is centered on this line, **consists** of two

Chapter 19 · Financial Statements for a Corporation   **623** accounts, **Capital Stock** and **Retained Earnings.** The **Capital Stock** account balance is from the work sheet's Balance Sheet section. The **Retained Earnings** account balance is from the statement of retained earnings. Again, the account names are listed at the left edge, and their individual balances are listed in the first amount column. *Total Stockholders' Equity* follows on the line below the last account name, indented about half an inch. The stockholders' equity total is entered in the second amount column.

On the line following total stockholders' equity, *Total Liabilities and Stockholders' Equity* is entered at the left edge. The total of the Liabilities section and the total of the Stockholders' Equity section are added. The total is entered in the second amount column. This total must agree with the total assets amount. If it does, double rule the balance sheet. If it does not, check the addition on the balance sheet. Also check that the accounts and amounts have been transferred from the work sheet accurately.

The Starting Line Sports Gear
Balance Sheet
December 31, 20--

Assets		
Cash in Bank	15 179 00	
Accounts Receivable	10 404 00	
Merchandise Inventory	81 385 00	
Supplies	1 839 00	
Prepaid Insurance	1 375 00	
Delivery Equipment	19 831 00	
Office Equipment	9 825 00	
Store Equipment	5 200 00	
Total Assets		145 038 00
Liabilities		
Accounts Payable	13 850 00	
Federal Corporate Income Tax Payable	155 00	
Employees' Federal Income Tax Payable	640 00	
Employees' State Income Tax Payable	80 00	
Social Security Tax Payable	248 00	
Medicare Tax Payable	58 00	
Federal Unemployment Tax Payable	18 36	
State Unemployment Tax Payable	114 73	
Sales Tax Payable	2 428 00	
Total Liabilities		17 592 09
Stockholders' Equity		
Capital Stock	75 000 00	
Retained Earnings	52 445 91	
Total Stockholders' Equity		127 445 91
Total Liabilities and Stockholders' Equity		145 038 00

Figure 19–6 The Starting Line's Balance Sheet

Analyzing Amounts on the Balance Sheet

How Can You Detect Trends?

When analyzing financial statements, you learned that while the dollar amounts provided on the statements are useful, the analysis can be expanded and made more meaningful by expressing the dollar amounts as percentages.

Percentage amounts are used in vertical analysis and also in horizontal analysis. **Horizontal analysis** is the comparison of the same items on financial statements for two or more accounting periods or dates, and the determination of changes from one period or date to the next. In horizontal analysis, each amount on the current statement is compared with its corresponding amount on the previous statement. A **base period** is a period, usually a year, that is used for comparison.

Look at the example of a comparative balance sheet in **Figure 19–7.** By comparing the amounts for the two years, you can see that **Cash in Bank** increased by 49.76 percent and **Accounts Payable** decreased by 50.73 percent. The accountant might use this information to assess why cash has increased or why accounts payable have decreased. Perhaps the business is receiving more sales in cash instead of on account. The business may also be purchasing less inventory on account. These are trends that would be of interest to management.

The Starting Line Sports Gear
Comparative Balance Sheet
December 31, Current Year and Previous Year

	Current Year	Previous Year	Increase (Decrease) Current over Previous Dollars	Percent
Assets				
Cash in Bank	$ 15,179.00	$ 10,135.28	$ 5,043.72	49.76 %
Accounts Receivable	10,404.00	8,220.00	2,184.00	26.57
Merchandise Inventory	81,385.00	84,921.00	(3,536.00)	(4.16)
Supplies	1,839.00	1,587.00	252.00	15.88
Prepaid Insurance	1,375.00	0.00	1,375.00	—
Delivery Equipment	19,831.00	12,462.00	7,369.00	59.13
Office Equipment	9,825.00	5,854.00	3,971.00	67.83
Store Equipment	5,200.00	3,500.00	1,700.00	48.57
Total Assets	$ 145,038.00	$ 126,679.28	$ 18,358.72	14.49 %
Liabilities				
Accounts Payable	$ 13,850.00	$ 28,113.14	$ (14,263.14)	(50.73) %
Fed. Corp. Inc. Tax Payable	155.00	140.00	15.00	10.71
Employees' Fed. Inc. Tax Pay.	640.00	685.00	(45.00)	6.57
Employees' State Inc. Tax Pay.	80.00	72.00	8.00	11.11
Social Security Tax Payable	248.00	241.00	7.00	2.90
Medicare Tax Payable	58.00	56.35	1.65	2.93
Federal Unemployment Tax Pay.	18.36	16.50	1.86	11.27
State Unemployment Tax Pay.	114.73	103.10	11.63	11.14
Sales Tax Payable	2,428.00	2,481.00	(53.00)	2.14
Total Liabilities	$ 17,592.09	$ 31,908.09	$ (14,316.00)	(44.87) %
Stockholders' Equity				
Capital Stock	$ 75,000.00	$ 75,000.00	$ 0.00	0.00 %
Retained Earnings	52,445.91	19,771.19	32,674.72	165.26
Total Stockholders' Equity	$ 127,445.91	$ 94,771.19	$ 32,674.72	34.48 %
Total Liab. and Stockhldrs' Equity	$ 145,038.00	$ 126,679.28	$ 18,358.72	14.49 %

Figure 19–7 Comparative Balance Sheet Showing Horizontal Analysis

The Statement of Cash Flows

What Does the Statement of Cash Flows Report?

As you learned in Chapter 9, the *statement of cash flows* reports how the activities of a business caused the cash balance to change during the accounting period. This information is vital for sound decision making. See **Figure 19–8** for The Starting Line's statement of cash flows.

Most businesses consider cash to be a major asset, and they need a sufficient amount to operate efficiently. Maintaining a positive cash flow is a primary goal of financial management. An **adequate** amount of available cash allows a business to pay its debts in a timely manner, take advantage of discounts, purchase equipment, and fund expansion. Creditors and investors use the statement of cash flows to evaluate a company's ability to pay its debts and pay dividends.

Cash inflows (receipts of cash) come into and **cash outflows** (payments of cash) go out of a business from different activities. The statement of cash flows classifies these activities as *operating, investing,* or *financing.*

Cash Flows from Operating Activities

Operating activities include all transactions that occurred during the accounting period as part of normal business operations. The information needed to complete this section is taken from The Starting Line's income statement (**Figure 19–3**) and comparative balance sheet (**Figure 19–7**).

Recall that revenue is recorded when it is earned and expenses are recorded when they are incurred, regardless of when items are actually paid. This is called the *accrual basis of accounting.*

To determine operating cash inflows and outflows for the accounting period, the accountant must convert income statement and balance sheet amounts to the *cash basis of accounting,* which records revenues *only when cash is received* and expenses *only when cash is paid out.* This is the reason that The Starting Line's net sales reported on the income statement is $317,720.00, but sales to customers reported on the statement of cash flows is $315,536.00.

Cash Flows from Investing Activities

Investing activities include loans the business makes, payments received for those loans, purchase and sale of plant assets, and investments. *Plant assets* are property that will be used in the business for more than one year.

During the current accounting period, The Starting Line purchased delivery equipment, office equipment, and store equipment for cash. When a plant asset is purchased for cash, the appropriate general ledger asset account is debited and **Cash in Bank** is credited. Since the purchase of plant assets is a cash outflow, the $13,040.00 is placed in parentheses on the statement of cash flows.

Figure 19–8 Statement of Cash Flows

The Starting Line Sports Gear
Statement of Cash Flows
For the Year Ended December 31, 20--

Cash Flows from Operating Activities		
Cash Receipts from:		
Sales to Customers	$ 315,536.00	
Total Cash Receipts from Operating Activities		$ 315,536.00
Cash Payments for:		
Purchases	(221,859.32)	
Accrued Payables	(67.86)	
Operating Expenses	(65,545.10)	
Federal Corporate Income Tax Expense	(9,980.00)	
Total Cash Payments from Operating Activities		(297,452.28)
Net Cash Flows from Operating Activities		$ 18,083.72
Cash Flows from Investing Activities		
Purchase of Plant Assets	(13,040.00)	
Net Cash Flows from Investing Activities		(13,040.00)
Cash Flows from Financing Activities		-0-
Net Increase in Cash		$ 5,043.72

Cash Flows from Financing Activities

Financing activities are the borrowing activities needed to finance the company operations and the repayment of these debts. The Starting Line had no financing activities during this accounting period.

The statement of cash flows indicates that cash increased by $5,043.72 during the period. This amount agrees with the increase in cash shown on the comparative balance sheet.

Typical cash inflows and outflows for the three activities of a business are shown here.

Activity	Cash Inflows (receipts)	Cash Outflows (payments)
Operating Activities	• Sales • Interest Income	• Merchandise Purchases • Operating Expenses • Interest Expense • Fed. Corp. Income Tax
Investing Activities	• Selling plant assets • Selling investments	• Purchasing plant assets • Purchasing investments
Financing Activities	• Long-term borrowing • Issuing stock	• Repaying loans (principal, not interest) • Paying dividends on stock

SECTION 19.3
Assessment

After You Read

Reinforce the Main Idea

Use a chart like this one to describe how stockholders and creditors use each of the listed financial statements.

	Stockholders	Creditors
Statement of Retained Earnings		
Balance Sheet		
Statement of Cash Flows		

Problem 4 Analyzing a Balance Sheet

Use the comparative balance sheet for The Starting Line in **Figure 19–7** to answer the following questions.

1. Which asset account has the larger percentage of increase in the two years? Larger percentage of decrease?
2. Which liability account has the larger percentage increase in the two years?
3. Did the overall value (total assets) of the corporation increase or decrease in the two years? What is the dollar amount? What is the percentage?
4. What is the percentage increase in retained earnings in the two years?
5. What conclusions might you draw based on the change in Accounts Receivable? Accounts Payable?
6. Can you provide a possible explanation for the difference in the balance of the **Prepaid Insurance** account?

Math for Accounting

Larry Campbell is the owner of Craftsman Furniture, a successful family-owned furniture store. As the accountant for Craftsman Furniture, it is your responsibility to run a vertical analysis on the income statement. Mr. Campbell asks you to prepare an Executive Summary of a three-year vertical analysis. Using the information provided, calculate the percentages for each year in your working papers and complete the columns of the draft report. Mr. Campbell will review the draft with you so he clearly understands the trends in sales for Craftsman Furniture.

	Year 1	Year 2	Year 3
Net Sales	$1,000,000	$1,500,000	$1,600,000
Gross Profit on Sales	600,000	750,000	780,000
Total Operating Expenses	25,000	21,000	23,000
Net Income	$ 575,000	$ 729,000	$ 757,000

628 Chapter 19 • Financial Statements for a Corporation

Chapter 19
Visual Summary

Concepts

Individuals become owners of a corporation by buying shares of stock. The ownership is known as *stockholder's equity*. Here is the entry to record ownership of a corporation:

	GENERAL JOURNAL			PAGE 25	
DATE	DESCRIPTION	POST. REF.	DEBIT	CREDIT	
1 20--					1
2 Date	Cash in Bank		X X X X X		2
3	Capital Stock			X X X X X	3
4					4

Analysis

The primary difference between financial statements for a sole proprietorship and a corporation involves how ownership is presented on a balance sheet.

	Sole Proprietor	Corporation
Balance Sheet Section	Owner's Equity	Stockholder's Equity
Account(s)	Owner's Capital—the owner's investment of cash and assets in the business	• Capital Stock—the investments by stockholders (owners) • Retained Earnings—earnings the corporation generated in past periods and retained instead of distributing to stockholders as dividends.

Procedures

Analyze the financial data contained on the statements.

Vertical Analysis helps the accountant determine the relationships among items on a financial statement and the changes in these relationships from one period to another.

Horizontal Analysis compares amounts for the same item on a financial statement for two or more accounting periods.

Chapter 19
Review and Activities

Answering the Essential Question

What are some of the unique accounting requirements of a merchandising corporation?

A corporation is responsible to its shareholders. How does a corporation that buys and sells goods provide accurate reports to its stockholders?

Vocabulary Check

1. **Vocabulary** Write a brief essay titled "Financial Statements for a Corporation" using as many content vocabulary terms as possible. The essay should clearly explain how the terms relate to the topic. Underline each vocabulary term used.

- capital stock
- contribute
- stockholders' equity
- retain
- retained earnings
- comparability
- potential
- full disclosure
- materiality
- relevance
- reliability
- expand
- net sales
- net purchases
- administrative expenses
- gross profit on sales
- operating expenses
- operating income
- selling expenses
- assign
- interpret
- vertical analysis
- statement of retained earnings
- consist
- base period
- horizontal analysis
- adequate
- cash inflows
- cash outflows
- investing activities
- operating activities
- financing activities

Concept Check

2. **Analyze** Explain how stockholders' equity and owner's equity are similar.
3. What four financial statements are prepared by a corporation?
4. Explain how an income statement for a merchandising business differs from an income statement for a service business.
5. What type of expense appears on the income statement of a corporation but not on that of a sole proprietorship or partnership?
6. **Analyze** Compare and contrast the cost of merchandise available for sale and the cost of merchandise sold.
7. **Evaluate** Explain the value and benefits of horizontal analysis.
8. **Math** Tran, an entrepreneur, wants his computer business to earn $42,000 in profits. His research shows that the average net profit for his type of business is 15%. If Tran's business earns the average net profit percentage, how much revenue must be generated to deliver that net profit?

Chapter 19
Standardized Test Practice

Multiple Choice

1. One way to increase gross profit on sales is to
   a. increase sales revenue.
   b. increase cost of merchandise sold.
   c. decrease sales revenue.
   d. decrease expenses.

2. The _____ account represents the increase in stockholders' equity from net income that is held by the corporation and not distributed to stockholders as a return on their investment.
   a. operating income
   b. retained earnings
   c. working capital
   d. net sales

3. When using _____, each dollar amount on a financial statement is also stated as a percentage of a base amount on the same statement.
   a. horizontal analysis
   b. vertical analysis
   c. statement of retained earnings
   d. comparability

4. You are given the following information: cost of merchandise sold, $404,000; operating expenses, $785,122; and net sales, $557,225. What is the company's gross profit on sales?
   a. $381,122
   b. $227,897
   c. $267,135
   d. $153,225

5. Beginning merchandise inventory plus net purchases minus ending inventory equals
   a. net income.
   b. gross profit on sales.
   c. total expenses.
   d. cost of goods sold.

Short Answer

6. What three financial statements are prepared by a merchandising corporation?

Extended Response

7. Why is the income statement prepared before the statement of retained earnings? Why is the statement of retained earnings prepared before the balance sheet?

Chapter 19
Computerized Accounting

Preparing Financial Statements

Making the Transition from a Manual to a Computerized System

MANUAL METHODS	COMPUTERIZED METHODS
• After all business transactions have been journalized and posted, prepare the trial balance.	• After all business transactions have been journalized and posted, print a working trial balance.
• Calculate, journalize, and post the adjusting entries.	• Journalize the adjusting entries.
• Prepare the income statement, statement of retained earnings, and balance sheet.	• Print the income statement, statement of retained earnings, and balance sheet.

Chapter 19
Problems

Problem 5 **Preparing an Income Statement**

The work sheet for Sunset Surfwear for the year ended December 31 is shown in your working papers.

Instructions Prepare an income statement for Sunset Surfwear in your working papers. Refer to **Figure 19–3** for guidance in setting up the income statement.

Analyze

Identify which general ledger account, Purchases Discounts or Purchases Returns & Allowances, had a higher amount this period.

Problem 6 **Preparing a Statement of Retained Earnings and a Balance Sheet**

Instructions Use the work sheet and the income statement from **Problem 19–5** to prepare a statement of retained earnings and a balance sheet for Sunset Surfwear. Use the accounting stationery provided in your working papers.

Analyze

Calculate the ending balance of Retained Earnings assuming Sunset Surfwear had a net loss of $6,492 instead of a net income for the period.

Chapter 19
Problems

Problem 7 Preparing Financial Statements

Instructions The partially completed work sheet for Shutterbug Cameras is included in your working papers.

1. Complete the work sheet.
2. Prepare an income statement.
3. Prepare a statement of retained earnings.
4. Prepare a balance sheet.

Analyze

Identify the biggest asset account, the biggest liability account, and the highest expense account shown on the financial statements.

Chapter 19 Problems

Problem 8 — Completing a Work Sheet and Financial Statements

The trial balance for Cycle Tech Bicycles, prepared on a ten-column work sheet, is included in your working papers.

Instructions

1. Complete the work sheet for the year ended December 31. Use the following information to make the adjustments.

   Ending merchandise inventory $25,191

   Ending supplies inventory 1,221

   Expired insurance 825

   Total federal corporate income taxes for the year 3,472

2. Prepare an income statement.
3. Prepare a statement of retained earnings.
4. Prepare a balance sheet.

Cycle Tech Bicycles
Work Sheet
For the Year Ended December 31, 20--

ACCT. NO.	ACCOUNT NAME	TRIAL BALANCE DEBIT	TRIAL BALANCE CREDIT
101	Cash in Bank	21931 00	
115	Accounts Receivable	1782 00	
125	Merchandise Inventory	24028 00	
130	Supplies	4159 00	
135	Prepaid Insurance	1800 00	
140	Store Equipment	24895 00	
145	Office Equipment	16113 00	
201	Accounts Payable		11224 00
210	Fed. Corp. Income Tax Payable		
211	Emplys' Fed. Inc. Tax Payable		
212	Emplys' State Inc. Tax Payable		522 00
213	Social Security Tax Payable		144 00
214	Medicare Tax Payable		413 00
215	Sales Tax Payable		134 00
216	Fed. Unemployment Tax Payable		1915 00
217	State Unemployment Tax Payable		54 00
301			271 00
305	Capital Stock		40000 00
310	Retained Earnings		11091 00
401	Income Summary		
405	Sales		127151 00
410	Sales Discounts	246 00	
501	Sales Returns and Allowances	1328 00	
505	Purchases	66107 00	
510	Transportation In	983 00	
515	Purchases Discounts		822 00
601	Purchases Ret. and Allowances		376 00
605	Advertising Expense	23800 00	
625	Bankcard Fees Expense	181 00	
630	Fed. Corp. Income Tax Expense	3340 00	
645	Insurance Expense		
650	Maintenance Expense	1950 00	
655	Miscellaneous Expense	1831 00	
657	Payroll Tax Expense	834 00	
660	Rent Expense	10800 00	
665	Salaries Expense	4734 00	
675	Supplies Expense		
	Utilities Expense	4695 00	
		194117 00	194117 00
	Net Income		

Analyze

Predict whether net income would be higher or lower if the ending value of Merchandise Inventory was actually $28,000.

Chapter 19
Problems

Problem 9 — Evaluating the Effect of an Error on the Income Statement

The accounting clerk for River's Edge Canoe & Kayak prepared the income statement for the year ended December 31. The accounting supervisor at River's Edge noticed that the balance of the **Transportation In** account was erroneously omitted from this statement. **Transportation In** has a balance of $562.

Instructions Use the income statement shown in your working papers to answer the following questions.

1. In which section of the income statement is the account **Transportation In** entered?
2. How is net purchases affected by this omission (understated or overstated)? By what amount?
3. How does the omission of the **Transportation In** balance affect gross profit on sales? By what amount?
4. What is the correct amount for the cost of merchandise sold for the period?
5. What is the correct amount for net income?

Analyze
Determine the effect that an overstatement of expenses would have on net income.

Real-World Applications & Connections

Career Wise

Billing Coordinator

Billing coordinators are responsible for billing and customer service issues with their clients. These professionals deal directly with their clients, so interpersonal skills are important in this career.

Research and Share

a. Use the Internet to research the detailed responsibilities and day-to-day activities of billing coordinators.

b. Make a list of the job requirements of this career. What would be the most important professional qualities to have to succeed as a billing coordinator?

Global Accounting

Accounting for Inflation

As economies go through ups and downs, one way to measure economic activity is by the rate of *inflation*. Inflation is the general increase in the costs of goods and services. If your money does not go as far as it used to, the cause may be inflation. Countries use different methods of accounting for inflation. *Deflation*, on the other hand, occurs when the supply of goods is greater than the demand, and prices decrease over a period of time.

Instructions

Explain why investors should know how inflation accounting is handled when reviewing financial statements.

A Matter of Ethics

Reporting a Mistake

You are an accounting clerk for a major league sports franchise. Your responsibilities do not include any end-of-period activities. You learn that a co-worker made a significant error on the work sheet. It will affect the financial statements. You wonder whether to report the error or assume that someone else will catch it.

Activity

a. What are the ethical issues?
b. What are the alternatives?
c. Who are the affected parties?
d. How do the alternatives affect the parties?
e. What would you do?

Analyzing Financial Reports

Statement of Cash Flows

It shows cash flows from operating activities, investing activities, and financing activities. It also shows the net cash provided (or used) by each activity. A company's *sources* of cash can be revealing. For long-term success, a company's cash inflows must be from operating activities. If its main source of cash was investing activities, it may have sold some of its plant assets. If its main source of cash was financing activities, it may have issued stock or borrowed money. A firm that must sell its assets or borrow money to stay in business cannot do so indefinitely.

Instructions

Use the Internet to locate and save or print a copy of Amazon.com's annual report to complete the following. Look at the statements of cash flow to answer these questions.

1. For each of the years shown, which activity *provided* the largest source of cash: operating, investing, or financing?

2. For each of the years shown, which activity *used* the largest amount of cash?

H.O.T. Audit

Inventory

The following account balances have been entered in the Income Statement section of the work sheet.

Purchases	$142,734
Transportation In	2,663
Purchases Discount	3,915
Purchases Returns & Allowances	1,021

The beginning merchandise inventory is $37,172, and the ending merchandise inventory is $49,334.

Instructions

1. Determine the cost of delivered merchandise.
2. Determine the amount of net purchases for the period.
3. Determine the cost of merchandise sold.

Mini Practice Set 4

Main Task
Use special journals to complete the accounting tasks for Cordova Electronics.

Summary of Steps
- Journalize and post transactions.
- Prove cash.
- Prepare a schedule of accounts receivable and a schedule of accounts payable.
- Prepare a trial balance.
- Create a clear and coherent oral presentation that analyzes the results of the accounting cycle.

Why It's Important
Special journals are used more than the general journal for routine transactions.

Recording Business Transactions in Special Journals

Cordova Electronics

Company Background: Pedro and Justina Cordova own and operate Cordova Electronics, a merchandising business organized as a corporation. The retail business is a small electronics store providing a unique service to the local community. This husband and wife team has been operating successfully for 10 years.

Keeping the Accounting Records for Cordova Electronics: Since Cordova Electronics has a large volume of business transactions, the business uses special journals and a general journal. The chart of accounts for Cordova Electronics appears on the next page. The previous accounting clerk, Manny Canseco, has journalized and posted the business transactions for May 1 through May 15. Those transactions are included in the accounting stationery in the working papers accompanying this textbook. The transactions that follow took place between May 16 and May 31.

Your Job Responsibilities: The forms for completing this activity are included in the working papers accompanying this textbook.

1. Record the remaining May transactions in the sales journal (page 18), cash receipts journal (page 15), purchases journal (page 12), cash payments journal (page 14), and general journal (page 7).

2. Post the individual amounts from the five journals to the accounts receivable and accounts payable subsidiary ledgers on a daily basis.

3. Post the individual amounts from the General columns of the cash receipts, purchases, cash payments, and general journals on a daily basis.

4. Foot, prove, total, and rule the special journals.

5. Post the column totals of the special journals to the general ledger accounts. Use the following order for posting: sales journal, cash receipts journal, purchases journal, and cash payments journal.

6. Prove cash. The balance shown on check stub 899 is $6,109.45.

continued

7. Prepare a schedule of accounts receivable and a schedule of accounts payable.
8. Prepare a trial balance.

CHART OF ACCOUNTS
Cordova Electronics

ASSETS
101 Cash in Bank
105 Accounts Receivable
110 Merchandise Inventory
115 Supplies
120 Prepaid Insurance
150 Store Equipment
155 Office Equipment

LIABILITIES
201 Accounts Payable
205 Sales Tax Payable
210 Employees' Federal Income Tax Payable
211 Employees' State Income Tax Payable
212 Social Security Tax Payable
213 Medicare Tax Payable
214 Federal Unemployment Tax Payable
215 State Unemployment Tax Payable

STOCKHOLDERS' EQUITY
301 Capital Stock
302 Retained Earnings
303 Income Summary

REVENUE
401 Sales
405 Sales Discounts
410 Sales Returns and Allowances

COST OF MERCHANDISE
501 Purchases
505 Transportation In
510 Purchases Discounts
515 Purchases Returns and Allowances

EXPENSES
605 Advertising Expense
610 Bankcard Fees Expense
615 Miscellaneous Expense
620 Payroll Tax Expense
625 Rent Expense
630 Salaries Expense
635 Utilities Expense

Accounts Receivable Subsidiary Ledger		Accounts Payable Subsidiary Ledger	
LOR	Sam Lorenzo	COM	Computer Systems Inc.
MAR	Marianne Martino	DES	Desktop Wholesalers
MCC	Mark McCormick	HIT	Hi-Tech Electronics Outlet
SCO	Sue Ellen Scott	LAS	Laser & Ink-Jet Products
TRO	Tom Trout	OFF	Office Suppliers Inc.

Business Transactions: Cordova Electronics had the following transactions May 16 through May 31.

Date		Transactions
May	16	Sold $120.00 in merchandise plus 5% sales tax to Sam Lorenzo on account, Sales Slip 607.
	17	Received $126.00 from Tom Trout to apply on his account, Receipt 356.
	17	Issued Check 892 for $800.00 to Desktop Wholesalers in payment of our account balance.
	18	Issued Debit Memo 38 to Laser & Ink Jet Products for our return of $75.00 in merchandise.
	19	Wrote Check 893 for $1,200.00 to Computer Systems Inc. to apply on our account.
	19	Paid Hi-Tech Electronics Outlet $1,750.00 in full payment of our account balance by issuing Check 894.
	19	Issued Check 895 to Office Suppliers, Inc. for $770.00 in full payment of our account balance.
	20	Sold $90.00 in office equipment for cash to an employee, Bob Bell, Receipt 357.
	20	Purchased $1,200.00 in store equipment on account from Desktop Wholesalers, DW87, n/30.
	20	Received $210.00 from Mark McCormick to apply on his account, Receipt 358.
	20	Sue Ellen Scott sent us a check for $308.70 in payment of her $315.00 account less a 2% discount, Receipt 359.
	21	Purchased $1,500.00 in merchandise on account from Hi-Tech Electronics Outlet, Invoice HT99, terms 2/10, n/30.
	21	Marianne Martino sent us a $94.50 check to apply on her account, Receipt 360.
	21	Sold $400.00 in merchandise plus 5% sales tax to Mark McCormick on account, Sales Slip 608, terms 2/10, n/30.
	22	Issued Check 896 for $1,200.00 to Desktop Wholesalers, in payment of our account.
	23	Sold $500.00 in merchandise to Sue Ellen Scott on account plus 5% sales tax, Sales Slip 609, terms 2/10, n/30.
	24	Mark McCormick returned merchandise purchased on account, $100.00, plus 5% sales tax, Credit Memo 55.
	25	Paid the annual insurance premium of $1,600.00 by issuing Check 897 to Surfside Insurance Co.
	26	Discovered that **Transportation In** account should have been debited last month for $50.00 instead of the **Purchases** account. Recorded the correcting entry, Memorandum 26.

continued

Date	Transactions (cont.)
May 27	Bought $1,400.00 in merchandise on account from Computer Systems, Inc., CS75, terms 2/10, n/30.
28	Sold $200.00 in merchandise on account plus 5% sales tax to Marianne Martino, Sales Slip 610, terms 2/10, n/30.
29	Purchased $150.00 in supplies on account from Office Suppliers Inc., Invoice 9489.
30	Issued Check 898 for $1,500.00 in payment for the monthly rent.
31	Recorded cash sales of $1,200.00 plus 5% sales tax, Tape 22.
31	Recorded bankcard sales of $900.00 plus 5% sales tax, Tape 22.
31	Recorded bank service charge of $25.00 and bankcard fees of $100.00, May bank statement (record compound entry).
31	Wrote Check 899 to pay the payroll of $2,500.00 (gross earnings) for the pay period ended May 31. The following amounts were withheld: employees' federal income taxes, $400.00; employees' state income taxes, $50.00; FICA taxes: $155.00 for social security and $36.25 for Medicare.
31	Recorded the employer's payroll taxes for the May 31 payroll: FICA tax rate, 6.2% for social security and 1.45% for Medicare; federal unemployment tax rate, 0.8%; and state unemployment tax rate, 5.4%.

Analyze

1. Compute the amount by which the Cash in Bank account changed during the month.

2. Identify the two accounts that have the greatest impact on the trial balance total.

3. Calculate the total percentage of tax withholdings for the May 31 payroll.

Chapter 20

Completing the Accounting Cycle for a Merchandising Corporation

Chapter Topics

20-1 Journalizing Closing Entries

20-2 Posting Closing Entries

Visual Summary

Review and Activities

Standardized Test Practice

Computerized Accounting

Problems

Real-World Applications & Connections

Essential Question

As you read this chapter, keep this question in mind:

Why is it important to "clean the slate" before a new accounting period?

Main Idea

A corporation's net income or net loss is closed to Retained Earnings. After posting the closing entries, accountants prepare a post-closing trial balance.

Chapter Objectives

Concepts	Analysis	Procedures
C1 Journalize closing entries for a merchandising corporation.	**A1** Post closing entries to the general ledger accounts.	**P1** Prepare a post-closing trial balance. **P2** Describe the steps in the accounting cycle.

Real-World Business Connection

99 Cents Only Stores

Do you think it is possible to buy a new iPod Nano for 99 cents? It was for the first nine customers at a recent grand opening of a 99 Cents Only Store. Promoting the newest of its more than 275 retail outlets, 99 Cents Only Stores tempted customers with all kinds of bargains for 99 cents or less: name-brand and private-label foods, beverages, health and beauty products, toys, and household goods. How can selling products at such a low price make the company a profit? This retailer focuses on close-out inventory and distressed products purchased at discount prices.

Connect to the Business

At the end of a fiscal period, companies like 99 Cents Only Stores prepare closing entries to transfer all temporary account balances to a permanent account. The general ledger is then ready for a new accounting period.

Analyze

What might happen if temporary accounts were not closed before the next accounting period begins?

Focus on the Photo

Smart shoppers depend on the 99 Cents Only Stores for all kinds of discounted merchandise, including fresh produce and other food items. **What accounts might be involved for preparing closing entries for this merchandising business?**

SECTION 20.1
Journalizing Closing Entries

Get ready to learn how to **journalize** and post the closing entries for a merchandising corporation. You will specifically discover how the closing process differs for merchandising corporations.

Section Vocabulary
- journalize
- temporary
- retain

Steps for Closing the Ledger

How Are Closing Entries for a Corporation Different from Closing Entries for a Sole Proprietorship?

In Chapter 10 you made four entries to close the **temporary** general ledger accounts of a sole proprietorship:

1. Close the temporary accounts with credit balances to **Income Summary.**
2. Close the temporary accounts with debit balances to **Income Summary.**
3. Close the balance of **Income Summary** to capital.
4. Close the withdrawals account to capital.

Only the first three closing entries are made to close the temporary accounts for a merchandising business organized as a corporation. Since a corporation does not have a withdrawals account, the fourth closing entry is not needed.

The portion of The Starting Line's work sheet in **Figure 20–1** on pages 647 and 648 shows the account balances that are closed. Let's look closely at each closing entry.

1. Close the accounts with balances in the Credit column of the Income Statement section of the work sheet (revenue and contra cost of merchandise accounts) to **Income Summary.** After this closing entry has been journalized and posted, the **Sales, Purchases Discounts,** and **Purchases Returns and Allowances** accounts have zero balances.

Income Summary

Debit	Credit
Adj. 3,536.00	Clos. 323,590.00

Sales

Debit	Credit
−	+
Clos. 320,450.00	Bal. 320,450.00

Purchases Discounts

Debit	Credit
−	+
Clos. 1,340.00	Bal. 1,340.00

Purchases Returns and Allowances

Debit	Credit
−	+
Clos. 1,800.00	Bal. 1,800.00

GENERAL JOURNAL   PAGE 23

	DATE	DESCRIPTION	POST. REF.	DEBIT	CREDIT	
1	20--	Closing Entries				1
2	Dec. 31	Sales		320 450 00		2
3		Purchases Discounts		1 340 00		3
4		Purchases Returns and Allow.		1 800 00		4
5		Income Summary			323 590 00	5
6						6

2. Close the accounts with balances in the Debit column of the Income Statement section of the work sheet (contra revenue, cost of merchandise, and expense accounts) to **Income Summary.** After this closing entry has been journalized and posted, the contra revenue, cost of merchandise, and expense accounts have zero balances.

Income Summary now has a credit balance of $32,674.72.

$	323,590.00	closing credit
−	3,536.00	adjustment debit
−	287,379.28	closing debit
$	32,674.72	credit balance

3. Close **Income Summary** to **Retained Earnings.**

Income Summary		Retained Earnings	
Debit	Credit	Debit	Credit
		−	+
Adj. 3,536.00	Clos. 323,590.00		Bal. 19,771.19
Clos. 287,379.28			Clos. 32,674.72
Clos. 32,674.72			Bal. 52,445.91

After the entry to close **Income Summary** to **Retained Earnings** has been journalized and posted, **Income Summary** has a zero balance. The balance of Retained Earnings is increased to $52,445.91.

The Starting Line
Work
For the Year Ended

	ACCT. NO.	ACCOUNT NAME	TRIAL BALANCE DEBIT	TRIAL BALANCE CREDIT	ADJUSTMENTS DEBIT	ADJUSTMENTS CREDIT
20	310	Income Summary			(a) 3 536 00	
21	401	Sales		320 450 00		
22	405	Sales Discounts	730 00			
23	410	Sales Returns and Allowances	2 000 00			
24	501	Purchases	206 700 00			
25	505	Transportation In	4 036 18			
26	510	Purchases Discounts		1 340 00		
27	515	Purchases Returns and Allow.		1 800 00		
28	601	Advertising Expense	2 450 00			
29	605	Bankcard Fees Expense	4 199 27			
30	630	Fed. Corporate Income Tax Exp.	9 840 00		(d) 155 00	
31	635	Insurance Expense			(c) 125 00	
32	650	Maintenance Expense	3 519 25			
33	655	Miscellaneous Expense	348 28			
34	657	Payroll Tax Expense	3 826 83			
35	660	Rent Expense	14 000 00			
36	665	Salaries Expense	29 374 60			
37	670	Supplies Expense			(b) 3 710 00	
38	680	Utilities Expense	2 364 87			
39			435 798 28	435 798 28	7 526 00	7 526 00
40		Net Income				
41						

Figure 20–1 Closing Entries Needed for a Corporation (continued)

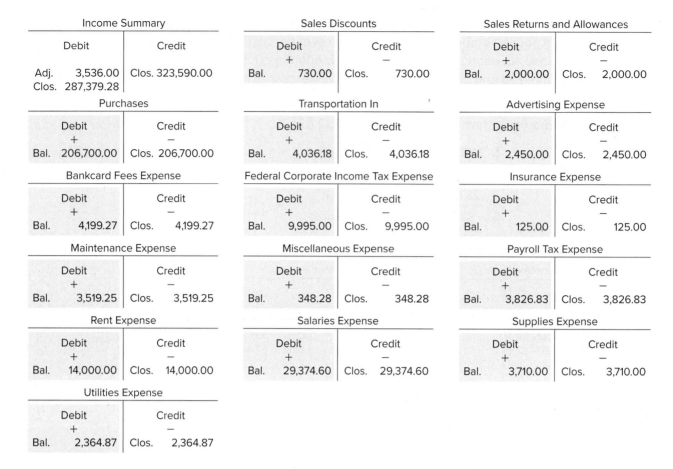

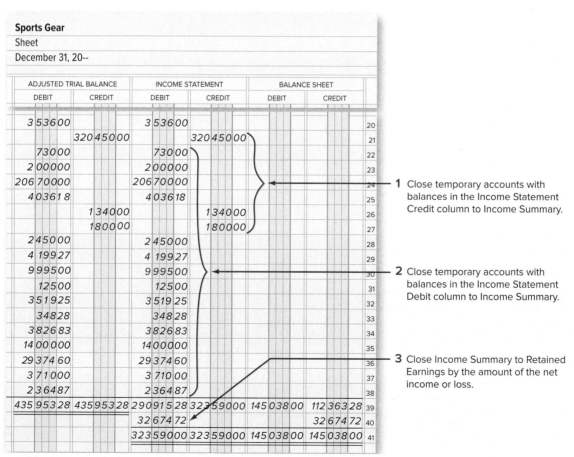

Figure 20–1 Closing Entries Needed for a Corporation

648 Chapter 20 • Completing the Accounting Cycle for a Merchandising Corporation

	GENERAL JOURNAL				PAGE 23	
DATE	DESCRIPTION	POST. REF.	DEBIT		CREDIT	
Dec. 31	Income Summary		287 379 28			6
	Sales Discounts				730 00	7
	Sales Returns and Allow.				2 000 00	8
	Purchases				206 700 00	9
	Transportation In				4 036 18	10
	Advertising Expense				2 450 00	11
	Bankcard Fees Expense				4 199 27	12
	Fed. Corp. Inc. Tax Expense				9 995 00	13
	Insurance Expense				125 00	14
	Maintenance Expense				3 519 25	15
	Miscellaneous Expense				348 28	16
	Payroll Tax Expense				3 826 83	17
	Rent Expense				14 000 00	18
	Salaries Expense				29 374 60	19
	Supplies Expense				3 710 00	20
	Utilities Expense				2 364 87	21
Dec. 31	Income Summary		32 674 72			22
	Retained Earnings				32 674 72	23

Reading Check

Explain

In your own words, explain the process of closing the ledger.

Closing Entry to Transfer a Net Loss

How Do You Close a Net Loss to Retained Earnings?

Sometimes businesses incur net losses. Suppose that after posting the first two entries closing the temporary accounts, the **Income Summary** account is as shown. Before it is closed, **Income Summary** has a debit balance of $5,000.00:

$ 3,612.00   adjustment debit
+ 65,178.00   closing debit
− 63,790.00   closing credit
$ 5,000.00   debit balance

Income Summary			
Debit		Credit	
Adj.	3,612.00	Clos.	63,790.00
Clos.	65,178.00		

To close **Income Summary**, credit it for $5,000 and debit **Retained Earnings** for $5,000. The net loss amount *decreases* the earnings the business **retains**. This closing entry is recorded in the general journal as follows:

	GENERAL JOURNAL			PAGE 23	
DATE	DESCRIPTION	POST. REF.	DEBIT	CREDIT	
	Closing Entries				1
Dec. 31	Retained Earnings		5 000 00		22
	Income Summary			5 000 00	23

SECTION 20.1
Assessment

After You Read

Reinforce the Main Idea

Using a diagram like this one, describe the step-by-step process for journalizing a merchandising corporation's closing entries. Add or remove answer boxes as needed.

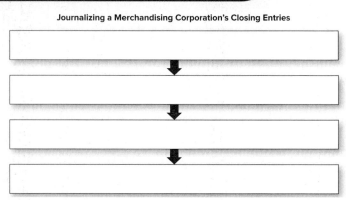

Journalizing a Merchandising Corporation's Closing Entries

Problem 1 Identifying Accounts Affected by Closing Entries

The following account names appear in the chart of accounts of Larkin's Department Store.

Accounts Receivable	Purchases
Bankcard Fees Expense	Purchases Discounts
Capital Stock	Purchases Returns and Allowances
Cash in Bank	Retained Earnings
Equipment	Sales
Fed. Corp. Income Tax Expense	Sales Discounts
Fed. Corp. Income Tax Payable	Sales Returns and Allowances
Income Summary	Sales Tax Payable
Insurance Expense	Supplies
Merchandise Inventory	Supplies Expense
Miscellaneous Expense	Transportation In
Prepaid Insurance	Utilities Expense

Instructions Use the form in your working papers to answer the following questions about each account. Assume that all accounts have normal balances.

1. Is the account affected by a closing entry?
2. During closing, is the account debited or credited?
3. During closing, is **Income Summary** debited or credited?

Math for Accounting

The closing debit entry to the **Income Summary** account was $263,000, and the closing credit entry was $300,000. If the **Income Summary** account had a $42,000 credit balance after these entries, was the inventory adjustment a debit or credit? What was the amount?

650 Chapter 20 • Completing the Accounting Cycle for a Merchandising Corporation

SECTION 20.2
Posting Closing Entries

Closing entries recorded in the general journal are posted to the general ledger.

Closing the General Ledger

How Do You Close a Merchandising Corporation's General Ledger?

Figure 20–2 shows the portion of The Starting Line's general ledger affected by the closing **process** after the closing entries have been posted. Write the term *Closing Entry* (abbreviated *Clos. Ent.* here) for each posting in the Description column of the general ledger account.

Section Vocabulary
- process
- adjusting
- consists
- analyze

ACCOUNT Retained Earnings   **ACCOUNT NO.** 305

DATE	DESCRIPTION	POST. REF.	DEBIT	CREDIT	BALANCE DEBIT	BALANCE CREDIT
20-- Dec. 1	Balance	✓				19 771 19
31	Clos. Ent.	G23		32 674 72		52 445 91

ACCOUNT Income Summary   **ACCOUNT NO.** 310

DATE	DESCRIPTION	POST. REF.	DEBIT	CREDIT	BALANCE DEBIT	BALANCE CREDIT
20-- Dec. 31	Adj. Ent	G22	3 536 00		3 536 00	
31	Clos. Ent.	G23		323 590 00		320 054 00
31	Clos. Ent.	G23	287 379 28			32 674 72
31	Clos. Ent.	G23	32 674 72			

ACCOUNT Sales   **ACCOUNT NO.** 401

DATE	DESCRIPTION	POST. REF.	DEBIT	CREDIT	BALANCE DEBIT	BALANCE CREDIT
20-- Dec. 1	Balance	✓				300 000 00
31		S12		10 750 00		310 750 00
31		CR13		9 700 00		320 450 00
31	Clos. Ent.	G23	320 450 00			

Figure 20–2 Partial General Ledger at the End of the Fiscal Period (continued)

ACCOUNT Sales Discounts   ACCOUNT NO. 405

DATE	DESCRIPTION	POST. REF.	DEBIT	CREDIT	BALANCE DEBIT	BALANCE CREDIT
20--						
Dec. 1	Balance	✓			700 00	
31		CR13	30 00		730 00	
31	Clos. Ent.	G23		730 00	—	

ACCOUNT Sales Returns and Allowances   ACCOUNT NO. 410

DATE	DESCRIPTION	POST. REF.	DEBIT	CREDIT	BALANCE DEBIT	BALANCE CREDIT
20--						
Dec. 1	Balance	✓			1 850 00	
4		G20	150 00		2 000 00	
31	Clos. Ent.	G23		2 000 00	—	

ACCOUNT Purchases   ACCOUNT NO. 501

DATE	DESCRIPTION	POST. REF.	DEBIT	CREDIT	BALANCE DEBIT	BALANCE CREDIT
20--						
Dec. 1	Balance	✓			190 000 00	
19		CP14	1 300 00		191 300 00	
31		P12	15 400 00		206 700 00	
31	Clos. Ent.	G23		206 700 00	—	

ACCOUNT Transportation In   ACCOUNT NO. 505

DATE	DESCRIPTION	POST. REF.	DEBIT	CREDIT	BALANCE DEBIT	BALANCE CREDIT
20--						
Dec. 1	Balance	✓			3 761 18	
24		CP14	275 00		4 036 18	
31	Clos. Ent.	G23		4 036 18	—	

ACCOUNT Purchases Discounts   ACCOUNT NO. 510

DATE	DESCRIPTION	POST. REF.	DEBIT	CREDIT	BALANCE DEBIT	BALANCE CREDIT
20--						
Dec. 1	Balance	✓				1 200 00
31		CP14		140 00		1 340 00
31	Clos. Ent.	G23	1 340 00			—

Figure 20–2 Partial General Ledger at the End of the Fiscal Period (continued)

ACCOUNT _Purchases Returns and Allowances_ ACCOUNT NO. _515_

DATE	DESCRIPTION	POST. REF.	DEBIT	CREDIT	BALANCE DEBIT	BALANCE CREDIT
20--						
Dec. 1	Balance	✓				1600 00
16		G21		200 00		1800 00
31	Clos. Ent.	G23	1800 00			—

ACCOUNT _Advertising Expense_ ACCOUNT NO. _601_

DATE	DESCRIPTION	POST. REF.	DEBIT	CREDIT	BALANCE DEBIT	BALANCE CREDIT
20--						
Dec. 1	Balance	✓			2450 00	
31	Clos. Ent.	G23		2450 00	—	

ACCOUNT _Bankcard Fees Expense_ ACCOUNT NO. _605_

DATE	DESCRIPTION	POST. REF.	DEBIT	CREDIT	BALANCE DEBIT	BALANCE CREDIT
20--						
Dec. 1	Balance	✓			4124 27	
31		CP14	75 00		4199 27	
31	Clos. Ent.	G23		4199 27	—	

ACCOUNT _Federal Corporate Income Tax Expense_ ACCOUNT NO. _630_

DATE	DESCRIPTION	POST. REF.	DEBIT	CREDIT	BALANCE DEBIT	BALANCE CREDIT
20--						
Dec. 1	Balance	✓			9840 00	
31	Adj. Ent.	G22	155 00		9995 00	
31	Clos. Ent.	G23		9995 00	—	

ACCOUNT _Insurance Expense_ ACCOUNT NO. _635_

DATE	DESCRIPTION	POST. REF.	DEBIT	CREDIT	BALANCE DEBIT	BALANCE CREDIT
20--						
Dec. 31	Adj. Ent.	G22	125 00		125 00	
31	Clos. Ent.	G23		125 00	—	

Figure 20–2 Partial General Ledger at the End of the Fiscal Period (continued)

ACCOUNT _Maintenance Expense_ ACCOUNT NO. _650_

DATE	DESCRIPTION	POST. REF.	DEBIT	CREDIT	BALANCE DEBIT	BALANCE CREDIT
20--						
Dec. 1	Balance	✓			3 519 25	
31	Clos. Ent.	G23		3 519 25		

ACCOUNT _Miscellaneous Expense_ ACCOUNT NO. _655_

DATE	DESCRIPTION	POST. REF.	DEBIT	CREDIT	BALANCE DEBIT	BALANCE CREDIT
20--						
Dec. 1	Balance	✓			328 28	
31		CP14	20 00		348 28	
31	Clos. Ent.	G23		348 28		

ACCOUNT _Payroll Tax Expense_ ACCOUNT NO. _657_

DATE	DESCRIPTION	POST. REF.	DEBIT	CREDIT	BALANCE DEBIT	BALANCE CREDIT
20--						
Dec. 1	Balance	✓			3 826 83	
31	Clos. Ent.	G23		3 826 83		

ACCOUNT _Rent Expense_ ACCOUNT NO. _660_

DATE	DESCRIPTION	POST. REF.	DEBIT	CREDIT	BALANCE DEBIT	BALANCE CREDIT
20--						
Dec. 1	Balance	✓			12 000 00	
31		CP14	2 000 00		14 000 00	
31	Clos. Ent.	G23		14 000 00		

ACCOUNT _Salaries Expense_ ACCOUNT NO. _665_

DATE	DESCRIPTION	POST. REF.	DEBIT	CREDIT	BALANCE DEBIT	BALANCE CREDIT
20--						
Dec. 1	Balance	✓			25 374 60	
31		CP14	4 000 00		29 374 60	
31	Clos. Ent.	G23		29 374 60		

Figure 20–2 Partial General Ledger at the End of the Fiscal Period (continued)

Figure 20-2 Partial General Ledger at the End of the Fiscal Period

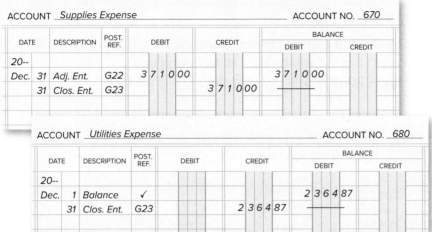

Reading Check

Recall

Where do you post closing entries that have been recorded in the general journal?

Preparing a Post-Closing Trial Balance

How Do You Prepare a Post-Closing Trial Balance?

A post-closing trial balance is prepared at the end of the accounting period to prove that the general ledger accounts are in balance after all **adjusting** and closing entries have been posted. **Figure 20-3** shows the post-closing trial balance for The Starting Line.

The Starting Line Sports Gear
Post-Closing Trial Balance
December 31, 20--

Account	Debit	Credit
Cash in Bank	15,179.00	
Accounts Receivable	10,404.00	
Merchandise Inventory	81,385.00	
Supplies	1,839.00	
Prepaid Insurance	1,375.00	
Delivery Equipment	19,831.00	
Office Equipment	9,825.00	
Store Equipment	5,200.00	
Accounts Payable		13,850.00
Fed. Corp. Income Tax Payable		155.00
Employees' Fed. Income Tax Payable		640.00
Employees' State Income Tax Payable		80.00
Social Security Tax Payable		248.00
Medicare Tax Payable		58.00
Fed. Unemployment Tax Payable		18.36
State Unemployment Tax Payable		114.73
Sales Tax Payable		2,428.00
Capital Stock		75,000.00
Retained Earnings		52,445.91
Totals	145,038.00	145,038.00

Figure 20-3 Post-Closing Trial Balance

Completing the Accounting Cycle for a Merchandising Business

What Is the Accounting Cycle for a Merchandising Business?

You have completed the study of the accounting cycle for a merchandising business organized as a corporation, which **consists** of the following steps:

1. Collect and verify source documents.
2. **Analyze** each business transaction.
3. Journalize each transaction.
4. Post to the general and subsidiary ledgers.
5. Prepare a trial balance.
6. Complete a work sheet.
7. Prepare the financial statements—income statement, statement of retained earnings, and balance sheet. Publicly held corporations also prepare a statement of cash flows.
8. Journalize and post the adjusting entries.
9. Journalize and post the closing entries.
10. Prepare a post-closing trial balance.

The accounting cycle for a service and a manufacturing business follows the same steps. Also, regardless of how a business is organized—sole proprietorship, partnership, or corporation—the basic steps of the accounting cycle are the same. **Figure 20–4** illustrates the accounting cycle.

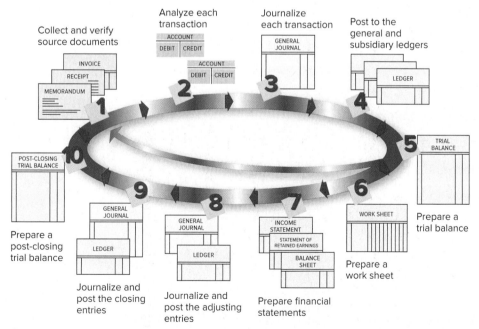

Figure 20–4 The Accounting Cycle

The accounting system used, whether manual or computerized, does not affect the steps in the accounting cycle. In a computerized accounting system, however, the computer performs many of the routine procedures such as posting.

SECTION 20.2 Assessment

After You Read

Reinforce the Main Idea

Using a diagram like this one, describe the step-by-step process for posting a merchandising corporation's closing entries. Add or remove answer boxes as needed.

Posting a Merchandising Corporation's Closing Entries

Problem 2 Analyzing a Source Document

Instructions Review the source document and prepare the journal entry to record this transaction in your working papers. Your Backpack Inc. uses a cash payments journal to record disbursements.

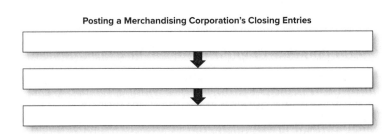

Problem 3 Organizing the Steps in the Accounting Cycle

Instructions List the following steps of the accounting cycle in their proper order. Use the form provided in your working papers or a separate sheet of paper.

Analyzing business transactions

Collecting and verifying source documents

Completing the work sheet

Journalizing business transactions

Journalizing and posting adjusting entries

Journalizing and posting closing entries

Posting journal entries to ledgers

Preparing financial statements

Preparing a post-closing trial balance

Preparing a trial balance

Math for Accounting

Totals on the post-closing trial balance were assets, $156,000, and liabilities, $75,000. If the ending balance of **Retained Earnings** was $45,000, what was the capital stock account's ending balance?

Chapter 20
Visual Summary

Concepts
Three steps to close the temporary accounts of a merchandising business:

1. Close all temporary accounts with credit balances to **Income Summary.**
2. Close all temporary accounts with debit balances to **Income Summary.**
3. Close the balance of **Income Summary** to the **Retained Earnings** account.

Analysis
The basic accounting cycle is the same whether

- the business is a service provider or merchandiser.
- the business is organized as a sole proprietorship, corporation, or partnership.
- the business uses a manual or computerized accounting system.

Procedures
A post-closing trial balance is prepared at the end of the accounting period.

The Starting Line Sports Gear
Post-Closing Trial Balance
December 31, 20--

Account	Debit	Credit
Cash in Bank	15 179 00	
Accounts Receivable	10 404 00	
Merchandise Inventory	81 385 00	
Supplies	1 839 00	
Prepaid Insurance	1 375 00	
Delivery Equipment	19 831 00	
Office Equipment	9 825 00	
Store Equipment	5 200 00	
Accounts Payable		13 850 00
Fed. Corp. Income Tax Payable		155 00
Employees' Fed. Income Tax Payable		640 00
Employees' State Income Tax Payable		80 00
Social Security Tax Payable		248 00
Medicare Tax Payable		58 00
Fed. Unemployment Tax Payable		18 36
State Unemployment Tax Payable		114 73
Sales Tax Payable		2 428 00
Capital Stock		75 000 00
Retained Earnings		52 445 91
Totals	145 038 00	145 038 00

Chapter 20
Review and Activities

Answering the Essential Question

Why is it important to "clean the slate" before a new accounting period?

Like sole proprietorships that you studied in Chapter 10, corporations need to "close the books" at the end of each accounting period. How do businesses get ready for a new accounting period?

Vocabulary Check

1. **Vocabulary** The following terms were introduced in Chapter 10 when we learned about closing the accounting cycle for a sole proprietorship business. How do they relate to a merchandising corporation? What differences exist between the closing entries for a sole proprietorship and the closing entries for a merchandising corporation? Use the following terms in your written explanation:

- permanent accounts
- temporary accounts
- closing entries
- journalize
- temporary
- retain
- process
- adjust
- analyze
- consist

Concept Check

2. **Analyze** Explain how the closing entries for corporations and sole proprietorships are similar and how they are different.
3. What account(s) are used to close temporary accounts with debit and credit balances?
4. How are temporary accounts with debit balances closed?
5. What is written in the Description column of the general ledger account as the closing entries are posted?
6. What is the purpose of preparing a post-closing trial balance?
7. **Analyze** Does the Income Summary account appear on the post-closing trial balance? Why or why not?
8. Illustrate the 10 steps in the accounting cycle.
9. **Math** Rudy's watch store just celebrated one year in business. Rudy wants to know the stock-turnover rate for six months to check his progress. Over 12 months, his sale totaled $156,000. His average inventory on hand totaled $68,000. What was the rate?
10. **English Language Arts** Research Supply and Demand Economics using the Internet or your school library, and explain how this system works to regulate the flow of goods and services within an economy. Create a poster presentation that illustrates the basic concepts of supply and demand within an economy. Be sure to cite the sources for your information.

Chapter 20
Standardized Test Practice

Multiple Choice

1. The procedure for transferring information from a journal to ledger accounts is called
    a. journalizing.
    b. adjusting.
    c. file maintenance.
    d. posting.
    e. none of these answers

2. Proving the accuracy of the adjusting and closing entries is best defined as
    a. preparing a post-closing trial balance.
    b. preparing the statements.
    c. closing the temporary accounts.
    d. adjusting the ledger accounts.
    e. none of these answers

3. To close the Income Summary account of a corporation with a net loss, the balance is closed into the
    a. Retained Earnings account with a debit.
    b. Retained Earnings account with a credit.
    c. account which caused the net loss with a debit.
    d. none of these answers

4. Which of the following accounts is closed at the end of a fiscal period?
    a. Membership Fees Income
    b. Accounts Receivable
    c. Delivery Equipment
    d. Retained Earnings

True or False

5. The Income Summary account appears on the post-closing trial balance.

6. Retained Earnings is credited when the business earns a profit.

Short Answer

7. Where can you find the information needed to prepare closing entries?

Extended Response

8. Explain the closing entries for a business organized as a corporation.

Chapter 20
Computerized Accounting

Closing a Fiscal Year and Preparing a Post-Closing Trial Balance

Making the Transition from a Manual to a Computerized System

MANUAL METHODS	COMPUTERIZED METHODS
• Journalize the entries to close the revenue, expense, Income Summary, and Withdrawals accounts.	• Closing the fiscal year is performed only if you want to clear income and expense accounts at the end of the fiscal year.
• Post the closing entries.	• If you have selected the closing option, the software journalizes closing entries.
• After the closing entries have been posted to the ledger, prepare a trial balance to verify the equality of debits and credits.	• A post-closing trial balance can be printed to verify that all revenue and expense accounts have been closed.
• All expense and revenue accounts should have zero balances.	

Chapter 20
Problems

Problem 4 Journalizing Closing Entries

The following amounts appeared in the Income Statement section of Sunset Surfwear's work sheet.

Instructions In your working papers, record the closing entries for the year ended December 31. Start with general journal page 13.

	ACCT. NO.	ACCOUNT NAME	INCOME STATEMENT DEBIT	INCOME STATEMENT CREDIT
13	310	Income Summary		7 000 00
14	401	Sales		90 000 00
15	405	Sales Discounts	1 000 00	
16	410	Sales Returns and Allowances	2 400 00	
17	501	Purchases	25 000 00	
18	505	Transportation In	3 000 00	
19	510	Purchases Discounts		500 00
20	515	Purchases Returns and Allowances		1 500 00
21	625	Federal Corporate Income Tax Exp.	5 532 00	
22	635	Insurance Expense	300 00	
23	650	Miscellaneous Expense	6 000 00	
24	655	Rent Expense	12 000 00	
25	665	Supplies Expense	450 00	
26	675	Utilities Expense	14 000 00	
27			69 682 00	99 000 00
28		Net Income	29 318 00	
29			99 000 00	99 000 00
30				

Analyze

Identify the effect of the closing entries on the Retained Earnings account.

Chapter 20 Problems

Problem 5 Journalizing and Posting Closing Entries

The following account balances appeared in the Income Statement section of the work sheet of Shutterbug Cameras.

Instructions

1. Journalize the closing entries for the year ended December 31. Start with page 14 of a general journal (in your working papers).
2. Post the closing entries to the general ledger accounts, which are included in your working papers.

	ACCT. NO.	ACCOUNT NAME	INCOME STATEMENT DEBIT	INCOME STATEMENT CREDIT
20	310	Income Summary	4 000 00	
21	401	Sales		150 000 00
22	410	Sales Returns and Allowances	5 000 00	
23	501	Purchases	90 000 00	
24	505	Transportation In	5 000 00	
25	510	Purchases Discounts		1 000 00
26	515	Purchases Returns and Allowances		1 500 00
27	620	Federal Corporate Income Tax Exp.	4 700 00	
28	645	Miscellaneous Expense	300 00	
29	650	Rent Expense	6 000 00	
30	660	Supplies Expense	1 630 00	
31	670	Utilities Expense	3 000 00	
32			119 630 00	152 500 00
33		Net Income	32 870 00	
34			152 500 00	152 500 00
35				

Analyze

Describe what would happen if the accountant for Shutterbug Cameras made a mistake and did not close the Transportation In account.

Chapter 20 Problems

Problem 6 Identifying Accounts for Closing Entries

A partial list of the accounts used by Cycle Tech Bicycles appears on page 607. All of the accounts have nonzero balances.

General Ledger
- 101 Cash in Bank
- 650 Miscellaneous Expense
- 125 Merchandise Inventory
- 665 Supplies Expense
- 215 Sales Tax Payable
- 505 Transportation In
- 305 Retained Earnings
- 401 Sales
- 405 Sales Discounts
- 310 Income Summary
- 501 Purchases
- 515 Purchases Returns and Allowances
- 601 Advertising Expense
- 625 Federal Corporate Income Tax Expense
- 657 Rent Expense
- 135 Prepaid Insurance

Instructions In your working papers, list all account numbers and names for accounts that will be debited when closed. Next, list all account numbers and names for accounts that will be credited when closed.

Analyze
Examine your list. Determine whether it contains all of the accounts. If it does not, explain why.

continued

Chapter 20 Problems

Problem 7 Completing End-of-Period Activities

The general ledger accounts for River's Edge Canoe & Kayak as of December 31, the end of the period, appear in the working papers.

Instructions In your working papers:

1. Prepare a trial balance on a ten-column work sheet.

2. Complete the work sheet. Use the following adjustment information.
   - Merchandise inventory, December 31 $20,000
   - Supplies inventory, December 31   900
   - Unexpired insurance, December 31   1,800
   - Total federal corporate income taxes for the year   2,965

3. Prepare an income statement from the work sheet information.

4. Prepare a statement of retained earnings.

5. Prepare a balance sheet.

6. Journalize and post the adjusting entries. Begin on general journal page 14.

7. Journalize and post the closing entries.

8. Prepare a post-closing trial balance.

Analyze
Conclude whether the company made a profit for the year.

Chapter 20 Problems

Problem 8 Preparing Adjusting and Closing Entries

In the middle of the end-of-period activities, the accountant for Buzz Newsstand was called away because of an illness in the family. Before leaving, the accountant prepared the work sheet and the financial statements. However, the business manager can locate only the trial balance shown here (before adjustments) and the income statement.

Instructions In your working papers:

1. Journalize the adjusting and closing entries on page 16 of the general journal.
2. Using the information provided, prepare a post-closing trial balance.

Buzz Newsstand
Trial Balance, Before Adjustments
December 31, 20--

Acct.	Account	Debit	Credit
101	Cash in Bank	12,035.00	
115	Accounts Receivable	6,106.00	
130	Merchandise Inventory	64,800.00	
135	Supplies	3,916.00	
140	Prepaid Insurance	5,400.00	
145	Delivery Truck	46,106.00	
201	Accounts Payable		4,690.00
204	Fed. Corp. Income Tax Pay.		—
215	Sales Tax Payable		416.00
301	Capital Stock		40,000.00
305	Retained Earnings		24,603.00
310	Income Summary		
401	Sales		299,156.00
410	Sales Returns and Allowances	9,500.00	
501	Purchases	168,624.00	
505	Transportation In	8,236.00	
510	Purchases Discounts		2,950.00
515	Purchases Returns and Allowances		2,108.00
601	Advertising Expense	4,000.00	
625	Fed. Corp. Income Tax Expense	12,500.00	
635	Insurance Expense	—	
650	Miscellaneous Expense	1,600.00	
660	Salaries Expense	26,900.00	
665	Supplies Expense	—	
675	Utilities Expense	4,200.00	
	Totals	373,923.00	373,923.00

Chapter 20
Problems

Buzz Newsstand
Income Statement
For the Year Ended December 31, 20--

Revenue:				
Sales			299 156 00	
Less: Sales Ret. and Allow.			9 500 00	
Net Sales				289 656 00
Cost of Merch. Sold:				
Merch. Inv., Jan 1, 20--			64 800 00	
Purchases	168 624 00			
Plus: Transportation In	8 236 00			
Cost of Del. Merch.		176 860 00		
Less: Purch. Discounts	2 950 00			
Purch. Ret. and Allow.	2 108 00	5 058 00	171 802 00	
Net Purchases				
Cost of Merch. Avail.			236 602 00	
Merch. Inv., Dec. 31, 20--			60 400 00	
Cost of Merch. Sold				176 202 00
Gross Profit on Sales				113 454 00
Operating Expenses:				
Advertising Expense			4 000 00	
Insurance Expense			1 800 00	
Miscellaneous Expense			1 600 00	
Salaries Expense			26 900 00	
Supplies Expense			2 744 00	
Utilities Expense			4 200 00	
Total Oper. Expenses				41 244 00
Operating Income				72 210 00
Less: Fed. Inc. Tax Exp.				14 913 00
Net Income				57 297 00

Analyze

Determine the impact of these adjusting entries on net income.

Real-World Applications & Connections

CASE STUDY

Merchandising Business: Department Store

You work in the accounting department for Pearl's, a trendy department store. The accounting manager asked you to review the work of an accounting intern who completed the closing process. Some temporary accounts in the post-closing trial balance still have balances.

INSTRUCTIONS

a. Explain how to tell which accounts should have zero balances after closing.

b. Explain how to determine which accounts have and have not been closed.

c. Suggest ways to explain to the intern why temporary accounts have zero balances after closing.

d. List human-relations skills you might use in your explanation to the intern.

21st century Skills

Direct Deposit

For many employees, the days of getting a paycheck and taking it to the bank are long gone. Many employers use direct deposit by which they place a worker's pay directly into his or her bank account. To set up direct deposit, the worker usually has to provide his or her bank account number and the bank's routing number to the employer. The worker may supply the employer with a voided check or deposit slip. With this, the payroll clerk is able to deposit money into the worker's account.

Discuss and Write

a. Discuss with a partner why you think businesses are willing to offer direct deposit to employees.

b. List three benefits direct deposit offers to workers. Would you like to have direct deposit? Why or why not?

Career Wise

General Ledger Accountant

The general ledger is the primary source of information from which companies generate their financial reports. The general ledger accountant is responsible for making sure that all information in the ledger, including all business transactions, is accurate and up to date, and properly accounted for. General ledger accountants may themselves be responsible for recording the day-to-day bookkeeping activities of an organization, or may work in a supervisory role, directing other accountants or clerks and reviewing their work. A general ledger accountant may also be referred to as a corporate accountant or a management accountant.

Research and Share

a. Use the Internet to locate three sources of information on general ledger accounting.

b. What level of education is required to be a general ledger accountant? What are the roles and responsibilities of a general ledger accountant?

H.O.T Audit

Closing Entries

Mr. Raheev Singh's accountant made the following closing entry:

GENERAL JOURNAL PAGE XX

	DATE	DESCRIPTION	POST. REF.	DEBIT	CREDIT	
1	20--					1
2	Date	Income Summary		17000 00		2
3		Purchases Discounts			8000 00	3
4		Purchases Ret. and Allow.			3200 00	4
5		Sales Discounts			2300 00	5
6		Sales Returns and Allow.			3500 00	6

Instructions

Review the closing entry. On a separate sheet of paper, prepare the closing entries that the accountant should have made.

Chapter 21

Accounting for Publicly Held Corporations

Chapter Topics

21-1 Publicly Held Corporations

21-2 Distribution of Corporate Earnings

21-3 Financial Reporting for a Publicly Held Corporation

Visual Summary

Review and Activities

Standardized Test Practice

Computerized Accounting Problems

Real-World Accounting Careers

Real-World Applications & Connections

Essential Question

As you read this chapter, keep this question in mind:

What is the relationship between a publicly held corporation and its stockholders?

Main Idea

Investors from the general public purchase stock of publicly held corporations. A corporation distributes a portion of its earnings to stockholders in the form of dividends. Corporate financial statements report stock issues and dividends.

Chapter Objectives

Concepts	Analysis	Procedures
C1 Describe the characteristics of the corporate form of business organization.	**A1** Prepare journal entries to record the issue of stock to investors. **A2** Prepare journal entries to record distribution of earnings to owners.	**P1** Prepare financial statements for publicly held corporations.

Real-World Business Connection

SanDisk

Did you ever wonder how your cell phone or digital camera stores all your photos and other data so conveniently? What would you do without your flash drive? These are the questions that the people at SanDisk, the inventor of digital flash memory, hope you never have to worry about. Since 1988, SanDisk has been making digital flash cards for devices such as cell phones and cameras, and USB flash drives for improved data storage.

Connect to the Business

SanDisk is a corporation that is traded on the New York Stock Exchange. When a company has stockholders, it has to keep profits high in order to maintain its interest among investors. This requires constant innovation and planning to achieve higher goals. It also requires a lot of planning on the part of accountants to report to stockholders and predict how the company will perform under current and future economic conditions.

Analyze

What factors determine whether or not a company will maintain a high value on its stock?

Focus on the Photo

SanDisk's exponential growth has been driven by its compact flash-drive technology that created a revolution in multi-media devices like cameras. ***What kinds of financial information might a publicly held corporation like SanDisk need to report?***

SECTION 21.1
Publicly Held Corporations

The United States has more sole proprietorships and partnerships than corporations. Corporations, however, account for more business activity than the other two forms of business combined. Many high profile companies like The Coca-Cola Company and General Motors are organized as corporations.

Characteristics of a Corporation

How Is a Corporation Different from Other Forms of Business Organization?

Get ready to shift your focus from merchandising corporations to publicly held corporations. A **closely held corporation** is a corporation owned by a few persons or by a family. The stock of a closely held corporation is *not* sold to the general public.

A **publicly held corporation** is one whose stock is widely held, has a large market, and is usually traded on a stock exchange such as the New York Stock Exchange. This chapter will teach you how to prepare journal entries that report stock entries and dividends for publicly held corporations.

In previous chapters The Starting Line could have been either a closely held or publicly held corporation. In this chapter The Starting Line will illustrate transactions for a publicly held corporation.

The corporation has several unique features:

- **Legal Permission to Operate**—To operate a business as a corporation, its incorporators (organizers) file an application with state officials for permission to operate. When the application has been approved, it becomes the corporation's *charter.* The charter indicates the purpose of the business and spells out the rules under which the business is to operate. The charter also states the type and amount of stock a corporation is authorized to issue.
- **Separate Legal Entity**—A corporation is a separate legal **entity** that is created and exists only by law. A corporation may enter into contracts, borrow money, and conduct business in the same manner as a person. It may acquire, own, and sell property in its name. It can also sue and be sued in the courts.
- **Stockholders**—The ownership of a corporation is divided into units called *shares of stock.* The owners of a corporation are called *stockholders.* Each stockholder receives a *stock certificate* as proof of ownership. The stock certificate lists the name of the stockholder, the number of shares issued, and the date the shares were issued.
- **Professional Management**—Stockholders own the corporation, but they do not manage it. The stockholders elect a **board of directors**, who govern and are responsible for the affairs of the corporation.

Section Vocabulary

- closely held corporation
- publicly held corporation
- board of directors
- authorized capital stock
- par value
- common stock
- proxy
- preferred stock
- Paid-in Capital in Excess of Par
- entity
- corporation

Capital Stock

How Do You Measure Ownership of a Corporation?

Stockholders' equity is the value of the stockholders' claims to the corporation's assets. As you learned in Chapter 19, corporations report stockholders' equity in two parts:

- the equity paid into the corporation by stockholders
- the equity earned by the corporation and retained in the business

The maximum number of shares a corporation may issue is called its **authorized capital stock**. The authorized number of shares is usually much higher than the number of shares the corporation plans to sell right away. This allows the corporation to sell additional shares at a later time.

State laws may require that an amount or value be assigned to each share of stock before the corporation sells it to the public. The amount assigned to each share is referred to as **par value**, the per-share dollar amount printed on the stock certificates. The par value is used to determine the amount credited to the capital stock account. Par values of $1, $5, and $25 are common. *Par value* is a fixed amount and is almost never the current value of a share of stock. A stock's current selling price on the stock market is called *market value*.

The corporate charter specifies the types of capital stock that a **corporation** may issue. The two main types of stock are *common* and *preferred*.

Common Stock

If the corporation issues only one class of capital stock, it is called **common stock**. The owners of common stock participate in the corporation as follows:

- Elect the board of directors and, through it, exercise control over the operations of the corporation. Stockholders are entitled to one vote for each share of stock they own. The election occurs at the stockholders' meeting, which is usually held once a year. If a stockholder cannot attend the meeting, he or she may send in a **proxy**, which gives the stockholder's voting rights to someone else.
- Share in the earnings of the corporation by receiving dividends declared by the board of directors.
- Are entitled to share in the assets of the corporation if it goes out of business.

> **Common Mistakes**
>
> **Par Value versus Market Value** The terms *par value* and *market value* can be easily confused. Par value is the legal capital per share that is set when the corporation is first established and actually is unrelated to "value." The market value per share is equal to the current share price. In most cases, the market value per share will exceed the par value.

Preferred Stock

To appeal to as many investors as possible, a corporation may also issue preferred stock. **Preferred stock** has certain privileges (or preferences) over common stock. Preferred stockholders participate as follows:

- Are entitled to receive dividends before common stockholders. The preferred stock dividend is stated in specific dollars, such as $6, or as a percentage of the stock's par value, such as 6 percent. The stock itself is then referred to as "preferred $6 stock" or "preferred 6% stock."
- Are given preference over common stockholders to distributions of the corporate assets should the company go out of business.

In return for these special privileges, the preferred stockholders give up two rights: to vote, and to participate in the control of the corporation. Usually, investors buy preferred stock to receive the stated dividend.

Reading Check

Define

In your own words, define common stock.

Issuing Common Stock

When a corporation issues common stock, the **Common Stock** account is credited for the par value of the stock. Let's look at some examples.

Issuing Common Stock at Par Value.

When The Starting Line was incorporated, it had the following transaction.

Business Transaction

On January 3 The Starting Line Sports Gear issued 10,000 shares of $10 par common stock at $10 per share. The Starting Line received $100,000 for the shares, Memorandum 3.

Journal Entry

GENERAL JOURNAL PAGE 46

DATE	DESCRIPTION	POST. REF.	DEBIT	CREDIT
20--				
Jan. 3	Cash in Bank		100 000 00	
	Common Stock			100 000 00
	Memorandum 3			

Issuing Common Stock in Excess of Par Value.

Investors are often willing to pay more than par value for the stock of a corporation. When a corporation sells its stock at a price that is above par, the excess over par is credited to a separate stockholders' equity account called **Paid-in Capital in Excess of Par**. **Paid-in Capital in Excess of Par** appears in the chart of accounts immediately following the **Common Stock** account. The amounts recorded in this account are not profits to the corporation. Instead, they represent

676 Chapter 21 • Accounting for Publicly Held Corporations part of the stockholders' investment in the corporation. This account follows the same rules of debit and credit as other stockholders' equity accounts.

In the second year of operations, The Starting Line completed the following two additional transactions.

Business Transaction

On January 4 The Starting Line issued 5,000 shares of $10 par common stock at $11.50 per share, Memorandum 147. The Starting Line received $57,500 for the shares.

Before recording the transaction, determine how much of the $57,500 is credited to **Common Stock** and how much is credited to **Paid-in Capital in Excess of Par.**

Credit to **Common Stock:**

5,000 shares at $10 par value	$50,000

Credit to **Paid-in Capital in Excess of Par:**

5,000 shares at $1.50 ($11.50 issue price − $10.00 par value) per share	+7,500
Total cash received	$57,500

Remember, the amount credited to the **Common Stock** account is the par value of the shares issued.

Journal Entry

GENERAL JOURNAL PAGE 99

	DATE		DESCRIPTION	POST. REF.	DEBIT	CREDIT	
1	20--						1
2	Jan.	4	Cash in Bank		57500 00		2
3			Common Stock			50000 00	3
4			Paid-in Cap. in Ex. of Par			7500 00	4
5			Memorandum 147				5
6							6

Issuing Preferred Stock

When preferred stock is issued, a corporation credits the **Preferred Stock** account for the stock's par value. Preferred stock is almost always issued at its par value.

The Starting Line was incorporated on January 3. It was authorized to issue 1,000 shares of preferred stock with a par value of $100 and a stated dividend of $6. On January 5, the company had the following transaction.

Business Transaction

On January 5 The Starting Line issued 250 shares of preferred $6 stock, $100 par, at $100 per share. The Starting Line received $25,000 for the shares, Memorandum 5.

Journal Entry

GENERAL JOURNAL   PAGE 47

DATE		DESCRIPTION	POST. REF.	DEBIT	CREDIT
20--					
Jan.	5	Cash in Bank		25000 00	
		Preferred Stock			25000 00
		Memorandum 5			

SECTION 21.1
Assessment

After You Read

Reinforce the Main Idea

Create a chart like this one to compare common stock and preferred stock. Add answer rows as needed.

Characteristic	Common Stock	Preferred Stock

Problem 1 Examining Capital Stock Transactions

Dublin Corporation was organized and authorized to issue 10,000 shares of $100 par, preferred 8% stock and 500,000 shares of $10 par common stock. The three transactions recorded in the following T accounts took place during the first month of operations.

Instructions In your working papers, describe each of the three transactions.

```
            Cash in Bank                            Preferred Stock
      Debit         |  Credit                  Debit    |    Credit
        +           |    −                       −      |      +
  (1) 300,000       |                                   |  (2) 200,000
  (2) 200,000       |                                   |

           Common Stock                    Paid-in Capital in Excess of Par
      Debit    |    Credit                  Debit    |    Credit
        −      |      +                       −      |      +
               |  (1) 300,000                        |  (3) 200,000
               |  (3) 500,000                        |
```

Math for Accounting

A corporation sells 3,000 shares of $25 par common stock and receives $97,500. How much did each share sell for? What was the per share amount of paid-in capital in excess of par?

SECTION 21.2
Distribution of Corporate Earnings

In Section 21.1 you learned to record stock issue transactions for a corporation. In this section you will learn how corporations **distribute** earnings to the stockholders. Corporations like General Motors have a long history of distributing a portion of earnings to their stockholders every year, making their stock an attractive investment.

Section Vocabulary
- dividend
- distribute

Dividend Accounts

What Are Corporate Dividends?

When an owner of a sole proprietorship or a partnership wishes to take money out of the business, a check is written on the checking account of the business. The amount of the check is recorded as a debit in the owner's withdrawals account, which reduces the owner's equity.

The owners (stockholders) of a publicly held corporation cannot withdraw cash whenever they want. Instead, they receive dividends. A **dividend** is a distribution of cash to stockholders. Dividends reduce retained earnings.

The corporation's board of directors *declares,* or authorizes, dividends. Before a dividend is declared, the corporation should have a sufficient amount of cash available to pay the dividend. In addition, since dividends decrease retained earnings, there must be an adequate balance in the **Retained Earnings** account. **Figure 21–1** illustrates the important dates in the dividend process.

A separate account named **Dividends** is used to record dividends declared. The **Dividends** account is a contra stockholders' equity account. At the end of the accounting period, the **Dividends** account is closed to the **Retained Earnings** account. The rules of debit and credit for **Dividends** are shown in this T account. Notice that they are the opposite of the rules for the stockholders' equity accounts.

```
                Dividends
        Debit           │   Credit
          +             │     −
    Increase Side       │ Decrease Side
    Normal Balance      │
```

Dividend amounts could be debited directly to the **Retained Earnings** account. Most corporations, however, prefer to use a separate account so that the dividend amounts can be easily determined.

A liability account, **Dividends Payable,** is used to record the amount of dividends that will be paid on the payment date. Like all liability accounts, **Dividends Payable** is increased by credits and decreased by debits. The normal balance of the **Dividends Payable** account is a credit balance.

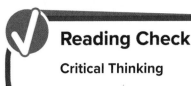

Reading Check

Critical Thinking

Why are dividends attractive to stockholders?

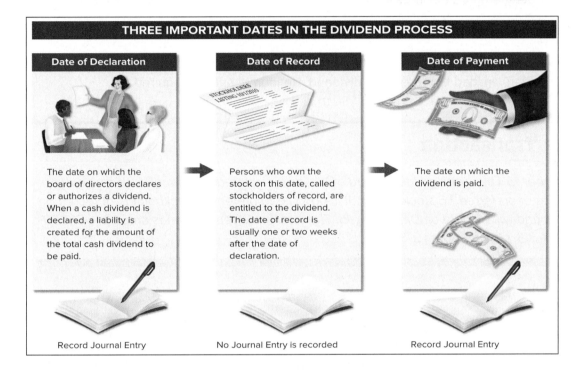

Figure 21–1 Important Dates in the Dividend Process

Dividend Transactions

How Do You Journalize Dividend Transactions?

A corporation authorized to issue two types of stock uses separate dividend and dividend payable accounts for each type. The Starting Line uses these:

Dividends—Preferred   **Dividends Payable—Preferred**

Dividends—Common   **Dividends Payable—Common**

Let's look at the transactions involved when dividends are declared by the board of directors of The Starting Line during its second year of operations.

Dividends on Preferred Stock

As mentioned, preferred stockholders have certain preferences over common stockholders; one is the right to receive dividends before common stockholders. The preferred stock dividend amount is predetermined, or *stated*. The stated dividend indicates the amount to be paid to preferred stockholders per year. The Starting Line issued 250 shares of preferred $6 stock; that is, it will pay a $6 dividend for each share of preferred stock annually or in semiannual or quarterly installments:

Annually	$6.00
Semiannually	$3.00 ($6.00 ÷ 2)
Quarterly	$1.50 ($6.00 ÷ 4)

Date of Declaration.

A journal entry records the dividend on preferred stock on the *date of declaration,* the date the board declares it.

Date of Record.

On November 29 the corporation checks its records and prepares a list of preferred stockholders entitled to receive the dividend. No journal entry is required on this date.

Business Transaction

On November 15 The Starting Line's board of directors declared an annual cash dividend on the 250 shares of preferred $6 stock issued. It is payable to preferred stockholders of record on November 29 and will be paid on December 15. The total preferred dividends amount is $1,500 (250 shares × $6), Memorandum 215.

Journal Entry

GENERAL JOURNAL   PAGE 72

	DATE	DESCRIPTION	POST. REF.	DEBIT	CREDIT
1	20--				
2	Nov. 15	Dividends—Preferred		1500 00	
3		Dividends Payable—Preferred			1500 00
4		Memorandum 215			

Date of Payment.

On December 15 a check for $1,500, the total dividend payable on preferred stock, is written and deposited in a special dividends checking account. Separate checks written on this account are made payable to each preferred stockholder entitled to receive the dividend.

Business Transaction

On December 15 The Starting Line issued Check 1373 for $1,500 in payment of the dividend on preferred stock declared November 15.

Journal Entry

GENERAL JOURNAL   PAGE 81

	DATE	DESCRIPTION	POST. REF.	DEBIT	CREDIT
1	20--				
2	Dec. 15	Dividends Payable—Preferred		1500 00	
3		Cash in Bank			1500 00
4		Check 1373			

Dividends on Common Stock

Dividends on common stock may be declared in one of two ways.

1. The board of directors may declare a dividend amount *per common share*. The Starting Line's board declared a 50¢ per common share dividend. It will pay a 50¢ dividend for each share of common stock.

2. The board may decide to declare the *total cash dividend* for both preferred and common stock. In this case the company first pays the preferred dividends and then divides the remainder equally among the common stockholders. For example, The Starting Line's board of directors declared a $9,000 total cash dividend. Preferred stockholders will receive $1,500 (250 shares × $6). It splits the remainder, $7,500 ($9,000 − $1,500), among the common stockholders. Each share of common stock receives a 50¢ dividend ($7,500 ÷ 15,000 shares).

Dividends on common stock, like preferred stock, can be paid annually, semiannually, or quarterly. Most publicly held corporations pay quarterly.

SECTION 21.2
Assessment

After You Read

Reinforce the Main Idea

Create a diagram like this one to describe which accounts (if any) are debited and credited for each event.

Event	Account Debited	Account Credited
Dividend on common stock—date of declaration		
Dividend on common stock—date of record		
Dividend on common stock—date of payment		

Problem 2 Distributing Corporate Earnings

During its first year of operation, Longhorn Corporation issued 17,500 shares of $10 par common stock. At the end of the year, the corporation had a net income of $350,000. The board of directors declared a cash dividend of $5 per share.

Instructions Answer these questions in your working papers.

1. How much of the net income did Longhorn distribute to the stockholders?
2. How much of the net income did the corporation retain?

Problem 3 Analyzing a Source Document

This memorandum contains data about the first quarter dividend on the common stock of Rob Williams Thrift Market Inc.

Instructions Prepare the journal entries to record the following in your working papers. Use general journal page 18.

- declaration of the dividend
- payment of the dividend, Check 221

Rob Williams Thrift Market Inc.
2347 Eastern Parkway
Orange, IA 50322-6922

MEMORANDUM 37

TO: Chief Financial Officer
FROM: Albert MacFish, Chairman of the Board
DATE: April 8, 20--
SUBJECT: 1st quarter dividend

On April 1, 20--, the board of directors declared a $1 per share dividend on the 5,679 shares of common stock issued. The date of record is April 15, 20--. The date of payment is April 30, 20--.

Math for Accounting

On January 28 the board of directors for Jelly Bean Works declared an annual cash dividend on 200 shares of preferred $10 stock. What is the total amount of the dividend? How is this dividend accounted for in the company's books?

SECTION 21.3
Financial Reporting for a Publicly Held Corporation

Many corporations prepare a statement of stockholders' equity instead of a statement of retained earnings. This section examines the information reported on the statement of stockholders' equity, how this statement is prepared, and how stockholders' equity is reported on a corporation's balance sheet.

> **Section Vocabulary**
> - statement of stockholders' equity
> - contrast

The Income Statement

Does a Publicly Held Corporation's Income Statement Differ from That of a Closely Held Corporation?

The income statement of a publicly held corporation is similar to that prepared by a closely held corporation. Remember, the federal corporate income taxes paid by a corporation are reported separately on the income statement.

The Statement of Stockholders' Equity

What Is a Statement of Stockholders' Equity?

In Chapter 19 you learned how to prepare a statement of retained earnings for a closely held corporation. That statement reports the changes in the **Retained Earnings** account during the period. It showed:

Retained Earnings, beginning balance
+ Net Income
Retained Earnings, ending balance

An increasing number of corporations prepare a statement of stockholders' equity rather than a statement of retained earnings. In **contrast** to the statement of retained earnings, the **statement of stockholders' equity** reports the changes in *all* stockholders' equity accounts during the period. It also provides information about the transactions affecting stockholders' equity during the period. The information reported on a statement of stockholders' equity includes

- the number of shares of each type of stock issued,
- the total amount received for those shares,
- the net income or net loss for the period, and
- dividends declared during the period.

Figure 21–2 shows the statement of stockholders' equity for The Starting Line at the end of its second year of operations. The information needed to prepare the statement comes from the work sheet and from the general ledger accounts.

The statement is prepared as follows:

Figure 21–2 Statement of Stockholders' Equity

The Starting Line Sports Gear
Statement of Stockholders' Equity
For the Year Ended December 31, 20--

	$100 Par Preferred $6 Stock	$10 Par Common Stock	Paid-in Capital in Excess of Par-Common	Retained Earnings	Totals
Balance, January 1, 20--		100 000 00		27 600 00	127 600 00
Issuance of 5,000 shares of common stock		50 000 00	7 500 00		57 500 00
Issuance of 250 shares of Preferred Stock, January 5, 20--	25 000 00				25 000 00
Net Income				13 525 00	13 525 00
Cash Dividends:					
Preferred Stock				(1 500 00)	(1 500 00)
Common Stock				(7 500 00)	(7 500 00)
Balance, December 31, 20--	25 000 00	150 000 00	7 500 00	32 125 00	214 625 00

- The names of the four stockholders' equity accounts appear at the top of the amount columns. There is also a Totals column at the far right.
- The first line shows the balance of each account at the beginning of the period.
- The next lines describe various transactions affecting the stockholders' equity accounts, such as issuance of stock, net income, and dividends declared. The transactions are described in the left column. The increase or decrease amounts are recorded in the individual account columns and in the Totals column. For example, the second line on the statement indicates that the company issued 5,000 shares of common stock during the period. The issuance of the 5,000 shares increased the balance of the **Common Stock** account by $50,000. Since the shares were issued at a price above par value, the **Paid-in Capital in Excess of Par-Common** account increased by $7,500.
- The final line shows the balance of each account at period-end.

As you review **Figure 21–2**, notice that the amounts that decrease account balances, like dividends declared, are enclosed in parentheses. The cash dividends declared are listed separately for preferred and common stock. Also notice that the Retained Earnings column of the statement of stockholders' equity contains the same information that is reported on a statement of retained earnings. Remember that net income increases retained earnings, and dividends declared decrease retained earnings. The statement of stockholders' equity is used to prepare the corporation's balance sheet.

Reading Check

List

What information is included on a statement of stockholders' equity?

The Balance Sheet

How Do You Prepare the Stockholders' Equity Section of the Balance Sheet?

The assets and liabilities sections of a balance sheet for a publicly held corporation are similar to those of a balance sheet for a closely held corporation. Notice in **Figure 21–3** that the **Dividends Payable** accounts are reported in the liabilities section. Since The Starting Line paid the dividends before the end of the year, it reports zero balances in these liability accounts.

The stockholders' equity section of a publicly held corporation's balance sheet is more detailed than that of a closely held corporation. As you can see in **Figure 21–4**, each type of stock issued by The Starting Line is listed separately under the heading "Paid-in Capital." Preferred stock is listed before common stock. Each listing describes the following:

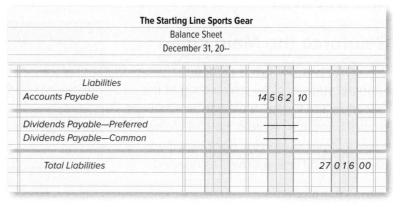

Figure 21–3 Reporting the Balances of the Dividends Payable Accounts on the Balance Sheet

- the par value,
- the number of shares authorized, and
- the number of shares issued.

The **Dividends** accounts are not listed in the stockholders' equity section because they are closed to **Retained Earnings** at the end of the year. Thus, the retained earnings amount shown on the balance sheet has been reduced by the dividends declared during the period.

The Statement of Cash Flows

Where Are Stock Issues and Dividends Reported on This Statement?

The Financing Activities section of the statement of cash flows reflects activities related to ownership of the corporation. A stock issue results in a cash inflow from a financing activity. Payment of dividends is a cash outflow.

The Starting Line Sports Gear
Balance Sheet
December 31, 20--

Stockholders' Equity			
Paid-in Capital:			
$6 Preferred Stock, $100 par, 1,000 shares authorized,			
250 shares issued	25 000 00		
Common stock, $10 par, 20,000 shares authorized,			
15,000 shares issued	150 000 00		
Paid-in Capital in Excess of Par	7 500 00		
Total Paid-in Capital		182 500 00	
Retained Earnings		32 125 00	
Total Stockholders' Equity			214 625 00
Total Liabilities and Stockholders' Equity			241 641 00

Figure 21–4 The Stockholders' Equity Section of the Balance Sheet

SECTION 21.3
Assessment

After You Read

Reinforce the Main Idea

Create a chart like this one to identify the financial statement on which each of the following appear: cash dividends, common stock, dividends payable, paid-in capital in excess of par, preferred stock, retained earnings. Add answer rows as needed.

Statement of Stockholders' Equity	Balance Sheet

Problem 4 Examining the Statement of Stockholders' Equity

The following transactions of Victor Jewelry Corporation took place during the period.

Instructions Use the form in your working papers to indicate which of the transactions is reported on the statement of stockholders' equity.

Transactions:

1. Paid accounts payable of $50,000.
2. Issued 2,000 shares of $10 par common stock, receiving $15 per share.
3. The board of directors declared a cash dividend of $12,000 for all common stockholders.
4. Bought equipment on account at a total cost of $125,000.
5. Paid the cash dividend declared in Transaction 3.
6. Issued 500 shares of $100 par, $7 preferred stock.
7. Paid the federal income tax installment of $5,000.
8. Earned a net income of $150,000 for the period.

Math for Accounting

Ron Kanai, president of Sunny Days, a Hawaiian-based sun care products manufacturer and retailer, is looking for a buyer. He believes an analysis of stock prices and dividends will appeal to potential corporate buyers. As the company's accounting manager, use spreadsheet, accounting, or graphics software to prepare the following information in a clustered column chart showing the stock price and dividends declared for the past two years. What does the chart say about Sunny Days' financial condition that could attract potential buyers?

	Year 1				Year 2			
	Q1	Q2	Q3	Q4	Q1	Q2	Q3	Q4
Stock Price	13.50	13.60	14.90	17.00	17.50	20.00	23.00	25.50
Dividend Declared	2.50	2.50	2.90	3.00	3.25	3.50	4.90	4.95

Chapter 21
Visual Summary

Concepts
The corporation has several unique characteristics:

A CORPORATION
has legal permission from the state to operate.
is recognized as a separate legal entity.
is owned by stockholders.
is operated by professional managers hired by a board of directors that is elected by the stockholders.

Analysis
Journal entries to record the issue of common stock:

GENERAL JOURNAL PAGE 85

	DATE	DESCRIPTION	POST. REF.	DEBIT	CREDIT	
1	20--					1
2	April 25	Cash in Bank		65 550 00		2
3		Common Stock			60 000 00	3
4		Paid in Cap. in Ex. of Par			5 550 00	4
5		Memorandum 145				5

Procedures
Journal Entries Required for Dividends:

Event	Importance	Journal Entry Required
Dividend declared	Corporation's board of directors votes to distribute earnings to owners.	Debit **Dividends**, Credit **Dividends Payable**
Dividend date of record	The corporation makes a list to show who owns stock on the date of record.	No journal entry is made.
Dividend paid	The corporation writes the dividend checks to the owners.	Debit **Dividends Payable**, Credit **Cash in Bank**
End of the year	The corporation closes the **Dividends** account.	Debit **Retained Earnings**, Credit **Dividends**

Chapter 21
Review and Activities

Answering the Essential Question

What is the relationship between a publicly held corporation and its stockholders?

In a sense, the stockholders "own" the corporation because they've spent money to be a part of the corporation. How does a corporation share its earnings with stockholders?

Vocabulary Check

1. **Vocabulary** Use and underline the content vocabulary words in a brief essay titled "Corporations and Capital Stock."

- board of directors
- closely held corporation
- entity
- publicly held corporation
- authorized capital stock
- common stock
- corporation
- par value
- preferred stock
- proxy
- Paid-in Capital in Excess of Par
- distribute
- dividend
- contrast
- statement of stockholders' equity

Concept Check

2. Name three characteristics of a corporation.
3. **Analyze** Explain the difference between a closely held corporation and a publicly held corporation.
4. When stock is sold at a price above its par value, what amount is credited to the capital stock account?
5. What is the classification of the Paid-in Capital in Excess of Par account? What is the normal balance?
6. **Analyze** Compare the methods used by sole proprietorships and corporations for taking money out of the business.
7. Name and explain the three dates important to the dividend process.
8. **Analyze** Compare and contrast the statement of retained earnings and the statement of stockholders' equity.
9. **Math** Tera is about to open a motorcycle shop. He must choose between a space for rent or a small building for sale. The rent is $1,200 per month, but he cannot make any improvements. The monthly payment for the building would be $800, but he must pay taxes and extra insurance at $300/month. What is the better monthly rate?

Chapter 21
Standardized Test Practice

Multiple Choice

1. The type of stock issued by a corporation when only one class of stock is issued is
    a. preferred stock.
    b. common stock.
    c. capital stock.
    d. a dividend.

2. Earnings distributed to stockholders are called
    a. retained earnings.
    b. revenue.
    c. capital gains.
    d. dividends.
    e. none of these answers

3. When 10,000 shares of $10 par-value common stock are issued at $16 per share, Paid-In Capital in Excess of Par, Common Stock is credited for
    a. $160,000.
    b. $60,000.
    c. $100,000.
    d. none of the above

4. When a dividend is paid in cash, the accounts debited and credited are
    a. Dividends Received and Dividends Payable.
    b. Cash and Dividends Payable.
    c. Dividends Payable and Cash.
    d. Common Stock and Cash.

5. A legal form that asks stockholders to transfer their voting rights is called a
    a. security.
    b. proxy.
    c. preemptive right.
    d. stock split.

Short Answer

6. What is the purpose of closing entries?
7. Where can you find the information needed to prepare the closing entries?

Extended Response

8. List the ten steps in the accounting cycle.

Chapter 21
Computerized Accounting

Customizing Financial Statements

Making the Transition from a Manual to a Computerized System

MANUAL METHODS
- Using accounting stationery, you may reorganize information or add details to financial statements.

COMPUTERIZED METHODS
- Accounting software gives you the option to select and sort data that appears on each financial statement.
- You may also customize information on a report.

Chapter 21 Problems

Problem 5 Distributing Corporate Earnings

During the first year of operations, Sunset Surfwear issued 18,500 shares of $10 par common stock. At the end of the year, the corporation had net income of $380,000. The board of directors declared a $5 cash dividend per share of common stock.

Instructions Answer the following questions in your working papers:

1. How much of the net income earned for the year was paid to the common stockholders?

2. How much of the net income was retained by the corporation?

Analyze

Compare the date of declaration and date of payment with the date of record. Explain what entry or action is required on each date.

Chapter 21 Problems

Problem 6 Journalizing the Issue of Stock

On June 1 InBeat CD Shop was incorporated and authorized to issue 1,000 shares of $100 par, preferred 9% stock, and 10,000 shares of $25 par common stock.

Instructions In your working papers, record the following transactions on general journal page 1.

Date		Transactions
June	1	Issued 200 shares of preferred 9% stock at $100 per share, Memorandum 3.
	2	Issued 3,000 shares of common stock at $25 per share, Memorandum 7.
	6	Received $31 per share for 2,000 shares of common stock issued, Memorandum 10.
	7	Issued 50 shares of preferred 9% stock at par, Memorandum 11.

Analyze

Identify the stock issued at a price above par. Calculate the total amount of capital that InBeat CD Shop received in the first week of June.

Chapter 21
Problems

Problem 7 Journalizing Common and Preferred Stock Dividend Transactions

Shutterbug Cameras issued 8,000 shares of $80 par, preferred 7% stock and 35,000 shares of $25 par common stock.

Instructions In your working papers:

1. Record the following transactions on general journal page 14.
2. Post the transactions to the general ledger accounts provided in your working papers. Net income was $296,490.

Date	Transactions
Oct. 15	The board of directors declared an annual cash dividend on the preferred 7% stock, payable on December 1, Memorandum 407.
Nov. 5	Declared an annual cash dividend of $1.25 on 35,000 shares of common stock, payable on December 17, Memorandum 415.
Dec. 1	Paid the dividend declared on October 15, Check 1163.
17	Paid the dividend declared on November 5, Check 1201.

Analyze

Calculate the amount of dividends payable on common stock and dividends payable on preferred stock at the end of December.

Chapter 21
Problems

Problem 8 — Preparing Corporate Financial Statements

River's Edge Canoe & Kayak is authorized to issue 10,000 shares of $100 par, preferred 9% stock and 500,000 shares of $5 par common stock. The following balances appeared in the Balance Sheet section of the company's work sheet for the year ended December 31.

Instructions

1. Prepare the statement of stockholders' equity.

    a. During the period the corporation issued 500 shares of 9% preferred stock at par and 25,000 shares of common stock at $9.

    b. The net income for the year was $425,000.

2. Prepare the balance sheet.

101	Cash in Bank	$506,010
115	Accounts Receivable	850,680
120	Notes Receivable	400,000
130	Merchandise Inventory	388,815
135	Prepaid Ins.	3,600
140	Supplies	10,500
145	Delivery Truck	139,298
150	Store Equipment	363,009
201	Accounts Payable	62,412
202	Dividends Payable—Preferred	22,500
203	Dividends Payable—Common	$150,000
210	Fed. Corp. Income Tax Payable	20,000
215	Sales Tax Payable	4,500
301	Preferred Stock	500,000
303	Common Stock	750,000
304	Paid-in Capital in Excess of Par	300,000
305	Retained Earnings	735,000
307	Dividends—Common	262,500
308	Dividends—Preferred	45,000

Analyze

Identify the total stockholders' equity on December 31.

Chapter 21
Problems

Problem 9 — Recording Stockholders' Equity Transactions

Buzz Newsstand is authorized to issue 100,000 shares of $5 par common stock and 5,000 shares of $100 par, preferred 8% stock. On January 1, the beginning of the period, the stockholders' equity accounts had the following balances:

301	Preferred Stock	$150,000
302	Paid-in Capital in Excess of Par—Preferred	11,250
303	Common Stock	225,000
304	Paid-in Capital in Excess of Par—Common	112,500
305	Retained Earnings	366,800
307	Dividends—Preferred	0
308	Dividends—Common	0

SOURCE DOCUMENT PROBLEM

Problem 21–9

Use the source documents in your working papers to complete this problem.

Instructions In your working papers:

1. Record the following transactions on general journal page 42. Close the **Dividends** and **Retained Earnings** accounts.
2. Prepare the stockholders' equity section of the balance sheet.

Date	Transactions
Mar. 15	The board of directors approved a total semiannual cash dividend of $62,250 for preferred and common stockholders. The dividend is payable to stockholders of record as of April 15 with payment on May 1, Memorandum 635.
Apr. 19	Issued 500 shares of preferred stock at $108, Memorandum 651.
May 1	Paid the dividends declared on March 15, Check 1256.
Sept. 1	The board of directors approved a total semiannual cash dividend of $79,250 for preferred and common stockholders. The dividend is payable to stockholders of record as of October 1 with payment on November 1, Memorandum 828.
Nov. 1	Paid the dividend declared on September 1, Check 2451.

Analyze

Compare the December 31 **Retained Earnings** balance with the beginning balance of $366,800. Why did the balance change?

🌍 Real-World Accounting Careers

Danielle Davidson

Armanino McKenna LLP

Q What do you do?

A As a senior auditor, my role is to direct the day-to-day activities in a client's audit. I typically will audit significant issues directly, but I also spend time training my staff and serving as a liaison between my team and client management.

Q What are your day-to-day responsibilities?

A Most of my time is spent at the client's location. I may prepare draft financial statements, lead my team through the steps of an audit or meet with the client to discuss the project. I also oversee the budget and plan for each project.

Q What factors have been key to your success?

A In my nine years with the company, I've gone from working part time at the front desk to a bookkeeping position to my current auditor role. By developing professional relationships and setting goals for myself, I was able to progress in my career.

Q What do you like most about your job?

A I like the flexibility and the fact that I get to wear many different hats. I may be meeting with managers, giving my staff evaluations, coordinating the team or contacting clients in addition to performing audits.

Career Facts

Real-World Skills

Ability to manage time wisely; project management skills; strong interpersonal skills

Training and Education

Bachelor's degree in accounting or finance; CPA license

Career Skills

Knowledge of basic financial statements, debits and credits, and budgets; proficiency with Microsoft Office software

Career Path

Start as a staff accountant or auditor, then move to senior auditor after two to four years

Tips from Robert Half International

Learning is a continual process — even after you've graduated from school. Build your skills on the job by volunteering for projects beyond your immediate area of responsibility. You can also join industry associations and take professional education courses to remain at the forefront of your field.

College and Career Readiness

How might flexibility and "wearing different hats" help you advance in your career? Write and create a brochure that explains to classmates how to be flexible, and how wearing different hats can help with career development.

Real-World Applications & Connections

Career Wise

Certification for Management Accounting Careers

When exploring accounting careers, it can be helpful to locate resources that are specific to an accounting field.

Activity

Use the Internet to research a web site such as the Institute of Management Accountants. Identify any accounting licensing and certification programs that might be required of certified management accountants.

Global Accounting

Tariffs and Duties

Thinking of boosting your company's sales by selling products in another country? Consider the impact of tariffs. A *tariff*, or *duty*, is a tax imposed on imports of specific products. For example, if your U.S.-based company sells computer hardware in Venezuela, a 5 percent tariff will be charged on each item.

Instructions

Assume that Venezuelan competitors sell Model AB computer monitors for $210. You typically charge $205. Will your price be competitive after considering the tariff? Why or why not?

A Matter of Ethics

Using Insider Information

You are an accountant for a large shipping company, like United Parcel Service (UPS). Through your work, you learn that your company is planning to buy a company that makes shipping containers. It's a small company whose stock is traded on the stock exchange. You think about purchasing several thousand shares of this company's stock, expecting to make a profit when your company purchases it.

Activity

Define the ethical issues involved. What are the alternatives? Who are the affected parties? How do the alternatives affect the parties? What would you do?

Analyzing Financial Reports

Evaluating Stockholders' Equity

The statement of stockholders' equity provides details about changes in the stockholders' equity accounts during the period. This information helps investors and analysts better understand the company's capital. Successful companies earn money for future operations (reported as retained earnings) through sales and rely less on raising new capital through the issuance of stock.

Instructions

Use the Internet to locate and save or print a copy of Amazon.com's annual report to complete the following. Use the statement of stockholders' equity to answer these questions.

1. Explain how retained earnings changed during the period?
2. How much did additional paid-in capital increase during this same time?

H.O.T. Audit

Balance Sheet

The balance sheet of Automart, Inc. includes the following amounts:

Preferred 6% Stock, $100 par value: $90,000

Common Stock, $10 par value: 150,000

Paid-in Capital in Excess of Par—Common: 40,000

Retained Earnings: 55,000

Instructions

Audit the balance sheet and report the number of shares of preferred and common stock issued as well as the selling price per share. Prepare a brief memorandum that contains your findings.

UNIT 5

Accounting for Special Procedures

A Look Back
Unit 4 described the accounting cycle for a merchandising corporation.

A Look Ahead
In Unit 5 you will learn special accounting procedures including cash funds, depreciation, uncollectable accounts, and inventories.

Keys to Success

Q How can joining related student organizations help you get the most out of your career and technology courses?

A Student organizations are more than just exclusive clubs for high achievers or a way to add credentials to your college application. Participation in these groups provides a great instructional tool to help you build skills for a particular career or subject matter, deepen your understanding of what you learn in class, and develop leadership abilities. Student organizations hold competitive events, conferences and activities, giving opportunities to meet others with similar interests and promoting lifelong responsibility for community service and professional development.

English Language Arts

Writing The Internet is a valuable source of information about career and technology student organizations. Go online to research the following: DECA, FBLA, SkillsUSA, TSA. Write a brief description of each group. *What is the emphasis of each organization? How might participation in these groups help your class work?*

Focus on the Photo
Joining a student group with others who share your interests can provide opportunities to participate in events that help you develop skills and even earn money for the organization. **What method might you use to analyze financial transactions for special procedures?**

Chapter 22 • Cash Funds 701

Chapter 22: Cash Funds

Chapter Topics

22-1 The Change Fund

22-2 The Petty Cash Fund

Visual Summary

Review and Activities

Standardized Test Practice

Computerized Accounting

Problems

Real-World Applications & Connections

Essential Question

As you read this chapter, keep this question in mind:

How do businesses monitor and protect cash?

Main Idea

Some businesses keep cash on hand to make change for customers. Businesses use petty cash funds because writing checks for small amounts is impractical, costly, and time consuming.

Chapter Objectives

Concepts

C1 Record the entry to establish a change fund.

C2 Prove the cash in the cash register drawers each business day.

C3 Open and replenish a petty cash fund.

Analysis

A1 Journalize opening a petty cash fund.

A2 Prepare a petty cash requisition to replenish the petty cash fund.

A3 Use a petty cash register to record petty cash disbursements.

Procedures

P1 Journalize replenishing a petty cash fund.

P2 Determine whether cash is short or over, and record the shortage or overage.

Real-World Business Connection

Papa Murphy's

The "take and bake" concept—buying a custom-made, uncooked pizza to take home and bake—began in 1981 under a company called Papa Aldo's Pizza. In 1995, Papa Murphy's was born when Papa Aldo's merged with a former take-and-bake competitor, Murphy's Pizza. Today, with more than 1,200 stores in at least 35 states and lots of growth potential, a Papa Murphy's franchise is an attractive business option for entrepreneurs.

Connect to the Business

Some Papa Murphy's franchisees have annual gross sales of more than $1 million, with most individual purchases totaling $20 or less, so you can imagine the daily cash inflow that a franchise owner faces. The stores keep cash on hand to make change for customers and pay for small expenses. Accounting for and protecting this cash is an important function in managing this kind of business.

Analyze

What documentation should Papa Murphy's franchisees keep for small cash payments?

Focus on the Photo

Businesses such as Papa Murphy's keep and take in significant amounts of cash every day. Employees must follow company rules for all financial transactions, including policies for cash control and record retention. ***How might keeping good records protect against theft or loss?***

SECTION 22.1
The Change Fund

In earlier chapters, you learned that businesses use checking accounts for depositing cash receipts and making cash payments. Merchandising businesses, such as Sears or Hollister, keep some cash on hand so that they can make changes for customers who pay for purchases with cash. In this chapter, you will learn how to manage, prepare, and record petty cash fund transactions.

Section Vocabulary
- change fund
- establishes
- attached

Establishing a Change Fund

What Is a Change Fund?

A **change fund** is an amount of money, consisting of varying denominations of bills and coins, that is used to make change in cash transactions. For example, a customer who pays for a $13.80 purchase with a $20 bill will receive $6.20 in change.

When a business first **establishes** a change fund, the amount needed for the fund is estimated. The size of the fund does not change unless the business finds that it needs more or less change than it had originally estimated. The change fund is established by writing a check for the amount of the fund. The check is made payable to the person in charge of the change fund. That person cashes the check and places the bills and coins in the cash register drawer. The transaction is recorded in an asset account called **Change Fund.** Let's look at The Starting Line's establishment of a change fund as an example.

Business Transaction

On May 1 the accountant for The Starting Line wrote Check 2150 for $100 to establish a change fund.

Analysis	Identify	1. The accounts affected are **Change Fund** and **Cash in Bank.**
	Classify	2. Both **Change Fund** and **Cash in Bank** are asset accounts.
	+/–	3. **Change Fund** is increased by $100. **Cash in Bank** is decreased by $100.

Debit-Credit Rule	4. Increases to asset accounts are recorded as debits. Debit **Change Fund** for $100.
	5. Decreases to asset accounts are recorded as credits. Credit **Cash in Bank** for $100.

T Accounts 6.

Change Fund		Cash in Bank	
Debit +	Credit –	Debit +	Credit –
100			100

continued

Journal Entry

7.

	GENERAL JOURNAL			PAGE 32	
DATE	DESCRIPTION	POST. REF.	DEBIT	CREDIT	
20--					1
May 1	Change Fund		100 00		2
	Cash in Bank				3
	Check 2150			100 00	4
					5

If the business needs to increase the amount in the change fund, the accountant writes a check for the amount of the cash increase. For example, suppose that The Starting Line needs to increase the change fund from $100 to $125. The amount of the check is $25. The journal entry debits **Change Fund** for $25 and credits **Cash in Bank** for $25. This brings the **Change Fund** account balance to $125.

Using the Change Fund

How Do You Use the Change Fund?

The amount of cash in the change fund is put into the cash register drawer at the beginning of the day. When a cash sale occurs, the salesclerk rings the sale on the cash register. The sale is automatically recorded on the cash register tape. At the end of the day, the cash in the cash register drawer is counted. A cash proof is prepared to verify that the amount of cash in the drawer equals the total cash sales for the day plus the change fund. The amount of cash in the change fund is set aside for use as change for the next day. The balance of the cash from the drawer is deposited in the checking account.

Let's look at an example. Suppose The Starting Line has $470 in the cash register drawer at the end of the day on May 15. The cash register tape shows that cash sales, including sales tax, total $370. **Figure 22–1** shows the cash proof. As you can see, the amount of cash in the drawer at the end of the day minus the amount of cash in the change fund equals the total sales shown on the cash register tape.

Most businesses require that salesclerks sign the cash proof to indicate that they have counted the cash in the drawer and verified its accuracy, both when they receive the drawer and turn it in. The supervisor also checks these amounts and signs the cash proof. The cash proof form is **attached** to the cash register tape, which is the source document for recording the cash sales for the day.

Reading Check

Explain

Which accounts are debited and credited when a change fund is established?

Figure 22–1 Cash Proof Form

Recording Cash Short and Over

What If Cash in the Drawer Does Not Match the Records?

Many cash transactions occur during each business day. Occasionally, a salesclerk makes an error and gives the incorrect amount of change to a customer. When this happens, the amount of cash in the cash register drawer, less the beginning change fund, does not agree with the cash sales amount recorded on the cash register tape. If the salesclerk gives a customer too much change, the amount of cash in the drawer at the end of the day is *short*, or less than it should be. If the salesclerk gives a customer too little change, the cash amount is *over*, or more than it should be.

The amount of cash either gained or lost because of errors is recorded in the **Cash Short & Over** account. **Cash Short & Over** is a temporary account. Because cash shortages are expenses to the business, they are recorded as debits to **Cash Short & Over**. Cash overages are revenue for the business; they are recorded as credits to the **Cash Short & Over** account. Note that the **Cash Short & Over** account does not have a normal balance.

Cash Short & Over	
Debit Cash Shortages	Credit Cash Overages

At the end of the accounting period, the balance of the **Cash Short & Over** account is closed to **Income Summary**. Let's look at an example of a shortage.

Business Transaction

On May 2 the cash register tape shows that cash sales are $520 and sales taxes are $26, for a total of $546. The actual cash in the cash register drawer, after subtracting the amount of the change fund, is $545.

T Accounts

Cash in Bank		Cash Short & Over	
Debit + 545	Credit −	Debit +	Credit

Sales		Sales Tax Payable	
Debit −	Credit + 520	Debit −	Credit + 26

Journal Entry

GENERAL JOURNAL PAGE 33

DATE	DESCRIPTION	POST. REF.	DEBIT	CREDIT
20--				
May 2	Cash in Bank		545 00	
	Cash Short & Over		1 00	
	Sales			
	Sales Tax Payable			520 00
	Cash Proof—Cash reg. tape			26 00

The journal entry to record a cash overage is similar to that for a cash shortage. The only difference is that **Cash Short & Over** is *credited* for the amount of the overage.

In the previous example, suppose that the amount of cash in the drawer is $547 after the amount of the change fund is subtracted. The journal entry would debit **Cash in Bank** for $547 and credit **Cash Short & Over** for $1. The credits to **Sales** and **Sales Tax Payable** remain the same.

SECTION 22.1
Assessment

After You Read

Reinforce the Main Idea

A business that uses a cash register prepares a cash proof each day. Create a diagram similar to this one to show the step-by-step method for proving cash and recording any discrepancy.

Proving Cash

Problem 1 Preparing a Cash Proof

The change fund for Messina's Grocery Store is $200 per cash register. On March 31 total cash sales from cash register 6 are $964 and sales taxes are $57.84. A count of cash shows $1,216.84 in the cash register drawer.

Instructions

1. Use the form in your working papers to prepare a cash proof. Sign your name as the salesclerk.

2. Record the March 31 cash sales on page 2 of a general journal.

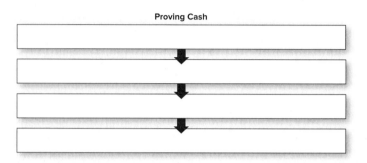

Mar 31
Tape 49

964.00 CA
57.84 ST

Problem 2 Recording a Cash Overage

The change fund for Visions Hair Salon is $300 per register. On April 14, at the end of the business day, the manager counts the funds in the register. Cash sales total $1,496 and sales taxes amount to $89.77 in register 2. The cash drawer contains $1,895.77.

Instructions

1. Use the form in your working papers to prepare a cash proof. Sign your name as the salesclerk.

2. Record the April 14 cash sales and overage on page 4 of the general journal.

Math for Accounting

The Children's Consignment Shop operates primarily with cash transactions. One of your responsibilities is to prove cash register cash. The cash register tape showed cash received from sales and sales taxes collected to be $620.40. The count of cash in the drawer was: 16 $20 bills, 18 $10 bills, 26 $5 bills, 44 $1 bills, 66 quarters, 30 dimes, 52 nickels, and 130 pennies. The change fund is $75 dollars. What was the net cash received? Was cash short, over, or in balance? If the total sales tax rate is 10%, what is the amount of cash sales and the sales tax payable?

708 Chapter 22 • Cash Funds

SECTION 22.2
The Petty Cash Fund

Businesses often make small cash payments for delivery fees, postage stamps, supplies, and other similar purchases.

Establishing the Petty Cash Fund

What Is a Petty Cash Fund?

Small, **incidental** cash payments are made from the **petty cash fund**. The word *petty* indicates that only small amounts of cash are paid out of this fund. When setting up a petty cash fund, each business determines the maximum amount that will be paid out by a petty cash disbursement. A **petty cash disbursement** is any payment made from the petty cash fund. All payments over the maximum amount are paid by check. The person responsible for maintaining the petty cash fund and for making cash disbursements is called the **petty cashier**.

Section Vocabulary
- petty cash fund
- petty cash disbursement
- petty cashier
- petty cash voucher
- petty cash requisition
- petty cash register
- incidental
- specified
- supplemental

To establish a petty cash fund, a business estimates the amount of cash needed in the fund for a certain period of time, usually a month. This estimate is based on the company's past experiences.

The Starting Line decides to establish a petty cash fund. Crystal Casteel, an office clerk, is appointed petty cashier. The petty cash fund will contain $100. Any payments under $10 will be paid from the petty cash fund. Any payments over $10 will be paid by check. The company accountant, Greta Keegan, prepares a check for $100, payable to "Petty Cashier Crystal Casteel," to establish the fund. The transaction is recorded in an asset account called **Petty Cash Fund.** Crystal cashes the check and places the money, consisting of small denominations of bills and coins, in a petty cash box. For internal control purposes, the petty cash box is kept in an office safe or a locked desk drawer. Crystal is responsible for the $100 in the petty cash fund. When members of the staff need small amounts of cash for items like stamps or office supplies, Crystal will disburse the cash and store the receipts in the petty cash box. Let's learn how to journalize transactions to open a petty cash fund.

Business Transaction

On May 1 Check 2151 for $100 was issued to establish the petty cash fund.

Analysis	Identify	1. The accounts affected are **Petty Cash Fund** and **Cash in Bank.**
	Classify	2. Both **Petty Cash Fund** and **Cash in Bank** are asset accounts.
	+/−	3. **Petty Cash Fund** is increased by $100. **Cash in Bank** is decreased by $100.

continued

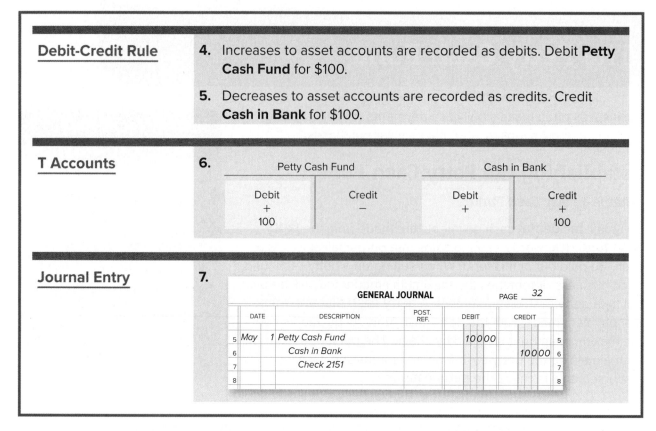

Sometimes petty cash disbursements occur more often than expected, and the petty cash fund is used up before the end of the **specified** time period. If this happens often, the company may decide to increase the fund's amount. To increase the petty cash fund, the accountant debits the **Petty Cash Fund** account and credits the **Cash in Bank** account for the amount of the increase.

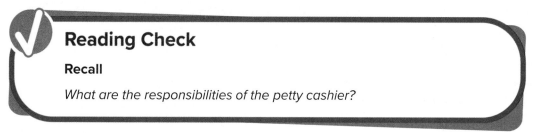

Reading Check

Recall

What are the responsibilities of the petty cashier?

Using the Petty Cash Fund

How Do You Use the Petty Cash Fund?

The petty cashier is responsible for making payments from the petty cash fund. Whenever a cash payment is made, a petty cash voucher is completed. A **petty cash voucher** is a proof of payment from the petty cash fund.

Figure 22–2 on page 709 shows a petty cash voucher. These vouchers are usually prenumbered. If they are not, the petty cashier numbers them when issuing them.

The petty cash voucher includes the following information:

1. the date of the payment
2. the person or business to whom the payment is made

3. the amount of the payment
4. the reason for the payment
5. the account to be debited
6. the signature of the person approving the payment (usually the petty cashier)
7. the signature of the person receiving the payment

PETTY CASH VOUCHER

No. 001
               1
Date May 2, 20--

2
Paid to Premier Office Supply Co.   3 $ 7.10

4
For Printer paper

5
Account Supplies

Approved by
6 *Crystal Casteel*

Payment received by
7 *John Marks*

Figure 22–2 Petty Cash Voucher

After the petty cash disbursement is made, the voucher is filed in the petty cash box until the fund is reimbursed.

Replenishing the Petty Cash Fund

To *replenish* a petty cash fund means to restore the fund to its original cash balance. As the cashier makes payments from the fund, the amount of cash in the petty cash box decreases. Some businesses set a minimum amount that must be kept. When the amount of cash in the petty cash box reaches the minimum or a low amount, the accountant replenishes the petty cash fund.

The Starting Line replenishes its petty cash fund once a month, when the balance reaches the minimum amount, or at the end of the accounting period. Replenishing the petty cash fund affects the general ledger accounts that the petty cash disbursements impact (such as **Supplies** and **Delivery Fees**), which must be updated.

Replenishing the petty cash fund requires reconciling the cash balance in the fund and then preparing a petty cash requisition form.

Reconciling the Petty Cash Fund

The petty cashier reconciles the petty cash fund to determine whether it is in balance. To reconcile the petty cash fund, the petty cashier first adds all paid petty cash vouchers. This total is then subtracted from the original cash balance of the petty cash fund. The difference is the *reconciled petty cash balance,* or the amount of

Common Mistakes

Entry to Replenish Petty Cash Fund
A common mistake made when the Petty Cash Fund is replenished is the debiting of the Petty Cash account rather than debiting the accounts affected by the petty cash payments. Remember, the accounts affected by the petty cash payment should be debited when the fund is replenished. The Petty Cash account is debited only when it is first established or when there is a need to increase the original amount.

money that *should be* in the petty cash box. If the count of the cash in the petty cash box agrees with the reconciled balance, the petty cash fund is in balance. If the two amounts do not agree, the petty cash fund is either short or over.

On May 31 the total of all the petty cash vouchers for the month was $87.75. As you recall, the original petty cash balance was $100. The reconciled petty cash balance is:

Original balance	$100.00
Total of paid petty cash vouchers	− 87.75
Reconciled petty cash balance	$ 12.25

The amount of cash in the petty cash box that Crystal counted was $12.25. The petty cash fund, therefore, is in balance.

Preparing a Petty Cash Requisition Form

After reconciling the petty cash fund, the petty cashier prepares a **petty cash requisition**, which is a form requesting money to replenish the petty cash fund. **Figure 22–3** shows a typical petty cash requisition. This form serves as the source of information for the check written to replenish the petty cash fund. The check stub then serves as the source document for the entry recorded in the general journal.

To prepare the petty cash requisition, the petty cashier first sorts the paid petty cash vouchers by account and totals the vouchers for each account. The cashier records the account title and total amount to be debited to it on the petty cash requisition. Review **Figure 22–3** again. During May The Starting Line made petty cash disbursements affecting the general ledger accounts **Supplies, Delivery Expense, Miscellaneous Expense,** and **Advertising Expense.**

PETTY CASH REQUISITION

Accounts for which payments were made:	Amount
Supplies	$24.45
Delivery Expense	19.00
Miscellaneous Expense	29.50
Advertising Expense	14.80
TOTAL CASH NEEDED TO REPLENISH FUND	$87.75

Requested by: *Crystal Casteel* Date 5/31/20--
Approved by: *Greta Keegan* Date 5/31/20--
Check No. 2341

Figure 22–3 Petty Cash Requisition

The total of all paid petty cash vouchers is the amount of cash needed to replenish the petty cash fund. After receiving the petty cash requisition from the petty cashier, the accountant writes a check for the total of the paid vouchers. The check is made payable to the petty cashier, who cashes the check and places the money in the petty cash box.

Using a Petty Cash Register

Some businesses use a **petty cash register** to record all disbursements made from the petty cash fund. The petty cash register is a **supplemental** record that summarizes the types of petty cash disbursements. It is not an accounting journal because amounts from it are not posted to general ledger accounts.

Recording Petty Cash Vouchers in a Petty Cash Register.

Not all businesses that have a petty cash fund use a petty cash register. Those who do might use a form similar to that shown in **Figure 22–4** on page 712. This illustration shows a typical petty cash register with vouchers recorded. It shows the month's disbursements following the fund's establishment.

The establishment of the petty cash fund on May 1 is noted on line 2 of the register. On each line that follows, each petty cash payment is identified by date, voucher number, and a brief explanation. The amount of each disbursement is entered in the Payments column *and* in the appropriate Distribution of Payments column. The register has three special amount columns: Supplies, Delivery Expense, and Miscellaneous Expense. The General Amount column is used for petty cash payments that do not belong in one of the three special amount columns. This column has two subdivisions, one for the account name and the other for the amount.

Totaling and Proving the Petty Cash Register.

Replenishing the petty cash fund requires totaling and proving the petty cash register. Refer to **Figure 22–4** on page 712 as you read the following steps on totaling and proving a petty cash register:

1. Enter the date the fund is being replenished in the Date column. Also enter the word *Totals* in the Description column.
2. Single rule the amount columns.
3. Foot each amount column.
4. Verify that the total of the Payments column is equal to the total of the Distribution of Payments columns.

Payments	Distribution of Payments	
	$24.45	Supplies
	19.00	Delivery Expense
	29.50	Miscellaneous Expense
	14.80	Advertising Expense
$87.75	= $87.75	

   Once verified, the totals are recorded below the footings.

5. Draw a double rule under the amount columns to show that the totals have been proved.
6. You will now enter the petty cash fund replenishment information. Skip one line, and then enter the reconciled petty cash balance (the amount of cash that *should be* in the petty cash box before it is replenished). For The Starting Line, that amount is $12.25.
7. On the next line, write the amount of the check written to replenish the petty cash fund.
8. Add the balance that *should be* in the petty cash fund and the amount of the check. The sum should equal the original amount of the petty cash fund.

Figure 22–4 A Typical Petty Cash Register

				PETTY CASH REGISTER					PAGE 1	
						DISTRIBUTION OF PAYMENTS				
DATE		VOU. NO.	DESCRIPTION	PAYMENTS	SUPPLIES	DELIVERY EXPENSE	MISC. EXPENSE	GENERAL		
								ACCOUNT NAME	AMOUNT	
20--										
May	1	—	Est. Petty Cash ($100)							
	2	1	Printer paper	7 10	7 10					
	3	2	Postage on incoming mail	2 50			2 50			
	4	3	Newspaper Ad.	9 80				Adv. Expense	9 80	
	5	4	Gas & Parking	9 50			9 50			
	7	5	Daily newspaper	3 50			3 50			
	8	6	Collect telegram	1 25			1 25			
	10	7	Pens and pencils	2 50	2 50					
	12	8	Dara's Delivery Service	9 50		9 50				
	16	9	Daily newspaper	3 50			3 50			
	18	10	Memo pads	8 45	8 45					
	20	11	Postage stamps	1 00			1 00			
	22	12	Ad in H.S. yearbook	5 00				Adv. Expense	5 00	
	26	13	File folders	6 40	6 40					
	29	14	Dara's Delivery Service	9 50		9 50				
	30	15	Gas & tolls	8 25			8 25			
	31		Totals	87 75	24 45	19 00	29 50		14 80	
			Reconciled bal. $ 12.25							
			Replen. check +187.75							
			Total $100.00							

As you can see, the petty cash register helps the petty cashier keep track of the petty cash disbursements by account. When the petty cash fund is replenished, the totals of the special columns and the amounts recorded in the General column are listed on the petty cash requisition form.

Using a Petty Cash Envelope

Small businesses sometimes use *petty cash envelopes* to record petty cash disbursements. A form very similar to the petty cash register is printed on the front of the petty cash envelope. Petty cash disbursements are recorded on the form on the envelope.

The paid petty cash vouchers are placed in the petty cash envelope. When the petty cash fund is replenished, the petty cash envelope, containing all paid vouchers for the period, is sealed and filed. A new envelope is used to record the next period's petty cash disbursements.

Reading Check

Critical Thinking

Why is the petty cash register not considered a journal?

Journalizing the Check to Replenish the Petty Cash Fund

The check stub and the petty cash requisition are the source documents for recording the journal entry for a check written to replenish the petty cash fund. Let's learn how to journalize transactions to replenish a petty cash fund.

Notice that this transaction does not affect the **Petty Cash Fund** account. Replenishing petty cash requires crediting **Cash in Bank** and debiting the accounts for which petty cash payments were made.

Business Transaction

On May 31 Check 2341 is written to replenish the petty cash fund.

Analysis

Identify
1. The accounts affected are **Supplies, Delivery Expense, Miscellaneous Expense, Advertising Expense,** and **Cash in Bank.**

Classify
2. **Supplies** is an asset account. **Delivery Expense, Miscellaneous Expense,** and **Advertising Expense** are expense accounts. **Cash in Bank** is an asset account.

+/−
3. **Supplies** is increased by $24.45. **Delivery Expense** is increased by $19. **Miscellaneous Expense** is increased by $29.50. **Advertising Expense** is increased by $14.80. **Cash in Bank** is decreased by $87.75.

Debit-Credit Rule

4. Increases in asset accounts are recorded as debits. Debit **Supplies** for $24.45. Increases in expense accounts are recorded as debits. Debit **Delivery Expense** for $19; and **Miscellaneous Expense** for $29.50. Debit **Advertising Expense** for $14.80.

5. Decreases in asset accounts are recorded as credits. Credit **Cash in Bank** for $87.75.

T Accounts

6.

Supplies		Delivery Expense	
Debit + 24.45	Credit −	Debit + 19.00	Credit −

Miscellaneous Expense		Advertising Expense	
Debit + 29.50	Credit −	Debit +	Credit −

Cash in Bank	
Debit +	Credit − 87.75

continued

Journal Entry 7.

	GENERAL JOURNAL			PAGE 41
DATE	DESCRIPTION	POST. REF.	DEBIT	CREDIT
20--				
May 31	Supplies		24 45	
	Delivery Expense		19 00	
	Miscellaneous Expense		29 50	
	Advertising Expense		14 80	
	Cash in Bank			
	Check 2341			87 75

Handling Cash Short and Over in the Petty Cash Fund

How Do You Handle Shortages and Overages in the Petty Cash Fund?

The petty cashier could occasionally make an error when paying cash from the petty cash fund. This will cause the amount of cash in the petty cash box not to agree with the reconciled petty cash balance. Any amounts of cash *gained* or *lost* through errors made by the petty cashier are recorded in the **Cash Short & Over** account.

Let's look at an example. At the end of June, Crystal Casteel, The Starting Line's petty cashier, classified and totaled the petty cash vouchers. The accounts affected by the petty cash disbursements follow:

Supplies	$15.75
Delivery Expense	20.45
Miscellaneous Expense	21.80
Advertising Expense	25.00
	83.00

	GENERAL JOURNAL			PAGE 43
DATE	DESCRIPTION	POST. REF.	DEBIT	CREDIT
20--				
June 30	Supplies		15 75	
	Delivery Expense		20 45	
	Miscellaneous Expense		21 80	
	Advertising Expense		25 00	
	Cash Short & Over		1 50	
	Cash in Bank			84 50

Figure 22–5 Journal Entry to Replenish the Petty Cash Fund

The June petty cash disbursements totaled $83. Crystal then reconciled the petty cash fund.

Original balance	$100
Total of paid petty cash vouchers	− 83
Reconciled petty cash balance	$ 17

Crystal counted the cash in the petty cash box and found only $15.50—a shortage of $1.50 ($17.00 − 15.50). Bringing the petty cash fund up to the original $100.00 requires $84.50 ($83.00 + $1.50). The $1.50 cash shortage is an expense and is debited to the **Cash Short & Over** account.

When Crystal prepared the petty cash requisition form, she listed the accounts to be debited for the petty cash disbursements. She also indicated that the **Cash Short & Over** account is to be debited for $1.50. The journal entry to record the replenishment is shown in **Figure 22–5.**

A petty cash *overage* is recorded in a similar manner. If, for example, the cash in the petty cash box is $17.75, a cash overage of 75¢ exists. **Cash Short & Over** is credited for that amount in the journal entry. Instead of needing $83.00 to replenish the fund, only $82.25 is required.

The cash shortage or overage is also reported in the petty cash register if a business uses one. **Figure 22–6** shows how the June cash shortage of $1.50 is recorded in the petty cash register.

PETTY CASH REGISTER PAGE 2

	DATE	VOU. NO.	DESCRIPTION	PAYMENTS	SUPPLIES	DELIVERY EXPENSE	MISC. EXPENSE	GENERAL ACCOUNT NAME	AMOUNT	
1	20--									1
2	June 1	16	Postage stamps	5 00			5 00			2
14	30	29	Button	1 25				Adv. Expense	1 25	14
15	30		Totals	83 00	15 75	20 45	21 80		25 00	15
16										16
17			Reconciled bal. $ 17.00							17
18			Cash short (1.50)							18
19			Replen. check 84.50							19
20			Total $100.00							20
21										21

Figure 22–6 Recording a Cash Shortage in the Petty Cash Register

SECTION 22.2
Assessment

After You Read

Reinforce the Main Idea

The journal entry to replenish the petty cash fund depends on whether the fund is in balance, over, or short. Use a diagram like this one to show the three possible journal entries using the accounts **Cash in Bank, Supplies, Delivery Expense, Miscellaneous Expense,** and **Cash Short and Over.**

Transaction Description	Account Debited	Account Credited

Problem 3 — Analyzing a Source Document

The petty cash clerk for Riddle's Card Shop prepared the accompanying petty cash requisition.

Instructions Review the document and prepare Check 973 to replenish the petty cash fund in your working papers.

PETTY CASH REQUISITION

Accounts for which payments were made:	Amount
Office Supplies	$12.40
Postage Expense	12.00
Misc. Expense	16.00
Cash Short and Over	(.89)
TOTAL CASH NEEDED TO REPLENISH FUND	$39.51

Requested by: __Brent Roy__ Date 2/28/20--
Approved by: __Hugh Morrison__ Date _____
Check No. 941

Math for Accounting

It is another busy day at Olde Time Swimming Hole, a private swimming facility for your neighborhood. You have a summer job running the concession stand, and you maintain the petty cash fund. To make your accounting easier, you record the cash sales and the petty cash transactions in one report at the end of the day. You always attach the cash register tape to prove the cash drawer. Using the following figures, determine the revenue and expenses for the day. How much money will remain in the cash drawer?

Opening cash drawer	$225
Petty cash fund	100
Cash sales	600
Pool supplies bought	55
Photocopy paper purchased	8

Chapter 22
Visual Summary

Concepts

At the end of each business day, a cash proof is prepared to reconcile the amount of cash in the register to the recorded cash sales. To prove cash, count the cash in the cash register drawer, subtract the change fund, and compare that amount with the amount of total cash sales, shown on the cash register tape.

Counted cash in cash register drawer **—** Change Fund **=** Total cash sales from cash register tape

Analysis

Here is the entry to establish a petty cash fund:

	DATE	DESCRIPTION	POST. REF.	DEBIT	CREDIT	
1	20--					1
2	Date	Petty Cash Fund		x x x x x		2
3		Cash in Bank			x x x x x	3
4						4

GENERAL JOURNAL   PAGE 85

Procedures

At the end of each business day, a cash proof is prepared to reconcile the cash in the register to the recorded cash sales.

- To record a cash shortage:

  Debit: **Cash in Bank**   Credit: **Sales**
  Debit: **Cash Short and Over**   Credit: **Sales Tax Payable**

- To record a cash overage:

  Debit: **Cash in Bank**   Credit: **Sales**
                                         Credit: **Sales Tax Payable**
                                         Credit: **Cash Short and Over**

Chapter 22
Review and Activities

Answering the Essential Question

How do businesses monitor and protect cash?

Special procedures are necessary for cash. Why does cash need to be so carefully protected?

Vocabulary Check

1. **Vocabulary** Write a brief essay titled "Two Cash Funds," using and underlining all of the content vocabulary terms.

   - change fund
   - establish
   - attach
   - incidental
   - petty cash disbursement
   - petty cash fund
   - petty cashier
   - petty cash voucher
   - specify
   - petty cash register
   - petty cash requisition
   - supplemental

Concept Check

2. What is a change fund, and which accounts are debited and credited when a change fund is established?
3. **Evaluate** How and when is cash proofed? Explain why proofing cash is important to a business.
4. Why would a business set up a petty cash fund?
5. Describe how to set up a petty cash fund.
6. **Analyze** What information is recorded in a petty cash register? Why might a company keep a petty cash register?
7. Why is a petty cash register not considered a journal?
8. **Synthesize** Describe a scenario that would result in a cash shortage.
9. Why is a cash shortage treated like an expense?
10. **Math** For her travel agency, Kari borrowed $25,000 at 6% interest for four years. Calculate the amount of interest she will pay.
11. **English Language Arts** Think of an item you wanted for which you had to save money in order to buy. Tell what sacrifices you made to save enough to buy the item, what things you had to give up and what extra work you did, if any, to earn the money. Was the thing you wanted worth the sacrifices you made? Give your answer in a one-page essay.

Chapter 22
Standardized Test Practice

Multiple Choice

1. An amount of cash kept on hand and used for making small payments is called
   a. revenue.
   b. cash.
   c. petty cash.
   d. prepaid interest.
   e. none of these answers

2. A form showing proof of a petty cash payment is a
   a. check.
   b. petty cash slip.
   c. petty cash check stub.
   d. journal.

3. A petty cash on hand amount that is more than a recorded amount is called
   a. cash over.
   b. cash short.
   c. cash credit.
   d. cash debit.

4. The entry to replenish the petty cash fund requires

Debit	Credit
a. Petty Cash Fund	Cash
b. Petty Cash Expense	Petty Cash Fund
c. Cash	Petty Cash Fund
d. Various Expense accounts	Cash

5. Michelle is a cashier for The Pet Store. When she counted the cash in her register drawer at the end of the day, the total was $959.74. According to the electronic register, she should have had a balance of $969.85. Was Michelle short or over in her drawer and by how much?
   a. $10.11 short
   b. $10.11 over
   c. $10.38 short
   d. $10.38 over

True or False

6. When a change fund is established, the Change Fund account is debited, and the Cash in Bank account is credited.

Short Answer

7. When does a cash shortage occur?

8. What information is contained on a petty cash voucher?

Chapter 22
Computerized Accounting

Maintaining Cash Funds

Making the Transition from a Manual to a Computerized System

MANUAL METHODS	COMPUTERIZED METHODS
• A check stub or deposit slip is the source document for journalizing cash payments.	• Checks and deposits are automatically entered in the check register when transactions are journalized and posted
• Add deposits/deduct the checks to keep a running balance in the check register.	• The checking account balance is updated as each transaction is posted.
• Compare the check register with the bank statement.	• Enter the bank statement balance.
• Add deposits in transit to the statement balance. Deduct outstanding checks.	• Verify outstanding checks and deposits not included on statement.
• Deduct bank charges from the check register.	• Record bank charges as a payment.
• Journalize and post bank charges.	

Chapter 22
Problems

Problem 4 — Establishing a Change Fund

On February 1 Sunset Surfwear issued Check 115 to establish a change fund of $150. At the end of the business day on February 2, the shop's cash register tape showed cash sales of $340 plus sales taxes of $17. An actual cash count of the money indicated that $505 was in the cash register drawer.

Instructions

1. Record the entry to establish the change fund on page 1 of the general journal.
2. Prepare a cash proof for Feb. 2. Sign your name on the Salesclerk line.
3. Record the cash sales for Feb. 2 on page 1 of the general journal.

Analyze

Determine whether the shortage or overage represents revenue or expense to the business.

Problem 5 — Establishing and Replenishing a Petty Cash Fund

InBeat CD Shop established a petty cash fund for $100.

Instructions

1. In your working papers, record the entry to establish the petty cash fund on page 6 of a general journal.
2. Record the entry for replenishing the petty cash fund on page 10 of the general journal. There was $5 cash in the petty cash fund box on February 28.

Date	Transactions
Feb. 1	Issued Check 112 for $100 to establish the petty cash fund.
28	Issued Check 146 to replenish the petty cash fund. Paid petty cash vouchers included Supplies, $40; Advertising Expense, $16; Maintenance Expense, $27; and Miscellaneous Expense, $12.

Analyze

Conclude whether cash was short or over.

Chapter 22
Problems

Problem 6 — Establishing and Replenishing a Petty Cash Fund

Shutterbug Cameras, a camera store, decided to establish a petty cash fund. On February 1 the accountant, Al Rosen, issued Check 1018 for $70 to establish the fund. The following disbursements were made.

Instructions

1. In your working papers, record the entry to establish the petty cash fund on page 9 of a general journal.
2. Make a list of the paid petty cash vouchers.
3. Classify the petty cash disbursements by account. Calculate the total amount paid out for each account.
4. Prepare a petty cash requisition, signing your name as the petty cashier. On February 28 there was $1.50 in the petty cash box.
5. Record the entry in the general journal (page 11) to replenish the petty cash fund on February 28. Use Check 1191.

Date		Transactions
Feb.	1	Purchased memo pads for the office, $2.75, Voucher 101 (**Supplies**).
	3	Prepared Voucher 102 for a newspaper ad, $7.50 (**Advertising Expense**).
	5	Prepared Voucher 103 for the postage on an outgoing package, $1.75 (**Miscellaneous Expense**).
	8	Paid Dandy Delivery Service $5.65, Voucher 104 (**Delivery Expense**).
	10	Prepared Voucher 105 for pens and pencils, $3.75 (**Supplies**).
	12	Paid $2.20 for postage stamps, Voucher 106 (**Miscellaneous Expense**).
	15	Paid Dandy Delivery Service $6.75, Voucher 107 (**Delivery Expense**).
	20	Paid the news carrier $4.25 for delivery of the daily newspaper, Voucher 108 (**Miscellaneous Expense**).
	22	Bought typing paper for $7.50, Voucher 109 (**Supplies**).
	25	Paid $4.50 to Dandy Delivery Service, Voucher 110 (**Delivery Expense**).
	27	Prepared Voucher 111 for an advertisement, $10 (**Advertising Expense**).
	28	Paid $4.40 for postage stamps, Voucher 112 (**Miscellaneous Expense**).
	28	Prepared Voucher 113 for an advertisement, $7.50 (**Advertising Expense**).

Analyze

Calculate the total petty cash disbursements for Delivery Expense during February.

Chapter 22
Problems

Problem 7 Using a Petty Cash Register

Cycle Tech Bicycles decided to establish a petty cash fund. On February 1 the accountant issued Check 3724 for $120 to establish the fund. The following disbursements were made.

SOURCE DOCUMENT PROBLEM

Problem 22–7

Use the source documents in your working papers to complete this problem.

Instructions

1. In your working papers, record the entry to establish the petty cash fund in a general journal, page 5.

2. Enter the information about the establishment of the petty cash fund on line 1 of a petty cash register, page 1.

3. Record the petty cash disbursements in the petty cash register.

4. Foot, total, and prove the petty cash register on February 28.

5. Record the petty cash fund replenishment information in the explanation column below the totals. On February 28 there was $4.35 in the petty cash box.

6. Prepare a petty cash requisition form. Use the form provided in your working papers and sign your name as petty cashier.

7. Record the issuance of Check 3875 to replenish the petty cash fund in the general journal, page 8.

Date		Transactions
Feb.	2	Prepared Voucher 1 for a $9.25 newspaper advertisement (**Advertising Expense**).
	5	Prepared Voucher 2 for pens and pencils, $5 (**Supplies**).
	9	Paid $12.50 for flowers for an employee's birthday, Voucher 3 (**Miscellaneous Expense**).
	12	Bought cash register tape for $3.95, Voucher 4 **(Supplies).**
	19	Prepared Voucher 5 for $15 to pay National Express for parts delivered (**Delivery Expense**).
	20	Paid $3.90 for postage stamps, Voucher 6 (**Miscellaneous Expense**).
	22	Paid $16 to have the show window cleaned, Voucher 7 (**Miscellaneous Expense**).
	24	Bought an $11 advertisement in the local newspaper, Voucher 8 (**Advertising Expense**).
	25	Bought stationery for $10, Voucher 9 (**Supplies**).
	26	Prepared Voucher 10 to National Express for packages delivered, $8.25 (**Delivery Expense**).
	27	Prepared Voucher 11 incorrectly and voided it.
	27	Purchased memo pads for the office for $3, Voucher 12 (**Supplies**).

continued

Chapter 22
Problems

Date	Transactions (cont.)
Feb. 28	Paid the news carrier a $4 tip for the daily newspaper delivery, Voucher 13 (**Miscellaneous Expense**)
28	Paid $6.80 for postage stamps, Voucher 14 (**Miscellaneous Expense**).
28	Prepared Voucher 15 for a $10 newspaper advertisement (**Advertising Expense**).

Analyze
Identify the number of payments that were made from the petty cash fund during the month of February.

Problem 8 Handling a Petty Cash Fund

River's Edge Canoe & Kayak petty cash fund was established on February 1 for $100, by writing check 1763. The accounts for which petty cash disbursements are likely to be made include **Supplies, Gas Expense, Advertising Expense, Delivery Expense,** and **Miscellaneous Expense.**

Instructions

1. In your working papers, record the entry to establish the petty cash fund on page 12 of a general journal.
2. Record the establishment of the fund on the first line of the petty cash register, page 1.
3. Record each petty cash disbursement in the petty cash register.
4. Foot, prove, total, and rule the petty cash register on February 28.
5. Reconcile the petty cash fund. The amount in the petty cash box is $1.50.
6. Prepare a petty cash requisition. Sign your name as petty cashier.
7. Record the entry to replenish the petty cash fund by issuing Check 1798 in the general journal, page 15.
8. Record the replenishment information in the petty cash register.
9. The accountant believes the petty cash fund should be increased by $25. Record the issuance of Check 1799 on February 28 on page 16 of the general journal.

continued

Chapter 22
Problems

Date		Transactions
Feb.	1	Bought an $8 advertisement in the local newspaper, Voucher 101.
	2	Prepared Voucher 102 for $7.50 to pay Mercury Messenger for packages delivered.
	3	Bought adding machine tape for 75¢, Voucher 103.
	5	Paid $9.50 for flowers for an employee's birthday, Voucher 104.
	7	Prepared Voucher 105 for $5.25 for printer toner.
	9	Paid $4.40 for postage stamps, Voucher 106.
	12	Bought $9 worth of gasoline, Voucher 107.
	15	Paid $8.50 to have the shop's windows washed, Voucher 108.
	18	Bought memo pads, pencils, and pens for office use, $6.30, Voucher 109.
	20	Prepared Voucher 110 for $7.50 to pay Mercury Messenger for packages delivered.
	23	Prepared Voucher 111 incorrectly and voided it.
	23	Bought stationery for $8, Voucher 112.
	27	Paid the news carrier $4.75 for the daily newspaper, Voucher 113.
	28	Prepared Voucher 114 for $7.50 to pay Mercury Messenger for packages delivered.
	28	Bought gasoline, $5.80, Voucher 115.
	28	Prepared Voucher 116 for a newspaper advertisement, $5.

Analyze

List the petty cash disbursements that are charged to the Miscellaneous Expense account.

Chapter 22
Problems

Problem 9 Locating Errors in a Petty Cash Register

On February 1 a petty cash fund of $150 was established for Buzz Newsstand, Check 948. The petty cashier writes a voucher for each petty cash disbursement. The vouchers are entered in a petty cash register, which is included in your working papers. When the petty cash register was totaled on February 28, the accounting clerk discovered that the footings of the Distribution of Payments columns did not equal the total of the Payments column.

Instructions

1. Compare the petty cash disbursement information in your working papers with the entries in the petty cash register.
2. Correct any errors you find in the petty cash register by drawing a line through the incorrect item and writing the correction above it.
3. Total all columns after the corrections are made.
4. Record the replenishment information on the register. The amount in the petty cash box on February 28 was $8.10.

Analyze
Explain how you determined the amount of the check to replenish the petty cash fund on February 28.

Real-World Applications & Connections

CASE STUDY

Merchandising Business: Health Foods

The Healthy Alternative sells vitamins and natural health-care products. It has two electronic cash registers that track inventory and record sales directly to the computerized accounting system. The store owner is concerned because the actual cash on hand is usually short when it is compared to the sales records.

INSTRUCTIONS

1. You work for a local CPA firm that is auditing The Healthy Alternative's accounting records. What advice would you give the owner about cash controls and protection?

2. Explain to the owner why it is important for the cash records to match the accounting records.

21st century Skills

Special Savings Programs

Many employers offer savings programs to help employees save for end-of-year purchases. The employee makes regular, automatic withdrawals from each paycheck to a company-managed savings account. The employee can withdraw the total amount at the end of the year.

Discuss and Write

a. As a class, discuss why some people set aside money for the end of the year.

b. Write a list of items for which you would like to save.

Career Wise

Management Accountants and Senior Management Accountants

One way to enhance, or further an accounting career would be to seek professional development through lifelong learning. Lifelong learning goals might include taking a college course, a training seminar, or even pursuing an advanced degree. For example, management accountants may gain experience to pursue senior management accounting positions.

Activity

Conduct research for both management accountants and senior management accountants. How are the educational requirement, roles, and responsibilities similar? How are they different? What experience would a management accountant need to grow into a senior management accountant position?

Spotlight on Personal Finance

Your Spending Records

You should track your cash daily to be aware of the cash you have at the start and end of each day. If the amounts differ, you should have receipts for expenses or estimate where the remainder went such as for cold drinks or snacks. If you cannot account for the difference, you should pay more attention to spending cash.

Activity

For three days, track your personal expenses like a petty cash fund. Record how much you spend on each item on an envelope or sheet of paper. At the end of each day, is your cash over or short?

H.O.T Audit

Petty Cash Requisition

Review the Petty Cash Requisition form below and the journal that was prepared to replenish the petty cash fund on June 30. The petty cash fund was established for $50.00. There was $2.00 in the petty cash fund box on June 30.

PETTY CASH REQUISITION

Accounts for which Payments were made:	Amount
Supplies	$12.00
Delivery Expense	10.00
Miscellaneous Expense	11.00
Advertising Expense	13.00
Cash Short and Over	2.00

TOTAL CASH NEEDED TO REPLENISH FUND $48.00

GENERAL JOURNAL
20—

June 30 Supplies	$12.00	
Delivery Expense	10.00	
Misc Expense	11.00	
Advertising Expense	13.00	
Cash Short and Over	2.00	
Cash		$48.00
Check# 500		

Chapter 23
Plant Assets and Depreciation

Chapter Topics

23-1 Plant Assets and Equipment

23-2 Calculating Depreciation

23-3 Accounting for Depreciation Expense at the End of a Year

Visual Summary

Review and Activities

Standardized Test Practice

Computerized Accounting

Problems

Real-World Accounting Careers

Real-World Applications & Connections

Essential Question

As you read this chapter, keep this question in mind:

How does the value of an asset change as it gets older?

Main Idea

A business uses plant assets for more than one accounting period, so it spreads the cost of these assets over a number of years. A business must also calculate depreciation of certain plant assets.

Chapter Objectives

Concepts	Analysis	Procedures
C1 Identify plant assets.	**A1** Calculate partial-year depreciation of plant assets.	**P1** Record depreciation of plant assets.
C2 Explain the need to depreciate plant assets.	**A2** Determine the book value of a plant asset.	**P2** Prepare depreciation schedules.
C3 Calculate annual depreciation of plant assets.		

Real-World Business Connection

John Deere (Deere & Company)

"Nothing Runs Like a Deere®" is an effective marketing slogan for the world's largest manufacturer of agricultural and forestry equipment, and a major provider of products and services for construction, lawn and turf care, landscaping and irrigation. The *Fortune* 500 company, employing more than 50,000 people in 35 countries, was founded in 1837 when Illinois blacksmith John Deere developed the first commercially successful cast-steel plow.

Connect to the Business

A manufacturer such as John Deere uses plant assets—examples are land, buildings, and manufacturing equipment—to generate revenue over many years. Depreciation allows the company to match the cost of assets to revenue over several accounting periods.

Analyze

Assume John Deere buys a truck to transport goods to dealerships across the country and uses it for 10 years. What costs are associated with buying the truck? Will the truck have any value when it needs to be replaced?

Focus on the Photo

Agricultural equipment like this John Deere combine makes it possible for American farmers to plant and harvest crops such as soybeans. **What other assets might a farming business purchase in order to generate revenue?**

SECTION 23.1
Plant Assets and Equipment

Businesses own many different types of assets. One category of asset requires special treatment in the accounting records. These assets, such as office equipment and buildings, have two things in common:

- They are expected to produce benefits for the business for more than one year.
- They are purchased for use in operating the business, not for resale.

Rush Trucking owns many assets in this category, including computer equipment, office equipment, and trucks. Let's explore how businesses account for these types of assets and learn how to calculate and record the annual and partial-year depreciation of plant assets.

Section Vocabulary

- plant assets
- depreciation
- disposal value
- straight-line depreciation
- consumed

Current and Plant Assets

How Do You Match the Cost of Assets to the Revenue They Help Generate?

Throughout this textbook you have learned about various assets that a business uses in its operation. These assets can be classified as either current assets or **plant assets**.

Current Assets	Plant Assets
Assets that are either consumed or converted to cash during the normal operating cycle of the business, usually one year. Examples are: - cash - accounts receivable (cash to be collected from customers within a short period of time) - merchandise (sold within a short period of time)	Long-lived assets that are used in the production or sale of other assets or services over several accounting periods. Examples are: - land - buildings - delivery equipment - store equipment - office equipment

In Chapter 18 you learned about assets such as supplies and prepaid insurance. As these assets are used, their costs are converted to expenses. This conforms to the *matching principle* of accounting, which states that during an accounting period, expenses must be matched with the revenue earned. Since current assets are **consumed** within one accounting period, the costs of current assets can be easily matched to the revenue for the period.

Reading Check

Recall

How are current assets and plant assets similar and different?

Estimating Depreciation of a Plant Asset

What Four Factors Are Used to Estimate Depreciation?

Plant assets are used over a number of accounting periods. To follow the matching principle, the cost of a plant asset is spread over, or allocated to, the periods in which the asset will be used to produce revenue.

Allocating, or spreading the cost of a plant asset over that asset's useful life is called **depreciation**. For accounting purposes businesses depreciate all plant assets except land. The cost of land is not depreciated because land is considered to have an unlimited useful life. In this chapter you will learn how to calculate and record the depreciation of plant assets.

Common Mistakes

Depreciation A common mistake is to depreciate land. Land is not depreciated because its service life never ends.

For example, suppose that a plant asset costs $40,000 and has a useful life of 10 years. The cost of the asset is depreciated over 10 years. A portion of the $40,000 is transferred to an expense account each year. At the end of 10 years, the cost of this plant asset will have been recognized as an expense.

It is important to remember that depreciation is an *estimate*. No one can predict with certainty the useful life or the disposal value of an asset.

Four factors are used to calculate depreciation of a plant asset:

- its cost
- its estimated useful life
- its estimated disposal value
- the depreciation method used

Plant Asset Cost

The cost of a plant asset is the price the business paid to purchase it plus any sales taxes, delivery charges, and installation charges. The total cost is the amount debited to the plant asset account (for example, **Delivery Equipment**) at the time of purchase.

Estimated Useful Life of a Plant Asset

The *estimated useful life* of a plant asset is the number of years it is expected to be used before it wears out, becomes outdated, or is no longer needed by the business. The number of years a plant asset can be used varies from one asset to another. A delivery truck might have a useful life of six years. A building, on the other hand, might have a useful life of 30 years.

The useful life, also called the *service life*, might not be as long as the asset's total productive life. For example, the productive life of a computer can be eight years or more. Some companies, however, trade in old computers for new ones every two years. In this case, these computers have a two-year useful life. This means the cost of these computers (less their expected trade-in value) is charged to depreciation expense over a two-year period.

In estimating useful life, the accountant considers past experiences with the same type of asset. The Internal Revenue Service (IRS) also publishes guidelines on the estimated useful lives for many types of assets.

Reading Check

Recall

What are the four factors that are used to calculate depreciation on a plant asset?

Estimated Disposal Value of a Plant Asset

At some point a plant asset will be replaced or discarded. Usually this occurs while the asset still has some monetary value. For example, if a business buys a new delivery truck, the old delivery truck can often be traded in to reduce the price of the new truck.

The estimated amount that a plant asset will be worth at the time of its replacement is called the **disposal value**. The disposal value assigned to a plant asset is an estimate that is based on previous experience. The IRS also publishes guidelines on disposal values.

Depreciation Methods

Several methods for computing depreciation expense are acceptable. Here you will learn a simple, widely used depreciation method called the *straight-line method*. **Straight-line depreciation** equally distributes the depreciation expense over the asset's estimated useful life. Other methods of computing depreciation include units-of-production and accelerated methods.

- *Units-of-production method* estimates useful life measured in units of *use* rather than units of *time*.
- *Accelerated depreciation methods* are based on the theory that an asset loses more value in the early years of its useful life than in the later years. Two types of accelerated depreciation are the *sum-of-the-years'-digits* method and the *declining-balance* method.

Depreciation for Tax Reporting

The federal income tax law has rules for depreciating assets. These rules include the accelerated cost recovery system (ACRS). It is called *accelerated* because it allows the business to recognize depreciation expense over a shorter period of time. The ACRS method does not take disposal value into consideration. Congress modified ACRS in 1986 resulting in *MACRS* (pronounced makers), the *modified accelerated cost recovery system.* This system is used for tax accounting purposes only. MACRS is not acceptable for financial reporting, because it often allocates costs over an arbitrary period that is less than the asset's useful life, and it fails to estimate disposal value. It is intended to be an incentive for businesses to invest in plant assets. The higher depreciation expense results in a lower income tax liability.

SECTION 23.1
Assessment

After You Read

Reinforce the Main Idea

Use a diagram like this one to show the four factors that are used to calculate a plant asset's depreciation.

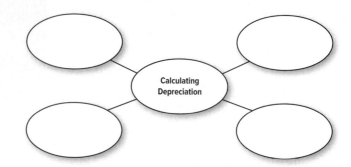

Problem 1 Classifying Asset Accounts

Listed here are the assets of New England Sports Equipment Inc.

- Accounts Receivable
- Building
- Cash in Bank
- Change Fund
- Delivery Equipment
- Land
- Merchandise Inventory
- Office Equipment
- Office Furniture
- Petty Cash Fund
- Prepaid Insurance
- Store Equipment
- Supplies

Instructions In your working papers, indicate whether each asset listed is a current asset or a plant asset by placing a check mark in the correct column. The first account is completed as an example.

Asset	Current Asset	Plant Asset
Accounts Receivable	✓	

Math for Accounting

You work for Island Tropics, a trendy clothing store. It recently purchased a computer system for $20,000. The computers have an estimated useful life of five years. The $20,000 cost can be depreciated over the useful life of the asset as an expense on the tax return of the business. What is the amount that can be deducted each year as an expense if the computer's estimated disposal value is $1,000 and the straight-line depreciation method is used?

SECTION 23.2
Calculating Depreciation

The Starting Line Sports Gear purchased a delivery truck on January 5 for $16,500 cash. The truck has an estimated disposal value of $1,500 and an estimated useful life of five years.

Section Vocabulary
- accumulated depreciation
- book value

Calculating Depreciation

How Do You Calculate Plant Asset Depreciation?

To calculate depreciation you need to know the cost of the truck, its estimated useful life, and its estimated disposal value.

First calculate the amount to be depreciated:

Original Cost	−	Estimated Disposal Value	=	Amount to Be Depreciated
$16,500	−	$1,500	=	$15,000

The estimated disposal value represents the part of the asset's cost that the business expects to recover. Therefore, the estimated disposal value should not be treated as an expense.

Straight-Line Depreciation

Calculate the annual depreciation expense using the straight-line method:

Amount to Be Depreciated	÷	Estimated Useful Life	=	Annual Depreciation Expense
$15,000	÷	5	=	$3,000

The annual depreciation expense for the delivery truck is $3,000.

For straight-line depreciation, the depreciation rate is 1 divided by the years of useful life. In this example the depreciation rate is 20% per year (1 ÷ 5 years).

Note that the $3,000 depreciation expense is for a full year. Suppose that The Starting Line purchased the delivery truck on April 5 instead of January 5. During the first year, the delivery truck will be used for only nine months. Therefore, the depreciation expense is calculated for nine months.

Common Mistakes

Partial-Year Depreciation You may think April 5 to the end of the year is eight months since December is the twelfth month and April is the fourth month. April 5 to the end of the year is actually nine months, every month except January, February, and March.

Annual Depreciation Expense	×	Fraction of Year	=	Partial-Year Depreciation Expense
$3,000	×	9/12	=	$2,250

Chapter 23 • Plant Assets and Depreciation 739

Declining-Balance Depreciation

With the declining-balance method, the annual depreciation expense is the asset's book value multiplied by the declining-balance rate. The declining-balance rate can vary, but it is usually double the straight-line rate. For a five-year asset, the rate is 40% (2 × [1 ÷ 5]).

Disposal value is not used in the computation. However, after the last year of depreciation, the total depreciation expense "stops" at the asset's disposal value. The annual depreciation expense for the truck's first year is $6,600 ($16,500 × 40%).

Plant Asset Records

Where Do You Record Plant Asset Values?

Businesses maintain records for each plant asset and the depreciation taken for that asset. The record in **Figure 23–1** provides detailed information about the delivery truck, including:

1. the date of purchase
2. the original cost
3. the estimated useful life
4. annual depreciation
5. accumulated depreciation
6. book value at the end of each year

The lower part of the plant asset record contains the depreciation schedule. The amount of depreciation expense accumulates from one year to the next. **Accumulated depreciation** is the total amount of depreciation for a plant asset that has been recorded up to a specific point in time. The accumulated depreciation at the end of the third year is $9,000.

The far right column of the depreciation schedule shows the **book value** of the plant asset, the original cost less accumulated depreciation.

At the end of the third year, the book value of the delivery truck is $7,500 ($16,500 − $9,000). Note that the delivery truck's book value at the end of five years is $1,500. This is the truck's estimated disposal value. Under the straight-line method, it cannot be depreciated below its estimated disposal value.

PLANT ASSET RECORD

ITEM: Delivery Truck
SERIAL NUMBER: 2911-50041
PURCHASED FROM: Winding Creek Auto
ESTIMATED LIFE: 5 years **3**
GENERAL LEDGER ACCOUNT: Delivery Equipment
MANUFACTURER: VanPower
EST. DISPOSAL VALUE: $1,500.00
LOCATION: Company Garage

DEPRECIATION METHOD: Straight-line
DEPRECIATION PER YEAR: $3,000.00 **4**

5

DATE	EXPLANATION	ASSET			ACCUMULATED DEPRECIATION			BOOK VALUE
		DEBIT	CREDIT	BALANCE	DEBIT	CREDIT	BALANCE	
1 1/5/2008	Purchased	16,500 **2**		16,500				**6** 16,500
12/31/2008						3,000	3,000	13,500
12/31/2009						3,000	6,000	10,500
12/31/2010						3,000	9,000	7,500
12/31/2011						3,000	12,000	4,500
12/31/2012						3,000	15,000	1,500

Figure 23–1 Plant Asset Record

SECTION 23.2
Assessment

After You Read

Reinforce the Main Idea

Using a diagram like this one, show the step-by-step procedure to record the first year's depreciation using the straight-line method.

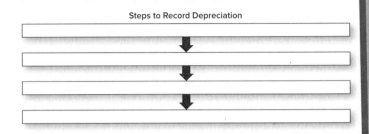

Steps to Record Depreciation

Problem 2 Calculating Depreciation Expense

Instructions For each of the following plant assets:

1. Calculate the amount to be depreciated,
2. Calculate the annual depreciation expense using the straight-line method,
3. Calculate the depreciation expense for the first year.

Use the form provided in your working papers.

Plant Asset	Months Owned First Year	Original Cost	Estimated Disposal Value	Estimated Useful Life
1. Cash register	8	$ 450	$ 30	7 years
2. Computer	2	6,500	1,500	5 years
3. Conference table	6	1,900	100	25 years
4. Delivery truck	3	36,400	6,400	5 years
5. Desk	11	3,180	300	20 years

Problem 3 Completing a Plant Asset Record

Use the following information to complete the blank asset record found in your working papers. The company uses the straight-line method for depreciating all plant assets. Item Purchased: Xerox Copier—Serial No. X42599757, $12,500, from K&C Office Equipment.

Purchased: 10/1/2011
Estimated Life: 5 years
Location: Executive Offices
Estimated Disposal Value: $1,700

Math for Accounting

Office furniture for Teen Counseling Center is estimated at a total value of $45,000 with a 10% disposal value. The estimated useful life is four years. Calculate the annual depreciation expense for the office furniture using the straight-line method.

SECTION 23.3
Accounting for Depreciation Expense at the End of a Year

After depreciation on plant assets has been calculated, adjustments are made to record depreciation for the period. These adjustments bring the general ledger into agreement with the plant asset records.

Adjusting for Depreciation Expense

How Do You Adjust for Depreciation Expense?

When a plant asset is purchased, the accountant sets up a depreciation schedule for the asset like the one in **Figure 23–1** on page 738. The amount of depreciation expense for each plant asset is recorded in the accounting records at the end of the year. The information to record the adjustments for depreciation comes from the plant asset records.

Many businesses prepare a summary of depreciation expense for each type of plant asset. For example, a business may have 10 delivery trucks. Each truck has its own plant asset record. At the end of the year, the depreciation expense for all 10 trucks is totaled. This total is entered on a summary form under the name of the asset account, in this case, **Delivery Equipment. Figure 23–2** shows The Starting Line's depreciation summary form for its plant assets.

The Starting Line's total depreciation expense for the year is $33,000. This amount includes the depreciation expense for all plant assets. The accumulated depreciation for all of The Starting Line's plant assets is $100,250.

Making the Depreciation Expense Adjustment

The adjustment for depreciation affects two accounts: **Depreciation Expense** and **Accumulated Depreciation.**

The Depreciation Expense Account.

Depreciation Expense is an expense account. During the year the account has a zero balance because the adjustment for depreciation is recorded at the end of

2008 SUMMARY OF DEPRECIATION EXPENSE
December 31, 2008

Asset	Cost	Depreciation Expense	Depreciation to Date
Building	50,000	2,500	8,125
Delivery Equipment	16,500	3,000	9,000
Office Equipment	50,000	2,500	8,125
Store Equipment	250,000	25,000	75,000
Totals	366,500	33,000	100,250

Figure 23–2 Depreciation Summary Form

the period. **Depreciation Expense** is reported on the income statement. At the end of the year, **Depreciation Expense** is closed to **Income Summary.**

Businesses have a depreciation expense account for each type of plant asset. Some examples are:

- **Depreciation Expense—Delivery Equipment** (trucks, vans, automobiles)
- **Depreciation Expense—Office Furniture** (desks, chairs, filing cabinets)

The Accumulated Depreciation Account.

The balance of **Accumulated Depreciation** represents the total amount of depreciation expensed since the business purchased the asset. Each type of plant asset has an accumulated depreciation account. Typical account names are:

- **Accumulated Depreciation—Delivery Equipment**
- **Accumulated Depreciation—Building**

Accumulated Depreciation is classified as a contra asset account. Recall that the balance of a contra account reduces the balance of its related account. In the case of an accumulated depreciation account, the related account is a plant asset account. For example, if the asset account is **Delivery Equipment,** the contra asset account is **Accumulated Depreciation—Delivery Equipment.**

Accumulated Depreciation	
Debit	Credit
−	+
Decrease Side	Increase Side
	Normal Balance

The debit and credit rules for an accumulated depreciation account are opposite those for an asset account. The balance of an accumulated depreciation account is reported on the balance sheet as a decrease to its related plant asset account.

Account Name

Asset	Expense	Contra Asset
Del. Equip.	Depr. Exp.—Del. Equip.	Accum. Depr.—Del. Equip.
Office Equip.	Depr. Exp.—Office Equip.	Accum. Depr.—Office Equip.
Store Equip.	Depr. Exp.—Store Equip.	Accum. Depr.—Store Equip.

The Adjustment.

Let's learn how to record depreciation of plant assets. Look at The Starting Line's depreciation schedule in **Figure 23–1** on page 738. The delivery truck annual depreciation expense is $3,000.

continued

Adjustment

On December 31 the accounting clerk for The Starting Line records the depreciation for the delivery truck.

Analysis	Identify	1. The accounts affected are **Depreciation Expense—Delivery Equipment** and **Accumulated Depreciation—Delivery Equipment**.
	Classify	2. **Depreciation Expense—Delivery Equipment** is an expense account. **Accumulated Depreciation—Delivery Equipment** is a contra asset account.
	+/−	3. Both **Depreciation Expense—Delivery Equipment** and **Accumulated Depreciation—Delivery Equipment** are increased by $3,000.
Debit-Credit Rule		4. Increases to expense accounts are recorded as debits. Debit **Depreciation Expense—Delivery Equipment** for $3,000.
		5. Increases to contra asset accounts are recorded as credits. Credit **Accumulated Depreciation—Delivery Equipment** for $3,000.
T Accounts		6.

Depreciation Expense—Delivery Equipment

Debit + 3,000	Credit −

Accumulated Depreciation—Delivery Equipment

Debit −	Credit + 3,000

Figure 23–3 Work Sheet with Depreciation Adjustments

The Starting
Work
For the Year Ended

ACCT. NO.	ACCOUNT NAME	TRIAL BALANCE DEBIT	TRIAL BALANCE CREDIT	ADJUSTMENTS DEBIT	ADJUSTMENTS CREDIT
140	Delivery Equipment	16 500 00			
142	Accum. Depr.—Delivery Equip.		6 000 00		(e) 3 000 00
145	Office Equipment	5 000 00			
147	Accum. Depr.—Office Equipment		5 625 00		(f) 2 500 00
615	Depr. Expense—Delivery Equip.	—		(e) 3 000 00	
620	Depr. Expense—Office Equip.	—		(f) 2 500 00	

744 Chapter 23 · Plant Assets and Depreciation

Accountants make similar adjustments to record depreciation for other plant assets, such as buildings and office equipment.

Analysis of the Accumulated Depreciation Account.

Suppose this is the end of the third year of the estimated useful life of The Starting Line's delivery truck. For each year the same adjustment was made to record the depreciation of the delivery truck:

- a debit to **Depreciation Expense—Delivery Equipment**
- a credit to **Accumulated Depreciation—Delivery Equipment**

After the books have been closed each year, the **Depreciation Expense—Delivery Equipment** account has a zero balance. (Remember that expense accounts are closed at the end of each year.) In contrast the **Accumulated Depreciation—Delivery Equipment** account shows the total amount of depreciation expensed since the asset was purchased. At the end of the third year, the total is $9,000.

Recording Depreciation Adjustments on a Work Sheet

After preparing the adjustment for depreciation, the accountant enters it in the Adjustments section of the work sheet.

Refer to **Figure 23–3**. Locate the accumulated depreciation accounts in the Trial Balance Credit column ($6,000 and $5,625). Note that the depreciation expense accounts do not have balances in the Trial Balance section. Adjustments (e) and (f) are entered in the Adjustments section to show the depreciation adjustments

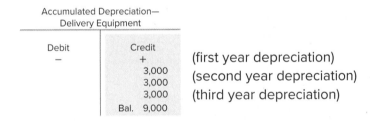

Figure 23–3 Work Sheet with Depreciation Adjustments (continued)

for the year. Note that the depreciation expense accounts are debited and the accumulated depreciation accounts are credited. Also note that no adjustments are made to the asset accounts.

Each amount is extended to the other work sheet columns.

	Column Extended to	
Account	Adjusted Trial Balance	Financial Statement
Del. Equip.	Debit (unchanged)	Balance Sheet
Accum. Depr.—Del. Equip.	Credit (increased)	Balance Sheet
Office Equip.	Debit (unchanged)	Balance Sheet
Accum. Depr.—Office Equip.	Credit (increased)	Balance Sheet
Depr. Exp.—Del. Equip.	Debit (increased)	Income Statement
Depr. Exp.—Office Equip.	Debit (increased)	Income Statement

Reporting Depreciation Expense and Accumulated Depreciation on Financial Statements

Figure 23–4 on page 744 shows placement of the depreciation expense accounts on the partial income statement of The Starting Line.

Figure 23–5 on page 744 shows placement of the plant asset and related accumulated depreciation accounts. Both types of accounts appear in the Assets section of the balance sheet. Notice that the accumulated depreciation account is listed immediately below the related plant asset account.

Figure 23–4 Income Statement

The Starting Line Sports Gear		
Income Statement		
For the Year Ended December 31, 20--		
Operating Expenses		
Depreciation Expense—Delivery Equip.	3 0 0 0 00	
Depreciation Expense—Office Equip.	2 5 0 0 00	
Total Operating Expenses		94 3 5 1 00
Operating Income		51 3 4 2 00

For **Delivery Equipment:**

- The original cost is entered in the first amount column on the first line ($16,500).
- The accumulated depreciation is entered in the first amount column on the second line ($9,000).
- The difference between cost and accumulated depreciation is entered in the second amount column on the second line ($7,500).

The book value of the delivery equipment is $7,500. The book value of each plant asset reported on the balance sheet should be the same as that shown on the plant asset record.

Figure 23–5 Balance Sheet

The Starting Line Sports Gear			
Balance Sheet			
December 31, 20--			
Assets			
Delivery Equipment		16 500 00	
Less: Accum. Depr.—Delivery Equip.		9 000 00	7 500 00
Office Equipment		50 000 00	
Less: Accum. Depr.—Office Equip.		8 125 00	41 875 00

Reading Check

Recall

Why does the Depreciation Expense account close with a zero balance at the end of the year?

Adjusting and Closing Entries for Depreciation Expense

What Are the Adjusting and Closing Entries for Depreciation?

After the accountant has completed the work sheet and prepared the financial statements, the adjustments for depreciation expense are recorded in the general journal. The information for the journal entries is taken directly from the Adjustments section of the work sheet.

Adjustment

Record the December 31 adjusting journal entries for depreciation.

Analysis	Identify	1. The accounts affected are **Depreciation Expense—Delivery Equipment, Depreciation Expense—Office Equipment, Accumulated Depreciation—Delivery Equipment,** and **Accumulated Depreciation—Office Equipment.**
	Classify	2. **Depreciation Expense—Delivery Equipment** and **Depreciation Expense—Office Equipment** are expense accounts. **Accumulated Depreciation—Delivery Equipment** and **Accumulated Depreciation—Office Equipment** are contra asset accounts.
	+/−	3. **Depreciation Expense—Delivery Equipment** is increased by $3,000. **Depreciation Expense—Office Equipment** is increased by $2,500. **Accumulated Depreciation—Delivery Equipment** is increased by $3,000. **Accumulated Depreciation—Office Equipment** is increased by $2,500.

continued

Debit-Credit Rule

4. Increases in expense accounts are recorded as debits. Debit **Depreciation Expense—Delivery Equipment** for $3,000 and **Depreciation Expense—Office Equipment** for $2,500.

5. Increases in contra asset accounts are recorded as credits. Credit **Accumulated Depreciation—Delivery Equipment** for $3,000 and **Accumulated Depreciation—Office Equipment** for $2,500.

T Accounts

6.

Depreciation Expense—Delivery Equipment		Accumulated Depreciation—Delivery Equipment	
Debit + 3,000	Credit −	Debit −	Credit + 3,000

Depreciation Expense—Office Equipment		Accumulated Depreciation—Office Equipment	
Debit + 2,500	Credit −	Debit −	Credit + 2,500

Journal Entry

7.

GENERAL JOURNAL   PAGE 21

	DATE	DESCRIPTION	POST. REF.	DEBIT	CREDIT	
1	20--					1
2	Dec. 31	Depr. Exp.—Del. Equip.		3 000 00		2
3		Accum. Depr.—Del. Equip.			3 000 00	3
4	31	Depr. Exp.—Office Equip.		2 500 00		4
5		Accum. Depr.—Office Equip.			2 500 00	5
6						6

After adjusting entries have been journalized and posted, the next step in the accounting cycle is to close the ledger. In the second closing entry, you'll remember, accounts with debit balances in the Income Statement Debit column of the work sheet are closed to **Income Summary.** This closing entry includes the depreciation expense accounts.

After this closing entry has been posted to the general ledger, the balances of the depreciation expense accounts are reduced to zero.

Closing

Second Closing Entry—Depreciation accounts only.

Analysis

Identify
1. The accounts affected are **Depreciation Expense—Delivery Equipment, Depreciation Expense—Office Equipment,** and **Income Summary.**

Classify
2. **Depreciation Expense—Delivery Equipment** and **Depreciation Expense—Office Equipment** are expense accounts. **Income Summary** is a temporary capital account.

+/−
3. **Depreciation Expense—Delivery Equipment** is decreased by $3,000 and **Depreciation Expense—Office Equipment** is decreased by $2,500; the total decrease is $5,500. The $5,500 is transferred to the **Income Summary** account.

Debit-Credit Rule

4. To transfer the expenses to the **Income Summary** account, debit **Income Summary** for $5,500.

5. Decreases in expense accounts are recorded as credits. Credit **Depreciation Expense—Delivery Equipment** for $3,000 and **Depreciation Expense—Office Equipment** for $2,500.

T Accounts

6.

Income Summary	
Debit	Credit
5,500	

Depreciation Expense—Delivery Equipment	
Debit +	Credit −
3,000	Clos. 3,000

Depreciation Expense—Office Equipment	
Debit +	Credit −
2,500	Clos. 2,500

Journal Entry

7.

GENERAL JOURNAL PAGE 22

	DATE	DESCRIPTION	POST. REF.	DEBIT	CREDIT	
1	20--					1
2	Dec. 31	Income Summary		5 500 00		2
3		Depr. Exp.—Del. Equip.			3 000 00	3
4		Depr. Exp.—Office Equip.			2 500 00	4
5						5

SECTION 23.3
Assessment

After You Read

Reinforce the Main Idea

Use a diagram like this one to show the last five steps of the accounting cycle as they relate to depreciation.

End-of-Period Accounting Cycle Steps

Step Number	How This Step Relates to Depreciation

Problem 4 Analyzing a Source Document

Instructions In your working papers:

1. Record the purchase in the general journal, page 4.

2. Record the partial depreciation expense for the first year in the general journal, page 7.

   a. Disposal value—$200
   b. Estimated useful life—5 years
   c. Fiscal year end—December 31.
   d. Method—straight-line

3. Record the annual depreciation, one year later, in the general journal, page 13.

All Purpose Office Equipment
996 Lake Drive, Sacramento, CA 94203

INVOICE NO. 14492
DATE: June 4, 20--
ORDER NO.: 22688
SHIPPED BY: UPS
TERMS: Cash

TO Universal Auto Supply
1422 Central Blvd.
Sacramento, CA 94203

QTY.	ITEM	UNIT PRICE	TOTAL
1	Xerox Copy Machine Serial Number 24X612987	$3,200.00	$3,200.00
	Sales Tax		256.00
	Total		$3,456.00

Problem 5 Preparing a Depreciation Schedule and Journalizing the Depreciation Adjusting Entry

Quade Corporation bought a copy machine on January 7 of the current year for $2,360. It has an estimated useful life of five years and an estimated disposal value of $100.

Instructions In your working papers:

1. Prepare a depreciation schedule for the copy machine using the straight-line method of depreciation. (Use the form provided in your working papers.)

2. Journalize the adjustment for the copy machine's depreciation at the end of the first year.

3. Journalize the closing entry for the expense account affected by the adjusting entry.

4. What is the book value of the asset after five years? Is this the same as the disposal value?

continued

750 Chapter 23 • Plant Assets and Depreciation

End of	Purchase Price	Depreciation Expense	Accumulated Depreciation	Book Value
1st Year	5,000			

Math for Accounting

You are preparing a depreciation schedule for a new piece of equipment you purchased for your business, Supreme T-Shirt Outfitters, for $5,000. It has a five-year estimated useful life and a $500 estimated disposal value. Calculate its annual depreciation using the straight-line method. Next calculate the accumulated depreciation and the equipment's book value at the end of each of the five years. Use a table like this one. Add rows for years 2 through 5.

Chapter 23
Visual Summary

Concepts
To match the cost of an asset with the revenue it is used to generate during a year, annual straight-line depreciation is calculated:

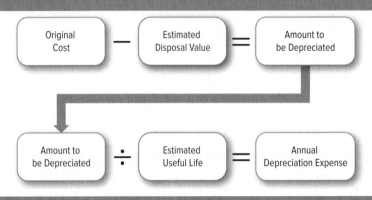

Analysis
To calculate depreciation for part of a year:

Annual Depreciation Expense × **Fraction of Year** = **Partial Year Depreciation Expense**

Procedures
To record depreciation:

Depreciation Expense—Description of Asset	
Debit + xx	Credit −

Accumulated Depreciation—Description of Asset	
Debit −	Credit + xx

Chapter 23
Review and Activities

Answering the Essential Question

How does the value of an asset change as it gets older?

Machinery, such as a car, is worth more when it's new than it is after it's been used for a year. Why? How do companies calculate the value of equipment that is getting older?

Vocabulary Check

1. **Vocabulary** All of the content vocabulary terms relate to the depreciation of a company's assets. Use the terms to describe depreciation. You can either write a short essay or make a diagram or chart to provide the information.
   - consume
   - plant assets
   - depreciation
   - disposal value
   - straight-line depreciation
   - accumulated depreciation
   - book value

Concept Check

2. **Analyze** Explain the meaning of the term *plant assets*. Compare and contrast *plant assets* and *current assets*.
3. Depreciation is the application of what accounting principle?
4. What four factors affect the depreciation calculation?
5. How is the annual depreciation expense for a plant asset calculated under the straight-line method?
6. **Analyze** How does calculating depreciation for part of a year depend on the yearly calculation?
7. **Synthesize** What is accumulated depreciation, and how does it relate to the book value of an asset?
8. Which two accounts are affected by an adjusting entry for depreciation?
9. What is the purpose of a depreciation schedule?
10. When is the depreciation of a plant asset recorded?
11. **Math** National Bank recently purchased six new computers. Each computer cost $2,245. The estimated life of each computer is 5 years with a trade-in value of $125 at the end of the 5 years. What is the annual depreciation for all six computers.
12. **English Language Arts** Think of an item for which it would be useful to know the depreciation schedule. Tell why it would be useful to know how long the item would last from the date you first purchase it and how much it will cost you on a yearly basis to own. Tell why the item is worth less each year that you own it.

Chapter 23
Standardized Test Practice

Multiple Choice

1. A truck used in daily operations of a corporation would be considered a(n)

    a. long-term asset.

    b. plant asset.

    c. intangible asset.

    d. either A or B.

    e. none of these answers

2. A method that allocates an equal portion of the total depreciation for a plant asset (cost minus salvage) to each accounting period in its service life is

    a. accelerated depreciation.

    b. units-of-production.

    c. straight-line depreciation.

    d. sum-of-the-years'-digits depreciation.

    e. declining-balance depreciation.

3. The Crimson Cartage Company purchased a new truck at a cost of $42,000 on July 1. The truck is estimated to have a useful life of 6 years and a salvage value of $6,000. Using the straight-line method, how much depreciation expense will be recorded for the truck during the first year ended December 31?

    a. $3,000

    b. $3,500

    c. $4,000

    d. $6,000

    e. $7,000

4. The adjusting entry for the depreciation of a plant asset such as equipment involves a credit to

    a. equipment expense.

    b. accumulated equipment.

    c. depreciation expense.

    d. accumulated depreciation.

5. The entry to record depreciation for the fax machine at the end of the fiscal period

    a. Debit Accumulated Depreciation—Maintenance Equipment, credit Depreciation Expense—Maintenance Equipment

    b. Debit Depreciation Expense—Maintenance Equipment, credit Accumulated Depreciation—Maintenance Equipment

    c. Debit Accumulated Depreciation—Office Equipment, credit Depreciation Expense—Office Equipment

    d. Debit Depreciation Expense—Office Equipment, credit Accumulated Depreciation—Office Equipment

True or False

6. The estimated value of a plant asset at the end of its useful life is its book value.

Short Answer

7. Why is the cost of land not depreciated?

8. What four factors affect the depreciation calculation?

# Chapter 23
Computerized Accounting

Recording Depreciation

Making the Transition from a Manual to a Computerized System

MANUAL METHODS	COMPUTERIZED METHODS
• Estimated annual depreciation is calculated when a fixed asset is purchased.	• Estimated annual depreciation is calculated when a fixed asset is purchased.
• Each accounting period, an adjusting entry is journalized and posted for each plant asset account in the general ledger.	• A depreciation journal entry is recorded for the month's amount. The journal entry can be automatically posted each month as a recurring entry.
• Recording depreciation affects two accounts, **Accumulated Depreciation** and **Depreciation Expense**.	• Amounts are posted to the **Depreciation Expense** and **Accumulated Depreciation** accounts. Their balances are automatically updated.
• The **Depreciation Expense** account is closed at the end of the fiscal year.	

Chapter 23
Problems

Problem 6 Opening a Plant Asset Record

On July 10 Sunset Surfwear purchased a scanner from Taunton Equipment for $1,500. Taunton Equipment charged Sunset Surfwear $200 to install the scanner. The scanner has an estimated useful life of four years and an estimated disposal value of $260.

Instructions Prepare a plant asset record, including the depreciation schedule for the new scanner. Use the form provided in your working papers.

  Serial number: TMC46312
  General ledger account: **Office Equipment**
  Location: Main Street store
  Depreciation method: Straight-line
  Manufacturer: Brothers Company

Analyze
Describe how you determined the book value at the end of each period.

Problem 7 Recording Adjusting Entries for Depreciation

The following adjustments for depreciation were entered on the work sheet for InBeat CD Shop for the year ended December 31.

	Adjustments	
	Debit	Credit
Store Equipment		
Accumulated Depreciation—Store Equipment		(e) 3,800
Office Equipment		
Accumulated Depreciation—Office Equipment		(f) 1,400
Depreciation Expense—Store Equipment	(e) 3,800	
Depreciation Expense—Office Equipment	(f) 1,400	

Instructions Record the adjusting entries on general journal page 11.

Analyze
Explain how the adjustment affects the book value of the store equipment.

Chapter 23
Problems

Problem 8 Reporting Depreciation Expense on the Work Sheet and Financial Statements

The trial balance of Shutterbug Cameras appears on the work sheet included in your working papers. All adjustments, except those for depreciation, are already recorded on the work sheet.

Instructions

1. Record the following adjustments for depreciation expense on the work sheet.
   a. Depreciation for office equipment is $2,500.
   b. Depreciation for store equipment is $1,200.
2. Complete the work sheet.
3. Prepare an income statement, statement of retained earnings, and balance sheet for Shutterbug Cameras for the year ended December 31.

Analyze
Calculate the total current assets.

Problem 9 Calculating and Recording Depreciation Expense

Cycle Tech Bicycles purchased manufacturing equipment on August 1 for $410,000. The equipment has an estimated useful life of 25 years and an estimated disposal value of $20,000. Cycle Tech uses the straight-line method of depreciation. The partial depreciation schedule found in your working papers is set up for the equipment.

Instructions

1. Calculate annual depreciation, accumulated depreciation, and book value for each of the first two years. The fiscal year ends December 31. Use the form provided in your working papers.
2. Calculate the depreciation adjustment to be entered on the work sheet at the end of the first year. Use T accounts to show the accounts debited and credited.
3. Journalize the adjustment for depreciation at the end of the first year, general journal page 21.

Analyze
Determine the balance of the accumulated depreciation account at the end of the first and second years.

758 Chapter 23 • Plant Assets and Depreciation

Chapter 23 Problems

Problem 10   Calculating and Recording Adjustments

The December 31 trial balance of Rivers Edge Canoe & Kayak is included in your working papers.

Instructions

1. Calculate and record end-of-period adjustments on the work sheet.

   a. Ending **Merchandise Inventory** is $15,000.

   b. Supplies on hand total $1,450.

   c. The amount of the expired insurance premium is $5,000.

   d. Use the following information to calculate the estimated annual depreciation expense using the straight-line method.

Plant Asset	Cost	Estimated Disposal Value	Estimated Useful Life
Store Equipment	$ 13,000	$ 1,000	10 years
Delivery Truck	32,000	2,000	10 years
Building	160,000	10,000	25 years

   e. The total federal income tax expense for the year is $4,250.

2. Complete the work sheet.

3. Journalize and post the adjusting entries on page 16 of the general journal.

4. Journalize and post the closing entries on page 16 of the general journal.

Analyze
Calculate the total depreciation expense for the year.

Problem 11   Examining Depreciation Adjustments

On May 2 Buzz Newsstand purchased a new machine for $2,700. It has an estimated disposal value of $100 and an estimated useful life of eight years. Buzz uses the straight-line method.

On December 31 the adjustment for depreciation for the first year was entered: **Accumulated Depreciation—Store Equipment** was credited for $325, and **Depreciation Expense—Store Equipment** was debited for $325.

Instructions Answer the following questions regarding this adjustment:

1. What is wrong with the adjustment for depreciation made on December 31? What is the correct entry?

continued

Chapter 23
Problems

2. One year from now, another adjustment for this machine will be entered on the work sheet. Assume the error from the previous year is not corrected. What amount should be entered in the Adjustments section for annual depreciation expense?

Analyze

Determine whether the current period net income will be too high or too low if the original error is not corrected.

Real-World Accounting Careers

Angela York

Protiviti Inc.

Q What do you do?

A I assist publicly held companies in complying with the Sarbanes-Oxley Act of 2002 by working with them on their internal audits. I work primarily with FORTUNE® 1000 and FORTUNE® 500 companies in the banking and professional services industries.

Q What are your day-to-day responsibilities?

A I go to the client's site and work with on-site personnel, carefully review records and files, verify the accuracy of documents, and report my findings to management. I also oversee other consultants and handle budgeting for each project.

Q What factors have been key to your success?

A Project lengths vary; I may work for a client for just a few weeks or up to a year. So the ability to adapt to a new environment, quickly shift gears, and work autonomously are all key factors.

Q What do you like most about your job?

A In internal audit, you really see the value you add for a company. At the end of a project, I can see how what I've done improves a client's operating efficiencies, saves money or ensures accurate records.

Career Facts

Real-World Skills
Adaptability and flexibility; professionalism; time management skills; strong communication abilities

Training and Education
Bachelor's degree in accounting; CPA or CIA credential strongly encouraged; MBA not necessary but beneficial

Career Skills
Knowledge of accounting principles; business writing and research skills; strong analytical and deductive skills

Career Path
Start out as a consultant and progress to senior consultant

Tips from Robert Half International
Prepare for the common job-interview request, "Tell me about yourself" by rehearsing a 15-second "sound bite" that describes your strongest skills and relevant experience in two or three sentences.

College and Career Readiness
Angela York mentions that she works for FORTUNE 1000 and FORTUNE 500 companies. What kinds of companies are these? Use online resources to learn more about these two types of companies, and then define them in your own words.

Real-World Applications & Connections

Career Wise

Auditor

An auditor can be a great career for those who enjoy mathematics and analyzing data. An auditor must examine lots of financial information and be able to understand and analyze that information. An auditor should be very focused and able to pay attention to detail.

Education Requirements

Auditors have the same educational requirements as CPAs, including state accounting board and American Institute of Public Accounting requirements. Auditors in most States also have to take continuing education throughout their career to maintain their professional license.

Roles

In addition to performing audits, auditors also analyze business statements and budgets, and help clients plan investments, keep up to date with the latest technology, and reduce taxes.

Responsibilities

Auditors review financial statements to make sure they follow Generally Accepted Auditing Standards (GAAS). They inform investors whether the financial statements represent a "fair" presentation of an organization's finances.

Activity

The Generally Accepted Auditing Standards are a good tool for understanding what an auditor must do. These standards are set out in the Code of Professional Conduct by the American Institute of CPAs. The standards can be read here: http://pcaobus.org/Standards/Auditing/Pages/AU150.aspx. Review the general standards, standards of fieldwork, and standards of reporting, and on a separate piece of paper, write what each standard means in your own words.

Global Accounting

Plant Assets

Global companies may use the *revaluation model* allowed by International Accounting Standard 16 to record plant assets. After initially recording an asset at cost, this model revalues an asset at its fair market value less subsequent depreciation, if fair market value can be measured reliably.

Instructions

Define *fair market value*. How would revaluation affect the income statement?

Analyzing Financial Reports

Vertical Analysis

Vertical analysis makes it easy to compare financial statements with industry standards, which are average percentages for companies in the same industry.

Instructions

Use the Internet to locate and save or print a copy of Amazon.com's annual report for these tasks.

1. Complete a vertical analysis of Amazon.com's cost of sales and gross profit for the year reflected in the report.

2. Complete a vertical analysis of Amazon.com's assets for the year reflected in the report.

3. Analyze the impact of accounts receivable and accounts payable on the balance sheet and cash flow statements.

4. Analyze the impact of salary expenses and payroll tax expense on total expenses and net income.

H.O.T. Audit

Adjustments for Depreciation

Brookside Sports Equipment Inc. purchased a new cash register for $3,900 on August 3 from Best Office Equipment. The cash register has an estimated useful life of ten years and an estimated disposal value of $300. Review the journal entry below and answer the following questions.

1. What is wrong with the adjustment for depreciation made on December 31?

2. Because of the error, is the book value of the cash register too high or too low?

GENERAL JOURNAL PAGE 61

	DATE	DESCRIPTION	POST. REF.	DEBIT	CREDIT	
1	20--					1
2	Dec. 31	Depreciation Exp.—Store Equip.		360 00		2
3		Accum. Depr.—Store Equip.			360 00	3
4						4

Chapter 24

Uncollectible Accounts Receivable

Chapter Topics

24-1 The Direct Write-Off Method

24-2 The Allowance Method

24-3 Estimating Uncollectible Accounts Receivable

Visual Summary

Review and Activities

Standardized Test Practice

Computerized Accounting

Problems

Real-World Applications & Connections

Essential Question

As you read this chapter, keep this question in mind:

What does a business do when a debt cannot be collected?

Main Idea

A business using the direct write-off method to write off uncollectible accounts converts an account receivable to an expense when it become clear the customer will not pay the bill. A business using the allowance method estimates the bad debt expense and writes off that amount.

Chapter Objectives

Concepts

C1 Explain methods used to write off uncollectible accounts.

C2 Determine uncollectible accounts receivable.

C3 Use the direct write-off method for uncollectible accounts.

Analysis

A1 Calculate bad debts expense.

A2 Make an adjusting entry for uncollectible accounts.

A3 Use the allowance method to record uncollectible accounts.

Procedures

P1 Record the collection of an account previously written off.

P2 Describe two methods to estimate uncollectible accounts expense.

Real-World Business Connection

YouTube

The video-sharing Web site YouTube is more than just a company—it is such a powerful cultural phenomenon that people worldwide use the name as a verb. ("Let's YouTube our home video!") In 2005, the first video was posted to the site; today, users upload more than 24 hours of video every minute. As with Google (which bought the company in 2006), it is difficult to remember a time when YouTube was not part of the Internet.

Connect to the Business

Many experts wondered how YouTube would ever sell enough advertising to make money; after all, server space for all those videos is expensive. In recent years, however, virtually every major advertising campaign has incorporated YouTube videos. The company is selling ads in more than 20 countries, with space on the home page pulling in the priciest rates.

Analyze

Why is it important for YouTube to have good collection policies for the money it is owed?

Focus on the Photo

YouTube is not just an entertaining home video site—it has become a source of revenue for advertisers and a number of content providers. **How might businesses collect revenue that they are owed?**

SECTION 24.1
The Direct Write-Off Method

Many businesses sell goods or services on account. Some charge customers cannot or will not pay the amounts they owe. This chapter will teach you what methods businesses use to write off uncollectible accounts.

Section Vocabulary
- uncollectible account
- direct write-off method
- prospective

Extending Credit

Why Do Businesses Extend Credit?

Selling goods and services on credit is a standard practice for businesses of all sizes and types. They offer credit because they expect to sell more than they would by accepting only cash.

Retail stores typically ask customers seeking to purchase on account to complete a credit application. Before a business extends credit, it should check each **prospective** customer's credit rating to help determine the customer's ability to pay the amounts charged on account. Businesses get these credit ratings from the national consumer-credit-reporting companies Equifax, Experian, and TransUnion. Wholesalers and manufacturers use reports from wholesale credit bureaus and national credit-rating organizations such as Dun & Bradstreet.

An **uncollectible account,** or *bad debt,* is an account receivable that the business cannot collect. The business eventually removes the account receivable from its records, and the amount becomes an expense to the business. Businesses account for bad debts by using the *direct write-off method* or the *allowance method.*

The Direct Write-Off Method

What Is the Direct Write-Off Method?

The *direct write-off method* is used primarily by small businesses and those with few charge customers. Under the **direct write-off method,** when the business determines that the amount owed is not going to be paid, the uncollectible account is removed from the accounting records. **Uncollectible Accounts Expense** is debited and **Accounts Receivable** (both controlling and subsidiary) is credited. The direct write-off method is the only method a business can use for income tax purposes.

Writing Off an Uncollectible Account

On June 4 2012 The Starting Line Sports Gear sold football equipment on account to Robert Galvin for $250 plus $15 sales tax. It recorded the transaction as a $265 debit to **Accounts Receivable** (controlling) and the subsidiary account **Accounts Receivable—Robert Galvin** and as a credit to **Sales** for $250 and **Sales Tax Payable** for $15.

For over a year, The Starting Line tried to collect this account. The length of time is one consideration used to determine uncollectible accounts receivable. It is now apparent that Robert Galvin is not going to pay the $265.

In effect, a business must decrease total assets and increase total expenses when a customer fails to pay a debt. The Starting Line must decrease its **Accounts Receivable** account and increase its expense related to uncollectible accounts.

Business Transaction

On August 25, 2013, The Starting Line wrote off as uncollectible Galvin's account for $265, Memorandum 170.

Analysis

Identify
1. The accounts affected are **Uncollectible Accounts Expense, Accounts Receivable** (controlling), and **Accounts Receivable— Robert Galvin** (subsidiary).

Classify
2. **Uncollectible Accounts Expense** is an expense account. **Accounts Receivable** (controlling) and **Accounts Receivable— Robert Galvin** (subsidiary) are asset accounts.

+/−
3. **Uncollectible Accounts Expense** is increased by $265. **Accounts Receivable** (controlling) and **Accounts Receivable— Robert Galvin** (subsidiary) are decreased by $265.

Debit-Credit Rule

4. Increases to expense accounts are recorded as debits. Debit **Uncollectible Accounts Expense** for $265.

5. Decreases to asset accounts are recorded as credits. Credit **Accounts Receivable** (controlling) and **Accounts Receivable— Robert Galvin** (subsidiary) for $265.

T Accounts

6.

Uncollectible Accounts Expense		Accounts Receivable	
Debit + 265	Credit −	Debit +	Credit − 265

Accounts Receivable Subsidiary Ledger
Robert Galvin

Debit +	Credit − 265

Journal Entry

7.

GENERAL JOURNAL   PAGE 13

	DATE	DESCRIPTION	POST. REF.	DEBIT	CREDIT	
1	2013					1
2	Aug. 25	Uncollectible Accounts Expense		265 00		2
3		Accts. Rec./Robert Galvin			265 00	3
4		Memorandum 170				4
5						5

Chapter 24 • Uncollectible Accounts Receivable   767

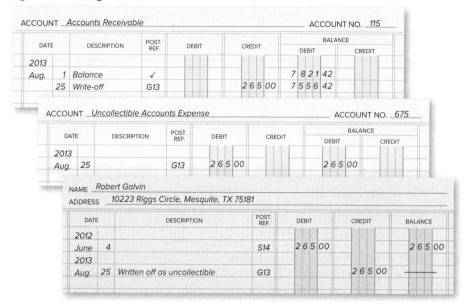

Figure 24–1 Ledger Accounts After an Uncollectible Write-Off

Figure 24–1 shows how this transaction is posted to the general ledger and the accounts receivable subsidiary ledger.

Notice the explanation entered in Robert Galvin's account. When an account is written off as uncollectible, it is important to note on the subsidiary ledger that the account was not paid off but was written off.

Collecting a Written-Off Account

Occasionally, a charge customer whose account was written off as uncollectible later pays the amount owed. When this happens:

- First, reinstate the customer's account, or reenter it in the accounting records.
- Second, record the cash receipt.

Business Transaction

On September 5 The Starting Line received $265 from Robert Galvin, whose account was written off as uncollectible on August 25, Memorandum 176 and Receipt 1109. First reinstate the account receivable.

Analysis	Identify	1. The accounts affected are **Accounts Receivable** (controlling), **Accounts Receivable—Robert Galvin** (subsidiary), and **Uncollectible Accounts Expense**.
	Classify	2. **Accounts Receivable** (controlling) and **Accounts Receivable—Robert Galvin** (subsidiary) are asset accounts. **Uncollectible Accounts Expense** is an expense account.
	+/−	3. **Accounts Receivable** (controlling) and **Accounts Receivable—Robert Galvin** (subsidiary) are increased by $265. **Uncollectible Accounts Expense** is decreased by $265.

continued

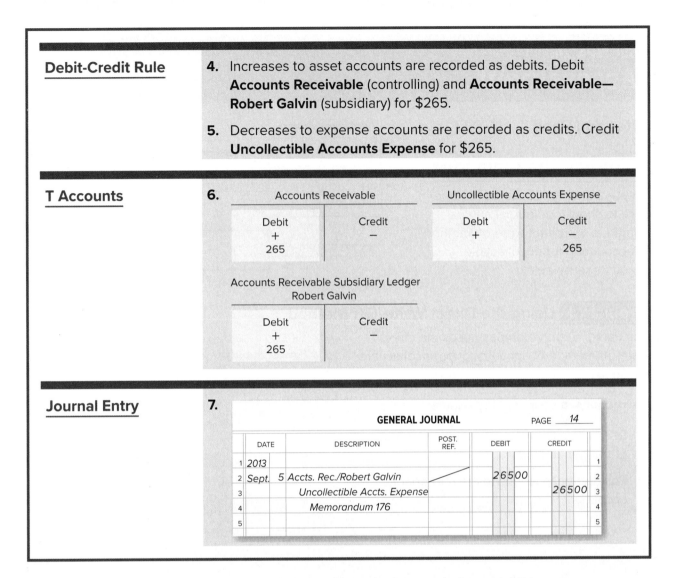

After this transaction has been posted, the cash receipt is recorded as a debit to **Cash in Bank** for $265 and a credit to **Accounts Receivable** (controlling) for $265. Also, the subsidiary account **Accounts Receivable—Robert Galvin** is credited for $265. Receipt 1109 is the source document for this entry.

Robert Galvin's subsidiary ledger account in **Figure 24–2** contains data about the sale, write-off, reinstatement, and cash receipt.

NAME *Robert Galvin*
ADDRESS *10223 Riggs Circle, Mesquite, TX 75181*

DATE		DESCRIPTION	POST. REF.	DEBIT	CREDIT	BALANCE
2012						
June	4		S14	265 00		265 00
2013						
Aug.	25	Written off as uncollectible	G13		265 00	—
Sept.	5	Reinstated	G14	265 00		265 00
	5		G14		265 00	

Figure 24–2 Robert Galvin Account

SECTION 24.1 Assessment

After You Read

Reinforce the Main Idea

Create a diagram similar to the one here to show four journal entries related to the direct write-off method of accounting for uncollectible accounts receivable.

Transaction Description → Journal Entry

Problem 1 Using the Direct Write-Off Method

The Parker Supply Company uses the direct write-off method of accounting for uncollectible accounts.

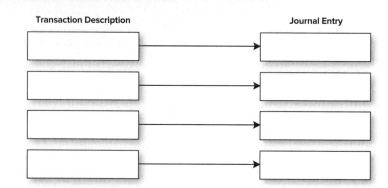

Parker Supply Company MEMORANDUM 78

TO: Accounting Clerk
FROM: Jon Herbert, Collection Manager
DATE: November 30, 20--
SUBJECT: Account write-off

It has been determined by the collection department that Account No. 4698214, Sonya Dickson, in the amount of $630, is uncollectible. This memo is your authorization to write this amount off as uncollectible.
  Thank you.

Instructions In your working papers:

1. Record the following transactions on page 21 of the general journal.
2. Post the transactions to the appropriate accounts.

Date		Transactions
Apr. 10	1.	Sold merchandise on account to Sonya Dickson, $600 plus $30 sales tax, Sales Slip 928.
Nov. 30	2.	Wrote off Sonya Dickson's account as uncollectible, $630, Memorandum 78.
Dec. 30	3.	Received $630 from Sonya Dickson in full payment of her account. This account was written off on November 30, Memorandum 89 and Receipt 277.

Math for Accounting

You are the accounts receivable clerk for Fast and Friendly Shipping. The balance of the **Accounts Receivable** account is $25,000. For six months you have been trying to collect the amounts owed by three companies: ABC Company, $450; XYZ Company, $500; and Nice Try Company, $350. These accounts are still unpaid. Your supervisor asked you to write off these accounts using the direct write-off method. What is the balance of **Accounts Receivable** after you write off these accounts as uncollectible?

SECTION 24.2
The Allowance Method

In Section 24.1 you learned that the direct write-off method is used by

- small businesses,
- businesses with few credit customers, and
- all businesses for tax purposes.

Section Vocabulary

- allowance method
- book value of accounts receivable

In contrast, businesses that have many credit customers use the allowance method of accounting for uncollectible accounts. This method allows businesses to match revenue with the expenses incurred to earn that revenue. In this section you will learn about the allowance method and continue to journalize transactions involving uncollectible accounts.

Matching Uncollectible Accounts Expense with Revenue

Why Do Businesses Use the Matching Principle to Report Uncollectible Accounts?

When the direct write-off method of accounting for uncollectible accounts is used, an unpaid account is written off when the business determines that it will not be paid. Under the direct write-off method, it often happens that the sale is recorded in one period, and the uncollectible accounts expense is recorded in the following period. This violates the matching principle.

One of the fundamental principles of accounting is that *revenue should be matched with the expenses incurred in generating that revenue.* This means that expenses incurred to earn revenue should be deducted in the same period that the revenue is recorded. The uncollectible accounts expense should be reported in the year in which the sale takes place. However, the uncollectible account expense is usually not determined with certainty until some future period. That is, a credit sale in the year 2012 may not be determined to be uncollectible until some time in 2013. In order to conform to the matching principle, the credit sales are recorded in the year 2012 and an *estimate* of the uncollectible accounts expense is also recorded in 2012.

The Starting
Work
For the Year Ended

	ACCT. NO.	ACCOUNT NAME	TRIAL BALANCE DEBIT	TRIAL BALANCE CREDIT	ADJUSTMENTS DEBIT	ADJUSTMENTS CREDIT
1	101	Cash in Bank	12 650 00			
2	115	Accounts Receivable	44 893 00			
3	117	Allowance for Uncollectible Accts.		125 00		(b)1 350 00
23	670	Supplies Expense	600 00			
24	675	Uncollectible Accounts Expense			(b)1 350 00	
25						

Figure 24–3 Recording the Adjustment for Uncollectible Accounts on the Work Sheet

The estimate of uncollectible accounts expense is recorded as an end-of-period adjustment. The adjusting entry meets two objectives:

1. **Accounts Receivable** is reduced to the amount the business can reasonably expect to receive.
2. The estimated uncollectible accounts expense is charged to the current period.

Reading Check

In Your Own Words

Explain why businesses estimate uncollectible accounts expense for the year in which the corresponding sale takes place.

The Allowance Method

How Do You Use the Allowance Method to Report Uncollectible Accounts Expense?

The **allowance method** of accounting for uncollectible accounts matches the estimated uncollectible accounts expense with sales made during the same period. At the end of the period, the accountant must calculate the uncollectible accounts expense that will result from the sales made during the period. The estimated uncollectible accounts expense is recorded as an adjustment on the work sheet. The two accounts affected by this adjustment are **Uncollectible Accounts Expense** and **Allowance for Uncollectible Accounts.**

When the adjustment is made, the business does not know exactly which charge customers will not pay the amounts they owe. Therefore, the estimated uncollectible amount cannot be credited to **Accounts Receivable** (neither the controlling nor the subsidiary). Since the **Accounts Receivable** account cannot be used to record the estimated uncollectible amount, another account is opened. This account is **Allowance for Uncollectible Accounts.**

Allowance for Uncollectible Accounts is used to summarize the *estimated* uncollectible accounts receivable of the business. It is classified as a contra asset account.

Common Mistakes

Allowance for Uncollectible Accounts Because the allowance for uncollectible accounts has a normal credit balance, it might be easy to misclassify this account as a liability, which also has a normal credit balance. Instead, though, a contra asset represents a reduction in a related asset.

Allowance for Uncollectible Accounts	
Debit	Credit
−	+
Decrease Side	Increase Side Normal Balance

Figure 24–3 Recording the Adjustment for Uncollectible Accounts on the Work Sheet (continued)

Line
Sheet
December 31, 20--

ADJUSTED TRIAL BALANCE		INCOME STATEMENT		BALANCE SHEET		
DEBIT	CREDIT	DEBIT	CREDIT	DEBIT	CREDIT	
1265000				1265000		1
4489300				4489300		2
	147500				147500	3
60000		60000				23
135000		135000				24
						25

Adjustment

On December 31 The Starting Line estimates its uncollectible accounts expense for the year ended December 31 to be $1,350. (Various methods are used to estimate uncollectible accounts expense. You will learn about two of these methods in Section 24.3.)

Analysis Identify
1. The accounts affected are **Uncollectible Accounts Expense** and **Allowance for Uncollectible Accounts.**

Classify
2. **Uncollectible Accounts Expense** is an expense account. **Allowance for Uncollectible Accounts** is a contra asset account.

+/−
3. **Uncollectible Accounts Expense** is increased by $1,350. **Allowance for Uncollectible Accounts** is increased by $1,350.

Debit-Credit Rule
4. Increases to expense accounts are recorded as debits. Debit **Uncollectible Accounts Expense** for $1,350.

5. Increases to contra asset accounts are recorded as credits. Credit **Allowance for Uncollectible Accounts** for $1,350.

T Accounts
6.

Uncollectible Accounts Expense

Debit	Credit
+	−
1,350	

Allowance for Uncollectible Accounts

Debit	Credit
−	+
	1,350

Journal Entry
7.

GENERAL JOURNAL   PAGE 15

	DATE	DESCRIPTION	POST. REF.	DEBIT	CREDIT	
1	20--	*Adjusting Entries*				1
2	Dec. 31	Uncollectible Accts. Expense		135000		2
3		Allow. for Uncollectible Accts.			135000	3
4						4
5						5

Allowance for Uncollectible Accounts appears on the balance sheet as a deduction from **Accounts Receivable.** By using **Allowance for Uncollectible Accounts:**

- The balance of **Accounts Receivable** still equals the total of the customer accounts in the subsidiary ledger.
- The balance of **Allowance for Uncollectible Accounts** represents the amount the business *estimates* to be uncollectible.
- The difference between **Accounts Receivable** and **Allowance for Uncollectible Accounts** represents the book value of accounts receivable.

The **book value of accounts receivable** is the amount the business can reasonably expect to collect from its accounts receivable. **Figure 24–3** on pages 771 and 773 illustrates how this adjustment is recorded and extended on the work sheet.

Notice that **Allowance for Uncollectible Accounts** has a $125 balance in the Trial Balance Credit column. This balance is carried over from previous years. If the previous years' uncollectible accounts exactly equaled the estimate, **Allowance for Uncollectible Accounts** would have a zero balance. This seldom happens.

Also notice that the new balance is extended first to the Adjusted Trial Balance Credit column and then to the Balance Sheet Credit column.

Reporting Estimated Uncollectible Amounts on the Financial Statements

The **Uncollectible Accounts Expense** account appears on The Starting Line's income statement as an expense. See **Figure 24–4** for the placement of the **Uncollectible Accounts Expense** account in its partial income statement.

Figure 24–4 Reporting Uncollectible Accounts on the Income Statement

The Starting Line Sports Gear			
Income Statement			
For the Year Ended December 31, 20--			
Operating Expenses			
Supplies Expense		600 00	
Uncollectible Accounts Expense		1 350 00	
Total Operating Expenses			38 345 00
Operating Income			24 698 00

On the balance sheet, **Allowance for Uncollectible Accounts** is listed immediately below **Accounts Receivable** in the Assets section. See The Starting Line's partial balance sheet in **Figure 24–5.**

Notice that the balances of **Accounts Receivable** and **Allowance for Uncollectible Accounts** are entered in the first amount column. The difference between the two balances—the book value of accounts receivable—is entered in the second amount column.

Figure 24–5 Reporting Allowance for Uncollectible Accounts on the Balance Sheet

The Starting Line Sports Gear			
Balance Sheet			
December 31, 20--			
Assets			
Cash in Bank			12 650 00
Accounts Receivable		44 893 00	
Less: Allowance for Uncollectible Accounts		1 475 00	43 418 00

Journalizing the Adjusting Entry for Uncollectible Accounts

After the work sheet has been completed and the financial statements have been prepared, the adjusting entries are journalized. The information for the adjusting entries is found in the Adjustments section of the work sheet as shown in **Figure 24–3** on pages 771 and 773.

Figure 24–6 Journalizing and Posting the Adjusting Entry for Uncollectible Accounts

	GENERAL JOURNAL			PAGE 14	
DATE	DESCRIPTION	POST. REF.	DEBIT	CREDIT	
20--	*Adjusting Entries*				1
Dec. 31	Uncollectible Accounts Expense	675	1 350 00		2
	Allowance for Uncollectible Accounts	117		1 350 00	3
					4

ACCOUNT Allowance for Uncollectible Accounts   ACCOUNT NO. 117

DATE	DESCRIPTION	POST. REF.	DEBIT	CREDIT	BALANCE DEBIT	BALANCE CREDIT
20--						
Dec. 1	Balance	✓				125 00
31	Adjusting Entry	G14		1 350 00		1 475 00

ACCOUNT Uncollectible Accounts Expense   ACCOUNT NO. 675

DATE	DESCRIPTION	POST. REF.	DEBIT	CREDIT	BALANCE DEBIT	BALANCE CREDIT
20--						
Dec. 31	Adjusting Entry	G14	1 350 00		1 350 00	

See **Figure 24–6** for the way the adjusting entry for the estimated uncollectible accounts expense is recorded in the general journal and posted to the appropriate general ledger accounts.

At the end of the period, the balance of the **Uncollectible Accounts Expense** account is closed, along with the balances of the other expense accounts, to **Income Summary**. **Uncollectible Accounts Expense** has a zero balance at the beginning of the next period. The balance of **Allowance for Uncollectible Accounts** is not affected by the closing entries. It is a permanent account, and its balance at the beginning of the next period remains $1,475.

Writing Off Uncollectible Accounts Receivable

When it becomes clear that a charge customer is not going to pay the amount owed, the accountant removes the uncollectible account from the accounting records.

Allowance for Uncollectible Accounts acts as a reservoir; that is, at the end of the period, the adjusting entry "fills it up." The account balance is saved until it is needed some time in the future. When a charge customer's account finally proves uncollectible, the business dips into that reservoir to write off the account. In other words **Allowance for Uncollectible Accounts** is reduced when a specific account is written off. Let's look at an example.

Business Transaction

On April 18, after many attempts to collect the amount owed, The Starting Line decides to write off the account of Megan Sullivan for $150, Memorandum 236.

Analysis	Identify	1. The accounts affected are **Allowance for Uncollectible Accounts, Accounts Receivable** (controlling), and **Accounts Receivable—Megan Sullivan** (subsidiary).
	Classify	2. **Allowance for Uncollectible Accounts** is a contra asset account. **Accounts Receivable** (controlling) and **Accounts Receivable—Megan Sullivan** (subsidiary) are asset accounts.
	+/−	3. **Allowance for Uncollectible Accounts** is decreased by $150. **Accounts Receivable** (controlling) and **Accounts Receivable—Megan Sullivan** (subsidiary) are decreased by $150.

continued

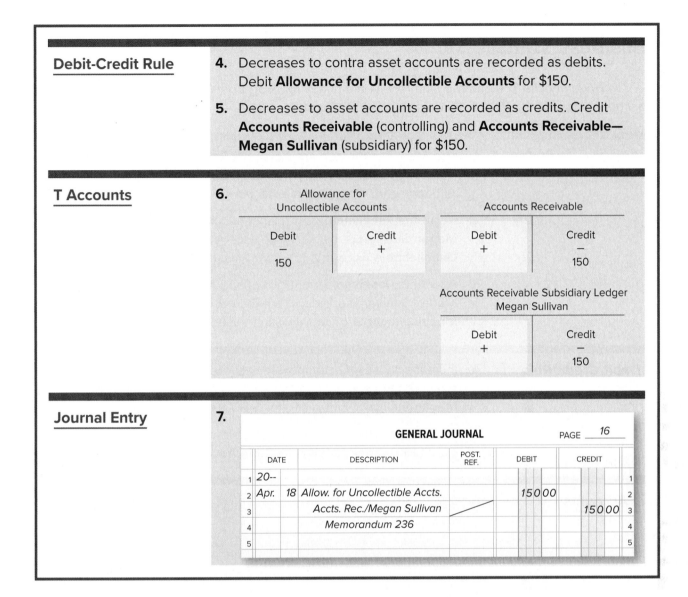

Under the allowance method, the write-off of a specific account does not affect an expense account. Recall that the expense was already recorded as an adjusting entry.

The adjusting entry recorded the entire estimated expense for the period, not just this one customer's bad debt.

Collecting an Account Written Off by the Allowance Method

A charge customer whose account was written off as uncollectible might later pay the amount owed. When this happens:

- First, reinstate the customer's account.
- Second, record the cash receipt.

Business Transaction

On November 19 The Starting Line received a check for $150 from Megan Sullivan, whose account was written off April 18, Memorandum 294 and Receipt 2243.

Analysis

Identify
1. The accounts affected are **Accounts Receivable** (controlling), **Accounts Receivable—Megan Sullivan** (subsidiary), and **Allowance for Uncollectible Accounts.**

Classify
2. **Accounts Receivable** (controlling) and **Accounts Receivable—Megan Sullivan** (subsidiary) are asset accounts. **Allowance for Uncollectible Accounts** is a contra asset account.

+/−
3. **Accounts Receivable** (controlling) and **Accounts Receivable—Megan Sullivan** (subsidiary) are increased by $150. **Allowance for Uncollectible Accounts** is increased by $150.

Debit-Credit Rule

4. Increases to asset accounts are recorded as debits. Debit **Accounts Receivable** (controlling) and **Accounts Receivable—Megan Sullivan** (subsidiary) for $150.

5. Increases to contra asset accounts are recorded as credits. Credit **Allowance for Uncollectible Accounts** for $150.

T Accounts

6.

Accounts Receivable		Allowance for Uncollectible Accounts	
Debit + 150	Credit −	Debit −	Credit + 150

Accounts Receivable Subsidiary Ledger
Megan Sullivan

Debit + 150	Credit −

Journal Entry

7.

GENERAL JOURNAL   PAGE 18

	DATE	DESCRIPTION	POST. REF.	DEBIT	CREDIT	
1	20--					1
2	Nov. 19	Accts. Rec./Megan Sullivan		150 00		2
3		Allow. for Uncollectible Accts.			150 00	3
4		Memorandum 294				4
5						5

After this transaction to reinstate the customer's account is posted, the cash receipt transaction is journalized and posted. **Figure 24–7** shows Megan Sullivan's account after the cash receipt transaction is posted. The account shows that:

1. The account had been declared uncollectible and was written off.
2. The account was reinstated.
3. The account was collected in full.

NAME Megan Sullivan
ADDRESS 883 Bisbee Drive, Dallas, TX 75211

DATE		DESCRIPTION	POST. REF.	DEBIT	CREDIT	BALANCE
20--						
Jan.	1	Balance	✓			150 00
Apr.	18	Written off as uncollectible	G16		150 00	—
Nov.	19	Reinstated	G18	150 00		150 00
	19		G18		150 00	—

Figure 24–7 Megan Sullivan Account

SECTION 24.2
Assessment

After You Read

Reinforce the Main Idea

Use a diagram like the one shown here to describe what happens when payment is received for an account receivable that had been written off.

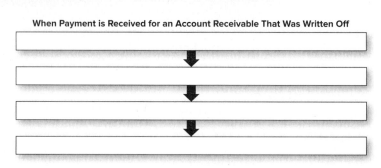

Problem 2 Writing Off Accounts Using the Allowance Method

Taylor Furniture Company Inc. uses the allowance method to account for uncollectible accounts.

Instructions In your working papers:

1. Record the following transactions in the general journal on page 24.
2. Post the transactions to the appropriate accounts.
3. Prepare the Assets section of the balance sheet for Taylor Furniture Company Inc. using the partial general ledger in your working papers. The balance of other asset accounts are **Merchandise Inventory,** $42,000; **Supplies,** $1,500; and **Prepaid Insurance,** $1,200.

Date	Transactions
May 4	1. Using the allowance method, wrote off the account of Jack Bowers for $1,050 as uncollectible, Memorandum 241.
Nov. 18	2. Received $1,050 from Jack Bowers in full payment of his account, which was written off May 4, Memorandum 321 and Receipt 1078.
Dec. 31	3. The adjusting entry for the estimated uncollectible accounts expense for the year ended December 31 was $1,850.

Math for Accounting

A review of the accounting records for Mary Sawyer's business, Secret Garden, revealed a disturbing trend. Her uncollectible accounts continue to increase. You suspect that Mary is far too nice about extending "in store" credit. You strongly recommend that she change her credit policies and collect on the outstanding accounts for this year. However Mary does not seem to understand the big picture, and she requested an illustration. Use a line graph to chart the uncollectibles for the past five years using the following figures.

Year 1 $1,500 Year 3 $2,300 Year 5 $2,800
Year 2 $1,875 Year 4 $2,422

SECTION 24.3
Estimating Uncollectible Accounts Receivable

As you learned in the previous section, businesses estimate the uncollectible accounts expense at the end of the period using the allowance method. In this section you will learn about the percentage of net sales and the aging of accounts receivable methods. These two ways to estimate uncollectible accounts expenses are based on judgment and past experience.

Section Vocabulary

- percentage of net sales method
- aging of accounts receivable method
- approximately
- assumes

Percentage of Net Sales Method

How Do You Use Net Sales to Compute Uncollectible Amounts?

When using the **percentage of net sales method** for estimating uncollectible accounts expense, the business assumes that a certain percentage of each year's net sales will be uncollectible. To find the adjustment for uncollectible accounts expense:

1. Determine the percentage.
2. Calculate net sales.
3. Multiply net sales by the percentage.
4. Enter the amount calculated above on the work sheet.

Let's see how this method works. First, the percentage is determined. As you can see, in recent years The Starting Line's actual uncollectible accounts have been **approximately** 2 percent of net sales. On this basis, The Starting Line's accountant believes that the 2 percent figure should be used to estimate uncollectible accounts expense.

Year	Net Sales	Uncollectible Accounts	Percentage
2008	$ 59,000	$1,062	(1.8%)
2009	65,000	1,430	(2.2%)
2010	67,000	1,273	(1.9%)
Totals	$191,000	$3,765	(2.0%)

Second, the amount of net sales is calculated. Remember that net sales equals sales minus sales discounts and sales returns and allowances.

As shown, net sales for The Starting Line is $67,500.

Sales		$74,500
Less: Sales Discounts	$3,000	
Sales Returns and Allowances	4,000	− 7,000
Net Sales		
		$67,500

Third, the uncollectible accounts expense for the current year is determined by multiplying net sales by the percentage. The Starting Line's uncollectible accounts expense for the period is estimated to be $1,350 ($67,500 × .02).

Under this method the amount calculated is recorded as the adjustment on the work sheet and later is entered into the accounting records by journalizing the adjusting entry. At the beginning of the next period, **Allowance for Uncollectible Accounts** will have a credit balance of $1,475 ($1,350 adjustment plus $125 existing balance in the account).

Reading Check

Explain

List the steps to find the adjustment for uncollectible accounts expense using the percentage of net sales method.

Aging of Accounts Receivable Method

How Do You Use Accounts Receivable to Compute Uncollectible Amounts?

The **aging of accounts receivable method** classifies the accounts receivable according to the number of days each account is past due. This method **assumes** that the longer an account is overdue, the less likely it is to be collected. Use these steps to find the adjustment for uncollectible accounts expense:

1. *Age,* or classify and group, each account according to the number of days it is past due.

2. Use past experience to determine the percentage of each group that will be uncollectible.

3. Multiply the uncollectible amount for each group by the percentage for that group.

4. Add the results for all groups.

5. Enter on the work sheet the total estimated uncollectible amount (calculated above) adjusted by any balance in **Allowance for Uncollectible Accounts.**

Let's look at an example. First, each customer account is *aged,* or classified and grouped according to the number of days it is overdue. The computer printout in **Figure 24–8** is a schedule of The Starting Line's aged accounts receivable that has grouped each customer account.

Next, The Starting Line's accountant estimates what percentage of each group will be uncollectible based on past experience. The percentages range from 2 to 80 percent.

782 Chapter 24 • Uncollectible Accounts Receivable

Figure 24–8 Computer-Generated Analysis of Accounts Receivable

The Starting Line Sports Gear
ANALYSIS OF ACCOUNTS RECEIVABLE
December 31, 20--

Account ID	Customer's Name	Total Amount Owed	Not Yet Due	Days Past Due 1–30	31–60	61–90	Over 90
DIM	Joe Dimaio	$ 300.00	$ 300.00				
GAL	Robert Galvin	50.00		$ 50.00			
KLE	Casey Klein	800.00					$ 800.00
MON	Anita Montero	200.00	200.00				
RAH	Shashi Rahim	175.00			$ 175.00		
RAM	Gabriel Ramos	1,000.00	1,000.00				
SUL	Megan Sullivan	40.00				$ 40.00	
TAM	Tammy's Fitness	225.00	225.00				
WON	Kim Wong	306.50		306.50			
YOU	Lara Young	750.00	750.00				
	TOTALS	$ 3,846.50	$ 2,475.00	$ 356.50	$ 175.00	$ 40.00	$ 800.00

Figure 24–9 Accounts Receivable Aging Schedule

The Starting Line Sports Gear
AGING OF ACCOUNTS RECEIVABLE
Estimated Uncollectible Amount
December 31, 20--

Age Group	Amount	Estimated Percentage Uncollectible	Estimated Uncollectible Amount
Not yet due	$ 2,475.00	2%	$ 49.50
1–30 days past due	356.50	4%	14.26
31–60 days past due	175.00	10%	17.50
61–90 days past due	40.00	20%	8.00
Over 90 days past due	800.00	80%	640.00
Total	$ 3,846.50		$ 729.26

Third, the total uncollectible amount for each group is multiplied by the percentage for that group. The resulting amounts are the estimated uncollectible amounts for each group. The computer printout in **Figure 24–9** illustrates how the estimated uncollectible amount is determined from the accounts receivable analysis in **Figure 24–8.** As you can see, The Starting Line estimates that a total of $729.26 of its accounts will be uncollectible. This total represents the end-of-period balance of **Allowance for Uncollectible Accounts.**

Finally, an adjustment is entered on the work sheet. This adjustment will bring the balance of **Allowance for Uncollectible Accounts** to $729.26, the estimated amount.

Suppose the balance of **Allowance for Uncollectible Accounts** reported in the Trial Balance Credit column is $49.80. To determine the adjustment, subtract the balance of **Allowance for Uncollectible Accounts** from the total estimated uncollectible amount. The adjustment amount is $679.46 ($729.26 − $49.80).

After the adjusting entry has been journalized and posted, the balance of **Allowance for Uncollectible Accounts** is $729.26 (the balance as determined by the aging schedule).

Uncollectible Accounts Expense		Allowance for Uncollectible Accounts	
Debit + Adj. 679.46	Credit −	Debit −	Credit + Bal. 49.80 Adj. 679.46 Bal. 729.26

SECTION 24.3
Assessment

After You Read

Reinforce the Main Idea

Use a diagram like this one to compare and contrast the two methods for estimating uncollectible accounts discussed in Section 24.3.

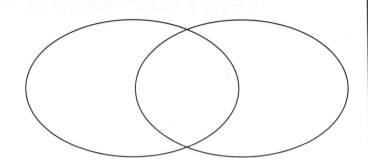

Problem 3 Estimating Uncollectible Accounts Expense Using the Percentage of Net Sales Method

Following are the end-of-period account balances for several stationery and office supply companies. Each company uses the percentage of net sales method to estimate its uncollectible accounts expense. The percentage used by each company is also listed.

Instructions Using the form provided in your working papers:

1. Calculate the amount of the adjustment for uncollectible accounts expense using the percentage of net sales method.
2. Record the adjusting entry for the estimated uncollectible accounts expense for Davis Inc.

	Sales	Sales Discounts	Sales Returns and Allowances	Percentage of Net Sales Uncollectible
Andrews Co.	$142,360	$1,423	$ 936	2
The Book Nook	209,100	3,180	1,139	1
Cable Inc.	173,270	1,730	1,540	1½
Davis Inc.	65,460	650	690	2
Ever-Sharp Co.	95,085	900	1,035	1¼

Math for Accounting

Hernando's Card Shop is planning to expand. Before it expands, the owner wants to review the uncollectible accounts. Hernando asks you to calculate the percentage of uncollectible accounts per year.

	Net Sales	Uncollectible Accounts
Year 1	$23,000	$ 800
Year 2	28,750	1,200
Year 3	46,000	1,345
Year 4	52,000	1,150

Chapter 24
Visual Summary

Concepts
Write off a specific uncollectible account using the direct write-off method:

Uncollectible Accounts Expense	
Debit + xxx	Credit −

Accounts Receivable (controlling/subsidiary)	
Debit +	Credit − xxx

Analysis
Write off a specific account using the allowance method:

Allowance for Uncollectible Accounts	
Debit − xxx	Credit +

Accounts Receivable (controlling/subsidiary)	
Debit +	Credit − xxx

Procedures
Two common methods used to estimate uncollectible Accounts Receivable are the *percentage of net sales* method and the *aging of accounts receivable method*.

Percentage of Net Sales Method:

Net Sales × Percentage of Net Sales Estimated Uncollectable = Uncollectible Accounts Expense

Aging of Accounts Receivable Method:

Age Group	Amount	×	Estimated Percentage Uncollectible	=	Estimated Uncollectible Amount
Not yet due			xx		xxxx
1-30 days past due	xxxxx		xx		xxxx
31-60 days past due	xxxxx		xx		xxxx
61-90 days past due	xxxxx		xx		xxxx
91-180 days past due	xxxxx		xx		xxxx
Over 180 days past due	xxxxx		xx		xxxx
Total					xxxx

Allowance for uncollectible accounts, end of period ⬆

Chapter 24
Review and Activities

Answering the Essential Question

What does a business do when a debt cannot be collected?

When a company extends credit, it assumes the risk that the debt might not be paid. How can businesses protect themselves from accounts that can't be collected?

Vocabulary Check

1. **Vocabulary** Use the content vocabulary words to write instructions for dealing with uncollectible accounts receivable.

- direct write-off method
- uncollectible account
- What is the root word in *prospective*
- allowance method
- book value of accounts receivable
- percentage of net sales method
- Use *approximate*
- aging of account receivable method
- What is the definition of *assume*

Concept Check

2. **Analyze** Explain the difference between the direct write-off method and the allowance method.
3. Which method is likely to be used by a business with many charge customers? Which is likely to be used by a business that sells mainly on a cash basis?
4. How does each method identify the amount of the uncollectible expense?
5. In the direct write-off method of recording an uncollectible account receivable, which accounts are debited and credited?
6. **Synthesize** What can businesses do to prevent bad debts?
7. What is the account classification of Allowance for Uncollectible Accounts?
8. When a specific amount is written off using the allowance method, which accounts are debited and credited?
9. What is the first journal entry made when a customer pays an amount that was previously written off? Name the accounts debited and credited for both methods.
10. In the second journal entry, which accounts are debited and credited?
11. **Analyze** Compare and contrast the percentage of net sales method and the aging of accounts method.
12. **Math** The Tenth National Bank has contracted with Cheryl Adams to provide business writing skills training for its employees. Because she will be giving this training 35 times over the next year at this company, Cheryl has agreed to discount her fee by 7%. She usually charges $85 per hour for classroom time and $55 per hour for preparation time. Each training session will involve 4 hours in the classroom and 3 hours of preparation. How much will Cheryl have invoiced the bank by the end of the year?

Chapter 24
Standardized Test Practice

Multiple Choice

1. The _____ of accounting for uncollectible accounts matches potential bad debts expense with the sales from the same period.

   a. direct write-off method

   b. matching principle

   c. aging of accounts receivable method

   d. allowance method

2. Under the allowance method, the entry to record the estimated bad debts expense

   a. has no effect on net income.

   b. increases net income.

   c. has no effect on total assets.

   d. reduces total assets.

3. Cameron Corporation ages accounts receivable to estimate uncollectibles. The aging schedule estimates $2,340 of uncollectible accounts. Prior to adjustment, allowance for uncollectible accounts has a debit balance of $300. The expense reported on the income statement for uncollectibles will be

   a. $2,640.

   b. $2,040.

   c. $2,340.

   d. $300.

4. Vasquez Construction uses the percentage of sales method to estimate uncollectibles. Net credit sales for the current year amount to $500,000 and management estimates 2% will be uncollectible. Allowance for doubtful accounts prior to adjustment has a debit balance of $890. The amount of expense reported on the income statement will be

   a. $10,000.

   b. $10,890.

   c. $9,110.

   d. $890.

5. SPH, Inc. uses the direct write-off method for uncollectible accounts. The entry to record the $64 invoice that Dave Case did not pay is

   a. debit Accounts Receivable/Dave Case, credit Uncollectible Accounts Expense.

   b. debit Allowance for Uncollectible Accounts, credit Accounts Receivable/Dave Case.

   c. debit Uncollectible Accounts Expense, credit Accounts Receivable/Dave Case.

   d. debit Uncollectible Accounts Expense, credit Cash.

True or False

6. Uncollectible accounts are debts the business plans not to pay.

7. An uncollectible account that is written off and then paid must be reinstated in the accounting records.

Short Answer

8. Explain which accounting principle is violated when the direct write-off method of accounting for uncollectible accounts is used.

Chapter 24
Computerized Accounting

Writing Off Uncollectible Accounts Receivable

Making the Transition from a Manual to a Computerized System

MANUAL METHODS	COMPUTERIZED METHODS
• Examine each customer account and organize it based on its due date. • Using accounting stationery, list each customer's account in the appropriate past due columns.	• A preformatted report feature allows the accountant to generate an aging schedule. The software pulls data from information stored in the accounts receivable ledgers.
• Determine the method to be used to write off uncollectible accounts. • Record a general journal entry to record the writeoff and post it to the general ledger and subsidiary ledger. • Calculate new balances.	• Determine the method to be used to write off uncollectible accounts. • Outstanding invoices due are tied to the customer's ID and assigned to the account, **Allowance for Uncollectible Accounts.**

Chapter 24
Problems

Problem 4   Using the Direct Write-Off Method

Sunset Surfwear uses the direct write-off method of accounting for uncollectible accounts.

Instructions

1. In your working papers, record the following transactions on page 14 in the general journal.
2. Post the transactions to the appropriate accounts.

Date		Transactions
June	1	Wrote off the $288.75 account of Alex Hamilton as uncollectible, Memorandum 223.
	4	Wrote off the $243.60 account of Helen Jun as uncollectible, Memorandum 249.
	14	Wrote off the $57.75 account of Nate Moulder as uncollectible, Memorandum 255.
	22	Received $288.75 from Alex Hamilton in full payment of his account, Memorandum 298 and Receipt 944.
	29	Wrote off the $100.80 account of Martha Adams as uncollectible, Memorandum 329.

Analyze

Describe the impact on **Accounts Receivable** and **Uncollectible Accounts Expense** when an account is written off.

Problem 5   Calculating and Recording Estimated Uncollectible Accounts Expense

InBeat CD Shop uses the percentage of net sales method of accounting for uncollectible accounts. At the end of the period, the following account balances appeared on InBeat's trial balance:

Accts. Rec.	$110,000	Allow. for Uncoll. Acct	$ 4,000
Sales	900,000	Sales Ret. and Allow.	50,000
Sales Disc.	10,000	Uncoll. Accts. Expense	—

continued

Chapter 24
Problems

Instructions

1. In your working papers, calculate the amount of the adjustment for uncollectible accounts for the period ended June 30. Management estimates that uncollectible accounts will be 1 percent of net sales.
2. Journalize the adjusting entry on page 8 in the general journal.
3. Post the adjusting entry to the general ledger accounts.

Analyze
Determine the book value of accounts receivable.

Problem 6 Writing Off Accounts Under the Allowance Method

Shutterbug Cameras uses the allowance method of accounting for uncollectible accounts.

SOURCE DOCUMENT PROBLEM

Problem 24–6

Use the source documents in your working papers to record the transactions for this problem.

Instructions

1. In your working papers, record the following transactions on page 9 in the general journal.
2. Post the transactions to the appropriate accounts.
3. Prepare the Assets section of the balance sheet for Shutterbug Cameras using the partial general ledger in the working papers. The balances of other asset accounts are **Merchandise Inventory,** $33,000; **Supplies,** $2,000; and **Prepaid Insurance,** $1,200.

Date		Transactions
June	2	Wrote off the $593.25 account of Kalla Booth as uncollectible, Memorandum 329.
	9	Wrote off the $840 account of Click Studios as uncollectible, Memorandum 343.
	10	Received $131.25 from Jimmy Thompson in full payment of his account, which was written off on November 10, Memorandum 349 and Receipt 210.
	12	Wrote off the $945 account of FastForward Productions as uncollectible, Memorandum 474.
	30	Recorded the adjusting entry for estimated uncollectible accounts for the period. The uncollectible accounts expense estimate is based on 2 percent of the net sales of $150,000.
	30	Recorded the closing entry for **Uncollectible Accounts Expense.**

continued

Chapter 24
Problems

Analyze

Calculate the book value of the accounts receivable after the closing entries have been posted.

Problem 7 Estimating Uncollectible Accounts Expense

Cycle Tech Bicycles uses the allowance method of accounting for uncollectible accounts. The company estimates the uncollectible amount by aging its accounts receivable accounts.

Instructions

1. Complete the analysis of accounts receivable that is included in your working papers.

2. Calculate the estimated uncollectible amount. Use the form provided in your working papers.

3. Journalize the June 30 adjusting entry for uncollectible accounts expense on page 11 of the general journal. Before the adjusting entry, **Allowance for Uncollectible Accounts** had a credit balance of $142.

4. Post the adjusting entry to the general ledger accounts.

Analyze

Calculate the book value of accounts receivable.

continued

Chapter 24
Problems

Problem 8 — Reporting Uncollectible Amounts on the Financial Statements

The work sheet for River's Edge Canoe & Kayak is included in your working papers. The trial balance is complete, and all of the adjustments except the one required for uncollectible accounts have been entered.

Instructions

1. Record the adjustment for uncollectible accounts expense in the Adjustments section of the work sheet. Uncollectible accounts are estimated to be 1.5 percent of net sales. Label the adjustment (a).
2. Complete the work sheet.
3. Prepare an income statement, a statement of retained earnings, and a balance sheet.
4. Record the adjusting entries on page 18 of the general journal.
5. Post the adjusting entries. Record and post the closing entries.

Analyze

Compute the balance of **Allowance for Uncollectible Accounts** before and after the adjusting entry.

Chapter 24
Problems

Problem 9 — Using the Allowance Method for Write-Offs

Buzz Newsstand uses the allowance method for uncollectible accounts.

Instructions In your working papers, journalize the following transactions on page 13 in the general journal. Post the transactions to the account of Lee Adkins.

Date	Transactions
Jan. 14	Wrote off the $194.50 account of Lee Adkins as uncollectible, Memorandum 498.
June 25	Received a check for $30 from Lee Adkins on account, Memorandum 767 and Receipt 98.
Dec. 10	Received notice that Lee Adkins declared bankruptcy. Received 40 percent of the balance not paid, Receipt 288 and Memorandum 941.

Analyze

Conclude which account you would debit when writing off an account using the direct write-off method instead of the allowance method.

Real-World Applications & Connections

CASE STUDY

Merchandising Business: Formal Wear

The Black Tie rents formal attire. It accepts cash, debit cards, credit cards, and personal checks. To reserve an outfit requires a down payment. The balance, due 10 days after the return, is recorded as an account receivable. Lately the store has had problems collecting accounts receivable. Customers say they do not receive an invoice for the balance until it is past due.

INSTRUCTIONS

1. Develop a plan for quality that will reduce uncollectible accounts. Your plan must be attainable and measurable.
2. Determine how to implement your plan. Will you need to change procedures, modify the accounts receivable system, modify the customer file system, hire another clerk, or retrain personnel?

21st Century Skills

Budgeting for IT Productivity

Most of today's business leaders agree that an organization that does not invest wisely in information technology (IT) cannot survive. IT increases productivity. However, while investing in IT is necessary, it is important to match productivity gains to technology costs. Productivity is the primary measure of technology's economic impact. It is often difficult to measure technology's impact, but as with every other budgetary item, expenditures must align with organizational goals. When the costs for IT software and hardware are added up, decisions need to be made concerning the difference between investments in IT and the performance it generates.

Activity

Use the Internet to research ways IT costs influence businesses' investments. Write a few paragraphs describing how IT affects productivity. Think about the impact of integrating IT into accounting. Discuss how integrating technology into accounting assists companies in decision making.

Career Wise

Certified Management Accountant

A certified management accountant specializes in preparing and presenting financial information in a way that best enables management to make informed business decisions. While many accountants focus on preparing financial information thoroughly and according to regulations, management accountants focus on what that financial information means to the company and its operations. Many management accountants get their start in finance accounting. Management accountants perform many duties, including forecasting, planning, variance analysis, and monitoring costs.

Research and Share

a. Use the Internet to locate three sources of information on certified management accounting.

b. What level of education is required to be a certified management accountant? What organization offers certification for this profession?

Spotlight on Personal Finance

Credit Check

To obtain credit, you must complete a credit application. The business will check your income, banking record, current debt amount, debt payment record, age, employment history, current employer, address, and so on.

Activity

Imagine that you just graduated from high school, have a part-time job, and plan to work your way through college. What are some steps you can take to establish a good credit rating?

H.O.T. Audit

Methods of Writing Off Uncollectible Accounts

Petty Matero, owner of Matero Department Store, uses the allowance method to account for uncollectible accounts. The March 30 trial balance shows a debit balance of $300 in the Uncollectible Accounts Expense account. You discovered that Jan Anderson's account for $300 was written off on March 1 using the direct write-off method instead of the allowance method.

Activity

What entry was recorded in the general journal on March 1? Which entry should have been recorded in the general journal on March 1?

Chapter 25

Inventories

Chapter Topics

25-1 Determining the Quantity of Inventories

25-2 Determining the Cost of Inventories

25-3 Choosing an Inventory Costing Method

Visual Summary

Review and Activities

Standardized Test Practice

Computerized Accounting

Problems

Real-World Accounting Careers

Real-World Applications & Connections

Essential Question

As you read this chapter, keep this question in mind:

How do merchandising businesses keep track of the value of their inventories?

Main Idea

Two methods of tracking merchandise are the perpetual inventory system and the periodic inventory system. Businesses can choose one of four methods to assign cost values to inventories. Businesses must apply consistency and conservatism when reporting merchandise inventory on the financial statements.

Chapter Objectives

Concepts	Analysis	Procedures
C1 Explain the importance of maintaining accurate inventory records.	**A1** Take a physical inventory count and record inventories.	**P1** Assign a value to merchandise inventory using the lower-of-cost-or-market rule.
C2 Explain the difference between a periodic and a perpetual inventory system.	**A2** Determine the cost of merchandise inventory using the specific identification; first-in, first-out; last-in, first-out; and weighted average cost methods.	**P2** Explain the accounting principles of consistency and conservatism.

Real-World Business Connection

Ford Motor Company

100 years after Henry Ford revolutionized large-scale manufacturing with the moving assembly line, Ford Motor Company is working to bring important changes to the American auto industry while reducing the public's dependence on non-renewable energy sources. Ford's plug-in hybrid electric vehicle research program is partnering with regional energy companies, with the goal of enabling plug-in electric vehicles to use electric grids, so that drivers can choose when and for how long to recharge.

Connect to the Business

The vehicle inventory management system at Ford includes a wireless Real Time Locating System, which uses low-power radio tags at factories to locate vehicles, and WhereNet, an inventory management software system that allows the company to locate within seconds a specific vehicle parked in a lot full of thousands of other vehicles.

Analyze

What might happen if Ford could not effectively evaluate the numbers of different vehicle models it should carry in inventory?

Focus on the Photo

There are several ways to track inventory and assign a value to it at the end of a fiscal period, including Ford's innovative wireless system that instantly identifies cars in its inventory. **What are some other ways a business might take inventory?**

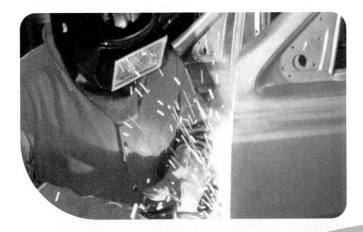

SECTION 25.1
Determining the Quantity of Inventories

In this chapter, you will learn the methods businesses use to track inventory. Maintaining control over inventory is essential. To maintain this control and determine the most efficient inventory levels, a business establishes a system of inventory-tracking procedures. A manufacturer controls inventories of raw materials, work in process, and finished goods. Wholesalers and retailers **maintain** control over merchandise inventory. Can you imagine the problems that a store like Wal-Mart would encounter if it did not have an inventory control system?

Section Vocabulary

- perpetual inventory system
- point-of-sale terminal (POS)
- online
- periodic inventory system
- maintain

Merchandise Inventory

How Is Tracking Merchandise Helpful?

Earlier you learned that *merchandise* refers to goods a business purchases for resale to customers. The **Merchandise Inventory** account shows the cost of goods purchased for resale. By tracking merchandise, a business knows how much merchandise is sold and which items are selling well. **Merchandise Inventory** is the only account reported on both the balance sheet (as a current asset) and the income statement (to calculate the cost of merchandise sold).

Methods of Tracking Inventory

How Do You Keep Track of Inventory?

Businesses can choose between two methods to track merchandise: the *perpetual inventory system* and the *periodic inventory system*.

Perpetual Inventory System

Most large businesses and smaller ones with automated accounting systems use the **perpetual inventory system**, which keeps a constant, up-to-date record of merchandise on hand at any point in time. When a business uses a perpetual inventory system, management can obtain the quantity on hand and cost of any item at any point in time. The business can use the information to determine when to reorder items. This avoids loss of sales due to inadequate inventory. **Figure 25–1** shows an example of a computer printout for a perpetual inventory system.

Wilton Outdoor Center
DAILY INVENTORY REPORT
Department 47
October 13, 20--

Stock No.	Item	Unit	Quantity	Unit Cost	Total Value
7651	Kilmer Rods	Each	8	$ 31.80	$ 254.40
7560	Tyon Rods	Each	12	36.40	436.80
7762	Peterson Rods	Each	11	29.75	327.25
7785	K & R Rods	Each	6	26.30	157.80
7208	Weber Reels	Each	5	35.20	176.00
7338	Pro Reels	Each	8	41.40	331.20
7193	Artcraft Reels	Each	4	47.10	188.40
7525	#7 Fishing Hook	Box	26	4.86	126.36
7937	#9 Fishing Hook	Box	31	5.24	162.44
	Total				$ 2,160.65

Figure 25–1 Computer Printout from a Perpetual Inventory System

The perpetual inventory system records an entry in the **Merchandise Inventory** account every time a purchase or sale occurs. After each sale the system also enters the cost of merchandise sold in the accounting records. The perpetual inventory system uses a **Cost of Merchandise Sold** account.

Computers update a perpetual inventory system through electronic cash registers, or **point-of-sale terminals (POS)**, that are **online** (linked to a central computer system). Such machines read bar codes that identify the item being sold. The computer records the sale and automatically updates the inventory information. Merchandise purchases are also entered into the computer to update the inventory records.

Businesses that have not yet automated their operations can also use a perpetual inventory system. They use stock cards or sheets like the one in **Figure 25–2** to record the amount of increase or decrease for every purchase and sale. For example, a business records the quantity of merchandise purchased in the stock card's *In* column and the quantity of sold items in the *Out* column.

STOCK CARD

STOCK NO.	ITEM
C 1297	Altmore Disc Player
SUPPLIER	SUPPLIER'S CATALOGUE NO.
Star Electric	91246
UNIT	MINIMUM	MAXIMUM
Each	15	60

DATE	EXPLANATION	IN	OUT	BALANCE
7/1/20--	Balance on Hand			48
7/8/20--	Shipping Order 21928		6	42
7/12/20--	Shipping Order 22201		10	32
7/20/20--	Shipping Order 22456		8	24
7/24/20--	Shipping Order 22719		12	12
7/24/20--	Purchase Req. 19426			
7/31/20--	Receiving Report 21563	48		60

Figure 25–2 Stock Card Used in a Perpetual Inventory System

Periodic Inventory System

In the **periodic inventory system**, inventory records are updated only after a physical count of merchandise on hand is made. This system does not adjust inventory records for every purchase and sale. Instead, an adjusting entry is made at the end of the accounting period. See Chapter 18, pages 574–575 for a review of how to calculate the inventory adjustment. Small businesses that maintain manual accounting records generally use this system.

Physical Inventory Count

In either inventory tracking system, a business takes a physical count of its merchandise at least once a year. This is called *taking inventory* and is part of the system of internal controls. Businesses using the periodic inventory system must use the physical count to update their accounting records.

The process of identifying and counting all items of merchandise for businesses with many items in inventory is very time consuming. Therefore, inventory is usually counted when it is at its lowest level. Seasonal businesses take physical counts and record inventories at the

INVENTORY SHEET

DATE Jan. 5, 20--   CLERK Arlene Stone   PAGE 8

STOCK NO.	ITEM	UNIT	QUANTITY	UNIT COST	TOTAL VALUE
1901	Needles	Pkg	24	1 14	27 36
2132	Thread	Spool	12	65	7 80
2136	Thread	Spool	18	55	9 90
3245	Zipper	Each	18	1 50	27 00
1917	Pins	Box	24	79	18 96
4971	Buttons	Pkg	12	89	10 68
4993	Tape Measure	Each	15	1 49	22 35
				TOTAL FOR THIS SHEET	789 14

Figure 25–3 Inventory Sheet

end of the peak selling period after they have sold most of the merchandise. A ski shop, for instance, probably takes inventory in May or June.

The total number of a particular item on hand is recorded on an inventory card or an inventory sheet. **Figure 25–3** shows a typical inventory sheet. It lists each item's stock number, description, quantity on hand, unit cost, and the total cost of its inventory on hand.

SECTION 25.1
Assessment

After You Read

Reinforce the Main Idea

Identify three facts about each inventory tracking system covered in this section. Create a chart similar to this one.

Tracking System	Fact #1	Fact #2	Fact #3

Problem 1 Preparing Inventory Reports

In your working papers, complete the manual inventory sheet for Carole's Gift Shop. Use today's date and your name as the clerk.

Stock No.	Item	Unit	Quantity	Unit Cost
1790	Greeting Cards	Doz.	32	6.00
2217	Plush Toys	Each	20	2.50
1900	Balloons	Doz.	12	.50
1201	Wrapping Paper	Each	30	1.12
1205	Ribbon	Spool	25	.75
3495	Novelty Buttons	Doz.	12	2.50
2722	Music Boxes	Doz.	6	60.00
4200	Party Supplies	Doz.	10	6.50
1907	Gift Boxes	Doz.	5	2.75
1742	Vases	Doz.	2	12.50

Math for Accounting

Katie's Ceramic Emporium received 9 dozen ceramic elephants for a total cost of $306.72. The invoice also includes 17 dozen songbird figurines in various colors, with a cost of $822.12. Based on the information on the invoice, calculate the unit cost of the elephants and the songbirds.

SECTION 25.2
Determining the Cost of Inventories

Once a business determines the quantity of merchandise it has on hand, it calculates the cost of that merchandise. There are four inventory costing methods used to assign costs to merchandise.

Inventory Costs

Why Is It Difficult to Determine Inventory Costs?

When merchandise is purchased, it is recorded in the accounting records at cost. However, assigning a cost to each item in inventory can be complicated. A business may purchase the same item many times within a single inventory period, and the cost may change from one purchase to the next. The challenge is to decide which cost applies to each item.

Methods of Assigning Costs to Inventories

How Do You Set the Cost of Inventory?

Businesses use one of four methods to determine inventory cost:

- specific identification
- first-in, first-out (FIFO)
- last-in, first-out (LIFO)
- weighted average cost

The Specific Identification Costing Method

Under the **specific identification method**, the exact cost of each item is determined and assigned to that item. The actual cost of each item is obtained from the invoice.

Businesses that sell a small number of items with high unit prices most often use the specific identification method. Appliance stores, automobile dealerships, and furniture stores often use this method.

The Entertainment Store uses the periodic inventory system. It started the year with a beginning inventory of 15 DVD players. During the period the store purchased an additional 50 DVD players. When a physical inventory count was taken on May 31, there were 12 players still on hand. The cost of the DVD players was calculated as follows.

Section Vocabulary

- specific identification method
- first-in, first-out method (FIFO)
- last-in, first-out method (LIFO)
- weighted average cost method

Knowledge Extension

When the prices of beginning inventory and net purchases are not available, a business can use the gross profit method to estimate the cost of its ending inventory. As the name implies, the *gross profit method* uses the percentage of gross profit to estimate the cost of the inventory. The gross profit percentage is gross profit divided by net sales. Usually the gross profit percentage is based on financial statements from several previous periods. Once the gross profit percentage is determined, it is a matter of working backward through the income statement to estimate the cost of ending inventory.

Date	Description	Units	Cost		Total
June 1	Beginning inventory	15	$250	=	$ 3,750
Aug. 4	Purchase	20	250	=	5,000
Dec. 8	Purchase	10	253	=	2,530
Feb. 27	Purchase	10	258	=	2,580
May 1	Purchase	10	260	=	2,600
	Total	65			$16,460

Using the specific identification method, the accountant checks the invoices to find the actual cost of each of the 12 players still on hand at the end of May. The accountant found the following:

4 purchased @ $253 each	=	$1,012
5 purchased @ $258 each	=	1,290
3 purchased @ $260 each	=	780
12 players (ending inventory)		$3,082

The cost of ending inventory is $3,082. Once the cost of ending inventory has been calculated, the cost of merchandise sold can be computed:

Merchandise available for sale
− Cost of ending inventory
Cost of merchandise sold

In this example the cost of merchandise sold using the specific identification method is $13,378:

	Units	Cost
Cost of players available for sale	65	$16,460
Less ending inventory	12	3,082
Cost of merchandise sold	53	$13,378

Reading Check

Critical Thinking

Why do all businesses not use the specific identification costing method?

The First-In, First-Out Costing Method

The **first-in, first-out method (FIFO)** of assigning cost assumes that the first items purchased (first in) are the first items sold (first out). The FIFO method assumes that the items purchased most recently are the ones on hand at the end of the period. The physical flow of most merchandise is first-in, first-out. For example, think about milk that is stocked by a supermarket. Since milk is perishable, the supermarket stocks the shelves with the milk it purchased first. As that milk is sold, later purchases of milk are added at the back of the shelves.

Let's apply the FIFO costing method to our DVD player example.

Date	Description	Units	Cost		Total
June 1	Beginning inventory	15	$250	=	$ 3,750
Aug. 4	Purchase	20	250	=	5,000
Dec. 8	Purchase	10	253	=	2,530
Feb. 27	Purchase	10	258	=	2,580
May 1	Purchase	10	260	=	2,600
	Total	65			$16,460

Under the FIFO method, the items purchased first are assumed to be the items sold first. In other words the 53 DVD players sold are assumed to be as follows:

	Units
Beginning inventory	15
Aug. 4	20
Dec. 8	10
Feb. 27	8
Total sold	53

> **Common Mistakes**
>
> **FIFO, LIFO, and ending inventory**
> It might be easy to confuse FIFO and LIFO when calculating ending inventory. These acronyms describe more directly the calculation of cost of goods sold rather than ending inventory. For example FIFO (first-in, first-out) directly suggests which inventory units are sold—the first ones—and therefore are used to calculate the cost of goods sold. The inventory units not sold are the last ones in and are used to calculate the ending inventory.

The items remaining in inventory are:

	Units
Feb. 27 (10 bought − 8 sold)	2
May 1	10
Total	12

The cost of the ending inventory using the FIFO method is:

10	units @ $260 each	=	$2,600
2	units @ $258 each	=	516
12	units		$3,116

The cost of merchandise sold using the FIFO method is $13,344:

	Units	
Cost of players available for sale	65	$16,460
Less ending inventory	12	3,116
Cost of merchandise sold	53	$13,344

The Last-In, First-Out Costing Method

The **last-in, first-out method (LIFO)** of assigning inventory cost assumes that the last items purchased (last in) are the first items sold (first out). The LIFO method assumes that the items purchased first are still on hand at the end of the period. The earliest costs, therefore, are the ones used to assign a cost to the inventory. The physical flow of a stone and gravel company is last-in, first-out. When new gravel is purchased and delivered, it is deposited on top of the existing gravel. As gravel is taken from the top of the pile, the first gravel used is the last gravel delivered.

Let's return to the DVD player example and apply the LIFO costing method.

Date	Description	Units	Cost		Total
June 1	Beginning inventory	15	$250	=	$ 3,750
Aug. 4	Purchase	20	250	=	5,000
Dec. 8	Purchase	10	253	=	2,530
Feb. 27	Purchase	10	258	=	2,580
May 1	Purchase	10	260	=	2,600
	Total	65			$16,460

Using the LIFO method, the 12 players remaining in stock are from the beginning inventory of 15 units. The cost of the ending inventory is $3,000 (12 units @ $250 each).

The cost of merchandise sold using the LIFO method is $13,460:

	Units	
Cost of players available for sale	65	$16,460
Less ending inventory	12	3,000
Cost of merchandise sold	53	$13,460

Reading Check

Compare and Contrast

How are LIFO and FIFO similar? How are they different?

The Weighted Average Cost Method

A fourth method of assigning inventory costs is the weighted average cost method. The **weighted average cost method** assigns the average cost to each unit in inventory. The average cost is calculated by:

- adding the number of units on hand at the beginning of the period and the number of units purchased
- adding the cost of the units on hand at the beginning of the period and the cost of the units purchased
- dividing the total cost by the total number of units

The average cost per unit is used to determine the cost of the ending inventory. Again, we will use the DVD player example to apply the weighted average cost method.

Date	Description	Units	Cost		Total
June 1	Beginning inventory	15	$250	=	$ 3,750
Aug. 4	Purchase	20	250	=	5,000
Dec. 8	Purchase	10	253	=	2,530
Feb. 27	Purchase	10	258	=	2,580
May 1	Purchase	10	260	=	2,600
	Total	65			$16,460

The average cost per player is $253.23 ($16,460 ÷ 65 units).

The cost of the ending merchandise inventory using the weighted average cost method is $3,038.76 (12 units × $253.23).

The cost of merchandise sold using the weighted average cost method is $13,421.24:

	Units	
Cost of players available for sale	65	$16,460.00
Less ending inventory	12	3,038.76
Cost of merchandise sold	53	$13,421.24

The following table illustrates the different inventory costing methods for The Entertainment Store's inventory of DVD players. After the company selects an inventory costing method, it is applied consistently as you will see in the next section.

Method	Cost of Players Sold	Ending Inventory
Specific identification	$13,378.00	$3,082.00
FIFO	13,344.00	3,116.00
LIFO	13,460.00	3,000.00
Weighted average	13,421.24	3,038.76

SECTION 25.2
Assessment

After You Read

Reinforce the Main Idea

Identify two facts about each inventory costing method covered in this section. Create a chart similar to this one.

Costing Method	Fact #1	Fact #2

Problem 2 Determining Inventory Costs

The following items were purchased by Kudos Leather Goods during the month of April:

April 2	34 wallets @ $12.95 each
April 8	24 wallets @ $13.10 each
April 18	15 wallets @ $13.25 each
April 26	20 wallets @ $13.27 each

On April 1 the business had in inventory 19 wallets valued at $12.90 each. On April 30 the business had 36 wallets in inventory; of these wallets, 8 were purchased on April 2, 15 were purchased on April 8, 3 were purchased on April 18, and 10 were purchased on April 26.

Instructions In your working papers, calculate the cost of the ending inventory using:

a. the specific identification method

b. the FIFO method

c. the LIFO method

d. the weighted average method

Math for Accounting

Foxfire Golf Club Pro Shop uses the first-in, first-out method for inventory costing. At the beginning of the year, Foxfire had 30 golf club sets on hand. The golf club sets were purchased for $800 each. An additional 12 golf club sets were purchased during the year at $875 each. When inventory was taken at the end of the season, 5 golf sets were still on hand. Using the FIFO method of inventory valuation, what was the total cost of merchandise sold?

SECTION 25.3
Choosing an Inventory Costing Method

A business may use any one of the four inventory costing methods. Careful consideration is given to this choice because it affects the gross profit reported by the business.

Consistency and Inventory Costing

How Do You Apply the Consistency Principle?

When a business applies the same accounting methods in the same way from one period to the next, the business is applying the GAAP **consistency principle**. Once a business chooses an inventory costing method, the business must use it consistently. This helps owners and creditors compare financial reports from one period to another.

Businesses are permitted to change costing methods but must declare the reasons for changing and how the change will affect the financial statements. In addition the business must get permission for the change from the Internal Revenue Service.

Section Vocabulary
- consistency principle
- lower-of-cost-or-market rule
- market value
- conservatism principle

Comparison of the Four Inventory Costing Methods

How Does the Inventory Costing Method Affect the Reported Gross Profit?

When deciding which inventory costing method to use, the owner or manager compares the four methods and selects the one that is likely to be the most beneficial to the company. The owner or manager considers the present economic conditions and the future economic outlook. He or she will also consider whether prices and demand for the product will remain stable, increase, or decrease.

The cost of ending inventory affects the cost of merchandise sold, which in turn affects the income or loss reported on the income statement. The following table compares The Entertainment Store's gross profit on sales using the four inventory costing methods. The company sold all of the DVD players for $320 each. Total sales are $16,960 (53 units × $320).

	Specific Identification	First-In, First-Out	Last-In, First-Out	Weighted Average
Sales	$16,960.00	$16,960.00	$16,960.00	$16,960.00
Less: Cost of merchandise sold	13,378.00	13,344.00	13,460.00	13,421.24
Gross profit on sales	$ 3,582.00	$ 3,616.00	$ 3,500.00	$ 3,538.76

Over the year, the price that The Entertainment Store paid to purchase each player increased from $250 to $260. As the table on page 810 shows, in a period of rising prices, the LIFO method results in the lowest gross profit on sales. The FIFO method results in the highest gross profit on sales.

Businesses pay income taxes on income earned. As this example indicates, the inventory costing method used by a business can increase or decrease its taxes.

Reading Check

Explain

Why must businesses consistently use the same inventory method?

Conservatism and the Lower-of-Cost-or-Market Rule

How Do You Apply the Conservatism Principle?

Merchandise inventory appears on the income statement and the balance sheet. Cost is the most common basis for reporting inventory. However, inventory might be worth less than its cost. For example, some merchandise items may deteriorate or become obsolete. If their value becomes less than the recorded cost, the difference is a loss to the business.

The **lower-of-cost-or-market rule** requires that the cost of the ending inventory that appears on the financial statements is the lower of its cost (calculated using one of the four inventory methods) or its *market value*. **Market value** of an item is the current price that is charged for a similar item of merchandise in the market. That is, it is the price that a retailer like The Entertainment Store would pay a wholesaler or manufacturer for a specific item. Market value is the cost at which the inventory item could be replaced at the date of the financial statements.

Let's look at an example. Assume that The Entertainment Store determines that the current market value of the DVD players is $248 each. At market the players are worth $2,976 ($248 × 12 units). Assume also that The Entertainment Store uses the FIFO inventory costing method. Under this method the cost of the ending inventory was $3,116.

Following the lower-of-cost-or-market rule, The Entertainment Store will report inventory at $2,976.

Lower of	
Cost (FIFO)	$3,116
or	
Market	2,976

The GAAP **conservatism principle** of accounting states that it is best to present amounts that are least likely to result in an overstatement of income or assets. To be conservative is to take the safe route. The lower-of-cost-or-market rule is conservative for two reasons:

1. Decreases in inventory value (losses) are recognized when they occur, but increases in inventory value are not recorded.

2. Inventory as reported on the balance sheet is never more, but may be less, than the actual cost of the inventory.

SECTION 25.3 Assessment

After You Read

Reinforce the Main Idea

Create a chart like this one to describe the two accounting principles covered in this section. Also describe how each principle applies to merchandise inventory.

Accounting Principle	Description	Application to Inventory

Problem 3 Analyzing a Source Document

Read the following memorandum and complete the assigned task.

Toys & Things

MEMORANDUM

TO: Accounting Clerk
FROM: Accounting Manager
DATE: June 30, 20--
SUBJECT: Change in Inventory Method

We have received approval to change from the LIFO to the FIFO method of determining our inventory costs. Please calculate the cost of the Walk-A-Long Dolls using the FIFO method. There are 36 dolls in inventory.

Walk-A-Long Dolls Beginning inventory   8@ $15.45
                   Purchases   6/11   12@ $15.95
                               6/17   10@ $16.25
                               6/22   6@ $16.40

Instructions

1. What is the new value of the ending inventory?
2. Assume that all 36 dolls were sold for $21.95. What is the gross profit for this item?

Math for Accounting

The management of Baby Steps Children's Store wants to report the largest gross profit on sales. Using the graph, compare the gross profit on sales for the four inventory costing methods. Which method results in the largest gross profit on sales?

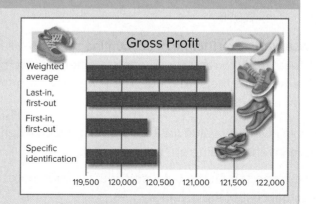

Gross Profit

- Weighted average
- Last-in, first-out
- First-in, first-out
- Specific identification

119,500 120,000 120,500 121,000 121,500 122,000

Chapter 25
Visual Summary

Concepts
Differences Between a Periodic and Perpetual Inventory System

Periodic Inventory System	Perpetual Inventory System
Used by small businesses with manual accounting records.	Used by most large businesses and smaller ones with automated accounting systems.
Requires a physical count of the merchandise on hand to update inventory records.	Provides current inventory records at any point in time.
How inventory accounting records are updated: • An adjusting entry is made to the **Merchandise Inventory** account at the end of the accounting period. • The cost of merchandise sold is calculated at the end of the accounting period.	How inventory accounting records are updated: • When a sale is made, the decrease to inventory is handled electronically using point-of-sale terminals that scan in the product's bar code. • When merchandise is purchased, the increase to inventory is entered directly into the inventory system.

Analysis
Inventory Costing Methods

Specific identification method	Tracks the exact cost of each item.
First-in, first-out method (FIFO)	Assumes that the first items purchased are the first items sold.
Last-in, first-out method (LIFO)	Assumes that the last items purchased are the first items sold.
Weighted average cost method	Uses an average cost for each inventory item

Procedures
In reporting its inventory, a business must follow two accounting principles:

Consistency: This principle requires that once a business chooses an inventory costing method, it must use that method consistently. If it changes methods, it must obtain permission from the Internal Revenue Service and then report the reasons for the change and how it affects the financial statements.	**Conservatism:** This accounting principle requires that when they can choose among procedures, accountants must choose the safer, or more conservative, route by presenting accounts that are least likely to result in an overstatement of income or assets.

Chapter 25
Review and Activities

Answering the Essential Question

How do merchandising businesses keep track of the value of their inventories?

Usually a business's inventory is its largest current asset. Knowing how much product a business has on hand is critical for determining the value of this asset.

Vocabulary Check

1. **Vocabulary** Create a fill-in-the-blank sentence for each of these Content Vocabulary terms. The sentence should contain enough information to help determine the missing word.

   - maintain
   - perpetual inventory system
   - online
   - periodic inventory system
   - point-of-sale terminal (POS)
   - specific identification method
   - first-in, first-out method (FIFO)
   - last-in, first-out method (LIFO)
   - weighted average cost method
   - consistency principle
   - conservatism principle
   - lower-of-cost-or-market rule
   - market value

Concept Check

2. **Analyze** Why does a company need a control system for its merchandise inventory? What are the benefits of determining the most efficient inventory levels?
3. **Analyze** Explain the difference between a periodic and a perpetual inventory system. How does each use inventory records, stock records, and point-of-sale terminals?
4. When is the best time for a business to take a physical inventory?
5. **Apply** What is meant by the phrase "with the FIFO method, the inventory cost is based on the most recent costs"?
6. Why does the inventory costing method used affect gross profit on sales and net income?
7. In the lower-of-cost-or-market rule, what does *cost* mean? What does *market* mean?
8. **Synthesize** Use your definitions above to explain the lower-of-cost-or-market rule.
9. How does the consistency principle help the owners, creditors, and the general public?
10. What approach to presenting income does the conservatism principle require?
11. **Math** An inventory record for Wholesale Hardware Suppliers shows a receipt of 100 chainsaws on April 1. Another shipment of 125 arrived on April 15. On May 10, 100 chainsaws were shipped out. On June 2, 50 arrived and on June 17, 75 were shipped out. find the number on hand as of June 17.

Chapter 25
Standardized Test Practice

Multiple Choice

1. The inventory system in which a constant, up-to-date record of merchandise on hand is maintained is called the

   a. first-in, first-out system.

   b. last-in, first-out system.

   c. perpetual inventory system.

   d. periodic inventory system.

   Use the following information to answer questions 2-4.

   Johnson, Inc. had the following inventory data: Inventory item T25 had 1,600 units at a unit price of $14 in inventory on January 1. The first purchase during the year was for 1,000 units at $15. The second purchase was for 1,000 units at $18. The December 31 inventory consisted of 1,200 units. The market price is $15.

2. The total cost of the ending inventory using FIFO is

   a. $16,800.

   b. $18,000.

   c. $18,468.

   d. $21,000.

3. The weighted average unit price of the ending inventory is

   a. $14.50.

   b. $15.39.

   c. $15.67.

   d. $16.00.

4. Assume the weighted average cost method is used. Calculate the ending inventory using the lower-of-cost-or-market rule.

   a. $16,000

   b. $18,000

   c. $18,468

   d. $21,000

True or False

5. The FIFO method of inventory assumes that old goods are sold first and that goods which are on hand are valued at current prices.

6. The weighted average cost method tracks the exact cost of each item.

Short Answer

7. What advantage is there to using a perpetual inventory system?

Extended Response

8. What is the lower-of-cost-or-market rule, and why is it used?

Chapter 25
Computerized Accounting

Inventory Costing

Determining the Cost of Inventories in a Perpetual Inventory System

MANUAL METHODS	COMPUTERIZED METHODS
• Enter purchases of merchandise onto stock cards.	• When the accounting system is set up, the costing method must be determined.
• Enter sales of items onto stock cards.	• Purchases of merchandise update not only the general ledger but also update the inventory records.
• At period end, calculate the number and value of remaining inventory items.	• As sales are made, the system updates the inventory records.
	• Inventory quantities and values are available when needed.

Chapter 25 Problems

Problem 4 Calculating the Cost of Ending Inventory

Sunset Surfwear sells wet suits. On January 2 there were 21 wet suits at a total cost of $4,809 in inventory.

Date	Description	Wet Suits	Cost	Total
Jan. 2	Beginning inventory	21	$229	$ 4,809
Jan. 3	Purchase	10	235	2,350
Mar. 17	Purchase	6	238	1,428
July 27	Purchase	12	240	2,880
Sept. 27	Purchase	10	241	2,410
Nov. 29	Purchase	6	244	1,464
	Total	65		$ 15,341

At the end of the year, there were 17 wet suits in ending inventory. Of these, 1 was purchased on July 27, 10 were purchased on September 27, and 6 were purchased on November 29.

Instructions Assign a cost to the ending inventory using the following:

a. the specific identification method

b. the FIFO method

c. the LIFO method

d. the weighted average cost method

Analyze

Determine which inventory method resulted in the lowest cost of ending inventory.

Chapter 25
Problems

Problem 5 — Completing an Inventory Sheet

InBeat CD Shop assigns a cost to its inventory using the lower-of-cost-or-market rule. In your working papers, there is a partial inventory record. As an example, the first line of the inventory record has been completed.

INVENTORY RECORD

Item No.	Item	Ending Inventory	Cost per Unit	Current Market Value	Price to be Used	Total Cost
0247	Blank CDs	24	2.67	2.88	2.67	64.08

Instructions Complete the inventory record. Do the following:

1. Select the lower-of-cost-or-market value. Enter that amount in the Price to be Used column.

2. Calculate the total cost of each item by multiplying the units in ending inventory by the Price to be Used column.

3. Add the amounts in the Total Cost column to determine the total cost of the ending inventory.

Analyze
Identify how many items used market value rather than actual cost.

Chapter 25
Problems

Problem 6 Calculating Gross Profit on Sales

Using the four inventory costing methods, Shutterbug Cameras summarized the cost of its ending inventory as follows:

Specific Identification	First-In, First-Out	Last-In, First-Out	Weighted Average Cost
$21,476.00	$21,581.40	$21,410.93	$21,447.36

Shutterbug Cameras also reported the following amounts:

Net sales	$53,874.92
Purchases available for sale	57,621.31

Instructions Using the preceding information, determine the cost of merchandise sold and the gross profit on sales for each of the inventory costing methods.

Analyze
Conclude which method resulted in the largest gross profit on sales.

Problem 7 Reporting Ending Inventory on the Income Statement

Cycle Tech Bicycles operates on a fiscal year beginning January 1. At the beginning of the year, the shop had in stock six Model #8274, 10-speed bicycles, valued at $2,364 (6 bicycles @ $394 each). During the year the business made the following purchases:

Date	Bicycles	Cost		Total
Jan. 20	4	$399	=	$ 1,596
Mar. 5	5	415	=	2,075
Apr. 23	7	419	=	2,933
Aug. 14	4	423	=	1,692
Oct. 3	6	430	=	2,580
Nov. 17	3	435	=	1,305
Total Purchases	29			$12,181

There were seven bicycles in inventory at the end of the period. During the year the bicycles sold for $675 each.

continued

Chapter 25
Problems

Instructions

1. Calculate the cost of the ending inventory using the FIFO, LIFO, and weighted average cost methods.

2. Using the costs calculated in (1), determine the cost of merchandise sold for each inventory costing method.

3. Prepare a partial income statement for each inventory costing method showing sales and the calculation of gross profit on sales. Assume that the sales and purchases are net amounts.

Analyze
Identify the method that resulted in the lowest cost of merchandise sold.

Problem 8 Calculating Cost of Merchandise Sold and Gross Profit on Sales

Buzz Newsstand started the month of May with the following inventory of disposable cameras:

Stock No.	Brand	Units on Hand	Unit Cost	Selling Price
3845	Lenox	4	$9.60	$ 17.95
4931	Lancaster	6	8.40	17.29
9265	Paterson	3	8.10	16.88
4850	McMahon	5	7.60	15.95

SOURCE DOCUMENT PROBLEM

Problem 25–8

Use the source documents in your working papers to complete this problem.

Buzz Newsstand uses the FIFO method to calculate the cost of its merchandise inventory. The May 31 physical inventory count indicated:

| Lenox | 4 cameras | Paterson | 7 cameras |
| Lancaster | 5 cameras | McMahon | 5 cameras |

Instructions

1. How many units of each of the four cameras were sold during May?

2. Using the chart provided in your working papers, calculate the gross profit on sales for each type of camera.

continued

Chapter 25
Problems

Date		Transactions
May	2	Purchased 10 Lancaster cameras at $8.45 each.
	4	Purchased 5 McMahon cameras at $7.80 each.
	9	Purchased 6 Lenox cameras at $9.95 each plus a $4 transportation charge.
	14	Purchased 5 Paterson cameras at $8.25 each.
	17	Purchased 8 Lancaster cameras at $8.60 each.
	19	Purchased 4 Lenox cameras at $10.10 each plus a $5 transportation charge.
	27	Purchased 8 Paterson cameras at $8.30 each.
	29	Purchased 4 Lancaster cameras at $8.85 each.

Analyze
Conclude which type of camera was sold the most.

Real-World Accounting Careers

Dan Crowley

San Antonio Water System

Q What do you do?

A My job is to oversee the short- and long-term financial planning of the organization, which includes the budget process and the calculation and implementation of public utility rates.

Q What are your day-to-day responsibilities?

A I report directly to the CFO and interact with other executives and managers within the organization so they are fully informed of the budget, rates and financial modeling requirements. I also supervise a group of seven employees who make sure the budget is up-to-date and accurately represents the needs of the organization.

Q What factors have been key to your success?

A One is education. Another is the ability to think strategically as well as tactically. Teamwork is a third key to success in this position.

Q What do you like most about your job?

A No two days are the same. Whether I'm working on the budget or doing financial modeling or managing people, I get to perform a wide variety of tasks. I use a wide range of skills—some days, it's pure accounting; others, it may be economics, political science or personnel management.

Career Facts

Real-World Skills
A willingness to take on different challenges; public speaking abilities; managerial skills

Training and Education
Undergraduate degree in accounting or economics and MBA; CPA certificate; eight to 10 years of experience

Career Skills
Knowledge of accounting and economics; ability to perform financial analysis; comprehensive understanding of business operations

Career Path
Work as both an accountant and a financial analyst, then assume roles of increasing responsibility

Tips from Robert Half International
Your resume should always be submitted along with a strong cover letter. The letter should expand upon the key points in your resume, highlight your skills and experience that are most relevant to the job opening, and explain how you can benefit the prospective employer.

College and Career Readiness

What qualities do you think would be important to be a supervisor of other employees? Interview a local business owner and learn more about effective leadership qualities. Summarize your findings in a one-page report.

Real-World Applications & Connections

Career Wise

Compare and Contrast: Forensic Accountant and Auditor

Forensic accountants and auditors both are accountants that analyze existing financial records, but they use the information they gather for different purposes. Conduct research to learn more about each career.

Compare and contrast

Make a list of the similarities and differences between each of these professions.

Choose

Decide which job out of the two would better suit you. Write a paragraph explaining why you made the choice you did.

Activity

Review the information you wrote in response to the activities above and then write two descriptions. Write one description of the type of person you think would make a good forensic accountant and another description of the type of person you think would make a good auditor. Describe the qualities that make each person good at each job.

Global Accounting

Inventory Measurement

Many countries use International Accounting Standards (IAS) as their national financial reporting requirements. IAS 2 allows inventory valuation using the specific identification, FIFO, and weighted average methods. It does not allow use of the LIFO method, although countries such as the United States, Japan, and Mexico permit the use of LIFO according to their own national financial reporting standards.

Research and Write

a. Use the Internet to research the International Accounting Standards Board's stance regarding LIFO.

b. Explain why you think IAS do not permit the use of LIFO.

Analyzing Financial Reports

Inventory Levels

Knowing how much inventory is on hand helps managers make decisions about how much merchandise to purchase. Too much inventory means the business has not sold what it purchased and its money is tied up in inventory. A business that holds too little inventory may run out often and need to make frequent purchases which increases costs. It also can lose business when customers go to a competitor to purchase an out-of-stock item. Higher costs and out-of-stock situations mean lower profits.

Instructions

Use the Internet to locate and save or print a copy of Amazon.com's annual report to complete the following. Use the balance sheet and letter to shareowners to answer the following questions.

1. How did Amazon.com's merchandise inventory change over the period reflected in the report?
2. What could have this change?

H.O.T. Audit

Gross Profit Method

On the night of April 29, the inventory of Midtown Craft Store was destroyed by fire. The company's accounting records were saved. For insurance purposes, the owner will estimate the value of inventory that was destroyed using the gross profit method. By reviewing past income statements, the owner determined that the average gross profit percentage for the past five years was 39 percent. Using the information available, the owner prepared the partial income statement below. To estimate ending inventory and complete the income statement, follow these steps.

1. The gross profit is estimated by multiplying net sales by the gross profit percentage: $250,000 × .39 = $97,500.

2. The cost of merchandise sold is determined by subtracting the estimated gross profit from net sales: $250,000 − $97,500 = $152,500.

3. Ending merchandise inventory is the difference between merchandise available for sale ($199,000) and the cost of merchandise sold ($152,500). What is the estimated cost of inventory that was destroyed on April 29? Analyze the effects of the gross profit method on net income and gross profit. Why is this method not acceptable for determining year-end inventory?

Midtown Craft Store
Income Statement
For the Period January 1 through April 29, 20--

Net Sales		250 000 00
Cost of Merchandise Sold:		
Merchandise Inventory, Jan. 1	53 000 00	
Net Purchases	146 000 00	
Merchandise Available for Sale	199 000 00	
Less: Merchandise Inventory, April 29		
Cost of Merchandise Sold		
Gross Profit (39% of Net Sales)		

Chapter 26
Notes Payable and Receivable

Chapter Topics

26-1 Promissory Notes

26-2 Notes Payable

26-3 Notes Receivable

Visual Summary

Review and Activities

Standardized Test Practice

Computerized Accounting

Problems

Real-World Applications & Connections

Essential Question

As you read this chapter, keep this question in mind:

How do businesses borrow and lend money?

Main Idea

Businesses issue two types of notes: interest-bearing notes and non-interest-bearing notes. Businesses record the receipt of a note receivable as well as the payment of the note.

Chapter Objectives

Concepts

C1 Explain how businesses use promissory notes.

C2 Calculate and record notes payable and notes receivable.

Analysis

A1 Explain the difference between interest-bearing and non-interest-bearing notes.

Procedures

P1 Journalize transactions involving notes payable.

P2 Journalize transactions involving notes receivable.

Real-World Business Connection

Wells Fargo

Have you noticed that your money seems to disappear as soon as you earn it? Do topics such as credit, investing, and loan-shopping fill you with dread? Wells Fargo, one of America's biggest and most reputable banking and financial services providers, wants to help. Wells Fargo's Hands on Banking® (www.handsonbanking.org) is an interactive program designed to answer those big, complex personal finance questions and help young people become confident money managers.

Connect to the Business

When companies need money for large purchases or other investments, many sign a note promising in writing to pay a certain amount of money at a specific time. Some notes are accepted by individuals or other companies, but commonly it is a financial institution such as Wells Fargo doing the lending. Interest is calculated depending on amount, interest rate, and time.

Analyze

When Wells Fargo lends money to a company, what factors do you think it considers?

Focus on the Photo

Wells Fargo has provided financial services for businesses and individual customers for more than a century. Individuals buy homes, automobiles, and finance educations with loans provided by this bank. ***What kinds of business purchases might require a loan from a bank?***

SECTION 26.1
Promissory Notes

Many people sign a note to pay for the purchase of a vehicle over a certain period of time. The note may be with a company like Ford Motor Credit or a financial institution. In this chapter, you will learn how businesses borrow and lend money.

A Promise to Pay

What Is a Promissory Note?

A **promissory note**, often shortened to *note,* is a written promise to pay a certain amount of money at a specific time. Promissory notes are formal documents that are evidence of credit granted or received. Promissory notes are used in many transactions, including paying for products and services, and lending and borrowing money. Sellers sometimes ask for a note to replace an account receivable when a customer requests additional time to pay a past-due account.

Laws require a promissory note to contain certain information as shown in **Figure 26–1**.

Section Vocabulary

- promissory note
- note payable
- note receivable
- principal
- face value
- term
- issue date
- payee
- interest rate
- maturity date
- maker
- interest
- maturity value

Notes Payable and Notes Receivable

A **note payable** is a promissory note that a business issues to a creditor when it borrows or buys on credit. A **note receivable** is a promissory note that a business accepts from a credit customer.

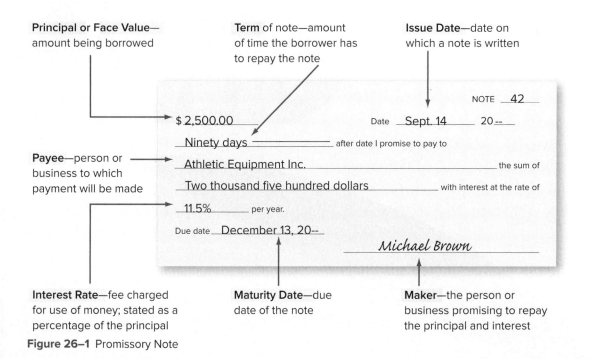

Figure 26–1 Promissory Note

The Maturity Date of a Note

When a note is signed, the maker of the note agrees to repay the amount of the note within a certain period of time, usually stated in days, months, or years. This time period is the **term** of the note. Both the term and the **issue date** (date on which the note is signed) are needed to determine the **maturity date** (due date) of a note.

In the note in **Figure 26–1** on page 828, Michael Brown, manager of The Starting Line Sports Gear, agreed to pay Athletic Equipment Inc. the principal plus interest 90 days from September 14. To determine the maturity date:

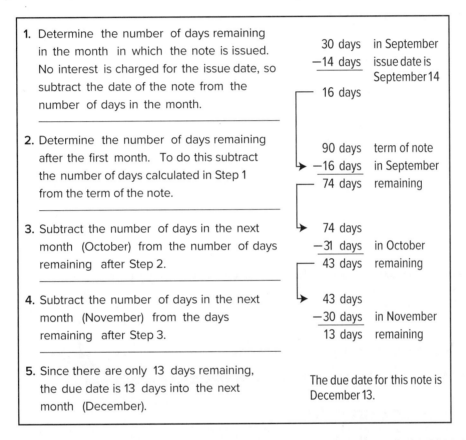

Some businesses and banks use time calendars to calculate a note's maturity date. **Figure 26–2** shows an example of a time calendar. The time calendar has two sets of days: (1) the day of the month (left and right columns), and (2) the day of the year, by month (middle columns).

To calculate a maturity date using the time calendar, follow these steps:

1. Locate the issue date of the note (for example, 14) in the Day of month column. Move across the month columns to the issue month (September). In our example September 14 is the 257th day of the year.

2. Add the number of days in the term of the note (90) to the day of the year. The sum of the two numbers is 347 (257 + 90).

3. Find the number 347 in the month columns. The 347th day of the year is in December. The maturity month is December. Move across to the Day of month column. The 347th day of the year corresponds to the 13th day of the month. The due date of the note is December 13.

Figure 26–2 Time Calendar

Day of month	Jan.	Feb.	Mar.	Apr.	May	June	July	Aug.	Sept.	Oct.	Nov.	Dec.	Day of month
1	1	32	60	91	121	152	182	213	244	274	305	335	1
2	2	33	61	92	122	153	183	214	245	275	306	336	2
3	3	34	62	93	123	154	184	215	246	276	307	337	3
4	4	35	63	94	124	155	185	216	247	277	308	338	4
5	5	36	64	95	125	156	186	217	248	278	309	339	5
6	6	37	65	96	126	157	187	218	249	279	310	340	6
7	7	38	66	97	127	158	188	219	250	280	311	341	7
8	8	39	67	98	128	159	189	220	251	281	312	342	8
9	9	40	68	99	129	160	190	221	252	282	313	343	9
10	10	41	69	100	130	161	191	222	253	283	314	344	10
11	11	42	70	101	131	162	192	223	254	284	315	345	11
12	12	43	71	102	132	163	193	224	255	285	316	346	12
13	13	44	72	103	133	164	194	225	256	286	317	347	13
14	14	45	73	104	134	165	195	226	257	287	318	348	14
15	15	46	74	105	135	166	196	227	258	288	319	349	15
16	16	47	75	106	136	167	197	228	259	289	320	350	16
17	17	48	76	107	137	168	198	229	260	290	321	351	17
18	18	49	77	108	138	169	199	230	261	291	322	352	18
19	19	50	78	109	139	170	200	231	262	292	323	353	19
20	20	51	79	110	140	171	201	232	263	293	324	354	20
21	21	52	80	111	141	172	202	233	264	294	325	355	21
22	22	53	81	112	142	173	203	234	265	295	326	356	22
23	23	54	82	113	143	174	204	235	266	296	327	357	23
24	24	55	83	114	144	175	205	236	267	297	328	358	24
25	25	56	84	115	145	176	206	237	268	298	329	359	25
26	26	57	85	116	146	177	207	238	269	299	330	360	26
27	27	58	86	117	147	178	208	239	270	300	331	361	27
28	28	59	87	118	148	179	209	240	271	301	332	362	28
29	29	...	88	119	149	180	210	241	272	302	333	363	29
30	30	...	89	120	150	181	211	242	273	303	334	364	30
311	311	...	90	...	1511	...	212	243	...	304	...	365	311

NOTE: For leap years, after February 28, the number of the day is one greater than that given in the table.

Reading Check

In Your Own Words

Explain who the maker and the payee of a note are.

Calculation of Interest on a Note

How Do You Calculate Interest on a Note?

Interest is the fee charged for the use of money. The **interest rate** is the interest stated as a percentage of the principal. The interest on a promissory note is based on three factors: *principal, interest rate,* and *term of the note.*

Calculating Interest Using a Formula

The formula used to calculate interest follows:

Interest = Principal × Interest Rate × Time

Interest rates are usually stated on an annual basis, that is, on a borrowing period of one year. To find the interest on a one-year promissory note, multiply the principal by the interest rate. The interest on an 11.5%, one-year $2,500 promissory note is $287.50 ($2,500 × .115 = $287.50).

If the term of a promissory note is less than one year, the time in the calculation is expressed as a fraction of one year. The fraction may be stated in days or months.

For example, on September 14 The Starting Line signed a note for $2,500 at 11.5% interest for 90 days. Since the term of the note is expressed in days, 365 days is used as the denominator of the time fraction.

The interest is calculated as follows:

Principal	×	Interest Rate	×	Time	=	Interest
$2,500	×	.115	×	90/365	=	$70.89

The interest on the note shown in **Figure 26–1** on page 826 is $70.89.

On the maturity date, The Starting Line will repay the maturity value of the note. **Maturity value** is the amount due at the due date. In our example the maturity value is $2,570.89 ($2,500.00 + $70.89).

If the term of this note had been three months instead of 90 days, the denominator of the time fraction would be 12. The interest would be calculated as follows:

Principal	×	Interest Rate	×	Time	=	Interest
$2,500	×	.115	×	3/12	=	$71.88

The maturity value would be $2,571.88 ($2,500.00 + $71.88).

Calculating Interest Using an Interest Table

To calculate interest, businesses and banks often use an interest table similar to the one in **Figure 26–3.** We use The Starting Line's note to illustrate.

- Find the term of the note in the Day column, 90.
- Follow the row across until you reach the column for the interest rate, 11.5%. Where the Day row and the Interest column meet is a factor, 2.835616. The factor is based on a principal amount of $100.
- Divide the principal of the note by 100. The result is 25 ($2,500 ÷ 100).
- Multiply the result by the factor to find the interest. The interest is $70.89 (25 × 2.835616).

In this example the interest calculated using both the equation and the interest table are the same. Sometimes small differences occur due to rounding.

SIMPLE INTEREST ON $100 (365 DAY BASIS)

	11.50 %		11.75 %		12.00 %		12.25 %		12.50 %		12.75 %
DAY	INTEREST	DAY	INTEREST	DAY	INTEREST	DAY	INTEREST	DAY	INTEREST	DAY	INTEREST
30	0.945205	30	0.965753	30	0.986301	30	1.006849	30	1.027397	30	1.047945
60	1.890411	60	1.931507	60	1.972603	60	2.013699	60	2.054795	60	2.095890
90	2.835616	90	2.897260	90	2.958904	90	3.020548	90	3.082192	90	3.143836
120	3.780822	120	3.863014	120	3.945205	120	4.027397	120	4.109589	120	4.191781
150	4.726027	150	4.828767	150	4.931507	150	5.034247	150	5.136986	150	5.239726
180	5.671233	180	5.794521	180	5.917808	180	6.041096	180	6.164384	180	6.287671
210	6.616438	210	6.760274	210	6.904110	210	7.047945	210	7.191781	210	7.335616
240	7.561644	240	7.726027	240	7.890411	240	8.054795	240	8.219178	240	8.383562
270	8.506849	270	8.691781	270	8.876712	270	9.061644	270	9.246575	270	9.431507
300	9.452055	300	9.657534	300	9.863014	300	10.068493	300	10.273973	300	10.479452
330	10.397260	330	10.623288	330	10.849315	330	11.075342	330	11.301370	330	11.527397
360	11.342466	360	11.589041	360	11.835616	360	12.082192	360	12.328767	360	12.575342
365	11.500000	365	11.750000	365	12.000000	365	12.250000	365	12.500000	365	12.750000
366	11.531507	366	11.782192	366	12.032877	366	12.283562	366	12.534247	366	12.784932

Figure 26–3 Interest Table

SECTION 26.1
Assessment

After You Read

Reinforce the Main Idea

Using a chart like this, describe the step-by-step procedure for determining the maturity value of a promissory note.

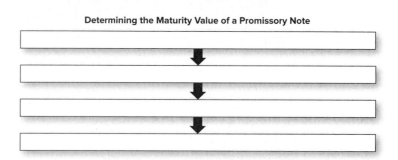

Determining the Maturity Value of a Promissory Note

Problem 1 Calculating Interest and Finding Maturity Values

Instructions Using the formula, compute the interest and maturity values for each of the following notes. Record your answers in your working papers. Use the interest table to check your computations.

	Principal	Interest Rate	Term
1.	$ 4,000	11.5%	60 days
2.	10,000	11.75%	90 days
3.	6,500	12.75%	60 days
4.	900	12.25%	120 days

Problem 2 Calculating Interest

Instructions Calculate the interest for each of the following notes. Record your answers in your working papers.

	Principal	Interest Rate	Term
1.	$ 600	15%	90 days
2.	3,500	12%	60 days
3.	9,600	9%	4 months
4.	2,500	10%	180 days

Math for Accounting

Marty Herick is the owner of CyberAction, a new computer-game store. Marty has just signed a promissory note with Excelsior Bank. He plans to use the loan to purchase and update his computer-game inventory. Using the formula, what is the interest on the $20,000, 90-day note with a 10.5% interest rate? What is the maturity value?

832 Chapter 26 • Notes Payable and Receivable

SECTION 26.2
Notes Payable

In this section you will journalize transactions involving notes payable. Recall that a note payable is a promissory note issued to a creditor. For example, a business may issue a note payable to borrow money from a bank. Notes that a business issues are recorded in the **Notes Payable** account. **Notes Payable** is a liability account; its normal balance is a credit. When the due date of a note extends beyond one year, the note is classified as a *long-term liability*. **Long-term liabilities** are debts that become due after one year.

Businesses frequently issue two types of notes: interest-bearing notes and non-interest-bearing notes. We consider both types of notes in this section.

Section Vocabulary

- long-term liabilities
- interest-bearing note payable
- non-interest-bearing note payable
- bank discount
- proceeds
- other expense

Interest-Bearing Notes Payable

What Is an Interest-Bearing Note Payable?

A note that requires the principal plus interest to be paid on the maturity date is called an **interest-bearing note payable**. The note issued by The Starting Line (in Section 26.1) is an interest-bearing note. Its maturity value is $2,570.89 ($2,500.00 principal + $70.89 interest).

Recording the Issuance of an Interest-Bearing Note Payable

Let's record The Starting Line's interest-bearing note payable as an example.

Business Transaction

On April 3 The Starting Line borrowed $7,000 from State Street Bank and issued a 90-day, 12% note payable to the bank, Note 6.

Analysis	Identify	1. The accounts affected are **Cash in Bank** and **Notes Payable.**
	Classify	2. **Cash in Bank** is an asset account. **Notes Payable** is a liability account.
	+/−	3. **Cash in Bank** is increased by $7,000. Notes Payable is increased by $7,000.
Debit-Credit Rule		4. Increases to asset accounts are recorded as debits. Debit **Cash in Bank** for $7,000.
		5. Increases to liability accounts are recorded as credits. Credit **Notes Payable** for $7,000.

continued

T Accounts

6.
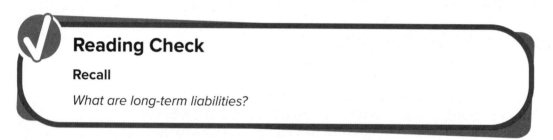

Journal Entry

7.

	GENERAL JOURNAL			PAGE 12
DATE	DESCRIPTION	POST. REF.	DEBIT	CREDIT
20--				
Apr. 3	Cash in Bank		7 000 00	
	Notes Payable			7 000 00
	Note 6			

✓ Reading Check

Recall

What are long-term liabilities?

Recording the Payment of an Interest-Bearing Note Payable

The maturity date of The Starting Line's note payable to State Street Bank is July 2. You can verify this by using the time calendar in **Figure 26–2** on page 830. The interest is $207.12, calculated as follows:

$$\text{Principal} \times \text{Interest Rate} \times \text{Time} = \text{Interest}$$
$$\$7,000 \times .12 \times 90/365 = \$207.12$$

The maturity value of the note is $7,207.12 ($7,000.00 principal + $207.12 interest).

Business Transaction

On July 2 The Starting Line issued Check 3892 for $7,207.12 payable to State Street Bank in payment of the note payable issued April 3.

Analysis	Identify	1. The accounts affected are **Notes Payable, Interest Expense,** and **Cash in Bank.**
	Classify	2. **Notes Payable** is a liability account. **Interest Expense** is an expense account. **Cash in Bank** is an asset account.
	+/−	3. **Notes Payable** is decreased by $7,000. **Interest Expense** is increased by $207.12. **Cash in Bank** is decreased by $7,207.12.

continued

Debit-Credit Rule	4. Decreases to liability accounts are recorded as debits. Debit **Notes Payable** for $7,000. Increases to expense accounts are recorded as debits. Debit **Interest Expense** for $207.12.
	5. Decreases to asset accounts are recorded as credits. Credit **Cash in Bank** for $7,207.12.

T Accounts 6.

Notes Payable		Cash in Bank	
Debit −	Credit +	Debit +	Credit −
7,000			7,207.12

Interest Expense	
Debit +	Credit −
207.12	

Journal Entry 7.

GENERAL JOURNAL   PAGE 22

	DATE	DESCRIPTION	POST. REF.	DEBIT	CREDIT	
1	20--					1
2	July 2	Notes Payable		7 000 00		2
3		Interest Expense		207 12		3
4		Cash in Bank			7 207 12	4
5		Check 3892				5
6						6

Non-Interest-Bearing Notes Payable

How Is Interest Paid on a Non-Interest-Bearing Note?

Sometimes a bank requires a borrower to pay the interest on a note in advance. On the issue date, the bank deducts the interest from the face value of the note. This reduces the amount of money the borrower receives. When interest is deducted in advance from the face value of the note, the note is called a **non-interest-bearing note payable**. The note is "non-interest-bearing" because no interest rate is stated on the note. The interest deducted in advance is called the **bank discount**. The interest rate used to calculate the bank discount is called the *discount rate*. The cash received by the borrower is called the **proceeds**. The proceeds equal the face value of the note minus the bank discount.

For a non-interest-bearing note payable, the maturity value is the same as the face value. This is because the interest is deducted from the face value on the issue date. **Figure 26–4** shows an example of a non-interest-bearing note payable.

Figure 26–4 Non-Interest-Bearing Note Payable

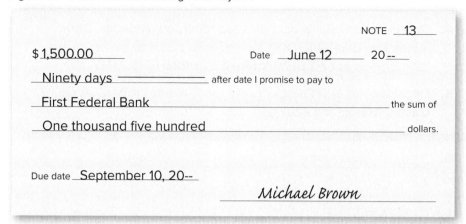

Calculating Non-Interest-Bearing Notes Payable

Let's calculate the proceeds of the non-interest-bearing note payable shown in **Figure 26–4**. The note was discounted at a rate of 12% by First Federal Bank, Note 13.

The first step in calculating the proceeds on a non-interest-bearing note is to calculate the bank discount. This is the interest on the note. (Notice that the formula is similar to the one used to compute interest on an interest-bearing note.)

Common Mistakes

Interest-Bearing versus Interest
The term non-interest-bearing might imply that the note has no interest charge. This is not the case. The term non-interest-bearing refers to the fact that no interest is stated on the note, whereas with an interest-bearing note, the rate of interest is stated. See **Figure 26–1** on page 826, and **Figure 26–4** on this page.

Face Value	×	Discount Rate	×	Time	=	Bank Discount
$1,500	×	.12	×	90/365	=	$44.38

The bank discount is subtracted from the face value of the note to determine the proceeds. The proceeds are $1,455.62 ($1,500.00 − $44.38).

Recording the Issuance of a Non-Interest-Bearing Note Payable

The bank discount is recorded in a contra liability account called **Discount on Notes Payable.** The normal balance of **Discount on Notes Payable** is a debit. The bank discount is the future interest expense on the note. However, the bank discount is not recorded in an expense account until the note matures and the interest expense has been incurred.

Now that we calculated the discount, let's record the issuance of the non-interest-bearing note for The Starting Line.

Business Transaction

On June 12 The Starting Line signed a $1,500, 90-day non-interest-bearing note payable that First Federal Bank discounted at a rate of 12%, Note 13.

Analysis Identify
1. The accounts affected are **Cash in Bank, Discount on Notes Payable,** and **Notes Payable.**

Classify
2. **Cash in Bank** is an asset account. **Discount on Notes Payable** is a contra liability account. **Notes Payable** is a liability account.

+/−
3. **Cash in Bank** is increased by $1,455.62. **Discount on Notes Payable** is increased by $44.38. **Notes Payable** is increased by $1,500.00.

Debit-Credit Rule
4. Increases to asset accounts are recorded as debits. Debit **Cash in Bank** for $1,455.62. Increases to contra liability accounts are recorded as debits. Debit **Discount on Notes Payable** for $44.38.

5. Increases to liability accounts are recorded as credits. Credit **Notes Payable** for $1,500.00.

T Accounts
6.

Cash in Bank		Notes Payable	
Debit + 1,455.62	Credit −	Debit −	Credit + 1,500.00

Discount on Notes Payable	
Debit + 44.38	Credit −

Journal Entry
7.

GENERAL JOURNAL   PAGE 20

DATE	DESCRIPTION	POST. REF.	DEBIT	CREDIT
20--				
June 12	Cash in Bank		1455 62	
	Discount on Notes Payable		44 38	
	Notes Payable			1500 00
	Note 13			

Businesses report the **Discount on Notes Payable** account on the balance sheet as a deduction from **Notes Payable**. The difference between the **Notes Payable** account and the **Discount on Notes Payable** account is the book value of notes payable. **Figure 26–5** shows the Liabilities section of the balance sheet for The Starting Line on June 30. It shows that the book value of notes payable is $1,455.62 ($1,500 − $44.38).

Figure 26–5 Reporting Non-Interest-Bearing Notes Payable on the Balance Sheet

The Starting Line Sports Gear		
Balance Sheet		
June 30, 20--		
Liabilities		
Notes Payable	1 500 00	
Less: Discount on Notes Payable	44 38	1 455 62

Reading Check

Compare and Contrast

How is an interestbearing note payable similar to a non-interestbearing note payable? How are they different?

Recording the Payment of a Non-Interest-Bearing Note Payable

When the non-interest-bearing note payable matures and is due, The Starting Line will

- pay First Federal Bank $1,500, the face value of the note, and
- record the interest expense by transferring the bank discount to interest expense.

We will look at each of these individually and as a compound journal entry.

Business Transaction

On September 10 The Starting Line issued Check 4241 for $1,500 to First Federal Bank in payment of the June 12 non-interest-bearing note payable.

Analysis	Identify	1. The accounts affected are **Notes Payable** and **Cash in Bank**.
	Classify	2. **Notes Payable** is a liability account. **Cash in Bank** is an asset account.
	+/−	3. **Notes Payable** is decreased by $1,500. **Cash in Bank** is decreased by $1,500.

continued

Debit-Credit Rule	4.	Decreases to liability accounts are recorded as debits. Debit **Notes Payable** for $1,500.
	5.	Decreases to asset accounts are recorded as credits. Credit **Cash in Bank** for $1,500.

T Accounts

6.

Notes Payable		Cash in Bank	
Debit −	Credit +	Debit +	Credit −
1,500			1,500

Journal Entry

7.

GENERAL JOURNAL PAGE 42

	DATE	DESCRIPTION	POST. REF.	DEBIT	CREDIT	
1	20--					1
2	Sept. 10	Notes Payable		1500 00		2
3		Cash in Bank			1500 00	3
4		Check 4241				4
5						5

When a non-interest-bearing note payable matures, the amount of the bank discount is recognized as an expense. The bank discount is transferred from the **Discount on Notes Payable** account to the **Interest Expense** account. As the following T accounts demonstrate, **Interest Expense** is debited for $44.38 and **Discount on Notes Payable** is credited for $44.38. When this transaction is recorded, the balance of the **Discount on Notes Payable** account is reduced to zero.

Interest Expense		Discount on Notes Payable	
Debit +	Credit −	Debit +	Credit −
9/10 44.38		6/12 44.38	9/10 44.38

You could record two separate journal entries:

1. the payment of the non-interest-bearing note payable (in the cash payments journal), then
2. the interest expense (in the general journal)

It is simpler, however, to prepare one compound entry in the general journal as shown.

	GENERAL JOURNAL			PAGE 43	
DATE	DESCRIPTION	POST. REF.	DEBIT	CREDIT	
20--					1
Sept. 10	Notes Payable		1500 00		2
	Interest Expense		44 38		3
	Cash in Bank			1500 00	4
	Discount on Notes Payable			44 38	5
	Check 4241				6

The **Interest Expense** account is classified as an other expense account. An **other expense** is a nonoperating expense. This means that the expense does not result from the normal operations of the business. Other expenses appear in a separate section on the income statement, as deductions from operating income.

SECTION 26.2
Assessment

After You Read

Reinforce the Main Idea

Create a table similar to this one to list three facts about the types of notes covered in this section.

Type of Note	Fact #1	Fact #2	Fact #3

Problem 3 Recording the Issuance of an Interest-Bearing Note Payable

On June 12 Frank's Lobster Pound issued a $9,000, 120-day, 12% note payable to American Bank of Commerce.

1. Which account is debited? What is the debit amount?
2. Which account is credited? What is the credit amount?
3. What is the classification of each account?
4. What is the maturity value of the note?

Problem 4 Recording the Issuance of a Non-Interest-Bearing Note Payable

On October 14 Canton Car Care Center issued a $10,000, 60-day, 12% non-interest-bearing note payable to Canton National Bank.

1. Which accounts are debited and which are credited? What are the debit and credit amounts?
2. Compute the bank discount. What is the amount of the proceeds?

Math for Accounting

Franklin Enterprises can borrow $10,000 for 30 days at 5% at the Jefferson City Bank or $10,000 for 45 days at 4.5% at Lincoln National Bank. Answer the following questions.

1. Which bank note results in the least amount of interest expense?
2. How much in interest expense can be saved?

SECTION 26.3
Notes Receivable

In this section you will journalize transactions involving notes receivable. If you have ever loaned someone money and asked the person to repay the loan by a specific date, you understand the basic concept of a note receivable. Sometimes such a loan includes payment of a specified amount of interest; other times no interest is expected.

Section Vocabulary
- other revenue

Recording the Receipt of a Note Receivable

How Do You Convert an Account Receivable to a Note Receivable?

When a customer needs additional time to pay an account receivable, he or she may be asked to sign a promissory note. The note replaces the account receivable. Promissory notes that a business accepts from customers are called *notes receivable*.

Notes Receivable is an asset account, and its normal balance is a debit. A note receivable is due on a specific date and carries an interest charge for the term of the note.

The interest earned on a note receivable is recorded in the **Interest Income** account. **Interest Income** is an other revenue account. **Other revenue**, also known as *nonoperating revenue* accounts, track revenue that a business receives from activities other than its normal operations. Other revenue appears in a separate section on the income statement, as an increase to operating income.

Business Transaction

On March 1 The Starting Line sold $1,750 of merchandise on account to Joe Dimaio. That transaction was recorded in The Starting Line's sales journal. Joe cannot pay his account by the due date. On April 8 The Starting Line received a 60-day, 12.5% note dated April 6 for $1,750 from Joe Dimaio to settle the account receivable, Note 4.

Analysis	Identify	1. The accounts affected are **Notes Receivable, Accounts Receivable** (controlling), and **Accounts Receivable—Joe Dimaio** (subsidiary).
	Classify	2. **Notes Receivable, Accounts Receivable** (controlling), and **Accounts Receivable—Joe Dimaio** (subsidiary) are asset accounts.
	+/−	3. **Notes Receivable** is increased by $1,750. **Accounts Receivable** (controlling) and **Accounts Receivable—Joe Dimaio** (subsidiary) are decreased by $1,750.

continued

Debit-Credit Rule	4. Increases to asset accounts are recorded as debits. Debit **Notes Receivable** for $1,750.
	5. Decreases to asset accounts are recorded as credits. Credit **Accounts Receivable** (controlling) for $1,750. Also credit **Accounts Receivable—Joe Dimaio** (subsidiary) for $1,750.

T Accounts 6.

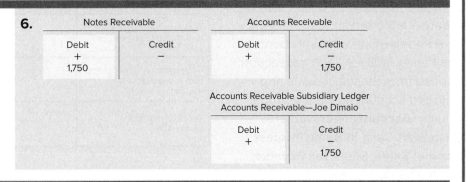

Journal Entry 7.

	GENERAL JOURNAL		PAGE 13	
DATE	DESCRIPTION	POST. REF.	DEBIT	CREDIT
20--				
Apr. 8	Notes Receivable		1 750 00	
	Accts. Rec./Joe Dimaio			1 750 00
	Note 4			

Recording the Payment of a Note Receivable

How Do You Record Payment of a Note?

The note from Joe Dimaio is due on June 5. The maturity value of the note is $1,785.96 ($1,750.00 principal + $35.96 interest).

Principal	×	Interest Rate	×	Time	=	Interest
$1,750	×	.125	×	60/365	=	$35.96

Business Transaction

On June 7 The Starting Line received a check dated June 5 for $1,785.96 from Joe Dimaio in payment of the $1,750 note of April 6 plus interest of $35.96, Receipt 996.

Journal Entry

	GENERAL JOURNAL		PAGE 18	
DATE	DESCRIPTION	POST. REF.	DEBIT	CREDIT
20--				
June 7	Cash in Bank		1 785 96	
	Notes Receivable			1 750 00
	Interest Income			35 96
	Receipt 996			

SECTION 26.3
Assessment

After You Read

Reinforce the Main Idea

Create a table similar to the one here to determine which accounts to debit and credit for each transaction. Choose from these accounts and write the account title in the proper column: **Accounts Receivable/Customer; Cash in Bank; Interest Income; Sales; Sales Tax Payable;** and **Notes Receivable**

Transaction	Account(s) Debited	Account(s) Credited
Sold merchandise on account to a charge customer plus sales tax		
Received an interest-bearing note in payment of the account receivable		
Received payment for the note		

Problem 5 Analyzing a Source Document

Instructions Examine the note illustrated here. In your working papers, make the appropriate journal entry on page 14 of the general journal for Eli's Catering Company. The note was discounted at a rate of 12% by First Federal Bank.

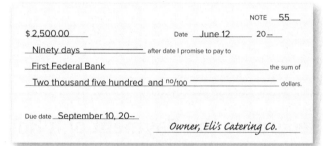

Math for Accounting

Your accounting manager has just finished a graph illustrating the possible interest-bearing notes available from the region's banks. Review the graph and give your boss your recommendation of which bank will provide the best loan value.

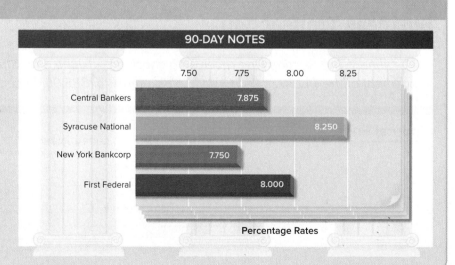

Chapter 26
Visual Summary

Concepts
How to calculate interest on a note payable or note receivable:

Interest = Principal × Interest Rate × Time

Analysis
When Interest is Paid on Interest-Bearing and Non-Interest-Bearing Notes Payable

| Interest-bearing notes are paid at maturity. | Non-interest-bearing notes are paid at issue. |

Procedures

To record the *receipt* of a note receivable converted from an account receivable:

Notes Receivable		Accounts Receivable (controlling/subsidiary)	
Debit + xxx	Credit −	Debit +	Credit − xxx

To record the *payment* of a note receivable:

Cash in Bank		Notes Receivable		Interest Income	
Debit + xxx	Credit −	Debit +	Credit − xxx	Debit −	Credit + xxx

Chapter 26
Review and Activities

Answering the Essential Question

How do businesses borrow and lend money?

Most people are familiar with the concepts of borrowing and lending, and these are important features of business. How can borrowing and lending affect the way companies do business?

Vocabulary Check

1. **Vocabulary** Arrange the Content Vocabulary in meaningful groups. Explain why you have grouped terms together. How are they related? For example, are they part of the same thing or is one the result of another?

- face value
- interest rate
- issue date
- maker
- maturity date
- note payable
- note receivable
- payee
- principal
- promissory note
- term
- interest
- maturity value
- interest-bearing note payable
- long-term liabilities
- bank discount
- non-interest-bearing note payable
- proceeds
- other expense
- other revenue

Concept Check

2. Name the two parties to a promissory note. Which party issues the note? Which party receives the note?
3. **Synthesize** Describe a situation in which a business might (a) receive a promissory note and (b) issue a promissory note.
4. What type of account is Notes Payable, and what is its normal balance? What type of account is Notes Receivable, and what is its normal balance?
5. **Analyze** Explain how interest-bearing and non-interest-bearing notes are different.
6. **Analyze** Explain how interest and bank discount are different.
7. What accounts are affected by the *issuance* of an interest-bearing note payable, and how are they affected? What accounts are affected by the *payment* of an interest-bearing note payable, and how are they affected?
8. What accounts are affected by the *receipt* of a note receivable, and how are they affected? What accounts are affected by the *payment* of a note receivable, and how are they affected?
9. **English Language Arts** Research how interest rates are computed. List the effects that rising and falling interest rates have on the economy. Then find a news article that explains the impact interest rates are currently having on some aspect of the economy. Create a PowerPoint presentation to share with your class.

Chapter 26
Standardized Test Practice

Multiple Choice

1. To journalize a 90-day, 10% note

    a. Debit Cash, credit Notes Receivable.

    b. Debit Cash, credit Notes Payable.

    c. Debit Accounts Receivable, credit Cash.

    d. Debit Cash, credit Accounts Receivable.

    Use the following information for questions 2 & 3.
    October 15, Morton Co. accepts a 60-day, 11% note from Anderson Imports for an extension of time on its account, $990.00 Notes Receivable No. 5.

2. The credit for this transaction would be made to

    a. Accounts Payable.

    b. Accounts Receivable.

    c. Notes Payable.

    d. Notes Receivable.

3. The effect of this transaction on the customer's account in the accounts receivable ledger is

    a. to decrease the account balance.

    b. to increase the account balance.

    c. no change in the account balance.

    d. not known.

4. Find the interest and maturity value for a 60-day note with principal of $1,500 and interest at 8 percent.

    a. $120.00 interest; $1,620.00 maturity value

    b. $32.88 interest; $1,532.88 maturity value

    c. $3.29 interest; $1,503.29 maturity value

    d. $19.74 interest; $1,519.74 maturity value

5. Qupre, Inc. signed a 90-day, 9.75% note with First State Bank for $1,200 on August 1. The maturity date for the note is

    a. October 30.

    b. November 1.

    c. October 28.

    d. November 2.

True or False

6. Promissory notes accepted by a business from charge customers are called notes payable.

7. The maturity value of an interest-bearing note equals the principal (face value) plus the interest.

Short Answer

8. How is the interest on a note calculated if the term is less than one year?

Chapter 26
Computerized Accounting

Notes Receivable and Payable

Making the Transition from a Manual to a Computerized System

MANUAL METHODS

- Using the general journal, record the receipt or issuance of the note.
- Post the entry to the appropriate accounts in the general ledger and subsidiary ledgers.
- Journalize and post the entry to record the receipt or payment of cash and interest.
- Calculate new balances for all accounts affected.

COMPUTERIZED METHODS

- Using the general journal, record the receipt or issuance of the note. The entry is automatically posted to the appropriate accounts.
- Record receipts or payment of notes.

Chapter 26
Problems

Problem 6 Recording Transactions for Interest-Bearing Notes Payable

Instructions In your working papers, record the following transactions in a cash receipts journal (page 22) and a cash payments journal (page 26).

Date	Transactions
Jan. 14	Sunset Surfwear borrowed $1,500 from First One Bank by issuing a 90-day, 12% interest-bearing note payable, Note 78.
Apr. 14	Issued Check 168 for $1,544.38 to First One Bank in payment of the $1,500 note issued on January 14, plus interest of $44.38.
May 31	Borrowed $12,400 from Merchant's Bank and Trust by issuing a 90-day, 12.5% interest-bearing note, Note 79.
Aug. 29	Paid Merchant's Bank and Trust the maturity value of the note issued on May 31, $12,782.19, Check 284.

Analyze
Calculate the amount of interest paid on notes in January.

Problem 7 Recording Transactions for Non-Interest-Bearing Notes Payable

Instructions In your working papers, record the following transactions in a cash receipts journal (page 14) and a cash payments journal (page 16).

Date	Transactions
June 10	InBeat CD Shop borrowed $6,000 from BankOne by issuing a 60-day, non-interest-bearing note payable (proceeds, $5,901.37) that the bank discounted at 10%, Note 67.
Aug. 9	Issued Check 205 for $6,000 in payment of the note issued June 10 and recorded the interest expense.
30	Borrowed $16,000 from Citizens Bank by issuing a 120-day, non-interest-bearing note payable less the 10.5% bank discount of $552.33, Note 68.
Dec. 28	Issued Check 398 in payment of the note issued on August 30 and recorded the interest expense.

continued

Chapter 26
Problems

Analyze
Explain why the account **Discount on Notes Payable** is used.

Problem 8 — Recording Notes Payable and Notes Receivable

Instructions In your working papers, record the following transactions in a cash receipts journal (page 47), cash payments journal (page 56), and general journal (page 19) for Cycle Tech Bicycles.

Date		Transactions
Mar.	19	Borrowed $9,000 from Desert Palms Savings and Loan by issuing a 90-day, 12% interest-bearing note payable, Note 87.
June	1	Received a 120-day, 13% note receivable for $1,900 from Greg Kellogg as a time extension on his account receivable, Note 6.
	17	Paid Desert Palms Savings and Loan the maturity value of the note issued on March 19, Check 2784.
Sept.	29	Received a check from Greg Kellogg for the maturity value of the note dated June 1, Receipt 628.
Oct.	6	Borrowed $2,700 from Jonesboro Bank and Trust by issuing a 60-day, non-interest-bearing note payable discounted at 11.5%, Note 88.
Dec.	5	Prepared a check for the note issued on October 6 and recorded the interest expense, Check 3954.

Analyze
If the October 6 note had been an interest-bearing note, how much cash would Cycle Tech Bicycles have received from the bank?

Problem 9 — Recording Notes Payable and Notes Receivable

The following is a partial list of accounts used by River's Edge Canoe & Kayak.

101	Cash in Bank	205	Notes Payable
115	Accounts Receivable	207	Discount on Notes Payable
120	Notes Receivable	415	Interest Income
201	Accounts Payable	640	Interest Expense

continued

Chapter 26
Problems

Instructions In your working papers, record the following transactions in a cash receipts journal (page 67), cash payments journal (page 73), and general journal (page 27).

Date	Transactions (cont.)
May 7	Borrowed $4,000 from Union Bank by issuing a 60-day, 9.5% non-interest-bearing note, Note 284.
15	Issued a $3,000, 90-day, 9% interest-bearing note to Trailhead Canoes in place of the amount owed on account, Note 285.
21	Received a 120-day, 10% note for $1,200 from Cathy Wilcox for an extension of time on her account, Note 94.
July 6	Issued Check 4711 in payment of the non-interest-bearing note given to Union Bank on May 7.
Aug. 13	Issued Check 5044 for the maturity value of the note issued to Trailhead Canoes on May 15.
Sept. 18	Received a check from Cathy Wilcox for the maturity value of the note dated May 21, Receipt 5921.

Analyze

Compare Notes 284 and 285. Which of the two notes was most advantageous to River's Edge Canoe & Kayak?

Problem 10 Renewing a Note Receivable

Occasionally, on the maturity date, a note may be renewed instead of being paid. When this occurs, (1) the interest on the first note is paid, (2) the first note is canceled, and (3) a new note for the same principal amount is issued, usually at a higher interest rate. Buzz Newsstand had the following transactions.

Instructions In your working papers, record the following transactions on general journal page 24.

SOURCE DOCUMENT PROBLEM

Problem 26–10

Use the source documents in your working papers to record the transactions for this problem.

Date	Transactions
Mar. 14	Sold merchandise on account to Saba Nadal for $1,800, plus sales tax of $108.00, terms 30 days, Sales Slip 388.
Apr. 13	Accepted a 60-day, 9% note for $1,908.00 from Saba Nadal in place of the account receivable, Note 416.

continued

Chapter 26
Problems

Date	Transactions
June 12	Received the interest due from Saba Nadal for the note dated April 13 and agreed to renew the note at 10% for 90 days, Receipt 1387 and Note 417.
Sept. 10	Received a check from Saba Nadal for the maturity value of the note issued June 12, Receipt 1555.

Analyze
Calculate the total amount of interest earned in March.

Real-World Applications & Connections

CASE STUDY

Data Mining

A business such as The Starting Line might benefit from data mining technology. Data mining gives companies the ability to assess, process, maintain, evaluate, and disseminate financial information to assist in decision making. Data mining is the process of extracting patterns from data and using it to understand a business's performance. For example, a company like The Starting Line could use data mining to learn customer buying habits, such as what days of the week or months of the year certain merchandise is being purchased.

ACTIVITY Use the Internet to research the technology of data mining. Write a one-page paper explaining how a company like The Starting Line can use data mining to learn more about its customers' spending habits. How could data mining help The Starting Line sell more goods? Share your ideas with the class.

21st Century Skills

The Technology Sector

Digital technology has the ability to improve our personal and business lives in many ways. Technology is one of the fastest growing sectors of our economy. The increasing demand for consumer electronics, such as iPods and digital cameras, the popularity of cell phones, and the opening up of new markets around the world have fueled the impressive growth of this sector. For the individual investor, buying shares in a technology sector mutual fund is a way to capitalize on this growth. However, investing in a single sector of the economy carries with it greater risk than does a more diversified approach to investing.

Research and Write

a. Use the Internet to find more information about technology sector mutual funds.

b. List one or two mutual funds as examples. Then write a paragraph describing the investment strategies, goals, and objectives of these funds.

Career Wise

Compare and Contrast: Management Accounting and CPA

Even though they are both accountants, a management accountant has different education requirements, roles, and responsibilities than a CPA. Review what you have learned about each career.

Activity

1. After reviewing what you have learned about management accountants and CPAs, make a list of the similarities between each of these professions.

2. Next, make a list of the differences between each of these professions.

3. Decide which job out of the two would better suit you. Write a paragraph explaining why you made the choice you did.

4. Write one description of the type of person you think would make a good management accountant and another description of the type of person you think would make a good CPA. Describe the qualities that make each person good at each job.

Spotlight on Personal Finance

Your Vehicle Loan

If you want to buy a vehicle but cannot pay cash, you need to borrow from a financial institution. To do so, you will be required to sign a legally binding note to make monthly payments for a required period of time.

Activity

Assume you want to buy a preowned vehicle but do not have all of the cash needed, and prefer not to ask your parents for it. Write a plan considering all aspects of the purchase.

H.O.T. Audit

Notes Payable

On May 1 Artist's Loft borrowed $8,000 from American National Bank. The art supply company issued an interest-bearing 90-day, 12% note payable to the bank. The journal entry that was made debited Cash in Bank for $7,230.40 and credited Notes Payable for $7,230.40. Determine whether this entry was correct. If not, why?

Mini Practice Set 5

Main Task

Complete the accounting cycle within an assigned time frame.

Summary of Steps

- Record transactions in various journals.
- Post individual transactions to various ledgers.
- Prove the special journals and post the column totals.
- Prove cash.
- Prepare schedules of accounts receivable and accounts payable.
- Complete the work sheet.
- Prepare financial statements.
- Journalize and post the adjusting and closing entries.
- Prepare a post-closing trial balance.
- Create a clear and coherent oral presentation that analyzes the results of the accounting cycle.

Why It's Important

Whether you are taking a test or working at a job, you will have time constraints.

Completing the Accounting Cycle for a Merchandising Corporation

Go Fly a Kite Inc.

Company Background: The Ramspart family owns and operates a wholesale-retail merchandising business organized as a corporation. The business, called Go Fly a Kite Inc., sells a variety of kites and paper airplanes to regional and local toy store businesses.

Your teacher will assign a due date for this project. Working with Go Fly a Kite Inc. will give you an opportunity to complete the accounting cycle within an assigned time frame.

Keeping the Accounting Records for Go Fly a Kite Inc.: Go Fly a Kite Inc. uses special journals and a general journal to record its business activity.

Max Martin, Go Fly a Kite's previous accounting clerk, has already journalized and posted the transactions for December 1 through December 15. The transactions recorded thus far are included in the accounting stationery in your working papers. The transactions for December 16 through December 31 are shown on the following pages.

Your Job Responsibilities: The forms for completing this activity are included in the working papers. As the accountant for Go Fly a Kite, you are to complete these tasks:

1. Record the remaining December transactions in the sales, cash receipts, purchases, cash payments, and general journals.

2. Post the individual amounts from the five journals to the accounts receivable and accounts payable subsidiary ledgers daily.

3. Post the individual amounts from the General columns of the cash receipts, purchases, cash payments, and general journal daily.

continued

4. Foot, prove, total, and rule the special journals at the end of the month.

5. Post the column totals of the special journals to the general ledger. Use this order for posting: sales, cash receipts, purchases, and cash payments.

6. Prove cash. The balance shown on check stub 619 is $22,752.83.

7. Prepare a schedule of accounts receivable and a schedule of accounts payable.

8. Prepare a trial balance on a ten-column work sheet for the year ended December 31.

9. Complete the work sheet. Use this December 31 adjustment information:

Merchandise Inventory	$24,850.43
Supplies inventory	120.00
Unexpired insurance	660.00
Total federal income taxes	4,500.00

10. Prepare the income statement from the work sheet information.

11. Prepare a statement of retained earnings.

12. Prepare a balance sheet.

13. Journalize and post the adjusting entries.

14. Journalize and post the closing entries.

15. Prepare a post-closing trial balance.

continued

CHART OF ACCOUNTS
Go Fly a Kite Inc.

ASSETS
101 Cash in Bank
105 Accounts Receivable
110 Merchandise Inventory
115 Supplies
120 Prepaid Insurance
125 Office Equipment
130 Store Equipment

LIABILITIES
201 Accounts Payable
205 Federal Corp. Income Tax Payable
210 Sales Tax Payable

STOCKHOLDERS' EQUITY
301 Capital Stock
305 Retained Earnings
310 Income Summary

REVENUE
401 Sales
405 Sales Discounts
410 Sales Returns and Allowances

COST OF MERCHANDISE
501 Purchases
505 Transportation In
510 Purchases Discounts
515 Purchases Returns and Allowances

EXPENSES
605 Advertising Expense
610 Bankcard Fees Expense
615 Insurance Expense
620 Miscellaneous Expense
625 Rent Expense
630 Salaries Expense
635 Supplies Expense
640 Utilities Expense
650 Federal Corp. Income Tax Expense

Accounts Receivable Subsidiary Ledger		Accounts Payable Subsidiary Ledger	
BES	Best Toys	BRA	Brad's Kites Ltd.
LAR	Lars' Specialties	CRE	Creative Kites Inc.
SER	Serendipity Shop	EAS	Easy Glide Co.
SMA	Small Town Toys	RED	Reddi-Bright Manufacturing
TOY	The Toy Store	STA	Stars Kites Outlet
		TAY	Taylor Office Suppliers

Business Transactions: Go Fly a Kite Inc. had the following transactions December 16 through December 31.

Date		Transactions
Dec.	16	Received Invoice 410 from Reddi-Bright Manufacturing for merchandise purchased on account, $1,475.00.
	16	Paid the quarterly federal income tax installment of $1,050.00, Check 610.
	16	Issued Check 611 for $2,548.00 to Brad's Kites Ltd. in payment of Invoice 112 for $2,600.00 less discount of $52.00.
	17	Paid the monthly salaries by issuing Check 612 for $4,750.00.
	17	Purchased $80.00 of supplies from Taylor Office Suppliers on account, Invoice 830.
	17	Received a check for $1,965.60 from Best Toys in payment of Sales Slip 479 for $2,003.40 less a cash discount of $37.80, Receipt 358.
	19	Sold merchandise on account to Best Toys, $2,600.00 plus $156.00 sales tax, Sales Slip 484.
	19	Prepared Receipt 359 for a $1,716.00 check received from Lars' Specialties in payment of Sales Slip 480 for $1,749.00 less a $33.00 cash discount.
	20	Purchased merchandise on account from Brad's Kites Ltd., Invoice 215, $1,560.00.
	20	Wrote Check 613 to Creative Kites Inc. to apply on account, $375.00.
	21	Prepared Credit Memorandum 44 for $106.00 for the return of $100.00 in merchandise by Best Toys, plus sales tax of $6.00.
	23	Sold merchandise to Lars' Specialties on account, $1,580.00 plus sales tax of $94.80, Sales Slip 485.
	23	Received a check from Serendipity Shop to apply on account, Receipt 360 for $300.00.
	23	Paid Easy Glide Co. for Invoice 326 for $1,890.00 less a $37.80 discount, Check 614 for $1,852.20.
	26	Received from The Toy Store a check for $1,102.40 in payment of Sales Slip 483 for $1,123.60 less a cash discount of $21.20, Receipt 361.
	26	Returned defective merchandise purchased on account from Brad's Kites Ltd., $150.00, Debit Memorandum 28.
	26	Received Invoice 335 from Easy Glide Co. for merchandise purchased on account totaling $1,630.00.
	28	Wrote Check 615 for $120.00 to the *Daily Examiner* for a monthly advertisement.

continued

Date	Transactions (cont.)
Dec. 28	Small Town Toys sent a check for $450.00 to apply on account, Receipt 362.
29	Paid Stars Kites Outlet $1,625.00 on account, Check 616.
29	Sold to The Toy Store $1,990.00 of merchandise on account, plus $119.40 sales tax, Sales Slip 486.
30	Issued Check 617 to Reddi-Bright Manufacturing for $700.00 to apply on account.
30	Sold merchandise totaling $560.00 plus $33.60 sales tax to the Serendipity Shop on account, Sales Slip 487.
31	Recorded the bank service charge of $10.00 and the bankcard fee of $150.00, from December bank statement.
31	Paid transportation charges of $51.60 for merchandise shipped from Easy Glide Co., Check 618.
31	Recorded cash sales of $3,995.10 plus $239.71 in sales tax, Tape 41.
31	Recorded bankcard sales of $1,736.27 plus sales tax of $104.18, Tape 41.

Analyze

1. Identify where Go Fly a Kite's outstanding customer balances are kept, and report which customer has the largest balance outstanding and the amount owed.

2. Locate the cash receipts journal, and identify the amount of cash to be debited to the general ledger Cash in Bank account.

3. Calculate Go Fly a Kite's working capital at December 31.

UNIT 6

Additional Accounting Topics

A Look Back
Unit 5 described accounting for special procedures, including cash funds, plant assets and depreciation, uncollectible accounts, inventories, and notes.

A Look Ahead
In Unit 6, you will learn about partnerships, including how to record financial information for a partnership from inception to liquidation, and the role of ethics in accounting.

Keys to Success

Q Why are writing skills important for an accounting career?

A Accounting is a numbers-oriented profession, but the ability to clearly communicate in written form is a learned skill that is crucial to professional success. Before you can start working with financial reports, you will need a well-written résumé and cover letter to impress potential employers. Once on the job, the ability to write reports and correspondence in clear, well-structured language, with correct spelling, punctuation and usage, helps you persuade others and build credibility with clients.

English Language Arts

Writing Accountants frequently write business correspondence to enhance their careers. Assume you own a private accounting firm that prepares tax returns for individuals and small businesses. Write a one-page letter introducing your business to potential clients. *What kind of business image will you project through your writing? How will you structure your letter?*

Focus on the Photo
Demonstrating responsibility, a strong work ethic, and good character is important in life and in business. ***What do ethics and social responsibility have to do with accounting?***

Chapter 27 • Introduction to Partnerships **861**

Chapter 27: Introduction to Partnerships

Chapter Topics

27-1 Partnership Characteristics and Partners' Equity

27-2 Division of Income and Loss

Visual Summary

Review and Activities

Standardized Test Practice

Computerized Accounting

Problems

Real-World Applications & Connections

Essential Question

As you read this chapter, keep this question in mind:

How do partnerships operate?

Main Idea

The partnership form of business organization has unique features, including unlimited liability and mutual agency. Partners divide profits and losses in several different ways.

Chapter Objectives

Concepts

C1 Identify the characteristics of a partnership.

C2 Identify the various accounting functions involved with a partnership.

Analysis

A1 Account for investments in a partnership.

A2 Account for partners' withdrawals.

Procedures

P1 Allocate profits and losses to the partners by different methods.

Real-World Business Connection

One Smooth Stone

Companies often have reasons for planning important business events, such as a national sales meeting or seminar. If a business is planning such an event, chances are it leaves the details up to a company like One Smooth Stone.

"The Stonies" as these event producers affectionately refer to themselves, take care of details such as hiring entertainment, creating staging and lighting designs, or finding a celebrity spokesperson. "We grasp the business situation quickly, make accurate and strategic recommendations, and do it all with a 'no problem' attitude and a smile."

Connect to the Business

One Smooth Stone is owned by several "partners" who collaborate and make business decisions as a team. In a company whose business is very detailed, this is not always easy. Planning the business so that each partner has a specific role can help ensure successful collaboration.

Analyze

What traits would a partner need in order to contribute to a smooth and successful partnership?

Focus on the Photo

To produce successful and exciting events, the partners of One Smooth Stone must know how to plan ahead creatively and collaboratively. **What issues should business partners agree upon before going into business with each other?**

SECTION 27.1
Partnership Characteristics and Partners' Equity

A *partnership* is an association of two or more persons as co-owners to operate a business for profit. Any type of business may be organized as a partnership, but it is more common for those providing **professional** services such as accounting and legal firms. In this chapter, you will identify the unique properties of partnership businesses such as unlimited liability and mutual agency. You will also explore how partners divide business profits and losses.

> **Section Vocabulary**
> - partnership agreement
> - mutual agency
> - professional
> - unique
> - ceases

Characteristics of a Partnership

What Are the Characteristics of a Partnership?

The partnership organizational form has certain **unique** features.

Ease of Formation

No special legal requirements must be met to form a partnership.

A Voluntary Arrangement.

A partnership is formed when two or more persons agree to operate as partners. No one can be forced into a partnership or required to continue as a partner.

The Partnership Agreement.

A partnership may be formed when two or more individuals verbally agree to operate a business as co-owners. However, it is advisable to have a **partnership agreement** in writing that states the terms under which the partnership will operate. It should include (1) each partner's name and address; (2) the name, location, and nature of the partnership; (3) the agreement date and the length of time the partnership is to exist; (4) each partner's investment; (5) each partner's duties, rights, and responsibilities; (6) the amount of withdrawals allowed each partner; (7) the procedure for sharing profits and losses; and (8) the procedures to follow when the partnership **ceases** to exist. Each partner should sign the agreement.

Unlimited Liability

Each partner is *personally* liable for the partnership's debts. This means that if the assets of the partnership cannot pay its creditors, the partners' personal assets may be used to pay those debts.

Limited Life

A partnership may end for a number of reasons, including any partner's death, withdrawal, bankruptcy, or incapacity. It may also end upon the completion of the project for which the partnership was formed, or at the expiration of the time set by the partners. For example, two architects may agree to combine their talents to design and oversee the construction of a building. When the building is completed, the partnership is dissolved.

Mutual Agency

Each partner is an agent of the partnership. In other words any partner has the legal right, in the name of the firm, to enter into agreements that are binding on all other partners. This is known as **mutual agency**.

Co-ownership of Partnership Property

When a partner invests assets in the partnership, he or she gives up all personal rights of ownership. The partners co-own all partnership assets.

Advantages and Disadvantages of a Partnership

A partnership combines the abilities, experiences, and resources of two or more individuals. It is easy to form and requires only the partners' agreement. Partners can usually make decisions without meeting formally. A partnership must have a legal purpose but has few other restrictions. Finally, it does not pay federal or state income taxes because each partner pays personal income taxes on his or her share of the net income of the business.

Of course, a partnership has disadvantages. It has a limited life, each partner is personally liable for the partnership's debts, and all partners may be held responsible for one partner's decisions. A partner cannot transfer his or her partnership interest without the other partners' consent. The partners need to be able to work together without major disagreements.

Reading Check

List

What terms are included in the partnership agreement?

Accounting for Partners' Equity

How Do You Account for a Partnership?

Accounting for owner's equity for a partnership is basically the same as for a sole proprietorship. A sole proprietorship has only one capital account, but a partnership has two or more capital accounts. A separate capital account is set up for each partner to record that partner's investment in the business. Each partner also has a separate withdrawals account.

Recording Partner Investments

When a partnership is formed, the value of cash and other assets invested by each partner is listed in the partnership agreement. Separate entries then record each partner's investment in the business. Let's look at an example.

On January 1 Rachel Wesley and Alex Tatsuno agree to form a partnership to operate a business. The name of the partnership is Boardwalk Bike & Skate Rentals. Each partner agrees to invest the following assets in the new business.

	Wesley	Tatsuno
Cash		$12,000
Office supplies		1,000
Office equipment (market value)		12,000
Building (market value)	$30,000	
Land	15,000	
Total assets invested	$45,000	$25,000

When assets other than cash are invested in a partnership, the asset accounts are debited for the market value of the assets.

Business Transaction

On January 1 Rachel Wesley and Alex Tatsuno contributed cash and other assets to form a partnership, Memorandum 1.

Journal Entry

GENERAL JOURNAL PAGE 1

DATE	DESCRIPTION	POST. REF.	DEBIT	CREDIT
20--				
Jan. 1	Building		30 000 00	
	Land		15 000 00	
	Rachel Wesley, Capital			45 000 00
	Memorandum 1			
1	Cash in Bank		12 000 00	
	Office Supplies		1 000 00	
	Office Equipment		12 000 00	
	Alex Tatsuno, Capital			25 000 00
	Memorandum 1			

Recording Additional Partner Investments

Any additional investments by the partners are recorded in a similar manner. For example, in April each partner agreed to invest $5,000 cash in the business. **Cash in Bank** is debited for $10,000 and the two partners' capital accounts are credited for $5,000 each.

Recording Partner Withdrawals

Because partners do not receive a salary, they may withdraw cash or other assets for personal use during the year. The withdrawal amounts must follow the terms of the partnership agreement. The amount is debited to the partner's withdrawals account and is credited to the appropriate asset account.

Business Transaction

On May 12 Rachel Wesley withdrew $5,000 cash for personal use, Check 123, and Alex Tatsuno withdrew $3,000 cash for personal use, Check 124.

Journal Entry

GENERAL JOURNAL   PAGE 36

	DATE		DESCRIPTION	POST. REF.	DEBIT	CREDIT	
1	20--						1
2	May	12	Rachel Wesley, Withdrawals		$5 000 00		2
3			Cash in Bank			$5 000 00	3
4			Check 123				4
5			Alex Tatsuno, Withdrawals		$3 000 00		5
6			Cash in Bank			$3 000 00	6
7			Check 124				7

SECTION 27.1 Assessment

After You Read

Reinforce the Main Idea

The partnership form of business organization is different from the other forms you have studied. Make a chart like this one and fill in the blanks with a short definition of each partnership characteristic.

CHARACTERISTICS OF A PARTNERSHIP	
Ease of Formation	
Unlimited Liability	
Limited Life	
Mutual Agency	
Co-ownership of Partnership Property	

Problem 1 Recording Partners' Investments

On June 1 Matthew Deck and Jennifer Rusk agree to combine their sole proprietorships into a new business, Dreamscapes Catering, organized as a partnership. The partnership will take over all assets of the two proprietorships. The assets invested by Deck and Rusk follow.

Instructions Prepare the journal entries required to record the investment by each partner. Use page 1 of the general journal in your working papers. The source document is Memorandum 1.

Investments	Deck	Rusk
Cash	$ 1,200	$ 2,300
Accounts receivable	2,000	7,000
Merchandise	8,000	5,000
Equipment	5,000	12,000
Van	12,000	—
Office Furniture	1,000	500

Math for Accounting

A partnership has two partners. Assume that assets total $123,400, liabilities are $18,700, and partner A has a capital balance of $63,234. What is the capital balance of partner B?

SECTION 27.2
Division of Income and Loss

At the end of each accounting period, the net income or net loss from partnership operations is divided among the partners. Partners may divide the income or loss among themselves in any way they choose. The specific method should be defined in the partnership agreement. If it is not, the law provides that net income or net loss be divided equally among the partners.

Dividing Profits and Losses

What Methods Can Be Used to Distribute Partnership Profits or Losses?

There are many ways for partners to divide profits and losses. The division of profits and losses is generally based on the services and capital contributed by the partners to the partnership. For example, if the partners share equally in the work of the business, but one partner has invested more capital, it seems only fair that the one who has invested more should profit more.

When the profits of the accounting period are divided, each partner's capital account is increased. If the business incurs a net loss for the accounting period, the capital account of each partner is decreased.

A number of methods can be used to distribute partnership profits or losses. We consider three of these methods:

- equal basis
- fractional share basis
- capital investment basis

We will look at two examples for Boardwalk Bike & Skate Rentals for each method: a net income of $24,000 and a net loss of $12,000.

Dividing Profits and Losses Equally

What Is the Journal Entry to Divide Income or Loss Equally?

The easiest way to divide a partnership's net income or net loss is on an equal basis. This method is often used when all partners invest equal amounts of capital and share equally in the work of the business.

The partnership agreement for Boardwalk Bike & Skate Rentals states that the profits or losses are to be divided between the two partners equally. During the first year of operation, the partnership earned a net income of $24,000. Each partner's share is $12,000 ($24,000 ÷ 2).

Remember that the balance of the **Income Summary** account is the net income or net loss for the period. In the third closing entry, **Income Summary** is closed to capital. On December 31 the balance of **Income Summary**, a $24,000 credit, is divided equally between the two partners.

Chapter 27 • Introduction to Partnerships

	GENERAL JOURNAL			PAGE 66	
DATE	DESCRIPTION	POST. REF.	DEBIT	CREDIT	
20--	Closing Entries				1
Dec. 31	Income Summary		24 000 00		2
	Rachel Wesley, Capital			12 000 00	3
	Alex Tatsuno, Capital			12 000 00	4

After the journal entry has been posted, the balance of Rachel Wesley's capital account is $62,000 and the balance of Alex Tatsuno's capital account is $42,000.

Rachel Wesley, Capital		Alex Tatsuno, Capital	
Debit −	Credit + Bal. 50,000 Clos. 12,000 Bal. 62,000	Debit −	Credit + Bal. 30,000 Clos. 12,000 Bal. 42,000

If the partnership incurs a net loss of $12,000 for the period, each partner's share of the net loss is $6,000 ($12,000 ÷ 2).

On December 31 the balance of **Income Summary**, a $12,000 debit, is divided equally between the two partners.

	GENERAL JOURNAL			PAGE 66	
DATE	DESCRIPTION	POST. REF.	DEBIT	CREDIT	
20--	Closing Entries				1
Dec. 31	Rachel Wesley, Capital		6 000 00		2
	Alex Tatsuno, Capital		6 000 00		3
	Income Summary			12 000 00	4

After the journal entry has been posted, the balance of Rachel Wesley's capital account is $44,000 and the balance of Alex Tatsuno's capital account is $24,000.

Rachel Wesley, Capital		Alex Tatsuno, Capital	
Debit − Clos. 6,000	Credit + Bal. 50,000 Bal. 44,000	Debit − Clos. 6,000	Credit + Bal. 30,000 Bal. 24,000

Reading Check

Recall

How are profits divided when there is no written agreement?

Dividing Profits and Losses on a Fractional Share Basis

How Do You Compute and Record Profit or Loss Based on Fractions?

Another way to divide net income or net loss is to assign each partner a stated fraction of the total. The size of the fraction usually depends on (1) the amount of each partner's investment and (2) the value of each partner's services to the business.

Suppose that the Boardwalk Bike & Skate Rentals partnership agreement states that the net income or net loss is to be divided between the two partners on the following basis:

- Wesley, two-thirds
- Tatsuno, one-third

When partners agree to share net income or net loss on a fractional share basis, that basis is often stated as a ratio. For example, Rachel Wesley's two-thirds share and Alex Tatsuno's one-third share could be expressed as a 2:1 ratio (2 to 1). To turn a ratio into a fraction, add the figures and use the total as the denominator of the fraction. In this particular example, the ratio is converted to fractions as follows:

$$2:1 = (2 + 1 = 3) = \tfrac{2}{3} \text{ and } \tfrac{1}{3}$$

The division of the $24,000 net income based on a 2:1 ratio is calculated as follows:

Wesley's share: $24,000 × ⅔ = $16,000
Tatsuno's share: $24,000 × ⅓ = $8,000

The December 31 closing entry follows:

GENERAL JOURNAL   PAGE 66

DATE	DESCRIPTION	POST. REF.	DEBIT	CREDIT
20--	Closing Entries			
Dec. 31	Income Summary		24 000 00	
	Rachel Wesley, Capital			16 000 00
	Alex Tatsuno, Capital			8 000 00

After the journal entry has been posted, the balance of Rachel Wesley's capital account is $66,000 and the balance of Alex Tatsuno's capital account is $38,000. Refer to the following T account for each of the partners.

Rachel Wesley, Capital	
Debit	Credit
−	+
	Bal. 50,000
	Clos. 16,000
	Bal. 66,000

Alex Tatsuno, Capital	
Debit	Credit
−	+
	Bal. 30,000
	Clos. 8,000
	Bal. 38,000

If the partnership incurs a net loss of $12,000 for the period, the December 31 distribution to the partners based on a 2:1 ratio is calculated as follows:

Wesley's share: ($12,000) × ⅔ = ($8,000)

Tatsuno's share: ($12,000) × ⅓ = ($4,000)

On December 31 the partnership makes the following closing entry:

	DATE	DESCRIPTION	POST. REF.	DEBIT	CREDIT	
1	20--	Closing Entries				1
2	Dec. 31	Rachel Wesley, Capital		8 000 00		2
3		Alex Tatsuno, Capital		4 000 00		3
4		Income Summary			12 000 00	4

GENERAL JOURNAL   PAGE 66

Rachel Wesley, Capital

Debit	Credit
−	+
Clos. 8,000	Bal. 50,000
	Bal. 42,000

Alex Tatsuno, Capital

Debit	Credit
−	+
Clos. 4,000	Bal. 30,000
	Bal. 26,000

After the journal entry has been posted, the balance of Rachel Wesley's capital account is $42,000 and the balance of Alex Tatsuno's capital account is $26,000.

Reading Check

Explain

What factors do partners usually consider when deciding on how the profits and losses of the partnership will be divided?

Dividing Profits and Losses Based on Capital Investments

How Do You Use Capital Investment to Divide Profit or Loss?

Net income and net loss can also be divided on the basis of the amount of capital contributed by the individual partners. To do this, compute the percentage of each partner's capital investment as follows:

$$\frac{\text{Individual Partner's Investment}}{\text{Total Partnership Investment}} = \text{Partner's Percentage}$$

Then multiply the net income or net loss by each partner's percentage.

Rachel Wesley and Alex Tatsuno made the following capital investments:

Rachel Wesley	$50,000
Alex Tatsuno	$30,000
Total investment	$80,000

Wesley's Percentage Tatsuno's Percentage

$$\frac{50{,}000}{80{,}000} = 62.5\% \qquad \frac{30{,}000}{80{,}000} = 37.5\%$$

The share of the $24,000 net income for each partner is calculated as follows:

Wesley's share $24,000 × .625 = $15,000
Tatsuno's share $24,000 × .375 = $9,000

On December 31 the partnership makes the following closing entry:

GENERAL JOURNAL PAGE 66

	DATE	DESCRIPTION	POST. REF.	DEBIT	CREDIT	
1	20--	Closing Entries				1
2	Dec. 31	Income Summary		24 000 00		2
3		Rachel Wesley, Capital			15 000 00	3
4		Alex Tatsuno, Capital			9 000 00	4
5						5

Rachel Wesley, Capital

Debit	Credit
−	+
	Bal. 50,000
	Clos. 15,000
	Bal. 65,000

Alex Tatsuno, Capital

Debit	Credit
−	+
	Bal. 30,000
	Clos. 9,000
	Bal. 39,000

After the journal entry has been posted, the balance of Rachel Wesley's capital account is $65,000 and the balance of Alex Tatsuno's capital account is $39,000.

If the partnership incurs a net loss of $12,000 for the period, the December 31 distribution is calculated as follows:

Wesley's share   ($12,000) × .625 = ($7,500)
Tatsuno's share ($12,000) × .375 = ($4,500)

On December 31 the closing entry is as follows:

GENERAL JOURNAL PAGE 66

	DATE	DESCRIPTION	POST. REF.	DEBIT	CREDIT	
1	20--	Closing Entries				1
2	Dec. 31	Rachel Wesley, Capital		7 500 00		2
3		Alex Tatsuno, Capital		4 500 00		3
4		Income Summary			12 000 00	4
5						5

After the journal entry has been posted, the balance of Rachel Wesley's capital account is $42,500, and the balance of Alex Tatsuno's capital account is $25,500.

Rachel Wesley, Capital		Alex Tatsuno, Capital	
Debit	Credit	Debit	Credit
−	+	−	+
Clos. 7,500	Bal. 50,000	Clos. 4,500	Bal. 30,000
	Bal. 42,500		Bal. 25,500

SECTION 27.2
Assessment

After You Read

Reinforce the Main Idea

Identify two facts about the methods of dividing partnership profits and losses covered in this section. Use a chart like this one to organize the information.

Method of Dividing Profits and Losses	Fact #1	Fact #2

Problem 2 Determining Partners' Fractional Shares

Following are the ratios used by several partnerships to divide net income or net loss.

Instructions Determine the fractions that are used to calculate each partner's share of net income or net loss. Use the form provided in your working papers.

1. 3:1
2. 5:3:1
3. 3:2:2:1
4. 2:1:1
5. 2:1

Problem 3 Analyzing a Source Document

The deposit slips for Mr. Walter's and Ms. Yount's investments in the partnership of Walter and Yount Tax Service are presented here.

Instructions Prepare a journal entry to record the investments by the partners. Use the journal provided in your working papers.

Math for Accounting

You and your best friend, Mary Jo Perry, operate an ice cream kiosk at the local outdoor shopping mall. You agree to divide profits and losses based on the capital investments made by each partner. Calculate the share of profit and loss due to each partner.

Your investment:	$15,000
Mary Jo Perry's investment:	25,000
Net income:	32,000

1. What is the percentage of investment by each partner?
2. What is your share of the profit?
3. If there is a loss, which partner will have the largest share of the loss?

Chapter 27
Visual Summary

Concepts
Accounting for a partnership involves the following accounting functions:

Record Original Investment	Record Subsequent Investment	Record Withdrawals
Record Partner 1's original investment in Partner 1's capital account.	Record Partner 1's subsequent investment in Partner 1's capital account.	Record Partner 1's withdrawals in Partner 1's withdrawals account.
Record Partner 2's original investment in Partner 2's capital account.	Record Partner 2's subsequent investment in Partner 2's capital account.	Record Partner 2's withdrawals in Partner 2's withdrawals account.

Analysis
The formation of a partnership involves recording the value of cash and other assets that each partner invests. To record a partner's investment in a partnership:

Cash in Bank (or other asset)	
Debit + xxx	Credit −

Individual Partner's Capital	
Debit −	Credit + xxx

Procedures
The terms of the partnership agreement should determine how partners may withdraw cash or assets from the business. To record a partner's cash withdrawal from the partnership:

Individual Partner's Withdrawals	
Debit + xxx	Credit −

Cash in Bank	
Debit +	Credit − xxx

Chapter 27
Review and Activities

Answering the Essential Question

How do partnerships operate?

Many businesses are started when two or more parties decide to work together for a common goal. What other kinds of partnerships are you familiar with?

Vocabulary Check

1. **Vocabulary** Write a paragraph using the content vocabulary terms. The paragraph should clearly explain how the terms relate to one another and to accounting. Underline each vocabulary term used.

- cease
- partnership agreement
- professional
- unique
- mutual agency

Concept Check

2. List five characteristics of a partnership.
3. List three advantages and three disadvantages to the partnership form of business.
4. What information is usually included in a partnership agreement?
5. **Analyze** Explain the differences and similarities between a partnership and a sole proprietorship.
6. **Analyze** Explain the differences and similarities between accounting for a partnership and accounting for a sole proprietorship.
7. When assets other than cash are invested in a partnership, at what amount are the assets recorded?
8. What is the journal entry to record a partner's investment of cash?
9. What determines the amount of cash or other assets a partner may withdraw from the business for personal use?
10. What factors do partners usually consider when deciding on how the profits and losses of the partnership will be divided?
11. **Evaluate** Name three methods partners might use to divide profits and losses. Describe the benefits of each.
12. **English Language Arts** Work with one of your classmates to act out in front of the class the following scenario: Two partners run a small business together. They are trying to decide whether or not to incorporate the business. One partner wants to incorporate and the other does not. You and your partner may choose the type of business. Each of you should research your position and give good reasons as to whether or not the business should incorporate. Make sure to write a script so that your arguments follow a complete structure.

Chapter 27
Standardized Test Practice

Multiple Choice

1. The withdrawals shows each partner's
   a. withdrawal of cash only.
   b. withdrawal of all assets.
   c. additional investments.
   d. all of the above.
   e. none of these answers

2. The right of all partners to contract on behalf of a partnership is called
   a. mutual agency.
   b. a partnership.
   c. a partner's agreement.
   d. voting rights.
   e. none of these answers

3. In a partnership, which transaction would occur if Partner A withdraws cash for personal use to purchase an automobile?
   a. Debit Salary Expense and credit Cash
   b. Debit Cash and credit Partner A, Withdrawal
   c. Debit Partner A, Withdrawal and credit Cash
   d. Debit Capital and credit Cash

4. If a partner invests a noncash asset in the partnership, the amount to be debited and credited would be
   a. the asset's cost.
   b. the asset's book value.
   c. the asset's current market value.
   d. an amount determined by the investing partner.

5. Rewrite: 3/5 as a decimal.
   a. 1.67
   b. .4
   c. .6
   d. .0167

True or False

6. The actions of one partner acting on behalf of the partnership are binding on all partners.

Short Answer

7. Name three methods partners might use to divide profits and losses.

8. What does it mean to share a profit at a ratio of 3:4?

Chapter 27
Computerized Accounting

Starting a New Company
Making the Transition from a Manual to a Computerized System

MANUAL METHODS	COMPUTERIZED METHODS
• Determine the form of business ownership. • Create a chart of accounts. • Open a general ledger account for each account. • Record the journal entry for the initial investment by the owner(s). • Post the opening entry to the appropriate ledger accounts.	• Use the new company setup guide offered with most computerized accounting systems. • Follow the steps that will prompt you to enter general information about the business, as well as specific information about the accounting records.

Chapter 27
Problems

Problem 4 Dividing Partnership Earnings

Listed here are the net income and the method of dividing net income or net loss for several partnerships.

Instructions Use the form provided in your working papers to determine each partner's share of the net income.

Net Income	Method of Dividing Partnership Earnings
1. $45,000	Equally: 2 partners
2. $89,700	Fractional share: 2/3, 1/3
3. $22,000	Fractional share: 3:1
4. $32,000	Fractional share: 2/5, 2/5, 1/5
5. $92,700	Equally: 3 partners

Analyze
Identify the accounts affected when net income is distributed to the partners.

Problem 5 Calculating the Percentage of a Partner's Capital Investment

Listed here are an individual partner's investments for several partnerships.

Instructions Using the form provided in your working papers, calculate Partner A's percentage ownership in each partnership.

Partnership	Partner A's Investment	Total Partnership Investment
1.	$60,000	$180,000
2.	25,000	125,000
3.	11,250	28,125
4.	30,000	40,000

Analyze
Calculate the percentage of ownership for the other partner(s) in partnerships 1 and 4.

Problem 6 Recording Investments of Partners

On May 1 Jason Pua and Roy Nelson formed the partnership called JR Landscaping. Jason contributed $8,100 cash and all of his landscaping equipment. Jason had

continued

Chapter 27
Problems

purchased the equipment for $2,500 last year. The market value of this equipment is now $1,800. Roy contributed $1,000 cash and his truck to the partnership. The truck currently has a market value of $5,600.

Instructions Prepare the journal entries required to record the investment that each partner contributed in creating the partnership. Use page 1 of the general journal in your working papers. The source document for this assignment is Memorandum 1.

Analyze
Compare the balance of Jason Pua's capital account with the capital account of Roy Nelson.

Problem 7 Sharing Losses Based on Capital Balances

Mariela DeJesus and Natasha Faircloth started the partnership In Shape Fitness. Mariela contributed a capital balance of $35,000. Natasha's capital contribution totaled $45,000. Their partnership agreement specified sharing net profits and net losses on the basis of their capital balances. The net loss for their first year of operation for In Shape Fitness was $28,500.

Instructions Prepare the journal entry required to divide the net loss between the partners. Use the journal in your working papers to record your entry.

Analyze
If Mariela took a $1,500 withdrawal, what was her capital account balance on December 31 after the closing entries were posted?

Problem 8 Partners' Withdrawals

Toni Graff, Ahmad Nu, and Lindsay Pane are partners in Travel Essentials. Their partnership agreement stated that withdrawals would be 10 percent for each partner based on the partnership's net profits or net losses recorded for the year. In the first year of operation, the partnership reported $25,000 in net profit.

Instructions Prepare the entry to record the withdrawals of Graff, Nu, and Pane based on the partnership agreement. Use the journal paper provided in your working papers.

Analyze
Explain how the $25,000 in profit would be shared among Graff, Nu, and Pane.

Chapter 27
Problems

Problem 9 | Preparing Closing Entries for a Partnership

Barbara Scott and Martin Towers are partners in the firm of Ten Column Accounting Services. Their partnership agreement states that Scott and Towers share net income or net loss in a 3:2 ratio.

At the end of the period, December 31, the business had a net loss of $9,700. During the period Scott withdrew $6,600 and Towers withdrew $5,400.

Instructions In your working papers, do the following:

1. Journalize the closing entries to divide the net loss between the partners and to close the withdrawals accounts. Use general journal page 14.

2. Post the closing entries to the general ledger accounts.

Analyze
Calculate the reductions in the Scott and Towers capital accounts.

Chapter 27
Problems

Problem 10 — Evaluating Methods of Dividing Partnership Earnings

Jo Garrity, Maureen O'Riley, and David White decided to form a partnership called OnTime Copy Shop. The partners will invest the following assets in the business:

	Garrity	O'Riley	White
Cash	0	$22,000	$40,000
Supplies	0	5,000	2,000
Equipment	0	7,500	4,000
Building	$50,000	0	0
Land	15,000	0	0

They are considering the following plans for the division of net income or net loss:

1. Equally
2. Garrity, 20%; O'Riley, 40%; White, 40%
3. In the same ratio as the beginning balances of their capital accounts

Instructions Assume that the business had a net income of $17,500 in its first year of operations. Calculate the division of net income under each of the three plans.

Analyze
Identify which method O'Riley would prefer and why.

Real-World Applications & Connections

Career Wise

General Ledger Accountant

General ledger accounting can be a great career for those who are interested in the day-to-day accounting tasks of a company. General ledger accountants keep the books for a business and make sure everything is properly done in case of an audit.

Education Requirements

A general ledger accountant must have a college degree. If you want to work at a larger company it must be a four-year college degree. For small organizations an associate's degree from a two-year college may be sufficient.

Roles

A general ledger accountant ensures the accuracy and completeness of the company's accounts. They can do this themselves, or they may supervise others in accounting or administrative positions.

Responsibilities

Responsibilities will vary depending on the size of the organization. At a small firm, the accountant may be required to take care of all accounting activities while in a larger firm the accountant may be responsible for only one specialized area.

Activity

Explore the career of general ledger accountant by searching for general ledger accountant jobs on the Internet or using other career search sources. Pick out three job listings and analyze the qualifications. What skills would make you a good fit for this job? What skills should you work on developing to get this job in the future?

Global Accounting

Joint Ventures

One way a company can increase sales is to enter a foreign market in a *joint venture* with a local firm. The two companies share capital, technology, risks, rewards, and control. The local company offers cultural knowledge and business contacts. Joint ventures require lower investment than other entry methods, but control is diluted and they can be difficult to manage.

Research and Write

1. Use the Internet to research joint ventures.
2. Brainstorm factors to consider before forming a joint venture.

A Matter of Ethics

Partner Loyalty

You and your best friend Harold have started a jewelry business. He manages the business while you make the jewelry. Your designs are very successful, but because of Harold's mismanagement, the business is not making any money. He suggests "doctoring" the financial statements and using them to raise capital from other investors. You are afraid that the business will not grow as long as Harold is running it, but he is a good friend and has some good ideas.

Activity

Determine the ethical issues. What are the alternatives? Who are the affected parties? How do the alternatives affect the parties? What would you do?

Analyzing Financial Reports

Partners' Equity

The net income or net loss from a partnership represents a return on the equity the partners invested in the business. Return on equity states a company's net profit or net loss as a percentage of equity. To compute it, divide net income by equity. Partners may invest different amounts of equity, so each partner's return may be different.

Instructions

Use the beginning balances in the T accounts for Wesley and Tatsuno on page 873 to complete these tasks.

1. If total net income is $24,000, calculate the return on total partnership equity.

2. Using an equal split of net income, calculate the return on each partner's equity.

3. Which partner earned the better return? Explain why. Why is this percentage different than the return on total equity?

H.O.T. Audit

Sharing Profits

The partnership agreement states that partners Mary, Yoto, and Sally are to share profits in the ratio of 3:2:1. Allan, the chief accountant, divided the net income of $86,000 as follows:

Mary's share = $43,000

Yoto's share = $28,000

Sally's share = $15,000

Instructions

Determine if the profits have been calculated properly. If not, recalculate the partners' shares on a separate sheet of paper.

Chapter 28

Financial Statements and Liquidation of a Partnership

Chapter Topics

28-1 Financial Statements for a Partnership

28-2 Liquidation of a Partnership

Visual Summary

Review and Activities

Standardized Test Practice

Computerized Accounting

Problems

Real-World Applications & Connections

Essential Question

As you read this chapter, keep this question in mind:

What happens when a partnership ends?

Main Idea

Financial statements for a partnership report the details of each partner's capital. In a liquidation the assets are sold, creditors are paid, and any remaining cash is distributed by the partners.

Chapter Objectives

Concepts

C1 Prepare an income statement for a partnership.

C2 Prepare a statement of changes in partners' equity.

C3 Prepare the Partners' Equity section of a balance sheet.

Analysis

A1 Account for partnership liquidation losses.

A2 Account for partnership liquidation gains.

Procedures

P1 Prepare the final entry to liquidate a partnership.

Real-World Business Connection

Goodman & Company, LLP

Formed in 1932, Virginia Beach, Virginia-based Goodman & Company, LLP, is among the U.S.'s top 30 public accounting firms. Companies from a variety of industries—construction, financial institutions, government contracting, real estate, nonprofit and health care are just a few—hire Goodman & Company for services ranging from accounting and auditing to human resource management and training, administering retirement plans, and business valuation services.

Connect to the Business

Goodman & Company has more than 80 partners and 500 employees. As with any partnership, the firm's statement of changes in partners' equity reports changes in the equity position of the owners, with a separate column for each partner. Some partnerships equally distribute income among partners, while others use ratios for allocation.

Analyze

How do you think income is allocated among partners in a company like Goodman & Company?

Focus on the Photo

Goodman & Company's numerous partners include companies with significant assets in the expanding field of health care. *If a partner's company went out of business, what might happen to the company's assets?*

SECTION 28.1
Financial Statements for a Partnership

At the end of the year, Deloitte & Touche LLP, one of the Big Four accounting firms, prepares its financial statements that include an income statement, a statement of changes in partners' equity, and a balance sheet. In this chapter, you will learn how partnerships prepare and report their financial statements.

Section Vocabulary
- statement of changes in partners' equity

The Income Statement

What Does a Partnership Income Statement Report?

A partnership's income statement is prepared in the same way as that for any business. If the partnership is a service business, expenses are subtracted from revenue to determine the net income or loss. It is not required, but the division of net income or net loss among the partners may be shown on the income statement.

The partnership agreement of Boardwalk Bike & Skate Rentals states that net income or net loss will be divided equally between the partners. At the end of the year, the firm had a net income of $24,000. Each partner's share is $12,000. **Figure 28–1** illustrates how to report the division of net income on the income statement.

Boardwalk Bike & Skate Rentals		
Income Statement		
For the Year Ended December 31, 20--		
Net Income		24 000 00
Division of Net Income:		
Rachel Wesley	12 000 00	
Alex Tatsuno	12 000 00	
Net Income		24 000 00

Figure 28–1 Reporting the Division of Net Income on the Income Statement

Reading Check

Critical Thinking

The Income Statement above shows an equal distribution of income. How would the income change if the distribution was on a 2:1 fractional basis?

890 Chapter 28 • Financial Statements and Liquidation of a Partnership

Boardwalk Bike & Skate Rentals
Statement of Changes in Partners' Equity
For the Year Ended December 31, 20--

	Wesley	Tatsuno	Totals
Beginning Capital, Jan. 1, 20–			
Add: Investments	50 000 00	30 000 00	80 000 00
Net Income	12 000 00	12 000 00	24 000 00
Subtotal	62 000 00	42 000 00	104 000 00
Less: Withdrawals	5 000 00	3 000 00	8 000 00
Ending Capital, Dec. 31, 20–	57 000 00	39 000 00	96 000 00

Figure 28–2 Statement of Changes in Partners' Equity

The Statement of Changes in Partners' Equity

What Is Unique About Reporting Equity in a Partnership?

The **statement of changes in partners' equity** reports the change in each partner's capital account resulting from business operations, investments, and withdrawals. It is similar to the statement of changes in owner's equity for a sole proprietorship, except it has a separate column for each partner. See the statement of changes in partners' equity for Boardwalk Bike & Skate Rentals in **Figure 28–2**. Notice that the net income is divided between the partners.

Reading Check

Recall

What is the purpose of the statement of changes in partners' equity?

The Balance Sheet

What Does a Partnership Balance Sheet Report?

The owners' equity section of the balance sheet for a partnership is called the *Partners' Equity* section. This section lists each partner's capital account separately. (See **Figure 28–3**.) The capital account amounts on the balance sheet are the ending capital amounts from the statement of changes in partners' equity.

Boardwalk Bike & Skate Rentals
Balance Sheet
December 31, 20--

Partners' Equity		
Rachel Wesley, Capital	57 000 00	
Alex Tatsuno, Capital	39 000 00	
Total Partners' Equity		96 000 00
Total Liabilities and Partners' Equity		126 000 00

Figure 28–3 The Partners' Equity Section of the Balance Sheet

SECTION 28.1
Assessment

After You Read

Reinforce the Main Idea

The financial statements for a partnership are the same as those for other forms of business except for items related to owners' equity. Make a chart like this one to describe how partnership financial statements are different from those of a sole proprietorship.

Differences in Partnership Financial Statements	
STATEMENT	DIFFERENCE
Income Statement	
Statement of Changes in Partners' Equity	
Balance Sheet	

Problem 1 Preparing the Income Statement and Balance Sheet for a Partnership

Information related to the operations of the Goldman and Jones partnership follows.

Instructions In your working papers, prepare the Division of Net Income section of the income statement and the Partners' Equity section of the balance sheet.

1. Goldman and Jones share profits in the ratio of 2:3.
2. Net income for the year ended December 31, 20--, was $35,000.
3. The January 1 capital balance for Goldman was $23,000.
4. The January 1 capital balance for Jones was $47,000.

Problem 2 Analyzing a Source Document

Instructions In your working papers or on a separate sheet of paper, determine each partner's share of the gain.

N *Buie Norman & Company*
CERTIFIED PUBLIC ACCOUNTANTS

Minutes of the Partners' Meeting
January 12, 20--

The partnership sold equipment for a gain of $26,400. The partners agreed to share the gain as follows:

Larry Bass, CPA	6/22
John Buie, CPA	6/22
Teri Anderson, CPA	3/22
Robert Norman, CPA	3/22
Paula Dunham, CPA	2/22
John Ruppe, CPA	2/22

Math for Accounting

On January 1 Don had a capital balance of $23,000 and Ruth's capital balance was $34,000. Compute each partner's capital balance on December 31 given the following:

- Net income of $28,000 is divided equally between the two partners.
- Don withdrew $3,600.
- Ruth withdrew $4,000.

SECTION 28.2
Liquidation of a Partnership

A partnership does not have an unlimited life. Partnerships end because of death or incapacity of a partner, upon the completion of a project, or at the end of a specified time period. In this section you will learn about the procedures followed to end a partnership.

Section Vocabulary
- dissolution
- liquidation
- terminations

Ending a Partnership

What Is the Difference Between Dissolving and Liquidating?

A partnership can be either *dissolved* or *liquidated*. Both are legal **terminations** of the partnership. A **dissolution** occurs when the partners change, but the partnership continues operations. For example, when a new partner is admitted, the partnership dissolves and a new partnership begins.

A **liquidation** occurs when the business ceases to exist. The liquidation converts all partnership assets to cash and pays all partnership debts. The individual partners then are paid any remaining cash. The process involves four steps:

1. Sell all noncash assets for cash.
2. Add all gains (or deduct all losses) resulting from the sale of noncash assets to or from the capital accounts of the partners based on the partnership agreement.
3. Pay all partnership creditors.
4. After the creditors have been paid, distribute any cash remaining to the partners based on the final balance in their capital accounts.

Liquidating Boardwalk Bike & Skate Rentals

How Do You Liquidate a Partnership?

Let's explore the liquidation of the Boardwalk Bike & Skate Rentals partnership. **Figure 28–4** shows the balance sheet on the date the partners decided to end Boardwalk Bike & Skate Rentals.

Rachel Wesley and Alex Tatsuno share profits and losses equally and agree to end the partnership based on the April 15 balance sheet. The liquidation requires converting all partnership assets to cash, paying all debts, and distributing the remaining cash to the partners. Ending the partnership required five transactions over a period of weeks.

Reading Check

Recall

What are some reasons why partnerships may end?

Boardwalk Bike & Skate Rentals
Balance Sheet
April 15, 20--

Assets			
Cash in Bank		18 000 00	
Accounts Receivable		33 000 00	
Merchandise Inventory		35 000 00	
Equipment	48 000 00		
Less: Accumulated Depreciation	(8 000 00)		
		40 000 00	
Total Assets			126 000 00
Liabilities			
Note Payable		10 000 00	
Accounts Payable		20 000 00	
Total Liabilities			30 000 00
Partners' Equity			
Rachel Wesley, Capital		57 000 00	
Alex Tatsuno, Capital		39 000 00	
Total Partners' Equity			96 000 00
Total Liabilities and Partners' Equity			126 000 00

Figure 28–4 Boardwalk Bike & Skate Rentals Balance Sheet

Sale of Partnership Accounts Receivable at a Loss

On April 20 Boardwalk sold its $33,000 in accounts receivable to a finance broker for $29,000. The receivables sale resulted in a $4,000 ($29,000 − $33,000) loss to the partnership. Wesley and Tatsuno divided the loss equally. The following entry records the company's sale of the receivables and loss distribution:

GENERAL JOURNAL PAGE 109

DATE	DESCRIPTION	POST. REF.	DEBIT	CREDIT
20–				
Apr. 20	Cash in Bank		29 000 00	
	Rachel Wesley, Capital		2 000 00	
	Alex Tatsuno, Capital		2 000 00	
	Accounts Receivable			33 000 00
	Receipt 341			

The individual balances in the accounts receivable subsidiary ledger are credited for a total of $33,000.

Sale of Partnership Merchandise at a Loss

On April 29 Boardwalk sold its merchandise inventory, which cost $35,000, for $32,000 to a discounter. The sale resulted in a $3,000 ($32,000 − $35,000) loss to the partnership. The partners divided the loss equally. The following entry records the sale of the inventory and the distribution of the loss to the partners:

	GENERAL JOURNAL			PAGE 110	
DATE	DESCRIPTION	POST. REF.	DEBIT	CREDIT	
20–					
Apr. 29	Cash in Bank		32 000 00		
	Rachel Wesley, Capital		1 500 00		
	Alex Tatsuno, Capital		1 500 00		
	Merchandise Inventory			35 000 00	
	Receipt 363				

Sale of Partnership Equipment at a Gain

The partnership owns equipment that cost $48,000 and has accumulated depreciation of $8,000. The book value of the equipment is $40,000 ($48,000 − $8,000).

On May 5 Boardwalk sold the equipment to a dealer for $42,000 resulting in a gain of $2,000 ($42,000 − $40,000). The gain was divided equally between the partners. The entry to record the sale of the equipment and distribution of the gain is as follows:

	GENERAL JOURNAL			PAGE 111	
DATE	DESCRIPTION	POST. REF.	DEBIT	CREDIT	
20–					
May 5	Accum. Depr.–Equipment		8 000 00		
	Cash in Bank		42 000 00		
	Rachel Wesley, Capital			1 000 00	
	Alex Tatsuno, Capital			1 000 00	
	Equipment			48 000 00	
	Receipt 371				

Reading Check

Compare and Contrast

How does the sale of a partnership's merchandise during regular business operations differ from the sale of merchandise during a liquidation?

Payment of Partnership Liabilities

On May 11 the partnership mailed Check 234 to pay the bank note and Checks 235 through 238 to pay the accounts payable balances. The entry to record the transactions follows.

	GENERAL JOURNAL			PAGE 112	
DATE	DESCRIPTION	POST. REF.	DEBIT	CREDIT	
20–					1
May 11	Notes Payable		10 000 00		2
	Accounts Payable		20 000 00		3
	Cash in Bank			30 000 00	4
	Checks 234–238				5
					6

The individual balances in the accounts payable subsidiary ledger are debited for a total of $20,000. This entry brings the **Accounts Payable** account to zero.

Only three accounts remain in the ledger after all noncash assets have been sold and the debts of the partnership have been paid: **Cash in Bank; Rachel Wesley, Capital;** and **Alex Tatsuno, Capital.**

```
        Cash in Bank                          Rachel Wesley, Capital
    Debit           Credit               Debit              Credit
Bal. 18,000         30,000               2,000          Bal. 57,000
     29,000                               1,500               1,000
     32,000
     42,000
Bal. 91,000                                              Bal. 54,500


                                         Alex Tatsuno, Capital
                                      Debit              Credit
                                      2,000          Bal. 39,000
                                      1,500               1,000

                                                     Bal. 36,500
```

Now cash can be distributed to the partners based on the final balances in their capital accounts: Wesley, $54,500 and Tatsuno, $36,500.

Final Distribution of Cash

On May 15 the partnership is ended by distributing the balance of **Cash in Bank** to Wesley and Tatsuno. The final transaction to end the partnership is recorded as follows:

	GENERAL JOURNAL			PAGE 113	
DATE	DESCRIPTION	POST. REF.	DEBIT	CREDIT	
20–					1
May 15	Rachel Wesley, Capital		54 500 00		2
	Alex Tatsuno, Capital		36 500 00		3
	Cash in Bank			91 000 00	4
	Checks 239 and 240 to				5
	liquidate the partnership				6
					7

SECTION 28.2
Assessment

After You Read

Reinforce the Main Idea

Use a diagram like this one to describe the actions needed to liquidate a partnership. Add answer boxes as needed.

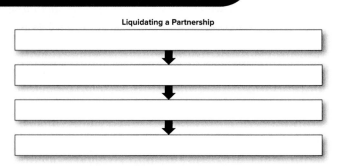

Liquidating a Partnership

Problem 3 Recording a Loss and a Gain on the Sale of Noncash Assets by a Partnership

Partners Gunther and Pertee share profits equally. In the process of ending the business, they sell all of the partnership's noncash assets. The transactions for the sales follow.

Instructions Record the transactions in general journal form in your working papers. Use general journal page 275.

Date		Transactions
Sept.	4	The merchandise inventory, which cost $45,000, was sold for $38,000.
	15	The office equipment, with a book value of $27,000, was sold for $29,000.

Math for Accounting

Max & Tex Novelty Store, a partnership, is ending. You are to help liquidate the partnership. Listed below are the remaining assets and liabilities of the business. Max & Tex share profits and losses equally.

Assets		Liabilities	
Cash in Bank	$20,000	Accounts Payable	$15,500
Accounts Receivable	$12,000	**Partners' Equity**	
Merchandise Inventory	$42,000	Marilyn Max, Capital	$59,250
Equipment	$60,000	Tom Tex, Capital	$59,250

Liquidation Transactions:

1. Customers will pay their accounts in full before the liquidation date.
2. Merchandise inventory was sold for $40,000 resulting in a $2,000 loss to the partnership.
3. Equipment was sold to a local retailer for $70,000 resulting in a gain of $10,000.
4. Accounts payable will be paid in full.

Instructions How much cash will each partner receive at final liquidation?

Chapter 28
Visual Summary

Concepts
Prepare the Partners' Equity section of a balance sheet.

ABC Partnership			
Balance Sheet			
July 31, 20--			
Partners' Equity			
Partner A, Capital		13 500 00	
Partner B, Capital		9 000 00	
Partner C, Capital		25 000 00	
Total Partners' Equity			47 500 00
Total Liabilities and Partners' Equity			60 500 00

Analysis
The sale of accounts receivable at a loss is recorded as follows, with the loss divided and debited to each partner according to the partnership agreement:

GENERAL JOURNAL   PAGE 100

DATE	DESCRIPTION	POST. REF.	DEBIT	CREDIT
20–				
May 1	Cash in Bank		52 000 00	
	Philip Hart, Capital		5 000 00	
	Donna Sanchez, Capital		5 000 00	
	Accounts Receivable			62 000 00
	Receipt 341			

Procedures
The final transaction to end the partnership is recorded as follows:

GENERAL JOURNAL   PAGE 110

DATE	DESCRIPTION	POST. REF.	DEBIT	CREDIT
20–				
May 15	Philip Hart, Capital		75 500 00	
	Donna Sanchez, Capital		45 000 00	
	Cash in Bank			120 000 00
	Checks 325 and 326			
	to liquidate the partnership			

Chapter 28
Review and Activities

Answering the Essential Question

What happens when a partnership ends?

When partners enter into a partnership agreement, all parties accept the chance of loss or gain. How are debts and assets handled when the partnership agreement ends?

Vocabulary Check

1. **Vocabulary** Create multiple-choice test questions for each Content Vocabulary term. Trade papers with a partner and answer his or her questions.

   - statement of changes in partners' equity
   - dissolution
   - liquidation

Concept Check

2. What does a partnership income statement report?
3. **Analyze** Explain how the income statement for a partnership may differ from that of a sole proprietorship.
4. Describe what a statement of partners' equity shows.
5. **Analyze** Explain how the statement of changes in partners' equity differs from a statement of changes in owners' equity.
6. What does a partnership balance sheet report?
7. **Analyze** Explain how the balance sheet for a partnership differs from that of a sole proprietorship.
8. When selling partnership accounts receivable at a loss, what accounts are affected?
9. What happens to the partners' capital accounts when noncash assets are sold at a loss?
10. What happens to the partners' capital accounts when noncash assets are sold at a gain?
11. What basis is used to distribute gains and losses from the sale of noncash assets?
12. When liquidating a partnership, what is the final transaction?
13. **Math** Calculate the amount each partner would get in the following scenarios: (1) Two partners share the earnings of $12,000 on a 3:2 basis. (2) Three partners share a loss of $16,000 on a 2:1:1 basis. (3) Two partners share liquidation of cash of $35,000 on a 4:3 basis.
14. **English Language Arts** Analyze the average annual salaries for two or three of the major accounting professions. Define the roles and duties of each profession, tell what responsibilities each usually has, and what skills and education are required for each. Explain why you think there is a disparity or similarity in income between them. Outline your findings in a PowerPoint or slide presentation.

Chapter 28
Standardized Test Practice

Multiple Choice

1. When there is a net loss, the
    a. Capital account is increased by the amount of the net loss.
    b. Drawing (Withdrawals) account is increased by the amount of the net loss.
    c. Capital account is decreased by the amount of the net loss.
    d. Drawing (Withdrawals) account is decreased by the amount of the net loss.

2. A partnership may be terminated by the
    a. partners' mutual agreement.
    b. death of a partner.
    c. admission of a new partner.
    d. all of the above.
    e. none of these answers

3. A partnership's capital accounts are changed if
    a. net income or net loss is recorded.
    b. partners invest additional cash in the business.
    c. partners withdraw cash from the business.
    d. all of the above.
    e. none of these answers

4. When both a partnership and the partnership business end, this is called a
    a. mutual understanding.
    b. dissolution.
    c. termination.
    d. liquidation.

5. The _____ shows the picture of a firm's financial position at a point in time.
    a. income statement
    b. balance sheet
    c. distribution of net income
    d. none of the above

True or False

6. In a liquidation, partners are paid before any partnership assets are distributed to the creditors.

Short Answer

7. What three financial statements are prepared for a partnership at the end of a period?

8. When selling partnership accounts receivable at a loss, what accounts are affected?

Chapter 28
Computerized Accounting

Setting Up the General Ledger

Making the Transition from a Manual to a Computerized System

MANUAL METHODS

- Using accounting stationery, open general ledger accounts based on the expected transactions of the business.
- Record the balance of each account. Ideally, end-of-period balances should be used. This ensures that the accounting records were in balance when they were transferred.

COMPUTERIZED METHODS

- Create a chart of accounts, selecting from predefined charts offered, or by entering select accounts based on the expected transactions for the business.
- Enter the beginning balance for each account.

Chapter 28
Problems

Problem 4 — Preparing an Income Statement and Balance Sheet for a Partnership

Joy Webster and Diana Ruiz have been in business since the beginning of the year. Now at the end of the year, they would like to know how the partnership has done in its first year of operation. Joy and Diana stated in the partnership agreement that they will share profits equally. The net income for the year was $5,780. The January 1 capital balances for both partners were $0. Joy invested $6,000 cash in the partnership throughout the year and withdrew $1,800. Diana invested $5,500 cash in the partnership and withdrew $1,200.

Instructions In your working papers, prepare the Division of Net Income section of the income statement and the Partners' Equity section of the balance sheet for the partnership.

Analyze
Identify the ending balances in the partners' capital accounts.

Problem 5 — Liquidating the Partnership with Losses on the Sale of Noncash Assets

Guice and Ward decide to end their partnership on September 21. They share profits and losses equally. The account balances of the partnership as of that date follow.

Cash	$6,500
Inventory	8,800
Equipment (book value)	2,700
Accounts Payable	4,000
Guice, Capital	6,800
Ward, Capital	7,200

The partnership sold the inventory for $5,000 and the equipment for $2,000. All of the accounts payable will be paid in full with the cash.

Instructions In your working papers, prepare the journal entries to record the liquidation of this partnership. Use general journal page 85.

Analyze
Conclude whether the partners' capital accounts increased or decreased as a result of the liquidation (before the final distribution). Explain why.

Chapter 28
Problems

Problem 6 — Recording a Gain or a Loss on the Sale of Noncash Assets by a Partnership

Hudson and Franklin are in the process of liquidating their partnership. They share profits and losses in a 3:1 ratio. They have sold all of the partnership's noncash assets. The transactions for the sales follow.

Date		Transactions
May 29	1.	The equipment, with a book value of $14,500, was sold for $12,775.
June 2	2.	The $7,800 in accounts receivable was sold to a finance broker for $6,900.
4	3.	The merchandise inventory, which cost $2,800, was sold for $1,900.

Instructions In your working papers, record the journal entries for the sale of noncash assets. Use general journal page 120.

Analyze

Explain why losses are distributed to the partners' capital accounts before any cash is paid out to the partners.

Problem 7 — Preparing a Statement of Changes in Partners' Equity

On January 1 Carol Farmer and Jim Romans formed a partnership, Research Consultants. Each partner invested $50,000 in cash on that date. The partnership agreement stated that the partners would share net income or loss equally.

During the year Farmer invested an additional $2,000 and withdrew $7,500 for personal use. Romans invested an additional $1,500 and withdrew $8,500. Net income for the first year was $33,176.

Instructions In your working papers, prepare a statement of changes in partners' equity for the year ended December 31.

Analyze

Identify the amount of partners' equity for Carol Farmer that will appear on the December 31 balance sheet.

Chapter 28
Problems

Problem 8 — Liquidating the Partnership

On October 15 Martinez and Royka decide to end their partnership. Their assets consist of $10,000 in cash and inventory that cost $105,000 and was sold for $95,000. Their only liability is a note payable for $10,000 that will be paid in full. Their capital balances are Martinez, $60,000 and Royka, $45,000. They share profits and losses equally.

Instructions In your working papers, prepare the journal entries to record the liquidation of the partnership.

Analyze
Calculate how much cash each partner would receive at the liquidation if the inventory was sold for $80,000.

Chapter 28
Problems

Problem 9 — Completing End-of-Period Activities for a Partnership

Richard Smooth and Carrie Overhill are partners in the firm of R&C Roofing. They agreed to divide net income or loss on the following basis: Smooth, ¾; Overhill, ¼.

The completed work sheet for R&C Roofing for the year ended December 31 appears in your working papers.

Instructions In your working papers:

1. Prepare an income statement for the partnership.
2. Prepare a statement of changes in partners' equity.
3. Prepare a balance sheet.
4. Journalize the adjusting and closing entries, beginning on page 27 of the general journal.

Analyze

Calculate the total year-end partners' equity if neither Smooth nor Overhill had withdrawn any cash from the business.

Real-World Applications & Connections

CASE STUDY

Partnership: Building and Design

Barnes Construction is a homebuilder that wants to expand into commercial real estate. It has a client who owns 10 acres of land on which she wants to build an apartment complex. Barnes is looking for a partner to help design and build it.

Slater Architectural Design is looking for new projects. Barnes and Slater form a partnership to design and build the complex for $10,400,000. Based on estimated costs, the project should earn net profits of 18 percent.

Instructions

1. If Barnes and Slater agree to share revenue and expenses equally, how much profit will each partner earn?
2. To limit each partner's liability to this project only, list the information that should be in the partnership agreement.

21st century Skills

Online Video Marketing Study

Years ago, marketers who wanted to analyze their competition would need to study recordings of television ads. Today, however, most TV spots can be found online, either at the company's Web site or on video sharing sites such as YouTube. Instant access to the competition's marketing spots can be a powerful tool for marketers.

Research and Write

a. Imagine you have a job as a marketing executive for a luxury car manufacturer. Using the Internet, find online videos of three of your competitors' TV ads.

b. What are some possible drawbacks to using online video as a research tool?

Career Wise

Government Accountant

Government accounting can be a great career for those who are interested in tax law, as they will have to learn and keep up-to-date on the tax laws.

Education Requirements

A government accountant must have at least a bachelor's degree in accounting or a related subject. Some also get a special certificate qualifying them as a Certified Government Financial Manager.

Roles

Government accountants must create, maintain and analyze the financial records of government agencies or private businesses that are regulated by the government.

Responsibilities

Government accountants are responsible for creating the budgets and examining the revenue and expenses of publicly funded programs.

Activity

Conduct research to find tax laws and read them. Would you like working with tax law? Why or why not?

Analyzing Financial Reports

Evaluating Partnership Operating Results

You learned earlier how to use ratio analysis to evaluate the financial health of a business. Partnerships are also interested in these measures of financial health. Return on sales is a measure of how much profit the business is earning for each dollar of total revenue. The current ratio and quick ratio show a company's ability to repay its debts.

Instructions

Review the ratios on pages 237–238, then use the balance sheet on page 815 for the following tasks.

1. Calculate the return on sales for Boardwalk Bike & Skate Rentals, assuming a revenue of $192,000 and net income of $24,000 for the fiscal period just ended.

2. Analyze Boardwalk's liquidity based on the current ratio and the quick ratio.

3. Write one to two paragraphs describing Boardwalk's financial health based on your analysis of its financial information.

H.O.T Audit

Partnerships

Ann Martha, the accountant for the R&P partnership, made an entry in the partnership books after some of the noncash assets were sold at a loss of $13,000. The remainder of the assets were sold at a gain of $4,000.

The capital balances before the distribution of gains and losses were Roberts, $40,000 and Patrick, $30,000. Roberts and Patrick shared profits and losses based on: Roberts 2/3; Patrick, 1/3. Martha made the following entry to distribute the balance of cash after all the debts were paid.

Roberts, Capital	$34,000
Patrick, Capital	$27,000
Total	$61,000

Instructions

Review Martha's calculations. If incorrect, re-compute the distribution of cash.

Mini Practice Set 6

Main Task
Complete the accounting cycle for Paint Works for its first month of business.

Summary of Steps
- Set up new general ledger accounts.
- Analyze the transactions.
- Journalize and post the transactions.
- Prepare a trial balance and a work sheet.
- Prepare the financial statements.
- Journalize and post the closing entries.
- Prepare a post-closing trial balance.
- Create a clear and coherent oral presentation that analyzes the results of the accounting cycle.

Why It's Important
You might work for a partnership someday. Maybe you will even open your own business with some partners. While you create the best product or service, you also need to know what is happening with your company's finances.

Completing the Accounting Cycle for a Partnership

Paint Works

Company Background: Paint Works is an interior and exterior painting company organized as a partnership. The partners are Laura Andersen, Sean Woo, and David Ingram. The business earns revenue from consultation and painting fees.

The partnership divides income and losses as follows: Laura and Sean each receive 33 percent, and David receives 34 percent.

Your Job Responsibilities: As the accountant for Paint Works, use the accounting stationery in your working papers to complete the following activities.

1. Open a general ledger account for each account in the chart of accounts.
2. Analyze each business transaction.
3. Enter each business transaction in the general journal. Begin on journal page 1.
4. Post each journal entry to the appropriate accounts in the general ledger.
5. Prepare a trial balance and then complete the work sheet.
6. Prepare an income statement.
7. Prepare a statement of partners' equity.
8. Prepare a balance sheet.
9. Journalize and post the closing entries.
10. Prepare a post-closing trial balance.

CHART OF ACCOUNTS
Paint Works

ASSETS
101 Cash in Bank
105 Accts. Rec.–Mountain View City School District
120 Computer Equipment
130 Office Supplies
135 Office Equipment
140 Painting Supplies
145 Painting Equipment

LIABILITIES
205 Accts. Pay.–Custom Color
210 Accts. Pay.–J & J Hardware and Lumber
215 Accts. Pay.–Paint Palace

PARTNERS' EQUITY
301 Laura Andersen, Capital
302 Laura Andersen, Withdrawals
303 David Ingram, Capital
304 David Ingram, Withdrawals
305 Sean Woo, Capital
306 Sean Woo, Withdrawals
310 Income Summary

REVENUE
401 Painting Fees
405 Consultation Fees

EXPENSES
505 Advertising Expense
510 Miscellaneous Expense
515 Rent Expense
520 Utilities Expense

Business Transactions: Paint Works began business operations on February 1 of this year. During the month of February, the business completed the transactions that follow.

Date	Transactions
Feb. 1	Laura Andersen, David Ingram, and Sean Woo agreed to form a partnership and invest the following assets, Memorandum 1:

Contributed Assets	Andersen	Ingram	Woo
Cash	$1,500	$1,000	$1,200
Computer Equipment	—	2,800	—
Office Equipment	100	—	—
Painting Supplies	150	—	225
Painting Equipment	1,375	—	1,675
Total assets invested	$3,125	$3,800	$3,100

Date	Transactions
1	Signed a one-year rental agreement for a warehouse and small office at $1500 per month. The first month's rent was paid, Check 1101.
1	Put an ad in the Call an Expert circular for $25, Check 1102.
1	Paid $100 to the utility company to start electric service, Check 1103.
1	Paid telephone company $175 to begin phone service and run dedicated line for computer, Check 1104.
2	Awarded a contract to paint a kitchen and family room for the McGuires. Collected $250 deposit, Receipt 1.

continued

Chapter 28 • Financial Statements and Liquidation of a Partnership

Date		Transactions
Feb.	2	Bought paint and border stencils from Custom Color for $200 on account, Invoice 742.
	2	Issued check 1105 for $55 for a business license to operate in the city of Mountain View, Alabama (Miscellaneous Expense).
	4	Purchased $115 in office supplies, Check 1106.
	5	Finished painting the kitchen and family room for the McGuires. Collected final payment of $450, Receipt 2.
	6	Joined Mountain View Chamber of Commerce by purchasing membership of $45 (Miscellaneous Expense), Check 1107.
	8	Bought painting equipment on account for $375 from Paint Palace, Invoice 1162.
	10	Received $60 for color and painting consultation, Receipt 3.
	12	Finished painting the cafeteria at the elementary school. Issued Invoice 101 to Mountain View City School District for $835.
	14	Sent Check 1108 for $200 to Custom Color on account.
	15	Andersen, Ingram, and Woo each made a withdrawal of $650 for personal use, Checks 1109, 1110, and 1111, respectively.
	15	Awarded a contract to paint exterior of the Wicker and Hartel law office building. Received deposit of $1,000, Receipt 4.
	16	Spent $135 for new paint brushes (Painting Supplies), Check 1112.
	16	Bought paint from Custom Color for $395 on account, Invoice 750.
	17	Received and paid Invoice 303 from Mountain View Realtors for ad in real estate brochure for $77, Check 1113.
	18	Received $125 for painting consultation, Receipt 5.
	19	Wrote Check 1114 for $85 to repair computer printer (Miscellaneous Expense).
	21	Paid Paint Palace $375 on account, Check 1115.
	22	Finished painting the Wicker and Hartel law office building. Collected final payment of $2,000, Receipt 6.
	24	Bought $90 of lumber (Painting Supplies) on account from J & J Hardware and Lumber, Invoice 207.
	25	Received $575 for painting and making minor repairs to a garage, Receipt 7.
	28	Andersen, Ingram, and Woo each made a withdrawal of $650 for personal use, Checks 1116, 1117, and 1118, respectively.

Analyze

For each partner, calculate the rate of return on equity for February, the first month of operations. Use the Partners' original investments of equity in the formula: Return on Equity = Net Income ÷ Equity. How does each partner's return compare to the overall return for the partnership?

Chapter 29
Ethics in Accounting

Chapter Topics

29-1 The Nature of Ethics

29-2 Ethics in the Accounting Profession

    Visual Summary

    Review and Activities

    Problems

    Real-World Applications & Connections

Essential Question

As you read this chapter, keep this question in mind:

How do ethics affect the success of a business?

Main Idea

The accounting profession requires its members to follow a code of ethics.

Chapter Objectives

Concepts

C1 Explain the meaning of ethics.

C2 Describe the components of business ethics.

C3 Identify the role of the accountant in business ethics.

Analysis

A1 Discuss how ethical behavior benefits individuals, businesses, and society.

A2 Explain the key principles an accountant is expected to follow.

Procedures

P1 Identify the accounting organizations that establish codes of ethics for the profession.

P2 Describe the Sarbanes-Oxley Act.

Real-World Business Connection

PricewaterhouseCoopers

The success or failure of business depends on ethical corporate accounting practices. PricewaterhouseCoopers, one of the world's largest professional services firms, is among the "Big Four" network of accounting firms that perform the majority of audits, or review of financial statements, of publicly traded companies, as well as of many private companies. In a recent year, PricewaterhouseCoopers employed approximately 163,000 people and had revenues of $26.2 billion.

Connect to the Business

A company enters into an agreement for auditors from PricewaterhouseCoopers to independently review and improve the company's operations and internal control processes. Auditors use strict standards to make sure that the business sticks to its agreements with suppliers, government, and customers. Ethical accounting practices protect assets and make the best use of a company's resources.

Analyze

Internal auditors evaluate procedures used to safeguard the resources of a business. What are some of these resources?

Focus on the Photo

As a world-class firm, PricewaterhouseCoopers adheres to strict standards of accounting and confidentiality when tallying votes for the Academy Awards, thus preserving the integrity of results that affect revenue for studios and individuals. **How would you define ethics in relation to business?**

SECTION 29.1
The Nature of Ethics

Have you ever had to make a difficult decision about a particular course of action or behavior? Did you consider how you would feel about the action or how others would view your choice? Did you consider how your behavior would affect others? If so, you have probably **encountered** an ethical dilemma.

Ethics

What Is Meant by Ethics?

Ethics is the study of our notions of right and wrong. In its broadest sense, ethics deals with human conduct in relation to what is morally good and bad. The term *ethics* often refers to a set of basic principles. Life is complex and individuals must face a variety of situations. A person's ethics, or basic principles, can provide guidelines for action when facing ethical dilemmas.

The basic values found in a system of personal ethics exist in systems of *business ethics* as well. Business ethics are not very different from ethics in general.

How Are Business Ethics Determined?

Business ethics refer to the **policies** and practices that reflect a company's core values such as honesty, trust, respect, and fairness. The ethics of a business can be seen by the way it treats employees and customers and how it attends to shareholder value, community service, supplier relationships, and regulatory law. Review the following components to understand how a business ethics program can be established.

The Law As a Guide.

A discussion of ethical performance should consider the relationship between law and ethics. While law and ethics both define proper and improper behavior, law attempts to formally define the general public's ideas about what makes something right or wrong. The law describes a minimum acceptable level of correct behavior.

In contrast, ethical concepts have more subtle implications and are more complex than written rules of law. For example, although a business may be well within the law to advertise certain products to teenagers, it may not be the "right" or "ethical" thing to do. A business may use law as a guide, but it must consider ethical principles and standards as well.

Statements of Company Values.

If you have heard the phrase, "actions speak louder than words," then you already know one of the best methods of maintaining a strong business ethics

Section Vocabulary

- ethics
- business ethics
- code of ethics
- ethics officer
- encountered
- policies
- conflicts
- obvious
- voluntarily
- principles
- implemented

program. A person learns standards and values by observing what others do, *not* by what they say.

In addition to having management and employees set good examples of ethical behavior, a business must state a well-defined framework of ethical concerns and core values. A **code of ethics** is a formal policy of rules and guidelines that describes the standards of conduct that a company expects from all its employees. The following topics are often addressed within a code of ethics:

- **Conflicts** of interest
- Product quality and testing
- Customer relations
- Employee relations
- Suppliers and consultants
- Expense reports
- Security
- Political contributions
- Environmental actions
- International business
- Workplace safety
- Technology
- Whistle-blowing

Training and Outreach.

After a code of ethics has been created, it is important to communicate that policy properly. Distribution of the written code, along with formal training, helps employees understand the importance of these policies and gives them realistic and concrete examples of what to do when they face an ethical dilemma.

Ethics Committees.

Ethics committees and *ethics officers* play an important role in developing and enforcing ethical processes. An **ethics officer** is the employee directly responsible for creating business conduct programs, evaluating performance, and enforcing standards of conduct. As ethical dilemmas surface, these officials help settle disputed issues and resolve problems. Once a hearing has been held and all issues have been resolved, an enforcement phase may follow.

Enforcement.

Effective enforcement is necessary to achieve an ethical work environment. A company's code of ethics might contain penalties for violations that include performance appraisal notes, probation, suspension, demotion, and termination.

Reading Check

Compare and Contrast

How are laws and ethical concepts similar? How are they different?

What Is the Accountant's Role?

Accountants play a major role in the operation, management, and development of business. In this role, they can face many ethical dilemmas. For example, if an accountant's manager gives instructions to record the physical inventory at its original costs when it is **obvious** that the inventory's value has decreased, what should the accountant do? If an auditor finds questionable accounting practices within a client's financial reporting, but knows the client is a major source of revenue for his or her firm, what action should he or she take?

As you can see, the actions and behaviors of accountants play a critical role in the maintenance of public trust in the business community. The accountant's opinions, practices, and behaviors directly impact how the company is viewed and how the profession, in turn, is assessed.

In general, the accountant should focus on the following guidelines:

- Avoid harm to stockholders.
- Optimize the interests of the public.
- Adhere to universal standards of what is right.
- Respect the human rights of all people.

Specific responsibilities of the accounting profession are expressed in the various codes of ethics created by major organizations. These codes of ethics are addressed in Section 29.2.

Ethical Behavior

What Are the Benefits of Acting Ethically?

While it may seem obvious that ethical behavior naturally benefits individuals, business, and society, the following discussion addresses specific benefits to each group.

Individuals

Acting ethically produces benefits that can affect the course of your life and the lives of others. These benefits include increased self-esteem, contentment, and self-respect. Ethical behavior also paves the way for achieving life goals. Your honesty, integrity, and accountability will allow you to find a satisfying career with the type of organization that shares your core values.

As you act ethically, you will also enjoy the benefit of acceptance in a society that rewards and honors appropriate behavior. As you focus more on the common good rather than selfish interests, you will earn trust and respect from friends, co-workers, employers, and society in general. Finally, ethical behavior will dramatically improve the quality of your decisions.

Businesses

A DePaul University study found that companies that made commitments to an ethics code provided more than twice the value to shareholders than those that did not. A study by Walker Information, a shareholder research firm, supports this finding, reporting that "good corporate behavior" leads to positive business outcomes.

Society

Society can be viewed as the sum of all social relationships between humans. Therefore, it is easy to understand why our individual actions contribute to the overall nature of our society. We cannot expect our society to become more ethical unless individuals and businesses commit to ethical behaviors.

As businesses act in ethical ways, new wealth for society is created. In the realm of ethical financial reporting, the public can be confident in the data provided and make informed investment decisions. In turn, greater capital funding is available for growth and productivity, yielding strong and healthy economies.

SECTION 29.1
Assessment

After You Read

Reinforce the Main Idea

Identify five components of a business ethics program that were discussed in this section. Use a chart like this one to list two examples of each component.

Component of a Business Ethics Program	Example #1	Example #2

Problem 1 Reporting Ethics Violations

Instructions Results from Walker Information's study on business integrity state that 65 percent of employees who know about an ethical violation choose not to report it. What weaknesses in a company's ethical environment might be responsible for such a finding? Discuss the components of a strong business ethics plan that management could apply to resolve this situation.

Problem 2 Exploring the Difference Between Ethics and Law

Instructions Several historically significant compliance or regulatory issues have led to the formation of regulatory agencies or laws. Adhering to legal regulations is the first step in behaving ethically. Explain why ethical behavior extends beyond the law. Describe one situation in which a company might act within the boundaries of the law yet still engage in an unethical act.

Math for Accounting

Assume that you are the chief financial officer for a major manufacturer of infant car seats. The lead product designer has informed management that the installation of a new latch system would improve product safety by 25 percent. The new latch system will add $4 to production costs per car seat. Current production costs are $52.50 per seat, and the product sells for $69.50. Management wants to compare the effect on net income if the product price increases by (a) $1, (b) $2, or (c) $5.

1. If the company sells 8,000 car seats in a year, what is the effect on net income for each proposed price increase scenario? Assume fixed operating costs are $115,000 annually.

2. As an officer of the company, what factors (other than pricing) do you think should be considered when making product design changes?

SECTION 29.2
Ethics in the Accounting Profession

Professional organizations are often separated from other organizations by a code of conduct or a code of ethics. To build public trust, most accountants **voluntarily** join a professional organization and accept the standards of conduct expected by the organization.

However, ethical behavior requires commitment beyond abiding by a few rules of conduct. No ethics code or written rules can apply to all situations that might arise on the job. Day-to-day job experiences often test an individual's personal judgment and personal ethics.

Written codes of ethics are generally designed to encourage ideal behavior. However, the codes must be both realistic and enforceable. Professional organizations are charged with enforcing the rules of conduct outlined in their codes.

Section Vocabulary
- integrity
- objectivity
- independence
- competence
- confidentiality
- voluntarily
- principles
- implemented

Key Principles

What Is the Foundation for Ethics in Accounting?

Certain **principles** provide the framework for rules of conduct that an accountant is expected to follow. These principles include *integrity, objectivity, independence, competence,* and *confidentiality.*

Integrity

The principle of **integrity** requires that accountants choose what is right and just over what is wrong. Guidance for behaving with integrity suggests that the accountant ask the following questions: Is this what a person of integrity would do? Have I made a right and just decision? Have I maintained the spirit of ethical conduct?

How do you explain the concept of integrity as related to the accounting profession? Making false or misleading entries in a client's books violates the accounting principle of integrity. Failing to correct false and misleading financial statements also violates integrity. Integrity helps accountants establish the trust necessary for others to rely on their professional judgment.

Objectivity

Every year publicly traded corporations must submit financial statements to the Securities and Exchange Commission. Certified public accountants (CPAs) audit these financial statements.

CPAs who perform audits are known as *independent auditors.* They do not work for the companies they audit. For example, suppose that Paul Corporation needs to give audited financial statements to the SEC. Paul Corporation hires

Findlay & Partners, a public accounting firm. Five CPAs who work for Findlay & Partners go to the offices of Paul Corporation and perform the audit. Paul Corporation is the *client* and Findlay & Partners is the *independent auditor.* The work of the independent auditor is critical. Many different users of financial information depend on audited financial statements when making decisions.

The principle of **objectivity** requires that accountants be impartial, honest, and free of conflicts of interest. When forming professional judgments, accountants should not be influenced by personal interests or relationships with others or behave in a way that would give even the *appearance* of improper behavior.

Independence

CPAs who audit public companies must maintain a position of *independence*. **Independence** in this sense means that the CPA does not have a financial interest in or a loan from the company he or she is auditing. An investment in the company or a loan from the company would interfere with the independence required of an auditor. Serving as a director, officer, or employee of the company would also demonstrate a lack of independence.

An accountant must act in such a way that maintains the general public's confidence in the services provided by the profession. Even if the accountant's financial interest is minor, the relationship could be suspect, and the accountant would be considered to lack independence. Even non-CPA employees in a public accounting firm may not accept more than token gifts from clients.

Competence

Competence refers to the knowledge, skills, and experience needed to complete a task. Accountants must perform only those services that they are competent to provide. Accountants are expected to maintain an appropriate level of competence through continuing education and to continually improve the quality of professional services. Services should be rendered promptly, carefully, thoroughly, and in accordance with appropriate technical and ethical standards. Accountants must be committed to learning and professional improvement throughout their professional lives.

Confidentiality

Accountants learn about everything from individual salaries to the business strategies of their clients. **Confidentiality** as related to the accounting profession is the requirement that accountants who acquire information in the course of work protect it and not disclose it without the appropriate legal or professional responsibility to do so. Above all, information should not be used for personal gain.

> **Reading Check**
>
> **Define**
>
> *In your own words, define one of the five principles that frame the rules of conduct for an accountant.*

Codes of Ethics

What Accounting Organizations Provide Guidelines?

Several accounting organizations provide guidelines to assist in ethical decision making.

American Institute of Certified Public Accountants

AICPA is the national organization of certified public accountants. The preamble to the AICPA Principles of Professional Conduct states that membership is voluntary and by accepting membership, a CPA assumes an obligation of self-discipline above and beyond the requirements of laws and regulations.

> **Knowledge Extension**
>
> Congress passed the Securities Exchange Act of 1934, which created the Security and Exchange Commission (SEC), "to protect investors, maintain fair, orderly, and efficient markets, and facilitate capital formation."

The AICPA Code also expresses the profession's recognition of its responsibilities to the public, to clients, and to colleagues. It calls for an unswerving commitment to honorable behavior, even at the sacrifice of personal advantage.

Institute of Management Accountants

The IMA is the leading professional organization devoted exclusively to management accountants and financial managers. Members of the IMA have a responsibility to maintain professional competence, uphold professional standards of confidentiality, avoid conflicts of interest, and communicate information fairly and objectively.

Institute of Internal Auditors

The IIA expects its members to demonstrate competence and to follow the principles of integrity, objectivity, and confidentiality. The IIA's code of ethics is necessary and appropriate because of the public trust placed in internal auditors.

> **Knowledge Extension**
>
> State Boards of Accountancy are generally appointed by the State Governor. State Boards issue rules to enforce state accountancy laws, which follow accounting and auditing standards established nationally (mostly by the AICPA).

Sarbanes-Oxley Act

How Does Federal Legislation Affect Accounting?

In the past decade, the need for stricter accounting regulations to protect investors arose from several corporate financial scandals. Certain major corporations were accused of misrepresenting their financial position. Public accounting firms that audited them failed to identify and prevent it.

Congress decided that the accounting profession and publicly held corporations needed to abide by legal standards of conduct. The most significant changes to corporate governance and accounting practice since the 1930s were **implemented** when the Sarbanes-Oxley (SOX) Act was signed into law in 2002. The SEC enforces the application of the regulations contained within SOX. SOX requires that CEOs, financial officers, accountants, and auditing firms comply with the new regulations and procedures. It also established an accounting board to oversee and investigate the audits and auditors of public companies.

SECTION 29.2
Assessment

After You Read

Reinforce the Main Idea

Identify five key principles for accountants that were discussed in this section. Use a chart like this one to describe two situations in which the accountant would need to apply each principle.

Principle	Situation #1	Situation #2

Problem 3 Promoting Principles of Conduct

Instructions Read the following scenario. Identify and discuss behaviors that you believe might violate the key principles of conduct for accountants.

In your new position as accounting manager with Triple B Markets, you supervise three staff accountants: Jennifer, Marcus, and Ing. You become aware of the following situations:

- Jennifer, a new employee, has never worked as an accountant but scored high marks in her college math classes.
- Marcus often records payments of utilities as assets instead of expenses because he wants to show a higher net income for the business.
- You overheard Ing talking with a friend on the phone about the payroll details of the company.

Math for Accounting

Assume that the cost per CPA employee to maintain professional competence is as follows: continuing professional education course, $900; travel, $350; lodging, $135; meals, $75; and lost work time, $480. If the firm has 37 CPA employees, what is the total cost of the commitment to quality work? What is the cost per employee? What is the possible cost of not maintaining professional competence?

Chapter 29
Visual Summary

Concepts
A business ethics program can be established with the following components:

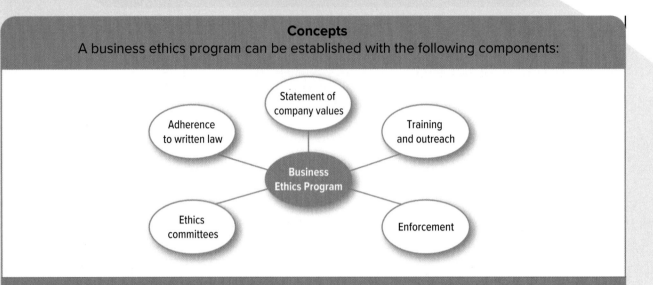

Analysis
Accountants are expected to follow these key principles:

Integrity	Choosing what is right and just over what is wrong.
Objectivity	Being impartial, honest and free from conflicts of interest, including personal interests or relationships that could influence professional judgment.
Independence	Having no financial interest in the company being audited.
Competence	Possessing the knowledge, skills and experience to perform assigned duties.
Confidentiality	Protecting information gained during the course of work, not disclosing that information without the appropriate legal or professional responsibility to do so, and not using that information for personal gain.

Procedures
The final transaction to end the partnership is recorded as follows:

GENERAL JOURNAL PAGE 110

	DATE	DESCRIPTION	POST. REF.	DEBIT	CREDIT	
1	20—					1
2	May 15	Philip Hart, Capital		75 500 00		2
3		Donna Sanchez, Capital		45 000 00		3
4		Cash in Bank			120 000 00	4
5		Checks 325 and 326				5
6		to liquidate the partnership				6

Chapter 29
Review and Activities

Answering the Essential Question

How do ethics affect the success of a business?

A company's reputation for "doing the right thing"—acting ethically—can affect its success. When you make consumer choices, do you look for companies that have good reputations?

Vocabulary Check

1. **Vocabulary** Write a short essay titled "Ethics in Accounting" using the content vocabulary terms. The essay should clearly explain how the terms relate to one another and to accounting. Underline each vocabulary term used.

   - business ethics
   - encounter
   - ethics
   - policies
   - code of ethics
   - conflict
   - ethics officer
   - obvious
   - integrity
   - principles
   - voluntarily
   - confidentiality
   - competence
   - independence
   - objectivity
   - implement

Concept Check

2. **Synthesize** How can a person's ethics help an individual face the dilemmas that are part of life's complexities? Describe a time when your own ethics helped you make the right choice.
3. How can a company communicate a code of ethics to its employees?
4. Name five topics often addressed in a code of ethics.
5. List four goals of an ethically trained accountant. What role does the accountant play in maintaining public trust in the business community?
6. Describe the accounting regulations signed into law in 2002. What role does the SEC play in regulating the accounting industry?
7. **Evaluate** What are the benefits of acting ethically, both for an individual and for a business?
8. **Evaluate** Are the concepts of competence and confidentiality important to the accounting profession? Explain the reasons for your answer.
9. What is the purpose of a code of ethics for a professional organization?
10. **English Language Arts** Find a current news story that deals with an ethical issue in business. Explain in a one-page paper what the accountant's role would have been in the situation described in the article, and the ethical way an accountant should handle such a situation. If appropriate for the article, discuss how any negative repercussions could have been avoided. Be sure to evaluate the validity and reliability of the source you use. Cite any source material you used.

Chapter 29
Problems

Problem 4 | Researching Ethics in the News

Instructions Use the library, newspaper, or Internet to find three recent examples of businesses involved in a historically significant compliance or regulatory issue or accused of engaging in unethical practices. For each example describe the company's practices, any legal charges involved, whether the issue led to the formation of regulatory agencies or laws, and why you think the business chose this course of action.

Problem 5 | Creating a Business Ethics Program

Instructions As the ethics officer for a national chain of coffeehouses, you must prepare a one- or two-page outline for a business ethics plan that covers the main issues you think are relevant to the company's operations. Include outline headings for a code of ethics, enforcement measures, and how you will communicate the ethics plan to the company's employees and managers.

Problem 6 | Making Ethical Decisions

Instructions Sean McGee works as a graphics designer for a large design firm. In his job he utilizes dozens of expensive software programs that the company purchased for its employees to use to complete client projects. Because Sean often takes work home to meet deadlines, he copied all software programs and installed them on his personal home computer. In addition to the work he does for his employer, he takes freelance jobs for extra cash. What are the ethical issues? Who are the affected parties? Do you think Sean has made an ethically correct decision? Why or why not? What would you have done?

Problem 7 | Making Ethical Decisions

Instructions Randy Simpson and Kyung Won worked as recruiters for a national job placement company. The company provided them a database of potential companies that might use a placement service as well as qualified applicants seeking jobs. After working together for several years, Randy and Kyung decided to open their own recruiting agency. When they worked for the national job placement company, both had signed an agreement not to take any company property with them if they left.

When Randy and Kyung were setting up their new offices, Kyung noticed Randy downloading files from a CD-ROM to his computer. When questioned, Randy said he had copied his list of clients in addition to names of potential new clients and qualified applicants. Kyung was worried that this act violated the agreement they had signed. Randy maintained that he had built solid relationships with his clients, and that he was entitled to take their contact information. As for the names of applicants, he argued, "These people need jobs. If I can help make that happen, I should." What are the ethical issues? What would you do if you were Kyung?

Chapter 29
Problems

Problem 8 — Examining the Impact of Unethical Decisions

Instructions Assume that the accountant for a drug manufacturing company made the decision to record the company's ending drug inventory at an amount higher than its actual worth. How does this decision affect gross profit and net income for that accounting period? In what ways might this decision affect shareholder value? How might this affect confidence in the company's financial statements in the future?

Problem 9 — Finding Out What Ethical Principles Mean to Your Classmates

Instructions Make a list of the five ethical principles discussed in this chapter. Ask your friends how they think the principles might influence their everyday life. Report your results in class.

Problem 10 — Finding Out What Ethical Principles Mean to Adults

Instructions Ask adults how they think the five principles of ethics influence professional behavior. Share your results in class.

Problem 11 — Applying a Code of Ethics to Personal Behavior

Instructions Write a code of ethics to guide student behavior in school organizations. Interview students outside your class to get ideas.

Problem 12 — Analyzing the Preamble to the Principles Section of the AICPA Code of Professional Conduct

Instructions Review the AICPA Preamble in the text. List two key points covered in each of the two paragraphs. Are the points important to your everyday behavior? Why or why not?

Real-World Applications & Connections

Compare and Contrast: International Accountant and Government Accountant

International accountants and government accountants are both specialized types of accountants, but they are also both very different types of jobs. Review what you have learned about each career.

Compare
After reviewing what you have learned about international accountants and government accountants, make a list of the similarities between each of these professions.

Contrast
After reviewing what you have learned about international accountants and government accountants, make a list of the differences between each of these professions.

Choose
After reviewing what you have learned about international accountants and government accountants, decide which job out of the two would better suit you. Write a paragraph explaining why you made the choice you did.

Activity
Review the information you wrote in response to the activities above and then write two descriptions. Write one description of the type of person you think would make a good international accountant and another description of the type of person you think would make a good government accountant. Describe the qualities that make each person good at each job.

Organization of Petroleum Exporting Countries (OPEC)

OPEC is an alliance of 11 oil producing nations that seeks to stabilize oil prices and help producers achieve a reasonable rate of return. OPEC members discuss oil production levels, pricing, demand, and environmental issues. OPEC can increase its oil production in order to prevent price increases or reduce oil production to hedge against falling prices.

Research and Share
a. Use the Internet to locate the Web site for OPEC.

b. List the ways in which OPEC impacts the oil market. Share your answer with a partner.

Ethics in the Real World

As you read in this chapter, accountants must abide by laws and regulations in order to manage business operations and transactions in acccounting. During the past decade, companies such as Enron, WorldCom, Tyco, Arthur Anderson, and Adelphia Communications failed because of ethics violations.

Activity

Select one of the companies mentioned above. Using relevant and reliable sources, research the ethics violations that led to the company's fall and the laws and regulations that apply. Detail your findings in a two-page paper. Discuss state regulation of the accounting industry. What role does the SEC play in regulating the accounting industry?

Corporate Responsibility

An ethical corporation makes a commitment to manage its roles in society—as a producer, employer, customer, and citizen—in a responsible manner. Implementing policies and practices in these areas is generally considered sound management. Many companies publish their commitment to corporate responsibility in their annual reports.

Activity

Use the Internet to locate and save or print a copy of Amazon.com's most current annual report for these tasks.

1. Complete a vertical analysis of Amazon.com's cost of sales and gross profit for the current year.
2. Complete a vertical analysis of Amazon.com's assets using the classifications above for the current year.

Ethics

Imagine one of your clients is a company that owns a hospital. As you review the hospital's various sources of income, you discover the hospital received a large donation from a pharmaceutical company. The hospital's code of ethics states that contributions from suppliers, in this case a pharmaceutical company, are not permitted.

Activity

Write a memo that gives your recommendation for action in this situation. In your memo, discuss the reasons you think the hospital's code of ethics does not allow contributions from pharmaceutical companies.

APPENDIXES

Appendix A: The Accrual Basis of Accounting A-2

Appendix B: Federal Personal Income Tax A-12

Appendix C: Using the Numeric Keypad A-18

Appendix D: Advanced Accounting Concepts A-22

Appendix E: Additional Reinforcement Problems A-27

Appendix F: Answers to Section Assessment Problems A-50

Appendix A
The Accrual Basis Of Accounting

Before you Read

Main Idea
Accrual accounting recognizes when revenue and expenses are earned and incurred rather than when they are collected and paid.

Read to Learn...
- how to record accrual entries. (p. A–2)
- how to record deferral entries. (p. A–6)
- how to use reversing entries. (p. A–8)

Key Terms
- cash basis of accounting
- accrual basis of accounting
- accruals
- accrued revenue
- accrued expense
- deferrals
- unearned revenue
- prepaid expense
- reversing entry

Some small businesses keep their financial records on a cash basis. The **cash basis of accounting** recognizes and records revenues when cash is received and expenses only when cash is paid out. Often, however, cash is received and payments are made in an accounting period other than the period in which the original transaction took place. For example, a landscaping service performs work for a client in December, but the customer may not be required to pay until January of the next year. However, if the revenue is not recorded for the service performed in December, the financial statements may not present fairly the business earnings for the year just ended. To overcome this, the accounting profession requires that most businesses use what is called the **accrual basis of accounting**. In accrual accounting, revenues are recognized and recorded when earned, and expenses are recognized and recorded when they are actually incurred. The system for accrual accounting focuses on when earned and incurred rather than on when collected and paid.

The accrual basis of accounting requires that accountants recognize revenue when the business makes a sale or performs a service, regardless of when the company receives the cash. Expenses are recognized as incurred, whether or not the business has paid out the cash. To accrue revenue means to record an amount in anticipation of future receipts and to accrue an expense means to record an amount in anticipation of a future payment. Accrual accounting involves two types of revenue and expenses: accruals and deferrals.

Accruals

How Do You Record Accruals?

Accruals relate to *transactions which have not yet been recorded* in the accounts. The word *accrue* means to accumulate or amass. Accruals for unrecorded revenue and expenses are required when:

1. *A revenue has been earned but not yet received.* For example, interest earned on a note receivable in one accounting period but not collected until the next accounting period.

2. *An expense has been incurred but not yet paid.* For example, you may earn wages from April 27 to April 30, but your employer may not pay your check to you until May 5.

Accruals are revenues or expenses that are gradually earned or incurred over time. In order to report a company's financial position fairly, accruals should be recognized in the accounting period in which they belong.

Accrued Revenue

Accrued revenue is revenue that has been earned but not received or recorded. Examples of unrecorded accrued revenues include unpaid fees for services completed and rent earned but not yet received. The accrued revenue example that follows is for unpaid interest earned on a note receivable.

Adjustment

On December 5, 2010, Rapid Growth Co. accepted a $5,000, 90-day, 9% note receivable from a charge customer, Janie Pippa. Pippa issued the note as an extension on her charge account. Rapid Growth Co. recorded the transaction as a debit of $5,000 to **Notes Receivable** and a credit of $5,000 to **Accounts Receivable** and Janie Pippa's account in the subsidiary ledger. On December 31, 26 days of interest had been earned (accrued) on the note receivable: $5,000 \times .09 \times 26/365 = \$32.05$. An adjusting entry is required on December 31 to record the amount of interest income earned from December 5 to December 31.

Analysis — Identify, Classify, +/−

1. For the adjusting entry, the accounts **Interest Receivable** and **Interest Income** are affected.
2. **Interest Receivable** is an asset account. **Interest Income** is a revenue account.
3. **Interest Receivable** is increased and **Interest Income** is increased.

Debit-Credit Rule

4. Increases to assets are recorded as debits. Debit **Interest Receivable** for $32.05.
5. Increases to revenues are recorded as credits. Credit **Interest Income** for $32.05.

T Accounts

6.

Interest Receivable		Interest Income	
Debit + Adj. 32.05	Credit −	Debit −	Credit + Adj. 32.05

Journal Entry

7.

GENERAL JOURNAL   PAGE 18

	DATE		DESCRIPTION	POST. REF.	DEBIT	CREDIT	
1	2010		Adjusting Entries				1
2	Dec.	31	Interest Receivable		32 05		2
3			Interest Income			32 05	3
4							4

On March 5, 2011, Rapid received a check from Janie Pippa for $5,110.96, the maturity value of the note ($5,000.00 principal + $110.96 interest). Of the total interest, $32.05 had been earned and recorded in the previous accounting period. The remaining $78.91 should be recorded as income earned in the *current* fiscal year.

In the entry for the receipt of Janie's payment, **Cash in Bank** is debited for $5,110.96, the total amount of cash received. **Notes Receivable** is credited for $5,000.00. **Interest Receivable** is credited for $32.05 since the amount owed has now been received. **Interest Income** is credited for $78.91, the amount of the interest earned that applies to the current accounting period. The entry for this transaction is shown below.

	GENERAL JOURNAL			PAGE 7
DATE	DESCRIPTION	POST. REF.	DEBIT	CREDIT
2011				
Mar. 5	Cash in Bank		5 110 96	
	Notes Receivable			5 000 00
	Interest Receivable			32 05
	Interest Income			78 91
	Receipt 416			

Accrued Expenses

Most expenses are recorded when they are paid or when a liability is incurred. There are, however, some business expenses that have not been paid—but accrue daily—and are not recorded until an adjusting entry is made at the end of the fiscal period. Some examples of these **accrued expenses** are salaries or wages earned by working employees but not yet paid to them, property taxes owed but not yet paid, and interest accrued on unpaid notes payable. The example that follows shows how to calculate interest due and payable.

Adjustment

On November 1, 2010, Rapid Growth Co. signed a $2,000, 90-day, 12% note payable for the purchase of equipment from Taylor Equipment. Rapid Growth Co. recorded the transaction as a debit of $2,000 to **Equipment** and a credit of $2,000 to **Notes Payable.** On December 31, 60 days of interest was owed (accrued) on the note payable: $2,000 × .12 × 60/365 = $39.45. An adjusting entry is required on December 31 to record the amount of interest expense incurred from November 1 to December 31.

Analysis	Identify	1. For the adjusting entry, the accounts **Interest Expense** and **Interest Payable** are affected.
	Classify	2. **Interest Expense** is an expense account. **Interest Payable** is a liability account.
	+/−	3. **Interest Expense** is increased and **Interest Payable** is increased.

A-4   Appendix A • The Accrual Basis of Accounting

Debit-Credit Rule	4. Increases to expenses are recorded as debits. Debit **Interest Expense** for $39.45.
	5. Increases to liabilities are recorded as credits. Credit **Interest Payable** for $39.45.

T Accounts

6.

Interest Expense			Interest Payable	
Debit +	Credit −		Debit −	Credit +
Adj. 39.45				Adj. 39.45

Journal Entry

7.

GENERAL JOURNAL PAGE 18

DATE	DESCRIPTION	POST. REF.	DEBIT	CREDIT
Dec. 31	Interest Expense		39 45	
	Interest Payable			39 45

On March 5 Rapid sent a check to Taylor Equipment for $2,059.18, the maturity value of the note ($2,000.00 principal + $59.18 interest). Of the total interest expense, $39.45 was already owed as of December 31. That amount was recorded as an adjusting entry in the previous accounting period. The remaining $19.73 should be recorded as an expense for the *current* accounting period.

In the entry for the payment of the note, **Notes Payable** is debited for $2,000.00. **Interest Expense** is debited for $19.73, the amount of interest expense incurred that applies to the current accounting period. **Interest Payable** is debited for $39.45 since the amount that was accrued has now been paid. **Cash in Bank** is credited for $2,059.18, the total amount of cash paid. The entry for this transaction is shown below.

GENERAL JOURNAL PAGE 7

DATE	DESCRIPTION	POST. REF.	DEBIT	CREDIT
Mar. 5	Notes Payable		2 000 00	
	Interest Expense		19 73	
	Interest Payable		39 45	
	Cash in Bank			2 059 18
	Check # 1597			

To acquire cash with which to provide services, communities impose a tax on real property. Property taxes are usually paid twice a year, with each payment covering a six-month period or one-half the annual tax.

Underwood Merchandisers operates in a building on its own land. The value of the real property is $300,000. The total annual property taxes are $7,800. Underwood is located in a state where property taxes must be paid in May and November. The May payment covers the first six months of the calendar year, and the November payment covers the last six months of the calendar year.

Underwood's fiscal period ends on April 30. As a result, at the end of its fiscal period, Underwood owes property taxes of $2,600 ($7,800 × 4/12 or $3,900 × 4/6) for the months of January, February, March and April. These taxes, however, are not due to be paid until May.

On April 30 Underwood makes the following adjustment to the accounting records to reflect the correct amount of property tax for the period.

	GENERAL JOURNAL			PAGE 26	
DATE	DESCRIPTION	POST. REF.	DEBIT	CREDIT	
20--	Adjusting Entries				1
Apr. 30	Property Tax Expense		2600 00		2
	Property Tax Payable			2600 00	3

On May 15 Underwood wrote a check for the semiannual property tax payment of $3,900. Of this amount, $2,600 is an expense that was already charged to the prior fiscal period, and $1,300 is an expense charged to the current accounting period. The entry in May to record this payment is shown in the general journal below.

	GENERAL JOURNAL			PAGE 3	
DATE	DESCRIPTION	POST. REF.	DEBIT	CREDIT	
20--					1
May 15	Property Tax Expense		1300 00		2
	Property Tax Payable		2600 00		3
	Cash in Bank			3900 00	4
	Check 401				5

Deferrals

How Do You Record Deferrals?

Deferrals are for *transactions that have been recorded* in the accounts. The word *defer* means to wait or delay. Deferrals are for receipts that are not yet earned and for payments that are not yet expenses.

1. *Cash may have been received but not yet earned.* For example, rent received in advance when some of the rental time is in one accounting period and the balance is in the next accounting period.

2. *Cash may have been paid but the expenses not incurred.* For example, when premiums on insurance are paid covering part of one accounting period and the balance covers the next accounting period.

Adjusting entries are required to: (1) recognize unrecorded accrued revenues and expenses or (2) allocate recorded deferred revenues and expenses to the appropriate accounting period.

Unearned Revenue

Unearned revenue is revenue received before it is actually earned. Examples include cash received in advance for rental properties, for season tickets to various events, for insurance premiums, and for work to be performed in the future. Because the business has an obligation to deliver the merchandise or perform the service for which it has already received payment, unearned revenue represents a current liability for the business.

Boyce Real Estate rents space in its building to a tax accountant. On November 1 the accountant paid Boyce $1,200 in advance for rent covering the months of November through April. The entry to record this transaction is shown below.

	DATE		DESCRIPTION	POST. REF.	DEBIT	CREDIT	
	20--						1
1	Nov.	1	Cash in Bank		1200 00		2
2			Unearned Rental Income			1200 00	3
3			Receipt 181				4

GENERAL JOURNAL PAGE 55

As the business delivers the merchandise or performs the service, it earns a part of the advance payment. The earned portion must be transferred from the liability account to a revenue account through an adjusting entry made at the end of the period.

When Boyce's fiscal period ends on December 31, the company will have earned rental income of $400 for two of the six months paid in advance ($1,200 × 1/3). The adjusting entry to record the rental income is shown below.

GENERAL JOURNAL PAGE 65

	DATE		DESCRIPTION	POST. REF.	DEBIT	CREDIT	
1	20--		Adjusting Entries				1
2	Dec.	31	Unearned Rental Income		400 00		2
3			Rental Income			400 00	3
4							4

Prepaid Expenses

A **prepaid expense** is an expense paid in advance. Examples include the purchase of office supplies, premiums paid on insurance policies, rent paid in advance on the use of property or equipment, and bank discounts on non-interest-bearing notes payable. Examples of adjustments for prepaid expenses appear in Chapter 18 of this textbook.

Reversing Entries

What Are Reversing Entries?

After the financial statements have been prepared and the books are closed, many accountants find it helpful to reverse some of the adjusting entries before starting to record routine transactions for the new accounting period. A **reversing entry** is an optional general journal entry that is the exact opposite of a specific adjusting entry. Reversing entries are made on the first day of the new accounting period and simplify the recordkeeping in the new period.

We will use the note receivable accrual on page A–3 to see how reversing entries work. But first, we will review it *without* using reversing entries. The adjusting entry recorded accrued revenue of $32.05 in 2010 as a debit to **Interest Receivable** and a credit to **Interest Income**:

Interest Receivable		Interest Income	
Debit	Credit	Debit	Credit
+	–	–	+
Adj. 32.05			Adj. 32.05

Refer to the general journal entry on page A–4. When Janie paid the note on March 5, 2011, the total amount of interest received was $110.96. Since the interest owed was received, the accountant credited **Interest Receivable** for $32.05. The remaining interest of $78.91 ($110.96 2 $32.05) was recorded as interest income. Notice that the accountant had to allocate the total interest to **Interest Receivable** and **Interest Income** based on the adjusting entry made in the prior period. These two accounts appear as follows after posting:

Interest Receivable		Interest Income	
Debit	Credit	Debit	Credit
+	–	–	+
Adj. 32.05	3/5 32.05		3/5 78.91
Bal. 0			Bal. 78.91

Now let's use a reversing entry instead. The adjusting entry is recorded the same way whether reversing entries are used or not:

GENERAL JOURNAL   PAGE __18__

DATE		DESCRIPTION	POST. REF.	DEBIT	CREDIT
2010		Adjusting Entries			
Dec.	31	Interest Receivable		32 05	
		Interest Income			32 05

The accountant records the following reversing entry on January 1, 2011. Note that it is the exact opposite of the December 31 adjusting entry.

A-8   Appendix A • The Accrual Basis of Accounting

	GENERAL JOURNAL			PAGE 1	
DATE	DESCRIPTION	POST. REF.	DEBIT	CREDIT	
2011	Reversing Entries				1
Jan. 1	Interest Income		32 05		2
	Interest Receivable			32 05	3

The accountant writes the words *Reversing Entries* in the Description column of the General Journal before recording the first reversing entry. This heading eliminates the need for an explanation to be written after each reversing entry. The date of each reversing entry is the first day of the new accounting period. After recording reversing entries, the accountant posts them to the general ledger accounts.

The **Interest Income** account is a temporary account that is closed at the end of each accounting period. It begins the new period with a zero balance. After the reversing entry is posted in the new period, **Interest Income** has a $32.05 debit balance. This is not the normal balance for this account. As you recall, income accounts have a normal credit balance. Here are the accounts after the reversing entry is posted:

Interest Income			Interest Receivable	
Debit −	Credit +		Debit +	Credit −
Rev. 32.05			Adj. 32.05	Rev. 32.05
Bal. 32.05			Bal. 0	

When the note is paid, the accountant records it in the normal manner:

	GENERAL JOURNAL			PAGE 7	
DATE	DESCRIPTION	POST. REF.	DEBIT	CREDIT	
2011					1
Mar. 5	Cash in Bank		5110 96		2
	Notes Receivable			5000 00	3
	Interest Income			110 96	4
	Receipt 416				5

Notice that the accountant does *not* need to allocate the interest between the receivable and income accounts based on the adjusting entry. Instead, all interest is credited to **Interest Income**. Here is **Interest Income** after posting:

Interest Income	
Debit −	Credit +
Rev. 32.05	3/5 110.96
	Bal. 78.91

The Interest Income account balance is $78.91, the same as it would have been if reversing entries had not been used. The reversing entry simplifies the entry *made in the new period* to record the note's interest. Reversing entries save time and help prevent errors in the new period.

Appendix A • The Accrual Basis of Accounting **A-9**

Appendix A
Problems

Problem A–1 Identifying Accruals and Deferrals

Instructions Use the form in your working papers that is similar to the one below. For each item listed below, indicate by placing a check mark in the correct column whether the item is a prepaid expense, unearned revenue, an accrued expense, or accrued revenue. The first item has been completed as an example.

1. Rent received in advance for a two-year rental of a computer.
2. A two-year premium paid on a fire insurance policy.
3. Advance tuition collected by a university.
4. Interest on an interest-bearing note payable due next year.
5. Cash received for a three-year subscription to a magazine.
6. Fees due for the completed designs for three of five buildings.
7. Property taxes incurred for the last three months of the fiscal period.
8. Salaries owed but not yet paid.
9. Interest on an interest-bearing note receivable that matures in the next accounting period.
10. Office supplies purchased.
11. Cash received for season tickets for home football games.

Item	Prepaid Expense	Unearned Revenue	Unearned Revenue	Accrued Revenue
1.		✓		

Problem A–2 Recording Adjusting Entries

The Gore Paper Company uses the accrual basis of accounting. Its fiscal period ends on June 30. The following account balances appear in the company's general ledger as of year-end.

Cash in Bank	$34,616	Office Supplies Expense	$0
Interest Receivable	0	Salaries Expense	75,000
Office Supplies	8,335	Rental Income	0
Salaries Payable	0	Interest Income	0
Unearned Rental Income	1,000		

Instructions Record the adjusting entries on the general journal in your working papers.

1. The office supplies on hand as of June 30 are valued at $935.
2. Gore's 5-day weekly payroll totals $1,500. Salaries have been earned but not yet paid or recorded for June 28–30.
3. On June 30, 30 days of interest had accrued on an $8,000, 90-day, 9% note receivable from a charge customer.
4. Of the $1,000 recorded in the **Unearned Rental Revenue** account, $250 had been earned as of June 30.

Problem A–3 Recording Transactions for Notes Payable

The Villareal Corporation uses the accrual basis of accounting. Its fiscal period ends on December 31. On November 15, the Villareal Corporation borrowed $9,500 from the Second National Bank by issuing a 120-day, 9.5% interest-bearing note payable.

Instructions Record the following transactions on the general journal.

1. The issuance of the note payable (Note 7).
2. The adjusting entry to record the amount of accrued interest payable at year-end.
3. The payment of the note on the maturity date (Check 411).

Problem A–4 Identifying Accruals and Deferrals

The Cassenada Dance Company uses the accrual basis of accounting. Located in Seattle, Washington, the company's fiscal year-end is July 31. Property taxes are paid in June and December. Total annual property taxes on the studio and parking lot are $12,000.

Instructions

1. Record the payment made on June 30 for the first six months' property tax in the general journal.
2. Record the adjusting entry necessary at the fiscal year-end, July 31.

Appendix B
Federal Personal Income Tax

Before you Read

Main Idea
Every income-earning individual must file an income tax return to report taxable income and to calculate income tax owed.

Read to Learn...
- how the federal tax system was created. (p. A–12)
- the purpose of some common federal tax forms. (p. A–12)
- how a federal income tax return is prepared. (p. A–13)
- how to file a return online. (p. A–14)

Key Terms
- voluntary compliance
- Form W-4
- Form 1099
- Form 1040EZ
- adjusted gross income
- taxable income
- e-file

In order to become an informed and contributing member of society, every individual should have a basic knowledge of the federal income tax system. For the system to work fairly and efficiently, everyone must pay their fair share in an accurate and timely manner.

Creation of the Federal Income Tax System

How Was the Federal Tax System Created?

At its founding in 1776, this country did not have a federal personal income tax system. Individual states taxed their residents to raise revenue for needed expenditures. The first personal income tax was created in 1862 to finance the Civil War. A commission levied taxes on a small percentage of people to generate needed revenue. However, this tax was abolished in 1872, soon after the war ended. A federal personal income tax was again proposed in 1895 but was declared unconstitutional.

In 1913 Congress ratified the 16th Amendment to the Constitution allowing the federal government to levy a tax on individuals based on personal income. The income tax system was established under the principle of **voluntary compliance**. This means that all individual taxpayers are responsible for filing their tax return. All taxpayers must report their taxable income, accurately calculate the income tax owed, and file the necessary forms on time.

The Internal Revenue Service (IRS) also has responsibilities under our present tax system. The IRS must administer the tax laws fairly and process tax returns in an efficient manner.

Tax Forms and Procedures

What Are Three Common Tax-Related Forms?

For the federal tax system to run efficiently, all income-earning individuals must prepare and file certain tax forms. For personal income tax, the earning period is from January 1 through December 31. Every individual has from January 1 until April 15 each year to file a tax return and pay the government any taxes owed on income earned in the previous year.

For most taxpayers the most common and largest form of income is the gross earnings paid to them by employers. Other common forms of taxable income include interest earned on savings accounts and investments, stock dividends, rental income, and profits from business operations run as sole proprietorships or partnerships.

Federal Withholding Tax Estimation

As you have learned, the employer withholds a certain amount from each employee's paycheck for taxes. This system of "pay as you earn" was enacted during the early 1940s.

Since the exact amount earned for the year will not be determined until the end of December, the amount of taxes withheld from each employee's paycheck is an *estimate* of the amount the employee will owe at the end of the year. As you learned in Chapter 12, all employees fill out a **Form W-4** when they begin work. This indicates to the employer how much to withhold for federal income tax from each paycheck. For more information on completing Form W-4, refer back to Chapter 12, pages 318–319.

Earned Income

Individual taxpayers receive two common federal forms to report income for the year. The first is Form W-2. This form is prepared by employers and reports to employees and the federal government the total gross wages for the year and the amount withheld from employees' wages to pay the federal income tax. This form is given to the employees in the month of January so the information is available for employees to complete their individual tax returns. To review the Form W-2, refer to Chapter 13, page 361.

Interest Income

The other common form used to report income is **Form 1099**. This form reports interest earned in savings accounts and is prepared by banks or other institutions. Interest from all 1099 forms is added together and the total interest earned for the year is reported on your federal tax return.

The Federal Income Tax Return

What Tax Form Do Most Students File?

The most common tax return prepared by students is Form 1040EZ.

Preparing Form 1040EZ

To file **Form 1040EZ**, several requirements must be met including:

1. Taxable income from wages and tips must be less than $100,000.
2. Taxable interest cannot be more than $1,500.

A completed Form 1040EZ is presented in **Figure B–1** on pages A–15 and A–16. To complete this form, follow these steps:

1. Enter your name, address, and social security number.
2. Check the box on the next line if you want $3 to go to the Presidential Election Campaign Fund.
3. Add the amount(s) in box 1 of your Form(s) W-2 and put that amount on line 1 of the form. This should be the total gross earnings from all of your jobs this year.
4. Add the total interest received, as reported on Form(s) 1099, and enter the total amount of interest on line 2.

5. Add lines 1 and 2 and put the total on line 4. This is your total income for the year, called **adjusted gross income**.

6. Since you are a student, you are probably claimed as a dependent by your parents or guardians. On line 5, check YES.

7. Go to the worksheet for dependents on the back of the form and complete A through G. Enter the amount from G on line 5 on the front of the form.

8. Subtract line 5 from line 4 and enter the amount on line 6. This is your **taxable income**, the amount of income that is taxed.

9. Add the amounts in box 2 of your Form(s) W-2 and enter the amount on line 7. This is the amount *already* deducted from your wages to pay your federal tax liability.

10. Enter the amount from line 7 on line 9.

11. Take the amount entered on line 6, and look up the amount of your actual tax liability on the tax table in your working papers. Enter this amount on line 10.

12. If line 9 is greater than line 10, then you have paid in more than you owe and are entitled to a **refund**. Enter the amount of the overpayment on line 11a.

13. If line 10 is greater than line 9, then you owe the federal government more money. Enter the amount you owe on line 12.

If you owe additional money, you must include a check for the entire amount when you file the tax return. Sign the return and enter the date and your occupation on the designated lines.

Before your file your tax return, Form 1040EZ, make sure you:

1. Check all calculations.
2. Attach Copy B (one copy) of each of your W-2s, which are supplied to you by your employer(s).
3. File the return by April 15.

Filing a Tax Return Electronically

How Do I File My Return Online?

Most taxpayers are eligible to file their tax returns online. The Internal Revenue Service (IRS) offers the *e-file* option as a quick, easy, and accurate way to file tax returns. Detailed information about *e-file* is available on the IRS Web site.

Paid income tax preparers can use *e-file* if they are an Authorized IRS *e-file* Provider. If you prepare your own tax return, you can use *e-file* with a personal computer connected to the Internet. In most states, you can also file your state return at the same time.

You have several options to prepare and *e-file* your own tax return:

- purchase commercially available software,
- download software from an Internet site, or
- prepare and file your return online.

The IRS does not provide *e-file* software or offer online filing. A number of companies, tested and approved by the IRS, offer it. Some charge a fee and others do not. The IRS Web site provides additional information. Like any product or service, you must shop around and choose the product that is right for you.

After you transmit your return, you are notified whether it has been accepted or rejected. If it is not accepted, you are provided with customer support to correct the return and resubmit it.

If you owe money, you can authorize an electronic funds withdrawal from your bank account or use a credit card. If you are due to receive a refund, it can be

Figure B–1 Completed Form 1040EZ

deposited directly into your bank account. Refunds are received much faster using this system compared to mailing your tax return.

The IRS also offers a service called *Tele-Tax*. You can call the IRS 24 hours a day and hear recorded messages on over 150 common tax topics. These topics can also be accessed on the IRS Web site.

Form 1040EZ (20--)   Page **2**

Use this form if

- Your filing status is single or married filing jointly. If you are not sure about your filing status, see page 11.
- You (and your spouse if married filing jointly) were under age 65 and not blind at the end of 20--. If you were born on January 1, 19--, you are considered to be age 65 at the end of 20--.
- You do not claim any dependents. For information on dependents, use TeleTax topic 354 (see page 6).
- Your taxable income (line 6) is less than $100,000.
- You do not claim any adjustments to income. For information on adjustments to income, use TeleTax topics 451-458 (see page 6).
- The only tax credit you can claim is the earned income credit. For information on credits, use TeleTax topics 601-608 and 610 (see page 6).
- You had only wages, salaries, tips, taxable scholarship or fellowship grants, unemployment compensation, or Alaska Permanent Fund dividends, and your taxable interest was not over $1,500. But if you earned tips, including allocated tips, that are not included in box 5 and box 7 of your Form W-2, you may not be able to use Form 1040EZ (see page 12). If you are planning to use Form 1040EZ for a child who received Alaska Permanent Fund dividends, see page 13.
- You did not receive any advance earned income credit payments.

If you cannot use this form, use TeleTax topic 352 (see page 6).

Filling in your return

For tips on how to avoid common mistakes, see page 20.

If you received a scholarship or fellowship grant or tax-exempt interest income, such as on municipal bonds, see the booklet before filling in the form. Also, see the booklet if you received a Form 1099-INT showing federal income tax withheld or if federal income tax was withheld from your unemployment compensation or Alaska Permanent Fund dividends.

Remember, you must report all wages, salaries, and tips even if you do not get a Form W-2 from your employer. You must also report all your taxable interest, including interest from banks, savings and loans, credit unions, etc., even if you do not get a Form 1099-INT.

Worksheet for dependents who checked "Yes" on line 5

(keep a copy for your records)

Use this worksheet to figure the amount to enter on line 5 if someone can claim you (or your spouse if married filing jointly) as a dependent, even if that person chooses not to do so. To find out if someone can claim you as a dependent, use TeleTax topic 354 (see page 6).

- **A.** Amount, if any, from line 1 on front   __5,954.00__
  + 250.00   Enter total ▶   **A.** __6204.00__
- **B.** Minimum standard deduction . . . . **B.** __800.00__
- **C.** Enter the **larger** of line A or line B here . . . . **C.** __6204.00__
- **D.** Maximum standard deduction. **Single,** enter $4,850; if **married filing jointly,** enter $9,700 . . . . **D.** __4850.00__
- **E.** Enter the **smaller** of line C or line D here. This is your standard deduction. . . . . **E.** __4850.00__
- **F.** Exemption amount.
  - If single, enter -0-.
  - If married filing jointly and—
    —both you and your spouse can be claimed as dependents, enter -0-.
    —only one of you can be claimed as a dependent, enter $3,100.   **F.** __-0-__
- **G.** Add lines E and F. Enter the total here and on line 5 on the front **G.** __4850.00__

If you checked "**No**" on line **5** because no one can claim you (or your spouse if married filing jointly) as a dependent, enter on line 5 the amount shown below that applies to you.

- Single, enter $7,950. This is the total of your standard deduction ($4,850) and your exemption ($3,100).
- Married filing jointly, enter $15,900. This is the total of your standard deduction ($9,700), your exemption ($3,100), and your spouse's exemption ($3,100).

Mailing return

Mail your return by **April 15, 20--**. Use the envelope that came with your booklet. If you do not have that envelope or if you moved during the year, see the back cover for the address to use.

Form **1040EZ** (20--)

Figure B-1 Completed Form 1040EZ (continued)

Appendix B
Problems

Problem B-1 Preparing Form 1040EZ

Michael Feld is a junior at Lewiston High School. He lives with his parents at 274 West Polomia Drive in Hanksville, Ohio, 03856. Mike's social security number is 036-23-8825. Mike wants to contribute $3 to the Presidential Election Campaign Fund.

Mike worked full-time last summer at Harriet's Sub Shop as a counter clerk. He received a Form W-2, which stated that he had total gross wages of $6,201.00 and $192.00 was deducted for federal income tax.

Mike also had a savings account. The Kingsman National Bank sent him Form 1099, which stated he had earned interest of $113.00 on his account.

Mike lives at home with his parents and is claimed as a dependent on their tax return.

Instructions Using this information, prepare Form 1040EZ for Mike Feld. A blank Form 1040EZ is provided in your working papers. Use the tax chart in your working papers to calculate the federal income tax.

Problem B-2 Preparing Form 1040EZ

Alicia DeSantis is a senior at Georgetown High School. She lives at 825 Patterson Street in Montville, Texas, 09436. Her social security number is 043-73-8217. She worked part-time during the school year at Hilltop Deli and full-time during the summer as a lifeguard and swimming instructor at Montville's town pool.

Alicia wants to contribute $3 to the Presidential Election Campaign Fund. Alicia lives with her mother and is claimed as a dependent on her mother's tax return.

Her Form W-2 from the deli reported that she earned $2,143.60 during the year and $112.00 was deducted for federal income tax. The town of Montville sent her a W-2 that reported she earned $4,815.00 during the summer and $206.00 was deducted for federal income tax.

Alicia opened a savings account at the local bank that sent her a Form 1099 that stated she earned $118.50 in interest last year. She also had another savings account set up for college that paid her $201.70 in interest.

Instructions Prepare Form 1040EZ for Alicia. A blank Form 1040EZ is provided in your working papers. Use the tax chart in your working papers to calculate the tax owed.

Appendix C
Using the Numeric Keypad

Before you Read

Main Idea

Using a numeric keypad by touch will help you enter numbers quickly and accurately.

Read to Learn...
- about the key locations. (p. A–18)
- how to enter numbers by touch. (p. A–19)

Key Term
- home keys

Electronic calculators and computer keyboards have ten-key numeric keypads. When you key numerical data, your ability to use the numeric keypad by touch will make your task easier and faster.

Locating Keys

How Are Keys Arranged on a Numeric Keypad?

The ten-key numeric keypad is usually arranged in four rows of three keys. The locations of the 1 to 9 keys are the same on all equipment. The locations of the 0 (zero), decimal, Enter, and other function keys vary. Some typical arrangements are illustrated here.

On the ten-key numeric keypad, the 4-5-6 keys are called the **home keys**. These keys are the "starting point" from which you operate the other number keys. The index finger of your right hand should rest on the 4 key, your middle finger on the 5 key, and your ring finger on the 6 key. Most keypads have a special "help" on the home keys so you can easily locate them: the 5 key may have a raised dot or the surfaces of the 4-5-6 keys may be concave (indented).

From the home keys, you reach up or down to tap other keys. The index finger is also used for the 7 and 1 keys. The middle finger is used for the 8 and 2 keys. The ring finger is used for the 9 and 3 keys.

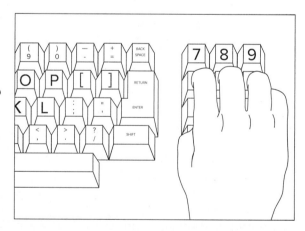

The fingers used for the 0 (zero), decimal, and Enter keys depend on the keypad's arrangement. On some keypads, the thumb taps the 0 (zero) key and the ring finger taps the decimal key. Depending on its location, the Enter key on a computer may be operated by the little finger or the thumb.

A-18   Appendix C • Using the Numeric Keypad

On an electronic calculator, numbers are entered by using function keys: plus (+), minus (−), multiply (×), divide (÷), and so on. A computer keypad has different function keys for multiply (*) and divide (/). On an electronic calculator, the asterisk (*) is used for the total.

The fingers used to operate these keys depend on the keys' location on the keyboard. Locate these keys on your keypad and determine the correct fingers to use to operate them.

Entering Numbers by Touch

How Do You Use the Keys by Touch?

Throughout the remaining pages of this appendix, you will demonstrate use of the numeric keypad by touch. That is, you will enter numbers on the keypad without looking at your fingers.

Your practice consists of adding columns of numbers. If you are using an electronic calculator, tap the plus (1) key after you have entered a number in a column. After entering all the numbers in a column, tap the total (*) key. If you are using a computer spreadsheet program, tap the Enter key after you have entered a number in a column. This will force a line break; each time you strike the Enter key, the cursor will move to the next line. After entering all the numbers in a column, use the software function or formula to add the numbers.

Using the 4-5-6 Keys

1. Demonstrate use of the numeric keypad by touch. Locate the 4-5-6 keys (the home keys) on your keypad. Also locate the Enter key if you are using a computer or the plus key if you are using a calculator.

2. Place your index finger on the 4 key, your middle finger on the 5 key, and your ring finger on the 6 key.

3. To enter a number, tap the number keys, one at a time, in the same order as you read the digits from left to right. Always keep your fingers on the home keys.

4. When you have entered the last digit, tap the Enter key (computer) or the plus key (calculator).

5. Using the following problems, practice entering columns of numbers at a comfortable pace until you feel confident about each key's location.

6. After entering all of the numbers in a column, add the numbers using the software (computer) or tap the total key once (calculator).

1.	2.	3.	4.	5.	6.
444	555	666	456	554	664
555	666	454	654	445	445
666	444	545	465	564	566
456	654	446	556	664	645
564	546	646	656	565	465
646	465	546	465	655	654

Using the 1, 7 and 0 Keys

1. Demonstrate use of the numeric keypad by touch. Locate the 1, 7 and 0 (zero) keys on your keypad.

2. Place your fingers on the home keys.

3. Practice the reach from the home keys to each new key. Reach down to the 1 key and up to the 7 key with your index finger. Be sure to return your finger to the home keys after tapping the 1 and 7 keys. Strike the 0 (zero) key with your thumb.

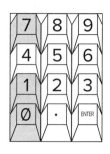

4. Using the following problems, practice adding columns of numbers containing the new keys. Practice at a comfortable pace until you feel confident about each key's location. Be sure to keep your fingers in home-key position.

1.	2.	3.	4.	5.	6.
444	014	140	107	011	141
471	107	701	074	170	117
174	740	701	104	710	417
741	101	704	007	004	047
710	114	471	411	471	104
407	441	117	047	174	114

Using the 3 and 9 Keys

1. Locate the 3 and 9 keys on your keypad.

2. Place your fingers on the home keys.

3. Practice the reach from the home keys to each new key. Reach down to the 3 key and up to the 9 key with your ring finger. Be sure to return your finger to the home keys after tapping the 3 and 9 keys.

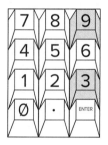

4. Using the following problems, practice adding columns of numbers containing the new keys. Practice at a comfortable pace until you feel confident about each key's location. Be sure to keep your fingers in home-key position.

1.	2.	3.	4.	5.	6.
666	669	339	966	939	699
999	663	363	393	363	936
333	936	336	966	393	939
963	396	936	633	639	336
639	936	636	393	369	696
399	363	996	993	369	939

Using the 2 and 8 Keys

1. Locate the 2 and 8 keys on your keypad.

2. Place your fingers on the home keys.

3. Practice the reach from the home keys to each new key. Reach down to the 2 key and up to the 8 key with your middle finger. Be sure to return your finger to the home keys after tapping the 2 and 8 keys.

4. Using the following problems, practice adding columns of numbers containing the new keys. Practice at a comfortable pace until you feel confident about each key's location. Be sure to keep your fingers in home-key position.

1.	2.	3.	4.	5.	6.
555	228	885	285	582	828
888	852	285	258	558	825
222	522	825	525	582	852
582	252	588	858	825	258
822	528	258	582	525	885
522	855	852	825	582	282

Using the Decimal Key

1. Locate the decimal key on your keypad.
2. Place your fingers on the home keys.
3. Depending on the arrangement of keys on your numeric keypad, you may use your thumb, your middle finger, or your ring finger to tap the decimal key. Practice the reach from the home keys to the decimal key. Be sure to return your finger to the home keys after tapping the decimal key.

4. Using the following problems, practice adding columns of numbers containing the decimal key. Practice at a comfortable pace until you feel confident about the key's location. Be sure to keep your fingers in home-key position.

1.	2.	3.	4.	5.	6.
.777	.978	.998	8.78	7.88	8.79
.888	.987	.879	8.89	7.87	7.98
.999	.878	.787	8.87	8.97	9.89
.789	.987	.878	7.88	9.77	9.87
.897	.789	.797	9.87	7.97	7.89
.978	.797	.899	7.98	8.79	9.78

7.	8.	9.	10.	11.	12.
468.	48.2	.8	284.0	41.87	154.88
.489	2,537.	5,827.	100.	4,057.4	888.
214.2	852.	.024	8.45	89.45	.0082
7.12	3.978	18.73	56.0	2.25	200.08
6,394.4	257.0	85.00	23.00	20.0	632.48
.58	.2684	1.045	.89	36.248	64.1

13.	14.	15.	16.	17.	18.
267.50	425.21	1.25	467.54	65.27	9.78
4.19	414.50	0.18	95.14	102.38	5.94
87.64	1,684.84	585.56	6,926.95	8,216.58	652.25
654.84	49.95	7.50	35.00	1,852.84	3,782.70
1,750.67	720.65	11.60	7.13	4.60	39.25
141.82	77.61	23.55	154.95	79.15	36.87

Appendix D
Advanced Accounting Concepts

Before you Read

Main Idea

The accuracy of financial statements to fairly report the financial position of a business is essential to both management and investors.

Read to Learn...
- how to prepare and issue the annual report (p. A–22)
- about the role of management in financial reporting (p. A–23)
- about the role of auditors in financial reporting (p. A–24)
- about the Sarbanes-Oxley Act (p. A–24)
- about aspects of financial reporting (p. A–25)

Key Terms
- notes to the financial statements
- unqualified opinion
- qualified/adverse opinion
- revenue expenditure
- capital expenditure

Preparing and Issuing Financial Statements

The information presented on financial statements allows management to make more informed business decisions and enables investors and creditors to evaluate the business's financial operations. Each year, large corporations that issue stock to the general public must report the financial performance of the business to all stockholders. This document is called the *annual report*.

Preparing and Issuing the Annual Report

Why Must Stockholders Be Informed?

In order for stockholders to assess the performance of a business, financial statements and other important items of information must be provided. Investors need to be accurately informed as to the financial status of their investment.

Purpose of Annual Reports

Based on the annual report, investors and creditors and other interested parties can evaluate the present and future success of the business. Stockholders are not aware of daily or weekly business activity throughout the year and need a yearly report to see how everything is going. At school you receive progress reports and report cards which inform you and others of your success in the class. You use this information to make decisions regarding your future goals for the class. The annual report is the report card for businesses. It reports the financial successes and failures for all interested parties. Management, investors, and other parties make decisions based on this information.

Financial Statements Included in the Annual Report

An annual report will include the four financial statements you learned in Chapter 19. These are the *income statement*, the *statement of retained earnings*, the *balance sheet* and the *statement of cash flows*. Three of these statements report the financial activity over the entire year and the balance sheet reports the financial position of the business on a specific date—the last day of the period. These statements provide a financial picture of the business.

Reading Check

Recall

How do public corporations keep their investors informed about the financial status of their investments?

Analysis Included in the Annual Report

The annual report will often also include a letter to the stockholders from an official of the corporation highlighting the past year's performance and plans and new initiatives for the coming year. This is management's opportunity to present the corporation in a most favorable light.

Management may also include its interpretation and analysis of the company's financial status as well as comments on how the financial statements were prepared. These **notes to the financial statements** explain the accounting policies and procedures that were utilized, which may help investors analyze the financial information.

Role of Management in Financial Reporting

What Are The Responsibilities of Management?

All financial statements are prepared by management and must reflect an accurate financial picture of the business.

Objectives of Financial Reporting

The financial information concerning a business may be analyzed for a variety of reasons. These reasons include ability to make rational investment and credit decisions, potential for positive cash flows, and resources the business can utilize. One financial statement provides a limited amount of financial information. For complete analysis, all four statements should be examined and compared.

Management Responsibilities

It is the responsibility of management to report the financial position of the business in the most accurate way possible. Management must follow all GAAP (generally accepted accounting principles) guidelines in recording and maintaining financial records and in preparing financial statements. Management must also be free of bias financial statements and must be prepared with a high degree of integrity, and objectivity, along with high ethical standards.

Requirements and Procedures

The financial information for the present period should be compared to identical information from previous periods. An important accounting concept, *comparability,* refers to financial information that can be compared from one accounting period to another or from one business to another. When financial statements are prepared in this manner they are called *comparative financial statements*. This information can also be compared to industry averages for similar business operations on a local, regional, or national level.

Another responsibility of management is to present an analysis between certain amounts included on the financial statements. This process is called *ratio analysis*. It is comparing two amounts on a financial statement and the evaluation of the relationship between these amounts. Ratio analysis is used to evaluate a business's liquidity, profitability, and financial position. There are several types of ratio analysis and they are valuable management tools for future planning. For example, your hardware store had a merchandise inventory turnover rate of 3.4 for the past year. You discover the industry average turnover rate for similar stores was 4.2. You should now explore to see why your ratio is lower than average. There are several reasons why it may be lower but the ratio analysis identified merchandise inventory as an area for you to examine.

Management must also prepare financial statements using the *conservation principle* which states that amounts reported should be the least likely to result in an overstatement of income or assets.

Reading Check

Explain

Why must management be free of bias and have a high degree of integrity when preparing financial statements?

Role of Auditors in Financial Reporting

How Do We Know the Financial Information is Accurate?

In order to assure accurate information is being reported, the financial statements are audited by a group of independent accountants called *certified public accountants* (CPAs).

Need for Independent Experts

The purpose of an audit by CPAs is to assure that the management of the business has followed all accounting procedures and guidelines, (GAAP) in preparing the financial statements. As independent experts, auditors examine the financial statements and issue a professional, independent opinion. This opinion is usually expressed as a *"fair presentation"* meaning management has followed GAAP in preparing the statements. This endorsement by the auditors is shared with stockholders in the annual report.

Types of Audit Opinions

When auditors determine the financial statements are a *"fair presentation"* this is referred to as an **unqualified opinion**. When auditors feel the statements may not fairly represent the financial status of the business, they will identify the areas of concern in the annual report and issue a **qualified or adverse opinion**.

The Sarbanes-Oxley Act

How is accounting accuracy being addressed?

Creditability in financial reporting is essential for confidence in the U.S. capital markets. With accounting scandals in *WorldCom, Tyco International* and *Adelphia Communications* and the total collapse of *Enron* in 2001, Congress passed the Sarbanes-Oxley Act in 2002. This law provides for increasing regulations of auditors, identifying the types of accounting services auditors may provide, establishing higher accountability of corporate executives, and creating harsher penalties for violations. Enforced by the SEC, it is hoped this legislation would reduce the chances of corporate and accounting fraud and restore confidence in the preparation and issuance of financial statements by corporations. Unfortunately, with the demise of Enron, one of the largest public accounting firms, Arthur Anderson, was also dismantled.

Aspects of Financial Reporting

Can accounting procedures effect financial statements?

When preparing financial statements, corporations are often faced with accounting decisions that may have an effect on the reported financial position of the business. Even though GAAP guidelines are being followed, some accounting procedures involve choices. The following are two examples of how financial statements can be manipulated by various accounting procedures. By applying different acceptable accounting procedures (GAAP), different amounts may result. However, disclosure of your procedures must be made known.

Cost Allocation

When fixed assets are acquired, the business must determine the method of depreciation that will be applied. Depreciation is the allocation of the cost of the asset over its useful life. Each method will result in a different cost per year and also a different asset value for the asset. An asset being depreciated using the sum-of-the-years-digits method would have a different asset value than the same asset depreciated using the declining-balance method. Thus businesses have a certain degree of control as to the allocated value of fixed assets and the amount of depreciation recorded each year.

Management can allocate cost using various methods for depreciation of plant assets, depletion for natural resources, and amortization of intangible assets. All of these procedures directly affect the amounts recorded on financial statements.

Allocation of Expenditures

Revenue expenditures are expenses that are used up before the end of the current period. Examples are rent, utilities, maintenance and salaries. These amounts appear on the income statement for the period. **Capital expenditures** are expenses that will also benefit future accounting periods. These could include the rebuilding of a machine or redesign of an office. These would be recorded as increases to assets. If these expenses are not allocated or classified in the proper manner, both the income statement and the balance sheet will be affected. Misallocation of these expenses will have an effect on the present financial statements and also future statements.

Appendix D
Problems

Problem D-1

There have been several large accounting scandals in the past few years. These have caused harm to investors, employees, creditors, and seriously hurt the reputation of the corporation. Although accounting procedures were a major aspect of their illegal practices, all had a different central problem.

Instructions Pick one of the following corporations that had serious financial scandal in the past few years.

Adelphia Communications   Arthur Anderson   Enron   Tyco International WorldCom

Research the corporation using articles, news releases, books, or the Internet and formulate questions to analyze the issue. Be sure to evaluate the validity and reliability of the source material you use.

a. What business was the corporation operating?

b. Who were the individuals held responsible for the scandal?

c. Write a brief summary of the nature of the scandal.

d. How could this scandal have been prevented?

Problem D-2

The Sarbanes-Oxley Act was passed by Congress in 2002. Research the passage of this law and answer the following questions.

a. What is the legal name of this law?

b. What did it change regarding corporate executives in relation to financial statements?

c. What do you think it means when it states that "internal controls have been strengthened"?

d. Who enforces SOX?

Appendix E
Additional Reinforcement Problems

Chapter 3, Problem 3A: Determining the Effects of Business Transactions on the Accounting Equation

Pamela Wong started her own business called Thunder Graphics Desktop Publishing.

Instructions Use the form provided in your working papers. For each of the following transactions:

1. Identify the accounts affected.
2. Write the amount of the increase (+) or decrease (−) in the space provided on the form.
3. Determine the new balance for each account.

Trans. No.	Transactions
1	Pamela Wong, the owner, opened a checking account for the business by depositing $48,000 of her personal funds.
2	Paid the monthly rent of $1,500.
3	Bought office furniture on account for $1,000.
4	Pamela Wong invested $3,000 of office equipment in the business.
5	Paid cash for a new computer for the business, $5,000.
6	Paid for an advertisement in the local newspaper, $200.
7	Completed graphic desktop publishing services for a client and sent a bill for $800.
8	Paid $700 on account for the office furniture bought earlier.
9	Received $500 on account from a client.
10	Pamela Wong withdrew $1,000 for personal use.
11	Received $400 cash for desktop publishing services completed for a client.

Analyze

After analyzing all of the transactions, complete the following tasks.

(a) Compute the amount of total equipment assets.
(b) Compute the amount of total assets.
(c) Compute the total liabilities and owner's equity.
(d) Conclude whether the accounting equation is in balance.

Chapter 4, Problem 4A: Analyzing Transactions Affecting Assets, Liabilities, and Owner's Equity

Thunder Graphics Desktop Publishing had the following transactions. A partial list of the accounts used to record and report business transactions appear on the next page.

---

101 Cash in Bank
110 Accounts Receivable—
     Roger McFall
125 Office Equipment
130 Office Furniture
135 Computer Equipment

201 Accounts Payable—
     Computer Warehouse, Inc.
205 Accounts Payable—
     Pro Computer Company
301 Pamela Wong, Capital

Instructions On the forms provided in your working papers:

1. Prepare a T account for each account listed above.
2. Analyze and record each of the following business transactions in the appropriate T accounts. Identify each transaction by number.
3. After recording all transactions, compute and record the account balance and identify the normal side of each T account.
4. Add the balances of those accounts with normal debit balances.
5. Add the balances of those accounts with normal credit balances.

Trans. No.	Transactions
1	Pamela Wong invested $30,000 into the business.
2	Invested office equipment, valued at $650, into the business.
3	Bought a computer on account from Computer Warehouse, Inc. for $9,360.
4	Bought new office equipment for $1,550, Check 100.
5	Bought new office furniture on account from Computer Warehouse, Inc. for $1,250.
6	Sold the older office equipment on account to Roger McFall for $650.
7	Paid $3,000 on account to Computer Warehouse Inc., Check 101.
8	Received $400 on account from Roger McFall.
9	Paid $3,500 on account to Computer Warehouse Inc., Check 102.
10	Bought computer on account from Pro Computer Company for $775.

continued

Analyze

Appraise whether the accounting equation is in balance.

Chapter 5, Problem 5A — Analyzing Transactions Affecting Revenue, Expenses, and Withdrawals

Pamela Wong owns and operates Thunder Graphics Desktop Publishing. A partial list of the chart of accounts appears in your working papers.

Instructions On the forms provided in your working papers:

1. Prepare a T account for each account listed.
2. Analyze and record each of the following business transactions in the appropriate T accounts. Identify each transaction by number.
3. After recording all transactions, compute a balance for each account.
4. Test for the equality of debits and credits.

Trans. No.	Transactions
1	Pamela Wong invested $25,000 in the business.
2	Bought a new computer on account from Computer Warehouse, Inc. for $1,400.
3	Ms. Wong invested office equipment, valued at $650, in the business.
4	Paid the rent for the month, $750, Check 207.
5	Wrote Check 208 for $145 for minor repairs to the equipment.
6	Completed graphics design work on account for Adams, Bell, and Cox, Inc., $850.
7	Paid the $175 utility bill, Check 209.
8	Deposited the daily design receipts in the bank, $1,200.
9	Sent Check 210 for $700 to Computer Warehouse, Inc. as payment on account.
10	Received $425 from Adams, Bell, and Cox, Inc.
11	Ms. Wong withdrew $150 for personal needs.
12	Paid the $85 telephone bill, Check 211.

Analyze

Calculate the total expenses the business recorded in the above transactions.

Chapter 6, Problem 6A — Recording General Journal Transactions

Instructions On the form provided in your working papers, record the following transactions for the month of January on page 7 of the general journal for Thunder Graphics Desktop Publishing.

Date		Transactions
Jan.	1	Received $900 for completion of design work for Designers Boutique, Receipt 545.
	3	Purchased a desk, chairs, and table valued at $2,400 from Computer Warehouse, Inc. Made a down payment of $400 and agreed to pay the balance within 60 days, Check 801/Invoice CW402.
	7	Ran a $300 special feature ad in Solutions Software's monthly magazine on account, Invoice SS60.
	10	Completed a $3,500 print production job for Adams, Bell, and Cox, Inc. Received a down payment of $500 and agreed to receive the balance within 30 days, Receipt 546/ Invoice TG90.
	15	Pamela Wong withdrew $1,500 from the business for personal use, Check 802.
	17	Paid a carpenter $175 to install a new office door, Check 803.
	20	Purchased a $4,000 computer system from Pro Computer Company on account, Invoice 783.
	27	Paid Computer Warehouse, Inc. $1,000 to apply on our account, Check 804.
	29	Wrote Check 805 for $300 to Solutions Software in payment of an ad purchased on July 7.
	31	Pamela Wong invested a $200 file cabinet in the business, Memo 40.

Analyze
Calculate the total cash the business received during January.

Chapter 7, Problem 7A — Journalizing and Posting Transactions

The balances in the general ledger accounts of Thunder Graphics are shown below.

Instructions On the forms provided in your working papers:

1. Open the accounts in the general ledger with their beginning balances as of March 1 of the current year.

2. Record March transactions on page 15 of the general journal.

continued

101	Cash in Bank	$ 10,000
110	Accounts Receivable—Roger McFall	—
125	Office Equipment	2,500
135	Computer Equipment	3,000
201	Accts. Pay.—Computer Warehouse, Inc.	—
301	Pamela Wong, Capital	8,000
302	Pamela Wong, Withdrawals	4,000
401	Design Revenue	9,000
405	Print Production Revenue	4,000
505	Maintenance Expense	1,200
510	Miscellaneous Expense	300

3. Post each journal entry to the appropriate accounts in the ledger.
4. Prove the ledger by preparing a trial balance.

Date		Transactions
Mar.	1	Pamela Wong invested additional assets in the business: cash, $3,000, and office equipment, $500, Memorandum 75.
	5	Paid $600 to Premier Painters for painting the office, Check 912.
	7	Discovered that $1,500 in design revenue was incorrectly journalized and posted to the **Print Production Revenue** account last month, Memorandum 76.
	9	Purchased $3,600 computer system from Computer Warehouse, Inc. on account, Invoice CW204.
	12	Deposited $4,000 into the checking account. $3,000 was earned from design revenue and $1,000 from print production revenue, Receipts 300–310.
	15	Completed print production for Roger McFall. Sent him a bill for $600, Invoice TG601.
	20	Discovered that $50 in miscellaneous expense was incorrectly journalized and posted to the **Maintenance Expense** account, Memorandum 77.
	25	Pamela Wong withdrew $2,000 from the business, Check 913.
	28	Received $300 from Roger McFall to apply to his account, Receipt 311.
	31	Issued Check 914 for $1,800 to Computer Warehouse, Inc. to apply on our account.

Analyze

Compute the change in the cash account for March.

Chapter 8, Problem 8A — Preparing a Six-Column Work Sheet

The final balances in the general ledger of Thunder Graphics Desktop Publishing at the end of May are listed below.

101	Cash in Bank	$16,095
105	Accts. Rec.—Adams, Bell, and Cox Inc.	1,729
110	Accts. Rec.—Roger McFall	281
113	Accts. Rec.—Designers Boutique	—
115	Accts. Rec.—Pat Cooper	714
120	Office Supplies	228
125	Office Equipment	5,027
130	Office Furniture	2,940
135	Computer Equipment	8,200
201	Accts. Pay.—Computer Warehouse, Inc.	4,027
205	Accts. Pay.—Pro Computer Company	2,441
207	Accts. Pay.—Solutions Software	—
301	Pamela Wong, Capital	24,927
305	Pamela Wong, Withdrawals	2,400
310	Income Summary	—
401	Design Revenue	8,116
405	Print Production Revenue	4,050
501	Advertising Expense	735
505	Maintenance Expense	1,335
510	Miscellaneous Expense	285
520	Rent Expense	3,100
530	Utilities Expense	492

Instructions Use the form in your working papers to prepare a work sheet for the month ended May 30.

Analyze

Identify the net income or net loss for the month of May, and explain its effect on the Pamela Wong, Capital account.

Chapter 9, Problem 9A — Interpreting Financial Information

You are applying for a job with Thunder Graphics Desktop Publishing. The job includes preparing financial statements. In order to determine your ability to do the job, the owner, Pamela Wong, has given you the following information. At the beginning of the last fiscal period, the **Pamela Wong, Capital** account had a balance of $46,295. At the end of the period, the account showed a balance of

continued

$49,152. During the period, Ms. Wong made an additional investment of $2,000 and had withdrawals of $600. The revenue for the period was $12,948.

Analyze

Use the form provided in your working papers to compute the total expenses for the period.

Chapter 10, Problem 10A — Preparing Closing Entries

The work sheet of Thunder Graphics Desktop Publishing for the year ended December 31 is presented in your working papers.

Instructions Using the information in the work sheet, prepare the four journal entries to close the temporary capital accounts. Use general journal page 21 provided in your working papers.

Analyze

Compute the balance in the Pamela Wong, Capital account after the closing entries.

Chapter 11, Problem 11A — Recording Deposits in the Checkbook

On October 20, Pamela Wong, owner of Thunder Graphics Desktop Publishing, deposited the following in the checking account of the business.

Cash: $800

Checks: Boyden Company; drawn on National Bank of Commerce; Monroe; ABA No 45-1204; $50.29.

Gail's Supplies; drawn on Regent Bank; Rayville; ABA No 49-3401; $255.48.

Clinton Co.; drawn on Sun Federal Savings and Loan; Bastrop; ABA No 52-6429; $788.40.

On October 21, Pamela Wong received the electricity bill from Central Utilities for $314.69 for September 17 to October 18 service.

The October bank statement for Thunder Graphics listed a bank service charge of $21.80, dated October 23.

Instructions On the forms provided in your working papers:

1. Complete a deposit slip using the October 20 information.

2. Record the deposit on check stub 1068 using $2,468.14 as the balance brought forward.

3. Complete check stub 1068 and prepare Check 1068 to Central Utilities. Use October 22 as the date and sign your name as drawer.

continued

Appendix E • Additional Reinforcement Problems   A-33

4. Record this service charge in the checkbook on check stub 1069.
5. Journalize the entry for the bank service charge in the general journal. Use October 25 as the date.
6. Answer the following questions.
   a. What was the amount carried forward from check stub 1067?
   b. What was the checkbook balance before writing Check 1068?
   c. What was the total amount deposited on October 20?
   d. Identify the payee of Check 1068.
   e. What was the balance brought forward on check stub 1069?

Analyze

Assess the effect on net income if the bank service charge was not journalized.

Chapter 12, Problem 12A — Preparing a Payroll Register

Thunder Graphics Desktop Publishing has four employees. They are paid on a weekly basis with overtime paid for all hours worked over 40. The overtime rate is 1½ times the regular hourly rate of pay. The names of the employees and other information needed to prepare the payroll are as follows.

Employee No.	Employee	Status	Exemptions	Rate/Hour	Union
173	Don Hoffman	M	1	$6.95	No
168	Manual Gongas	S	0	7.10	Yes
167	Riley Sullivan	M	2	7.40	Yes
175	Marcy Jackson	S	1	6.95	Yes

During the week ended Oct. 15, Gongas worked 41 hours, Sullivan worked 39¼ hours, Jackson worked 42½ hours, and Hoffman worked 38¾ hours.

Instructions On the form provided in your working papers:

1. Prepare a payroll register. The date of payment is October 15. List employees in alphabetical order by last name.
   a. Compute FICA taxes at 6.2% for social security and 1.45% for Medicare.
   b. Use the tax table in Chapter 12 to determine federal income taxes. The state income tax is 1.5% of gross earnings.
   c. All employees have a deduction for hospital insurance that is $4.75 for single employees and $7.85 for married employees.
   d. Union members pay weekly dues of $3.25.

2. After the payroll information is entered in the payroll register, total the columns and check the accuracy of the totals.

continued

Analyze
Identify the employee with the highest gross pay and the employee with the highest net pay.

Chapter 13, Problem 13A — Recording Payroll Transactions

Thunder Graphics Desktop Publishing pays its employees each week. The payroll register for the week ended December 17 is shown in your working papers.

Instructions On the general journal form provided in your working papers:

1. Record the December 17 payment of the payroll on page 34, Check 831.

2. Compute the employer's payroll taxes (FICA tax rates, 6.2% social security, 1.45% Medicare; state unemployment tax rate, 5.4%; federal unemployment tax rate, 0.8%).

3. Record the journal entry for the employer's payroll taxes.

4. Record the payment of all FICA and employees' federal income taxes, Check 832. The previous account balances were: **Employees' Federal Income Tax Payable**, $189, **Social Security Tax Payable**, $191.48, and **Medicare Tax Payable**, $44.92.

5. Record the income tax payment to the state, Check 833. The balance in the **Employees' State Income Tax Payable** account before the December 17 payroll was $178.40.

Analyze
Calculate the company's total FICA liability for the week ended December 17.

Chapter 14, Problem 14A — Recording and Posting Sales and Cash Receipt Transactions

On January 1, T-Shirt Trends, a merchandising business, had the following balances in its general ledger and accounts receivable subsidiary ledger.

General Ledger		
101	Cash In Bank	$ 7,500
115	Accounts Receivable	10,920
215	Sales Tax Payable	1,500
401	Sales	18,620
405	Sales Discounts	1,400
410	Sales Returns and Allowances	300

continued

Accounts Receivable Subsidiary Ledger

MOR	Morales, Gabriela	$ 420
ROU	Roundabout Fashions	3,150
SAM	Samstead Fashions	5,250
SAN	Sandpiper Sports Club	2,100

Instructions On the forms provided in your working papers:

1. Open the general ledger accounts and record the January 1 balances.
2. Open the customer accounts in the accounts receivable subsidiary ledger and record the January 1 balances.
3. Record the January transactions. Use page 12 in the general journal.
4. Post the transactions to the general ledger and accounts receivable subsidiary ledger accounts.

Date		Transactions
Jan.	1	Received a check for $3,087 from Roundabout Fashions payment of their $3,150 account less a 2% cash discount of $63, Receipt 82.
	4	Gabriela Morales notified us that $40 in T-shirts were defective. Issued Credit Memorandum 20 for $42, which includes a 5% sales tax of $2.
	9	Sold $1,000 in merchandise plus a sales tax of $50 to Roundabout Fashions on account, Sales Slip 90.
	15	Sandpiper Sports Club sent us a check in payment of their $2,100 account less a 2% cash discount, Receipt 83.
	17	Granted Samstead Fashions $252 credit for the return of $240 in merchandise plus a $12 sales tax, Credit Memorandum 21.
	20	Received a check from Roundabout Fashions in payment of their $1050 account balance, Receipt 84.
Jan.	25	Samstead Fashions sent us a check for $2,000 to apply on their account, Receipt 85.
	31	Cash sales for the month were $5,000 plus $250 in sales tax, Tape 44.
	31	Bankcard transactions for the month were $6,000 plus $300 in sales tax, Tape 44.

Analyze

Break down January's sales into cash sales (including bankcard) and sales on account.

Chapter 15, Problem 15A — Recording Purchases and Cash Payment Transactions

Business transactions for the month of March for T-Shirt Trends are presented below. A partial general ledger appears in your working papers.

Instructions On the form provided in your working papers, record March transactions on page 11 of the general journal.

Date		Transactions
Mar.	2	Issued Check 5500 for $6,790 to Dancing Wind Clothing Manufacturers in payment of their $7,000 invoice less a 3% cash discount.
	4	Purchased $2,800 in merchandise on account from Wilmington Shirt Factory, Invoice WSF123, terms 2/10, n/30.
	6	Wrote Check 5501 for $180 to Mario's Trucking for delivery of merchandise shipped from Wilmington Shirt Factory, FOB shipping point.
	8	Received a credit allowance of $300 from Wilmington Shirt Factory for damaged merchandise, issued Debit Memo 44.
	10	Received a premium notice from Keystone Insurance Company for $3,400, issued Check 5502.
	12	Bought $175 in supplies from Carter Office Supply on account, Invoice 97, terms n/30.
	14	Paid Wilmington Shirt Factory the $2,500 balance owed less a 2% cash discount, Check 5503.
	19	Returned $50 in supplies to Carter Office Supply, issued Debit Memorandum 45.
	25	Paid employees gross earnings of $4,000. Rates for taxes withheld are: Employees' Federal Income Tax, 15%; State Income Tax, 6%; Social Security Tax, 6.2%; and Medicare Tax, 1.45%. Issued Check 5504 for the net amount.
	30	Wrote Check 5505 to Carter Office Supply for the amount due on our account.

Analyze

Determine the number of postings to the accounts payable subsidiary ledger based on the transactions provided.

Chapter 16, Problem 16A — Recording and Posting Sales, Cash Receipts, and General Journal Transactions

T-Shirt Trends, a retail merchandising store, uses special journals for recording its business transactions. On May 1, the store had the following balances in its general ledger and accounts receivable subsidiary ledger accounts.

General Ledger

101	Cash in Bank	$ 4,700
115	Accounts Receivable	7,250
135	Supplies	300
215	Sales Tax Payable	600
401	Sales	40,000
405	Sales Discounts	1,500
410	Sales Returns and Allowances	500

Accounts Receivable Subsidiary Ledger

HAL	Hal's Fitness Studios	$ 1,500
HIO	Dave Hioki	200
MOR	Gabriela Morales	100
ROU	Roundabout Fashions	2,000
SAM	Samstead Fashions	2,500
SAN	Sandpiper Sports Club	900
WIL	Virginia Williams	50

Instructions On the forms provided in your working papers:

1. Open the accounts and enter beginning balances in T-Shirt Trends' general ledger.
2. Open the customer accounts in the accounts receivable subsidiary ledger and record the beginning balances.
3. Record the following transactions in the sales journal (page 9), the cash receipts journal (page 10), and the general journal (page 2).
4. Post to the customer accounts daily from the journals.
5. Post the entries in the General Credit column of the cash receipts journal daily.
6. Post general journal entries daily.
7. Foot, prove, total, and rule the sales and cash receipts journals.
8. Post the column totals from the special journals.
9. Prepare a schedule of accounts receivable.

continued

Date		Transactions
May	1	Received a check for $50 from Virginia Williams in payment of her account, Receipt 202.
	3	Sandpiper Sports Club sent a check for $882 in payment of their $900 account less 2% cash discount, Receipt 203.
	5	Sold Virginia Williams $300 in merchandise plus 5% sales tax of $15, Sales Slip 404.
May	7	Received a check for $1,470 from Hal's Fitness Studios in payment of their $1,500 account balance less 2% cash discount, Receipt 204.
	9	Virginia Williams returned $100 in merchandise plus sales tax of $5, issued Credit Memo 50.
	11	Sold $700 in merchandise on account to Sandpiper Sports Club plus sales tax of $35, terms, 2/10, n/30, Sales Slip 405.
	13	Dave Hioki sent a check for $100 to apply on his account, Receipt 205.
	14	Received $25 from the sale of supplies to Hanson's Market, Receipt 206.

Analyze

Calculate the percentage of sales on account to total sales in May.

Chapter 17, Problem 17A — Recording Special Journal and General Journal Transactions

T-Shirt Trends, a retail merchandising business, uses special journals to record its business transactions.

Instructions On the forms provided in your working papers:

1. Record the July transactions in the sales journal (page 15), cash receipts journal (page 16), purchases journal (page 15), cash payments journal (page 16), and general journal (page 3).
2. Foot, prove, total, and rule the special journals.
3. Prove cash. The beginning cash in bank balance is $7,500. The check stub balance at the end of the month is $13,409.

continued

Date		Transactions
July	2	Sold $500 in merchandise on account to Dave Hioki plus 5% sales tax, Sales Slip 333.
	2	Purchased $1,400 in merchandise on account from Cherokee Productions, Inc., Invoice 7001, terms 2/10, n/30.
	3	Issued Check 535 for $1,960 in payment of Dancing Wind Clothing Manufacturers $2,000 invoice less a 2% discount.
	4	Received $1,764 from Hal's Fitness Studios in payment of our $1,800 invoice less a 2% discount, Receipt 200.
	5	Issued Credit Memo 27 to Gabriela Morales for the return of $100 in merchandise plus $5 sales tax.
	6	Purchased $2,000 in store equipment on account from Carter Office Supply, Invoice CO123, terms n/30.
	6	Received $2,450 from Sandpiper Sports Club in payment of our $2,500 invoice less a 2% discount, Receipt 201.
	7	Sold $1,000 in merchandise on account plus 5% sales tax to Virginia Williams, Sales Slip 334.
	7	Issued Debit Memo 14 for $300 to Wilmington Shirt Factory for credit taken on damaged merchandise.
	8	Wrote Check 536 to Chapel Hills Realtors in payment of monthly rent of $900.
	8	Sold old office equipment for $75 cash to Ann Martin, an employee, Receipt 202.
	9	Purchased $3,000 in merchandise on account from CompuRite Solutions, Invoice CR79, terms 2/10, n/30.
	10	Sold $3,500 in merchandise plus 5% sales tax on account to Roundabout Fashions, Sales Slip 335.
	12	Bought $150 in supplies from Palace Party Supplies, on account, Invoice 876.
	13	Issued Check 537 for $1,600 to Academy Insurance Company for annual business insurance premium.
	14	Wrote Check 538 to CompuRite Solutions for $2,940 in payment of Invoice CR79 for $3,000 less 2% discount.
	15	Cash sales amounted to $3,000 plus 5% sales tax, Tape 29.
	15	Bankcard sales were $5,000 plus 5% sales tax, Tape 29.
	18	Purchased $1,100 in merchandise on account from Sullivan Screen Printers, Invoice 423, terms 2/10, n/30.
	20	Purchased $750 in merchandise for cash from Cherokee Productions, Inc., Check 539.
	22	Paid the monthly utility bill, $150, Check 540.
	23	Purchased $4,500 in merchandise on account from Dancing Wind Clothing Manufacturers, Invoice 777, terms, 2/10, n/30.

continued

Date	Transactions
24	Received $525 from Dave Hioki to apply on his account, Receipt 203.
24	Wrote Check 541 to Sullivan Screen Printers for $1,078 in payment of Invoice 423 for $1,100 less a 2% discount.
26	Issued Debit Memo 15 for $200 to Dancing Wind Clothing Manufacturers for the return of merchandise.
31	Cash sales amounted to $2,200 plus $110 sales tax, Tape 30.
31	Bank card sales amounted to $1,500 plus $75.00 sales tax, Tape 30.
31	Wrote Check 542 to pay the payroll of $2,000 (Gross Earnings) for the pay period ended July 31. The following amounts were withheld: employees' federal income taxes, $300; FICA taxes, $124 for social security and $29 for Medicare; employees' state income taxes, $60.
31	Recorded the employer's payroll taxes (FICA tax rate, 6.2% for social security, and 1.45% for Medicare; federal unemployment tax rate, 0.8%; and state unemployment rate, 5.4%).

Analyze

Identify any transactions recorded in the purchases journal that were not for merchandise to resell.

Chapter 18, Problem 18A — Calculating Adjustments and Preparing the Ten-Column Work Sheet

A partially completed work sheet for T-Shirt Trends is shown in your working papers. Adjustments need to be made to the following items at year-end.

Data for Adjustments

Merchandise Inventory, December 31	$47,394
Supplies on hand, December 31	678
Insurance premium expired during the period	1,260
Additional federal corporate income tax owed	168

Instructions On the form provided in your working papers, complete the December 31 work sheet including adjustments for T-Shirt Trends.

Analyze

Identify which asset accounts are affected by adjustments.

Chapter 19, Problem 19A — Preparing Financial Statements

The completed work sheet for T-Shirt Trends for the year ended December 31 appears in your working papers.

Instructions On the forms provided in your working papers, prepare the following financial statements:

1. The income statement for the fiscal year ended December 31.
2. A statement of retained earnings.
3. A balance sheet.

Analyze

Identify the change in the Retained Earnings account during the year.

Chapter 20, Problem 20A — Journalizing Closing Entries

The account balances for T-Shirt Trends appear in your working papers as of December 31.

Instructions On the form provided in your working papers, journalize all the closing entries on page 17 in the general journal.

Analyze

Identify the balance in the Income Summary account before it was closed to Retained Earnings.

Chapter 21, Problem 21A — Recording Stockholders' Equity Transactions

T-Shirt Trends was organized as a publicly held corporation on September 1 and authorized to issue 10,000 shares of $100 par, $8 preferred stock, and 25,000 shares of $50 par common stock.

Instructions On the form provided in your working papers, record the following transactions on page 1 of the general journal.

continued

Date		Transactions
2010		
Sept.	1	Issued 4,000 shares of $8 preferred stock at $100 per share, Receipt 101.
	12	Issued 12,000 shares of common stock at $50 per share, Receipt 102.
Nov.	15	Received $52.50 per share for 2,500 shares of common stock issued, Receipt 215.
Dec.	16	Received $5,000 for 50 shares of $8 preferred stock issued, Receipt 298.
2011		
Mar.	24	Issued 1,000 shares of common stock at $55 per share, Receipt 389.
Aug.	15	Declared a dividend of $32,400 on 4,050 shares of $8 preferred stock outstanding, payable on September 15 to stockholders of record on September 1, Memorandum 316.
	15	Declared cash dividend of 25¢ per share on 15,500 shares of common stock outstanding, payable on September 30 to stockholders of record on September 1, Memorandum 317.
Sept.	15	Paid preferred stock dividend declared August 15, Check 1763.
	30	Paid common stock dividend declared August 15, Check 1802.
Dec.	31	Prepared the closing entries in the general journal to close **Income Summary** for the net income of $102,600 and to close the two Dividends accounts.

Analyze

Identify the effect on RIdentify the effect on Retained Earnings of the dividends declared on August 15.

Chapter 22, Problem 22A — Maintaining a Petty Cash Register

Maureen Miller operates T-Shirt Trends, a T-shirt specialty business. She established a petty cash fund on February 1 by writing Check 336 for $100. She also records all petty cash disbursements in a petty cash register. Petty cash disbursements are usually made for expenses such as office supplies, delivery expense, miscellaneous expenses, and advertising expense.

Instructions On the forms provided in your working papers:

1. Record the establishment of the petty cash fund on the first line of the petty cash register, page 1.
2. Record each of the following petty cash disbursements in the petty cash register.
3. Foot, prove, total, and rule the petty cash register at the end of the month.

continued

4. Reconcile the petty cash fund. An actual cash count of the fund shows a balance of $5.68.

5. Check 362 was issued to replenish the petty cash fund. Record the replenishment information in the petty cash register.

Date		Transactions
Feb.	1	Paid $1.20 for postage due on a letter, Voucher 101.
	3	Purchased a $9 ad in the *Towne Sentinel* paper, Voucher 102.
	5	Bought a ream of typing paper for $4.75, issued Voucher 103.
	6	Paid $6 for a mailgram, Voucher 104.
	8	Issued Voucher 105 for $9.90 to Express Couriers in payment for the delivery of packages.
	10	Paid the newscarrier $4.10 for daily newspapers and issued Voucher 106.
	13	Bought pens, pencils, and memo pads for $6.60, Voucher 107.
	15	Purchased a $9.50 advertisement in a local trade journal, Voucher 108.
	19	Issued Voucher 109 for $8.70 for office supplies.
	22	Express Couriers delivered financial data from a client, $9.90, Voucher 110.
	26	Issued Voucher 111 for $2.12 for additional postage due on mail.
	27	Bought an ad in the *Towne Sentinel* for $8.70 and issued Voucher 112.
	28	Purchased a typewriter ribbon for $5.50, issuing Voucher 113.
	28	Voided Voucher 114 due to an error.
	28	Issued Voucher 115 for $4.10 to the newscarrier.

Analyze

At the end of February, conclude whether cash was short or over, by what amount, and whether this should be recorded as a revenue or an expense.

Chapter 23, Problem 23A — **Calculating and Recording Depreciation Expense**

The T-Shirt Trends Company purchased a delivery truck on October 12, 2009, at a cost of $78,900. The delivery truck has an estimated useful life of three years and an estimated disposal value of $900.

Instructions On the forms provided in your working papers:

1. Prepare a depreciation schedule for the delivery truck using the straight-line method.

2. Record the following transactions on general journal page 41. Post the transactions to the ledger accounts provided.

continued

3. Answer the following questions.

   a. How did you determine the depreciation from October 12 to December 31?
   b. What was the book value of the truck on December 31, 2009?
   c. What was the book value of the truck on December 31, 2010?
   d. How did you determine the annual depreciation on December 31, 2012?

Date	Transactions
2009 Dec. 31	Recorded the adjusting entry for the year's depreciation expense on the delivery truck.
2010 Dec. 31	Recorded the adjusting entry for the year's depreciation expense on the delivery truck.

Analyze

Identify how much depreciation expense will be recorded over the useful life of the asset. Identify which account will show the depreciation recorded over the asset's useful life.

Chapter 24, Problem 24A — Calculating and Recording Uncollectible Accounts Expense

T-Shirt Trends uses the allowance method of accounting for uncollectible accounts. At the end of the fiscal period, the following accounts appeared on T-Shirt's trial balance.

Accounts Receivable	$18,647.25
Allowance for Uncollectible Accounts	1,060.50
Sales	452,612.00
Sales Discounts	5,431.35
Sales Returns and Allowances	13,578.36
Uncollectible Accounts Expense	0.00

Instructions On the forms provided in your working papers:

1. Determine the amount of the adjustment for uncollectible accounts for the fiscal period ended December 31. Management estimates that uncollectible accounts will be 1.25% of net sales.
2. Journalize the adjusting entry in the general journal, page 10.
3. Post the adjusting entry to the general ledger accounts.
4. Determine the book value of accounts receivable.

continued

Analyze

When an account is written off in the next fiscal period, identify the account to be debited.

Chapter 25, Problem 25A — Accounting for Inventories

The Shoooot-It Camera Shop uses a one-year fiscal period beginning January 1. At the beginning of the fiscal period, the shop had a beginning inventory of film valued at $867.90 (526 rolls at $1.65 per roll). During the year, the shop made the following purchases:

January 13	400 rolls of film @ $1.67	=	$ 668
March 2	600 rolls of film @ $1.73	=	1,038
April 14	300 rolls of film @ $1.76	=	528
July 8	700 rolls of film @ $1.79	=	1,253
September 3	200 rolls of film @ $1.81	=	362
November 19	300 rolls of film @ $1.83	=	549
Totals	2,500 rolls		$4,398

There were 621 rolls of film in inventory at the end of the fiscal year.

Instructions On the form provided in your working papers, calculate the cost of the ending inventory using the specific identification, fifo, lifo, and weighted average cost methods. For the specific identification method, 165 of the rolls were purchased on September 3 and 456 were purchased on July 8.

Analyze

Conclude which inventory valuation method gives the highest value to Shoooot-It's ending inventory.

Chapter 26, Problem 26A — Calculating Current and Future Interest

Frequently, notes that are issued or received in one fiscal period do not mature until the next fiscal period. As a result, the interest expense paid or the interest income received applies to two different fiscal periods. Listed below is the information for 10 different notes.

Instructions On the form provided in your working papers, determine the following for each note.

1. Determine the maturity date. Assume February has 28 days.
2. Determine what portion of the interest applies to the current year and what portion of the interest applies to the following year. December 31 is the end of the fiscal period.

continued

	Amount	Issue Date	Interest Rate	Term
1.	$ 1,100	December 10	9%	30 days
2.	700	November 21	12%	60 days
3.	17,100	October 10	10%	90 days
4.	4,000	December 5	15%	60 days
5.	15,000	November 10	6.5%	120 days
6.	3,000	September 8	7%	180 days
7.	6,600	November 17	10%	70 days
8.	840	October 1	9.5%	6 months
9.	1,200	December 1	10%	3 months
10.	2,700	August 1	8%	9 months

Analyze

Identify which notes will have more interest expense for the following year as compared to the current year.

Chapter 26, Problem 26B — Recording Non-Interest-Bearing Notes Payable

Your company, National Paper Supplies, frequently borrows money for short periods from First Bank by issuing promissory notes.

Instructions In your working papers, record the following transactions on page 6 of the general journal.

Date		Transactions
Jan.	15	Borrowed $12,000 from First Bank by issuing a 90-day, non-interest-bearing note payable that the bank discounted at 12%, Note 45.
Apr.	15	Issued Check 2561 for $12,000 in payment of the note issued Jan.15 and recorded the interest expense.
Oct.	31	Borrowed $18,000 from First Bank by issuing a 120-day, non-interest-bearing note payable that the bank discounted at 13%, Note 46.
Dec.	31	Made an adjusting entry to record the accrued interest expense for Note 46 at year-end.

continued

Analyze

Calculate the amount of interest expense incurred in the current year for Notes 45 and 46.

Chapter 27, Problem 27A — Recording Partners' Investments

On June 15, Walter Cohen and Janice Connor agreed to combine their individual sole proprietorships into one firm organized as a partnership called Cohen & Connor Legal Services. The partners were to invest all the assets of their two former businesses. Those assets are listed below.

	Cohen	Connor
Cash	$247,000	$517,500
Accounts Receivable	78,000	
Merchandise Inventory	123,000	
Equipment	154,500	

Instructions Prepare the general journal entries (page 1) required to record the partners' investments. The source document is Memorandum 1.

Analyze

Calculate the total of all the assets contributed to the new partnership.

Chapter 28, Problem 28A — Liquidation of a Partnership

The partnership of Joyce, Jane, and Jill is liquidating. The partnership account balances as of November 30 are:

Cash	$ 8,000	Accounts Payable	$ 8,000
Accounts Receivable	16,000	Joyce, Capital	10,800
Inventory	8,000	Jane, Capital	13,200
Equipment	3,000	Jill, Capital	3,000

Profits and losses are shared 4:5:1 to Joyce, Jane, and Jill respectively.

Date		Transactions
Dec.	1	The accounts receivable are sold for $15,000.
	8	The inventory is sold for $7,000.
	15	The equipment is sold for $5,000.
	23	The accounts payable are paid in full.
	24	The balance of cash is distributed to the partners.

Instructions Prepare the entries required to liquidate the partnership. Start with general journal page 213.

Analyze
Calculate the gain or loss on each asset that was sold.

Appendix F
Answers to Section Assessment Problems

Chapter 1

Problem 1-1
Perform an honest analysis of likes and dislikes. Use the chart on page 7 to help identify skills and traits you possess.

Problem 1-2
Use the personal career profile form (Figure 1-1) as a reference for the types of information. List personal information (values, interests, skills, etc.) and then select three possible careers.

Problem 1-3
The possibility for individuals with accounting degrees vary but include careers as: tax accountants, auditors, management accountants, controllers, or chief financial officers.

Problem 1-4
For the accountants mentioned in the section:

- Drew Taylor – Interested in working with numbers and music.
- Maya Cruz – Involved in environmental causes, studied accounting.

Jana Passeno – Interested in serving the public; good at decision making.

Problem 1-5
Answers will vary.

Problem 1-6
Answers will vary.

Chapter 2

Problem 2-1
Perform an honest analysis of personal traits. Try to relate the answers to personal activities and experiences that correspond with entrepreneurship.

Problem 2-2
The situations such as these have opposing viewpoints. The drop in sales could signal that a new salesperson is needed or some improvements are in order. Some owners, though, may be reluctant to invest more money in a business where sales are down. You might ask for new paint and awnings in return for the higher rent.

Problem 2-3
Greenwood Sky Divers is the name of the business entity. It has been in business for six years, and is a "going concern." The accounting period for Greenwood Sky Divers is one year.

Chapter 3

Problem 3-1
1. $10,000
2. $26,000
3. $3,000
4. $26,000
5. $6,000
6. $13,000
7. $16,000
8. $8,000
9. $21,000
10. $20,000
11. $35,000
12. $4,500

Problem 3-2
1. Cash in Bank increased $30,000. Jan Swift, Capital increased $30,000.
2. Office Furniture increased $700. Jan Swift, Capital increased $700.
3. Cash in Bank decreased $4,000. Computer Equipment increased $4,000.
4. Office Furniture increased $5,000. Accounts Payable increased $5,000.
5. Accounts Receivable increased by $700. Office Furniture decreased $700.
6. Cash in Bank decreased $2,000. Accounts Payable decreased $2,000.

Problem 3-3

1. Cash in Bank decreased by $50. Jan Swift, Capital decreased by $50.
2. Cash in Bank increased by $1,000. Jan Swift, Capital increased by $1,000.
3. Cash in Bank decreased by $600. Jan Swift, Capital decreased by $600.
4. Cash in Bank decreased by $800. Jan Swift, Capital decreased by $800.
5. Cash in Bank increased by $200. Accounts Receivable decreased by $200.

Chapter 4
Problem 4-1

Accounts Receivable – asset, debit, credit, debit
R. Lewis, Capital – owner's equity, credit, debit, credit
Office Equipment – asset, debit, credit, debit
Accounts Payable – liability, credit, debit, credit

Problem 4-2

1a. The asset Office Equipment is increased. Increases in assets are recorded as debits.
 b. The liability Accounts Payable is increased. Increases in liabilities are recorded as credits.
2a. The asset Office Furniture is increased. Increases in assets are recorded as debits.
 b. The owner's capital account Alice Roberts, Capital is increased. Increases in the owner's capital account are recorded as credits.
3a. The liability Accounts Payable is decreased. Decreases in liabilities are recorded as debits.
 b. The asset Cash in Bank is decreased. Decreases in assets are recorded as credits.

Chapter 5
Problem 5-1

Cash in Bank—asset, debit, credit, debit
Accounts Receivable—asset, debit, credit, debit
Airplanes—asset, debit, credit, debit
Accounts Payable—liability, credit, debit, credit
Caroline Palmer, Capital—owner's equity, credit, debit, credit
Caroline Palmer, Withdrawals—owner's equity, debit, credit, debit
Flying Fees—revenue, credit, debit, credit
Advertising Expense—expense, debit, credit, debit
Food Expense—expense, debit, credit, debit
Fuel and Oil Expense—expense, debit, credit, debit
Repairs Expense—expense, debit, credit, debit

Problem 5-2

1. Dr. Utilities Expense; Cr. Cash in Bank
2. Dr. Accounts Receivable; Cr. Service Fees
3. Dr. John Albers, Withdrawals; Cr. Cash in Bank
4. Dr. Advertising Expense; Cr. Cash in Bank

Chapter 6
Problem 6-1

1. JayMax Office Supply
2. Dario's Accounting Services
3. April 9, 20--
4. 479
5. Fax Machine
6. $299
7. Payable in 30 days

Problem 6-2

Step 1 D
Step 2 F
Step 3 E
Step 4 B
Step 5 A
Step 6 C

Problem 6-3

1. Passenger Van, asset, increase, debit; Cash in Bank, asset, decrease, credit
2. Utilities Expense, expense, increase, debit; Cash in Bank, asset, decrease, credit
3. Cash in Bank, asset, increase, debit; Day Care Fees, revenue, increase, credit

Chapter 7

Problem 7-1

Cash in Bank $10,000 Dr.

Accounts Receivable – Mark Cohen $2,000 Dr.

Accounts Payable – Jenco Industries $1,000 Cr.

Tom Torrie, Capital $35,000 Cr.

Admissions Revenue $0

Problem 7-2

Account Balances:

Cash in Bank $10,000 (Dr.)

David Serlo, Capital $10,000 (Cr.)

Problem 7-3

May 20: Dr. Office Equip. $1,500;
Cr. Computer Equip. $1,500; Memorandum 47

Problem 7-4

July 7: Dr. Advertising Expense $300;
Cr. Rent Expense $300; Memorandum 13

Chapter 8

Problem 8-1

Store Equipment, asset, debit

Rent Expense, expense, debit

Service Fees Revenue, revenue, credit

Accounts Payable–Rubino Supply, liability, credit

Scott Lee, Capital, owner's equity, credit

Advertising Expense, expense, debit

Accounts Receivable–John Langer, asset, debit

Scott Lee, Withdrawals, owner's equity, debit

Maintenance Expense, expense, debit

Office Supplies, asset, debit

Problem 8-2

1. Hailey Office Supply
2. Garmot Electrical Co.
3. June 15
4. $6.23
5. Two
6. The invoice number is 220

Problem 8-3

Store Equipment, Balance Sheet, debit

Rent Expense, Income Statement, debit

Service Fees Revenue, Income Statement, credit

Accounts Payable–Rubino Supply, Balance Sheet, credit

Scott Lee, Capital, Balance Sheet, credit

Advertising Expense, Income Statement, debit

Accounts Receivable–John Langer, Balance Sheet, debit

Scott Lee, Withdrawals, Balance Sheet, debit

Maintenance Expense, Income Statement, debit

Office Supplies, Balance Sheet, debit

Chapter 9

Problem 9-1

1. Stratford Learning Ctr.
2. $832
3. Sandra Miller
4. Payment on account
5. August 21
6. Rose Hughes
7. Roxbury, N.Y.

Problem 9-2

1. $62,500
2. $29,915
3. $29,470
4. $10,630
5. $7,060
6. $25,010

Problem 9-3

Return on Sales = 20.53%

Chapter 10

Problem 10-1

1. Dr. Ticket Revenue $6,000; Cr. Income Summary $6,000
2. Dr. Income Summary $3,100; Cr. Gas and Oil Expense $700, Miscellaneous Expense $600, Utilities Expense $1,800

Problem 10-2

1.,2. Dr. Utilities Expense $129; Cr. Cash in Bank $129

3.,4. Dr. Income Summary $129; Cr. Utilities Exp. $129

Problem 10-3

Accts. Rec.–Linda Brown, Balance Sheet, No, Yes

Advertising Expense, Income Statement, Yes, No

Cash in Bank, Balance Sheet, No, Yes

Exercise Class Revenue, Income Statement, Yes, No

Exercise Equipment, Balance Sheet, No, Yes

Income Summary, N/A, Yes, No

Laundry Equipment, Balance Sheet, No, Yes

Maintenance Expense, Income Statement, Yes, No

Membership fees, Income Statement, Yes, No

Miscellaneous Expense, Income Statement, Yes, No

Office Furniture, Balance Sheet, No, Yes

Rent Expense, Income Statement, Yes, No

Repair Tools, Balance Sheet, No, Yes

Ted Chapman, Capital, Balance Sheet, Yes, Yes

Ted Chapman, Withdrawals, Balance Sheet, Yes, No

Utilities Expense, Income Statement, Yes, No

Chapter 11

Problem 11-1

1., 2. Total Deposit $967.08

3. Balance $2,394.24

4. Balance $2,394.24

5. Balance $2,119.24

Problem 11-2

1. $100
2. $25
3. Miscellaneous Expense

Chapter 12

Problem 12-1

Lang, Richard: 38, $7.80, $296.40, $0, $296.40

Longas, Jane: 43, $7.25, $290, $32.63, $322.63

Quinn, Betty: 44 1/4, $8.30, $332.00, $52.91, $384.91

Sullivan, John: 39 1/2, $8.30, $327.85, $0, $327.85

Talbert, Kelly: 40, $7.50, $300.00, $0, $300.00

Trimbell, Gene: 42 1/2, $9.75, $390.00, $36.56, $426.56

Varney, Heidi: 34 1/4, $8.75, $299.69, $0, $299.69

Wallace, Kevin: 46, $9.25, $370.00, $83.25, $453.25

Problem 12-2

Cleary, Kevin: S, 0, $155.60, $9.65, $2.26, $16, $3.11, $31.02, $124.58

Halley, James: S, 1, $184.10, $11.41, $2.67, $12, $3.68, $29.76, $154.34

Hong, Kim: S, 0, $204.65, $12.69, $2.97, $23, $4.09, $42.75, $161.90

Jackson, Marvin: M, 1, $216.40, $13.42, $3.14, $6, $4.33, $26.89, $189.51

Sell, Richard: M, 2, $196.81, $12.20, $2.85, $0, $3.94, $18.99, $177.82

Totals: $957.56, $59.37, $13.89, $57, $19.15, $149.41, $808.15

Problem 12-3

Medicare $4.57

Social Security $19.53

Federal Income Tax $32.00

Problem 12-4

Check number 79, net pay $263.94

Chapter 13

Problem 13-1

1. $2,193.40
2. $31.80
3. $503.25
4. $33.75
5. $1,690.15

Problem 13-2

Social Security Tax Payable $299.87

Medicare Tax Payable $70.13

Federal Unemployment Tax Payable $38.69

State Unemployment Tax Payable $261.18

Problem 13-3

Employees' federal income tax, payroll entry

Employer's social security tax, payroll tax entry

U.S. savings bonds, payroll entry

Employer's Medicare tax, payroll tax entry

Federal unemployment tax, payroll tax entry

Employees' state income tax, payroll entry

Union dues, payroll entry

Employees' social security tax, payroll entry

State unemployment tax, payroll tax entry

Employees' Medicare tax, payroll entry

Problem 13-4

April 30: Dr. Social Security Tax Payable $318.55, Medicare Tax Payable $113.28, Employees' Federal

Income Tax Payable $286; Cr. Cash in Bank $717.83

April 30: Dr. State Income Tax Payable $205.60; Cr. Cash in Bank $205.60

Problem 13-5

Dr. Salaries Exp. $1,328.07, Cr. Soc. Sec. Tax Pay. $82.34, Cr. Medicare Tax Payable $19.26, Cr. Fed. Inc. Tax Pay. $184.00, Cr. State Inc. Tax. Pay. $26.56, Cr. Hosp. Ins. Prem. Pay. $20.00, Cr. Cash in Bank $995.91

Chapter 14
Problem 14-1

Ending account balances:

Cash in Bank $554

Accounts Receivable $3,000

Sales $3,554

Problem 14-2

Sept. 1: Dr. Accts. Rec./James Palmer $318; Cr. Sales $300, Sales Tax Pay. $18; Sales Slip 101

Sept. 4: Dr. Accts. Rec./Anna Rodriguez $636; Cr. Sales $600, Sales Tax Pay. $36; Sales Slip 102

Sept. 7: Dr. Sales Ret. + Allow. $300, Sales Tax Pay. $18; Cr. Accts. Rec./James Palmer $318; Credit Memo. 15

Sept. 19: Dr. Sales Ret. + Allow. $40, Sales Tax Pay. $2.40; Cr. Accts. Rec./Anna Rodriguez $42.40; Credit Memo. 16

Problem 14-3

May 15: Dr. Cash in Bank $1908; Cr. Sales $1800, Sales Tax Payable $108; Tape 40

Problem 14-4

Mar. 1: Dr. Cash in Bank $140.40; Cr. Sales $130, Sales Tax Payable $10.40; Sales Slip 49

Mar. 5: Dr. Accts. Rec./Kelly Wilson $324; Cr. Sales $300, Sales Tax Payable $24; Sales Slip 55

Mar. 17: Dr. Cash in Bank $810; Cr. Sales $750, Sales Tax Payable $60; Tape 65

Chapter 15
Problem 15-1

1. Case Construction Company
2. Westmoreland Paint and Supply Company
3. 7894
4. November 15, 20--
5. December 1, 20--
6. 25 gallons
7. 5
8. White, gray, brown, beige and peach.
9. $20.00
10. $500.00

Problem 15-2

Sept. 2: Dr. Purchases $900; Cr. Accts. Pay./Sunrise Novelty Supply $900; Invoice SN110

Sept 7: Dr. Accts. Pay./Sunrise Novelty Supply $50; Cr. Purchases Returns and Allowances $50; Debit Memorandum 18

Problem 15-3

Nov. 12: Dr. Accts. Pay./Randall's Café and Bookstore $16.00; Cr. Purchases Ret. and Allow. $16.00; Debit Memo 559

Problem 15-4

May 1: Dr. Purchases $10,500; Cr. Cash in Bank $10,500; Check 1150

May 5: Dr. Transportation In $325; Cr. Cash in Bank $325; Check 1151

May 7: Dr. Prepaid Insurance $2,500; Cr. Cash in Bank $2,500; Check 1152

Chapter 16
Problem 16-1

Ending debit balance:

Accounts Receivable $27,720

Ending credit balances:

Sales Tax Payable $2,020

Sales $37,000

Problem 16-2

1. Sales credit $820, Sales Tax Payable credit $41, Accounts Receivable debit $861
2. Ending Balance $1,161

Problem 16-3

Totals:

General credit $105

Sales credit $6,200

Sales Tax Payable credit $372

Accounts Receivable credit $5,580

Sales Discounts debit $110

Cash in Bank debit $12,147

Total debits/credits $12,257

Chapter 17

Problem 17-1

Column totals:

Accounts Payable $4,550

Purchases $3,600

General $950

Problem 17-2

Ending Cash in Bank balance $5,610.59

Problem 17-3

Nov. 2: Dr. Accts. Pay./Colonial Products Inc. $900; Cr. Purchases Discounts $27, Cash in Bank $873; Check 104

Chapter 18

Problem 18-1

1. More
2. $1,533
3. Merchandise Inventory
4. Income Summary

Problem 18-2

1. Amount $2,454.70
   Account debited, Supplies Expense
   Account credited, Supplies
2. Amount $740
   Account debited, Insurance Expense
   Account credited, Prepaid Insurance
3. Amount $105
   Account debited, Federal Corporate Income Tax Expense
   Account credited, Federal Corporate Income Tax Payable

Problem 18-3

1. $9,995
2. Balance Sheet
3. $3,710
4. $155

Problem 18-4

1. Rotary Supply Corporation
2. 6
3. $177.09
4. Accounts Payable—K & L Electrical
5. Purchases Returns and Allowances

Chapter 19

Problem 19-1

1. Kevin Cleary, Capital
2. Capital Stock
3. Capital Stock and Retained Earnings

Problem 19-2

Subtotal $70.25, sales tax $2.81

Total $73.06

Problem 19-3

1. $96,506
2. $16,703
3. $27,783
4. $17,148

Problem 19-4

1. Office Equipment, Merchandise Inventory.
2. Federal Unemployment Tax Payable.
3. It increased by $18,358.72, or 14.49%.
4. 165.26%
5. The increase in Accounts Receivable indicates that either the company is selling more product or is less efficient at collecting payments. The decrease in Accounts Payable indicates that the company is paying bills earlier. In order to see the whole picture, one would need to analyze the data in relation to the other financial statements.
6. The company prepaid for a new insurance contract.

Chapter 20

Problem 20-1

Accounts Receivable, no

Bankcard Fees Expense, yes, credited, debited

Capital Stock, no

Cash in Bank, no

Equipment, no

Fed. Corp. Income Tax Expense, yes, credited, debited

Fed Corp. Income Tax Payable, no

Income Summary, yes, depends (loss = credit), depends (loss = credit)

Insurance Expense, yes, credited, debited

Merchandise Inventory, no

Miscellaneous Expense, yes, credited, debited

Prepaid Insurance, no

Purchases, yes, credited, debited

Purchases Discounts, yes, debited, credited

Purchases Returns and Allowances, yes, debited, credited

Retained Earnings, yes, depends (loss = debit), depends (loss = credit)

Sales, yes, debited, credited

Sales Discounts, yes, credited, debited

Sales Returns and Allowances, yes, credited, debited

Sales Tax Payable, no

Supplies, no

Supplies Expense, yes, credited, debited

Transportation In, yes, credited, debited

Utilities Expense, yes, credited, debited

Problem 20-2

July 12: Dr. Rent Exp. $750; Cr. Cash in Bank $750

Problem 20-3

Collecting and verifying source documents

Analyzing business transactions

Journalizing business transactions

Posting journal entries to ledgers

Preparing a trial balance

Completing the work sheet

Preparing financial statements

Journalizing and posting adjusting entries

Journalizing and posting closing entries

Preparing a post-closing trial balance

Chapter 21

Problem 21-1

1. 30,000 shares of common stock were issued at par, $10 per share in cash.
2. 2,000 shares of preferred stock were issued at par, $100 per share in cash.
3. 50,000 shares of common stock, par $10, were issued at $14 per share.

Problem 21-2

1. $87,500
2. $262,500

Problem 21-3

Apr. 1: Dr. Dividends—Common $5,679; Cr. Dividends Payable—Common $5,679; Memorandum 37

Apr. 30: Dr. Dividends Payable—Common $5,679; Cr. Cash in Bank $5,679; Check 221

Problem 21-4

The transactions that are reported on the statement of stockholders' equity are 2, 3, 6, and 8.

Chapter 22

Problem 22-1

1. Net cash $1,016.84
   Cash short $5
2. Mar 31: Dr. Cash in Bank $1,016.84, Cash Short and Over $5; Cr. Sales $964, Sales Tax Payable $57.84

Problem 22-2

1. Net cash $1,595.77
   Cash over $10
2. Apr. 14: Dr. Cash in Bank $1,595.77; Cr. Sales 1,496, Sales Tax Pay. $89.77, Cash Short and Over $10

Problem 22-3

Check $39.51; new check stub balance $77,393.35

Chapter 23

Problem 23-1

Building, plant

Cash in Bank, current

Change Fund, current

Delivery Equipment, plant

Land, plant

Merchandise Inventory, current

Office Equipment, plant

Office Furniture, plant

Petty Cash Fund, current

Prepaid Insurance, current

Store Equipment, plant

Supplies, current

Problem 23-2

Cash register $420, $60, $40

Computer $5,000, $1,000, $166.67

Conference table $1,800, $72, $36

Delivery truck $30,000, $6,000, $1,500

Desk $2,880, $144, $132

Problem 23-3

Book Value:

10/01/2011, $12,500

12/31/2011, $11,960

12/31/2012, $9,800

12/31/2013, $7,640

12/31/2014, $5,480

12/31/2015, $3,320

09/31/2016, $1,700

Problem 23-4

1. June 4: Dr. Office Equipment $3,456; Cr. Cash in Bank $3,456; Invoice 14492

2. Dec. 31: Dr. Depreciation Expense $379.87; Cr. Accumulated Depreciation $379.87

3. Dec. 31: Dr. Depreciation Expense $651.20; Cr. Accumulated Depreciation $651.20

Problem 23-5

1. Book Value:
   Jan. 7, $2,360
   First year, $1,908
   Second year, $1,456
   Third year, $1,004
   Fourth year, $552
   Fifth year, $100

2. Dec. 31: Dr. Depr. Exp.—Office Equip. $452; Cr. Accum. Depr.—Office Equip. $452

3. Dec. 31: Dr. Income Summary $452; Cr. Depr. Exp.—Office Equip. $452

4. $100, Yes

Chapter 24

Problem 24-1

Journal Entries

Apr. 10: Dr. Accts. Rec./Sonya Dickson $630; Cr. Sales $600, Sales Tax Pay. $30; Sales Slip 928

Nov. 30: Dr. Uncollectible Accounts Expense $630; Cr. Accts. Rec./Sonya Dickson $630; Memorandum 78

Dec. 30: Dr. Accts. Rec./Sonya Dickson $630; Cr. Uncollectible Accounts Expense $630; Memorandum 89

Dec. 30: Dr. Cash in Bank $630; Cr. Accts. Rec./Sonya Dickson $630; Receipt 277

Ledger Balances

Ending debit balances:

Cash in Bank $10,058

Accounts Receivable $7,290

Uncollectible Accounts Expense $928

Accts. Rec.—Sonya Dickson $0

Ending credit balances:

Sales Tax Pay. $278

Sales $24,760

Problem 24-2

Journal Entries

May 4: Dr. Allowance for Uncollectible Accounts $1,050; Cr. Accts. Rec./Jack Bowers $1,050; Memorandum 241

Nov. 18: Dr. Accts. Rec./Jack Bowers $1,050; Cr. Allowance for Uncollectible Accounts $1,050; Memorandum 321

Nov. 18: Dr. Cash in Bank $1,050; Cr. Accts. Rec./Jack Bowers $1,050; Receipt 1078

Dec. 31: Dr. Uncollectible Accounts Expense $1,850; Cr. Allowance for Uncollectible Accounts $1,850

Ledger Balances

Ending debit balances:

Cash in Bank $10,470

Accounts Receivable $19,350

Uncollectible Account Expense $1,850

Accts. Rec.—Jack Bowers $0

Ending credit balances:

Allowance for Uncollectible Accounts $3,050

Problem 24-3

(1) Uncollectible Accounts Expense:
Andrews Co. $2,800.02
The Book Nook $2,047.81
Cable Inc. $2,550.00
Davis Inc. $1,282.40
Ever-Sharp Co. $1,164.38

(2) Dec. 31: Dr. Uncollectible Accounts Expense $1,282.40; Cr. Allowance for Uncollectible Accounts $1,282.40

Chapter 25

Problem 25-1

Preparing Inventory Reports

Total value of inventory $794.10

Problem 25-2

Determining Inventory Costs

a. $472.55

b. $477.25

c. $465.25

d. $470.52

Problem 25-3

1. Ending inventory $575.90
2. Gross profit $214.30

Chapter 26

Problem 26-1

1. Interest $75.62, Maturity value $4,075.62
2. Interest $289.73, Maturity value $10,289.73
3. Interest $136.23, Maturity value $6,636.23
4. Interest $36.25, Maturity value $936.25

Problem 26-2

1. Interest $22.19
2. Interest $69.04
3. Interest $288.00
4. Interest $123.29

Problem 26-3

1. Cash in Bank $9,000
2. Notes Payable $9,000
3. Cash in Bank, asset; Notes Payable, liability
4. $9,355.07

Problem 26-4

1. Cash in Bank is debited for $9,802.74; Discount on Notes Payable is debited for $197.26; Notes Payable is credited for $10,000.
2. Discount $197.26, Proceeds $9,802.74

Problem 26-5

June 12: Dr. Cash in Bank $2,426.03, Discount on Notes Payable $73.97; Cr. Notes Payable $2,500; Note 55

Chapter 27

Problem 27-1

June 1: Dr. Cash in Bank $1,200, Accts. Rec. $2,000, Merchandise Inventory $8,000, Equipment $5,000, Van $12,000, Office Furniture $1,000; Cr. Matthew Deck, Capital $29,200; Memorandum 1

June 1: Dr. Cash in Bank $2,300, Accts. Rec. $7,000, Merchandise Inventory $5,000, Equipment $12,000, Office Furniture $500; Cr. Jennifer Rusk, Capital $26,800; Memorandum 1

Problem 27-2

1. 3/4, 1/4
2. 5/9, 3/9, 1/9
3. 3/8, 2/8, 2/8, 1/8
4. 2/4, 1/4, 1/4
5. 2/3, 1/3

Problem 27-3

Nov. 13: Dr. Cash in Bank $20,000; Cr. Walter, Capital $8,000, Yount, Capital $12,000

Chapter 28
Problem 28-1

Division of Net Income:

Goldman $14,000

Jones $21,000

Partners' Equity:

Goldman $37,000

Jones $68,000

Problem 28-2

Shares of gain:

Bass $7,200

Buie $7,200

Anderson $3,600

Norman $3,600

Dunham $2,400

Ruppe $2,400

Problem 28-3

Sept. 4: Dr. Cash in Bank $38,000, Gunther, Capital $3,500, Pertee, Capital $3,500; Cr. Merchandise Inventory $45,000

Sept. 15: Dr. Cash in Bank $29,000; Cr. Gunther, Capital $1,000, Pertee, Capital $1,000, Office Equipment $27,000

Chapter 29
Problem 29-1

Weaknesses might include ethics programs that do not provide clear guidelines for action when faced with various situations. Some employees might not report violations out of fear of retribution. A business ethics plan that includes well-communicated methods for reporting violations (like a Web site, an ethics office, or ethics seminars) could help remedy this problem.

Problem 29-2

The law covers most major ideas about the public's definition of proper and improper behavior, but it does not offer guidance on every dilemma. Ethical concepts cover more subtle instances, where the proper course of action might not be so clear-cut. Use the following resources to find an example in which the actions of a business were legal but might be considered unethical: *BusinessWeek Online*, *The Wall Street Journal*, local newspapers, and business magazines.

Problem 29-3

The following behaviors are possible violations of principles of professional conduct. Marcus appears to be recording transactions in a misleading manner in order to increase the amount of profit reported for the business. This violates the accounting principle of integrity. Jennifer has not been appropriately trained and educated for accounting work. This violates the principle of competence. Ing's behavior violates confidentiality when he shares confidential financial information with anyone not approved to receive it.

Career Skills

Making Career Choices

A career differs from a job in that it is a series of progressively more responsible jobs in one field or a related field. You will need to learn some special skills to choose a career and to help you in your job search. Choosing a career and identifying career opportunities require careful thought and preparation.

Steps to Making a Career Decision

There are three core areas of career decision making: self-assessment, career exploration, and goal setting. These are the five basic steps to making a career decision:

1. Create a self-profile with these headings: lifestyle goals, values, interests, aptitudes, skills and abilities, personality traits, learning styles. Fill in information about yourself.
2. Identify possible career choices based on your self-assessment.
3. Gather information on each choice, including future trends.
4. Evaluate your choices based on your self-assessment.
5. Make your decision.

After you make your decision, create a career plan that explains how you will reach your goal. Include short-term, medium-term, and long-term goals. In making your choices, explore the future opportunities in this field or fields over the next several years. What impact will new technology and automation have on job opportunities in a rapidly evolving workplace environment? Remember, if you plan, you make your own career opportunities.

COLLEGE AND CAREER PORTFOLIO

A college and career portfolio is a collection of information about a person, including documents, projects, and work samples that show a person's skills, talents, and qualifications. It includes information needed for a job search or to apply for college. Turn to the end of this Career Skills Handbook for more information and instructions for creating your own college and career portfolio.

Career Research Resources

In order to gather information on various career opportunities, there are a variety of sources to research:

- **Libraries** Your school or public library offers books, magazines, pamphlets, videos, and other print, online, and multimedia reference materials on careers. The U.S. Department of Labor publishes the *Dictionary of Occupational Titles (DOT),* which describes about 20,000 jobs and their relationships with data, people, and things; the *Occupational Outlook Handbook (OOH),* with information on more than 200 occupations; and the *Guide for Occupational Exploration (GOE),* a reference that organizes the world of work into interest areas that are subdivided into work groups and subgroups.
- **The Internet** The Internet is a primary source of research on any topic. It is especially helpful in researching careers.
- **Career Consultations** Career consultation, an informational interview with a professional who works in a career that interests you, provides an opportunity to learn about the realities of a career.
- **On-the-Job Experience** On-the-job experience can be valuable in learning firsthand about a job or career. You can find out if your school has a work-experience program, or look into a company or organization's internship opportunities. Interning gives you direct work experience and often allows you to make valuable contacts.

The Job Search

To aid you in your actual job search, there are various sources to explore. You should contact and research all the sources that might produce a job lead, or information about a job. Keep a contact list as you proceed with your search. Some job search resources include:

- **Networking with family, friends, and acquaintances.** This means contacting people you know personally, including school counselors, former employers, and professional people.
- **Cooperative education and work-experience programs.** Many schools have programs in which students work part-time on a job related to one of their classes. Many also offer work-experience programs that are not limited to just one career area, such as marketing.
- **Newspaper ads.** Reading the Help Wanted advertisements in your local papers will provide a source of job leads, as well as teach you about the local job market.
- **Employment agencies.** Most cities have two types of employment agencies, public and private. These employment agencies match workers with jobs. Some private agencies may charge a fee, so be sure to know who is expected to pay the fee and what the fee is.
- **Company personnel offices.** Large and medium-sized companies have personnel offices to handle employment matters, including the hiring of new workers. You can check on job openings by contacting the office by telephone or by scheduling a personal visit.
- **Searching the Internet.** Cyberspace offers multiple opportunities for your job search. Many Web sites provide lists of companies offering employment. There are thousands of career-related Web sites, so find those that have jobs that interest you. Companies that interest you may have a Web site, which may provide information on their benefits and opportunities for employment.

 ## Applying for a Job

When you have contacted the sources of job leads and found some jobs that interest you, the next step is to apply for them. You will need to complete application forms, write letters of application, and prepare your own résumé. Before you apply for a job, you will need to have a work permit if you are under the age of 18 in most states. Some state and federal labor laws designate certain jobs as too dangerous for young workers. Laws also limit the number of hours of work allowed during a day, a week, or the school year. You will also need to have proper documentation, such as a green card if you are not a U.S. citizen.

Job Application

You can obtain the job application form directly at the place of business, by requesting it in writing, or over the Internet. It is best if you can fill the form out at home, but some businesses require that you fill it out at the place of work.

Fill out the job application forms neatly and accurately, using standard English, the formal style of speaking and writing you learned in school. You must be truthful and pay attention to detail in filling out the form.

Personal Fact Sheet

To be sure that the answers you write on a job or college application form are accurate, make a personal fact sheet before filling out the application:

- Your name, home address, and phone number
- Your Social Security number
- The job you are applying for
- The date you can begin work
- The days and hours you can work
- The pay you want
- Whether or not you have been convicted of a crime
- Your education

- Your previous work experience
- Your birth date
- Your driver's license number if you have one
- Your interests and hobbies, and awards you have won
- Your previous work experience, including dates
- Schools you have attended
- Places you have lived
- Accommodations you may need from the employer
- A list of references—people who will tell an employer that you will do a good job, such as relatives, students, former employers.

Letters of Recommendation

Letters of recommendation are helpful. You can request teachers, counselors, relatives, and other acquaintances who know you well to write these letters. They should be short, to the point, and give a brief overview of your important accomplishments or projects. The letter should describe your character and work ethic.

Letter of Application

Some employees prefer a letter of application, rather than an application form. This letter is like writing a sales pitch about yourself. You need to tell why you are the best person for the job, what special qualifications you have, and include all the information usually found on an application form. Write the letter in standard English, making certain that it is neat, accurate, and correct.

Résumé

The purpose of a résumé is to make an employer want to interview you. A résumé tells prospective employers what you are like and what you can do for them. A good résumé summarizes you at your best in a one- or two-page outline. It should include the following information:

1. **Identification** Include your name, address, telephone number, and e-mail address.
2. **Objective** Indicate the type of job you are looking for.
3. **Experience** List experience related to the specific job for which you are applying. List other work if you have not worked in a related field.
4. **Education** Include schools attended from high school on, the dates of attendance, and diplomas, degrees, licenses, or certifications earned. A professional certification is a designation earned by a person to assure qualification to perform a job or task. You may also include courses you are taking or have taken that are related to the job you are applying for.
5. **References** Include up to three references or indicate that they are available. Always ask people ahead of time if they are willing to be listed as references for you.

A résumé that you put online or send by e-mail is called an electronic résumé. Some Web sites allow you to post them on their sites without charge. Employers access these sites to find new employees. Your electronic résumé should follow the guidelines for a print résumé. It needs to be accurate. Stress your skills and sell yourself to prospective employers.

Cover Letter

If you are going to get the job you want, you need to write a great cover letter to accompany your résumé. Think of a cover letter as an introduction: a piece of paper that conveys a smile, a confident hello, and a nice, firm handshake. The cover letter is the first thing a potential employer sees, and it can make a powerful impression. The following are some tips for creating a cover letter that is professional and gets the attention you want:

- **Keep it short.** Your cover letter should be one page, no more.
- **Make it look professional.** Key your letter on a computer and print it on a laser printer. Use white or buff-colored paper. Key your name, address, phone number, and e-mail address at the top of the page.
- **Explain why you are writing.** Start your letter with one sentence describing where you heard of the opening. "Joan Wright

suggested I contact you regarding a position in your marketing department," or "I am writing to apply for the position you advertised in the Sun City Journal."

- **Introduce yourself.** Give a short description of your professional abilities and background. Refer to your attached résumé: "As you will see in the attached résumé, I am an experienced editor with a background in newspapers, magazines, and textbooks." Then highlight one or two specific accomplishments.

- **Sell yourself.** Your cover letter should leave the reader thinking, "This person is exactly who we are looking for." Focus on what you can do for the company. Relate your skills to the skills and responsibilities mentioned in the job listing. If the ad mentions solving problems, relate a problem you solved at school or work. If the ad mentions specific skills or knowledge required, mention your mastery of these in your letter. (Also be sure these skills are included on your résumé.)

- **Provide all requested information.** If the Help Wanted ad asked for "salary requirements" or "salary history," include this information in your cover letter. However, you do not have to give specific numbers. It is okay to say, "My wage is in the range of $10 to $15 per hour." If the employer does not ask for salary information, do not offer any.

- **Ask for an interview.** You have sold yourself, now wrap it up. Be confident, but not pushy. "If I would be an asset to your company, please call me at [insert your phone number]. I am available for an interview at your convenience." Finally, thank the person. "Thank you for your consideration. I look forward to hearing from you soon." Always close with a "Sincerely," followed by your full name and signature.

- **Check for errors.** Read and re-read your letter to make sure each sentence is correctly worded and there are no errors in spelling, punctuation, or grammar. Do not rely on your computer's spell checker or grammar checker. A spell check will not detect if you keyed "tot he" instead of "to the." It is a good idea to have someone else read your letter, too. He or she might notice an error you overlooked.

 Interview

Understanding how to prepare for and follow up on interviews is critical to your career success. At different times in your life, you may interview with a teacher or professor, a prospective employer, a supervisor, or a promotion or tenure committee. Just as having an excellent résumé is vital for opening the door, interview skills are critical for putting your best foot forward and seizing the opportunity to articulate why you are the best person for the job.

Research the Company

Your ability to convince an employer that you understand and are interested in the field you are interviewing to enter is important. Show that you have knowledge about the company and the industry. What products or services does the company offer? How is it doing? What is the competition? Use your research to demonstrate your understanding of the company.

Prepare Questions for the Interviewer

Prepare interview questions to ask the interviewer. Some examples include:

- "What would my responsibilities be?"
- "Could you describe my work environment?"
- "What are the chances to move up in the company?"
- "Do you offer training?"
- "What can you tell me about the people who work here?"

Dress Appropriately

Nonverbal communication is 90 percent of communication, so dressing appropriately is of the utmost importance. Wear clothing that is appropriate for the job for which you are applying. In most situations, you will be safe if you wear clean, pressed, conservative business clothes in neutral colors. Pay special attention to grooming. Keep makeup light and wear very little jewelry. Make certain your nails and hair are clean, trimmed, and neat. Do not carry a large purse, backpack, books, or coat. Simply carry a pad of paper, a pen, and extra copies of your résumé and letters of reference.

Exhibit Good Behavior

Conduct yourself properly during an interview. Go alone; be courteous and polite to everyone you meet. Relax and focus on your purpose: to make the best possible impression.

- Be on time.
- Be poised and relaxed.
- Avoid nervous habits.
- Avoid littering your speech with verbal clutter such as "you know," "um," and "like."
- Look your interviewer in the eye and speak with confidence.
- Use nonverbal techniques to reinforce your confidence, such as a firm handshake and poised demeanor.
- Convey maturity by exhibiting the ability to tolerate differences of opinion.
- Never call anyone by a first name unless you are asked to do so.
- Know the name, title, and the pronunciation of the interviewer's name.
- Do not sit down until the interviewer does.
- Do not talk too much about your personal life.
- Never bad-mouth your former employers.

Be Prepared for Common Interview Questions

You can never be sure exactly what will happen at an interview, but you can be prepared for common interview questions. There are some interview questions that are illegal. Interviewers should not ask you about your age, gender, color, race, or religion. Employers should not ask whether you are a parent, married or pregnant, or question your health or disabilities.

Take time to think about your answers now. You might even write them down to clarify your thinking. The key to all interview questions is to be honest, and to be positive. Focus your answers on skills and abilities that apply to the job you are seeking. Practice answering the following questions with a friend:

- "Tell me about yourself."
- "Why do you want to work at this company?"
- "What did you like/dislike about your last job?"
- "What is your biggest accomplishment?"
- "What is your greatest strength?"
- "What is your greatest weakness?"
- "Do you prefer to work with others or on your own?"
- "What are your career goals?" or "Where do you see yourself in five years?"
- "Tell me about a time that you had a lot of work to do in a short time. How did you manage the situation?"
- "Have you ever had to work closely with a person you didn't get along with? How did you handle the situation?"

 ## After the Interview

Be sure to thank the interviewer after the interview for his or her time and effort. Do not forget to follow up after the interview. Ask, "What is the next step?" If you are told to call in a few days, wait two or three days before calling back.

If the interview went well, the employer may call you to offer you the job. Find out the terms of the job offer, including job title and pay. Decide whether you want the job. If you decide not to accept the job, write a letter of rejection. Be courteous and thank the person for the opportunity and the offer. You may wish to give a brief general reason for not accepting the job. Leave the door open for possible employment in the future.

Follow Up With a Letter

Write a thank-you letter as soon as the interview is over. This shows your good manners, interest, and enthusiasm for the job. It also shows that you are organized. Make the letter neat and courteous. Thank the interviewer. Sell yourself again.

Accepting a New Job

If you decide to take the job, write a letter of acceptance. The letter should include some words of appreciation for the opportunity, written acceptance of the job offer, the terms of employment (salary, hours, benefits), and the starting date. Make sure the letter is neat and correct.

Starting a New Job

Your first day of work will be busy. Determine what the dress code is and dress appropriately. Learn to do each task assigned properly. Ask for help when you need it. Learn the rules and regulations of the workplace.

You will do some paperwork on your first day. Bring your personal fact sheet with you. You will need to fill out some forms. Form W-4 tells your employer how much money to withhold for taxes. You may also need to fill out Form I-9. This shows that you are allowed to work in the United States. You will need your Social Security number and proof that you are allowed to work in the United States. You can bring your U.S. passport, your Certificate of Naturalization, or your Certificate of U.S. Citizenship. If you are not a citizen of the United States, bring your green card. If you are under the age of 16 in some states, you need a different kind of work permit.

You might be requested to take a drug test as a requirement for employment in some states. This could be for the safety of you and your coworkers, especially when working with machinery or other equipment.

EMPLOYABILITY SKILLS

You will need employability skills to succeed in a rapidly evolving workplace environment. These skills include personal and interpersonal skills, such as functioning effectively as part of a team and demonstrating leadership skills, no matter what position you are in. There are also certain qualities and behaviors that are needed to be a good employee:

- Attend work regularly.
- Be prompt.
- Make the most productive use of your time.
- Be cooperative, responsible, and honest.
- Obey company rules.
- Have a positive attitude.
- Show enthusiasm and pride.
- Tolerate differences.
- Be open-minded.
- Show respect.
- Be flexible.
- Take initiative.
- Be willing to learn new skills.
- Listen attentively.
- Use an appropriate voice.
- Demonstrate planning and time-management skills.
- Keep your workplace clean and safe.
- Understand the legal and ethical responsibilities related to your job.
- Understand the relationship between health and achievement.
- Understand and avoid the implications of substance abuse.

Leaving a Job

If you are considering leaving your job or are being laid off, you are facing one of the most difficult aspects in your career. The first step in resigning is to prepare a short resignation letter to offer your supervisor at the conclusion of the meeting you set up with him or her. Keep the letter short and to the point. Express your appreciation for the opportunity you had with the company. Do not try to list all that was wrong with the job.

You want to leave on good terms. Do not forget to ask for a reference. Do not talk about your employer or any of your coworkers. Do not talk negatively about your employer when you apply for a new job.

If you are being laid off or face downsizing, it can make you feel angry or depressed. Try to view it as a career-change opportunity. If possible, negotiate a good severance package. Find out about any benefits to which you may be entitled. Perhaps the company will offer services for finding new employment.

Take Action!

It is time for action. Remember the networking and contact lists you created when you searched for this job. Reach out for support from friends, family, and other acquaintances. Consider joining a job-search club. Assess your skills. Upgrade them if necessary. Examine your attitude and your career choices. Decide the direction you wish to take and move on!

Build Your College and Career Portfolio

A college and career portfolio is a collection of information about a person, including documents, projects, and work samples that showcase a person's academic and professional skills, talents, accomplishments, and qualifications. It includes the information needed for a job search or for applying for college. Your portfolio can be a paper portfolio in a folder, a digital portfolio with electronic files, or a combination of both. You can use your college and career portfolio throughout your life to keep track of your academic and career goals and accomplishments.

- **Personal Fact Sheet** When you apply for a job, you will probably fill out an application that asks for information that may not be on your résumé. For that reason you should include a personal fact sheet in your college and career portfolio.
- **Evaluate Yourself** The information you know about yourself can help you choose a career that is right for you. Update your self-evaluation periodically to make sure you are on the right path.
- **Conduct Career Research** Create a section for your portfolio called Career Research. Include information about career clusters and careers that interest you and sources of information you find helpful. Also include notes from career interviews and career evaluations. Update the Career Research section of your portfolio as you continue to explore your career options.
- **Prepare a Career Plan** After you have made a career decision, you can make a career plan. Create a section for your portfolio called career plan. Your first step in making your career plan is setting a career goal. Then you can set the short-term goals, medium-term goals, and long-term goals that will lead you to your career goal. Include goals related to education or training and other experiential learning. Review, update, or create new career plans as you continue to explore your career options.
- **Résumé and Cover Letter** Your college and career portfolio should include your résumé and a sample cover letter that you can use when following up job leads. When you find a job that interests you, note the qualifications required. Then customize your cover letter and résumé so that they are tailored to the job. Relate the skills you have to the skills required for the job.
- **Develop References** You should supply references when you apply for a job. You may also need them when applying for college. Include a list of your references in your college and career portfolio. Include each person's name, title and company, address, phone number, and e-mail address. If your references will provide written letters of reference, include copies in your portfolio. People to ask include former managers, teachers, counselors, or other trusted adults in the community who can comment on your reliability and attitude.
- **Showcase Your Technology Skills** The best way to show an employer what you know about technology is by demonstrating your technology skills! As you research the career that interests you, take note of the hardware, software, and other technology tools that are current in the field. Then, learn to use the technologies and include examples that show your mastery of these tools in your college and career portfolio. Include a list of hardware and software that you know how to use.
- **Awards, Honors, and Certifications** If you have received awards or honors, include any relevant information about them in your college and career portfolio. Also, if you have any licenses or certifications related to your continuing education or job search, also include these in your portfolio.

Review Key Concepts

1. What are the five steps to making a career decision?
2. What three types of goals should a career plan include?
3. Why is a personal fact sheet useful?
4. What are employability skills?
5. What is the role of professional certifications in a career search?
6. What is the role of a career and college portfolio?
7. What are the functions of résumés and portfolios?
8. Why is it important to demonstrate leadership skills?
9. What are five positive work qualities?
10. What are three questions you should be prepared to answer in a job interview?

Critical Thinking

11. Compare and contrast the role of a résumé and a cover letter.
12. Analyze why your career choice might change as you get older.
13. Predict the consequences of choosing a career that conflicts with your personal values.
14. Evaluate how tracking employment trends and technology trends can help you manage your own career.
15. Explain why it is important to think critically, demonstrate strong communication skills, and function effectively as part of a team in order to be successful in the workplace.
16. Analyze the importance of time management and project management skills in your chosen career field. Explain your answer.

Challenge Yourself!

17. Imagine that you have been asked to work on a project team either at school or where you work.
    - Think about the leadership and teamwork skills that you would need to be a successful member of the project team.
    - Demonstrate your knowledge of leadership and teamwork skills by creating a checklist that outlines what these skills are.
    - Work with a partner to identify how you would demonstrate these skills and behaviors in a work or school environment. Relate the skills to the "Employability Skills" section on page C-6. For example, offering to perform a task that another team member cannot complete may demonstrate initiative and support for a fellow team member.
18. Research careers of personal interest to you. Look at career Web sites to find job opportunities and learn about the accompanying duties.
    - Find out what type of education, certification, job training, and experience are required to meet your career goals.
    - Create a five-year plan that breaks down your goals.
    - What do you need to do now in order to meet your goals? What will you need to do next year? How will you assess your progress?

Glossary

A

account A subdivision under assets, liabilities, or owner's equity that summarizes the changes and shows the balance for a specific item. (p. 59)

accountant A person who handles a broad range of responsibilities, makes business decisions, and prepares and interprets financial reports. (p. 13)

accounting clerk An entry-level job that can vary from specializing in one part of the system to doing a wide range of tasks. (p. 14)

accounting cycle Activities performed in an accounting period that help the business keep its records in an orderly fashion. (p. 146)

accounting equation The accounting relationship between assets and the two types of equities. Assets = liabilities + owner's equity. (p. 56)

accounting period The period of time covered by an accounting report. (p. 40)

accounting system A system designed to collect, document, and report on business transactions. (p. 37)

accounts payable The amount of money owed, or payable, to the creditors of a business. (p. 59)

accounts payable subsidiary ledger A separate ledger that contains accounts for all creditors; it is summarized in the **Accounts Payable** controlling account in the general ledger. (p. 465)

accounts receivable The amount of money owed to a business by its credit customers. (p. 59)

accounts receivable subsidiary ledger A separate ledger that contains accounts for each charge customer; it is summarized in the **Accounts Receivable** controlling account in the general ledger. (p. 427)

accrual basis of accounting A system that recognizes and records revenues and expenses when they are earned and incurred rather than when cash is collected and paid. (p. A-2)

accruals Transactions that have not yet been recorded in the accounts because a revenue has been earned but not yet received or an expense has been incurred but not yet paid. (p. A-2)

accrued expenses A business expense that has been incurred but not paid or recorded. (p. A-4)

accrued revenue Business revenue that has been earned but not received or recorded. (p. A-3)

accumulated Build up, accrue, amass. (p. 386)

accumulated depreciation The total amount of depreciation for a plant asset that has been recorded up to a specific point in time. (p. 740)

accumulated earnings The employee's year-to-date gross earnings. (p. 362)

accuracy Correct; free from error or mistake. (p. 229)

acknowledgment Recognition of a favorable act or achievement. (p. 8)

acquire To obtain or come into control of something. (p. 54)

adequate Sufficient for a specific need. (p. 626)

adjust Adapt, change or move slightly. (p. 545)

adjusted gross income The total income for the year, after adjustments, on a federal income tax return. (p. A-14)

adjusting entries Journal entries that update the general ledger accounts at the end of a period. (p. 587)

adjustment An amount that is added to or subtracted from an account balance to bring that balance up to date. (p. 572)

administrative expenses Costs related to the management of a business (for example, office expenses). (p. 616)

affected To influence or change someone or something. (p. 152)

aging of accounts receivable method A method of estimating the uncollectible accounts expense in which each customer's account is classified by age; the age classifications are multiplied by certain percentages; and the total estimated uncollectible amounts are added to determine the end-of-period balance of **Allowance for Uncollectible Accounts.** (p. 782)

allowance method A procedure for uncollectible accounts receivable; the business matches the estimated uncollectible accounts expense with the sales made during the same period. (p. 772)

analyze Examine carefully in order to understand. (p. 657)

annual Yearly; happening once a year or in the course of a year. (p. 393)

appropriate Suitable or right. (p. 380)

approximately Estimated; near or close. (p. 781)

assets Property or items of value owned by a business. (p. 55)

assigned Allocate, give, specify. (p. 618)

assumes Take for granted; suppose, presume. (p. 782)

assurance A firm promise, a guarantee. (p. 197)

attached Fasten, join, connect, put together. (p. 705)

attitudes Opinions and feelings about someone or something. (p. 31)

audit The review of a company's accounting systems and financial statements to confirm that it follows generally accepted accounting principles. (p. 16)

authorized capital stock The maximum number of shares of stock a corporation may issue. (p. 675)

automated teller machine (ATM) Computer terminal where account holders can conduct various banking activities, often outside the bank or at other locations. (p. 325)

automatically Involuntary, routine, habitual. (p. 545)

available Ready to be used. (p. 261)

B

balance sheet A report of the balances in the permanent accounts on a specific date. (p. 258)

bank discount The interest charge deducted in advance on a non-interest-bearing note payable. (p. 835)

bank service charge A fee the bank charges for maintaining bank records and processing bank statement items for the depositor. (p. 320)

bank statement An itemized record of all the transactions in a depositor's account over a given period, usually a month. (p. 318)

bankcard A bank-issued card honored by many businesses that can be used to withdraw cash and to make payments for goods and services at many businesses instead of writing checks; also called *debit card, ATM card, or check card.* (p. 325)

bankcard fee A fee charged for handling bankcard sales slips; usually based on the total amounts recorded on the sales slips processed. (p. 479)

base period A period that is used for comparison in financial statement analysis, usually a year. (p. 625)

beginning inventory The merchandise a business has on hand and available for sale at the beginning of a period. (p. 573)

benefit An advantage, subsidy, help. (p. 474)

blank endorsement A check endorsement that includes only the signature or stamp of the depositor. It does not specify the new owner of the check. (p. 313)

board of directors A group of people, elected by the stockholders, who govern and are responsible for the affairs of a corporation. (p. 674)

book value The original cost of a plant asset minus accumulated depreciation. (p. 740)

book value of accounts receivable The amount the business can reasonably expect to collect from its accounts receivable. (p. 774)

business entity The accounting assumption that a business exists independently of its owner's personal holdings. The accounting records and reports are maintained separately and contain financial information related only to the business. (p. 39)

business ethics The policies and practices that reflect a company's core values such as honesty, trust, respect, and fairness. (p. 916)

business transaction An economic event that causes a change—either an increase or a decrease—in assets, liabilities, or owner's equity. (p. 59)

C

calendar year Accounting period that begins on January 1 and ends on December 31. (p. 148)

canceled checks A check paid by the bank, deducted from the depositor's account, and returned with the bank statement to the account holder. (p. 319)

capital Money supplied by investors, banks, or owners of a business. (p. 32)

capital expenditure An expense that will benefit current and future accounting periods. (p. A-25)

capital Stock The account that represents the total amount of investment in the corporation by its stockholders (owners). (p. 606)

cash basis of accounting A system that recognizes and records revenue only when cash is received and expenses only when cash is paid out. (p. A-2)

cash discount The amount a customer can deduct from the total owed for purchased merchandise if payment is made within a certain time; also called *sales discount.* (p. 438)

cash inflows Receipts of cash. (p. 626)

cash outflows Payments of cash. (p. 626)

cash payments journal A special journal used to record all transactions in which cash is paid out or decreased. (p. 540)

cash receipt The cash received by a business in a single transaction. (p. 436)

cash receipts journal A special journal used to record all transactions in which cash is received. (p. 508)

cash sale A transaction in which the business receives full payment for the merchandise sold at the time of the sale. (p. 436)

ceases Stop, end, come to an end. (p. 864)

certified public accountant (CPA) A licensed professional who has met certain education and experience requirements and passed a national test. (p. 16)

change fund An amount of money, consisting of varying denominations of bills and coins, that is used to make change in cash transactions. (p. 704)

charge customer A customer to whom a sale on account is made. (p. 425)

chart of accounts A list of all accounts used by a business. (p. 86)

charter The legal permission, granted by a state, that gives a corporation certain rights and privileges and spells out the rules under which the corporation is to operate. (p. 35)

check A written order from a depositor telling the bank to pay a stated amount of cash to the person or business named on the check. (p. 310)

check 21 The Check Clearing for the 21st Century Act; it allows the conversion of a paper check to an electronic image that can be quickly processed between banks. (p. 323)

checking account An account that allows a person or business to deposit cash in a bank and then write checks and make ATM withdrawals and debit card purchases against the account balance. (p. 310)

closely held corporation A corporation, often owned by a few people or by a family, that does not offer its stock for sale to the general public. (p. 674)

closing entries Journal entries made to close, or reduce to zero, the balances in the temporary accounts and to transfer the net income or net loss for the period to the capital account. (p. 280)

code of ethics A formal policy of rules and guidelines that describes the standards of conduct that a company expects from all its employees. (p. 917)

commission An amount paid to an employee based on a percentage of the employee's sales. (p. 349)

common stock The stock issued by a corporation when it issues only one class of stock. (p. 675)

comparability Accounting characteristic that allows the financial information to be compared from one period to another period; also allows the comparison of financial information between businesses. (p. 609)

competence The principle that requires accountants to have the knowledge, skills, and experience needed to complete a task. (p. 922)

compound entry A journal entry with two or more debits or two or more credits. (p. 286)

compute To calculate; determine an answer using mathematics. (p. 225)

computerized accounting system A type of accounting system in which information is recorded by entering it into a computer; also known as an automated accounting system. (p. 37)

conduct To organize and carry out. (p. 68)

confidentiality The principle that requires accountants to protect and not disclose information acquired in the course of work unless they have the appropriate legal or professional responsibility to do so. (p. 922)

conflicts Divergence; opposing needs, wishes, or interests. (p. 917)

conservatism principle Accounting principle requiring that when there is a choice, accountants choose the safer or more conservative method that is least likely to result in an overstatement of income or assets. (p. 811)

consists To be composed or made up of. (pp. 623, 657)

consistency principle Accounting principle requiring a business to apply the same accounting methods in the same way from one period to the next. (p. 810)

constantly Unchanging, steady, unvarying. (p. 541)

consumed To use something up. (p. 734)

contra account An account whose balance is a decrease to its related account. (p. 431)

contrast A degree of difference between things having similar or comparable natures. (p. 685)

contributed To give or supply in common with others. (p. 606)

controlling account An account that serves as a control on the accuracy of the account balances in the subsidiary ledger; its balance must equal the total of all account balances in the subsidiary ledger. (p. 427)

Glossary G-3

converted To turn into something else. (p. 263)

corporation A business organization recognized by law to have a life of its own. (pp. 35, 675)

correcting entry An entry made to correct an error in a journal entry discovered after posting. (p. 199)

cost of merchandise The actual cost to the business of the merchandise sold to customers. (p. 463)

credit An agreement to pay for a purchase at a later time; an entry on the right side of an account. (p. 54)

credit cards A card issued by a business containing a customer's name and account number that facilitates the sale on account. (p. 425)

credit memorandum A form that lists the details of a sales return or sales allowance. (p. 430)

credit terms Terms that state the time allowed for payment for a sale on account. (p. 426)

creditor A business or person to whom money is owed. (p. 54)

current assets The assets that are either used up or converted to cash during the normal operating cycle of the business. (p. 263)

current liabilities The debts of the business that must be paid within the next accounting period. (p. 263)

current ratio The relationship between current assets and current liabilities; calculated by dividing current assets by current liabilities. (p. 263)

D

debit An entry on the left side of an account. (p. 87)

debit memorandum The form a business uses to notify its suppliers (creditors) of a return or to request an allowance. (p. 468)

deduction An amount that is subtracted from an employee's gross earnings as required by law and those an employee wishes to have withheld. (p. 353)

deferrals Transactions that have been recorded in the accounts because cash has been received but not yet earned or cash has been paid but the expenses not yet incurred. (p. A-6)

demonstrate To show something clearly; make obvious. (p. 93)

deposit slip A bank form used to list the cash and checks to be deposited. (p. 312)

depositor A person or business that has cash on deposit in a bank. (p. 311)

depreciation Allocating a plant asset's cost over its useful life. (p. 735)

detect To notice or discover; spot; become aware of. (p. 431)

direct deposit Depositing net pay directly into an employee's personal bank account through electronic funds transfer. (p. 361)

direct write-off method A procedure for uncollectible accounts receivable; the business removes the uncollectible account from its accounting records when it determines the amount is not going to be paid. (p. 766)

discount period The period of time within which an invoice must be paid if a discount is to be taken. (p. 463)

disposal value The estimated value of a plant asset at its replacement time; also called *salvage value*. (p. 736)

dissolution A legal change to the partnership such as a change in partners that does not generally affect the operations of the business. (p. 894)

distribute To give out; allocate; divide amount several or many. (p. 680)

dividend A distribution of cash to stockholders of a corporation; reduces the corporation's retained earnings. (p. 680)

drawee The bank on which a check is written. (p. 315)

drawer The person who signs a check. (p. 315)

due date The date by which an invoice must be paid. (p. 466)

dynamic Energetic; exciting. (p. 13)

E

e-file A system for filing tax returns with the IRS using a computer connected to the Internet. (p. A-14)

electronic badge readers An electronic device that scans the magnetic strip containing employee information on an employee's identification badge and then transfers the information directly to a computer. (p. 349)

electronic Federal Tax Payment System (EFTPS) System used by large businesses to deposit income tax payments by electronic funds transfer. (p. 391)

electronic funds transfer system (EFTS) A system that allows banks to transfer funds among accounts quickly and accurately without the exchange of checks. (p. 325)

employee's earnings record A record prepared for each employee that contains all payroll information related to the employee; it is kept on a quarterly basis. (p. 362)

enables To make possible, allow, facilitate. (p. 420)

encountered Come across; meet, especially by chance. (p. 916)

ending inventory The merchandise a business has on hand at the end of a period. (p. 573)

endorsement An authorized signature written or stamped on the back of a check that transfers ownership of the check. (p. 313)

ensure To make certain; guarantee. (p. 353)

entity An organization, such as a business, that has an identity separate from those of its members; something that has a separate and distinct existence and purpose. (p. 674)

entrepreneur A person who transforms ideas for products or services into real-world businesses. (p. 31)

equipped To provide with operational items. (p. 116)

equities The total financial claims to the assets of a business. (p. 56)

error A mistake, fault, slip. (p. 197)

establishes Set up, create, start, begin, launch. (p. 704)

estimates To roughly calculate or approximate a value, cost, or worth. (p. 580)

ethics The study of our notions of right and wrong; a set of basic principles. (p. 916)

ethics officer The employee directly responsible for creating business conduct programs, evaluating performance, and enforcing standards of conduct. (p. 917)

evaluate To determine value or worth after careful consideration. (p. 246)

eventually Finally, at last; in the end. (p. 59)

exceed To be greater or better than something else; surpass. (p. 88)

expanded To increase in size or scope. (p. 612)

expense The cost of goods or services used to operate a business. (p. 66)

external controls Measures and procedures provided outside the business to protect cash and other assets. (p. 310)

F

federal A form of government in which power is distributed between a central authority and a number of constituent territorial units, such as the United States. (p. 386)

federal Unemployment Tax Act (FUTA) Law that requires employers to pay unemployment taxes to the federal government. (p. 387)

financial accounting The type of accounting that focuses on reporting information to external users. (p. 38)

financial claim Legal right to an item. (p. 54)

financial reports Documents that present summarized information about the financial status of a business. (p. 37)

financial statements Reports prepared to summarize the changes resulting from business transactions that occur during an accounting period. (p. 246)

financing activities Business activities involving debt and equity transactions. (p. 627)

first-in first-out method (FIFO) An inventory costing method that assumes that the first items purchased (first in) were the first items sold (first out). (p. 805)

fiscal year An accounting period of twelve months. (p. 148)

FOB destination Shipping term specifying that the supplier pays the shipping cost to the buyer's destination. (p. 478)

FOB shipping point Shipping term specifying that the buyer pays the shipping charge from the supplier's shipping point. (p. 478)

footing A column total written in small penciled figures. (p. 501)

Form 940 Form that reports the employer's federal unemployment taxes. (p. 399)

Form 941 Employer's Quarterly Federal Tax Return; form used to report accumulated amounts of FICA and federal income tax withheld from employees' earnings for the quarter, as well as FICA tax owed by the employer. (p. 399)

Form 1040EZ A simplified federal income tax return; most students file this return. (p. A-13)

Form 1099 Form used by banks and other institutions to report interest earned on savings accounts and other income. (p. A-13)

Form 8109 Federal Deposit Tax Coupon; form sent with payment for FICA and federal income taxes or federal unemployment taxes for the calendar year. (p. 391)

Form W-2 Wage and Tax Statement; form that summarizes an employee's earnings and amounts withheld for federal, state, and local taxes. (p. 398)

Form W-3 Transmittal of Wage and Tax Statements; summarizes the information contained on the employees' Form W-2 and accompanies the Form W-2 sent to the federal government. (p. 398)

Form W-4 A form employees complete when they begin work; it indicates to the employer how much to withhold for federal income tax from each paycheck. (p. A-13)

for-profit businesses A business that operates to earn money for its owners. (p. 14)

401(k) plan A voluntary payroll deduction from gross earnings; the employee does not pay income tax on the amount contributed until the money is withdrawn from the plan, usually after age 59½. (p. 356)

foundation The base or basis on which something is built. (p. 5)

free enterprise system A system in which individuals are free to produce the goods and services they choose. (p. 30)

full disclosure Accounting principle requiring a financial report to include enough information to be complete. (p. 609)

fundamental Basic and necessary. (p. 37)

funds Money that is set apart for a specific use; financial resources. (pp. 54, 323)

G

GAAP See Generally accepted accounting principles. (p. 37)

general journal An all-purpose journal in which all the transactions of a business may be recorded. (p. 151)

general ledger A permanent record organized by account number. (p. 184)

generally accepted accounting principles (GAAP) The set of rules that all accountants use to prepare financial reports, issued by the Financial Accounting Standards Board (FASB). (p. 37)

generate To produce, make, or cause. (p. 66)

going concern The accounting assumption that a business is expected to survive and operate indefinitely. (p. 40)

gross earnings The total amount of money an employee earns in a pay period; also called *gross pay*. (p. 347)

gross profit on sales The amount of profit made during the fiscal period before expenses are deducted; it is found by subtracting the cost of merchandise sold from net sales. (p. 616)

H

home keys The 4, 5, 6 keys on a numeric keypad. (p. A-18)

horizontal analysis The comparison of the same item(s) on financial statements for two or more accounting periods or dates; used to determine changes from one period to another. (p. 625)

I

illustrates To make clear or explain by example or comparison. (p. 249)

imaged checks A copy of a canceled check, also called a substitute check; it is sent with the bank statement in place of the original canceled check. (p. 319)

impact The effect of something or someone on another; influence. (p. 189)

implemented Put into practice, apply, execute. (p. 924)

implies To suggest or indicate without direct statement. (p. 218)

impose To prescribe or demand. 500

incidental Minor, secondary, supplementary, subsidiary. (p. 709)

income statement A report of the net income or net loss for a specific period; sometimes called a *profit-and-loss statement or earnings statement*. (p. 247)

income Summary account The general ledger account used to summarize the revenue and expenses for the period. (p. 283)

incurs To bring upon oneself; to become liable or subject to. (p. 32)

independence The principle that requires accountants to have no financial interest in the company being audited. (p. 922)

indicated Show or prove; point out clearly; specify. (p. 422)

integrity The principle that requires accountants to choose what is right and just over what is wrong. (p. 921)

interest The fee charged for the use of money. (p. 830)

interest-bearing note payable A note that requires the face value plus interest to be paid on the maturity date. (p. 833)

internal controls Procedures within the business that are designed to protect cash and other assets and to keep reliable records. (p. 310)

interpret Explain, tell the meaning of; deduce. (p. 618)

intervals Gap, space, period, time between two events. (p. 393)

inventory The items of merchandise a business has in stock. (p. 422)

investing activities Business activities involving investments and plant assets. (p. 626)

investment Money or other property provided for the purpose of making a profit. (p. 60)

issued To hand out, provide, offer; to put forth or distribute. (p. 315)

J

journal A chronological record of the transactions of a business. (p. 148)

journalize A chronological record of the transactions of a business. (p. 646)

journalizing The process of recording business transactions in a journal. (p. 148)

L

last-in first-out method (LIFO) An inventory costing method that assumes that the last items purchased (last in) are the first items sold (first out). (p. 806)

ledger A group of accounts; also referred to as a *general ledger*. (p. 86)

ledger account forms The accounting stationery used to record financial information about specific accounts in a manual accounting system. (p. 185)

liabilities Amounts owed to creditors; the claims of creditors to the assets of a business. (p. 56)

lifestyle The way a person uses time, energy, and resources. (p. 8)

liquidation When a partnership ceases to exist, all partnership assets are converted to cash, all debts are paid, and any remaining cash is distributed to the individual partners. (p. 894)

liquidity ratio The measure of a business's ability to pay its current debts as they become due and to provide for unexpected needs of cash. (p. 263)

long-term liabilities Debts that are not required to be paid within the next accounting period. (p. 833)

loss The result when a business spends more money than it earns. (p. 30)

lower-of-cost-or-market rule The requirement that ending merchandise inventory be stated at the lesser of cost (calculated using one of the four inventory costing methods) or market value. (p. 811)

M

maintain Preserve, look after, uphold. (p. 800)

management accounting The type of accounting that focuses on reporting information to management; often referred to as accounting for internal users of accounting information. (p. 39)

manual Operated or done by hand, rather than automatically. (p. 164)

manual accounting system A type of accounting system in which information is processed by hand. (p. 37)

manufacturing business A business that buys raw materials, transforms them into finished products by using labor and machinery, and sells the finished products to individuals or other businesses. (p. 33)

market value The current price that is charged for a similar item of merchandise in the market. (p. 811)

matching principle Principle requiring that the expenses incurred in an accounting period are matched with revenue earned in the same period. (p. 227)

materiality An accounting guideline stating that information considered important (relative to the other information) should be included in financial reports. (p. 609)

maturity value The principal plus interest that must be paid on a promissory note's due date. (p. 831)

merchandise Goods bought to resell to customers. (p. 422)

merchandising business A business that buys finished goods and resells them to individuals or other businesses. (p. 32)

mutual agency The legal right of any partner to enter into agreements for the business that are binding on all other partners. (p. 865)

N

net income The amount by which total revenue exceeds total expenses for the accounting period. (p. 227)

net loss The amount by which total expenses exceed total revenue for the accounting period. (p. 230)

net pay The amount left after total deductions have been subtracted from gross earnings. (p. 360)

net purchases The total cost of all merchandise purchased during a period, less any purchases discounts, returns, and allowances. (p. 614)

net sales The amount of sales for the period less any sales discounts, returns, and allowances. (p. 613)

networking Making contacts with people to share information and advice. (p. 9)

non-interest-bearing note payable A note from which the interest is deducted in advance from the face value of the note; no interest rate is stated on the note. (p. 835)

not-for-profit organizations An organization that operates for purposes other than making a profit. (p. 15)

note payable A promissory note issued to a creditor. (p. 828)

note receivable A promissory note that a business accepts from a customer. (p. 828)

NSF check A check returned to the depositor by the bank because the drawer's checking account does not have sufficient funds to cover the amount; also called *dishonored* or *bounced check* (NSF stands for Not Sufficient Funds). (p. 323)

objectivity The principle that requires accountants to be impartial, honest, and free of conflicts of interest. (p. 922)

obtained To get or be given, gain, attain. (p. 460)

obvious Easily understood or seen. (p. 918)

occurs To happen, take place, come about. (p. 147)

on account The purchase of an item on credit. (p. 62)

online Describes a terminal or cash register that is linked to a centralized computer system. (p. 801)

operating activities Business activities involving normal business operations. (p. 626)

operating expenses The cash spent or assets consumed to earn revenue for a business; operating expenses do not include federal income tax expense. (p. 616)

operating income The excess of gross profit over operating expenses; taxable income. (p. 616)

other expense A nonoperating expense; an expense that does not result from the normal operations of the business. (p. 840)

other revenue Nonoperating revenue that a business receives from activities other than its normal operations. (p. 842)

outstanding checks A check that has been written but has not yet been presented to the bank for payment. (p. 320)

outstanding deposits A deposit that has been made and recorded in the checkbook but does not yet appear on the bank statement. (p. 320)

overtime rate Pay rate employers are required to pay certain employees covered by state and federal laws; the Fair Labor Standards Act of 1938 sets the rate at 1½ times the regular hourly rate after 40 hours per week. (p. 350)

owner's equity The owner's claims to the assets of the business. (p. 56)

packing slip A form that lists the items included in the shipment. (p. 462)

paid-in Capital in Excess of Par The account that represents the amount of cash received by a corporation over the stock's par value. (p. 676)

par value The dollar amount assigned to each share of stock before it is sold to the public; used to determine the amount credited to the capital stock account. (p. 675)

partnership A business owned by two or more persons, called partners, who agree to operate the business as co-owners. (p. 34)

partnership agreement A written agreement that states the terms under which the partnership will operate. (p. 864)

pay period The amount of time over which an employee is paid. (p. 346)

payee The person or business to whom a check is written or a note is payable. (p. 314)

payroll A list of the employees and the payments due to each employee for a specific pay period. (p. 346)

payroll clerk An employee responsible for preparing the payroll. (p. 346)

payroll register A form that summarizes information about employees' earnings for each pay period. (p. 359)

payroll Tax Expense The expense account used to record the employer's payroll taxes (FICA, FUTA, SUTA). (p. 387)

percentage A fraction or proportion expressed as a number out of 100. (p. 349)

percentage of net sales method A method of estimating uncollectible accounts expense in which a business assumes that a certain percentage of each year's net sales will be uncollectible. (p. 781)

period A length of time, stage; completion of an event. (p. 218)

periodic inventory system An inventory system that requires a physical count of the merchandise on hand to update inventory records. (p. 801)

permanent accounts Accounts that are continuous from one accounting period to the next; balances are carried forward to the next period (for example, assets, liabilities, and the owner's capital account). (p. 117)

perpetual inventory system An inventory system that keeps a constant, up-to-date record of the amount of merchandise on hand. (p. 800)

personal interest tests Tests that help individuals to identify their preferences and to match interests to potential careers. (p. 9)

personality A set of unique qualities that makes a person different from all other people. (p. 9)

petty cash disbursement Any payment made from the petty cash fund. (p. 709)

petty cash fund Cash kept on hand for making small, incidental cash payments. (p. 709)

petty cash register A record of all disbursements made from the petty cash fund. (p. 712)

petty cash requisition A form requesting money to replenish the petty cash fund. (p. 712)

petty cash voucher A form that provides proof of payment from the petty cash fund. (p. 710)

petty cashier The person responsible for maintaining the petty cash fund and for making petty cash disbursements. (p. 709)

physical inventory An actual count of all merchandise on hand and available for sale. (p. 573)

piece rate A specific amount of money paid by an employer for each item the employee produces. (p. 349)

plant assets Long-lived assets that are used in the production or sale of other assets or services over several accounting periods. (p. 734)

point-of-sale terminals (POS) An electronic cash register. (p. 801)

policies A general principle or plan used to make decisions or take action. (p. 916)

post-closing trial balance A list of the permanent general ledger account balances; it is prepared to prove the ledger after the closing entries are posted. (p. 293)

postdated check A check that has a future date instead of the actual date; it should not be deposited until the date on the check. (p. 324)

posting The process of transferring information from the journal to individual general ledger accounts. (p. 184)

potential Possible, but not yet actual or real. (p. 608)

preferred stock Stock whose owners have certain privileges over common stockholders. (p. 675)

premium The amount paid for insurance. (p. 474)

prepaid expense An expense paid in advance. (p. 578)

primary Main, chief, most important. (p. 460)

principles A basic law, truth, or belief that governs behavior. (p. 921)

procedure Way of doing something, especially by a series of steps. (p. 284)

proceeds The cash actually received by the borrower on a non-interest-bearing note payable. (p. 835)

process An organized series of actions that produce a result. (p. 651)

process An organized series of actions that produce a result. (p. 116)

processing stamp A stamp placed on a creditor's invoice that outlines the steps to be followed in processing the invoice for payment. (p. 462)

professional Expert, specialized; one engaged in a principal calling, employment, or vocation. 864

profit The amount earned above the amount of expense incurred to keep the business operating. (p. 30)

profitability ratios A ratio used to evaluate the earnings performance of a business during the accounting period. (p. 262)

promissory note A written promise to pay a certain amount of money at a specific time. (p. 828)

property Anything of value that a business or person owns and therefore controls. (p. 54)

proportion Size, number, or amount in relation to another thing. (p. 431)

prospective Possible or likely to come. (p. 766)

proving cash The process of verifying that cash recorded in the accounting records agrees with the amount entered in the checkbook. (p. 552)

proving the ledger Adding all debit balances and all credit balances of ledger accounts, and then comparing the two totals to see whether they are equal. (p. 197)

proxy A document that transfers a stockholder's voting rights to someone else. (p. 675)

public accounting firms A business that provides a variety of accounting services including the independent audit. (p. 16)

publicly held corporation A corporation whose stock is widely held, has a large market, and is usually traded on a stock exchange. (p. 674)

purchase order A written offer to a supplier to buy specified items. (p. 460)

purchase requisition A written request that a specified item or items be ordered. (p. 460)

purchases account The account used to record the cost of merchandise purchased during a period. (p. 463)

purchases allowance A price reduction given when a business keeps unsatisfactory merchandise it has bought. (p. 468)

purchases discount The buyer's cash discount for early payment of an invoice on account. (p. 462)

purchases journal A special journal used to record all transactions in which items are bought on account. (p. 532)

purchases return The return of merchandise bought on account to the supplier for full credit. (p. 468)

Q

quick ratio A measure of the relationship between short-term assets and current liabilities; calculated by dividing the total cash and receivables by the current liabilities. (p. 263)

R

ranging To vary between one extreme and another. (p. 380)

ratio analysis The process of evaluating the relationship between various amounts in the financial statements. (p. 262)

reconciling the bank statement The process of determining any differences between a bank statement balance and a checkbook balance. (p. 319)

relevance An accounting characteristic requiring that all information that would affect the decisions of financial statement users be disclosed in the financial reports. (p. 609)

reliability A characteristic requiring that accounting information be reasonably free of bias and error. (p. 609)

reliable Dependable, trustworthy, consistent. (p. 310)

report form A balance sheet format that lists classifications one under another. (p. 259)

requires To need; to demand as necessary. (p. 286)

resources Things that are useful to a person; sources of supply or support. (p. 8)

restrictive endorsement A check endorsement that transfers ownership to a specific owner and limits how the check may be handled (for example, *For Deposit Only*). (p. 313)

retailer A business that sells to the final user, the consumer. (p. 420)

retained A corporation's accumulated net income that is not distributed to stockholders. (p. 607)

retained earnings A corporation's accumulated net income that is not distributed to stockholders. (p. 607)

retains To keep in possession, hold on to. (p. 649)

return on sales A ratio that examines the portion of each sales dollar that represents profit; calculated by dividing net income by sales. (p. 262)

revenue Income earned from the sale of goods and services. (p. 66)

revenue expenditure An expense that is used up before the end of the current period. (p. A-25)

revenue recognition The GAAP principle that revenue is recorded on the date it is earned even if cash has not been received. (p. 126)

ruling Drawing a line; a single *rule* (line) drawn under a column of figures indicates that the entries above the rule are to be added or subtracted. If an amount is a total and no further processing is needed, a double rule is drawn under it. (p. 221)

S

salaries Expense The expense account used to record employees' earnings. (p. 380)

salary A fixed amount of money paid to an employee for each pay period. (p. 348)

sale on account The sale of merchandise that will be paid for at a later date. (p. 425)

sales A revenue account to record the amount of the merchandise sold. (p. 422)

sales allowance A price reduction granted for damaged goods kept by the customer. (p. 430)

sales discount See *cash discount*. (p. 438)

sales journal A special journal used to record only the sale of merchandise on account. (p. 498)

sales return Any merchandise returned for credit or a cash refund. (p. 430)

sales slip A form that lists the details of a sale. (p. 425)

sales tax A tax levied by a city or state on the retail sale of goods and services. (p. 426)

schedule of accounts payable A list of all creditors in the accounts payable ledger, the balance in each account, and the total amount owed to all creditors. (p. 551)

schedule of accounts receivable A list of each charge customer, the balance in the customer's account, and the total amount due from all customers. (p. 516)

selling expenses Expenses a business incurs to sell or market its merchandise or services. (p. 616)

sequence The following of one thing after another in a regular or fixed order; a series or succession. (p. 420)

series A group of related items that follow in order; chain, string. (p. 420)

service business A business that provides a needed service for a fee. (p. 32)

signature card A bank form containing the signature(s) of the person(s) authorized to write checks on a checking account. (p. 311)

significant Important; of a noticeably or measurably large amount. (p. 577)

similarly In a comparable way. (p. 280)

skills The activities that a person does well. (p. 6)

slide error Error that occurs when a decimal point is moved by mistake. (p. 198)

sole proprietorship A business owned by one person. (p. 34)

source document A paper prepared as evidence that a transaction occurred. (p. 147)

special endorsement A check endorsement that transfers ownership of the check to a specific individual or business. (p. 313)

special journals Journals that have amount columns for recording debits and credits to specific general ledger accounts. (p. 498)

specific Particular, definite, or named individually. (p. 40)

specific identification method An inventory costing method in which the exact cost of each item in inventory is determined and assigned; used most often by businesses that have a low unit volume of merchandise with high unit prices. (p. 804)

specified Particular, specific, precise, definite, individually named. (p. 710)

state Unemployment Tax Act (SUTA) Law that requires employers to pay unemployment taxes to individual states. (p. 387)

statement of cash flows A financial statement that summarizes the cash receipts and cash payments resulting from business activities during a period. (p. 262)

statement of changes in owner's equity A financial statement that summarizes changes in the owner's capital account as a result of business transactions during the period. (p. 252)

statement of changes in partners' equity A financial statement that reports the change in each partner's capital account resulting from business operations, investments, and withdrawals. (p. 891)

statement of retained earnings A financial statement that reports the changes in the Retained Earnings account during the period. (p. 622)

statement of stockholders' equity A financial statement that reports the changes that have taken place in all of the stockholders' equity accounts during the period. (p. 685)

stockholders' equity The value of the stockholders' claims to the corporation. (p. 606)

stop payment order A demand by the drawer, usually in writing, that the bank not honor a specific check. (p. 323)

straight-line depreciation A method that equally distributes the depreciation expense over an asset's estimated useful life. (p. 736)

subsidiary ledger A ledger with detailed data that is summarized to a controlling account in the general ledger. (p. 427)

sufficient Enough or adequate. (p. 33)

sum Total; whole or final amount. (p. 516)

summarize To reduce main points concisely. (p. 191)

supplemental Additional, extra, complementary. (p. 712)

T

task. A job, duty; work to be done, especially that assigned by another. (p. 124)

taxable income The amount of income that is taxed on a federal income tax return. (p. A-14)

temporary Brief, momentary, impermanent. (p. 646)

temporary accounts Accounts used to collect information that will be transferred to a permanent capital account at the end of a single accounting period (for example, revenue, expense, and the owner's withdrawals account). (p. 116)

terminations An ending, limit, or boundary; conclusion. (p. 894)

tickler file A file that contains a folder for each day of the month into which invoices are placed according to their due dates. (p. 466)

time card A record of the time an employee arrives at work, leaves work, and the total number of hours worked each day. (p. 348)

transfer To move something from one person or place to another. (p. 60)

transposition error Error that occurs when two digits within an amount are accidentally reversed, or transposed. (p. 198)

trial balance A list of all the general ledger account names and balances; it is prepared to prove the ledger. (p. 197)

uncollectible account An account receivable that the business cannot collect; also called a *bad debt*. (p. 766)

unearned revenue Revenue that is received before it is actually earned. (p. A-7)

unemployment taxes Taxes collected to provide funds for workers who are temporarily out of work; usually paid only by the employer (FUTA and SUTA). (p. 387)

unique One of a kind; exceptional, exclusive. (p. 864)

unqualified opinion A determination provided by an auditor that a company's financial statements are a "fair presentation." (p. 864)

values The principles a person lives by and the beliefs that are important to the person. (p. 6)

vary To change, differ, contrast. (p. 186)

vertical analysis A method of analysis that expresses financial statement items as percentages of a base amount. (p. 618)

voiding a check Making a check unusable by writing the word Void in ink across the front of the check. (p. 315)

volume Degree of loudness; intensity of a sound (p. 39)

voluntarily Willingly, done on purpose. (p. 921)

voluntary compliance The principle that all taxpayers are responsible for filing their tax returns accurately and on time. (p. A-12)

wage An amount of money paid to an employee at a specified rate per hour worked. (p. 348)

weighted average cost method An inventory costing method in which all purchases of an item are added to the beginning inventory of that item; the total cost is then divided by the total units to obtain the average cost per unit. (p. 807)

wholesaler A business that sells to retailers. (p. 420)

withdrawal The removal of cash or another asset from the business by the owner for personal use. (p. 68)

withholding allowance An allowance an individual claims on a Form W-4; it is used to calculate the amount of income tax withheld from an employee's paycheck. (p. 353)

work sheet A working paper used to collect information from the ledger accounts for use in completing end-of-period activities. (p. 218)

working capital The amount by which current assets exceed current liabilities. (p. 263)

Index A

A

ABA (American Bankers Association) number, 311
Accelerated cost recovery system (ACRS), 737
Accelerated depreciation methods, 736
Account balances, in general ledger, 195
Account names, on work sheet, 220–221
Account numbers, 86
Accountants, 13
Accounting
   accrual basis, A-2
   basic concepts and terminology, 54–56, 86–90, 116–121
   careers in, 13–18. *See also in Index B*
   cash basis, A-2
   double-entry, 87–90, 128
   financial, 38
   management, 39
   payroll, 413–416
   purpose of, 13
Accounting assumptions, 39–40
Accounting clerk, 14
Accounting cycle, 146
   analyzing business transactions, 148
   collecting/verifying source documents, 147–148
   journalizing adjusting entries, 572–573, 587–588
   journalizing business transactions, 148, 151–165
   journalizing closing entries, 280–289
   for merchandising business, 657
   posting, 189–194
   posting closing entries to general ledger, 291–293
   preparing a post-closing trial balance, 293–294
   preparing a trial balance, 197–201
   preparing financial statements, 246
   steps in, 146–148
   work sheets, 218–222, 225–230
Accounting equation, 56
   analyzing business transactions, 59–64
   effects of transactions on, 60, 89–90
Accounting period, 40, 148
   end of, 280
   end-of-period reports, 570
Accounting system, 37
Account(s), 59. *See also specific accounts*
   capital, 117
   cash received on, 438
   chart of, 86–87
   classifying, 59
   contra, 431
   controlling, 427
   four-column ledger account form, 185
   for merchandising businesses, 422
   normal balance, 88
   opening, 186
   permanent, 117–118
   rules of debit and credit, 87–90
   T, 87, 92–97
   temporary. *See* Temporary accounts
   zero balance in, 195
Accounts payable, 59
Accounts payable subsidiary ledger, 465
   posting to, 470–471
   proving, 551–552
Accounts payable subsidiary ledger form, 465–466
Accounts receivable, 59
   aging of, 782–784
   book value of, 774
   posting to, 431–433
   uncollectible. *See* Uncollectible accounts receivable
Accounts receivable subsidiary ledger, 426–427
   posting to, 500–501, 512–513
   proving, 516–517
Accrual basis of accounting, A-2
Accruals, A-2
Accrued expenses, A-4–A-6
Accrued revenue, A-3–A-4
Accumulated depreciation, 740
Accumulated earnings, 362
   employee's, 386
Accuracy, net income, 229
Acknowledgment, 8
Acquisition, 54
ACRS (accelerated cost recovery system), 737
Adjusted gross income, A-14
Adjusting entries, 587–590
   in computerized accounting, 596
   for depreciation expense, 742–749
   updating accounts through, 588–590
Adjustments, 572
   to Merchandise Inventory, 574–575
   to Prepaid Insurance, 578–581
   recording, 587–590
   to Supplies, 577–578
   to tax accounts, 580–581
   on ten-column work sheet, 570–576
Administrative expenses, 616
Adverse opinions (audits), A-34
Aging of accounts receivable method, 782–784
AICPA (American Institute of Certified Public Accountants), 923
Allowance method, 772–779
American Bankers Association (ABA) number, 311
American Institute of Certified Public Accountants (AICPA), 923
Analysis
   horizontal, 625
   ratio, 262–264, A-36
   vertical, 618
   what-if, 610
Annual payment, 393
Annual report, A-22–A-23
Asset accounts
   analyzing business transactions, 92–97
   rules of debit and credit, 88
Assets, 55
   on balance sheet, 259, 260
   contra, 743, 772
   current, 263, 734
   plant, 734–737, 740
   in T accounts, 92–97
ATM (automated teller machine), 325
ATM card, 325
Attitude, 31
Auditors, 921, A-24
Audits, 16
Authorized capital stock, 675
Automated bill paying, 325
Automated teller machine (ATM), 325

B

Bad debts, 766. *See also* Uncollectible accounts receivable
Balance sheet, 246, 258–261
   analysis of, 625
   in annual report, A-22
   assets section, 259, 260
   for corporations, 623–624
   heading, 258–259
   liabilities and owner's equity section, 259, 261
   for partnerships, 891
   proving, 261
   for publicly held corporations, 622, 687

on six-column work sheets, 225–227
for sole proprietorships, 258–261, 622
on ten-column work sheets, 587
Bank discount, 835
Bank service charges, 320–322, 545
Bank statement, 318–321, 332
Bank-by-phone service, 325
Bankcard fees, 479–480, 545
Bankcard sales, 437, 442–443, 511
Bankcards, 325, 437
Banking procedures, 310
    checking accounts, 310–315
    Electronic Funds Transfer System, 325, 326
    protecting cash, 310
Base period, 625
Beginning inventory, 573
Blank endorsement, 313
Board of directors, 674
Bonus, 349
Book of original entry, 148
Book value, 740, 774
Bounced check, 323
Business entities, 39
Business ethics, 916–917
Business operations, 32–33
Business organization, forms of, 33–35
Business transactions, 59
    and accounting equation, 59–64
    accounts for, 59
    analysis of, 92–97, 124–128, 148
    cash payment, 61–62
    credit, 62–64
    investments by owner, 60–61
    recording, 148
    revenue and expense, 66–68
    source documents for, 147–148
    withdrawals by owner, 68–69
Businesses, 30–35
    environment of, 30–31
    forms of organization, 33–35
    for-profit, 13–14
    not-for-profit, 14–15
    public accounting firms, 16–18
    types of, 32–33

C

Calendar year, 148
Canceled checks, 319
Capital, 32
Capital account, 117
Capital expenditures, A-25
Capital investments, division of profits and losses based on, 872–874
Capital stock, 606, 675–678
Careers
    in accounting, 13–18. *See also in Index B*
    choosing, 4–5
    researching, 9–10
    self-assessment for, 5–9
    setting goals for, 10–11
Cash
    protecting, 310
    proving, 318, 552–553
Cash basis of accounting, A-2
Cash discounts, 438–441
Cash flows. *See* Statement of cash flows
Cash funds
    change fund, 704–707
    maintaining, 723
    petty cash fund, 709–717
Cash inflows, 626

Cash outflows, 626
Cash payments, 61–62, 558
    for account purchases, 476–478
    controls over, 474
    recording, 474–480, 486, 558
    shipping charges, 478–479
Cash payments journal, 540–554
    posting from, 546–552
    proving, 549–554
    recording in, 541–546
Cash proof (change fund), 705
Cash receipt transactions, 436, 523
    cash discounts, 438–439
    cash transactions, 436–437
    computerized accounting for, 449
    journalizing and posting, 508–518
    for merchandising business, 436–444
    recording, 439–444
Cash receipts journal, 508–518
    completing, 513–514
    Internet sales, 518
    journalizing and posting to, 508–516
    proving, 513–514
    schedule of accounts receivable, 516–517
Cash received on account, 440
Cash refunds, 431
Cash register tape, as source document, 705
Cash registers, 436, 801
Cash sales, 436, 441–442
Cash short and over
    change fund, 706–707
    petty cash fund, 716–717
Cash transactions, 436–437
Cease, partnership, 864
Certified public accountant (CPA), 16
Change fund, 704–707
Charge customers, 425
    receipts for payments, 436, 437
    recording payments, 439–440
Chart of accounts, 86–87
    for merchandising corporations, 421–422
    setting up, 134
Charters, 35, 674
Check Clearing for the 21st Century Act (Check 21), 323
Check stubs, 147, 314
Checkbook, 311
Checking accounts, 310–315
    endorsing checks, 313
    making deposits to, 312–313
    opening, 311–312
    writing checks, 314–315
Checks, 310–311
    canceled, 319
    endorsing, 313
    imaged, 319
    issued, 315
    NSF, 323
    payroll, 361
    postdated, 324
    routing of, 326
    stopping payment on, 323
    voiding, 315
    writing, 314–315
Closely held corporation, 674
Closing entries, 280–289
    for capital, 286–288
    computerized preparation of, 299
    for corporations *vs.* sole proprietorships, 646–649
    for depreciation expense, 747–749
    for expenses, 285–286
    journalizing, 280–289, 646–649
    posting, 291–293, 651–655
    procedure, 284

for revenue, 284–285
for temporary accounts, 284–289
for withdrawals, 288–289
Closing process, 281–282
Closing the fiscal year, 663
Closing the general ledger, 646–649, 651–655
Code of ethics, 917, 923
Commission, 349
Common stock, 675
dividends, 680–683
issuing, 676–677
Comparability, 609, A-23
Comparative financial statements, A-24
Comparative income statement, 618
Competence, as principle of ethics, 922
Compound entry, 286
Compute net income or net loss, 225
Computerized accounting systems, 37, 103
adjusting entries, 596
chart of accounts, 134
closing entries, 299
closing fiscal year, 663
electronic spreadsheets, 74
financial statements, 633, 692
general journal entries, 171
inventory costing, 817
maintaining cash funds, 723
notes receivable and payable, 849
payroll tax liabilities, 405
post-closing trial balance, 663
posting to the general ledger, 206
preparing financial statements, 269, 633
preparing payroll, 368
preparing trial balance, 237
purchases and cash payments, 486, 558
reconciling bank statement, 332
recording depreciation, 756
recording purchases and cash payments, 486
sales and cash receipts, 449, 523
setting up the general ledger, 903
starting new company, 881
writing-off uncollectible accounts receivable, 790
Confidentiality, as principle of ethics, 922
Conflicts of interest, 917
Conservatism principle, 811, A-24
Consistency principle, 810
Consumed assets, 734
Contra account, 431
Contra assets, 743, 772
Contra cost of merchandise accounts, 469
Controlling account, 427
Converted, asset to cash, 263
Corporations, 35
accounting functions, 646–649, 676–677, 680–683
balance sheet, 623–625
characteristics of, 674
closely held, 674
financial statements of, 608–610
income statement, 612–619
publicly held, 674–678, 680–683, 685–687
recording ownership, 606
reporting stockholders' equity, 606–608
statement of cash flows, 626–627
statement of retained earnings, 622–623
Correcting entries, 199–201
Cost allocation, financial reporting and, A-25
Cost of merchandise accounts, 463
Cost of merchandise available for sale, 614
Cost of merchandise sold, 613–616
CPA (certified public accountant), 16
Credit, purchasing on, 54–55
Credit balance column, 186

Credit cards, 425, 437
Credit column, 186
Credit memorandum, 430
Credit terms, 426
Credit transactions, 62–64
Creditors, 54
Credits, 87, 128–129. See also Rules of debit and credit
Current assets, 263, 734
Current liabilities, 263
Current ratio, 263

D

Debit balance column, 186
Debit cards, 325
Debit column, 186
Debit memorandum, 468
Debit-credit rule, 87–90. See also Rules of debit and credit
Debits, 87, 128–129. See also Rules of debit and credit
Declining-balance depreciation, 736, 740
Deductions, payroll, 380–381
required, 353–356
voluntary, 356, 396
Deferrals, A-6–A-7
Deposit slip/ticket, 312–313
Depositor, 311
Depreciation, 735
accumulated, 740, 743–746
adjusting entries for, 747–749
adjusting for expense of, 742–748
calculating, 739–740
closing entries for, 747–749
on financial statements, 746–747
methods for, 736
of plant assets, 735–736
recording, 745–746, 756
for tax reporting, 737
Depreciation schedule, 736, 737, 742
Depreciation summary form, 742
Direct deposit, 325, 361
Direct write-off method, 766–769
Disclosure, full, 609
Discount period, 463
Dishonored check, 323
Disposal value, of plant assets, 736
Dissolution of partnerships, 894
Dividends, 680–683
Division of profits and losses (partnerships), 869–874
capital investment basis, 872–874
equal division, 869–870
fractional share basis, 871–872
Double rule, 259
Double-entry accounting, 87–90, 128
Drawee, 315
Drawer, 315
Due date, 466

E

E-file, A-14
EFTPS (Electronic Federal Tax Payment System), 391–392
EFTS (Electronic Funds Transfer System), 325, 326
Electronic badge readers, 349
Electronic Federal Tax Payment System (EFTPS), 391–392
Electronic Funds Transfer System (EFTS), 325, 326
Electronic funds withdrawal, A-15
Electronic spreadsheets, 74
Employees
earnings record of, 362–363
paying, 361

Index A   I-3

Employer's payroll taxes, 386–389
　journalizing, 387–389
　posting to general ledger, 389
　quarterly report, 399–400
Encountered, ethics, 916
Ending inventory, 573
　adjustment for, 574–575
　determining cost of. *See* Inventory costing
End-of-period reports, 570
Endorsement, check, 313
Entrepreneurs, 31
　traits of, 31
Equality of debits and credits, 128–129
Equities transactions
　analyzing, 92–97
　T accounts for, 92–97
Equity, 56, 865–866
　owner's, 56, 261
　partners', 891
　statement of stockholders' equity, 685–686
　stockholders', 606–607
Errors
　in cash proofs, 553
　in checkbooks, 321
　correcting, 164–165, 199–201
　finding, 128, 198–201
　in journal entries, 164–165
　in payroll, 362, 363
　slide, 198
　in subsidiary ledgers, 517
　transposition, 198
　in trial balances, 199–201
　when reconciling bank statement, 320
　when writing checks, 315
　on work sheets, 229
Estimated useful life (plant assets), 735–736
Ethics, 916–919
　accountant's role in, 917–918
　codes of, 917
　ethical behavior, 918–919
　key principles for, 921–922
　and Sarbanes-Oxley Act, 923–924
Ethics officer, 917
Evaluate profits or losses, 246
Expenditures, allocation of, A-25
Expense, 66
　accrued, A-4–A-6
　closing, 285–286
　depreciation, 742–747
　on income statement, 249
　operating, 616
　prepaid, 578, A-7
　uncollectible accounts, 771
Expense accounts
　rules of debit and credit for, 120
　as temporary accounts, 116
Expense transactions, 66–68
External controls, 310

F

Face value, 828
Fair presentation, A-24
FASB (Financial Accounting Standards Board), 37
Federal income tax
　EFTPS, 391–392
　paying, 391–392
　payroll deductions for, 353–354
Federal Insurance Contributions Act (FICA), 354, 386. *See also* FICA taxes
Federal Tax Deposit Coupon, 391, 394
Federal Unemployment Tax Act (FUTA), 387, 393–394

FICA (Federal Insurance Contributions Act), 354
FICA taxes, 354, 356
　employee's, 356
　employer's, 386
　paying, 391–392
FIFO. *See* First-in, first-out costing method
Financial accounting, 38
Financial Accounting Standards Board (FASB), 37
Financial claims, 55–56
Financial reporting, 37, A-23–A-25
Financial statements, 246. *See also specific financial statements, eg.:* Balance sheet
　in annual report, A-22
　comparative, A-23
　computerized preparation of, 246, 610, 633
　for corporations, 608–610, 612–619, 622–627
　customizing, 692
　depreciation on, 746–747
　end-of-period, 608
　manual preparation of, 246, 612–613
　notes to, A-34
　for partnerships, 890–891
　for publicly held corporations, 685–687
　for sole proprietorships, 246
Financing activities, 627
First-in, first-out (FIFO) costing method, 805–806
Fiscal year, 148
FOB destination, 478
FOB shipping point, 478
Footing, 501–502
Form 940, 399
Form 941, 399, 400
Form 1040EZ, A-13–A-14
Form 1099, A-13
Form 8109, 391
Form W-2, 398, A-13
Form W-3, 398
Form W-4, 353–354, A-13
For-profit businesses, 13–14
Foundation, 5
401(k) plan, 356
Four-column ledger account form, 185
Fractional division of profit and loss, 871–872
Free enterprise system, 30
Full disclosure, 609
Funds, 54
FUTA. *See* Federal Unemployment Tax Act

G

GAAP (generally accepted accounting principles), 37
　accounting period, 40
　business entity, 39
　conservatism, 811
　consistency, 810
　going concern, 40
　matching, 227, 612, 734, 771
　revenue recognition, 126, 428
General journal, 151
　adjusting entries in, 587–588
　closing entries in, 280–282
　correcting errors in, 164–165
　recording entries in, 151–165, 171
General ledger, 184
　account balances, 195
　closing, 646–649, 651–655
　posting to. *See* Posting
　setting up, 184–187, 903
Generally accepted accounting principles. *See* GAAP
Going concerns, 40
Gross earnings/pay, 347–350
Gross profit on sales, 616

H

Headings
    on financial statements, 247, 252, 258–259
    work sheet, 220
Home keys, A-18
Horizontal analysis, 625
Hourly wages, 348–349

I

IIA (Institute of Internal Auditors), 923
IMA (Institute of Management Accountants), 923
Imaged checks, 319
Implies, work sheet, 218
Income statement, 246–250, 612–613
    analysis of, 618–619
    in annual report, A-22
    comparative, 619
    for corporations, 612–619
    cost of merchandise sold, 613–616
    expenses section, 249
    gross profit on sales, 616
    for merchandising vs. service businesses, 612
    net income/loss sections, 249–250, 616–618
    operating expenses, 616
    for partnerships, 890–891
    publicly held corporations, 685
    revenue section, 247–249, 613
    on six-column work sheets, 225–227
Income Summary account, 283–289
Income taxes
    federal, 353–354, 391–392
    reporting depreciation, 737
    state and local, 356
    taxable income, A-14
Incurs expenses, 32
Independence, as principle of ethics, 922
Independent auditors, 921–922
Institute of Internal Auditors (IIA), 923
Institute of Management Accountants (IMA), 923
Insurance, 474–475
Integrity, as principle of ethics, 921
Interest, 830
    on checking accounts, 320
    on notes, 830–831, 833–835, A-2, A-4
    payable, A-4
    receivable, A-4
Interest rate, 828
Interest table, 831
Interest-bearing notes payable, 833–835
Interests, personal, 6, 9
Internal controls, 310, 363
Internal Revenue Service (IRS), 353, 398, 736, A-12, A-14, A-15, A-16
International sales, 423
Internet sales, 518
Inventories, 422
    beginning, 573
    ending, 573
    merchandise, 574–575, 800, 801
    methods of tracking, 800–802
    physical, 573
    recording, 801
Inventory costing, 804–808, 817
    choosing method for, 810–811
    consistency and, 810
    FIFO, 805–806
    LIFO, 806–807
    specific identification method, 804–805
    weighted average cost, 807–808
Inventory sheet, 801
Investing activities, 626
Investment, 60–61
Invoices, 147, 462
IRS. See Internal Revenue Service
Issue date, 828

J

Journalizing, 148
    cash payments, 540–554
    cash receipts, 508–518
    checks to replenish petty cash, 715–716
    closing entries, 280–289, 646–649
    correcting entries, 199–201
    employer's payroll taxes, 387–389
    in general journal, 151–165
    payroll, 380–383
    purchases, 532–534
    sales, 498–500, 502–503
    uncollectible accounts receivable, 766–769, 772–779
    work sheet adjustments, 587–588
Journals, 148. See also specific journals
    special, 498
    usefulness of, 187

K

Keeping the books, 86

L

Last-in, first-out (LIFO) costing method, 806–807
Ledger account form, 185
Ledger accounts, opening, 186
Ledgers, 86. See also specific ledgers
    subsidiary, 426–427
    usefulness of, 187
Liabilities, 56
    on balance sheet, 259, 261
    current, 263
    long term, 833
    payroll, 380–383, 396–398
Liability accounts
    analyzing business transactions, 94–97
    rules of debit and credit, 89–90
Lifestyle, 8
LIFO. See Last-in, first-out costing method
Liquidation of partnerships, 894–897
Liquidity ratio, 263
Local taxes, 356, 393
Long-term liabilities, 833
Losses, 30. See also Net loss
Lower-of-cost-or-market rule, 811

M

MACRS (modified accelerated cost recovery system), 737
Maintain control, merchandise inventory, 800
Maker, 828
Management
    financial reporting by, A-23–A-24
    income statement use by, 618–619
    journals and ledgers use by, 187
    and payroll accounting, 363
    work sheet use by, 227
Management accounting, 39
Manual accounting system, 37, 86
    accounts receivable subsidiary ledger, 427
    general ledger in, 185
Manufacturing businesses, 33
Market value, 811

Matching principle, 227, 612, 734, 771
Materiality, 609
Maturity date (notes), 828–830
Maturity value, 831
Medicare tax, 354–356
Memorandum, 147
   credit, 430
   debit, 468
Merchandise, 422
   cost of, 463
   purchasing, 460–463, 465–471, 475–476
Merchandise inventory, 574–575, 800, 801. *See also* Inventories
Merchandise Inventory account, 422
Merchandising businesses, 32, 420–423
   accounts for, 422
   cash receipt transactions, 436–444
   completing accounting cycle, 657
   income statement, 612
   international sales, 423
   operating cycle, 420
   primary source of income, 460
   purchasing process, 460–463
   sales transactions, 425–433
MICR number, on check, 312
Modified accelerated cost recovery system (MACRS), 737
Mutual agency, 865

N

Net income, 227
   closing entry for, 646–649
   division of, for partnership, 869–874
   on income statement, 249
   on work sheet, 227–230
Net loss, 230
   at closing, 288
   closing entry for, 649
   on income statement, 250
   on statement of changes in owner's equity, 255
   on work sheet, 230
Net pay, 360
Net purchases, 614
Net sales, 613
Networking, 9
Nonbank credit cards, 425
Non-interest-bearing notes payable, 835–840
Nonoperating revenue, 842
Normal balance, 88
Notes payable, 828, 849
   interest-bearing, 833–835
   maturity date, 829–830
   non-interest-bearing, 835–840
   promissory notes, 828–831
Notes receivable, 828, 842–843, 849
   maturity date, 829–830
   promissory notes, 828–831
Notes to the financial statements, A-23
Not-for-profit organizations, 14–15
NSF checks, 323
Numeric keypad, A-18–A-21

O

Objectivity, as principle of ethics, 921–922
On account, 62
   cash received, 440
   purchases, 466–468, 476–478, 533–534
   sales, 425–426, 499–500
Online banking, 325
Online POS terminals, 801
Operating activities, 626

Operating cycle, merchandising business, 420
Operating expenses, 616
Operating income, 616
Other expenses, 840
Other revenues, 842
Outstanding checks, 320
Outstanding deposits, 320
Overtime rate, 350
Owner's capital accounts
   analyzing business transactions, 93–94
   rules of debit and credit, 89–90
Owner's equity, 56, 261

P

Packing slip, 462
Paid-in Capital in Excess of Par, 676
Par value, 675
Partners' equity, 865–866, 891
Partnership agreement, 864
Partnerships, 34–35
   accounting functions, 864–865, 869–874
   characteristics of, 864–865
   dissolution of, 894
   division of profits and losses, 869–874
   financial statements for, 890–891
   liquidation of, 894–897
   partners' equity, 865–866
Pay period, 346
Payee, 314, 828
Paying employees, 361
Payroll, 346
   computerized preparation of, 368
   employee-paid withholdings. *See* Deductions
   employer's taxes, 386–389
   journalizing, 380–383, 544
   posting to general ledger, 383–384
Payroll accounting, 413–416
   deductions, 353–356, 380–381, 396
   employee earnings records, 362–363
   gross earnings/pay, 347–350
   managerial importance of, 363
   net pay, 360
   payroll registers, 359–360, 381–382
   preparing employee checks, 361
Payroll clerk, 346
Payroll deductions, 353–356, 380–381, 396
Payroll liabilities, 380–383, 396–398
Payroll registers, 359–360, 381–382
Payroll Tax Expense, 387–389
Payroll tax expense forms, 391, 398–400
Payroll tax liabilities, 391–396
Payroll taxes
   employer's FICA, 386
   FUTA and SUTA, 387
   journalizing, 387–389
   posting to general ledger, 389
Percentage, commission, 349
Percentage of net sales method, 781
Period, work sheet, 218
Periodic inventory system, 801
Permanent accounts, 117–118
Perpetual inventory system, 800–802
Personal interest tests, 9
Personality, 9, 10
Petty cash disbursements, 709
Petty cash envelope, 714
Petty cash fund, 709–717
   cash short and over in, 716–717
   establishing, 709–710
   using, 710–716

Petty cash register, 712–714
Petty cash requisition, 712
Petty cash voucher, 710
Petty cashier, 709
Physical inventory, 573
Piece rate, 349
Plant assets, 734–737
   depreciation of. *See* Depreciation
   recording values of, 740
Point-of-sale (POS) terminals, 801
Policies, 916
Post-closing trial balance, 280, 293, 656, 663
Postdated checks, 324
Posting, 185, 186, 189–194
   accounts payable subsidiary ledger, 470–471
   accounts receivable subsidiary ledger, 431–433
   cash payments journal, 546–552
   cash receipts journal, 512–516
   closing entries, 291–293, 651–655
   correcting entries, 201
   employer's payroll taxes, 389
   general ledger, 189–194, 207, 383–384, 389, 396–398
   payroll, 383–384
   payroll liabilities, 396–398
   purchases journal, 534–538
   sales journal, 500–501
Preferred stock, 675–676
   dividends, 681–682
   issuing, 678
Premium, insurance, 474
Prepaid expense, 578, A-7
Prepaid insurance, 474
Principal, 828
Proceeds (notes payable), 835
Processing stamp, 462
Professional services, 864
Profit, 30
Profitability ratios, 262
Promissory notes, 828–830. *See also* Notes payable; Notes receivable
   bank discount, 835
   calculating interest on, 830–831
Property, 54
Prospective customer's credit rating, 766
Protecting cash, 310
Proving
   accounts payable subsidiary ledger, 551–552
   balance sheet, 261
   cash, 318
   cash payments journal, 549–553
   cash receipts journal, 513–514
   ledger, 197
   purchases journal, 536–537
   sales journal, 501–502, 505
Proxy, 675
Public accounting firms, 16–18
Publicly held corporation, 674
   capital stock, 675–678
   dividends, 680–683
   financial reporting for, 685–687
Purchase orders, 460, 461
Purchase requisitions, 460, 461
Purchases, recording, 486, 558
Purchases account, 463
Purchases allowance, 468–471
Purchases discounts, 462–463
Purchases journal, 532–534
Purchases on account, 466–468
   in purchases journal, 533–534
   recording, 466–468, 476–478
Purchases return, 468–471
Purchasing process, 460–463

Q

Qualified opinions (audits), A-24
Quick ratio, 263–264

R

Ratio analysis, 262–264, A-24
Ratios
   current, 263
   liquidity, 263
   profitability, 262
   quick, 263–264
   return on sales, 262
Receipts, 147
Reconciling bank statements, 318–326, 332
Reinstatement, of written-off accounts receivable, 768–769, 777–779
Relevance, 609
Reliability, 609
Reliable records, 310
Report form, 245
Resources, 8
Restrictive endorsement, 313
Retailers, 420
Retained earnings, 607–608
   closing entries (net loss), 649
   statement of, 622–623
Return on sales, 262
Revenue, 66
   accrued, A-3–A-4
   on income statement, 247–249, 613
   unearned, A-7
Revenue accounts
   closing, 284–285
   rules of debit and credit, 87–90, 118–119
   as temporary accounts, 116
Revenue expenditures, A-25
Revenue realization, 428, 612
Revenue recognition, 126, 428, A-2
Revenue transactions, 66–68
Reversing entries, A-8–A-9
Routing checks, 326
Rules of debit and credit, 87–90, 118–119
   applying, 92–97, 124–128
   equality of debits and credits, 128–129
   for temporary accounts, 118–121
Ruling, 221
   purchases journal, 536–537
   special journals, 501–502
   trial balance on work sheet, 221, 222
   work sheet, 587

S

Salaries, 348, 349, 363
Salaries Expense, 347, 380
Sales, 523
   and cash receipt transactions, 444
   computerized accounting for, 449
   international, 423
   Internet, 518
Sales account, 422, 443
Sales allowances, 430–431
Sales credit column, 502
Sales discount, 438
Sales journal, 498–500
Sales on account, 428–431, 499–500
Sales returns, 430–431
Sales Returns and Allowances account, 431
Sales slips, 425–426

Index A   I-7

Sales tax, 426
Sales tax payable, posting, 503–504
Sales transactions
   and accounts receivable subsidiary ledger, 426–427
   for merchandising business, 425–433
   posting to accounts receivable subsidiary ledger, 431–433
   recording sales on account, 428–431
   sales on account, 425–426
Sarbanes-Oxley Act (2002), 923–924, A-25
Schedule of accounts payable, 551–552
Schedule of accounts receivable, 516–517
Securities and Exchange Commission (SEC), 921
Selling expenses, 616
Service businesses, 32, 612
Service charges, bank, 320–322, 545
Service life (plant assets), 735–736
Shareholders, 35
Shipping charges, 478–479
Signature card, 311
Six-column work sheets, 218–230
   account name section, 220–221
   balance sheet section, 225–227
   completing, 229–230
   heading, 220
   income statement section, 225–227
   showing net income/loss, 227–230
   trial balance section, 221–222
Skills, 6
Slide errors, 198
Social security tax, 354–356. *See also* FICA
Sole proprietorships, 34
   accounting functions, 116–121
   balance sheet, 258–261, 622
   closing entries, 646
   financial statements, 246
   income statement, 247–249
   post-closing trial balance, 293
   posting closing entries, 291–293
   preparing closing entries, 284–289
   ratio analysis, 262–264
   setting up accounting records, 214–215
   statement of cash flows, 261–262
   statement of changes in owner's equity, 252–255
Source documents, 147
   applying information from, 148
   bank statements, 318–325
   for cash payments journal, 540
   cash register tape, 705
   cash register tapes, 436
   collecting and verifying, 147–148
   credit memos, 430
   debit memos, 468
   sales slips, 425–426
Special endorsement, 313
Special journals, 498. *See also specific journals*
Specific identification method, 804–805
Spreadsheets, electronic, 74
State taxes, 356
   income, 393
   unemployment, 387, 395
State Unemployment Tax Act (SUTA), 387, 395
Statement of cash flows, 246
   in annual report, A-22
   for corporations, 626, 627
   for publicly held corporations, 687
   for sole proprietorships, 246
Statement of changes in owner's equity, 246, 252–255
Statement of changes in partners' equity, 891
Statement of retained earnings
   in annual report, A-22
   for corporations, 622–623
Statement of stockholders' equity, 685–686

Stock
   capital, 606, 607, 675–678
   common, 675–678, 683
   dividends, 680–683
   preferred, 675–676, 678
Stock certificate, 674
Stockholders, 35, 674
Stockholders' equity, 606, 675
   reporting, 606–608
   statement of, 685–686
Stop payment order, 323
Store credit cards, 425
Straight-line depreciation, 736, 739
Subsidiary ledgers, 426–427
   errors in, 517
   preparing schedules for, 516, 551–552
Substitute checks, 319
Sum-of-the-years'-digits method, 736
SUTA (State Unemployment Tax Act), 387, 395

T

T accounts, 87, 92–97
Tax reports, payroll, 398–400
Tax tables, 354, 355
Taxable income, A-14
Taxes
   EFTPS, 391–392
   employer's payroll, 386–389
   FICA, 354–356, 386
   income, 353–354, 391–392
   sales, 426
   social security, 354–356
   state and local, 356
   unemployment, 387, 393–395
Tele-Tax, A-16
Temporary accounts, 116–117
   analyzing transactions involving, 124–128
   closing entries for, 284–289
   rules of debit and credit for, 118–121
Ten-column work sheets, 570
   adjustments on, 570–575, 577–580
   balance sheet section, 587
   extending amounts, 587
   journalizing adjustments of, 587–588
   posting adjustments, 588–590
   trial balance section, 570–571, 583–587
Term of note, 828
Terminations, partnership, 894
Tickler file, 466
Time calendar, 829–830
Time card, 348
Traits
   of entrepreneurs, 31
   personal, 5–9
Transmittal of Wage and Tax statements (Form W-3), 398
Transposition errors, 198
Trial balance, 197–198
   computerized preparation of, 237
   finding/correcting errors in, 198–201
   manual preparation of, 197–201
   post-closing, 280, 293, 656, 663
   on six-column work sheet, 221–222
   ten-column work sheets, 570–571

U

Uncollectible accounts receivable, 766, 790
   allowance method, 772–779
   determining, 767
   direct write-off method, 766–769

I-8   Index A

estimating, 772, 781–784
jounalizing transactions involving, 766–769, 772–779
matching revenue with expense of, 771–772
reinstating, 768–769, 777–779
Unearned revenue, A-7
Unemployed workers, temporarily, 387
Unemployment taxes, 387, 393–395
Unique, partnership, 864
Units-of-production depreciation, 736
Unqualified opinions (audits), A-24

V

Values, personal, 6–8
Vertical analysis, 618
Voiding checks, 315
Voluntarily, 921
Voluntary compliance, A-12
Vouchers, petty cash, 710

W

Wage and Tax Statement
  Form W-2, 398, A-13
  Form W-3, 398
Wages, 348, 349, 363
Weighted average cost method, 807–808
What-if analysis, 610

Wholesalers, 420
Withdrawals
  closing entries for, 288–289
  electronic funds, A-15
  by owner, 68–69
Withdrawals accounts
  closing to capital, 288–289
  rules of debit and credit for, 121
  as temporary accounts, 116
Withholding allowance, 353–354
Work sheets
  adjustment for uncollectible accounts, 775–776
  six-column, 218–230
  ten-column, 570–575, 577–580, 583–590
Working capital, 263

Z

Zero balance, 195

Index B

A

A Matter of Ethics
    company property, 83
    ethics in the real world, 931
    gossip in the workplace, 143
    insufficient funds, 567
    meeting a deadline, 213
    money shuffling, 412
    padding a résumé, 27
    partner loyalty, 887
    reporting a mistake, 639
    showing favoritism, 495
    using insider information, 700
Accountemps, 379
Accounting Careers, 25, 48, 111, 141, 211, 275, 337, 410, 493, 565, 698, 761, 823. *See also* **Career Wise**
Analyzing Financial Reports
    calculating return on owner's equity, 307
    classifying the balance sheet, 603
    corporate responsibility, 931
    cost of sales, 567
    employee costs, 412
    evaluating partnership operating results, 909
    evaluating stockholders' equity, 700
    identifying corporate goals, 495
    inventory levels, 825
    partners' equity, 887
    return on sales, 277
    statement of cash flows, 639
    vertical analysis, 763
Avon (Mark cosmetics), 115

B

Businesses, 217, 569
    Accountemps, 379
    Donna Fenn's writing, 309
    Dutch Valley Foods, 419
    Education.com, 531
    1154 LILL Studio, 85
    5boronyc, 245
    Foot Locker, Inc., 569
    Ford Motor Company, 799
    GEICO, 459
    Goodman & Company, LLP, 889
    Hot Studio, 345
    Hulu, 53
    JCPenney, 497
    John Deere (Deere & Company), 733
    Mark Cosmetics (Avon), 115
    McFarlane Companies, 3
    The Night Agency, 145
    99 Cents Only Stores, 645
    Olympus, 29
    One Smooth Stone, 863
    Papa Murphy's, 703
    PricewaterhouseCoopers, 915
    San Antonio Spurs, 605
    SanDisk, 673
    The Solution People, 183
    Trainz.com, 279
    Wells Fargo, 827
    YouTube, 765

C

Career Wise. *See also* **Accounting Careers**
    accountant, 50
    auditor, 307, 762
    bank teller, 243
    billing coordinator, 638
    bookkeeper, 113
    certification for management accounting careers, 699
    certified management accountant, 797
    certified payroll professional, 377
    certified public accountant (CPA), 82
    compare and contrast
        forensic accountant and auditor, 824
        international accountant and government accountant, 930
        management accountant and CPA, 855
    forensic accountant, 411, 494
    general ledger accountant, 566, 671, 886
    government accountant, 529, 909
    human resources director, 27
    international accountant, 142, 212
    management accountant, 276
    retail sales professional, 457
    senior management accountant, 731
Case Study
    accounting and entrepreneurships, 49
    balance sheet and net income, 242
    career advice, 26
    data mining, 854
    merchandising business
        department store, 670
        fitness and athletic equipment, 456
        health foods, 730
        movie theater, 528
        training videos, 602
    partnerships: building and design, 908
    payroll: number of employees, 376
    service business
        entertainment, 338
        landscaping, 112
    setting up accounting records, 180

D

Dutch Valley Foods, 419

E

Education.com, 531
1154 LILL Studio, 85
Ethics. *See* **A Matter of Ethics**

F

Fenn, Donna, 309
Financial reports. *See* **Analyzing Financial Reports**
5boronyc, 245
Foot Locker, Inc., 569
Ford Motor Company, 799

G

GEICO, 459
Global Accounting
    accounting for inflation, 638
    the euro, 276
    global e-business, 212

international accounting standards, 50
international competitive advantage, 566
inventory measurement, 824
joint ventures, 886
offshoring, 411
Organization of Petroleum Exporting Countries (OPEC), 930
plant assets, 762
tariffs and duties, 699
time zones, 142, 242
utilizing e-procurement, 494
World Trade Organization, 338
Goodman & Company, LLP, 889

H

H.O.T. Audit
adjusting entries, 603
adjustments for depreciation, 763
balance sheet, 700
bank fees, 529
cash payments journal, 567
closing entries, 671
conducting an audit, 83
correcting entries, 495
correcting payroll, 377
errors in general journal entries, 181
ethics, 931
first audit, 112
gross profit method, 825
inventory, 639
methods of writing off uncollectible accounts, 797
notes payable, 855
partnerships, 907
petty cash requisition, 731
posting journal entries to general ledger accounts, 213
reconciling the bank statement, 339
sales tax, 457
sharing profits, 887
statement of changes in owner's equity, 277
T accounts, 143
work sheet, 306
Hot Studio, 345
Hulu, 53

J

JCPenney, 497
John Deere (Deere & Company), 733

M

Mark Cosmetics (Avon), 115
McFarlane Companies, 3

N

99 Cents Only Stores, 645
The Night Agency, 145

O

Olympus, 29
One Smooth Stone, 863

P

Papa Murphy's, 703
Personal finance. *See* **Spotlight on Personal Finance**
Petkeepers Ltd., 217
PricewaterhouseCoopers, 915

S

San Antonio Spurs, 605
SanDisk, 673
The Solution People, 183
Spotlight on Personal Finance
budget, 457
career, 27
closing activities you perform, 307
credit check, 797
earning power, 83
examining your currency, 529
finding the best loan, 213
maintaining a checking account, 50
monitoring supplies, 603
often overlooked bank fees, 143
personal financial records, 181
savings, 243
shopping for a checking account, 339
spending plan, 113
spending records, 731
vehicle loan, 855

T

Trainz.com, 279
21st Century Skills
allocating resources, 528
articles of incorporation, 456
budgeting for IT productivity, 796
data mining, 243
direct deposit, 670
ergonomics, 376
information and media literacy, 82
job stress, 306
keys to success, 26
online security, 339
online video marketing study, 908
public speaking, 412
punctuality, 180
Sarbanes-Oxley Act, 602
special savings programs, 730
teaching others, 113
the technology sector, 854
time management, 49

W

Wells Fargo, 827

Y

YouTube, 765